Gary Copenhaven
151 E. Shaw

Principles of

Heat Transfer

The Intext Series in Mechanical Engineering

Edward F. Obert, Consulting Editor

University of Wisconsin, Madison

Principles of HEAT TRANSFER

third edition

Frank Kreith
University of Colorado

INTEXT EDUCATIONAL PUBLISHERS
New York and London

Library of Congress Cataloging in Publication Data

Kreith, Frank.
Principles of heat transfer.

(Series in mechanical engineering)
Includes bibliographical references.
1. Heat—Transmission. I. Title.
QC320.K7 1973 536'.2 73-1784
ISBN 0-7002-2422-X

Intext Educational Publishers
257 Park Avenue South
New York, New York 10010

Contents

---

*Sections marked with an asterisk can be omitted in an undergraduate course without interrupting the continuity of the presentation.

6 Fundamentals of convection 309

7 Free convection 383

8 Forced convection inside tubes and ducts 415

9 Forced convection over exterior surfaces 457

Preface to the third edition

In preparing the third edition, I have attempted to improve the clarity of presentation, introduced numerical and computer methods used in industrial heat-transfer calculations, and eliminated material not absolutely necessary in an introductory text in heat transfer.

The numerical methods are presented in such a way that they can be handled by means of a hand calculator or a digital computer. All digital computer solutions are written in FORTRAN IV, and a table relating the symbols in the text to those used in the computer solutions, as well as a flow diagram, are included with each example. Emphasis has been placed on iterative schemes, primarily useful in the solution of transient and periodic heat-transfer problems, and on methods for solving simultaneous algebraic equations which find most of their applications in two- and three-dimensional steady-state conduction problems and radiation heat transfer. The iterative schemes are based on the Gauss-Seidel and the Liebmann methods. Matrix inversion is used to solve simultaneous algebraic equations. Explicit as well as implicit methods for solving transient heat-transfer problems are presented.

The chapter on radiation has been completely rewritten because many of my students have asked for more examples and greater clarity. Therefore, I have presented all the material first from the point of view of electromagnetic theory and then generalized it to handle heat transfer in enclosures of diffuse black, gray, and nongray surfaces. A section on heat transfer involving gaseous radiation has been included, but radiation to and from specularly reflecting surfaces has been omitted because to date all experimental data with industrial surfaces are in excellent agreement with calculations based on diffuse models.

In the chapters dealing with heat transfer by convection, I have updated the equations used for calculating the heat-transfer coefficients to correspond with correlations of the best available experimental data. I have also summarized the most important correlation equations in tabular form at the end of each chapter for the convenience of students and practicing engineers. Some of the new examples worked out in the body of the text are designed to illustrate current problems related to environmental and ecological topics.

The problems at the end of Chapters 1 through 5 have been arranged to correspond with the topics in each chapter, with easier problems followed by more difficult ones. New problems have been added throughout the text, and problems which students found either too difficult or ambiguous have been deleted.

I have also tried to make the new edition suitable for an introductory course in heat transfer at schools on the quarter system. For a shortened version of the course, I would suggest that the first four chapters be treated in detail, that the basic material on radiation and the network method of solution be covered, but that the sections on gaseous and solar radiation be omitted. The material on convection heat transfer in Chapters 6 through 9 can be condensed by first giving the student a physical understanding of laminar and turbulent transport processes, then discussing the use of dimensional analysis in the correlation of experimental data, and finally presenting the correlation equations necessary to calculate heat-transfer coefficients in engineering practice. Chapter 10 on heat transfer with change in phase can be handled similarly, but Section 10-4 on freezing and melting can be left out. Chapter 11 can be condensed by using either the log-mean-temperature or the effectiveness method for designing heat exchangers, but not both.

Although the time has not yet come when the engineering profession in the United States is ready to adopt the international system of units, I believe that this time is not too far away. In an effort to help this transition, I have not only included tables of conversion factors, but have also introduced SI units in key tables as well as in some homework problems. I hope that this modest effort will also help communications with engineers in the United Kingdom where the SI system of units is already used.

In preparing this edition I have had the benefit of seeing innovative methods of presentation used by authors of other heat-transfer texts, of receiving helpful suggestions and additional problems from students and teachers of heat transfer in all parts of the world, and of discussing specific points with my colleagues and friends. Dr. Jan Kreider of the University of Colorado proofread parts of the manuscript and wrote some of the computer programs. My wife Marion typed, cut, and pasted the final version of the manuscript. I am grateful to all of the people whose assistance and encouragement have given me the incentive to undertake and the perseverance to complete this revision.

Preface to the second edition

In the seven-year period since the first edition of this book was published, many changes have occurred in engineering technology. In preparing a new edition, I have tried to adapt it to the demands of the next seven years as I see them and to incorporate in it the most significant results of recent developments in the field of heat transfer.

The development and availability of digital computers today makes it almost mandatory that a competent engineer be familiar with numerical methods for solving heat-transfer problems. In the first edition I relied heavily on graphical methods for analyzing two- and three-dimensional steady-state conduction systems and solving unsteady heat-transfer problems. In order to adapt the new edition to the computer era, I have deleted some material on the graphical method of solving unsteady heat-conduction problems and have greatly expanded the section dealing with numerical methods. From a pedagogic point of view, however, I have found that a student grasps the numerical methods more easily if at least a cursory introduction of the graphical method is also presented. Solutions to the key illustrative problems in Chapter 4 are, therefore, presented both by the graphical method and by numerical methods which can easily be adapted to machine calculations.

Tremendous progress has been made in the last seven years in the field of boiling heat transfer. Not only has recent research given a better understanding of the mechanism of boiling heat transfer, but also the burnout heat flux can now be calculated with considerably more confidence than was possible seven years ago. Moreover, continuous evaporation in a duct has been investigated extensively and the mechanism is today much better understood. In order to bring the new edition up to date, I have incorporated recent results in the area of boiling heat transfer in the chapter dealing with Heat Transfer in Two-Phase Systems.

Heat-transfer problems related to entry from outer space into an atmosphere have taken on increasing importance in the past few years. Although this is an area in which a complete book could hardly do justice to all of the research of the past decade, a textbook which ignores this subject completely would not be in keeping with the times. Inasmuch as this is only an introductory text, I have incorporated a basic outline of the problems involved in entry heat transfer. Paramount among these is ablation cooling, and I have added a short section on this subject in Chapter 12 as well as some of the more important tools available for engineering calculations of heating rates under extreme conditions.

Since I believe that one of the most important features of an undergraduate textbook is the problems available to the students as exercises, I have increased considerably the number of problems at the end of each chapter. In this task I have been greatly aided by many of my colleagues who have permitted me to use some problems which they have found helpful in teaching heat transfer. Among these colleagues I would like to mention particularly the late Dr. W. Kays of Massachusetts Institute of Technology, Dr. D. Edwards at the University of California at Los Angeles, Dr. F. Landis of New York University, and Dr. P. J. Berenson of the Air Research Company.

In preparing the second edition, I have made a concerted effort to eliminate all of the small errors and misprints which plagued the first edition. In this task I have had the good fortune of being able to draw upon letters and suggestions from a number of people. I would particularly like to thank Dr. C. G. Downing of Oregon State University and Professor H. A. Johnson of the University of California at Berkeley for their help in eliminating the small misprints and errors in the first edition, and Mrs. E. Patton for typing the manuscript.

In closing, I would like to thank my colleagues and students who have encouraged me to undertake the task of preparing a second edition to this book by their response to the first edition.

Preface to the first edition

This book is an amplification of lecture notes used by the author in teaching a one-semester course in heat transfer for engineering students at the senior and first-year graduate level. The material has been organized as a text, but it can also serve as a convenient reference for practicing engineers interested in fundamental techniques of analyzing heat-transfer problems. It is assumed that the reader has an elementary knowledge of thermodynamics, fluid dynamics, dc circuit theory, calculus, and differential equations, although a certain amount of review of these topics has been included.

The purpose of the book is to present a basic introduction to the field of engineering heat transfer. The presentation endeavors to convey to the reader a physical understanding of the processes by which heat is transferred and to provide him with the tools necessary to obtain quantitative solutions to engineering problems involving one or more of the basic modes of heat flow. An effort has been made to present information from recent and authoritative sources, but the amount of empirical data included is no more than considered necessary to give the reader a sufficiently broad background to use the available literature effectively. No attempt has been made to present rigorous mathematical solutions which, although available for numerous problems, require a more advanced mathematical background than most engineering students acquire in their undergraduate curriculum.

Although the field of heat transfer is generally subdivided into conduction, radiation, and convection, in most practical situations heat is transferred by several of these modes simultaneously. The author deemed it desirable, therefore, to introduce a general method for handling heat-transfer problems in the first chapter. This method makes use of the similarity between the equations governing the flow of heat and the flow of electric current to develop an analogy between electrical and thermal systems. With the aid of this analogy, heat-transfer problems can be reduced to thermal networks which can be analyzed with simple and familiar principles of dc circuit theory. The thermal-circuit method of analysis also makes it possible to consider realistic problems throughout the book, and the reader can thereby acquire a "feel" for the order of magnitude of thermal resistances under various conditions. This is often of help in spotting errors in the solution to a particular problem when the numerical answer seems unreasonable in the light of past experience.

Interpreting a heat-transfer problem as a thermal circuit is not difficult once one has acquired a physical understanding of the analogies involved. However, the elements of the circuit can be evaluated quantitatively only after a detailed knowledge of the basic modes of heat transfer has been acquired. To this end, each heat-flow mechanism and its analysis is considered separately in subsequent chapters.

In the chapters dealing with conduction, heat transfer, numerical, graphical, and electrical-model-analogue methods of solution have been freely employed. Although numerical and graphical methods do not lend themselves to parameterizations, they yield rapid approximations to individual problems and are therefore widely used in industry. They also appeal to the students because the temperature field can be visualized.

A modified analogue-network method has been applied to the solution of radiant-heat transfer problems. This approach circumvents the need for matrix algebra in the solution of more complex problems and integrates radiant heat transfer into the overall scheme of analysis.

In the treatment of convection, the interrelationship between the flow of heat and the flow of the fluid has been emphasized. Prandtl's concept of the boundary layer has been presented and one problem, namely the flow over a flat plate, has been treated in detail. This problem is not only the simplest case to analyze, but is perhaps also the most important because many practical situations approximate flow over a plate, while others are described by equations which can be reduced by appropriate transformation to the flat-plate boundary-layer equations.

In as complex a field as convective heat transfer, the practicing engineer is forced to rely considerably on available experimental results. The derivation and interpretation of dimensionless parameters used to correlate empirical data can therefore not be overlooked. To acquaint the reader with the use of dimensional analysis in correlating heat-transfer data, the basic aspects of the Buckingham pi-theorem have been presented in the introductory chapter on convection and several pertinent examples have been worked out. The European method of deducing similarity parameters from differential equations is illustrated in the chapter on free convection.

The evaluation of convective-heat-transfer coefficients from empirical equations is taken up in separate chapters for free convection, forced convection inside conduits, and flow over tubes and other bodies. Heat transfer of boiling and condensing fluids is treated in the chapter on heat transfer with phase change, in which melting and freezing processes are also considered. The amount of empirical data presented has been limited to configurations of widest practical interest, but the effect of variable fluid properties on heat transfer and friction coefficients has been emphasized.

In the chapter dealing with the design and thermal analysis of heat exchangers, the concept of effectiveness has been emphasized. Basic material on the mean-temperature-difference method of analysis has been included because this technique is widely used in industry.

In Professor Andersen's chapter on mass transfer, the analogy between the transfer of mass, heat, and momentum has been emphasized. The material has been organized in such a manner that it can be presented immediately after, or simultaneously with, heat transfer by convection.

The Appendix contains a survey of thermal properties. The property tables are designed to supplement the discussion and to provide a handy source of data for solving the problems at the end of each chapter. Wherever answers are given, they have been obtained with the physical properties listed in the Appendix.

One of the most difficult and probably the most controversial task in preparing an introductory text for a field as broad as heat transfer is the selection of the material. The author has tried to avoid overspecialization as much as possible but has illustrated the basic principles by applying them to the solution of specific problems dealing with nuclear reactors, temperature-measuring techniques, solar radiation, high-speed flow, rocket-motor cooling systems, compact heat exchangers, and many other devices of current interest. Highly specialized areas such as regenerator theories, film cooling, and heat transfer at extreme temperatures and pressures have not been treated because the author believes that these problems do not fall within the scope of an introductory text. Wherever practical, however, a sufficiently complete bibliography has been included to enable the reader to pursue his special interest more fully.

This text makes no claim to originality. The author has merely attempted to collect pertinent material and present it in a teachable form. The material itself has been selected from the literature, and wherever possible the author has given credit to the original sources. In his approach the author has been greatly influenced by the philosophy of Dean L. M. K. Boelter and his coworkers, especially the late Mr. Earl Morrin.

The author acknowledges with pleasure the help and encouragement given him by his colleagues and students. Several members of the staff of the Mechanical Engineering Department at the University of California at Berkeley, particularly Professor H. A. Johnson, Dr. R. Drake (now at Princeton University), Professor R. V. Dunkel, and Dr. R. A. Seban, have offered suggestions and contributed some of the student exercises. Professor J. T. Anderson, Michigan State University, Professor W. M. Kays, Stanford University, Professor J. F. Lee, North Carolina State College,

Dr. P. J. Schneider, University of Minnesota, and Dr. L. B. Andersen, Lehigh University, have read the manuscript in its entirety and have contributed many valuable suggestions.

Dr. O. P. Bergelin, University of Delaware, Dr. A. J. Chabai, Lehigh University, and Dr. W. M. Rohsenow, Massachusetts Institute of Technology, have given technical advice. Mr. Kun Min checked the illustrative problems and prepared some of the drawings. Miss Joyce Broadhead and Mrs. Helen Farrell typed portions of the manuscript. Particular thanks are due to Marion Kreith for helping in many tangible and many more intangible ways.

1
Introduction

1-1. The relation of heat transfer to thermodynamics

Whenever a temperature gradient exists within a system, or when two systems at different temperatures are brought into contact, energy is transferred. The process by which the energy transport takes place is known as *heat transfer*. The thing in transit, called heat, cannot be measured or observed directly, but the effects it produces are amenable to observation and measurement. The flow of heat, as the performance of work, is a process by which the internal energy of a system is changed.

The branch of science which deals with the relation between heat and other forms of energy is called *thermodynamics*. Its principles, like all laws of nature, are based on observations and have been generalized into laws which are believed to hold for all processes occurring in nature, because no exceptions have ever been detected. The first of these principles, the first law of thermodynamics, states that energy can be neither created nor destroyed but only changed from one form to another. It governs all energy transformations quantitatively but places no restrictions on the direction of the transformation. It is known, however, from experience that no process is possible whose sole result is the net transfer of heat from a region of lower temperature to a region of higher temperature. This statement of experimental truth is known as the second law of thermodynamics.

All heat-transfer processes involve the transfer and conversion of energy. They must therefore obey the first as well as the second law of thermodynamics. At a first glance one might therefore be tempted to assume that the principles of heat transfer can be derived from the basic laws of thermodynamics. This, however, would be an erroneous conclusion because classical thermodynamics is restricted primarily to the study of equilibrium states, including mechanical and chemical as well as thermal equilibriums, and is therefore, by itself, of little help in determining quantitatively the transformations which occur from a lack of equilibrium in engineering processes. Since heat flow is the result of temperature nonequilibrium, its quantitative treatment must be based on other branches of science. The same reasoning applies to other types of transport processes such as mass transfer and diffusion.

Limitations of classical thermodynamics. Classical thermodynamics deals with the states of systems from a macroscopic view and makes no hypotheses about the structure of matter. To perform a thermodynamic analysis it is necessary to describe the state of a system in terms of gross characteristics, such as pressure, volume, and temperature, which can be measured directly and involve no special assumptions regarding the structure of matter. These variables or thermodynamic properties are of significance for the system as a whole only when they are uniform throughout it, i.e., when the system is in equilibrium. Thus, classical thermodynamics is not concerned with the details of a process but rather with equilibrium states and the relations among them. The processes employed in a thermodynamic analysis are idealized processes, devised merely to give information concerning equilibrium states.

From a thermodynamic viewpoint, the amount of heat transferred during a process simply equals the difference between the energy change of the system and the work done. It is evident that this type of analysis considers neither the mechanism of heat flow nor the time required to transfer the heat. It simply prescribes how much heat to supply to, or reject from, a system during a process between specified end states without taking care of whether or how this could be accomplished. The reason for this lack of information obtainable from a thermodynamic analysis is the absence of time as a variable. The question of how long it would take to transfer a specified amount of heat, although it is of great practical importance, does not usually enter into the thermodynamic analysis.

Engineering heat transfer. From an engineering viewpoint, the determination of the *rate of heat transfer at a specified temperature difference* is the key problem. To estimate the cost, the feasibility, and the size of equipment necessary to transfer a specified amount of heat in a given time, a detailed heat-transfer analysis must be made. The dimensions of boilers, heaters, refrigerators, and heat exchangers depend not only on the amount of heat to be transmitted, but rather on the rate at which the heat is to be transferred under given conditions. The successful operation of equipment components such as, for example, turbine blades or the walls of combustion chambers depends on the possibility of cooling certain metal parts by removing heat continuously at a rapid rate from a surface. Also, in the design of electric machines, transformers, and bearings, a heat-transfer analysis must be made to avoid conditions which will cause overheating and damage the equipment. These varied examples show that in almost every branch of engineering, heat-transfer problems are encountered which are not capable of solution by thermodynamic reasoning alone, but require an analysis based on the science of heat transfer.

In heat transfer, as in other branches of engineering, the successful solution of a problem requires assumptions and idealizations. It is almost impossible to describe physical phenomena exactly, and in order to express a problem in the form of an equation that can be solved it is necessary to make some approximations. In electric-circuit calculations, for example, it is usually assumed that the values of the resistances, capacitances, and inductances are independent of the current flowing through them. This assumption simplifies the analysis but may in certain cases limit the accuracy of the results severely.

It is important to keep the assumptions, idealizations, and approximations made in the course of an analysis in mind when the final results are interpreted. Sometimes insufficient information on physical properties makes it necessary to use engineering approximations to solve a problem. For example, in the design of machine parts for operation at elevated temperatures it may be necessary to estimate the proportional limit or the fatigue strength of the material from low-temperature data. To assure satisfactory operation of the part, the designer should apply a factor of safety to the results he obtains from his analysis. Similar approximations are also necessary in heat-transfer problems. Physical properties, such as the thermal conductivity or the viscosity, change with temperature, but if suitable average values are selected, the calculations can be considerably simplified without introducing an appreciable error in the final result. When heat is transferred from a fluid to a wall, as for example in a boiler, a scale forms under continued operation and reduces the rate of heat flow. To assure satisfactory operation over a long period of time, a factor of safety must be applied to provide for this contingency.

When it becomes necessary to make an assumption or approximation in the solution of a problem, the engineer must rely on his ingenuity and past experience. There are no simple guides to new and unexplored problems, and an assumption valid for one problem may be misleading in another. Experience has shown, however, that the first and foremost requirement for making sound engineering assumptions or approximations is a complete and thorough physical understanding of the problem at hand. In the field of heat transfer, this requires not only a familiarity with the laws and physical mechanisms of heat flow, but also with those of fluid mechanics, physics, and mathematics.

1-1. THE RELATION OF HEAT TRANSFER TO THERMODYNAMICS

1-2. Modes of heat flow

Heat transfer can be defined as the transmission of energy from one region to another as a result of a temperature difference between them. Since differences in temperatures exist all over the universe, the phenomena of heat flow are as universal as those associated with gravitational attractions. Unlike gravity, however, heat flow is governed not by a unique relationship, but rather by a combination of various independent laws of physics.

The literature of heat transfer generally recognizes three distinct modes of heat transmission: *conduction*, *radiation*, and *convection*. Strictly speaking, only conduction and radiation should be classified as heat-transfer processes, because only these two mechanisms depend for their operation on the mere existence of a temperature difference. The last of the three, convection, does not strictly comply with the definition of heat transfer because it depends for its operation on mechanical mass transport also. But since convection also accomplishes transmission of energy from regions of higher temperature to regions of lower temperature, the term "heat transfer by convection" has become generally accepted.

Each of these modes of heat transfer will be described and analyzed separately. Yet it should be emphasized that, in most situations occurring in nature, heat flows not by one, but by several of these mechanisms acting simultaneously. It is particularly important in engineering to be aware of the confluence of the various modes of heat transfer because, in practice, when one mechanism dominates quantitatively, useful approximate solutions are obtained by neglecting all but the dominant mechanism. However, a change of external conditions will often require that one or both of the previously neglected mechanisms be taken into account.

Conduction. Conduction is a process by which heat flows from a region of higher temperature to a region of lower temperature within a medium (solid, liquid, or gaseous) or between different mediums in direct physical contact. In conduction heat flow, the energy is transmitted by direct molecular communication without appreciable displacement of the molecules. According to the kinetic theory, the temperature of an element of matter is proportional to the mean kinetic energy of its constituent molecules. The energy possessed by an element of matter by virtue of the velocity and relative position of the molecules is called *internal energy*. Thus, the more rapidly the molecules are moving, the greater will be the temperature as well as the internal energy of an element of matter. When molecules in one region acquire a mean kinetic energy greater than that of molecules in an adjacent region, as manifested by a difference in temperature, the molecules possessing the greater energy will transmit part of their energy to the molecules in the lower-temperature region. The transfer of energy could take place by elastic impact (e.g., in fluids) or by diffusion of faster-moving electrons from the higher- to the lower-temperature regions (e.g., in metals). Irrespective of the exact mechanism, which is by no means fully understood, the observable effect of heat conduction is an equalization of temperature. However, if differences in temperature are maintained by addition and removal of heat at different points, a continuous flow of heat from the hotter to the cooler region will be established.

Conduction is the only mechanism by which heat can flow in opaque solids. Conduction is also important in fluids, but in nonsolid mediums it is usually combined with convection, and in some cases with radiation also.

Radiation. Radiation is a process by which heat flows from a high-temperature body to a body at a lower temperature when the bodies are separated in space, even when a vacuum exists between them. The term "radiation" is generally applied to all kinds of electromagnetic-wave phenomena, but in heat transfer only those phenomena which are the result of temperature and can transport energy through a transparent medium or through space are of interest. The energy transmitted in this manner is termed *radiant heat.*

All bodies emit radiant heat continuously. The intensity of the emissions depends on the temperature and the nature of the surface. Radiant energy travels at the speed of light (3×10^8 m/s) and resembles phenomenologically the radiation of light. In fact, according to the electromagnetic theory, light and thermal radiation differ only in their respective wavelengths.

Radiant heat is emitted by a body in the form of finite batches, or *quanta*, of energy. The motion of radiant heat in space is similar to the propagation of light and can be described by the wave theory. When the radiation waves encounter some other object, their energy is absorbed near its surface. Heat transfer by radiation becomes increasingly important as the temperature of an object increases. In engineering problems involving temperatures approximating those of the atmosphere, radiant heating may often be neglected.

Convection. Convection is a process of energy transport by the combined action of heat conduction, energy storage, and mixing motion. Convection is most important as the mechanism of energy transfer between a solid surface and a liquid or a gas.

The transfer of energy by convection from a surface whose temperature is above that of a surrounding fluid takes place in several steps. First, heat will flow by conduction from the surface to adjacent particles of fluid. The energy thus transferred will serve to increase the temperature and the internal energy of these fluid particles. Then the fluid particles will move to a region of lower temperature in the fluid where they will mix with, and transfer a part of their energy to, other fluid particles. The flow in this case is of fluid as well as energy. The energy is actually stored in the fluid particles and is carried as a result of their mass motion. This mechanism does not depend for its operation merely on a temperature difference and therefore does not strictly conform to the definition of heat transfer. The net effect, however, is a transport of energy, and since it occurs in the direction of a temperature gradient, is also classified as a mode of heat transfer and is referred to as *heat flow by convection.*

Convection heat transfer is classified according to the mode of motivating flow into *free convection* and *forced convection.* When the mixing motion takes place merely as a result of density differences caused by temperature gradients, we speak of *natural*, or free, convection. When the mixing motion is induced by some external agency, such as a pump or a blower, the process is called forced convection.

The effectiveness of heat transfer by convection depends largely upon the mixing motion of the fluid. Consequently a study of convective heat transfer is predicated on a knowledge of the characteristics of the fluid flow.

In the solution of heat-transfer problems, it is necessary not only to recognize the modes of heat transfer which play a role but also to determine whether a process is *steady* or *unsteady*. When the rate of heat flow in a system does not vary with time, i.e., when it is constant, the temperature at any point does not change and steady-state conditions prevail. Under steady-state conditions, the rate of heat influx at any point of the system must be exactly equal to the rate of heat efflux, and no change in internal energy can take place. The majority of engineering heat-transfer problems are concerned with steady-state systems. Typical examples are the flow of heat from the products of combustion to water in the tubes of a boiler, the cooling of an electric light bulb by the surrounding atmosphere, or the heat transfer from the hot to the cold fluid in a heat exchanger.

The heat flow in a system is *transient*, or unsteady, when the temperatures at various points in the system change with time. Since a change in temperature indicates a change of internal energy, we conclude that energy storage is part and parcel of unsteady heat flow. Unsteady-heat-flow problems are more complex than are those of steady state and can often be solved only by approximate methods. Unsteady-heat-flow problems are encountered during the warm-up periods of furnaces, boilers, and turbines or in the heat treatment and stress-relieving of metal castings.

A special case of unsteady heat flow occurs when a system is subjected to cyclic variations in the temperature of its environment. In such problems the temperature at a particular point in the system returns periodically to the same value; also, the rate of heat flow and the rate of energy storage undergo periodic variations. Problems of this type come under the classification of *periodic* or *quasi-steady-state heat transfer.* Typical examples are the variation of temperature of a building during any twenty-four-hour period or the heat flow through the cylinder walls of a reciprocating engine when the temperature of the gases within the cylinder changes periodically.

1-3. Basic laws of heat transfer

Any meaningful engineering analysis demands a quantitative answer. To perform such an analysis of heat-transfer problems we must investigate the physical laws and relations which govern the various mechanisms of heat flow. In this section we shall make a preliminary survey of the basic equations governing each of the three modes of heat transfer. Later we shall show how to combine these relations when several of the heat-flow mechanisms are operating concurrently, either in series or in parallel. Our preliminary aim is to obtain a broad perspective of the field without becoming involved in the details of any particular mechanism. We shall, therefore, consider only simple cases and postpone more complex problems for later chapters.

Conduction. The basic relation for heat transfer by conduction was proposed by the French scientist, J. B. J. Fourier, in 1822. It states that q_k, the *rate of heat flow by conduction* in a material, is equal to the product of the following three quantities:

1. k, the thermal conductivity of the material.
2. A, the area of the section through which heat flows by conduction, to be measured perpendicularly to the direction of heat flow.
3. dT/dx, the temperature gradient at the section, i.e., the rate of change of temperature T with respect to distance in the direction of heat flow x.

To write the heat conduction equation in mathematical form, we must adopt a sign convention. We specify that the direction of increasing distance x is to be the direction of positive heat flow. Then, since according to the second law of thermodynamics heat will automatically flow from points of higher temperature to points of lower temperature, heat flow will be positive when the temperature gradient is negative (Fig. 1-1). Accordingly, the elementary equation for one-dimensional conduction in the steady state is written

$$q_k = -kA \frac{dT}{dx} \tag{1-1}$$

For dimensional consistency of Eq. 1-1, the rate of heat flow q_k is

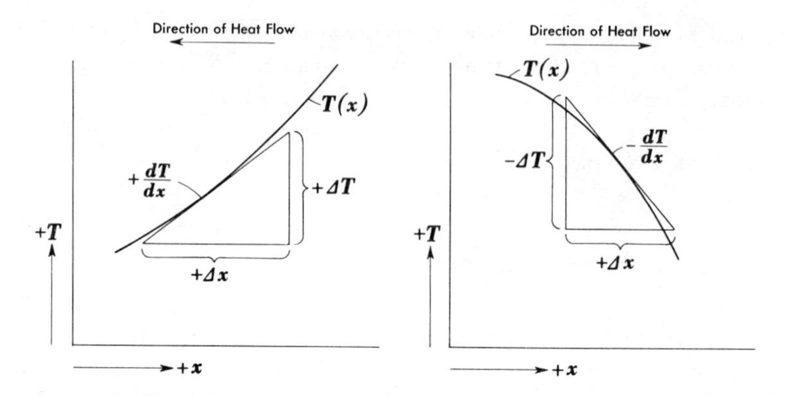

Fig. 1-1. Sketch illustrating sign convention for conduction heat flow.

expressed in Btu/hr, the area A in sq ft, and the temperature gradient dT/dx in F/ft. The thermal conductivity k is a property of the material and indicates the quantity of heat that will flow across a unit area if the temperature gradient is unity. The units for k used in this text are British thermal units per hour per square foot per unit temperature gradient in degrees Fahrenheit per foot, i.e.,

$$\frac{\text{Btu/hr sq ft}}{\text{F/ft}} \quad \text{or} \quad \frac{\text{Btu}}{\text{hr ft F}}$$ [1]

In the SI system (Système International d'Unités), the units of thermal conductivity are watts per square meter per unit temperature gradient in degree centigrade (or Kelvin) per meter, i.e.,

$$\frac{\text{watt/m}^2}{\text{K/m}} = \frac{\text{watt}}{\text{m K}}$$

and

$$1 \text{ w/m K} = 0.578 \text{ Btu/hr ft F}$$

Thermal conductivities of engineering materials at atmospheric pressure range from about 4×10^{-3} for gases through about 1×10^{-1} for liquids to 2.4×10^{2} Btu/hr ft F for copper. Orders of magnitudes of the thermal conductivity of various classes of material are shown in Table 1-1 and Fig. 1-2. Materials having a high thermal conductivity are called *conductors*, while materials of low thermal conductivity are referred to as *insulators*.

[1] Throughout the remainder of this book units of physical quantities will be given in conventionally abbreviated form without expanding them.

Table 1-1. Order of magnitude of thermal conductivity k.

Material	Btu/hr ft F	w/m K
Gases at atmospheric pressure	0.004 – 0.10	0.0069 – 0.17
Insulating materials	0.02 – 0.12	0.034 – 0.21
Nonmetallic liquids	0.05 – 0.40	0.086 – 0.69
Nonmetallic solids (brick, stone, cement)	0.02 – 1.5	0.034 – 2.6
Liquid metals	5.0 – 45	8.6 – 76
Alloys	8.0 – 70	14 – 120
Pure metals	30 – 240	52 – 410

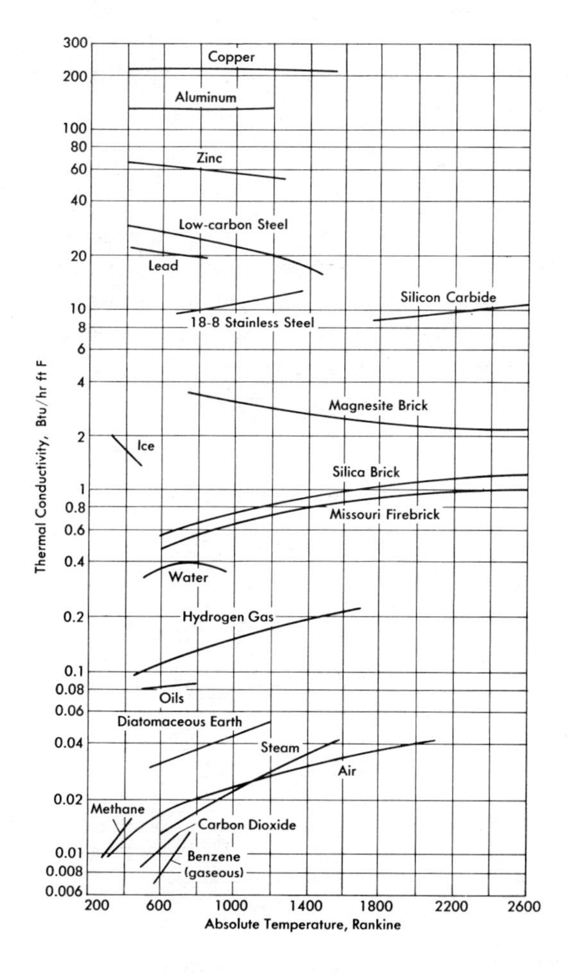

Fig. 1-2. Variation of thermal conductivity of solids, liquids, and gases with temperature.

In general, the thermal conductivity varies with temperature, but in many engineering problems the variation is sufficiently small to be neglected.

For the simple case of steady-state heat flow through a plane wall, the temperature gradient and the heat flow do not vary with time and the cross-sectional area along the heat-flow path is uniform. The variables in Eq. 1-1 can be separated and the resulting equation is

$$\frac{q_k}{A} \int_0^L dx = - \int_{T_{\text{hot}}}^{T_{\text{cold}}} k dT$$

The limits of integration can be checked by inspection of Fig. 1-3, where the temperature at the left-hand face ($x = 0$) is uniform at T_{hot} and the temperature at the right-hand face ($x = L$) is uniform at T_{cold}.

If k is independent of T, we obtain, after integration, the following expression for the rate of heat conduction through the wall

$$q_k = \frac{Ak}{L} (T_{\text{hot}} - T_{\text{cold}}) = \frac{\Delta T}{L/Ak} \tag{1-2}$$

In this equation ΔT, the temperature difference between the higher temperature T_{hot} and the lower temperature T_{cold}, is the driving potential which causes the flow of heat. L/Ak is equivalent to a *thermal resistance* R_k, which the wall offers to the flow of heat by conduction and we have

$$R_k = \frac{L}{Ak} \tag{1-3}$$

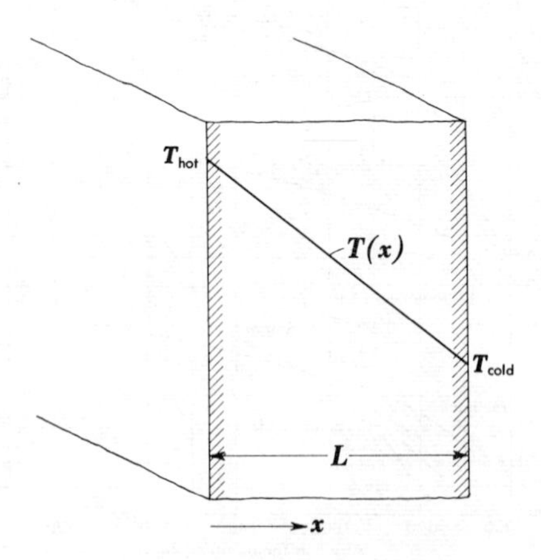

Fig. 1-3. Temperature distribution for steady-state conduction through a plane wall.

The reciprocal of the thermal resistance is referred to as the *thermal conductance*

$$K_k = \frac{Ak}{L} \tag{1-4}$$

and k/L, the thermal conductance per unit area, is called the *unit thermal conductance for conduction heat flow.* The subscript k indicates that the transfer mechanism is by conduction. The thermal conductance has the units of Btu/hr F temperature difference (watts/K in the SI system) and the thermal resistance has the units hr F/Btu. The concepts of resistance and conductance are helpful in the analysis of thermal systems where several modes of heat transfer occur simultaneously.

Radiation. The quantity of energy leaving a surface as radiant heat depends upon the *absolute temperature* and the nature of the surface. A perfect *radiator* or *blackbody*[2] emits radiant energy from its surface at a rate q_r given by

$$q_r = \sigma A_1 T_1^4 \text{ Btu/hr} \tag{1-5}$$

The heat flow rate q_r will be Btu/hr if A_1 is the surface area in sq ft, T_1 is the surface temperature in degrees Rankine (R), and σ is a dimensional constant with a value of 0.1714×10^{-8} Btu/hr sq ft R^4. In the SI units the heat flow rate q_r will be in watts if the surface area A_1 is in m^2, the absolute temperature in degrees Kelvin, and σ is 5.67×10^{-8} watts/m^2 K^4. The quantity σ is named the *Stefan-Boltzmann constant* after two Austrian scientists, J. Stefan, who in 1879 found Eq. 1-5 experimentally, and L. Boltzmann, who in 1884 derived it theoretically.

An inspection of Eq. 1-5 shows that any blackbody surface above a temperature of absolute zero radiates heat at a rate proportional to the *fourth power* of the *absolute temperature.* While the rate of emission is independent of the conditions of the surroundings, a *net* transfer of radiant heat requires a difference in the surface temperature of any two bodies between which the exchange is taking place. If the blackbody radiates to an enclosure which completely surrounds it and whose surface is also *black*, i.e., absorbs all the radiant energy incident upon it, the net rate of radiant heat transfer is given by

$$q_r = \sigma A_1 (T_1^4 - T_2^4) \tag{1-6}$$

where T_2 is the surface temperature of the enclosure in degrees Fahrenheit absolute.

---

[2] A detailed discussion of the meaning of these terms is presented in Chapter 5.

Real bodies do not meet the specifications of an ideal radiator but emit radiation at a lower rate than blackbodies. If they emit, at a temperature equal to that of a blackbody, a constant fraction of blackbody emission at each wavelength, they are called *gray bodies*. The net rate of heat transfer from a gray body at a temperature T_1 to a black surrounding body at T_2 is

$$q_r = \sigma A_1 \epsilon_1 (T_1^4 - T_2^4) \tag{1-7}$$

where ϵ_1 is the *emittance* of the gray surface and is equal to the ratio of emission from the gray surface to the emission from a perfect radiator at the same temperature.

If neither of two bodies is a perfect radiator and if the two bodies possess a given geometrical relationship to each other, the net heat transfer by radiation between them is given by

$$q_r = \sigma A_1 \mathfrak{F}_{1\text{-}2} (T_1^4 - T_2^4) \tag{1-8}$$

where $\mathfrak{F}_{1\text{-}2}$ is a modulus which modifies the equation for perfect radiators to account for the emittances and relative geometries of the actual bodies.

In many engineering problems, radiation is combined with other modes of heat transfer. The solution of such problems can often be simplified by using a thermal conductance K_r, or a thermal resistance R_r, for radiation. The definition of K_r is similar to that of K_k, the thermal conductance for conduction. If the heat transfer by radiation is written

$$q_r = K_r (T_1 - T_2') \tag{1-9}$$

the conductance, by comparison with Eq. 1-8, is given by

$$K_r = \frac{\sigma A_1 \mathfrak{F}_{1\text{-}2} (T_1^4 - T_2^4)}{T_1 - T_2'} \quad \text{Btu/hr F} \tag{1-10}$$

and the unit thermal conductance for radiation $\bar{h}_r$ by

$$\bar{h}_r = \frac{K_r}{A_1} = \frac{\sigma \mathfrak{F}_{1\text{-}2} (T_1^4 - T_2^4)}{T_1 - T_2'} \quad \text{Btu/hr sq ft F} \tag{1-11}$$

where T_2' is any convenient reference temperature whose choice is often dictated by the convection equation, which will be discussed next. Similarly, the thermal resistance for radiation is

$$R_r = \frac{T_1 - T_2'}{\sigma A_1 \mathfrak{F}_{1\text{-}2} (T_1^4 - T_2^4)} \quad \text{hr sq ft F/Btu} \tag{1-12}$$

Convection. The rate of heat transfer by convection between a surface and a fluid may be computed by the relation

$$q_c = \bar{h}_c A \Delta T \tag{1-13}$$

where q_c = rate of heat transfer by convection, Btu/hr;
A = heat transfer area, sq ft;
ΔT = difference between the surface temperature T_s and a temperature of the fluid T_∞ at some specified location (usually far away from the surface), F;
$\bar{h}_c$ = average unit thermal convective conductance (often called the surface coefficient of heat transfer or the convective heat-transfer coefficient), Btu/hr sq ft F.

In SI units the rate of heat transfer by convection will be in joules/second or watts if A is in m^2, ΔT in K, and $\bar{h}_c$ in watts/m^2 K; and 1 watt/m^2 K = 0.176 Btu/hr ft^2 F.

The relation expressed by Eq. 1-13 was originally proposed by the British scientist, Isaac Newton, in 1701. Engineers have used this equation for many years, even though it is a definition of $\bar{h}_c$ rather than a phenomenological law of convection. The evaluation of the convective heat-transfer coefficient is difficult because convection is a very complex phenomenon. The methods and techniques available for a quantitative evaluation of $\bar{h}_c$ will be presented in later chapters. At this point it is sufficient to note that the numerical value of $\bar{h}_c$ in a system depends on the geometry of the surface and the velocity, as well as on the physical properties of the fluid and often even on the temperature difference ΔT. In view of the fact that these quantities are not necessarily constant over a surface, the convective heat-transfer coefficient may also vary from point to point. For this reason we must distinguish between a *local* and an *average* convective heat-transfer coefficient. The local coefficient h_c is defined by

$$dq_c = h_c dA (T_s - T_\infty) \tag{1-14}$$

while the average coefficient $\bar{h}_c$ can be defined in terms of the local value by

$$\bar{h}_c = \frac{1}{A} \iint_A h_c dA \tag{1-15}$$

For most engineering applications, we shall be interested in average values. For general orientation, typical values of the order of magnitude of average convective heat-transfer coefficients encountered in engineering practice are presented in Table 1-2.

Using Eq. 1-13, we can define the thermal conductance K_c for con-

Table 1-2. Order of magnitude of convective heat transfer coefficients $\bar{h}_c$.

	Btu/hr sq ft F	w/m² K
Air, free convection	1 – 5	6 – 30
Superheated steam or air, forced convection	5 – 50	30 – 300
Oil, forced convection	10 – 300	60 – 1,800
Water, forced convection	50 – 2,000	300 – 6,000
Water, boiling	500 – 10,000	3,000 – 60,000
Steam, condensing	1,000 – 20,000	6,000 – 120,000

vective heat transfer as

$$K_c = \bar{h}_c A \tag{1-16}$$

and the thermal resistance to convective heat transfer R_c, which is equal to the reciprocal of the conductance, as

$$R_c = \frac{1}{\bar{h}_c A} \tag{1-17}$$

1-4. Combined heat-transfer mechanisms

In the preceding section the three mechanisms of heat transfer have been considered separately. In practice, however, heat is usually transferred in steps through a number of different series-connected sections, the transfer frequently occurring by two mechanisms in parallel for a given section in the system. The transfer of heat from the products of combustion in the combustion chamber of a rocket motor through a thin wall to a coolant flowing in an annulus over the outside of the wall will illustrate such a case (Fig. 1-4).

Products of combustion contain gases, such as CO, CO_2, and H_2O which emit and absorb radiation. In the first section of this system, heat is therefore transferred from the hot gas to the inner surface of the wall of the rocket motor by the mechanisms of convection and radiation acting in parallel. The total rate of heat flow q to the surface of the wall some distance from the nozzle is

$$\begin{aligned} q &= q_c + q_r \\ &= \bar{h}_c A(T_g - T_{sg}) + \bar{h}_r A(T_g - T_{sg}) \end{aligned} \tag{1-18}$$

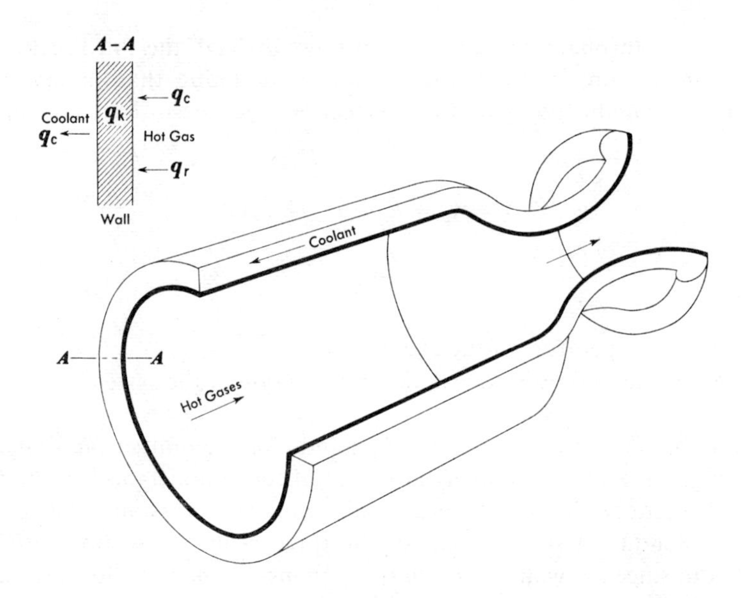

Fig. 1-4. Heat transfer in a rocket motor.

or

$$
\begin{aligned}
q &= (\bar{h}_c A + \bar{h}_r A)(T_g - T_{sg}) \\
&= (K_c + K_r)(T_g - T_{sg}) \\
&= \frac{T_g - T_{sg}}{R_1}
\end{aligned}
$$

where T_g = temperature of hot gas;
T_{sg} = temperature at inner surface of wall;
R_1 = combined or effective thermal resistance of the first section, $R_1 = 1/(\bar{h}_r + \bar{h}_c)A$.

In the steady state, heat is conducted through the shell, the second section of the system, at the same rate as to the surface and

$$
\begin{aligned}
q = q_k &= \frac{kA}{L}(T_{sg} - T_{sc}) \qquad (1\text{-}19) \\
&= K_k(T_{sg} - T_{sc}) \\
&= \frac{T_{sg} - T_{sc}}{R_2}
\end{aligned}
$$

where T_{sc} = surface temperature at wall on coolant side;
R_2 = thermal resistance of second section.

After passing through the wall, the heat flows through the third section of the system by convection to the coolant. Assuming that radiant heat transfer is negligible compared to convection, the rate of heat flow in the last step is

$$q = q_c = \bar{h}_c A (T_{sc} - T_c) \tag{1-20}$$
$$= \frac{T_{sc} - T_c}{R_3}$$

where T_c = temperature of the coolant;
R_3 = thermal resistance in the third section of the system.

It should be noted that the symbol $\bar{h}_c$ stands for the unit-surface conductance in general, but the numerical values of the conductances in the first and third sections of the system depend on many factors and will in general be different. Also the areas of the three heat-flow sections are not equal. But since the wall is very thin, the change in the heat-flow area is so small that it can be neglected in this system.

In practice, often only the temperatures of the hot gas and the coolant are known. The intermediate temperatures can be eliminated by algebraic addition of Eqs. 1-18, 1-19, and 1-20, or

$$q = \frac{T_g - T_c}{R_1 + R_2 + R_3} = \frac{\Delta T_{\text{total}}}{R_1 + R_2 + R_3} \tag{1-21}$$

where the thermal resistances of the three series-connected sections or heat-flow steps in the system are defined in Eqs. 1-18, 1-19, and 1-20.

In Eq. 1-21 the rate of heat flow is expressed only in terms of an overall temperature potential and the heat-transfer characteristics of individual sections in the heat-flow path. From these relations it is possible to evaluate quantitatively the importance of each individual thermal resistance in the path. An inspection of the order of magnitudes of the individual terms in the denominator often indicates means of simplifying a problem. When one or the other term dominates quantitatively, it is sometimes permissible to neglect the rest. As we gain facility in the techniques of determining individual thermal resistances and conductances, there will be numerous occasions where such approximations will be illustrated. There are, however, certain types of problems, notably in the design of heat exchangers, where it is convenient to simplify the writing of Eq. 1-21 by combining the individual resistances or conductances of the thermal system into one quantity, called the *overall unit conductance, the overall transmittance,* or the *overall coefficient of heat transfer, U*. The use of an overall coefficient is a convenience in notation, and it is important not to lose sight of the significance of the individual factors which determine the numerical value of U.

Writing Eq. 1-21 in terms of an overall coefficient gives

$$q = UA\Delta T_{\text{total}} \tag{1-22}$$

where

$$UA = \frac{1}{R_1 + R_2 + R_3} \tag{1-23}$$

The overall coefficient U may be based on any chosen area. To avoid misunderstandings, the area basis of an overall coefficient should therefore always be stated.

The numerical evaluation of the various resistances or conductances of a thermal system is generally the most difficult part of any engineering heat-transfer problem. In fact, the material to which most of the following chapters is devoted deals with the determination of individual resistances and conductances from external conditions which can either be measured or specified. Once the individual resistances or conductances have been evaluated, the overall coefficient of heat transfer can be obtained and, for steady-state conditions, the rate of heat transfer can be determined for a specified temperature difference. For heat flow along a path consisting of n thermal sections in series, the overall conductance UA is equal to the reciprocal of the sum of the resistances of the individual sections, or

$$UA = \frac{1}{R_1 + R_2 + \cdots + R_n} \tag{1-24}$$

where each resistance is the reciprocal of the sum of the conductances for that section.

The overall heat-transfer coefficient will be found useful primarily in problems involving thermal systems consisting of several series-connected sections. The analysis of heat flow at boundaries of complicated geometry and in unsteady-state conduction problems can be simplified by using a combined unit-thermal-surface conductance $\bar{h}$. The combined unit-thermal-surface conductance, or *unit-surface conductance* for short, combines the effects of heat flow by convection and radiation between a surface and a fluid and is defined by

$$\bar{h} = \bar{h}_c + \bar{h}_r \tag{1-25}$$

The unit-surface conductance specifies the average total rate of heat flow per unit area between a surface and a fluid per degree temperature difference. Its units are Btu/hr sq ft F (or watts/m^2 K in SI units).

1-5. Analogy between heat flow and electrical flow

Two systems are said to be analogous when both obey similar equations and also have similar boundary conditions. This means that the equation describing the behavior of one system can be transformed into the equation for the other system by simply changing the symbols of the variables. For example, the flow of heat through a thermal resistance is analogous to the flow of direct current through an electrical resistance because both types of flow obey similar equations. If we replace, in the heat-flow equation

$$q = \frac{\Delta T}{R} \tag{1-26}$$

the symbol for the temperature potential ΔT by the symbol for the electric potential, i.e., the voltage difference, ΔE and the symbol for the thermal resistance R by the symbol for the electrical resistance R_e, we obtain the equation for i, the flow rate of electricity, i.e., the current

$$i = \frac{\Delta E}{R_e} \tag{1-27}$$

Having once established the basic analogy, we can apply certain concepts from direct current theory to heat-transfer problems. For instance, an electric circuit has a corresponding thermal circuit, and vice versa. In the problem of the preceding section, the heat flow from the hot gases to the coolant can be visualized as analogous to the flow of current in a simple direct-current circuit. In the equation for the current flow, analogous to Eq. 1-21 for the heat flow, we find that

$$i = \frac{\Delta E}{R_{e1} + R_{e2} + R_{e3}} \tag{1-28}$$

where R_{e1} is the effective resistance of two parallel resistances. One of them is analogous to the thermal resistance encountered by the convection heat flow, the other to the thermal resistance met by the heat flow from the gas to the wall by radiation. The thermal circuit and the electrical circuit for this problem are shown in Fig. 1-5.

EXAMPLE 1-1. In the design of a heat exchanger for aircraft application, the maximum wall temperature is not to exceed 1000 F. For the conditions tabulated below, determine the maximum permissible thermal resistance per square foot area of the metal wall between hot gas on the one side and cold gas on the other.

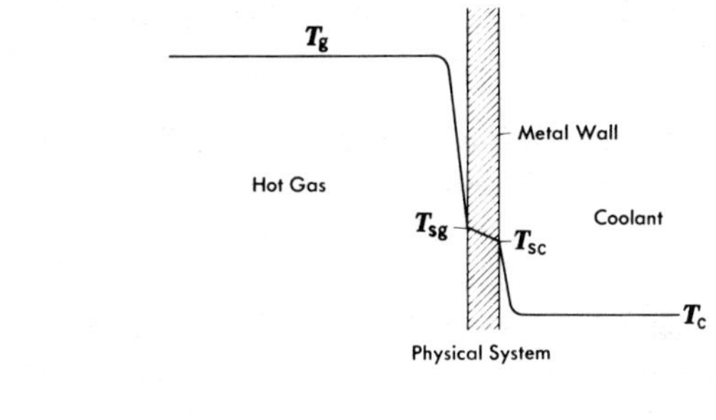

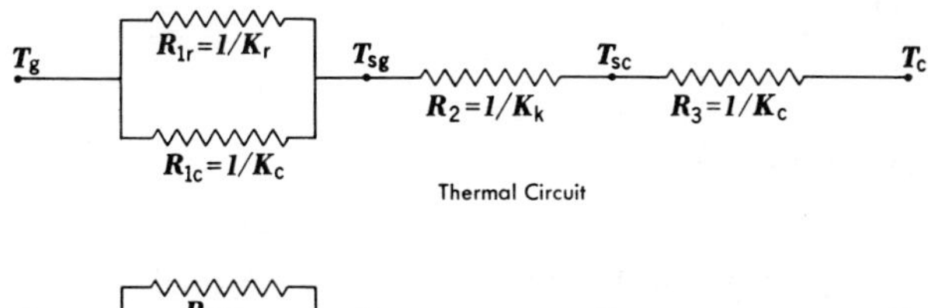

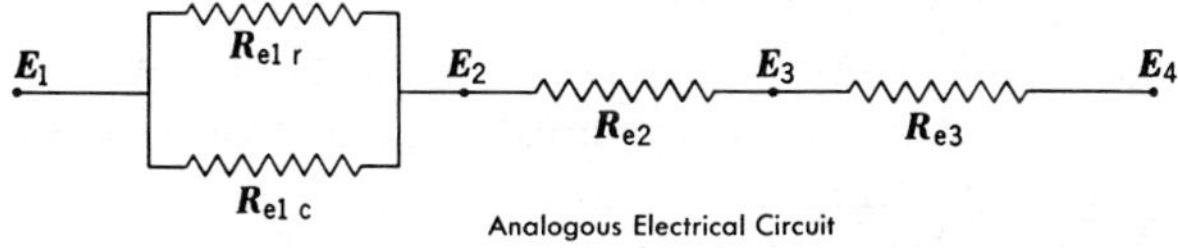

Fig. 1-5. Thermal circuit and analogous electric circuit for heat flow from a hot gas through a metal wall to a coolant.

Hot-gas temperature = 1900 F
Unit-surface conductance on hot side $\bar{h}_1$ = 40 Btu/hr sq ft F
Unit-surface conductance on cold side $\bar{h}_3$ = 50 Btu/hr sq ft F
Cold-gas temperature = 100 F

Solution: In the steady state we can write q/A from gas to hot side of wall = q/A from hot side of wall through wall to cold gas, or

$$\frac{q}{A} = \frac{T_g - T_{sg}}{AR_1} = \frac{T_{sg} - T_c}{A(R_2 + R_3)}$$

Substituting numerical values for the unit thermal resistances and temperatures yields

$$\frac{1900 - 1000}{1/40} = \frac{1000 - 100}{AR_2 + 1/50}$$

Solving for AR_2 gives

$$AR_2 = 0.005 \text{ hr sq ft F/Btu} \qquad \textit{Ans.}$$

1-5. Analogy between heat flow and electrical flow

A thermal resistance per unit area larger than 0.005 hr sq ft F/Btu would raise the inner wall above 1000 F.

The analogy between the flow of heat and the flow of electricity may be used as an aid in visualizing relations in a thermal system by relating them to a more familiar electrical system. The analogy between electrical and thermal systems illustrated in this section is by no means complete. Other useful analogies will be considered in later chapters in connection with problems in two-dimensional heat conduction and in transient thermal systems.

1-6. Units and dimensions

The proper use of units and dimensions will save time and avoid errors. *Dimensions* are our basic concepts of measurements such as length, time, temperature and mass. *Units* are the means of expressing dimension numerically, e.g., feet or meters for length, hours or seconds for time, or degrees F or degrees K for temperature.

There is a variety of systems of units in use. The three most common systems are shown in Table 1-3. The metric, the SI, and the British engineering system have only three basic units. The American engineering system, however, has four basically defined units and it is therefore necessary to use a conversion factor, g_c, to obtain consistent units. This peculiarity arises from Newton's second law of motion which states that force is proportional to the time rate of change of momentum. For a given mass this law may be written

$$F = \frac{1}{g_c} m a \tag{1-29}$$

Table 1-3. Common systems of units.

	Length	Time	Mass	Force	Energy
International (SI)	m	sec	kg	newton*	joule
British engineering	ft	sec	slug*	pound (force)	Btu, ft-lb$_f$
American engineering	ft	sec	pound (mass)	pound (force)	ft-lb$_f$, Btu or hp-hr

*Unit derived from basic units; all energy units are derived.

where F is the force,

m is the mass,

a is the acceleration, and

g_c is a constant whose numerical value and units depend on those selected for F, m, and a.

In the international system the unit of force, the newton, is defined as

$$1 \text{ newton} = \frac{1}{g_c} \times 1 \text{ kg} \times 1 \text{ m/sec}^2 \qquad \text{and thus} \qquad g_c = 1 \frac{\text{kg m}}{\text{newton sec}^2}$$

In the British engineering system the unit of mass, the slug, is derived from the three basic units: pound (force), feet, and seconds, and

$$1 \text{ lb}_f = \frac{1}{g_c} \times 1 \text{ slug} \times 1 \frac{\text{ft}}{\text{sec}^2} \qquad \text{and thus} \qquad g_c = 1 \frac{\text{slug ft}}{\text{lb (force) sec}^2}$$

In the American engineering system, which is used in this book, Eq. 1-29 becomes

$$1 \text{ lb}_f = \frac{1}{g_c} \times 1 \text{ (lb}_m) \times g \frac{\text{ft}}{\text{sec}^2}$$

where g is the acceleration of gravity in ft/sec^2,

lb_m is pound mass, and

lb_f is pound force.

The numerical value of the conversion factor constant g_c is determined by the acceleration imparted to a 1 lb mass by a 1 lb force, or

$$g_c = 32.174 \frac{\text{ft lb}_m}{\text{lb}_f \text{ sec}^2}$$

It should be noted that g and g_c are not similar quantities; the gravitational acceleration g depends on the location and on the altitude, whereas g_c is a constant. Note also that lb_f, the pound (force), and lb_m, the pound (mass), are not the same units, even though we use pound to express force, weight, or mass. The weight of a body, W, is defined as the force exerted on the body as a result of the gravitational field, or

$$W = \frac{g}{g_c} m \tag{1-30}$$

where W is the weight and has the dimension of force.

The common units of energy or heat are based on thermal phenomena and are defined as

1 Btu is the energy required to raise 1 lb_m of water 1 F at 68 F.
1 calorie (cal) is the energy required to raise 1 gm of water 1 C at 15 C.

In line with the definitions above,

$$1 \frac{\text{Btu}}{\text{lb}_\text{m}\ \text{F}} = 1 \frac{\text{cal}}{\text{gm C}}$$

and

$$1 \text{ Btu} = 252 \text{ cal}$$

In the SI system, with fundamental units of meter, kilogram, second, and degree Kelvin, both force and "thermal" energy are derived units. The joule (newton-meter) is the only energy unit in the SI system, and the watt (joule/sec) the unit of power. Thus, the specific heat in the SI system has the units joule/kg K and the appropriate conversion factors are

$$\begin{aligned} 1 \text{ Btu/lb}_\text{m}\text{ F} &= 4{,}184 \text{ J/kg K} \\ 1 \text{ Btu} &= 1054.35 \text{ J (or newton-meter)} \\ 1 \text{ Btu/hr} &= 0.293 \text{ w (or J/sec)} \end{aligned}$$

The SI system of units is intended to become a worldwide standard and, although it is not yet used in the United States, it is only a matter of time before it will be adopted universally.

It is often necessary to change from one set of units to another. To avoid mistakes simply treat the units are algebraic symbols and include the units of every conversion factor. The technique is illustrated below.

EXAMPLE 1-2. An empirical relation to determine the heat-transfer coefficient for air flow in a pipe is given by the relation

$$h = 0.10 \frac{V^{0.3}}{D^{0.7}}$$

where h = heat transfer coefficient in Btu/hr ft^2 F
V = velocity in ft/sec
D = inside diameter in ft

If h is to be expressed in watts/m^2 K, what should the constant in place of 0.10 be?

Solution:

$$h = \frac{0.1\, V^{0.3}}{D^{0.7}} \times \frac{\text{Btu}}{\text{hr ft}^2\,\text{F}} \times \frac{1054\ \text{J}}{\text{Btu}} \times \frac{1\ \text{hr}}{3600\ \text{sec}} \left(\frac{1\ \text{ft}}{12\ \text{in}}\right)^2 \times \left(\frac{1\ \text{in}}{0.0254\ \text{m}}\right)^2 \times \frac{1.8\ \text{F}}{1\ \text{K}} =$$

$$0.567\, \frac{V^{0.3}}{D^{0.7}}\, \frac{\text{watts}}{\text{m}^2\ \text{K}}$$

It is left as an exercise to determine the constant when V and D are also to be expressed in SI units (see Problem 1-31).

PROBLEMS

The problems below have been organized in the following manner: Problems 1-1 through 1-4 deal with conversion of units, 1-5 through 1-8 with conduction, 1-9 through 1-12 with radiation, 1-13 through 1-16 with convection, 1-17 through 1-26 with combined heat transfer mechanisms, and 1-27 through 1-30 with thermal circuits.

1-1. The heat transfer coefficient between a surface and a liquid is 10 Btu/hr sq ft F. How many calories per hour per square meter per degree Centigrade will be transferred in this system?

1-2. Convert the Stefan-Boltzmann constant in Eq. 1-5 to the SI system of units.

1-3. The thermal conductivity of asbestos at 86 F is given in Table A-2 of Appendix III in Btu/hr ft F. What is its value in watts per square centimeter per degree Centigrade per centimeter (w/cm C)? *Ans.* 1.50×10^{-3}

1-4. The thermal conductivity of silver at 212 F is 238 Btu/hr ft F. What is the conductivity in watts/m K?

1-5. If the weight, not the space, required for insulation of a plane wall is most significant, show analytically that the lightest insulation for a specified thermal resistance is that insulation which has the smallest product of density times thermal conductivity.

1-6. A furnace wall is to be constructed of bricks having standard dimensions 9 by $4\frac{1}{2}$ by 3 in. Two kinds of material are available: One has a maximum usable temperature of 1900 F and a thermal conductivity of 1 Btu/hr ft F, and the other has a maximum temperature limit of 1600 F and a thermal conductivity of 0.5 The bricks cost the same and can be laid in any manner, but we wish to design the most economical wall for a furnace with a temperature on the hot side of 1900 F and on the cold side of 400 F. If the maximum amount of heat transfer permissible is 300 Btu/hr for each square foot of area, determine the most economic arrangements for the available bricks.

1-7. To measure thermal conductivity, two similar 1-in.-thick specimens are placed in an apparatus shown in the accompanying sketch. Electric current is supplied to the 6- by 6-in. guarded heater, and a wattmeter shows that the power dissipation is 10 watts (w). Thermocouples attached to the warmer and to the cooler surfaces show temperatures of 120 F and 80 F respectively. Calculate the thermal conductivity of the material at the mean temperature in Btu/hr ft F and in watts/m K.

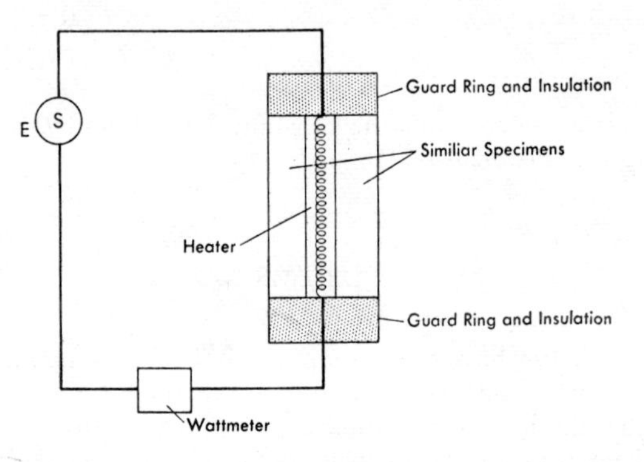

Prob. 1-7

1-8. To determine the value of the thermal conductivity of a structural material, a large 6-in.-thick slab of it was subjected to a uniform heat flux of 800 Btu/hr sq ft, while thermocouples imbedded in the wall 2 in. apart were read over a period of time. After the system had reached equilibrium, an operator recorded the readings of the thermocouples as shown below for two different environmental conditions:

Distance from Surface (in.)	Temperature (F)
Test 1	
0	100
2	150
4	206
6	270
Test 2	
0	200
2	265
4	335
6	406

From these data, determine an approximate expression for the thermal conductivity as a function of temperature between 100 and 400 F.

1-9. Determine the rate of radiant heat emission in Btu/hr sq ft from a blackbody at (a) 300 F; (b) 3000 F; (c) 3000 R; (d) 10,000 R.

Ans. (a) 5.72×10^2; (b) 2.46×10^5; (c) 1.39×10^5; (d) 1.71×10^7

1-10. Two large parallel planes, having surface conditions which approximate those of a blackbody, are maintained at 1500 and 500 F respectively. Determine the rate of heat transfer by radiation between the plates in Btu/hr sq ft and the radiation unit conductance in Btu/hr sq ft F and in watts/m^2 K.

1-11. A spherical vessel of 1-ft diameter is located in a large room whose walls are at 80 F. If the vessel is used to store liquid oxygen at -297 F and the surface of the storage vessel as well as the walls of the room are black, calculate the rate of heat transfer by radiation to the liquid oxygen in Btu/hr and in watts.

1-12. Repeat Problem 11, but assume that the surface of the storage vessel has an emittance of 0.1. Then determine the rate of evaporation of liquid oxygen in lb/hr, assuming that convection can be neglected.

1-13. The heat transfer coefficient for a 1-ft diameter sphere in still air is 1.2 Btu/hr sq ft F. If the air is at 80 F and the surface of the sphere at -297 F, determine the rate of heat transfer by convection.

1-14. Using Table 1-2 as a guide, prepare a similar table showing the order of magnitudes of the thermal resistances per unit area for convection between a surface and various fluids.

1-15. A thermocouple ($\frac{1}{32}$-in.-OD wire) is used to measure the temperature of quiescent gas in a furnace. The thermocouple reading is 300 F. It is known, however, that the rate of radiant heat flow per inch length from hotter furnace walls to the thermocouple wire is 0.1 Btu/hr-in. and the unit conductance between the wire and the gas is 1.2 Btu/hr sq ft F. With this information, *estimate* the true gas temperature. State your assumptions and indicate the equations used.

1-16. Water at a temperature of 120 F is to be evaporated slowly in a steady-flow system. The water is in a low-pressure container which is surrounded by steam. The steam is condensing at 225 F. The overall heat-transfer coefficient between the water and the steam is 200 Btu/hr sq ft F. Calculate the surface area of the container which would be required to evaporate water at a rate of 800 lb/hr. *Ans.* 39 sq ft

1-17. Heat is transferred through a plane wall from the inside of a room at 70 F to the outside air at 30 F. The unit-surface conductances at the inside and outside surfaces are 2 and 3 Btu/hr sq ft F, respectively. The thermal resistance of the wall per unit area is 3 hr sq ft F/Btu. Determine the temperature at the outer surface of the wall and the rate of heat flow through the wall per unit area.

1-18. Steam is condensing inside a pipe at 134 psia. The unit-surface conductance on the steam side is 500 Btu/hr sq ft F. The thermal resistance of the pipe per unit area is 0.001 hr sq ft F/Btu and the unit-surface conductance at the outside of the pipe is 4 Btu/hr sq ft F. (a) Estimate the percent of the overall thermal resistance offered by (1) the steam, (2) the pipe, and (3) the steam and the pipe. (b) Determine the temperature at the outer surface of the pipe if the pipe is suspended in a room at 70 F. The values of the unit conductances and the resistance are based on the outside area of the pipe.

1-19. A flat plate placed in the sunlight receives 200 Btu/hr sq ft of radiant heat from the sun and the atmosphere. If the air temperature is 80 F and the unit-surface conductance between the plate and the air is 2 Btu/hr sq ft F, determine the plate temperature. Neglect heat losses from the bottom of the plate.

1-20. How much Fiberglas insulation (k = 0.02 Btu/hr ft F) is needed to enable a guarantee that the outside temperature of a kitchen oven will not exceed 110 F? The maximum oven temperature to be maintained by the conventional type of thermostatic control is 550 F, the kitchen temperature may vary from 60 F to 90 F and the average heat-transfer coefficient between the oven surface and the kitchen is 2 Btu/hr sq ft F.

1-21. A heat-exchanger wall consists of a copper plate $\frac{3}{8}$ in. thick. The surface coefficients on the two sides of the plate are 480 and 1250 Btu/hr sq ft F, corresponding to fluid temperatures of 200 and 90 F, respectively. Assuming that the thermal conductivity of the wall is 220 Btu/hr ft F, (a) draw the thermal circuit, (b) compute the surface temperatures in deg F, and (c) calculate the heat flux in Btu/hr sq ft. *Ans.* (b) 119 F, 124 F; (c) 36,300 Btu/hr sq ft

1-22. A horizontal $\frac{1}{8}$-in.-thick flat copper plate, 2 ft long and 1 ft wide, is exposed in air at 80 F to radiation from the sun. If the total rate of incident solar radiation is 400 Btu/hr and the combined unit-surface conductances on the upper and lower surfaces are 4 and 3 Btu/hr sq ft F, respectively, determine the equilibrium temperature of the plate. *Ans.* 108.7 F

1-23. A submarine is to be designed to provide a comfortable temperature for the crew of no less than 70 F. The submarine can be idealized by a cylinder 30 ft in diameter and 200 ft in length. The combined unit-surface conductance on the interior is about 2.5 Btu/hr sq ft F, while on the outside the unit-surface conductance is estimated to vary from about 10 Btu/hr sq ft F (not moving) to 150 Btu/hr sq ft F (top speed). For the following wall constructions, determine the minimum size in kilowatts of the heating unit required if the sea water temperature varies from 34 to 55 F during operation: The walls of the submarine are (a) $\frac{1}{2}$-in. aluminum; (b) $\frac{3}{4}$-in. stainless steel with a 1-in. thick layer of Fiberglas insulation on the inside; and (c) of sandwich construction with a $\frac{3}{4}$-in.-thick layer of stainless steel, a 1-in.-thick layer of Fiberglas insulation, and a $\frac{1}{4}$-in. thickness of aluminum on the inside. What conclusions can you draw? *Ans.* (a) 525; (b) 57; (c) 57

1-24. A small gray sphere having an emissivity of 0.5 and a surface temperature of 1000 F is located in a blackbody enclosure having a temperature of 100 F. Calculate for this system: (a) the net rate of heat transfer by radiation per unit of surface area of the sphere, (b) the radiative thermal conductance in Btu/hr F if the surface area of the sphere is 0.1 sq ft, (c) the unit thermal resistance for radiation between the sphere and its surroundings, (d) the ratio of thermal resistance for radiation to thermal resistance for convection if the unit-surface conductance for convection between the sphere and its surroundings is 2.0 Btu/hr sq ft F, (e) the total rate of heat transfer from the sphere to the surroundings, and (f) the combined unit-thermal-surface conductance for the sphere.

1-25. A small oven with a surface area of 3 sq ft is located in a room in which the walls and the air are at a temperature of 80 F. The exterior surface of the oven is at 300 F and the net heat transfer by radiation between the oven's surface and the surroundings is 2000 Btu/hr. If the average unit-surface conductance for convection between the oven and the surrounding air is 2.0 Btu/hr sq ft F, calculate: (a) the net heat transfer between the oven and the surroundings in Btu/hr, (b) the thermal resistance at the surface for radiation and convection in hr F/Btu, and (c) the combined unit-thermal-surface conductance in Btu/hr sq ft F.

1-26. A simple solar heater consists of a flat plate of glass below which is located a shallow pan filled with water, so that the water is in contact with the glass plate above it. Solar radiation is passing through the glass at the rate of 156 Btu/hr sq ft. The water is at 200 F and the surrounding air at 80 F. If the heat-transfer coefficients between the water and the glass and the glass and the air are 5 Btu/hr sq ft F and 1.2 Btu/hr sq ft F, respectively, determine the time required to transfer 100 Btu/sq ft of surface to the water in the pan. The lower surface of the pan may be assumed insulated. *Ans.* 2.5 hr

1-27. Draw the thermal circuit, determine the rate of heat flow per unit area from a furnace wall, and estimate the exterior surface temperature under the following conditions:

1. Convective heat-transfer coefficient at the interior surface is 10 Btu/hr sq ft F.
2. Rate of heat flow by radiation from hot gases and particles (3500 F) to interior wall surface is 15,000 Btu/ hr sq ft.
3. Unit thermal conductance of wall (interior surface temperature about 1500 F) is 40 Btu/hr sq ft F.
4. Convection from outer surface.

1-28. The inner wall of a combustion chamber receives 50,000 Btu/hr sq ft by radiation from a gas at 5000 F. The convective unit conductance between the gas and the wall is 20 Btu/hr sq ft F. If the inner wall of the combustion chamber is at a temperature of 1000 F, determine the total unit thermal resistance in hr sq ft F/Btu. Also, draw the thermal circuit. *Ans.* 0.031

1-29. A composite refrigerator wall is composed of 2 in. of corkboard sandwiched between a $\frac{1}{2}$-in.-thick layer of oak and a $\frac{1}{32}$-in. thickness of aluminum lining on the inner surface. The average unit-connective-thermal conductances at the interior and exterior wall, respectively, are 2 and 1.5 Btu/hr sq ft F. (a) Calculate the individual resistance of this composite wall and the resistances at the surface. (b) Calculate the over-all conductance per unit area. (c) Draw the thermal circuit. (d) For an air temperature inside the refrigerator of 30 F and outside, of 90 F, calculate the rate of heat transfer per unit area.

Ans. (d) 7.4 Btu/hr sq ft

1-30. Draw the thermal circuit for heat transfer from the sun through a window to the air in a room. Identify each circuit element.

1-31. Determine the constant in Example 1-2 when V and D are also in SI units.

2
Steady one-dimensional heat conduction

2-1. Walls of simple geometrical configuration

In this section we shall consider steady-state heat conduction through simple systems in which the temperature and the heat flow are functions of a single coordinate.

Plane wall. The simplest case of one-dimensional heat flow, namely heat conduction through a plane wall, was treated in Sec. 1-3. We found that, for uniform temperatures over the hot and the cold surfaces, the rate of heat flow by conduction through a homogeneous material is given by

$$q_k = \frac{Ak}{L}(T_{\text{hot}} - T_{\text{cold}}) = \frac{\Delta T}{R_k} = K_k \Delta T \tag{2-1}$$

EXAMPLE 2-1. The interior surfaces of the walls in a large building are to be maintained at 70 F while the outer surface temperature is -10 F. The walls are 10 in. thick and constructed from a brick material having a thermal conductivity of 0.4 Btu/hr ft F. Calculate the heat loss for each square foot of wall surface per hour.

Solution: If we neglect the effect of the corners where the walls meet and the effect of mortared brick joints, Eq. 1-2 applies. Substituting the thermal conductivity and the pertinent dimensions in their proper units (e.g., $L = \frac{10}{12}$ ft) we obtain

$$\frac{q}{A} = \frac{(0.4)[70 - (-10)]}{10/12} = 38.4 \text{ Btu/hr sq ft}$$

Thus, 38.4 Btu will be lost from the building per hour through each square foot of wall surface area. *Ans.*

Hollow cylinders. Radial heat flow by conduction through a hollow circular cylinder is another one-dimensional conduction problem of considerable practical importance. Typical examples are conduction through pipes and through pipe insulation.

If the cylinder is homogeneous and sufficiently long that end effects may be neglected and the inner surface temperature is constant at T_i while the outer surface temperature is maintained uniformly at T_o, the rate of heat conduction is, from Eq. 1-1,

$$q_k = -kA\frac{dT}{dr} \tag{2-2}$$

where dT/dr = temperature gradient in the radial direction.

For the hollow cylinder (Fig. 2-1), the area is a function of the radius and

$$A = 2\pi rl \tag{2-3}$$

where r is the radius and l the length of the cylinder. The rate of heat flow

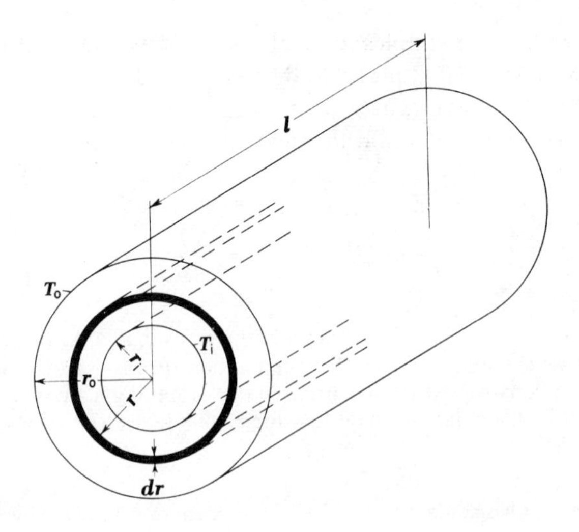

Fig. 2-1. Sketch illustrating nomenclature for conduction through a hollow cylinder.

by conduction can thus be expressed as

$$q_k = -k2\pi rl \frac{dT}{dr} \tag{2-4}$$

Separating the variables and integrating between T_o at r_o and T_i at r_i yields

$$T_i - T_o = \frac{q_k}{2\pi kl} \ln \frac{r_o}{r_i} \tag{2-5}$$

Solving Eq. 2-5 for q_k yields

$$q_k = \frac{T_i - T_o}{\dfrac{\ln (r_o/r_i)}{2\pi kl}} \tag{2-6}$$

the equation for calculating the rate of heat conduction through a hollow circular cylinder such as a pipe. An inspection of Eq. 2-6 shows that the rate of radial heat flow varies directly with the cylinder length l, the thermal conductivity k, the temperature difference between the inner and outer surfaces $T_i - T_o$, and inversely as the *natural logarithm*[1] of the ratio of the outside and inside radii r_o/r_i or the corresponding diameter ratio D_o/D_i. By analogy to the case of a plane wall and Ohm's law, the thermal resistance of the hollow cylinder is

$$R_k = \frac{\ln (r_o/r_i)}{2\pi kl} \tag{2-7}$$

The temperature distribution in the curved wall is obtained by integrating Eq. 2-4 from the inner radius r_i and the corresponding temperature T_i to an arbitrary radius r and the corresponding temperature T, or

$$\int_{r_i}^{r} \frac{q_k}{k(2\pi l)} \frac{dr}{r} = -\int_{T_i}^{T(r)} dT$$

which gives

$$T(r) = T_i - \frac{T_i - T_o}{\ln (r_o/r_i)} \ln \frac{r}{r_i}$$

---

[1]The natural logarithm of a number, ln, is 2.3026 times the logarithm to base 10.

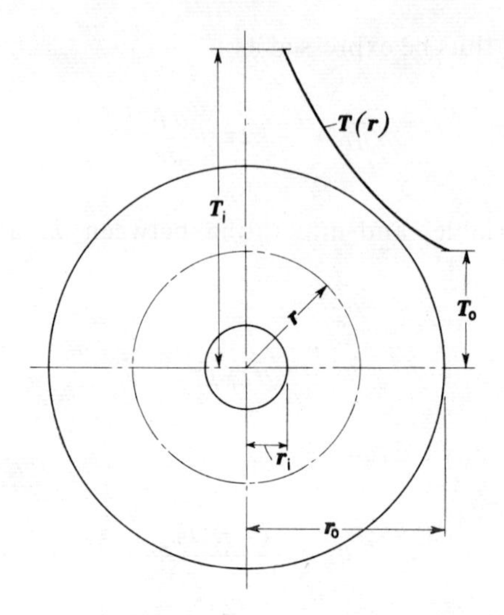

Fig. 2-2. Temperature distribution in a hollow cylinder.

Thus the temperature in a hollow circular cylinder is a logarithmic function of the radius r (Fig. 2-2), while for a plane wall the temperature distribution is linear.

For some applications it is helpful to have the equation for heat conduction through a curved wall in the same form as Eq. 2-1 for a plane wall. To obtain this form of equation we equate the right-hand sides of Eqs. 2-1 and 2-6 using, however, $L = (r_o - r_i)$, the thickness through which heat is conducted, and $A = \bar{A}$ in Eq. 2-1. This yields

$$\frac{k\bar{A}\Delta T}{r_o - r_i} = \frac{2\pi k l \Delta T}{\ln (r_o/r_i)}$$

from which $\bar{A}$ is

$$\bar{A} = \frac{2\pi(r_o - r_i)l}{\ln (r_o/r_i)} = \frac{A_o - A_i}{\ln (A_o/A_i)} \tag{2-8}$$

The area $\bar{A}$ defined by Eq. 2-8 is called the *logarithmic mean area.* The rate of heat conduction through a hollow circular cylinder can then be expressed as

$$q_k = \frac{T_i - T_o}{(r_o - r_i)/k\bar{A}} \tag{2-9}$$

For values of $A_o/A_i < 2$ (i.e., $r_o/r_i < 2$) the arithmetic mean area

$(A_o + A_i)/2$ is within 4 percent of the logarithmic mean area and may be used with satisfactory accuracy. For thicker walls this approximation is generally not acceptable.

EXAMPLE 2-2. Calculate the heat loss from 10 ft of 3-in. nominal-diameter pipe covered with $1\frac{1}{2}$ in. of an insulating material having a thermal conductivity of 0.040 Btu/hr ft F. Assume that the inner and outer surface temperatures of the insulation are 400 and 80 F, respectively.

Solution: The outside diameter of a nominal 3-in. pipe is 3.50 in. This is also the inside diameter of the insulation. The outside diameter of the insulation is 6.50 in. The logarithmic mean area is

$$\bar{A} = \frac{A_o - A_i}{\ln(A_o/A_i)} = \frac{10\pi(6.50 - 3.50)/12}{\ln(6.50/3.50)}$$

$$= \frac{7.85}{0.62} = 12.70 \text{ sq ft}$$

Since $r_o/r_i < 2$, the arithmetic mean area would be an acceptable approximation and

$$\frac{A_o + A_i}{2} = \frac{10\pi(6.50 + 3.50)}{(2)(12)} = 13.10 \text{ sq ft}$$

Applying Eq. 2-9, the rate of heat loss is

$$q_k = \frac{400 - 80}{(1.5/12)/(0.04)(12.7)} = 1300 \text{ Btu/hr} \qquad \textit{Ans.}$$

Spherical and parallelepiped shells. A sphere has the largest volume per outside surface area of any geometrical configuration. For this reason a hollow sphere is sometimes used in the chemical industry for low-temperature work, when heat losses are to be kept at a minimum. Conduction through a spherical shell is also a one-dimensional steady-state problem if the interior and exterior surface temperatures are uniform and constant. The rate of heat conduction for this case (Fig. 2-3) is

$$q_k = \frac{4\pi r_i r_o k(T_i - T_o)}{r_o - r_i} = k\sqrt{A_o A_i}\,\frac{T_i - T_o}{r_o - r_i} \tag{2-10}$$

if the material is homogeneous.

Equation 2-10 can also be used as an approximation for parallelepiped shells which have a small inner cavity surrounded by a thick wall. An example of such a system would be a small furnace surrounded by a large thickness of insulating material. For this type of geometry the heat

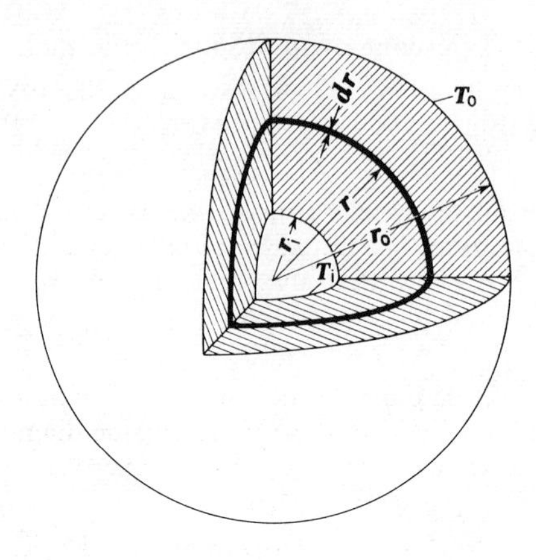

Fig. 2-3. Sketch illustrating nomenclature for conduction through spherical shell.

flow, especially in the corners, is not perpendicular to the bounding surfaces, and hence cannot strictly be considered one-dimensional. However, when the cavity is roughly cubic and the surrounding walls are thick ($A_o/A_i > 2$), the rate of heat flow can be estimated according to Schumann (1) by multiplying the geometric mean area in Eq. 2-10, $\sqrt{A_o A_i}$, by the semi-empirical correction factor 0.725. More accurate correction factors have been determined by Langmuir et al. (2) and are summarized in Ref. 8.

EXAMPLE 2-3. The working chamber of an electrically heated laboratory furnace is 6 by 8 by 12 in. and the walls, 6 in. thick on all sides, are made of a refractory brick (k = 0.2 Btu/hr ft F). If the temperature at the interior surface is to be maintained at 2000 F while the outside surface temperature is 300 F, estimate the power consumption in kilowatts (kw).

Solution: Under steady-state conditions the power consumption will equal the heat loss. The inner surface area A_i is

$$A_i = 2\,\frac{(6 \times 8) + (6 \times 12) + (8 \times 12)}{144} = 3 \text{ sq ft}$$

The outer surface area A_o is

$$A_o = 2\,\frac{(18 \times 20) + (18 \times 24) + (20 \times 24)}{144} = 17.7 \text{ sq ft}$$

Since $A_o/A_i > 2$, we can use Eq. 2-10 with the empirical correction factor 0.725, and the heat loss is

$$q_k = (0.2)(0.725)\sqrt{3 \times 17.7}\left(\frac{1700}{6/12}\right) = 3600 \text{ Btu/hr}$$

Since 1 Btu/hr = 2.93×10^{-4} kw, the power consumption is about 1.05 kw. *Ans.*

Effect of nonuniform thermal conductivity. It has already been mentioned that the thermal conductivity varies with temperature. The variation of thermal conductivity with temperature may be neglected when the temperature range under consideration is not too large or the temperature dependence of the conductivity is not too severe. On the other hand, when the temperature difference in a system causes substantial variations in the thermal conductivity, the temperature dependence must be taken into account.

For numerous materials, especially within a limited temperature range, the variation of the thermal conductivity with the temperature can be represented by the linear function

$$k(T) = k_0(1 + \beta_k T) \tag{2-11}$$

where k_0 is the thermal conductivity at $T = 0$ and β_k is a constant called the *temperature coefficient of thermal conductivity*. When the variation of thermal conductivity is available in the form of a curve showing how k varies with T, the temperature coefficient can be obtained approximately by drawing a straight line between the temperatures of interest and measuring its slope. Then k_0 is a hypothetical value of the thermal conductivity equal to the ordinate intercept at zero temperature. It is determined graphically by continuing the straight line representing the actual thermal conductivity over a limited temperature range through the axis of conductivity at zero temperature (Fig. 2-4).

With a linear approximation to the temperature variation of the thermal conductivity, the rate of heat flow by conduction through a plane wall is, from Eq. 2-1,

$$\frac{q_k}{A}\int_0^L dx = -\int_{T_{\text{hot}}}^{T_{\text{cold}}} k_0(1 + \beta_k T)\, dT$$

Integration gives

$$q_k = \frac{k_0 A}{L}\left[T_{\text{hot}} - T_{\text{cold}} + \frac{\beta_k}{2}\left(T_{\text{hot}}^2 - T_{\text{cold}}^2\right)\right]$$

which can be written more conveniently as

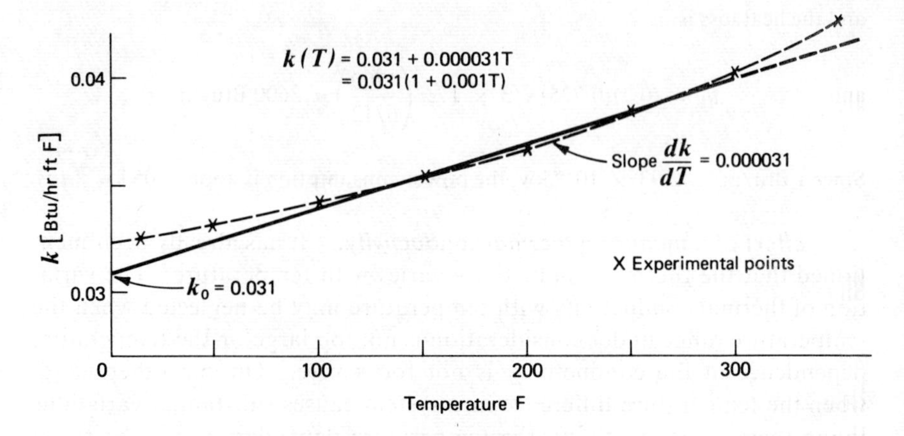

Fig. 2-4. Graphical determination of the temperature coefficient of thermal conductivity.

$$q_k = \frac{A(T_{hot} - T_{cold})}{L} k_0 \left(1 + \beta_k \frac{T_{hot} + T_{cold}}{2}\right) = \frac{\Delta T}{L/Ak_m} \tag{2-12}$$

where $k_m = k_0 \; [1 + \beta_k(T_{hot} + T_{cold})/2]$ represents a mean value of the thermal conductivity. For a linear variation of k with T, the thermal conductivity in Eq. 1-2 should therefore be evaluated at the arithmetic mean temperature $(T_{hot} + T_{cold})/2$.

EXAMPLE 2-4. The conductivity of an 85 percent magnesia insulating material is shown as a function of temperature in Fig. 2-4. (a) Determine β_k and k_0 for a linear approximation between 100 and 300 F. (b) Estimate the rate of heat flow per unit area between these temperatures for a slab of 3-in. thickness.

Solution: (a) By means of the graphical method illustrated in Fig. 2-4, the slope of the straight line connecting the thermal-conductivity curve between 100 and 300 F is found to be +0.000031. The ordinate intercept at 0 degrees (deg) is 0.031. Thus we have

$$k(T) = 0.031\,(1 + 0.001\,T) \qquad \text{for } 100 \text{ F} < T < 300 \text{ F}$$

The mean temperature is 200 F and the mean value of the thermal conductivity is

$$k_m = 0.031\,(1 + 0.001 \times 200) = 0.0372 \text{ Btu/hr ft F} \qquad \textit{Ans.}$$

(b) The rate of heat flow per unit area is, from Eq. 2-4,

$$\frac{q_k}{A} = \frac{\Delta T}{L/k_m} = \frac{200}{(3/12)/(0.0372)} = 29.8 \text{ Btu/hr sq ft} \qquad \textit{Ans.}$$

Radial heat conduction through hollow cylinders and spheres made of materials whose thermal conductivity varies linearly with temperature can be treated in a similar manner (see Problem 2-4).

2-2. *Composite structures*

The general method for analyzing problems of steady-state heat flow through composite structures has been presented in Sec. 1-4. In this section we shall consider some examples of composite structures in which the heat flow is one-dimensional, or at least approximately so. In order to make the treatment applicable to practical cases where the surface temperatures are generally not known, heat flow through thermal resistances at the boundaries will be included in the treatment. We shall assume that the system is exposed to a high-temperature medium, i.e., a *heat source*, of known and constant temperature on one side and to a low-temperature medium, i.e., a *heat sink*, of known and constant temperature on the other side. The surface conductances between the medium and the surface will be taken as constant over a given surface.

Composite walls. A composite wall, typical of the type used in a large furnace, is shown in Fig. 2-5. The inner layer, which is exposed to the high-temperature gases, is made of firebrick. The intermediate layer consists of an insulating brick and is followed by an outer layer of ordinary red brick. The temperature of the hot gases is T_i and the unit-surface conductance over the interior surface is $\bar{h}_i$. The atmosphere surrounding the furnace is at a temperature T_o and the unit-surface con-

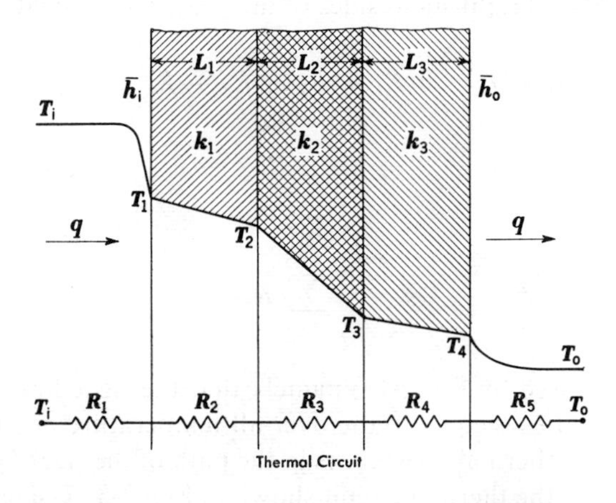

Fig. 2-5. Temperature distribution and thermal circuit for heat flow through a series composite plane wall.

ductance over the exterior surface is $\bar{h}_o$. Under these conditions there will be a continuous heat flow from the hot gases through the wall to the surroundings. Since the heat flow through a given area A is the same for any section, we obtain

$$q = \bar{h}_i A\,(T_i - T_1) = \frac{k_1 A}{L_1}(T_1 - T_2)$$

$$= \frac{k_2 A}{L_2}(T_2 - T_3) = \frac{k_3 A}{L_3}(T_3 - T_4) = \bar{h}_o A\,(T_4 - T_o) \qquad (2\text{-}13)$$

The symbols in Eq. 2-13 can be identified by inspection of Fig. 2-5. Equation 2-13 can be written in terms of the thermal resistances of the various sections as

$$q = \frac{T_i - T_1}{R_1} = \frac{T_1 - T_2}{R_2} = \frac{T_2 - T_3}{R_3} = \frac{T_3 - T_4}{R_4} = \frac{T_4 - T_o}{R_5} \qquad (2\text{-}14)$$

where the resistances may be determined from Eqs. 1-3 and 1-13 or by comparison of corresponding terms in Eqs. 2-13 and 2-14. Solving for the various temperature differences in Eq. 2-14 we obtain

$$\begin{aligned} T_i - T_1 &= qR_1 \\ T_1 - T_2 &= qR_2 \\ T_2 - T_3 &= qR_3 \\ T_3 - T_4 &= qR_4 \\ T_4 - T_o &= qR_5 \end{aligned} \qquad (2\text{-}15)$$

Adding the left- and right-hand sides of these equations yields

$$T_i - T_o = q(R_1 + R_2 + R_3 + R_4 + R_5) \qquad (2\text{-}16)$$

or

$$q = \frac{T_i - T_o}{\sum_{n=1}^{n=5} R_n} \qquad (2\text{-}17)$$

The result expressed by Eq. 2-17, namely that the heat flow through the five sections in series is equal to the over-all temperature potential divided by the sum of the thermal resistances in the path of the heat flow, can also be obtained from the thermal circuit shown in Fig. 2-5. Using the analogy between the flow of heat and the flow of electric current, Eq. 2-17 can be written directly.

EXAMPLE 2-5. A furnace wall consists of two layers, 9 in. of firebrick ($k = 0.8$ Btu/hr ft F) and 5 in. of insulating brick ($k = 0.1$ Btu/hr ft F). The temperature inside the furnace is 3000 F and the unit-surface conductance at the inside wall is 12 Btu/hr sq ft F. The temperature of the surrounding atmosphere is 80 F and the unit-surface conductance at the outer wall is 2 Btu/hr sq ft F. Neglecting the thermal resistance of the mortar joints, estimate (a) the rate of heat loss per square foot of wall and the temperatures at the (b) inner surface and (c) outer surface.

Solution: (a) The rate of heat flow is obtained from Eq. 2-17 as

$$\frac{q}{A} = \frac{3000 - 80}{\frac{1}{12} + \frac{9}{12}/0.8 + \frac{5}{12}/0.1 + \frac{1}{2}} = \frac{2920}{0.083 + 0.94 + 4.17 + 0.50}$$

$$= \frac{2920}{5.69} = 513 \text{ Btu/hr sq ft} \qquad \textit{Ans.}$$

It is of interest to note that the insulating brick, while representing only about one-third of the wall thickness, accounts for three-quarters of the total thermal resistance.

(b) Applying Eq. 2-14, the temperature drop between the furnace gases and the interior surface is $T_i - T_1 = qR_1 = (513)(0.083) = 43$ F. This relatively small temperature difference is in accordance with previous considerations indicating that the thermal resistance of the first section in the circuit is negligible. Thus, heat can flow without a large potential and the temperature at the interior wall is nearly equal to that of the gases, that is, $T_i = T_1 - 43 = 2957$ F *Ans.*

(c) The temperature of the outer surface, obtained in a like manner, is 336 F. *Ans.*

In numerous practical applications, combinations of series- and parallel-connected heat-flow paths are encountered. An example of such a case is illustrated by the composite wall shown in Fig. 2-6. An approximate solution can be obtained by assuming that the heat flow is essentially one-dimensional. The composite wall can then be divided into three sections. The thermal resistance of each section can be determined with the aid of the thermal circuit shown in Fig. 2-6. The intermediate layer consists of two separate thermal paths in parallel and its thermal conductance is the sum of the individual conductances. For a wall section of height $b_1 + b_2$ (Fig. 2-6) the conductance is

$$K_2 = \frac{k_2 b_1}{L_2} + \frac{k_1 b_2}{L_2} = \frac{1}{R_2}$$

per unit length of wall. Using Eq. 1-24, the overall unit conductance U from surface to surface is

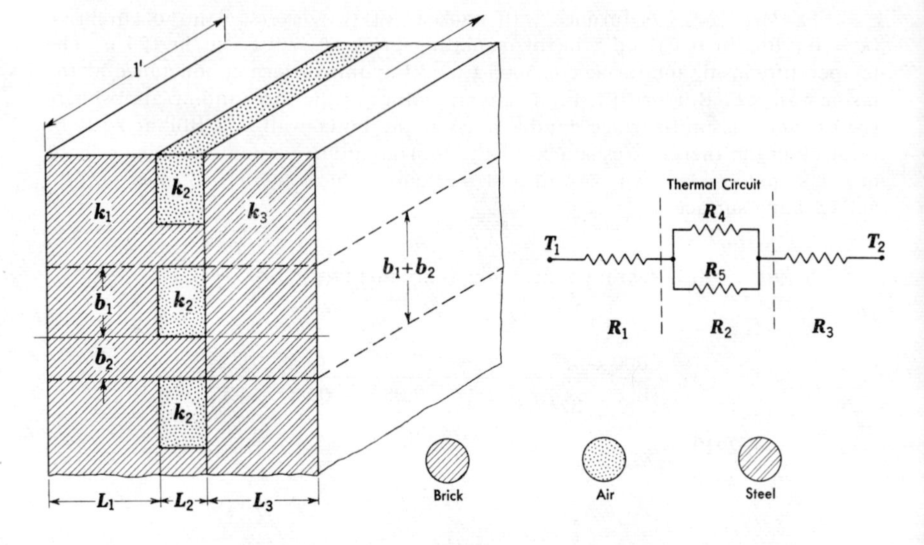

Fig. 2-6. Thermal circuit for a parallel-series composite wall. (L_1 = 1 in.; L_2 = 1/32 in.; L_3 = 1/4 in.; for Example 2-6, in which T_1 is at the center.)

$$U = \frac{1}{(b_1 + b_2)(R_1 + R_2 + R_3)} = \frac{1}{\dfrac{L_1}{k_1} + \dfrac{b_1 + b_2}{(k_1 b_2/L_2) + (k_2 b_1/L_2)} + \dfrac{L_3}{k_3}}$$

EXAMPLE 2-6. A layer of 2-in.-thick firebrick (k_b = 1.0 Btu/hr ft F) is placed between two $\frac{1}{4}$-in.-thick steel plates (k_s = 30 Btu/hr ft F). The faces of the brick adjacent to the plates are rough, having solid-to-solid contact only over 30 percent of the total area, with the average height of the asperities being $\frac{1}{32}$ in. If the outer steel-plate surface temperatures are 200 and 800 F respectively, specify the rate of heat flow per unit area.

Solution: The real system is first idealized by assuming that the asperities of the surface are distributed, as shown in Fig. 2-6. We note that the composite wall is symmetrical with respect to the center plane and therefore only consider one-half of the system. The overall heat-transfer coefficient for the composite wall is then

$$U = \frac{1/2}{R_1 + \dfrac{R_4 R_5}{R_4 + R_5} + R_3}$$

from an inspection of the thermal circuit.

The thermal resistance of the steel plate R_3 is, on the basis of a unit area, equal to

$$R_3 = \frac{L_3}{k_s} = \frac{1/4}{(12)(30)} = 0.694 \times 10^{-3} \text{ hr sq ft F/Btu}$$

The thermal resistance of the brick asperities R_4 is, on the basis of a unit area, equal to

$$R_4 = \frac{L_2}{0.3 k_b} = \frac{1/32}{(12)(0.3)(1.0)} = 8.7 \times 10^{-3} \text{ hr sq ft F/Btu}$$

Since the air is trapped in very small compartments, the effects of convection are small and it will be assumed that heat flows through the air by conduction. At a temperature of 300 F, the conductivity of air k_a is 0.02 Btu/hr ft F. Then R_5, the thermal resistance of the air trapped between the asperities, is, on the basis of unit area, equal to

$$R_5 = \frac{L_2}{0.7 k_a} = \frac{1/32}{(12)(0.7)(0.02)} = 187 \times 10^{-3} \text{ hr sq ft F/Btu}$$

The factors of 0.3 and 0.7 in R_4 and R_5 respectively, represent the percent of the total area for the two separate heat-flow paths.

The total thermal resistance for the two paths, R_4 and R_5 in parallel, is

$$R_2 = \frac{R_4 R_5}{R_4 + R_5} = \frac{(8.7)(187) \times 10^{-6}}{(8.7 + 187) \times 10^{-3}} = 8.3 \times 10^{-3} \text{ hr sq ft F/Btu}$$

The thermal resistance of *one half* of the solid brick, R_1, is

$$R_1 = \frac{1}{2} \frac{L_1}{k_b} = \frac{1}{2} \frac{2}{(12)(1.0)} = 83.5 \times 10^{-3} \text{ hr sq ft F/Btu}$$

and U, the overall heat-transfer coefficient, is

$$U = \frac{1/2 \times 10^3}{83.5 + 8.3 + 0.69} = 5.4 \text{ Btu/hr sq ft F}$$

An inspection of the values for the various thermal resistances show that the steel offers a negligible resistance, while the contact section, although only $\frac{1}{32}$ in. thick, contributes 10 percent to the total resistance. From Eq. 1-21, the rate of heat flow per unit area is

$$\frac{q}{A} = U \Delta T = 5.4(800 - 200) = 3250 \text{ Btu/hr sq ft} \qquad \textit{Ans.}$$

The thermal resistance between two surfaces is called *contact resistance*. The analysis of the contact resistance in the preceding problem

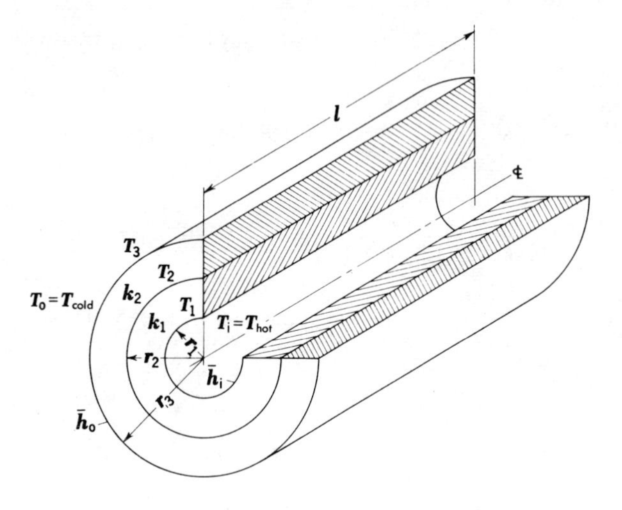

Fig. 2-7. Sketch illustrating nomenclature of composite cylinder wall.

is only approximate because, in addition to roughness, the contact resistance depends on the contact pressure. For more information on contact resistance, see Refs. 4, 5, 6, 17, and 18.

Concentric cylinders. Radial heat flow through concentric cylinders of different thermal conductivity is encountered in many industrial installations. An insulated pipe, with a hot fluid flowing inside and exposed to a colder medium on the outside, is typical of such problems (Fig. 2-7). If the pipe is relatively long, then the heat flow through the walls will be in a radial direction. In the steady state, the rate of heat flow through each section is the same and is represented by

$$q = 2\pi r_1 l \bar{h}_i (T_i - T_1) = \frac{T_{\text{hot}} - T_1}{R_1} \qquad \text{for the inner surface}$$

$$q = \frac{2\pi k_1 l}{\ln (r_2/r_1)} (T_1 - T_2) = \frac{T_1 - T_2}{R_2} \qquad \text{for the inner cylinder}$$

$$q = \frac{2\pi k_2 l}{\ln (r_3/r_2)} (T_2 - T_3) = \frac{T_2 - T_3}{R_3} \qquad \text{for the outer cylinder}$$

$$q = 2\pi r_3 l \bar{h}_o (T_3 - T_o) = \frac{T_3 - T_{\text{cold}}}{R_4} \qquad \text{for the outer surface}$$

In most practical applications the temperature of the fluid inside and the temperature of the medium surrounding the insulation are known or specified. The intermediate temperatures can be eliminated by addition of the temperature-difference terms and transposition. The resulting expression for the rate of heat flow through two concentric cylinders then becomes

$$q = \frac{T_i - T_o}{\dfrac{1}{2\pi r_1 l\bar{h}_i} + \dfrac{\ln(r_2/r_1)}{2\pi k_1 l} + \dfrac{\ln(r_3/r_2)}{2\pi k_2 l} + \dfrac{1}{2\pi r_3 l\bar{h}_o}} = \frac{T_{\text{hot}} - T_{\text{cold}}}{\sum_{n=1}^{n=4} R_n} \tag{2-18}$$

The overall heat-transfer coefficient U for this system can be based on any area, but its numerical value will depend on the area selected. Since the outer diameter is the easiest to measure in practice, $A_o = 2\pi r_3 l$ is usually chosen as the base area and the rate of heat flow is

$$q = UA_o(T_{\text{hot}} - T_{\text{cold}})$$

Then, by comparison with Eq. 2-18, the overall heat-transfer coefficient becomes

$$U = \frac{1}{\dfrac{r_3}{r_1\bar{h}_i} + \dfrac{r_3\ln(r_2/r_i)}{k_1} + \dfrac{r_3\ln(r_3/r_2)}{k_2} + \dfrac{1}{\bar{h}_o}} \tag{2-19}$$

EXAMPLE 2-7. Calculate the heat loss per linear foot from a 3-in.-steel sched. 40 pipe (3.07 in. ID, 3.500 in. OD, $k = 25$ Btu/hr ft F) covered with a $\frac{1}{2}$-in. thickness of asbestos insulation ($k = 0.11$ Btu/hr ft F). The pipe transports a fluid at 300 F with an inner unit-surface conductance of 40 Btu/hr sq ft F and is exposed to ambient air at 80 F with an average outer unit-surface conductance of 4.0 Btu/hr sq ft F.

Solution: Using Eq. 2-18, the rate of heat transfer for a length $l = 1$ ft is

$$\begin{aligned} q &= \frac{T_i - T_o}{R_1 + R_2 + R_3 + R_4} \\ &= \frac{220}{\dfrac{1}{\pi(3.07/12)40} + \dfrac{\ln(3.5/3.07)}{2\pi(25)} + \dfrac{\ln(4.5/3.5)}{2\pi(0.11)} + \dfrac{1}{\pi(4.5/12)(4)}} \\ &= \frac{220}{0.0312 + 0.00085 + 0.363 + 0.212} = 362 \text{ Btu/hr ft} \end{aligned}$$

Ans.

It is to be noted that the thermal resistance is concentrated in the insulation and in the low surface conductance at the outer surface, while the resistance of the metal wall is negligible. If the pipe were bare, the heat loss would be 1040 Btu/hr ft, or nearly thrice as large as with the insulation.

Critical thickness of insulation. The addition of insulation to the outside of small pipes or wires does not always reduce the heat transfer. We have previously noted that the radial rate of heat flow through a hollow cylinder is inversely proportional to the logarithm of the outer radius and the rate of heat dissipation from the outer surface is directly proportional to this radius. Thus, for a single-wall tube of fixed inner radius r_i, an increase in outer radius r_o (e.g., the insulation thickness) increases the thermal resistance due to conduction *logarithmically* and at the same time reduces the thermal resistance at the outer surface *linearly* with r_o. Since the total thermal resistance is proportional to the sum of these two resistances, the rate of heat flow may increase as insulation is added to a bare pipe or wire. If the insulation thickness is then further increased, the heat loss gradually drops below the loss for a bare surface. This principle is widely utilized in electrical engineering where lagging is provided for current-carrying wires and cables, not to reduce the heat loss, but to increase it. It is also of importance in refrigeration, where heat flow to the cold refrigerant should be kept at a minimum. In many such installations where small-diameter pipes are used, insulation on the outside surface would increase the rate of heat flow.

The relation between heat transfer and insulation thickness can be studied quantitatively with the aid of Eq. 2-18. In many practical situations the thermal resistance is concentrated in the insulation and at the outer surface. We shall therefore simplify Eq. 2-18 by assuming that T_i is the temperature at the inner surface of the insulation. This boundary condition applies to an insulated electric wire whose outer surface temperature T_i is fixed by the current density, the wire size, and the material. Then,

$$q = \frac{2\pi kl(T_i - T_o)}{\ln(r_o/r_i) + k/\bar{h}_o r_o} \tag{2-20}$$

where r_o is the outer radius, r_i the inner radius, and k the thermal conductivity of the insulation.

For a fixed value of r_i, the rate of heat flow is a function of r_o, i.e., $q = q(r_o)$, and will be a maximum at the value of r_o for which

$$\frac{dq}{dr_o} = \frac{-2\pi kl(T_i - T_o)[1/r_o - (k/\bar{h}_o r_o^2)]}{[\ln(r_o/r_i) + k/\bar{h}_o r_o]^2} = 0 \tag{2-21}$$

From Eq. 2-21 the radius for maximum heat transfer, called the *critical radius*, is $r_{oc} = k/\bar{h}_o$.

EXAMPLE 2-8. An electrical cable, $\frac{1}{2}$ in. OD, is to be insulated with rubber ($k = 0.09$ Btu/hr ft F). The cable is to be located in air ($\bar{h}_o = 1.5$ Btu/hr sq ft F) at 70 F. Investigate the effect of insulation thickness on the heat dissipation, assuming a cable surface temperature of 150 F.

Solution: Applying Eq. 2-20, the rate of heat dissipation per unit length is

$$q = \frac{2\pi(0.09)(150 - 70)}{\ln(r_o/\frac{1}{4}) + (0.09)(12/1.5r_o)} = \frac{45.2}{\ln 4r_o + (0.72/r_o)} \text{ Btu/hr ft}$$

if r_o is in inches. The first term of the denominator is proportional to the thermal resistance of the insulation, the second term to the surface resistance. In Fig. 2-8 each of these terms is plotted against the outer radius r_o. The dotted line, representing the sum of both terms, has a minimum at $r_{oc} = 0.72$ in. This is the critical radius at which the rate of heat dissipation reaches a maximum value of

$$q = \frac{45.2}{2.08} = 22.8 \text{ Btu/hr ft length}$$

If the wire were bare, the rate of heat dissipation would be 15.7 Btu/hr ft length, a reduction of 45 percent.

In a practical situation, the selection of the insulating thickness also requires a cost analysis, and usually a compromise between the desirability of dissipating as much heat as possible and the necessity of keeping cost down must be made. Such a compromise might be an insulation thickness of $\frac{1}{4}$ in. ($r_o = 0.5$ in.), which requires only about 50 percent of the material, yet dissipates heat at a rate equal to 98 percent of the maximum rate.

For cases when r_i is larger than $k/\bar{h}_o$, addition of insulation will always reduce the rate of heat transfer, and the optimum insulation thickness must be determined by a cost analysis that takes into account the cost

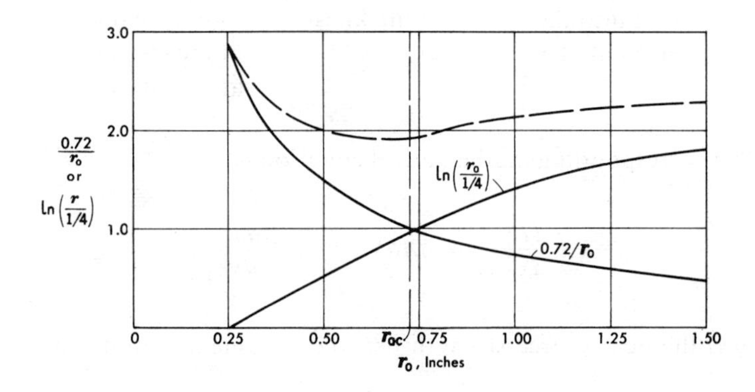

Fig. 2-8. Variation of thermal resistance with outside radius for an insulated electric wire.

and depreciation of the insulation, the cost and depreciation of the equipment required to make up for the energy lost as heat, and sometimes the space occupied by the insulation.

2-3. Systems with heat sources

Systems with heat sources (or sinks) are encountered in many branches of engineering. Typical examples are electric coils, resistance heaters, nuclear reactors, and the combustion of fuel in the fuel bed of a boiler furnace. The dissipation of heat from internal sources is also an important consideration in rating electric motors, generators, and transformers.

In this section we shall consider two simple cases: steady-state heat conduction in a flat plate and a circular cylinder with homogeneous internal heat generation. For a treatment of more complicated problems such as systems with nonuniform heat sources, constant local heat sources, or moving heat sources, see Refs. 3 and 8.

Flat plate with uniformly distributed heat sources. Consider a flat plate in which heat is generated uniformly. This plate could be a heating element such as a flat bus bar in which heat is generated by passing an electric current through it. If we assume that steady state exists, that the material is homogeneous, and that the plate is large enough that end effects may be neglected, an energy equation for a differential element (Fig. 2-9) can be expressed semantically as

Rate of heat conduction through the left face into the element at x	+	Rate of heat generation in the element of thickness dx	=	Rate of heat conduction through the right face out of the element at $x + dx$

The corresponding mathematical equation is

$$-kA\left.\frac{dT}{dx}\right|_{\text{at } x} + \dot{q}A\,dx = -kA\left.\frac{dT}{dx}\right|_{\text{at } x+dx} \tag{2-22}$$

where $\dot{q}$ is the heat-source-strength per unit volume and unit time. Since

$$-kA\left.\frac{dT}{dx}\right|_{x+dx} = -kA\left.\frac{dT}{dx}\right|_{x} + \frac{d}{dx}\left(-kA\left.\frac{dT}{dx}\right|_{x}\right)dx \tag{2-23}$$

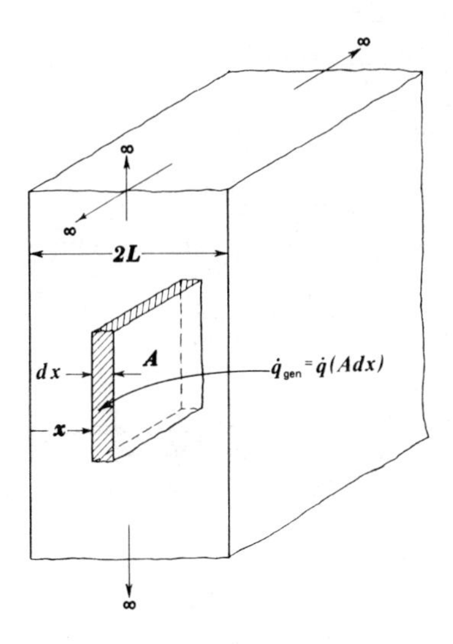

Fig. 2-9. Sketch illustrating nomenclature for heat conduction in a plane wall with internal heat generation.

Eq. 2-22 becomes

$$\dot{q} = -\frac{d}{dx}\left(k\,\frac{dT}{dx}\right) \tag{2-24}$$

If the thermal conductivity is constant and the heat generation uniform, Eq. 2-24 reduces to

$$-k\,\frac{d^2T}{dx^2} = \dot{q} \tag{2-25}$$

A solution to Eq. 2-25 is obtained by two successive integrations. The first yields the temperature gradient

$$\frac{dT}{dx} = -\frac{\dot{q}}{k}\,x + C_1 \tag{2-26}$$

and the second integration gives the temperature distribution

$$T = -\frac{\dot{q}}{2k}\,x^2 + C_1x + C_2 \tag{2-27}$$

where C_1 and C_2 are constants of integration whose values are determined by the boundary conditions. If we specify that the temperatures at both faces is T_o, then the boundary conditions are

$$T = T_o \quad \text{at } x = 0 \qquad \text{and} \qquad T = T_o \quad \text{at } x = 2L$$

For the solution to satisfy these conditions we substitute them successively into Eq. 2-27 to solve for C_1 and C_2. This yields

$$T_o = C_2$$

and

$$T_o = -\frac{\dot{q}}{2k}4L^2 + C_1 2L + T_o$$

Solving for C_1 we obtain

$$C_1 = \frac{\dot{q}L}{k}$$

Substituting these expressions for C_1 and C_2 in Eq. 2-27 yields the temperature distribution as

$$T = -\frac{\dot{q}}{2k}x^2 + \frac{\dot{q}L}{k}x + T_o \tag{2-28}$$

$$T - T_o = \frac{\dot{q}L^2}{2k}\left[2\left(\frac{x}{L}\right) - \left(\frac{x}{L}\right)^2\right]$$

Thus, the temperature distribution across the plate is a parabola with the apex at the median plane, $x = L$. The temperature difference between the center plane and the surface is

$$(T - T_o)_{\max} = \frac{\dot{q}L^2}{2k} \tag{2-29}$$

If the plate is immersed in a fluid at T_∞ and the surface conductance at both faces is $\bar{h}_o$, then the heat generated in one half of the plate must flow continuously through the adjacent face under steady-state conditions. This condition, expressed algebraically for a unit of area, is

$$\dot{q}L = -k\frac{\partial T}{\partial x}_{\text{at } x=0} = \bar{h}_o(T_o - T_\infty) \tag{2-30}$$

In Eq. 2-30 the first term represents the rate at which heat is generated in the plate, the second term the rate at which heat is conducted to the surface, and the third term the rate at which heat flows by convection and radiation from the surface to the surrounding medium. The temperature difference $T_o - T_\infty$ required to remove the heat from the surface is therefore

$$T_o - T_\infty = \frac{\dot{q}L}{\bar{h}_o} \tag{2-31}$$

EXAMPLE 2-9. A fluid ($T_\infty = 150$ F) of low electrical conductivity is heated by a long iron plate, $\frac{1}{2}$ in. thick and 3 in. wide. Heat is generated uniformly in the plate at a rate $\dot{q} = 100{,}000$ Btu/hr cu ft by passing electric current through it. Determine the unit-surface conductance required to maintain the temperature of the bar below 300 F.

Solution: Disregarding the heat dissipated from the edges, Eq. 2-29 applies and the temperature difference between the mid-plane and the surface is

$$(T - T_o)_{\max} = \frac{\dot{q}L^2}{2k} = \frac{(100{,}000)(\frac{1}{4}/12)^2}{(2)(25)} = 0.87 \text{ F}$$

The temperature drop in the iron is so low because its thermal conductivity is high ($k = 25$ Btu/hr ft F). From Eq. 2-31 we get

$$\bar{h}_o = \frac{\dot{q}L}{T_o - T_\infty} = \frac{(100{,}000)(\frac{1}{4}/12)}{150} = 14 \text{ Btu/hr sq ft F}$$

Thus the minimum unit-surface conductance which will keep the temperature in the heater below 300 F is 14 Btu/hr sq ft F. *Ans.*

Heat-transfer problems described in terms of simple differential equations, such as Eq. 2-25, are highly "idealized." In most real systems the thermal conductivity is a function of temperature, the heat sources are functions of the coordinates, and the boundary conditions are complex. In such cases it may not be possible to describe the problem in terms of an equation which yields an analytical solution, but it is always possible to obtain a numerical solution by means of a computer. The next problem illustrates the application of a computer to a relatively simple case for which the results of the computer can be compared with the answer obtained from an analytical solution. The same method can, with minor modifications, be applied to situations where the thermal conductivity is a function of temperature and the rate of heat generation is not uniform.

EXAMPLE 2-10. A large, one foot thick slab ($k = 25$ Btu/hr ft F) contains uniformly distributed heat sources ($\dot{q} = 100{,}000$ Btu/hr ft^3). The temperature at one face is 2000 F and heat is transferred to that surface [$q(0)$] at 1000 Btu/hr ft^2. Write a FORTRAN program to determine the steady state temperature distribution in the slab.

Solution: If end effects are neglected, the one-dimensional conduction equation applies, or

$$\frac{d^2 T}{dx^2} = -\frac{\dot{q}}{k}$$

The boundary conditions are:

$$T(0) = 2000 \text{ F}$$

$$\frac{dT}{dx}(0) = \frac{q(0)}{k} = -\frac{1000}{25} = -40 \text{ F/ft}$$

This equation can be solved by approximating the temperature and its derivative by a Taylor series:

$$T(x + \Delta x) \simeq T(x) + \frac{dT}{dx}(x)\,\Delta x$$

and from Eq. 2-22

$$\frac{dT}{dx}(x + \Delta x) \simeq \frac{dT}{dx}(x) + \left(-\frac{\dot{q}}{k}\right)\Delta x$$

The numerical technique to be used is called a "marching solution." The slab is divided into ten equal slices ($\Delta x = L/10$) and computations are carried out in a stepwise manner as follows:

1) The solution is "started" with the values of T and dT/dx at $x = 0$ which are known from the boundary conditions.

Table 2-1. Correspondence between physical variables and FORTRAN symbols for Example 2-10.

FORTRAN Variable	Equation Symbol	Description	Units
DTDX	dT/dx	Temperature gradient	F/ft
DX	Δx	Spatial increment	ft
FLUXO	$q(0)$	External surface heat flux	Btu/hr ft^2
K	k	Thermal conductivity	Btu/hr ft F
T	T	Temperature	F
QDOT	$\dot{Q}$	Heat generation rate	Btu/hr ft^3
X	x	Spatial coordinate	ft
XEND	L	Slab thickness	ft

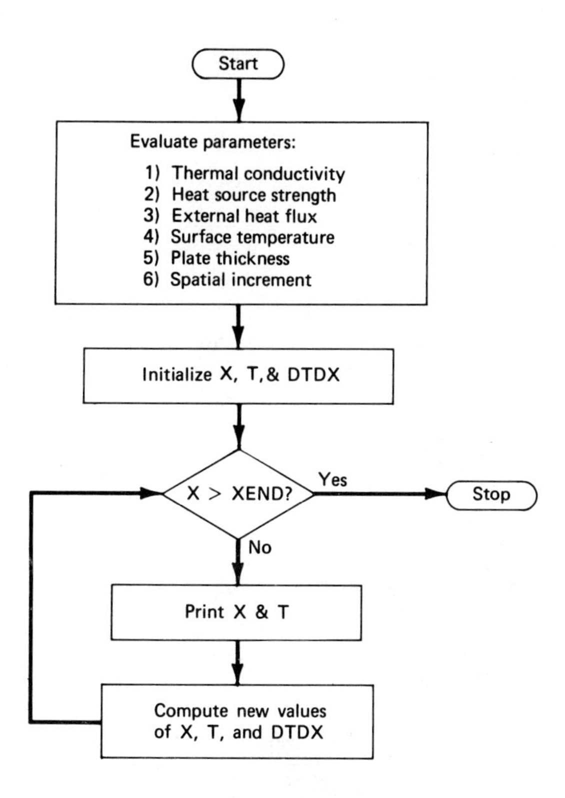

Fig. 2-10. Flow chart for Example 2-10.

2) The values at the first interior point are computed from the equations above.
3) In general, the values of T and dT/dx are known at a point x. Then values at the next point $x + \Delta x$ are computed from the two equations above.
4) This "marching" process is continued until $x = 1$ ft.

The symbols for the variables used in the program are related to the symbols used in the statement of the problem in Table 2-1, a flow diagram of the sequence of steps used in the computer solution is presented in Fig. 2-10, and the FORTRAN program for the solution of the problem is shown in Table 2-2, with explanatory remarks following the computer operations.

Table 2-2. FORTRAN program for Example 2-10.

```
      REAL K
```

The parameter K, normally an integer in FORTRAN, is declared to be a floating point constant. K is the thermal conductivity, called k in the text. (There are no small letters in FORTRAN.)

```
      K=25.
      QDOT=100000.
      FLUXO = 1000.
      XEND=1.
      DX=XEND/10.
```

The values for the parameters of the problem, i.e., K, thermal conductivity, QDOT, the internal heat generation rate per unit volume, FLUX0, the heat flux at the slab surface, XEND, the thickness of the slab, and DX, the spatial increment, are specified.

```
      T=2000.
      X=0.
      DTDX=-FLUXO/K
```

The initial values of the spatial coordinate X, the temperature T, and the temperature gradient DTDX are specified.

```
1     IF (X.GT.XEND) GO TO 2
```

Statement 1 determines whether the iteration is completed or whether another step is to be carried out. If the iteration is complete, i.e., if X > XEND, the program is diverted to statement 2 which terminates the program; otherwise the next 5 statements are executed.

```
      PRINT 3,X,T
      T=T+DTDX*DX
      DTDX=DTDX-(QDOT*DX)/K
      X=X+DX
      GO TO 1
```

The present values of X and T are printed as specified in the format statement 3. New values of T, DTDX, and X are computed and the computation is returned to statement 1.

```
2     CALL EXIT
3     FORMAT (2F7.1)
      END
```

Statement 2 terminates the run if called for by statement 1. The output (below) is printed out as specified in statement 3. The final card of the FORTRAN program is the END card.

```
    0.0  2000.0
     .1  1996.0
```

Table 2-2. Continued.

.2	1952.0
.3	1868.0
.4	1744.0
.5	1580.0
.6	1376.0
.7	1132.0
.8	848.0
.9	524.0
1.0	160.0

The program output consists of two columns; the first gives the values of the spatial coordinate X in feet; the second gives the values of the temperature in degrees F at that point.

Long solid cylinder with uniformly distributed heat sources. A long solid circular cylinder with uniform internal heat generation may be thought of as an idealization of a real system such as an electric coil, in which heat is generated as a result of the electric current in the wire, or a cylindrical fuel element of uranium 235, in which heat is generated by nuclear fission. The energy equation for an annular element (Fig. 2-11) formed between an inner cylinder of radius r and an outer cylinder of radius $r + dr$ is

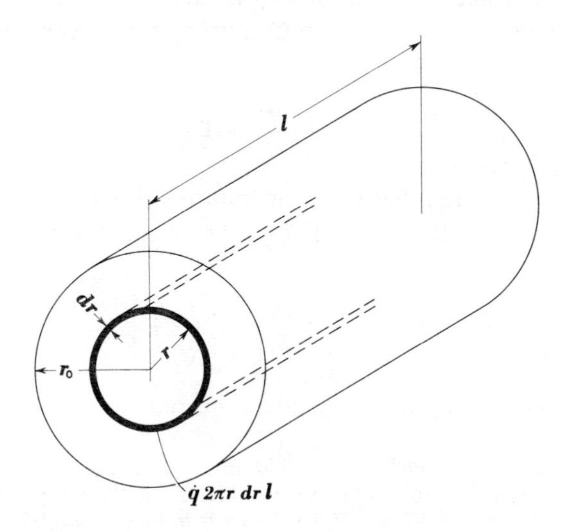

Fig. 2-11. Sketch illustrating nomenclature for heat conduction in a long circular cylinder with internal heat generation.

$$-kA_r \left.\frac{dT}{dr}\right|_r + \dot{q}l2\pi r dr = -kA_{r+dr} \left.\frac{dT}{dr}\right|_{r+dr} \tag{2-32}$$

where $A_r = 2\pi rl$ and $A_{r+dr} = 2\pi(r + dr)l$. Relating the temperature gradient at $r + dr$ to the temperature gradient at r, we obtain, after some simplifications,

$$\dot{q}r = -k\left(\frac{dT}{dr} + r\frac{d^2T}{dr^2}\right) \tag{2-33}$$

Integration of Eq. 2-33 can best be accomplished by noting that

$$\frac{d}{dr}\left(r\frac{dT}{dr}\right) = \frac{dT}{dr} + r\frac{d^2T}{dr^2}$$

and rewriting it in the form

$$\dot{q}r = -k\frac{d}{dr}\left(r\frac{dT}{dr}\right)$$

Integration then yields

$$\frac{\dot{q}r^2}{2} = -kr\frac{dT}{dr} + C_1 \tag{2-34}$$

from which we deduce that, to satisfy the boundary condition $dT/dr = 0$ at $r = 0$, the constant of integration C_1 must be zero. Another integration yields the temperature distribution

$$T = -\frac{\dot{q}r^2}{4k} + C_2$$

To satisfy the condition that the temperature at the outer surface, i.e., at $r = r_o$, be T_o, $C_2 = [(\dot{q}r_o^2/4k) + T_o]$. The temperature distribution is therefore

$$T = T_o + \frac{\dot{q}r_o^2}{4k}\left[1 - \left(\frac{r}{r_o}\right)^2\right] \tag{2-35}$$

with the maximum temperature equal to $(\dot{q}r_o^2/4k) + T_o$.

EXAMPLE 2-11. A sketch of a graphite-moderated reactor, typical of the type which will be used for commercial power production, is shown in Fig. 2-12. Heat is generated at the rate of 7.2×10^6 Btu/hr cu ft in long 2-in.-OD uranium rods[2] ($k = 17$ Btu/hr ft F) which are jacketed by an annulus in which water is flowing.

---

[2]This value of the thermal conductivity is taken from Ref. 11.

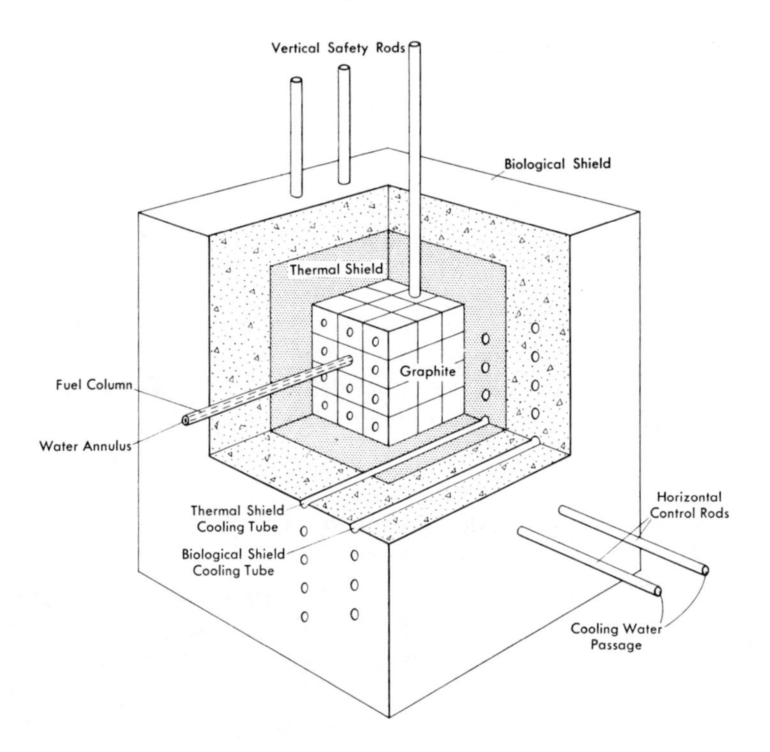

Fig. 2-12. Graphite-moderated reactor. (Reprinted from *General Electric Review*.)

For an average water temperature of 270 F and a unit-surface conductance of 10,000 Btu/hr sq ft F, determine the maximum temperature of the fuel rods.

Solution: Applying Eq. 2-34,

$$-k \left.\frac{dT}{dr}\right|_{r = r_o} = \frac{\dot{q} r_o}{2} = \frac{(7.2 \times 10^6 \text{ Btu/hr cu ft})(1/12 \text{ ft})}{2}$$

$$= 3 \times 10^5 \text{ Btu/hr sq ft}$$

The rate of heat flow by conduction at the outer surface equals the rate of heat flow by convection from the surface to the water

$$2\pi r_o \left(-k \frac{dT}{dr}\right)\Bigg|_{r_o} = 2\pi r_o \bar{h}_o (T_o - T_{\text{water}})$$

from which

$$T_o = \frac{-k(dT/dr)\,|_{ro}}{\bar{h}_o} + T_{\text{water}}$$

Upon substituting the data specified in the statement of the problem, we get

$$T_o = \frac{3 \times 10^5 \text{ Btu/hr sq ft}}{1 \times 10^4 \text{ Btu/hr sq ft F}} + 270 \text{ F} = 300 \text{ F}$$

and

$$T_{\max} = T_o + \frac{\dot{q} r_o^2}{4k} = 300 + 735 = 1035 \text{ F} \qquad \textit{Ans.}$$

The exact determination of the rate of heat conduction and temperature distribution in a nuclear pile is a complicated problem. The preceding problem only illustrates the basic idea, and for more complete information Refs. 12, 13, 14, 15, and 16 should be consulted.

2-4. Heat transfer from extended surfaces

The problems considered in this section are encountered in practice when a solid of relatively small cross-sectional area protrudes from a large body into a fluid at a different temperature. Such extended surfaces have wide industrial applications as fins attached to the walls of heat-transfer equipment for the purpose of increasing the rate of heating or cooling.

Fins of uniform cross section. As a simple illustration, consider a pin fin having the shape of a rod whose base is attached to a wall at surface temperature T_s (Fig. 2-13). The fin is cooled along its surface by a fluid at temperature T_∞. The fin has a uniform cross-sectional area A, is made of a material having a uniform conductivity k, and the heat-transfer coefficient between the surface of the fin and the fluid is $\bar{h}$. We shall assume that transverse temperature gradients are so small that the temperature at any cross section of the rod is uniform, i.e., $T = T(x)$ only. As shown in Ref. 10, even in a relatively thick fin the error in a one-dimensional solution is less than one percent.

To derive an equation for the temperature distribution, we make a heat balance for a small element of the fin. Heat flows by conduction into the left face of the element, while heat flows out of the element by conduction through the right face and by convection from the surface. Under steady-state conditions,

Rate of heat flow by conduction into element at x	=	rate of heat flow by conduction out of element at $(x + dx)$	+	rate of heat flow by convection from surface between x and $(x + dx)$

In symbolic form, this equation becomes

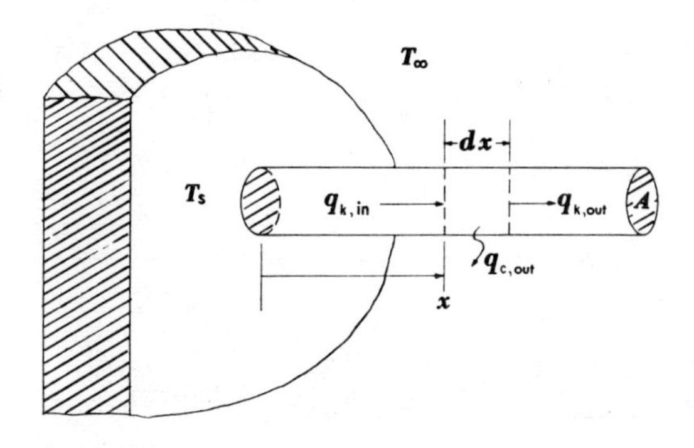

Fig. 2-13. Sketch and nomenclature for a pin fin protruding from a wall.

$$-kA\frac{dT}{dx} = \left[-kA\frac{dT}{dx} + \frac{d}{dx}\left(-kA\frac{dT}{dx}\right)dx\right] + \bar{h}Pdx(T - T_\infty) \tag{2-36}$$

where P is the perimeter of the rod and Pdx represents the surface area between sections x and $(x + dx)$ in contact with the surrounding fluid. If k and $\bar{h}$ are uniform equation 2-36 can be simplified to

$$\frac{d^2T}{dx^2} = m^2(T - T_\infty) \tag{2-37}$$

where $m^2 = \bar{h}P/kA$. Equation 2-37 is a standard form of an ordinary second-order linear differential equation whose general solution is

$$T - T_\infty = C_1e^{(m)0} + C_2e^{-mx} \tag{2-38}$$

where C_1 and C_2 are constants of integration whose values must be determined from the boundary conditions. One of the boundary conditions is $T = T_s$ at $x = 0$; the temperature at the base of the rod equals the temperature of the surface to which the rod is attached. To obtain a solution satisfying this condition, we substitute it in Eq. 2-38 and get

$$T_s - T_\infty = C_1e^{(m)0} + C_2e^{-(m)0} = C_1 + C_2 \tag{2-39}$$

To solve for C_1 and C_2 we need another equation, i.e., another boundary condition. The second boundary condition depends upon the nature of the problem. Since the appropriate selection of boundary conditions often causes considerable difficulty, we shall consider several cases which will assist the reader in gaining some facility in applying physical

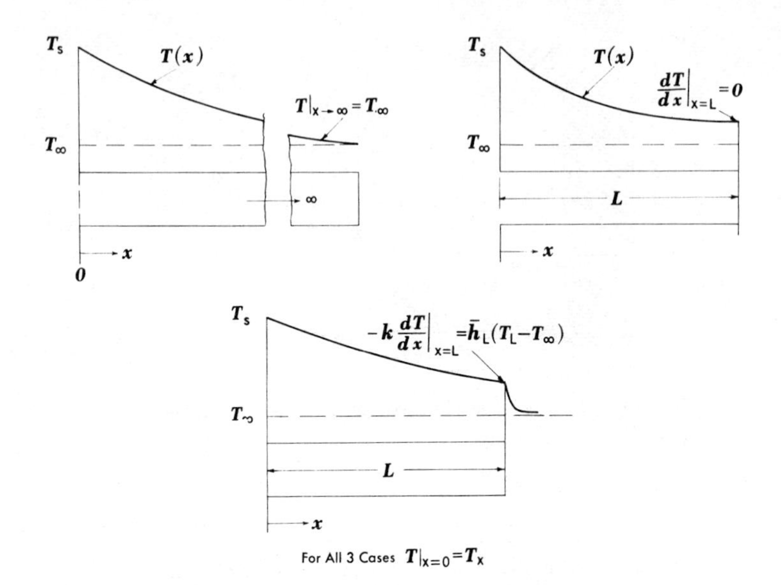

Fig. 2-14. Schematic illustration of three boundary conditions for a pin fin.

concepts to a mathematical analysis. Figure 2-14 illustrates schematically the conditions described by the three boundary conditions to be analyzed.

a) If the rod is infinitely long, its temperature will approach the temperature of the fluid as $x \to \infty$, or $T = T_\infty$ at $x \to \infty$. Substituting this condition in Eq. 2-38 yields

$$T - T_\infty = 0 = C_1 e^{m\infty} + C_2 e^{-m\infty} \tag{2-40}$$

Since the second term is zero, the boundary condition is satisfied only if $C_1 = 0$. Substituting 0 for C_1 in Eq. 2-39 gives

$$C_2 = T_s - T_\infty$$

and the temperature distribution becomes

$$T - T_\infty = (T_s - T_\infty)e^{-mx} \tag{2-41}$$

The heat-flow rate from the fin to the fluid can be obtained by two different methods. Since the heat flowing by conduction across the root of the fin must be transmitted by convection from the surface of the rod to the fluid,

$$q_{\text{fin}} = -kA \left.\frac{dT}{dx}\right|_{x=0} = \int_0^\infty \bar{h}P(T - T_\infty)\,dx \tag{2-42}$$

Differentiating Eq. 2-41 and substituting the result for $x = 0$ in Eq. 2-42 yields

$$q_{\text{fin}} = -kA[-m(T_s - T_\infty)e^{(-m)0}]_{x=0} = \sqrt{\bar{h}PkA}(T_s - T_\infty) \quad \text{(2-43)}$$

The same result is obtained by evaluating the convective heat flow from the surface of the rod

$$q_{\text{fin}} = \int_0^\infty \bar{h}P(T_s - T_\infty)e^{-mx}dx = -\frac{\bar{h}P}{m}(T_s - T_\infty)e^{-mx}\Big|_0^\infty$$

$$= \sqrt{\bar{h}PkA}\,(T_s - T_\infty)$$

Equations 2-41 and 2-43 are reasonable approximations of the temperature distribution and heat-flow rate in a finite fin if its length is very large compared to its cross-sectional area.

b) If the rod is of finite length, but the heat loss from the end of the rod is neglected, or if the end of the rod is insulated, the second boundary condition requires that the temperature gradient at $x = L$ be zero, or $dT/dx = 0$ at $x = L$. With these conditions

$$C_1 = \frac{T_s - T_\infty}{1 + e^{2mL}} \quad \text{and} \quad C_2 = \frac{T_s - T_\infty}{1 + e^{-2mL}}$$

The complete solution is therefore

$$T - T_\infty = (T_s - T_\infty)\left(\frac{e^{mx}}{1 + e^{2mL}} + \frac{e^{-mx}}{1 + e^{-2mL}}\right) \quad \text{(2-44)}$$

or, in simplified dimensionless form,[3]

$$\frac{T - T_\infty}{T_s - T_\infty} = \frac{\cosh m(L - x)}{\cosh (mL)} \quad \text{(2-44a)}$$

As L becomes infinite, Eq. 2-44 reduces to the previous solution, as would be expected.

The heat loss from the fin can be found from Eq. 2-42 by substituting the temperature gradient at the root

$$\left.\frac{dT}{dx}\right|_{x=0} = (T_s - T_\infty)m\left(\frac{1}{1 + e^{2mL}} - \frac{1}{1 + e^{-2mL}}\right)$$

---

[3]The reduction of Eq. 2-44 to 2-44a is left as an exercise for the reader. The hyperbolic cosine, cosh for short, is defined by $\cosh x = (e^x + e^{-x})/2$.

The above equation can be put into more convenient form by placing the final member of the right-hand side over a common denominator, and noting that

$$\frac{e^{mL} - e^{-mL}}{e^{mL} + e^{-mL}} = \tanh(mL)$$

where tanh is the hyperbolic tangent. The heat-flow rate from the rod is then found to be

$$q_{\text{rod}} = -kA \left.\frac{dT}{dx}\right|_{x=0} = \sqrt{P\bar{h}kA}\,(T_s - T_\infty) \tanh(mL) \qquad \text{(2-45)}$$

c) If the end of the rod loses heat by convection, the heat flowing by conduction to the face at $x = L$ must be equal to the convection heat flow from the end section of the rod to the fluid, or

$$-k \left.\frac{dT}{dx}\right|_{x=L} = \bar{h}_L(T_{x=L} - T_\infty)$$

The heat-transfer coefficient at the end face $\bar{h}_L$ is not necessarily equal to the value of $\bar{h}$ over the circumferential surface of the rod. Substituting for $T_{x=L}$ and $(dT/dx)_{x=L}$ from Eq. 2-38 we obtain

$$q_{x=L} = -k(C_1 m e^{mx} - C_2 m e^{-mx})_{x=L} = \bar{h}_L(T_{x=L} - T_\infty)$$

or

$$C_2 e^{-mL} - C_1 e^{mL} = \frac{\bar{h}_L}{km}(C_1 e^{mL} + C_2 e^{-mL}) \qquad \text{(2-46)}$$

Equations 2-39 and 2-46 can now be solved simultaneously to obtain the constants C_1 and C_2, just as in the previous cases. The algebra is somewhat more involved, but the principle is the same. It is left as an exercise for the reader to show that the dimensionless temperature distribution along the fin is

$$\frac{T - T_\infty}{T_s - T_\infty} = \frac{\cosh m(L - x) + (\bar{h}_L/mk)\sinh m(L - x)}{\cosh mL + (\bar{h}_L/mk)\sinh mL} \qquad \text{(2-47)}$$

and the heat-flow rate from the fin is

$$q_{\text{fin}} = \sqrt{P\bar{h}Ak}\,(T_s - T_\infty)\frac{\sinh mL + (\bar{h}_L/mk)\cosh mL}{\cosh mL + (\bar{h}_L/mk)\sinh mL} \qquad \text{(2-48)}$$

It is important to check the end results of more complicated cases by reducing them to the results available for simpler cases, because this often shows up errors which might otherwise remain unnoticed. Comparing Eqs. 2-47 and 2-48 with the corresponding results for case b (i.e., the end face insulated), we note that, for $\bar{h}_L = 0$, these equations indeed reduce to those obtained previously for $(dT/dx)_{\text{at } x=L} = 0$. Only the second terms in the numerator and the denominator contain $\bar{h}_L$. These terms indicate the influence of the heat loss from the end face of the rod and modify the results obtained when the end losses were neglected.

EXAMPLE 2-12. The temperature of steam flowing in a 3-in. steel pipe has been measured in the laboratory by means of a mercury-in-glass thermometer immersed in an oil-filled steel well. While a reliable pressure gauge in the line read 153 psia, the mercury thermometer indicated a temperature of 355 F. Reference to steam tables indicates that the saturation temperature of steam at 153 psia is 360 F. At first glance the thermometer reading, since it is below the saturation temperature, seems in error.

For the temperature-measuring station shown in Fig. 2-15, show that the thermometer reading is not inconsistent with the pressure reading and estimate the true temperature if the pipe-wall temperature is 200 F and the heat-transfer coefficient between the steam and the thermometer well is 50 Btu/hr sq ft F.

Solution: The thermometer well is essentially a hollow rod protruding into steam at temperature T_∞. Since heat flows from the steam along the well toward the cooler pipe walls, the thermometer does not indicate the true steam temperature, but rather the temperature at the bottom of the well. We can esti-

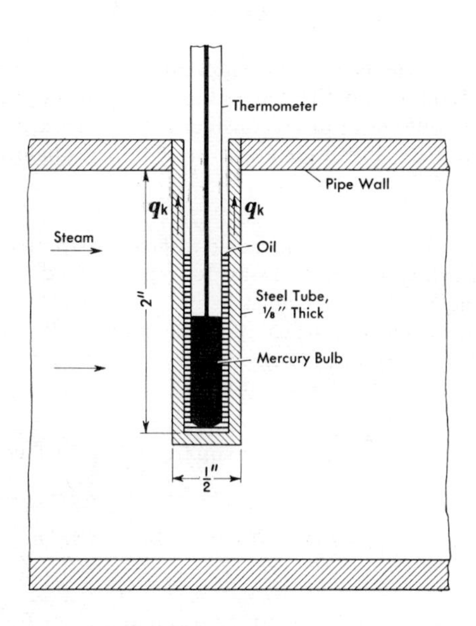

Fig. 2-15. Thermometer well for Example 2-12.

mate the steam temperature by treating the thermometer well as a rod. The conduction along the glass ($k = 0.5$ Btu/hr ft F) is neglected, since it is very small compared with the heat flow along the steel well. The cross-sectional area of the fin A is $(\pi/4)[(1/2)^2 - (1/4)^2] = 0.1475$ sq in. The perimeter P is $\frac{1}{2}\pi = 1.57$ in. The thermal conductivity of steel is 25 Btu/hr ft F. Thus

$$m = \sqrt{\frac{\bar{h}P}{kA}} = \sqrt{\frac{(50)(1.57)}{(25)(0.1475)} 12} = 16 \text{ ft}$$

As a first approximation, we could use Eq. 2-41 with $x = \frac{1}{6}$ ft. Then,

$$T - T_\infty = (T_s - T_\infty)e^{-16/6} = (T_s - T_\infty)0.07$$

Solving the above equation for T_∞ with $T_s = 200$ F, we obtain $T_\infty = 366$ F.

Although this first-order correction yields a reasonable value, the boundary conditions used to obtain Eq. 2-41 are not in physical correspondence with the conditions of the well. Since heat also flows from the end of the fin, Eq. 2-47 is more nearly true. For $x = L$, we get

$$T - T_\infty = (T_s - T_\infty)\frac{1}{\cosh mL + (\bar{h}_L/mk)\sinh mL}$$

Substituting numerical values, we find

$$T - T_\infty = (T_s - T_\infty)\frac{1}{7.18 + [50/(25)(16)](7.04)} = (T_s - T_\infty)(0.124)$$

The preceding equation yields a steam temperature T_∞ of 376 F. Therefore, the steam in the pipe is actually superheated and the error in the thermometer reading was 21 F. This error could be reduced considerably by increasing the length of the thermometer well to 3 in. and inserting it at a 45-deg angle to prevent it from touching the wall.

EXAMPLE 2-13. Estimate the increase in heat-dissipation rate which could be obtained from a cylinder wall by using four pin-shaped fins per square inch, each having a diameter of $\frac{3}{16}$ in. and a height of 1 in. Assume that the heat-transfer coefficient between the surface of the cylinder wall or a fin and the surrounding air is 25 Btu/hr sq ft F, the cylinder wall is at 600 F, and the air at 70 F. The wall and the fins are made of aluminum.

Solution: The heat dissipation per square inch of surface, without the fin, is

$$\frac{q}{A} = \frac{25}{144}(600 - 70) = 92 \text{ Btu/hr sq in.}$$

The heat dissipation for a single fin can be estimated from Eq. 2-45. The rate of heat transfer per fin q is equal to

$$\sqrt{P\bar{h}Ak}(T_s - T_\infty)\tanh mL$$

where

$$P = \left(\frac{3}{16}\right)\pi\left(\frac{1}{12}\right) = 0.0492 \text{ ft}$$

$$A = \left(\frac{3}{16}\right)^2\left(\frac{\pi}{4}\right)\left(\frac{1}{144}\right) = 0.000192 \text{ sq ft}$$

$$k = 120 \text{ Btu/hr ft F}$$

$$\sqrt{P\bar{h}Ak} = \sqrt{(0.049)\,(25)(0.000192)(120)} = 0.167$$

$$mL = \frac{1}{12}\sqrt{\frac{\bar{h}P}{kA}} = \frac{1}{12}\sqrt{\frac{(25)(0.0492)}{(120)(0.000192)}} = 0.192$$

so that

$$q_{\text{fin}} = (0.167)\,(530)\,(0.188) = 16.7 \text{ Btu/hr}$$

For four fins the heat-dissipation rate would be 66.8 Btu/hr. The rate of heat dissipation from the remaining wall surface would be approximately equal to the area not occupied by fins times the product of the heat-transfer coefficient and the temperature potential. On this basis, the total rate of heat transfer for the wall with fins is

$$\frac{q}{A} \simeq 67 + 92\,(1.0 - 0.11) = 149 \text{ Btu/hr sq in.}$$

Thus we see that the use of the fins can increase the heat-dissipation rate by over 50 percent. *Ans.*

The straight rectangular fin (Fig. 2-16) can be treated by the same methods as those used for the rod. If the width of the fin b is large compared with its thickness t, then the fin perimeter is

$$P = 2(b + t) \simeq 2b$$

The cross-sectional area of the fin is $A = bt$, and $m = \sqrt{\bar{h}P/kA} \simeq \sqrt{2\bar{h}/kt}$.

The development of the equations for the temperature distribution and the heat flow in a rectangular fin are identical to the previous cases, and the results may be applied directly. In many practical situations, however, the heat flow to or from a fin with convection at the end is calculated by an approximate method (10). Instead of using Eq. 2-48, the length of the fin is increased by one-half of its thickness and Eq. 2-45 for an insulated end is used. The corrected length in Eq. 2-45 is thus $[L + (t/2)]$. The error introduced by this approximation will be less than 8 percent as long as (ht/k) is less than 0.5.

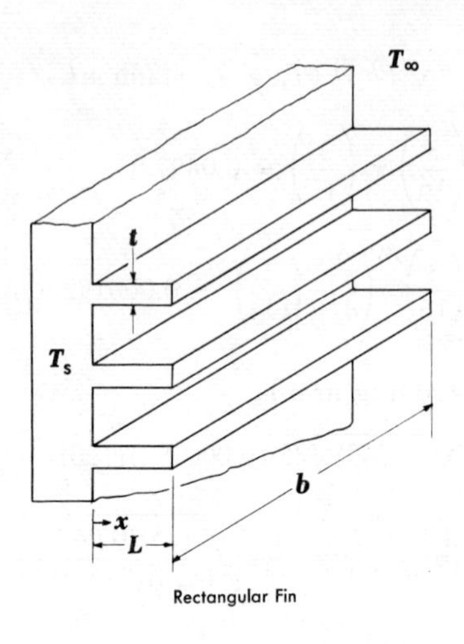

Fig. 2-16. Rectangular-fin nomenclature.

Selection and design of fins. The selection of a suitable fin geometry requires a compromise of the cost, the weight, the available space, and the pressure drop, as well as the heat-transfer characteristics of the extended surface. Harper and Brown (10) and Gardner (9) have analyzed a variety of fin geometries, and their papers are recommended for a general treatment of the problem. The following remarks are largely based on their work.

The heat-transfer effectiveness of a fin is measured by a parameter called the fin efficiency, η_f, which is defined as

$$\eta_f = \frac{\text{actual heat transferred by fin}}{\text{heat which would have been transferred if entire fin were at the base temperature}}$$

Using Eq. 2-45, the fin efficiency for a circular pin fin of diameter D and length L with an insulated end is

$$\eta_f = \frac{\tanh \sqrt{4L^2\bar{h}/kD}}{\sqrt{4L^2\bar{h}/kD}} \tag{2-49}$$

whereas for a fin of rectangular cross section (length L and thickness t) the fin efficiency with an insulated end is (see Ref. 9 and Problem 2-54)

$$\eta_f = \frac{\tanh \sqrt{\bar{h}PL^2/kA}}{\sqrt{\bar{h}PL^2/kA}} \tag{2-50}$$

If a rectangular fin is long, wide, and thin, $(P/A) = 2/t$. The heat loss from the end can be taken into account approximately by increasing L by $t/2$. The fin efficiency then becomes

$$\eta_f = \frac{\tanh[\sqrt{2\bar{h}/kt}(L + t/2)]}{\sqrt{2\bar{h}/kt}(L + t/2)}$$

It is often convenient to use the profile area of a fin, A_m, which for a rectangular shape is Lt whereas for a triangular cross section A_m is $Lt/2$, where t is the base thickness. In Fig. 2-17 the fin efficiency for rectangular and triangular fins are compared. One observes that rectangular fins have higher efficiencies than triangular fins of the same length, but the former also contain twice as much material as the latter. Fig. 2-18 shows the fin efficiency for circumferential fins of rectangular cross section (8, 9, 10).

For a plane surface of area A, the thermal resistance is $1/\bar{h}A$. The addition of fins increases the surface area, but at the same time it also introduces a conductive resistance over that portion of the original surface at which the fins are attached. The addition of fins will therefore not always increase the rate of heat transfer. In practice, the addition of fins is hardly ever justified unless $\bar{h}A/Pk$ is considerably less than unity.

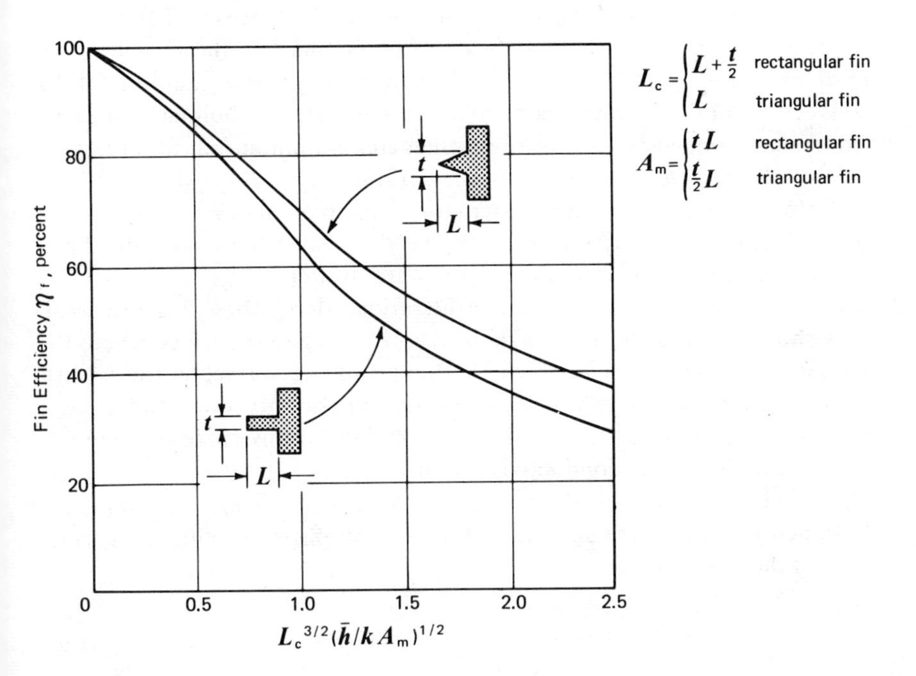

Fig. 2-17. Efficiencies of rectangular and triangular fins.

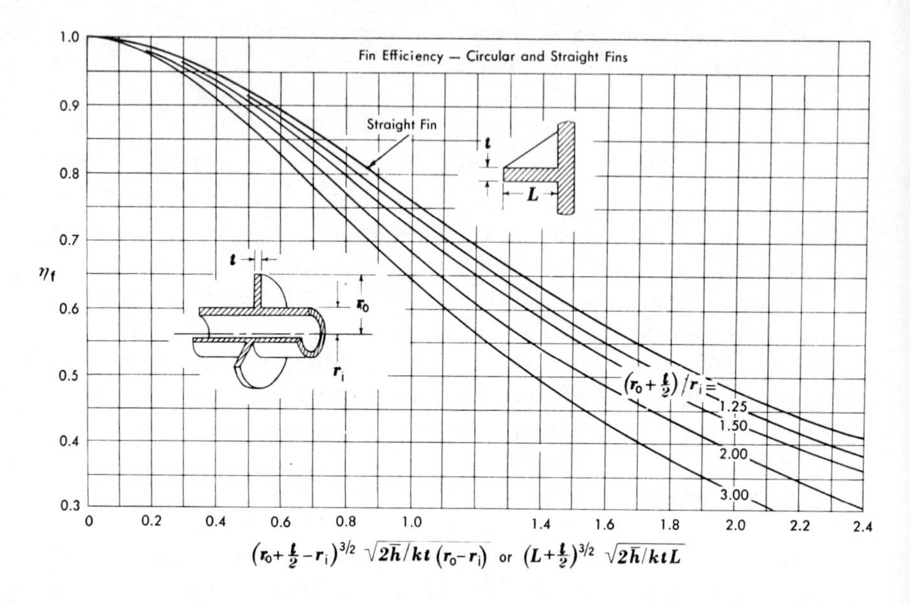

Fig. 2-18. Efficiency of circumferential fins of rectangular cross-sectional area.

Using the values of the average surface conductances in Table 1-2 as a guide, we can easily see that fins effectively increase the heat transfer to or from a gas, are less effective when the medium is a liquid in forced convection, but offer no advantage in heat transfer to boiling liquids or from condensing vapors. For a $\frac{1}{8}$-in.-diameter aluminum pin fin in a typical gas heater, $\bar{h}A/Pk = \bar{h}D/4k = [(20)(0.125/12)] / [(4)(115)] = 0.00045$, whereas in a water heater, for example, $\bar{h}A/Pk = [(1000) \times (0.125/12)] / [(4)(115)] = 0.022$. In the gas heater the fin would therefore be much more effective than in the water heater.

It is apparent from these considerations that, when fins are used, they should be placed on the side of the heat-exchange surface where the heat-transfer coefficient between the fluid and the surface is the lowest. Thin, slender, and closely spaced fins are superior to fewer and thicker fins from the heat-transfer standpoint. Obviously, fins made of materials having a high thermal conductivity are desirable.

To obtain the total efficiency of a surface with fins η_t we combine the unfinned portion of the surface at 100 percent efficiency with the surface area of the fins at η_f, or

$$A\eta_t = A - A_f + A_f\eta_f = A - A_f(1 - \eta_f) \tag{2-51}$$

where A = total heat-transfer area;
A_f = heat-transfer area of the fins.

The overall heat-transfer coefficient U, based on the total outer surface area, for heat transfer between two fluids separated by a wall with fins can then be expressed as

$$U = \frac{1}{\dfrac{1}{\eta_{to}\bar{h}_o} + R_{k_{\text{wall}}} + \dfrac{A_o}{\eta_{ti}A_i\bar{h}_i}} \tag{2-52}$$

where $R_{k_{\text{wall}}}$ = thermal resistance of the wall to which the fins are attached, in Btu/hr sq ft F outside surface;
A_o = total outer surface area, in sq ft;
A_i = total inner surface area, in sq ft;
η_{to} = total efficiency for outer surface;
η_{ti} = total efficiency for inner surface;
$\bar{h}_o$ = average unit conductance for outer surface, in Btu/hr sq ft F;
$\bar{h}_i$ = average unit conductance for inner surface, in Btu/hr sq ft F.

For tubes with fins on the outside only, the usual case in practice, η_{ti} is unity and $A_i = \pi D_i l$.

In the analysis presented in this chapter, details of the convection heat flow between the fin surface and the surrounding fluid have been omitted. A complete engineering analysis not only requires an evaluation of the fin performance, but must also take the interrelation between the fin geometry and the convection heat transfer into account. Problems on the convective heat transfer part of the design will be considered in later chapters.

PROBLEMS

The problems below have been organized in the following manner: Problems 2-1 through 2-7 deal with conduction through simple walls, 2-8 through 2-29 with conduction through composite walls, 2-30 through 2-41 with conduction in systems containing heat sources, and 2-42 through 2-57 with heat transfer to and from fins and extended surfaces.

2-1. The temperatures at the two faces of a 6-in.-thick plane concrete wall are maintained uniform at 50 and 100 F, respectively. Compare the rate of heat flow per unit area when the concrete is dry with the rate of heat flow when the concrete is wet (10 percent moisture).

2-2. A solution whose boiling point is 180 F boils on the outside of a 1-in. tube with a No. 14 BWG gauge wall. On the inside of the tube flows saturated steam at 60 psia. The surface heat-transfer coefficients are on the steam side 1500 and on the exterior surface 1100 Btu/hr sq ft F. Calculate the increase in the rate of heat transfer for a copper over a steel tube. *Ans.* 20%

2-3. The rate of heat flow per unit length q/L through a hollow cylinder of inside radius r_i and outside radius r_o is

$$q/L = \bar{A} k \Delta T/(r_o - r_i)$$

where $\bar{A} = 2\pi(r_o - r_i)/\ln(r_o/r_i)$. Determine the percent error in the rate of heat flow if the arithmetic mean area $\pi(r_o + r_i)$ is used instead of the logarithmic mean area $\bar{A}$ for ratios of inside to outside diameters (D_o/D_i) of 1.5, 2.0, and 3.0. Plot the results.

2-4. Show that the rate of heat conduction per unit length through a long hollow cylinder of inner radius r_i and outer radius r_o, made of a material whose thermal conductivity varies linearly with temperature, is given by the relation

$$q_k L = \frac{T_i - T_o}{(r_o - r_i)/k_m \bar{A}}$$

where T_i = temperature at the inner surface
T_o = temperature at the outer surface
$\bar{A} = 2\pi(r_o - r_i)/\ln(r_o/r_i)$
$k_m = k_o[1 + \beta_k(T_i + T_o)/2]$
L = length of cylinder

2-5. A long hollow cylinder is constructed from a material whose thermal conductivity is a function of temperature according to $k = 0.060 + 0.00060\,T$, where T is in deg F and k is in Btu/hr ft F. The inner and outer radii of the cylinder are 5 and 10 in., respectively. Under steady-state conditions, the temperature at the interior surface of the cylinder is 800 F and the temperature at the exterior surface is 200 F. (a) Calculate the rate of heat transfer per foot length, taking into account the variation in thermal conductivity with temperature. (b) If the surface heat-transfer coefficient on the exterior surface of the cylinder is 3 Btu/hr sq ft F, calculate the temperature of the air on the outside of the cylinder.
Ans. (a) 1959 Btu/hr; (b) 75.3 F

2-6. The following data are given for rock wool:

T(F)	100	200	300	400	500	600
k(Btu/hr ft F)	0.030	0.034	0.039	0.044	0.050	0.057

A 4-in.-thick layer of this rock wool is used to insulate an oven wall. If the temperature at the inner surface is 600 F and at the outer surface is 100 F, calculate the rate of heat flow per square foot and plot the temperature distribution when (a) an average value of k is used and (b) an equation is fitted to the above data for the thermal conductivity and this expression is used.

2-7. A 2-in.-thick slab has one side maintained at 200 F and the other side at 400 F. The temperature at the center plane of the material is 280 F, and the heat flow through the material is 3,500 Btu/hr ft. Obtain an expression for the thermal conductivity of the material as a function of temperature in the form $k = a + bT$, where T is the temperature in F.

2-8. A 1.0-ft-thick wall is made of a material which has a thermal conductivity of 0.5 Btu/hr ft F. The rate of heat transfer through this wall is to be reduced by placing a layer of insulating material with an average thermal conductivity of 0.2 Btu/hr ft F on one side. Assuming that the surface temperatures of the composite wall are to be 2,100 and 100 F, calculate the minimum thickness of insulating material which will insure that the rate of transfer per unit area will not exceed 600 Btu/hr ft^2.

2-9. In a manufacturing operation, a large sheet of plastic, $\frac{1}{2}$ in. thick, is to be glued to a 1-in.-thick sheet of cork board. To effect a bond, the glue is to be maintained at a temperature of 120 F for a considerable period of time. This is accomplished by a radiant heat source, applied uniformly over the surface of the plastic ($k = 1.3$ Btu/hr ft F). The exposed sides of the cork and of the plastic have a heat-transfer coefficient by convection of 2.0 Btu/hr sq ft F, and the room temperature during the operation is 75 F. Neglecting heat losses by radiation from the sheets, estimate the rate at which heat must be supplied to the surface of the plastic to obtain the required temperature at the interface. The thermal resistance of the glue may be neglected. Draw the thermal circuit for the system.

2-10. Steam having a quality of 98 percent at a pressure of 20 psia is flowing at a velocity of 3 fps through a $\frac{3}{4}$-in. steel pipe (1.05 in. OD, 0.824 in. ID). The heat-transfer coefficient at the inner surface, where condensation occurs, is 100 Btu/hr sq ft F. A dirt film at the inner surface adds a unit thermal resistance of 1.0 hr sq ft F/Btu. Estimate the rate of heat loss per foot length of pipe if (a) the pipe is bare, (b) the pipe is covered with a 2-in. layer of 85 percent magnesia insulation. For both cases assume that the unit-surface conductance at the outer surface is 2.0 Btu/hr sq ft F and that the environmental temperature is 70 F. Also estimate the change in quality per 10-ft length of pipe in both cases.

2-11. Estimate the rate of heat loss per unit length from a 2-in.-ID, $2\frac{3}{8}$-in.-OD steel pipe covered with asbestos insulation ($3\frac{3}{8}$ in. OD). Steam flows in the pipe. It has a quality of 99 percent and is at 300 F. The total thermal resistance at the inner wall is 0.015 hr sq ft F/Btu, the heat-transfer coefficient at the outer surface is 3.0 Btu/hr sq ft F, and the ambient temperature is 60 F.

2-12. Estimate the rate of heat flow per sq ft area through a furnace wall consisting of an 8-in.-thick inner layer of chrome brick, a center layer of kaolin insulating brick (4 in. thick) and an outer layer of masonry brick (4 in. thick). The unit-surface conductance at the inner surface is 15 Btu/hr sq ft F and the outer-surface temperature is 150 F. The temperature of the gases inside the furnace is 3000 F. What temperatures prevail in the steady state at the inner and outer surfaces of the center layer?

2-13. A 1-in.-OD, $\frac{3}{4}$-in.-ID copper pipe carries 10 gallons/minute (gpm) of brine at 20 F. The ambient air is at 70 F and has a dew point of 50 F. How much insulation with a conductivity of $k = 0.15$ Btu/hr ft F is needed to prevent condensation on the exterior of the insulation if $\bar{h}_c + \bar{h}_r = 3.0$ Btu/hr sq ft F on the outside?

2-14. A thin flat electrical heating element is covered on one side with a layer of asbestos ($\rho = 36$ lb/cu ft) 1 in. thick and on the other side with a carbon-steel plate $\frac{1}{8}$ in. thick. The heat-transfer coefficient for the fluid on the outer sides of this sandwich plate is 2 Btu/hr sq ft F and the fluid there is at 60 F. Calculate (a) the rate of heat dissipation at the heater, in watts per sq ft which will make the heater temperature 700 F under conditions of steady state, and (b) specify the temperature of the outside surface of the steel for these conditions.

2-15. A composite insulating wall is made of two layers of cork ($k = 0.025$ Btu/hr ft F) as shown in the accompanying sketch. If the spaces are filled with atmospheric air, determine the total unit thermal resistance of the wall and compare it with that of a solid wall of cork.

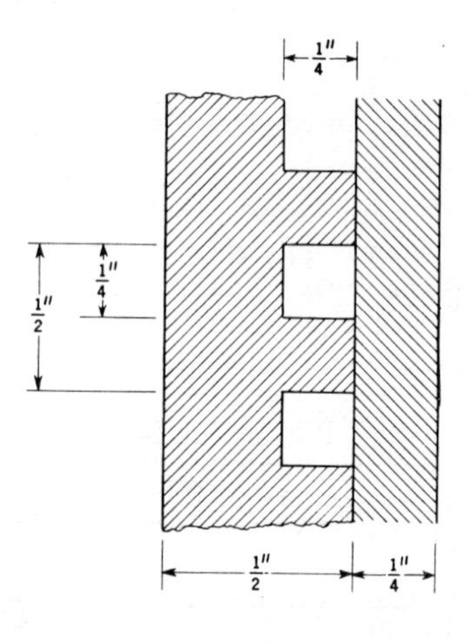

Prob. 2-15

2-16. A sheet of aluminum at 600 F is insulated with a 3-in. layer of powdered diatomaceous earth from air at 100 F. To keep this insulation in place, a layer of plastic is placed over it and bolted to the wall with $\frac{1}{4}$-in.-aluminum bolts on 6-in. centers. Calculate the percent increase in the thermal resistance per unit area of the insulation, including the bolts, if steel instead of aluminum bolts would be used.

2-17. (a) Find the thermal resistance per sq ft of a wall constructed of 2 by 4 wooden beams on 16-in. centers, with 1-in. boards on the exterior, $\frac{1}{2}$-in. sheet rock on the interior, and rock wool insulation in the space between the studs. (b) If the inside air temperature is 70 F, the outside 40 F, and the surface heat-transfer coefficient is 2 Btu/hr sq ft F on both sides, what is the rate of heat loss per square foot? (c) If a 20-ft cubical structure has walls and roof of this resistance, what is the necessary current rating of a 100-v electric heater if it is to maintain the specified conditions?

Data:		
	Two-by-four dimensions	= $1\frac{5}{8}$ in. x $3\frac{1}{2}$ in.
	One-inch board thickness	= $\frac{3}{4}$ in.
	Wood conductivity	= 0.10 Btu/hr ft F
	Sheet rock conductivity	= 0.30 Btu/hr ft F
	Rock wood conductivity	= 0.03 Btu/ft F

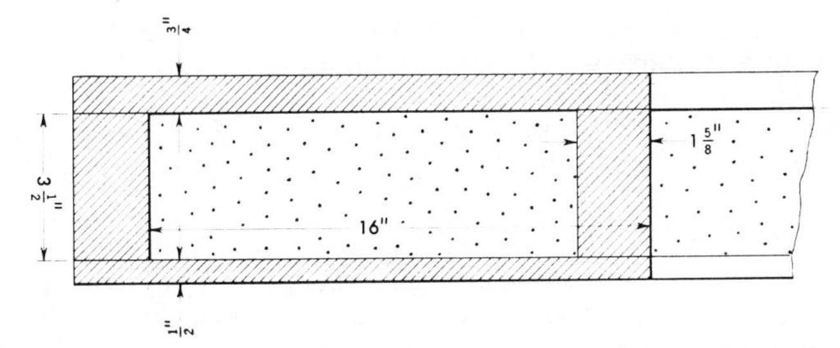

Prob. 2-17

2-18. A hollow sphere with inner and outer radii of R_1 and R_2, respectively, is covered with a layer of insulation having an outer radius of R_3. Derive an expression for the rate of heat transfer through the insulated sphere in terms of the radii, the thermal conductivities, the heat-transfer coefficients, and the temperatures of the interior and the surrounding medium of the sphere.

$$Ans.\quad q = 4\pi\Delta T \left(\frac{1}{h_1 R_1^2} + \frac{1}{h_3 R_3^2} + \frac{R_2 - R_1}{k_{12} R_1 R_2} + \frac{R_3 - R_2}{k_{23} R_2 R_3}\right)$$

2-19. A rocket motor has the shape of a sphere with one-sixth of it cut out to allow for the insertion of a nozzle. The chamber has a 14.5-in. ID. The wall consists of an inner layer of refractory $\frac{1}{4}$ in. thick ($k = 1$ Btu/hr ft F), followed by a 1-in.-thick layer of steel having a thermal conductivity given by the equation

$$k = 18(1 + 0.0011\,T) \qquad \text{Btu/hr ft F}$$

where T is in deg F. In steady-state operation the gases inside the chamber are at 5800 F, and the unit-surface conductance at the inner wall is 12 Btu/hr sq ft F. Determine the amount of heat transferred through the wall under steady-state operation in 10 sec if the outer surface temperature is 200 F. *Ans.* 806 Btu

2-20. The thermal conductivity of a material may be determined in the following manner: Saturated steam at 35 psia is condensed at the rate of 1.5 lb/hr inside of a hollow iron sphere which is $\frac{1}{2}$ in. thick and has an internal diameter of 20 in. The sphere is coated with the material whose thermal conductivity is to be evaluated. The thickness of the material to be tested is 4 in. and there are two thermocouples embedded in it, one at a distance of $\frac{1}{2}$ in. from the surface of the iron sphere and one, $\frac{1}{2}$ in. from the exterior surface of the system. If the inner

thermocouple indicates a temperature of 230 F and the outer thermocouple, a temperature of 135 F, calculate: (a) the thermal conductivity of the material surrounding the metal sphere, (b) the temperatures at the interior and exterior surfaces of the test material, and (c) the over-all heat-transfer coefficient based on the interior surface of the iron sphere, assuming that the thermal resistances at the surfaces, as well as at the interface between the two spherical shells, are negligible.
Ans. (a) 0.276 Btu/hr sq ft F; (b) 123 and 251 F; (c) .51 Btu/hr sq ft F

2-21. A cylindrical liquid oxygen (LOX) tank has a diameter of 4 ft, a length of 20 ft, and hemispherical ends. The boiling point of LOX is −297 F. An insulation is sought which will reduce the boil-off rate in the steady state to no more than 25 lb/hr. The heat of vaporization of LOX is 92 Btu/lb. If the thickness of this insulation is to be no more than 3 in., what would the value of its thermal conductivity have to be? *Ans.* ~0.005 Btu/hr ft F

2-22. The addition of insulation to a cylindrical surface, such as a wire, may sometimes increase the rate of heat dissipation to the surroundings. (a) For a No. 10 wire (0.102 in. diam), what is the optimum thickness of rubber insulation (k = 0.08 Btu/hr ft F) if the unit-surface conductance is 3 Btu/hr sq ft F? (b) If the current-carrying capacity of this wire is considered to be limited by the insulation temperature, what percent increase in capacity is realized by addition of the insulation?

2-23. A standard 4-in. steel pipe (ID = 4.026 in., OD = 4.500 in.) carries superheated steam at 1200 F in an enclosed space where a fire hazard exists, limiting the outer-surface temperature to 100 F. In order to minimize the insulation cost, two materials are to be used; first a high temperature insulation (relatively expensive) applied to the pipe and then magnesia (a less expensive material) on the outside. The maximum temperature of the magnesia is to be 600 F. The following constants are known:

Steam side coefficient	$\bar{h}$ = 100	Btu/hr sq ft F
High-temperature insulation conductivity	k = 0.06	Btu/hr ft F
Magnesia conductivity	k = 0.045	Btu/hr ft F
Outside heat-transfer coefficient	$\bar{h}$ = 2.0	Btu/hr sq ft F
Steel conductivity	k = 25	Btu/hr ft F
Ambient temperature	T_∞ = 70	F

(a) Specify the thickness for each insulating material. (b) Calculate the overall conductance based on the pipe OD. (c) What fraction of the total resistance is due to (1) steam side resistance, (2) steel pipe resistance, (3) insulation (combination of the two), and (4) outside resistance? (d) How much heat is transferred per hour, per foot length of pipe?

2-24. For the system outlined in Prob. 2-23, determine an expression for the critical radius of the insulation in terms of the thermal conductivity of the insulation and the surface coefficient between the exterior surface of the insulation and the surrounding fluid. Assume that the temperature difference, R_1, R_2, the heat-transfer coefficient on the interior, and the thermal conductivity of the material of the sphere between R_1 and R_2 are constant. HINT: Follow the example in Sec. 2-2 for the case of a cylindrical system. *Ans.* $R_{3,\text{crit}} = 2k_{23}/\bar{h}_3$

2-25. The interior of a refrigerator, having inside dimensions of $1\frac{1}{2}$-by $1\frac{1}{2}$-ft base area and 4-ft height, is to be maintained at 40 F. The walls of the refrigerator are constructed of two $\frac{1}{8}$-in. mild-steel sheets with 3 in. of glass-wool insulation between them. If the average heat-transfer coefficients at the inner and outer surfaces are 2.0 and 2.5 Btu/hr sq ft F respectively, *estimate* the rate at which heat must be removed from the interior to maintain the specified temperature in a kitchen at 85 F. What will be the temperature at the outer surface of the wall?

2-26. A nuclear reactor is shielded by a composite wall consisting of a 1-in.-thick layer of aluminum inside of a 5-ft thickness of concrete. The temperature at the inner surface of the aluminum is 1000 F, and there exists a thermal resistance equivalent to a layer of air 0.01 in. thick at the interface between the aluminum and the concrete. The thermal conductivity of the concrete varies with temperature according to $k = 0.5\,(1 + 0.002\,T)$, where T is in deg F and k is in Btu/hr ft F. The outer surface of the concrete may be assumed to be at 100 F. In order to lay a 1-in. pipe in the concrete wall, parallel to the reactor face, for a liquid metal which freezes at 200 F, it is necessary to determine the distance from the inner face at which the temperature reaches 200 F. Determine this distance first using an average value of k, and compare your approximate solution with the answer obtained by taking the temperature variation of the thermal conductivity into account.

2-27. Assuming one-dimensional heat flow, determine the rate of heat transfer per unit area through the composite wall shown in the sketch below.

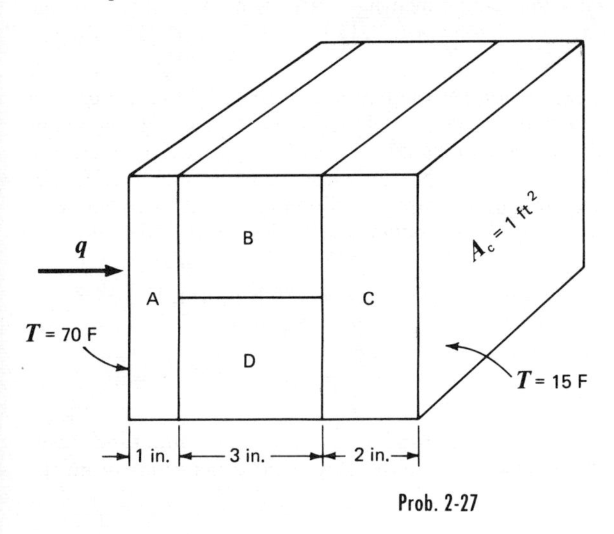

$k_A = 100$ Btu/hr ft F
$k_B = 20$ Btu/hr ft F
$k_C = 33$ Btu/hr ft F
$k_D = 45$ Btu/hr ft F
$A_B = A_D$

Prob. 2-27

2-28. A brick stack 200 ft tall with a 6-ft ID has a $4\frac{1}{2}$-in. fire-clay brick lining ($k = 0.60$ Btu/hr ft F) and a masonry brick outer wall which varies linearly from a 24-in. thickness at the base to 8 in. at the top. The heat-transfer coefficient from stack gas to wall is 10 and from the outer wall to air is 3 Btu/hr sq ft F. If the stack gas is 600 F and the air is 0 F, calculate accurately the heat loss from the stack, neglecting temperature drop up the stack.

2-29. One side of a 2-in.-thick aluminum plate is maintained at 500 F while the other side is covered with a 1-in.-thick layer of fiber glass, whose outside surface is maintained at 100 F. Determine the area of the slab required for the total heat flow through the aluminum-fiber glass combination to be 150,000 Btu/hr.

2-30. A plane wall 3 in. thick generates heat internally at the rate of 10^4 Btu/hr ft^3. One side of the wall is insulated, and the other side is exposed to an environment at 200 F. The convection heat-transfer coefficient between the wall and the environment is 100 Btu/hr ft^2 F. If the thermal conductivity of the wall is 10 Btu/hr ft F, calculate the maximum temperature in the wall.

2-31. A small dam which may be idealized by a large slab 4 ft thick is to be completely poured in a short period of time. The hydration of the concrete results in the equivalent of a distributed source of constant strength of 10 Btu/hr cu ft. If both dam surfaces are at 60 F, determine the maximum temperature to which the concrete will be subjected, assuming steady-state conditions. The thermal conductivity of the wet concrete may be taken as 0.7 Btu/hr ft F.

2-32. Two large steel plates at temperatures of 200 and 160 F are separated by a steel rod 1 ft long and 1 in. in diameter. The rod is welded to each plate. The space between the plates is filled with insulation which also insulates the circumference of the rod. Because of the voltage difference between the two plates, current flows through the rod dissipating electrical energy at a rate of 40 Btu/hr. Determine the maximum temperature in the rod and the heat flux at each end. Check your results by comparing the net heat-flow rate at the two ends with the total rate of heat generation. *Ans.* 218 F, −14.3 Btu/hr, 25.7 Btu/hr

2-33. (a) Derive an expression for the temperature rise of the center of a current-carrying wire relative to the surface as a function of the current, the diameter, and the electrical and thermal conductivities. (b) Compare the temperature differences between center and surface for No. 14 wires (0.064 in. diam) of copper and nichrome when both are carrying 15 amps. (c) Compare the surface temperature rise of these wires if the unit-surface conductance is 2 Btu/hr ft F for both.

Assume:		
	Thermal conductivity of copper	= 220 Btu/hr ft F
	of nichrome	= 8 Btu/hr ft F
	Electrical conductivity of copper	= 1.47×10^6 mho/in.
	of nichrome	= 3.76×10^3 mho/in.

2-34. Show that the temperature distribution in a long solid tube, insulated on the outside, cooled on the inside, with uniform heat generation within the solid, is given by

$$T(r) - T_o = -\frac{\dot{q}}{4k}(r^2 - r_o^2) + \frac{\dot{q}r_o^2}{2k}\ln\frac{r}{r_o}$$

2-35. The shield of a nuclear reactor can be idealized by a large 10-in.-thick flat plate having a thermal conductivity of 2 Btu/hr ft F. Radiation from the interior of the reactor penetrates the shield and produces heat generation in the shield which decreases exponentially from a value of 10 Btu/hr cu in. at the inner surface to a value of 1.0 Btu/hr cu in. at a distance 5 in. from the interior surface. If the exterior surface is kept cooling at 100 F by forced convection, determine the temperature at the inner surface of the shield. HINT: First set up the differential equation for a system in which the heat generation rate varies according to $q(x) = q(0)e^{-Cx}$.

2-36. Show that the temperature distribution in a sphere of radius r_o, made of a homogeneous material in which energy is released at a uniform rate per unit volume $\dot{q}$, is

$$T(r) = T_o + \frac{\dot{q}r_o^2}{6k}[1 - (r/r_o)^2]$$

2-37. In a cylindrical fuel rod of a nuclear reactor, heat is generated internally according to the equation

$$\dot{q} = \dot{q}_1[1 - (r/r_o)^2]$$

where $\dot{q}$ = local rate of heat generation per unit volume at r;
r_o = outside radius;
$\dot{q}_1$ = rate of heat generation per unit volume at the center line.

Calculate the temperature drop from the center line to the surface for a 1-in.-OD rod having a thermal conductivity of 15 Btu/hr ft F if the rate of heat removal from its surface is 500,000 Btu/hr sq ft.

2-38. Derive an expression for the temperature distribution in an infinitely long rod of uniform cross section within which there is uniform heat generation at the rate of 1 Btu/sec per in. length. Assume that the rod is attached to a surface at T_s and is exposed through a unit-surface conductance $\bar{h}$ to a fluid at T_∞.

2-39. Derive an expression for the temperature distribution in a plane wall in which there are uniformly distributed heat sources which vary according to the linear relation

$$\dot{q} = \dot{q}_w[1 - \beta(T - T_w]$$

where $\dot{q}_w$ is constant and equal to the heat generated per unit volume at the wall temperature T_w. Both sides of the plate are maintained at T_w and the plate thickness is $2L$.

2-40. A plane wall of thickness $2L$ has internal heat sources whose strength varies according to

$$\dot{q} = \dot{q}_0 \cos(ax)$$

where $\dot{q}_0$ is the heat generated per unit volume at the center of the wall ($x = 0$) and a is a constant. If both sides of the wall are maintained at a constant temperature of T_w, derive an expression for the total heat loss from the wall per unit surface area.

2-41. An electrical heater capable of generating 10,000 watts is to be designed. The heating element is to be a stainless steel wire, having an electrical resistivity of 32×10^{-6} ohms per inch length per square inch area. The operating temperature of the stainless steel is to be no more than 2300 F. The heat-transfer coefficient at the outer surface is expected to be no less than 300 Btu/hr sq ft F in a medium whose maximum temperature is 200 F. A transformer capable of delivering current at 9 and 12 v is available. Determine a suitable size for the wire, the current required, and discuss what effect a reduction in the heat-transfer coefficient could have. HINT: Demonstrate *first* that the temperature drop between the center and the surface of the wire is independent of the wire diameter, and determine its value.

2-42. The tip of a soldering iron consists of a $\frac{1}{4}$-in.-OD copper rod, 3 in. long. If the tip must be 400 F, what is the temperature of the base and the heat flow, in Btu/hr and in watts, into the base? Assume: $\bar{h}_c = 4.0$ Btu/hr sq ft F and $T_{air} = 70$ F.

2-43. Both ends of a $\frac{1}{4}$-in. copper U-shaped rod, as shown in the accompanying sketch, are rigidly affixed to a vertical wall, the temperature of which is maintained at 200 F. The developed length of the rod is 2 ft and it is exposed to air at 100 F. The combined radiative and convective unit-thermal conductance for this system is 6 Btu/hr sq ft F. (a) Calculate the temperature of the mid-point of the rod. (b) What will be the heat transfer from the rod?

Ans. (a) 120 F; (b) 33.6 Btu/hr

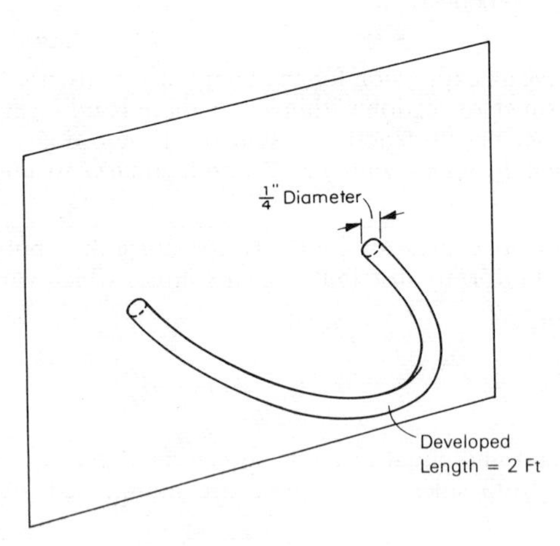

Prob. 2-43

2-44. Derive Eq. 2-47, showing all steps.

2-45. One end of a 1-ft-long steel rod is connected to a wall at 400 F. The other end is connected to a wall which is maintained at 200 F. Air is blown across the rod so that a heat transfer coefficient of 3 Btu/hr ft^2 F is maintained over the entire surface. If the diameter of the rod is 2 in., and the temperature of the air 100 F, what is the net rate of heat loss to the air in Btu per hour?

2-46. A circumferential fin of rectangular cross section $\frac{1}{4}$-in. long and $\frac{1}{8}$-in. thick, surrounds a 1-in.-diameter tube. The fin is constructed of mild steel and air blowing over the fin produces a heat transfer coefficient of 5 Btu/hr ft^2 F. If the temperatures of the base of the fin and of the air are 500 and 100 F, respectively, calculate the heat transfer rate from the fin.

2-47. Derive a differential equation for the temperature distribution in a straight triangular fin of length L and base width t. For convenience take the coordinate axis as shown in the sketch below and assume one-dimensional heat flow. If you are familiar with Bessel functions, solve the equations assuming you know the base temperature, the environmental temperature and the heat transfer coefficient, and present your results in dimensionless coordinates.

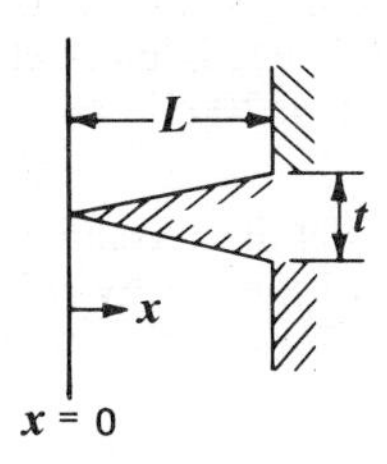

Prob. 2-47

2-48. A $\frac{1}{8}$-in.-thick aluminum plate has rectangular fins on one side, $\frac{1}{16} \times \frac{1}{2}$ in., spaced $\frac{1}{2}$ in. apart. The finned side is in contact with low-pressure air at 100 F and the average unit-surface conductance is 5 Btu/hr sq ft F. On the unfinned side water flows at 200 F and the unit-surface conductance is 50 Btu/hr sq ft F. Calculate (a) the efficiency of the fins, (b) the rate of heat transfer per unit area of wall, and (c) comment on the design if the water and air were interchanged. HINT: Show first that the fin efficiency is tanh mL/mL.

2-49. A turbine blade $2\frac{1}{2}$ in. long, cross-sectional area A of 0.005 sq ft, perimeter P of 0.400 ft is made of stainless steel (k = 15 Btu/hr ft F). The temperature of the root T_s is 900 F. The blade is exposed to a hot gas at 1600 F, and the unit-surface conductance $\bar{h}$ is 80 Btu/hr sq ft F. Determine the temperature distribution and the rate of heat flow at the root of the blade. Assume that the tip is insulated.

2-50. To determine the thermal conductivity of a long, solid 1-in.-diam rod, one half was inserted into a furnace while the other half was projecting into air at 80 F. After steady state had been reached, the temperatures at two points 3 in. apart were measured and found to be 258 F and 196 F respectively. The heat-transfer coefficient over the surface of the rod exposed to the air was estimated to be 4.0 Btu/hr sq ft F. What is the thermal conductivity of the rod?

2-51. Two long pieces of copper wire, $\frac{1}{16}$ in. in diameter, are to be soldered together end to end. If the air temperature is 80 F and the melting point of the solder is 450 F, what is the minimum rate of heat input required? Assume that the unit-surface conductance between the surface of the wire and the ambient air is 3.0 Btu/hr sq ft F.

2-52. Heat is transferred from water to air through a brass wall (k = 45 Btu/hr ft F). The addition of rectangular brass fins, 0.03 in. thick and 1 in. long, spaced 0.5 in. apart, is contemplated. Assuming a waterside heat-transfer coefficient of 30 Btu/hr sq ft F and an airside heat-transfer coefficient of 3 Btu/hr sq ft F, compare the gain in heat-transfer rate achieved by adding fins to (a) the waterside, (b) the airside, and (c) both sides. (Neglect temperature drop through the wall.)

2-53. The wall of a heat exchanger has a surface area on the liquid side of 20 sq ft (2 ft × 10 ft) with a unit-surface conductance of 45 Btu/hr sq ft F. On the other side of the heat-exchanger wall flows a gas, and the wall has 96 thin steel fins 0.02 in. wide and $\frac{1}{2}$ in. high (k = 25 Btu/hr ft F). The fins are 10 ft long and the unit-surface conductance on the gas side is 10 Btu/hr sq ft F. Assuming that the thermal resistance of the wall is negligible, determine the rate of heat transfer if the overall temperature difference is 100 F. *Ans.* 43,000 Btu/hr

2-54. Derive Eq. 2-50 in detail. Assume that tip losses are taken into account by extending the length of the pin by $t/2$, while the end of this extended pin is adiabatic ($L' = L + t/2, \partial T/\partial x = 0$ at $x = L'$).

2-55. The top of a 12-in. I beam is maintained at a temperature of 500 F, while the bottom is at 200 F. The thickness of the web is $\frac{1}{2}$ in. Air at 500 F is blowing along the side of the beam so that $\bar{h}$ = 7 Btu/hr sq ft F. The thermal conductivity of the steel may be assumed constant and equal to 25 Btu/hr ft F. Find the temperature distribution along the web from top to bottom and plot the result.

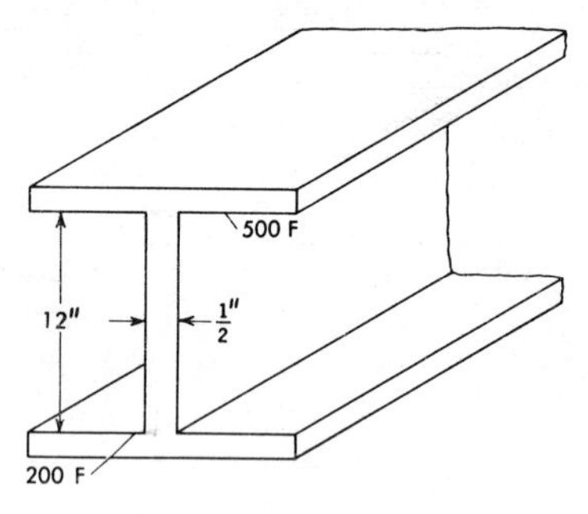

Prob. 2-55

2-56. The handle of a ladle used for pouring molten lead is 12 in. long. Originally the handle was made of $\frac{3}{4}$- by $\frac{1}{2}$-in. mild-steel bar stock. To reduce the grip temperature, it is proposed to form the handle of tubing $\frac{1}{16}$ in. thick to the same rectangular shape. If the average unit-surface conductance over the handle surface is 2.5 Btu/hr sq ft F, estimate the reduction of the temperature at the grip in 70 F air.

2-57. Derive Eq. 2-44a in detail.

REFERENCES

1. R. Schuman, Jr., *Metallurgical Engineering*, Vol. I, *Engineering Principles*. (Cambridge, Mass.: Addison-Wesley Publishing Company, 1952.)

2. I. Langmuir, E. Q. Adams, and S. F. Meikle, "Flow of Heat Through Furnace Walls," *Trans. Am. Electrochem. Soc.*, Vol. 24 (1913), pp. 53–84.

3. M. N. Ozisik, *Boundary Value: Problems of Heat Conduction* (Scranton, Pa.: International Textbook Comp., 1968.)

4. M. E. Barzelay, K. N. Tong, and G. F. Holloway, "Effect of Pressure on Thermal Conductance of Contact Joints," *NACA TN* 3245, May, 1955.

5. T. N. Cetinkale and M. Fishenden, "Thermal Conductance of Metal Surfaces in Contact," *General Discussion on Heat Transfer*, Conference of Institute of Mech. Eng. and ASME (1951), pp. 271–275.

6. E. Fried, "Study of Interface Thermal Conductance," General Electric Co. summary Rep. No. 65S D4394, March, 1965.

7. C. R. Wylie, Jr., *Advanced Engineering Mathematics*. (New York: McGraw-Hill Book Company, Inc., 1951.)

8. M. Jakob, *Heat Transfer*, Vol. I. (New York: John Wiley & Sons, Inc., 1949.)

9. K. A. Gardner, "Efficiency of Extended Surfaces," *Trans. ASME*, Vol. 67 (1945), pp. 621–631.

10. W. P. Harper and D. R. Brown, "Mathematical Equations for Heat Conduction in the Fins of Air-Cooled Engines," *NACA Report* 158, 1922.

11. H. A. Saller, "Uranium and its Alloys," *The Reactor Handbook*, Vol. 3, AECD-3647, p. 391. (USAEC, May, 1955.)

12. S. Glasstone, *Principles of Nuclear Reactor Engineering*. (Princeton, N.J.: D. Van Nostrand Company, Inc., 1955.)

13. J. R. Dietrich and D. Okrent, "Spatial Distribution of Heat Generation," *The Reactor Handbook*, Vol. 2, AECD-3646, pp. 87–121. (USAEC, May, 1955.)

14. A. S. Thompson and O. E. Rodgers, *Thermal Power from Nuclear Reactors*. (New York: John Wiley & Sons, Inc., 1956.)

15. C. F. Bonilla, *Nuclear Engineering*. (New York: McGraw-Hill Book Company, Inc., 1957.)

16. Forster and Wright, *Nuclear Engineering*, 2nd ed., (Boston: Allyn and Bacon, Inc., 1972.)

17. M. G. Cooper, B. B. Mikic, and M. M. Yovanovich, "Thermal Contact Conductances," Int. J. Heat and Mass Transfer, vol. 12 (1969), pp. 279–300.

18. R. T. Roca and B. B. Mikic, *Thermal Contact Resistance in a Non-ideal Joint*, Tech. Rep. No. 71821-77, Heat Transfer Laboratory, M.I.T., Cambridge, Mass., Nov. 1971.

3

Two- and three-dimensional steady-state conduction

3-1. *Methods of analysis*

In the preceding chapter we dealt with problems in which the temperature and the heat flow can be treated as functions of a single variable. Many practical problems fall into this category, but when the boundaries of a system are irregular or when the temperature along a boundary is nonuniform, a one-dimensional treatment may no longer be satisfactory. In such cases, the temperature is a function of two, and possibly even three, coordinates. The heat flow through a corner section where two or three walls meet, the heat conduction through the walls of a short, hollow cylinder, or the heat loss from a buried pipe are typical examples of this class of problems.

In this chapter we shall consider some methods for analyzing conduction in two- and three-dimensional systems. The emphasis will be placed on two-dimensional problems because they are less cumbersome to solve, yet they illustrate the basic methods of analysis for three-dimensional systems.

Heat conduction in two- and three-dimensional systems can be treated by *analytical*, *graphical*, *analogical*, and *numerical methods.* A complete treatment of analytical solutions requires a prior knowledge of Fourier series, Bessel functions, Legendre polynomials, Laplace transform methods, and complex variable theory. A number of excellent books dealing exclusively with mathematical solutions of heat-conduction problems are available (1, 2, 3, and 4), but since most of this material is too advanced for an introductory course, it will not be presented here. We shall consider only the analytical solution of one relatively simple problem to illustrate the analytical method of approach. Emphasis will be placed on numerical methods which are suitable for solution by means of digital computers.

3-2. *Derivation of the heat-conduction equation*

Consider a small element of material in a solid body. The element has the shape of a rectangular parallelepiped with its edges *dx*, *dy*, and *dz* parallel, respectively, to the *x*, *y*, and *z* axes as shown in Fig. 3-1. To obtain an equation for the temperature distribution we write an energy balance for the element semantically in the form

Rate of heat in-flow	+	Rate of heat generation by internal sources	=	Rate of heat out-flow	+	Rate of change in internal energy	(3-1)

or algebraically as

$$(q_x + q_y + q_z) + \dot{q}(dx\,dy\,dz) = (q_{x+dx} + q_{y+dy} + q_{z+dz}) + c\rho(dx\,dy\,dz)\frac{\partial T}{\partial \theta}$$

where the rate of heat generation per unit volume, $\dot{q}$, and the temperature,

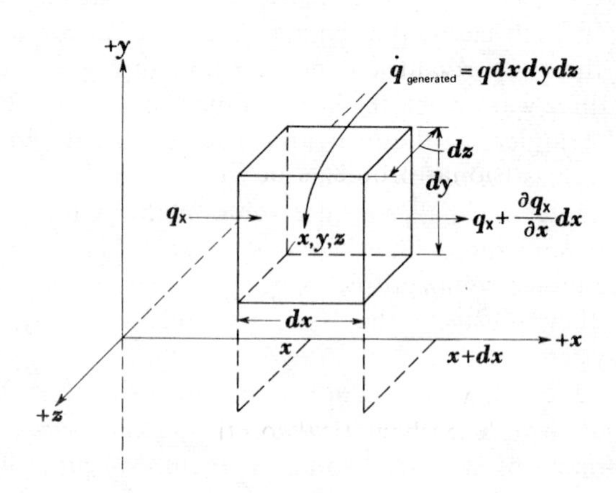

Fig. 3-1. Sketch illustrating nomenclature for the derivation of the general heat-conduction equation in Cartesian coordinates.

T, are in general functions of the three coordinates x, y, z as well as of time θ.

The rate of heat conduction into the element across the left face in the x direction, q_x, can according to Eq. 1-1 be written as

$$q_x = \left(-k\frac{\partial T}{\partial x}\right)dy\,dz$$

The temperature gradient is expressed as a partial derivative because T is not only a function of x, but also of y, z, and θ. The rate of heat conduction out of the element across the right face at $x + dx$, q_{x+dx} is

$$q_{x+dx} = \left[\left(-k\frac{\partial T}{\partial x}\right) + \frac{\partial}{\partial x}\left(-k\frac{\partial T}{\partial x}\right)dx\right]dy\,dz$$

Subtracting the heat-flow rate out of the element from the heat-flow rate into the element yields

$$q_x - q_{x+dx} = \frac{\partial\left(k\dfrac{\partial T}{\partial x}\right)}{\partial x}dx\,dy\,dz$$

and similarly for the y and z directions

$$q_y - q_{y+dy} = \frac{\partial\left(k\dfrac{\partial T}{\partial y}\right)}{\partial y}dx\,dy\,dz$$

$$q_z - q_{z+dz} = \frac{\partial\left(k\dfrac{\partial T}{\partial z}\right)}{\partial z}dx\,dy\,dz$$

Substituting these relations into the energy balance and dividing each term by $dx\,dy\,dz\,d\theta$ gives

$$\frac{\partial}{\partial x}\left(k\frac{\partial T}{\partial x}\right) + \frac{\partial}{\partial y}\left(k\frac{\partial T}{\partial y}\right) + \frac{\partial}{\partial z}\left(k\frac{\partial T}{\partial z}\right) + \dot{q} = c\rho\frac{\partial T}{\partial \theta} \tag{3-2}$$

if the system is homogeneous and the specific heat c and the density, ρ are independent of temperature. If also k is assumed to be uniform, Eq. 3-2 can be written

$$\frac{\partial^2 T}{\partial x^2} + \frac{\partial^2 T}{\partial y^2} + \frac{\partial^2 T}{\partial z^2} + \frac{\dot{q}}{k} = \frac{1}{a}\frac{\partial T}{\partial \theta} \tag{3-3}$$

where the constant $a = k/c\rho$ is called the *thermal diffusivity* and has the units, sq ft/hr, in the engineering system and sq m/sec in the SI system. Equation 3-3 is known as the general heat-conduction equation and governs the temperature distribution and the conduction heat flow in a solid having uniform physical properties.

If the system contains no heat sources, Eq. 3-3 reduces to the *Fourier equation*

$$\frac{\partial^2 T}{\partial x^2} + \frac{\partial^2 T}{\partial y^2} + \frac{\partial^2 T}{\partial z^2} = \frac{1}{a}\frac{\partial T}{\partial \theta} \tag{3-4}$$

If the system is steady, but heat sources are present, Eq. 3-3 becomes the *Poisson equation*

$$\frac{\partial^2 T}{\partial x^2} + \frac{\partial^2 T}{\partial y^2} + \frac{\partial^2 T}{\partial z^2} + \frac{\dot{q}}{k} = 0 \tag{3-5}$$

In the steady state the temperature distribution in a body free of heat sources must satisfy the *Laplace equation*

$$\frac{\partial^2 T}{\partial x^2} + \frac{\partial^2 T}{\partial y^2} + \frac{\partial^2 T}{\partial z^2} = 0 \tag{3-6}$$

For one-dimensional steady heat conduction, Eq. 3-6 becomes $d^2T/dx^2 =$

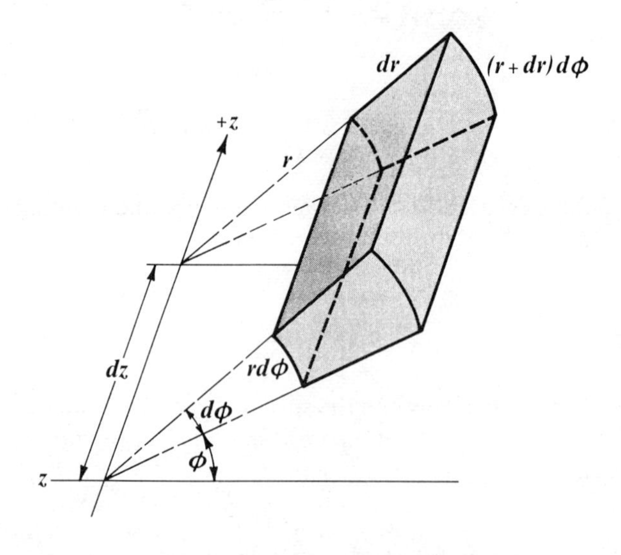

Fig. 3-2. Cylindrical coordinate system.

0, which yields, after integration, dT/dx = constant, as anticipated from Eq. 1-1 for steady one-dimensional heat conduction.

There are numerous problems in heat conduction which can be handled more conveniently in a cylindrical or spherical coordinate system. The general heat-conduction equation in the cylindrical coordinate system shown in Fig. 3-2 is

$$\frac{\partial^2 T}{\partial r^2} + \frac{1}{r}\frac{\partial T}{\partial r} + \frac{1}{r^2}\frac{\partial^2 T}{\partial \phi^2} + \frac{\partial^2 T}{\partial z^2} + \frac{\dot{q}}{k} = \frac{1}{a}\frac{\partial T}{\partial \theta} \tag{3-7}$$

The derivation of this equation is left as an exercise (see Problem 3-2). However, attention is directed to the fact that the area through which heat flows into the element in the positive r direction is $r\, d\phi\, dz$ and the area through which heat flows out of the element in the r direction is $(r + dr)d\phi dz$. This change in area results in the term $(1/r)(\partial T/\partial r)$ in Eq. 3-7.

In the spherical coordinate system shown in Fig. 3-3, the general heat-conduction equation becomes

$$\frac{1}{r^2}\frac{\partial}{\partial r}\left(r^2\frac{\partial T}{\partial r}\right) + \frac{1}{r^2 \sin\phi}\frac{\partial}{\partial \phi}\left(\sin\phi\frac{\partial T}{\partial \phi}\right) + \frac{1}{r^2\sin^2\phi}\frac{\partial^2 T}{\partial \psi^2} + \frac{\dot{q}}{k} = \frac{1}{a}\frac{\partial T}{\partial \theta} \tag{3-8}$$

3-3. *Analytical solution*

The objective of any heat transfer analysis is to predict either the rate of heat flow or the temperature distribution. In a two-dimensional

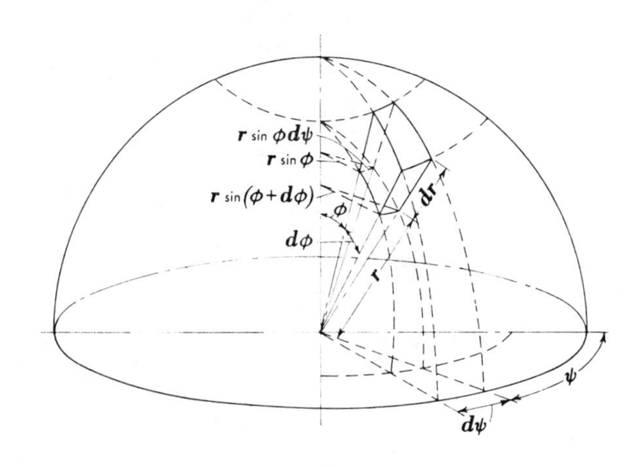

Fig. 3-3. Spherical coordinate system.

system without heat sources, the equation governing the temperature distribution in the steady state is

$$\frac{\partial^2 T}{\partial x^2} + \frac{\partial^2 T}{\partial y^2} = 0 \tag{3-9}$$

if the thermal conductivity is uniform. The solution of Eq. 3-9 will give $T(x, y)$, the temperature as a function of the two space-coordinates x and y. The rate of heat flow per unit area in the x and y directions, respectively, can then be obtained from Fourier's law

$$(q/A)_x = -k \frac{\partial T}{\partial y}$$

$$(q/A)_y = -k \frac{\partial T}{\partial x}$$

It should be noted that whereas the temperature is a scalar, the heat flux depends on the temperature gradient and is therefore a vector. The total rate of heat flow at a given point x, y, is the resultant of q_x and q_y at that point and is directed perpendicular to the isotherm as shown in Fig. 3-4. Thus, if the temperature distribution in a system is known, the rate of heat flow can easily be calculated. Usually, heat-transfer analyses concentrate therefore on determining the temperature field.

An analytical solution of a heat-conduction problem must satisfy the heat-conduction equation as well as the boundary conditions specified by the physical conditions of the particular problem. The classical approach to an exact solution of the Fourier equation is the separation-of-variables technique. We shall illustrate this approach by applying it

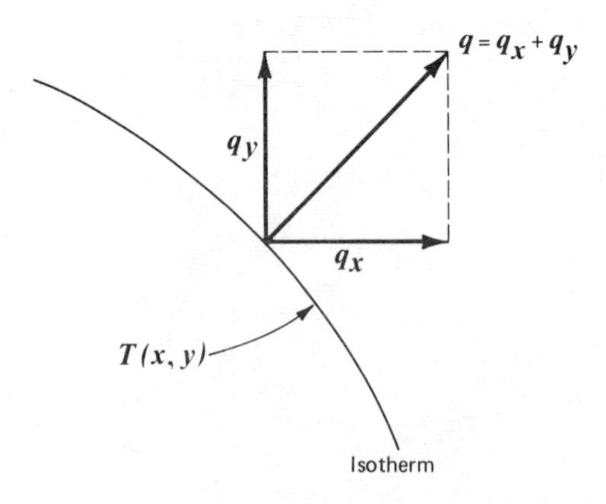

Fig. 3-4. Sketch showing heat flow in two dimensions.

to a relatively simple problem. Consider a thin rectangular plate, free of heat sources and insulated at the top and bottom surface (Fig. 3-5). Since $\partial T/\partial z$ is negligible the temperature is a function of x and y only. If the thermal conductivity is uniform, the temperature distribution must satisfy the equation

$$\frac{\partial^2 T}{\partial x^2} + \frac{\partial^2 T}{\partial y^2} = 0 \tag{3-9}$$

Equation 3-9 is a linear and homogeneous partial-differential equation which can be integrated by assuming a product solution for $T(x, y)$ of the form

$$T = XY \tag{3-10}$$

where $X = X(x)$, a function of x only, and $Y = Y(y)$, a function of y alone. Substituting Eq. 3-10 in Eq. 3-9 yields

$$-\frac{1}{X}\frac{d^2X}{dx^2} = \frac{1}{Y}\frac{d^2Y}{dy^2} \tag{3-11}$$

The variables are now separated. The left-hand side is a function of x only, while the right-hand side is a function of y alone. Since neither side can change as x and y vary, both must be equal to a constant, say λ^2. We have, therefore, the two total-differential equations

$$\frac{d^2X}{dx^2} + \lambda^2 X = 0 \tag{3-12}$$

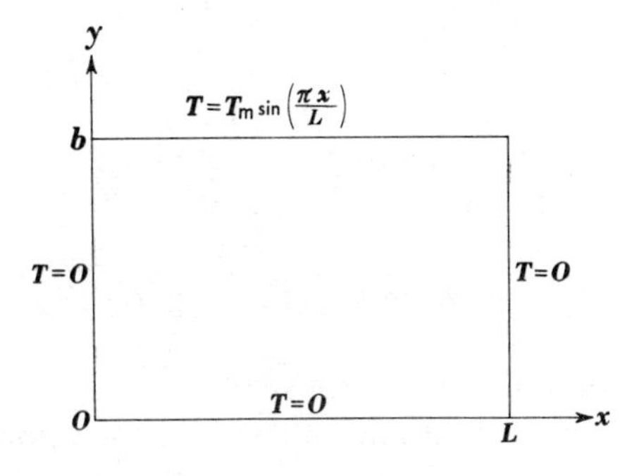

Fig. 3-5. Rectangular adiabatic plate.

and

$$\frac{d^2Y}{dy^2} - \lambda^2 Y = 0 \tag{3-13}$$

The general solution to Eq. 3-12 is

$$X = A\cos\lambda x + B\sin\lambda x,$$

the general solution to Eq. 3-13 is

$$Y = \mathrm{C}e^{-\lambda y} + De^{\lambda y},$$

and, therefore, from Eq. 3-10

$$T = XY = (A\cos\lambda x + B\sin\lambda x)(Ce^{-\lambda y} + De^{\lambda y}) \tag{3-14}$$

where A, B, C, and D are constants to be evaluated from the boundary conditions. As shown in Fig. 3-5, the boundary conditions to be satisfied are

$$T = 0 \quad \text{at } y = 0$$

$$T = 0 \quad \text{at } x = 0$$

$$T = 0 \quad \text{at } x = L$$

$$T = T_m \sin\frac{\pi x}{L} \quad \text{at } y = b$$

Substituting these conditions into Eq. 3-14 for T we get from the first condition

$$(\mathrm{A}\cos\lambda x + B\sin\lambda x)(C + D) = 0$$

from the second condition

$$A(Ce^{-\lambda y} + De^{\lambda y}) = 0$$

and from the third condition

$$(A\cos\lambda L + B\sin\lambda L)(Ce^{-\lambda y} + De^{\lambda y}) = 0$$

The first condition can be satisfied only if $C = -D$, and the second if $A = 0$. Using these results in the third condition, we obtain

$$(B\sin\lambda L)\,C(e^{-\lambda y} - e^{\lambda y}) = 2BC\sin\lambda L\sinh\lambda y = 0$$

To satisfy this condition, $\sin \lambda L$ must be zero or $\lambda = n\pi/L$, where n = 1, 2, 3, etc.[1] There exists therefore a different solution for each integer n and each solution has a separate integration constant C_n. Summing these solutions, we get

$$T = \sum_{n=1}^{\infty} C_n \sin \frac{n\pi x}{L} \sinh \frac{n\pi y}{L} \tag{3-15}$$

The last boundary condition demands that, at $y = b$,

$$\sum_{n=1}^{\infty} C_n \sin \frac{n\pi x}{L} \sinh \frac{n\pi b}{L} = T_m \sin \frac{\pi x}{L} \tag{3-16}$$

so that only the first term in the series solution with $C_1 = T_m/\sinh(\pi b/L)$ is needed. The solution therefore becomes

$$T(x, y) = T_m \frac{\sinh(\pi y/L)}{\sinh(\pi b/L)} \sin \frac{\pi x}{L} \tag{3-17}$$

The corresponding temperature field is shown in Fig. 3-6. The solid lines are isotherms and the dotted lines are heat-flow lines. It should be noted that lines indicating the direction of heat flow are perpendicular to the isotherms.

When the boundary conditions are not as simple as in the illustrative problem, the solution is obtained in the form of an infinite series. For example, if the temperature at the edge $y = b$ is a function of x, say

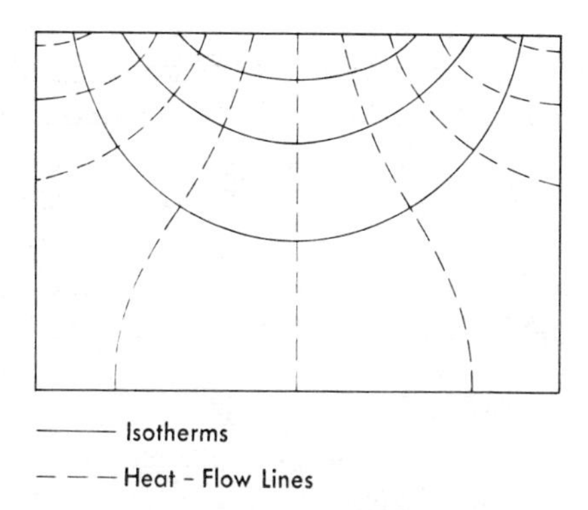

Fig. 3-6. Isotherms and heat-flow lines in the rectangular adiabatic plate shown in Fig. 3-5.

[1]The value $n = 0$ is excluded because it would give the trivial solution $T = 0$.

$T(x,b) = F(x)$, then the solution, as shown in Ref. 1, is the infinite series

$$T = \frac{2}{L}\sum_{n=1}^{\infty}\frac{\sinh(n\pi/L)y}{\sinh n\pi(b/L)}\sin\frac{\pi n}{L}x\int_0^L F(x)\sin\frac{n\pi}{L}x\,dx \tag{3-18}$$

which is quite laborious to evaluate quantitatively.

The separation-of-variables method can be extended to three-dimensional cases by assuming $T = XYZ$, substituting this expression for T in Eq. 3-6, separating the variables, and integrating the resulting total-differential equations subject to the given boundary conditions. Examples of three-dimensional problems are presented in Refs. 2, 3, and 4.

Analytical solutions are useful, but there are few practical problems with geometries and boundary conditions which can be solved analytically; and even when a solution has been obtained, it is often too complicated to justify the time and effort required to evaluate it quantitatively. We will, therefore, in the remainder of this chapter, consider methods of solution which have greater versatility.

3-4. Graphical method

The graphical method can yield rapidly a reasonably good estimate of the temperature distribution in geometrically complex two-dimensional systems with isothermal and insulated boundaries. The object of a graphical solution is to construct a network consisting of isotherms (lines of constant temperature) and lines of constant heat flow. The heat-flow lines are analogous to streamlines in a potential fluid flow, i.e. they are tangent to the direction of heat flow at any point. Consequently, no heat can flow across heat-flow lines and a constant amount of heat flows between any two of them. The isotherms are analogous to constant potential lines and heat flows perpendicular to them. Thus, lines of constant temperatures and lines of constant heat flux intersect at right angles. To obtain the temperature distribution one first prepares a scale model and then draws freehand, by trial and error, isotherms and heat-flow lines until they form a network of curvilinear squares. The procedure is illustrated in Fig. 3-7 for a corner section of unit depth with faces ABC at temperature T_1, faces FED at temperature T_2, and faces CD and AF insulated. Fig. 3-7a shows the scale model and Fig. 3-7b presents the curvilinear network of isotherms and heat-flow lines. It should be noted that the heat-flow lines emanating from isothermal boundaries are perpendicular to the boundary, except when they come from a corner. Flow lines leading to or from a corner of an isothermal boundary bisect the angle between the surfaces forming the corner.

A graphical solution, just as an analytical solution of a heat-con-

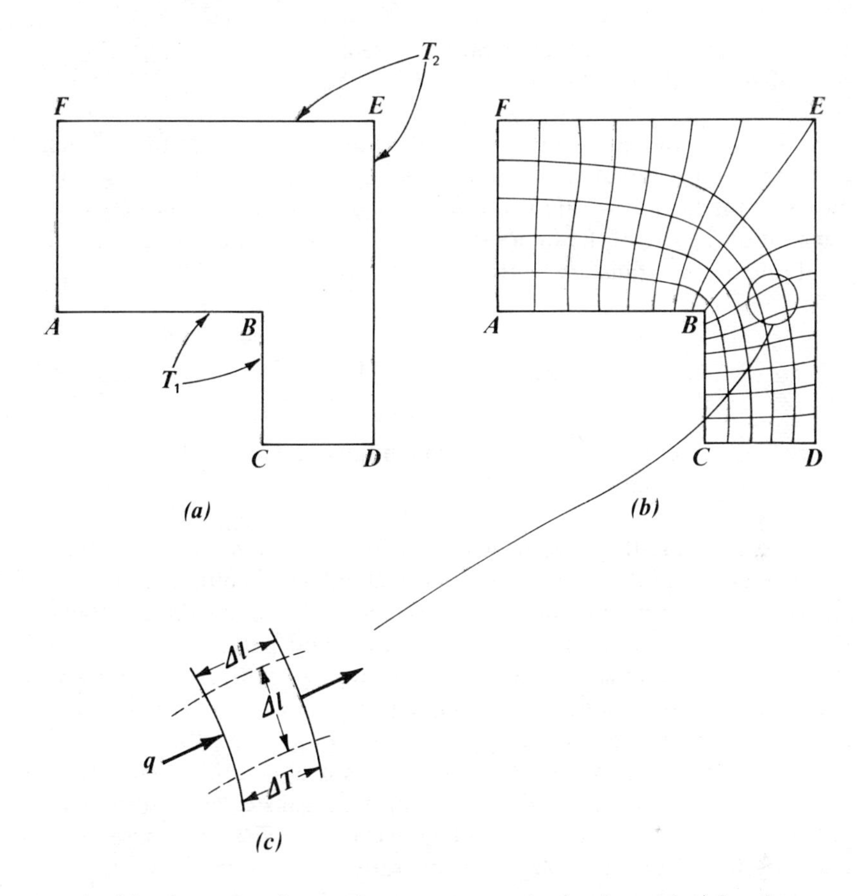

Fig. 3-7. Construction of a curvilinear square network: a) scale model, b) flux plot, c) typical curvilinear square.

duction problem described by the Laplace equation and the associated boundary conditions, is unique. Therefore, *any* curvilinear network, irrespective of the size of the squares, which satisfies the boundary conditions represents the correct solution. Taking any curvilinear square, e.g. Fig. 3-7c, the rate of heat flow is given by Fourier's law,

$$\Delta q = -k(\Delta l \cdot 1)\frac{\Delta T}{\Delta l} = -k\Delta T \tag{3-19}$$

This heat flow will remain the same across any square within any one heat-flow lane from the boundary at T_1 to the boundary at T_2. The ΔT across any one element in the heat-flow lane is therefore

$$\Delta T = \frac{T_2 - T_1}{N} \tag{3-20}$$

where N is the number of temperature increments
between the two boundaries at T_1 and T_2.

The total rate of heat flow from the boundary at T_2 to the boundary at T_1 equals the sum of the heat flow through all the lanes. According to Eq. 3-19 the heat-flow rate is the same through all lanes since it is independent of the size of the squares in a network of curvilinear squares. The total rate of heat transfer can therefore be written

$$q = \sum_{m=1}^{m=M} \Delta q_m = \frac{M}{N} k(T_2 - T_1) \tag{3-21}$$

where Δq_m is the rate of heat flow through the mth lane, and
M is the number of heat-flow lanes.

Thus, to calculate the rate of heat transfer we only need to construct a network of curvilinear squares in the scale model and count the number of temperature increments and heat-flow lanes. Although the accuracy of the method depends a good deal on the skill and patience of the person sketching the curvilinear square network, even a crude sketch can give a reasonably good estimate of the temperature distribution which, if desired, can be refined by the numerical method described in Sec. 3-6.

In any two-dimensional system in which heat is transferred from one surface at T_1 to another at T_2 the rate of heat transfer per unit depth depends only on the temperature difference $(T_1 - T_2)$, the thermal conductivity k, and the ratio (M/N). This ratio depends on the shape of the system and is called the shape factor S. The rate of heat transfer can thus be written

$$q = kS\Delta T \tag{3-22}$$

Values of S have been calculated for several shapes of practical significance (7,8,9) and are summarized in Table 3-1. Corner sections where two or three plane walls meet can also be handled by the shape-factor method. When all the interior walls are greater than one-fifth of the wall thickness, the appropriate shape factors are given in Table 3-1. Shape factors for rectangular enclosures with smaller inside dimensions are presented in Ref. 8.

EXAMPLE 3-1. Determine (a) the shape factor and (b) the rate of heat conduction through the corner section of Fig. 3-7a. The thermal conductivity of the material is 0.5 Btu/hr ft F, the face ABC is at 600 F, the face DEF is at 100 F, and the faces CD and AF are insulated.

Table 3-1. Conduction shape factor S for various systems.

$[q_k = Sk\,(T_1 - T_2)]$

Description of System	Symbolic Sketch	Shape Factor S†
Conduction through a homogeneous medium of thermal conductivity k between an isothermal surface and a sphere buried a distance z below		$\dfrac{2\pi D}{1 - (D/4z)}$
Conduction through a homogeneous medium of thermal conductivity k between an isothermal surface and a horizontal cylinder of length L buried with its axis a distance z below the surface		$\dfrac{2\pi L}{\cosh^{-1}(2z/D)}$ if $z/L \ll 1$.
Conduction through a homogeneous medium of thermal conductivity k between an isothermal surface and an infinitely long cylinder buried a distance z below (per unit length of cylinder)		$\dfrac{2\pi}{\cosh^{-1}(2z/D)}$
Conduction through a homogeneous medium of thermal conductivity k between an isothermal surface and a vertical circular cylinder of length L		$\dfrac{2\pi L}{\ln(4L/D)}$ if $D/L \ll 1$.
Horizontal thin circular disc buried far below an isothermal surface in a homogeneous material of thermal conductivity k		$\dfrac{4.45D}{1 - \dfrac{D}{5.67z}}$
Conduction through a homogeneous material of thermal conductivity k between two long parallel cylinders a distance L apart (per unit length of cylinders)		$2\pi/\cosh^{-1}\left(\dfrac{L^2 - 1 + r^2}{2L\,r}\right) + \cosh^{-1}\left(\dfrac{L + 1 - r}{2L}\right)$ where $r = r_1/r_2$ and $L = L/r_2$.
Conduction through two plane sections and the edge section of two walls of thermal conductivity k—inner and outer surface temperatures uniform‡		$\dfrac{al}{\Delta x} + \dfrac{bl}{\Delta x} + 0.54l$
Conduction through the corner section c of three homogeneous walls of thermal conductivity k—inner and outer surface temperatures uniform‡		$0.15\,\Delta x$ if Δx is small compared to the lengths of walls.

† All dimensions should be in feet.
‡ These shape factors apply only to enclosures whose inside dimensions are greater than one-fifth the wall thickness Δx. For enclosures having smaller inside dimensions see Ref. 8.

Solution: (a) The flux plot is shown in Fig. 3-7b. The faces AF and CD are insulated, and each of them forms, therefore, one boundary of a flow lane.

From the final plot we obtain the number of flow lanes ($N = 15$) and the number of curvilinear squares per flow lane ($M = 5$). Thus, the shape factor is

$$S = \frac{N}{M} = \frac{15}{5} = 3 \qquad \textit{Ans.}$$

(b) From Eq. 3-22, the rate of heat flow per unit thickness is

$$q = Sk(T_2 - T_1) = (3)(0.5)(600 - 100) = 750 \text{ Btu/hr ft} \qquad \textit{Ans.}$$

EXAMPLE 3-2. A long 6-in.-OD pipe is buried with its center line 30 in. below the surface in soil having an average thermal conductivity of 0.20 Btu/hr ft F. (a) Determine, by means of a flux plot, the rate of heat loss per foot length of pipe if the surface temperature of the pipe is 200 F and the surface of the soil is at 40 F. (b) Compare the result with that obtained by using the appropriate shape factor from Table 3-1.

Solution: (a) The flux plot for this problem is shown in Fig. 3-8. Because of the symmetry, only one half of the heat-flow field has been plotted. There are 18 heat-flow lanes leading from the pipe to the surface, and each flow lane consists

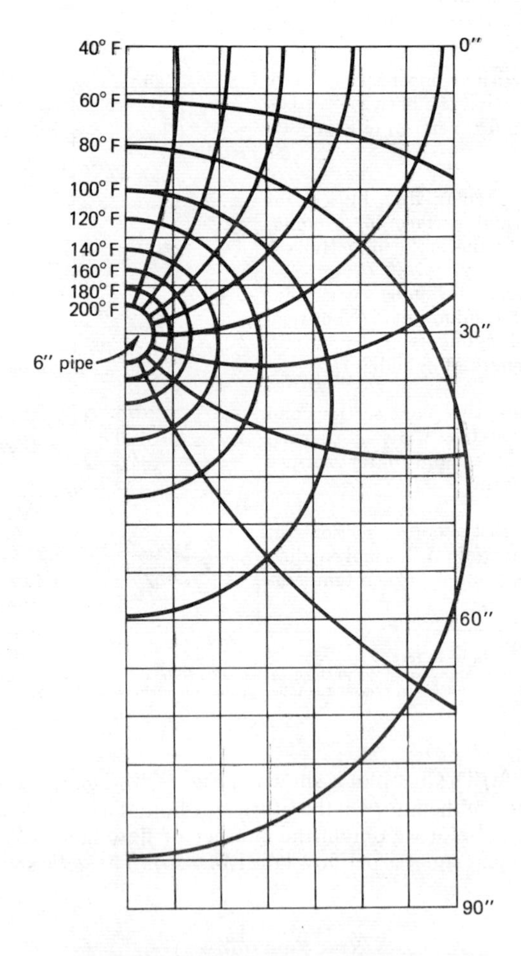

Fig. 3-8. Potential field for Example 3-2.

of 8 curvilinear squares. The shape factor is therefore $S = 18/8 = 2.25$ and the rate of heat flow per foot is

$$q = (0.2)(2.25)(200 - 40) = 72 \text{ Btu/hr ft} \qquad \textit{Ans.}$$

(b) From Table 3-1,

$$S = \frac{2\pi}{\cosh^{-1}(60/6)} = \frac{2\pi}{3.0} = 2.1$$

and the rate of heat flow per foot is

$$q = (0.2)(2.1)(160) = 67 \text{ Btu/hr ft} \qquad \textit{Ans.}$$

The reason for the difference between the two answers is that the potential field in Fig. 3-8 has been drawn approximately with a finite number of flow lines and isotherms.

3-5. *Analogical method*

When two or more phenomena can be described mathematically by the same equation, the phenomena are said to be mathematically analogous and the variables in one system are called the *analogues* of the corresponding variables in any other systems. A simple example of such a case is the two-dimensional Laplace equation. Not only does it apply to a temperature field, but, if the symbol $T(x, y)$ is replaced by $E(x, y)$ (the potential in an electric field), the equation governing the voltage distribution in an electrical field is obtained. An examination of the respective equations

$$\frac{\partial^2 T}{\partial x^2} + \frac{\partial^2 T}{\partial y^2} = 0$$

and

$$\frac{\partial^2 E}{\partial x^2} + \frac{\partial^2 E}{\partial y^2} = 0$$

shows that the electrical potential $E(x, y)$ can be regarded as the analogue of the thermal potential T. In other words, constant-voltage lines in an electric field correspond to constant-temperature lines in a heat-flow field, and lines of electric-current flow correspond to heat-flow lines. A similar correspondence can be established between a two-dimensional potential fluid-flow field in the steady state and either of the two aforementioned phenomena. Table 3-2 illustrates the analogy.

Table 3-2. Electrical and flow analogies.

Type of Flow Field	Potential Lines	Flow Lines
Heat	Constant temperature or isotherms	Heat-flow lines
Incompressible inviscid fluid	Constant velocity potential	Streamlines
Electricity	Constant voltage potential	Lines of force or electric current

A simple way to solve two-dimensional heat conduction problems is to use an electrical analog to construct a network of curvilinear squares and thus determine the shape factor. A convenient tool for this task is the Analog Field Plotter (12). It makes use of a thin sheet ($\sim$0.004 in.) of electrically conducting paper of relatively high resistivity which can be cut to a shape geometrically similar to that of the heat conductor. An electrical-current-flow pattern can be set up in the paper by means of suitably attached and energized electrodes. The resultant pattern of constant-voltage lines is detected and plotted directly on the paper by means of a searching stylus attached to a voltmeter. Boundary conditions corresponding to a constant-temperature potential in the heat flow are obtained in the electrical field by applying copper wires, or highly conductive areas of silver paint, to the surface of the paper and attaching them to a direct-voltage source. Insulated surfaces in the heat-flow field correspond to plain edges of the conducting paper. By selecting equal increments of voltage, adjacent lines become analogous to isotherms separated by the same temperature difference.

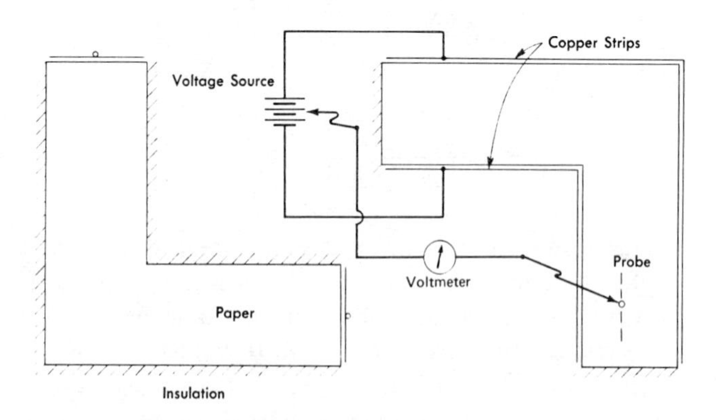

Fig. 3-9. Arrangement of the Analog Field Plotter for solving Example 3-1.

Since the heat-flow, or current-flow, lines are everywhere perpendicular to the potential lines, they can usually be sketched in freehand, so that the resulting network forms curvilinear squares. The flow lines can also be traced by simply reversing the insulating and conducting portions of the boundary. Figure 3-9 illustrates this method for the corner section considered in Example 3-1.

Electrical-geometrical analogues which can account for finite thermal resistances at the boundaries and for variable thermal conductivities have been developed by Kayan (13,14) and Ellerbrook et al. (15).

3-6. Numerical method

In many practical situations the geometry of the system and the boundary conditions are too complex to yield analytical or analogical solutions. Such problems can, however, be solved by numerical methods. These methods are based on finite difference techniques which are ideally suited for solution by means of high-speed, digital computers. However, before a numerical method can be applied to a heat-transfer problem, or any other physical problem described by a differential equation, some preliminary steps are necessary. The purpose of these preliminary steps is to approximate the differential equation and the boundary conditions by a set of algebraic equations. This is accomplished by replacing the continuous domain by a pattern of discrete points within the domain and introducing finite-difference approximations between the points.

To solve a heat-conduction problem numerically, we subdivide the system into a number of small but finite subvolumes and assign a reference number to each. Then we assume that each subvolume is at the temperature corresponding to its center and replace the physical system by a network of fictitious heat-conducting rods between the centers, or *nodal points*, of the subvolumes. Now, if a thermal conductance corresponding to the conductance of the material between nodal points is assigned to each rod, the heat flow in the rod network will approximate the heat flow in the continuous system. If N points are selected, a set of N algebraic equations is obtained. It can be solved by matrix inversion or a numerical method for the values of unknown at the N points.

If the absence of heat sources or sinks within the system, the rate of heat flow toward each nodal point must equal the rate of heat flow away from it in the steady state. To satisfy this condition we set up heat balances for each nodal point and thus obtain as many algebraic equations as there are nodal points in the system. To carry out a numerical solution, we first estimate nodal-point temperatures, and then correct them in successive steps until the rate of heat inflow equals the rate of heat outflow at every point in the system. The details of the numerical method will be illustrated in the following examples. A one-dimensional heat-flow problem has been selected as the first example because it illustrates the basic concepts without introducing unnecessary conceptual difficulties.

One-dimensional example. A circular pin fin is 0.25 ft long and 0.05 ft in diameter. Its base is attached to a wall at 300 F, while its surface is exposed to a gas at 100 F through an average unit-surface conductance of 20 Btu/hr sq ft F. The fin is made of a stainless steel with a thermal conductivity of 10 Btu/hr ft F. Assuming a uniform temperature at any cross-section, the temperature distribution along the fin and the rate of heat dissipation are to be determined numerically, and the results are to be compared with the analytic solution obtained in Sec. 2-4.

First Step. The first step is to subdivide the system and to replace the material between nodal points by fictitious rods having the same conductance as the material. We choose the uniform linear network shown in Fig. 3-10 with six equally spaced nodal points. This selection yields four complete internal subvolumes with half subvolumes at the base and at the tip of the fin. Increasing the number of nodal points will improve the accuracy of the solution but will also increase the amount of work and time required.

Second Step. The second step consists of writing heat-balance equations for each of the nodal points. The same heat-balance equation holds for all interior points, but the nodal points at the base and at the tip (i.e., the boundary conditions) require separate analyses. For an interior point m (Fig. 3-11a), the heat-balance to be satisfied is

$$q_{k,(m-1)\rightarrow m} + q_{c,\infty\rightarrow m} + q_{k,(m+1)\rightarrow m} = 0 \tag{3-23}$$

where $q_{k,(m-1)\rightarrow m} = K_{k,(m-1,m)}(T_{m-1} - T_m)$, the rate of heat flow by conduction from nodal point $(m - 1)$ to point m;

$q_{c,\infty\rightarrow m} = K_{c,(\infty,m)}(T_\infty - T_m)$, the rate of heat flow by convection

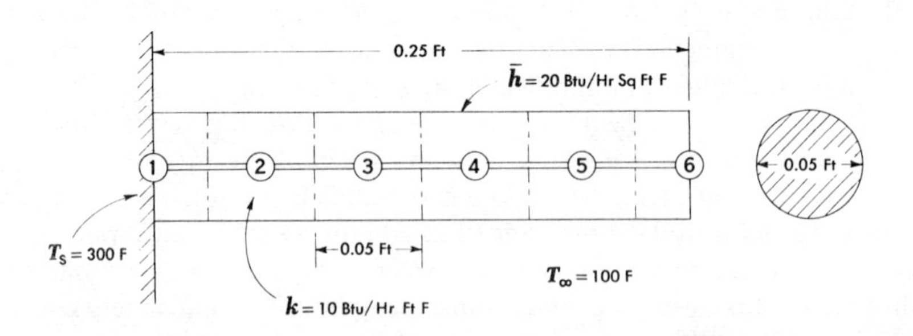

Fig. 3-10. Circular pin fin subdivided for numerical solution.

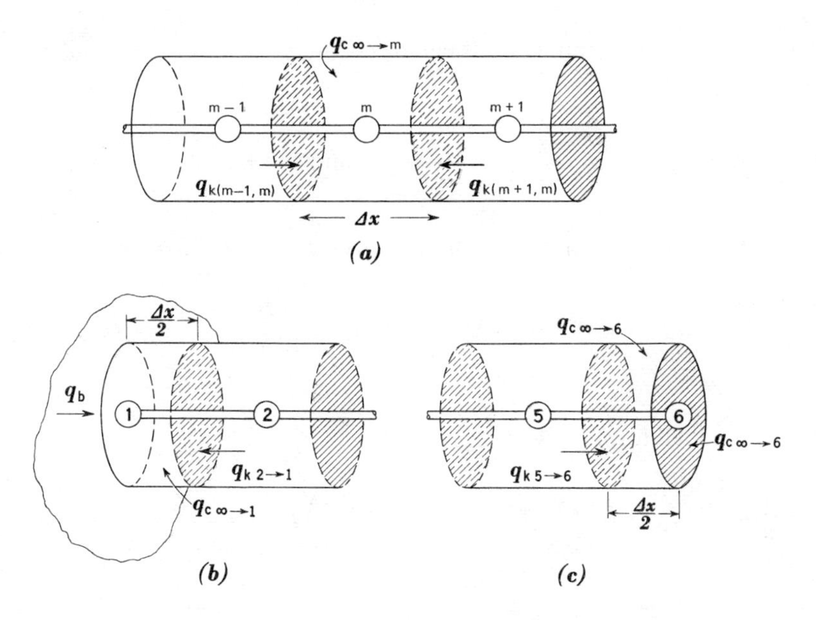

Fig. 3-11. Sketches illustrating heat balances.

from the surrounding gas to the surface of the sub-volume of nodal point m;

$q_{k,(m+1)\to m} = K_{k,(m+1,m)}(T_{m+1} - T_m)$, the rate of heat conduction from $(m + 1)$ to m.

The physical basis of Eq. 3-23 is analogous to Kirchhoff's first law for electrical circuits. The equation simply states that the algebraic sum of all the heat flows (or currents) at a junction point of a network equals zero in the steady state. The arrows indicate that the direction of positive heat flow is toward the nodal point m. This, of course, does not imply that heat flows only toward point m. If $T_{(m-1)}$, $T_{(m+1)}$ or T_∞ is less than T_m, any one of the three terms may be negative and a negative value for q_m indicates that heat is transferred away from point m.

Since we do not know at this stage what temperatures will make the heat flow *to* nodal point m equal to the heat flow *from* it, we set the left hand side of Eq. 3-23 equal to some residual Q_m, or

$$q_{k,(m-1)\to m} + q_{c,\infty\to m} + q_{k,(m+1)\to m} = Q_m \tag{3-24}$$

The residual in Eq. 3-24, Q_m, can be interpreted physically as the rate of internal energy change at point m which must be zero in the steady state.

Assuming straight-line temperature distributions along the fictitious rods between nodal points, a reasonable assumption when the points are close together, the conductances can be evaluated from Eqs. 1-4 and 1-16.[2] For the circular fin we have

$$K_{k,(m-1,m)} = K_{k,(m+1,m)} = \frac{k\pi D^2}{4\Delta x} \qquad \text{and} \qquad K_{c,(\infty,m)} = \bar{h}\pi D\Delta x$$

With these expressions for the conductances, Eq. 3-24 can be written

$$\frac{k\pi D^2}{4\Delta x}(T_{(m-1)} - T_m) + \bar{h}\pi D\Delta x(T_\infty - T_m) + \frac{k\pi D^2}{4\Delta x}(T_{(m+1)} - T_m) = Q_m \tag{3-25}$$

or in simplified form as

$$T_{(m-1)} + T_{(m+1)} + \frac{4\bar{h}\Delta x^2}{kD}T_\infty - \left(2 + \frac{4\bar{h}\Delta x^2}{kD}\right)T_m = \frac{Q_3}{k\pi D^2/4\Delta x} = Q'_m \tag{3-26}$$

In this form the residuals have the dimensions of temperature and their values are an indication of the accuracy of the temperature field. The term involving T_∞ drops out if all temperatures are measured above or below T_∞, and T_∞ is taken as zero.

At the base of the fin (Fig. 3-11b) the temperature T_1 is equal to the wall temperature and remains constant. The residual equation with T_∞ = 0 becomes therefore

$$\frac{4\Delta x}{k\pi D^2}q_b + T_2 - \left(1 + \frac{2\bar{h}\Delta x^2}{kD}\right)T_1 = \frac{Q_1}{k\pi D^2/4\Delta x} = Q'_1 \tag{3-27}$$

where q_b is rate of heat conduction from the wall to the base of the fin.

Table 3-3. Summary of residual equations.

Point	Equation for Q'	
1	$2.5\,q_b + T_2 - 240$	$= Q'_1$
2	$200 + T_3 - 2.4\,T_2$	$= Q'_2$
3	$T_2 + T_4 - 2.4\,T_3$	$= Q'_3$
4	$T_3 + T_5 - 2.4\,T_4$	$= Q'_4$
5	$T_6 + T_4 - 2.4\,T_5$	$= Q'_5$
6	$T_5 - 1.3\,T_6$	$= Q'_6$

[2]In the residual Eq. 3-24, terms of the order of Δx^4 are neglected (see Ref. 19). Equation 3-24 is, therefore, an acceptable approximation only when Δx^4 is small.

At the tip of the fin, heat is transferred by convection (Fig. 3-11c) and the residual equationfor point 6 is (see Prob. 3-35)

$$T_5 - \left(1 + \frac{2\bar{h}\Delta x^2}{kD} + \frac{\bar{h}\Delta x}{k}\right)T_6 = \frac{Q_6}{k\pi D^2/4\Delta x} = Q_6' \qquad \text{(3-28)}$$

Third Step. The third step consists of collecting the residual equations for all the nodal points and evaluating the coefficients. For the circular fin these equations are summarized in Table 3-3.

Fourth Step. The only remaining problem is finding the values of T_2, T_3, T_4, T_5, and T_6 which satisfy the residual equations and eliminate the residuals. Since T_1 is fixed by the boundary condition, we have a set of five equations in five unknowns. This set of equations could be solved algebraically, but in more complex problems with many nodal points, a simultaneous solution can become very time-consuming. We will therefore illustrate the use of numerical methods which can be adapted to

Table 3-4. Sequence of steps in a relaxation solution.

1. Assume values for the temperatures at the various nodes. Use physical insight and limits imposed by the boundary conditions to guess initial temperatures as near to their true values as possible.
2. Calculate the residual at each node, using the assumed temperatures.
3. Relax the largest residual to zero or close to zero by changing the temperature at that nodal point an appropriate amount. Observe that a positive value of a residual requires an increase in the nodal temperature to eliminate the residual, and vice-versa.
4. Change the residuals of the adjacent nodal points to take account of the temperature change in Step 3. The summary of residual equations will be useful to determine which nodal points are involved. For example, an increase of one degree in T_3 will decrease Q_3' by 2.4 degrees and at the same time increase Q_2' and Q_4' by one degree. The interrelation between temperature changes and residual can be handled most expeditiously by constructing a table showing the effects of a change of one degree in nodal temperature on the residuals of the surrounding nodes. Such a table illustrating the relaxation pattern for the fin problem is shown in Table 3-5.
5. Continue to relax residuals until they all are as close to zero as desired. Note that residuals may be over-relaxed or under-relaxed to speed up the convergence in hand calculations. This is usually not advantageous in machine calculations.
6. Check your answer by re-calculating the residuals using the final temperatures from Step 5. If a mistake is discovered, do not repeat the calculations, but continue the relaxation process using the temperatures from Step 5 and the corrected residuals. This step is omitted in machine calculations (they never make a mistake).

complex problems and to machine calculations. The first of these numerical methods to be illustrated is the so-called relaxation method. It is useful only for hand calculations, but it illustrates the numerical approach in a simple manner. The relaxation process is carried out in several steps listed in Table 3-4.

The relaxation solution for the fin is shown in Table 3-5. Because

Table 3-5. Relaxation solution for fin of Fig. 3-10.

	T_1 fixed	Q'_1	T_2	Q'_2	T_3	Q'_3	T_4	Q'_4	T_5	Q'_5	T_6	Q'_6
Initial Temperature Distribution, $(T - T_\infty)$ (Assumed)	200		120		70		30		20		10	
Initial Residuals (From equations in Table 3-3)		—		−18		−18		+18		−8		+7
Decrease T_2 by 8		—	112	+1.2		−26						
Decrease T_3 by 11				−9.8	59	+0.4		+7				
Decrease T_2 by 4		—	108	−0.2		−3.6						
Increase T_4 by 3						−0.6	33	−0.2		−5		
Increase T_6 by 5										0	15	+0.5
Final Residuals				−0.2		−0.6		−0.2		0		+0.5
$(T - T_\infty)$ relaxation	200		108		59		33		20		15	
$(T - T_\infty)$ analytical (Eq. 2-47)	200		107		57		32		19		15	

T_1 is fixed, no residual is calculated for this nodal point. However, after the temperatures have been evaluated, the rate of heat flow through the base of the fin can be determined by substituting the numerical value of T_2 into the residual equation for point 1. The sequence of steps of the relaxation process for the fin problem are shown in Table 3-6 and the reader should verify them by himself. For comparison, the temperature distribution $T(x) - T_\infty$ calculated from Eq. 2-47 is shown on the last line. The agreement is within 2 F.

The rate of heat flow to the fin is

$$q_b = \frac{240 - T_2}{2.55} = \frac{240 - 108}{2.55} = 51.7 \text{ Btu/hr}$$

from the first residual equation in Table 3-3, whereas the rate of heat flow calculated from the analytic solution, i.e., Eq. 2-48, is 49.5 Btu/hr, a difference of about 4 percent.

The steps followed in the hand calculations used in the solution of the preceding example could also be programmed for a digital computer. However, in programming for a digital computer the rapidity of convergence and the amount of memory storage required must be minimized. The efficiency of the computation process can be improved if instead of relaxing the residuals one by one, the temperatures required to make all of the residuals equal to zero are determined simultaneously by an iterative scheme (21). One such scheme, call the Gauss-Seidel method, starts with the same residual equations as the relaxation method (see Table 3-3), but assumes at the outlet that all the residuals are equal to zero. Thus, the system of equations to be solved for the pin fin shown in Fig. 3-10 consists of four equations for interior nodal points of the form

$$T_{m-1} + T_{m+1} - 2.4T_m = 0 \qquad (m = 2, 3, 4, 5)$$

and the equation for the nodal point at the end

$$T_{m-1} - 1.3T_m = 0 \qquad (m = 6)$$

Table 3-6. Relaxation pattern for fin problem.

	$\Delta Q_1'$	$\Delta Q_2'$	$\Delta Q_3'$	$\Delta Q_4'$	$\Delta Q_5'$	$\Delta Q_6'$
$\Delta T_2 = 1$	+1	−2.4	+1	....	....	....
$\Delta T_3 = 1$	...	+1	−2.4	+1	....	....
$\Delta T_4 = 1$	...	....	+1	−2.4	+1	....
$\Delta T_5 = 1$	...	....	....	+1	−2.4	+1
$\Delta T_6 = 1$	...	....	....	....	+1	−1.3
$\Delta T_2 = \Delta T_3 = \ldots = \Delta T_6 = 1$	+1.0	−1.4	−0.4	−0.4	−0.4	−0.3

In addition, the temperature at the base ($m = 1$) is prescribed as

$$T_m = 200 \qquad (m = 1)$$

The solution of the preceding set of equations by means of a digital computer is illustrated below, using FORTRAN IV computer language (22). The variables used in the computer program are related to those used in Eqs. 3-23 to 3-28 in Table 3-7. A flow diagram showing the sequence of operations in the computer solution is shown in Fig. 3-12. The FORTRAN program used for solving the problem is presented in Table 3-8 (see pages 106–7). Explanatory remarks are included to clarify the most important steps in the program.

Two-dimensional systems. The numerical method can readily be extended to two- and also to three-dimensional systems. Consider a two-dimensional system such as a solid of constant thickness b (Fig. 3-13). Subdivide the system into squares so that each subvolume has the dimensions $\Delta l\, \Delta l b$, and select the center of each subvolume as a nodal point. Then denote the location of each nodal point by two number subscripts, m and n. The first of these locates the row of the nodal point counted from the top, the second the column counted from the left in a Cartesian coordinate system. The first subscript locates the nodal point along the

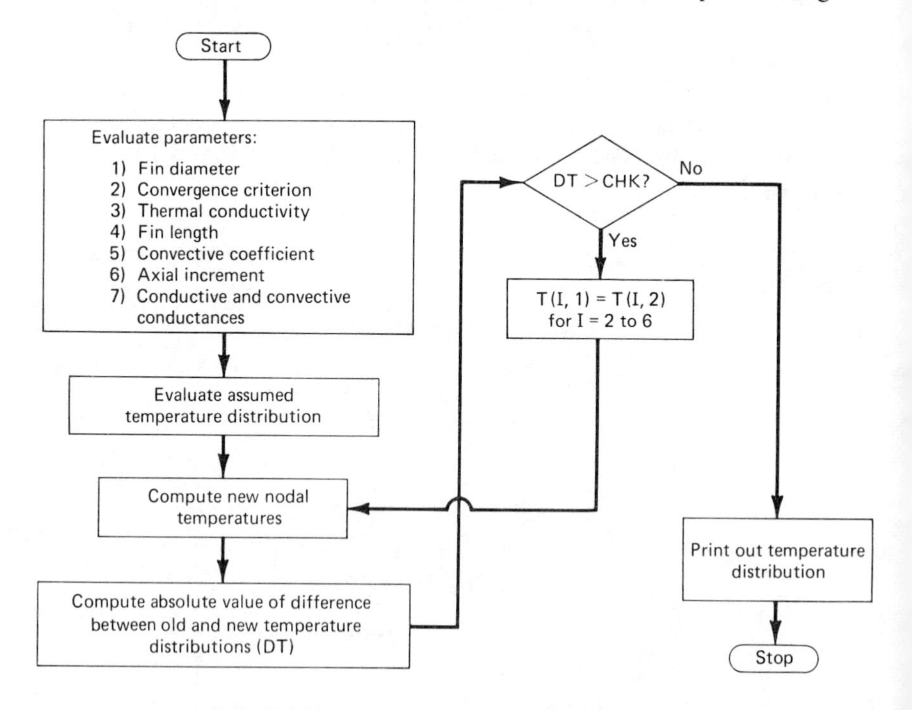

Fig. 3-12. Flow diagram for pin fin temperature distribution.

Table 3-7. Variables in FORTRAN program for pin fin temperature distribution related to those in Equations 3-23 to 3-28.

FORTRAN Variable	Equation Symbol in Text	Description	Units
CHK	–	Convergence criterion	F
D	D	Pin diameter	ft
DT (I)	–	$\vert T(I,1) - T(I,2)\vert$	F
DX	Δx	Axial increment	ft
H	$\bar{h}$	Average unit surface conductance	Btu/hr ft^2 F
K	k	Thermal conductivity	Btu/hr ft F
KK	$k\pi D^2/4\Delta x$	Axial conductance	Btu/hr F
KC	$\bar{h}\pi D\Delta x$	Convective conductance	Btu/hr F
KCE	$\bar{h}\pi D^2/4$	End conductance	Btu/hr F
L	–	Fin length	ft
PI	π	3.14159 ...	—
T (1, 1), etc.	T_1, etc.	Nodal temperatures	F

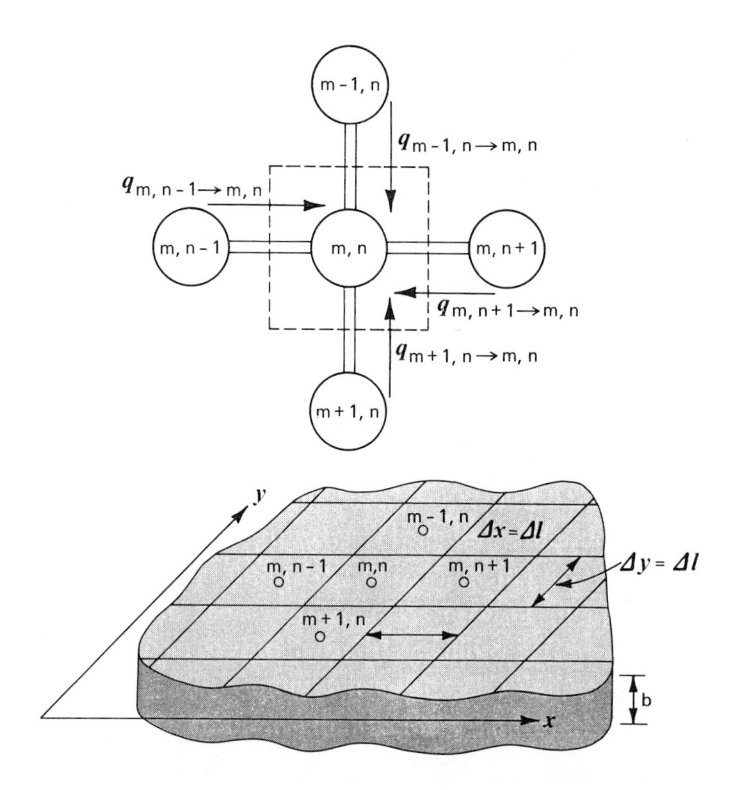

Fig. 3-13. Sketch showing nodal points for two-dimensional relaxation solution.

Table 3-8. FORTRAN IV program for pin fin temperature distribution (see Fig. 3-10).

```
      DIMENSION T(6,2),DT(6)
      REAL K,KK,KC,KCE,L
```

The array T has 12 locations reserved. The first subscript denotes the nodal location, the second either the "old" (subscript 1) or the newly computed value (subscript 2). The array DT stores absolute values of the differences between "old" and "new" values of T to be used in the convergence check portion of the program. The parameters, K, KK, KC, KCE, L, normally integers in FORTRAN, are declared to be floating point.

```
      D=.05
      CHK=.1
      K=10.
      L=.25
      H=20.
      DX=L/5.
      PI=4.*ATAN(1.)
      KK=K*PI*D*D/(4.*DX)
      KC=H*PI*D*DX
      KCE=H*PI*D*D/4.
```

The parameters of the program are evaluated. See Table 3-7 for their precise definition.

```
      DO 1 I=2,6
      T(I,1)=100.
1     CONTINUE
      T(1,1)=200.
```

The preceding four statements evaluate the initial guess of the temperature distribution. At the base of the fin T = 200 F; elsewhere initially a uniform value of 100 F is specified.

```
2     DO 3 I=2,5
      T(I,2)=(T(I-1,1)+T(I+1,1))/(2.+(KC/KK))
3     CONTINUE
      T(6,2)=T(5,1)/(1.+(KC/KK)+(KCE/KK))
      T(1,2)=200.
```

The preceding five statements compute the new values of T at each node using the heat balance equations as developed earlier.

Table 3-8. Continued.

```
      DO 4 I=2,6
      DT(I)=ABS(T(I,1)-T(I,2))
      IF (DT(I).GT.CHK) GO TO 6
4     CONTINUE
```

The absolute value of the difference between old and new temperatures is computed at each node. If any DT is larger than the convergence criterion the computation is diverted to statement 6 below. If all DT values are less than CHK, the temperature distribution is printed out and the program is terminated (next 2 statements).

```
      PRINT 5,CHK,(T(I,2),I=1,6)
      CALL EXIT
6     DO 7 I=2,6
      T(I,1)=T(I,2)
7     CONTINUE
      GO TO 2
```

The immediately preceding four statements replace the old values of T with the new values and return the computation to statement 2 for an additional iteration.

```
5     FORMAT (1H1,*NODAL TEMPERATURES, T(1) THRU T(6), TO AN ACCURACY
     1OF*,4.2,* DEG F*,//,6(5X,F5.1,/))
      END
```

Statement 5 specifies the format of the printout of the temperature distribution. The END card is the final statement of the program. Nodal temperatures, T(1) through T(6), to an accuracy of .10 degree F are given below.

```
200.0
107.6
 58.3
 32.1
 18.6
 15.2
```

The output of the program is the temperature distribution in the fin relative to the ambient temperature. (See Table 3-6 for relaxation solution.)

y-axis, the second along the x-axis. This convention corresponds to the numbering system used in computer programming. A steady-state heat balance on an interior point can be written as

$$q_{m-1,n\to m,n} + q_{m,n-1\to m,n} + q_{m+1,n\to m,n} + q_{m,n+1\to m,n} = 0 \tag{3-29}$$

Replacing the material between nodal points by fictitious rods having the same conductance K as the actual material (see Fig. 3-13), Eq. 3-29 can be expressed as a residual equation

$$K_{k,(m-1,n)-(m,n)}(T_{m-1,n} - T_{m,n}) + K_{k,(m,n-1)-(m,n)}(T_{m,n-1} - T_{m,n})$$
$$+ K_{k,(m+1,n)-(m,n)}(T_{m+1,n} - T_{m,n}) + K_{k,(m,n+1)-(m,n)}(T_{m,n+1} - T_{m,n}) = Q_{m,n} \tag{3-30}$$

where $Q_{m,n}$ is the residual which must be zero in the steady state. Since all the internal conductances are equal to

$$K_{k,(m-1,n)-(m,n)} = \frac{k(\Delta 1\, b)}{\Delta 1} = kb = K_{k,(m,n-1)-(m,n)} \quad \text{etc.}$$

the residual equation can be written as

$$Q'_{m,n} = \frac{Q_{m,n}}{kb} = T_{m-1,n} + T_{m,n-1} + T_{m+1,n} + T_{m,n+1} - 4T_{m,n} \tag{3-31}$$

In the form of Eq. 3-31 the residuals have the dimension of temperature and the values of the residuals reflect the accuracy of the temperature field at any step in the computations.

In subdividing a two-dimensional system it is usually convenient to select a square network of nodes. Interior points are then surrounded by full squares of material, but the nodal points at the edges have less material associated with them. Thus, while Eq. 3-31 applies to all interior points, the residual equations for nodal points at a boundary depend on the geometry as well as the particular boundary conditions. For example, consider a nodal point at a surface of a solid exposed to a convection boundary condition as shown in Fig. 3-14. Heat is transferred to the subvolume represented by the nodal point n,m through the left face by conduction, and the right face by convection. Conduction is also the heat transfer mode from the nodal points above and below, but the area is only $(\Delta 1 \cdot b/2)$ for these faces. Therefore, the energy balance for the node n,m at the boundary is

$$\frac{kb\Delta 1}{2}\,\frac{T_{m-1,n} - T_{m,n}}{\Delta 1} + kb\Delta 1\,\frac{T_{m,n-1} - T_{m,n}}{\Delta 1}$$

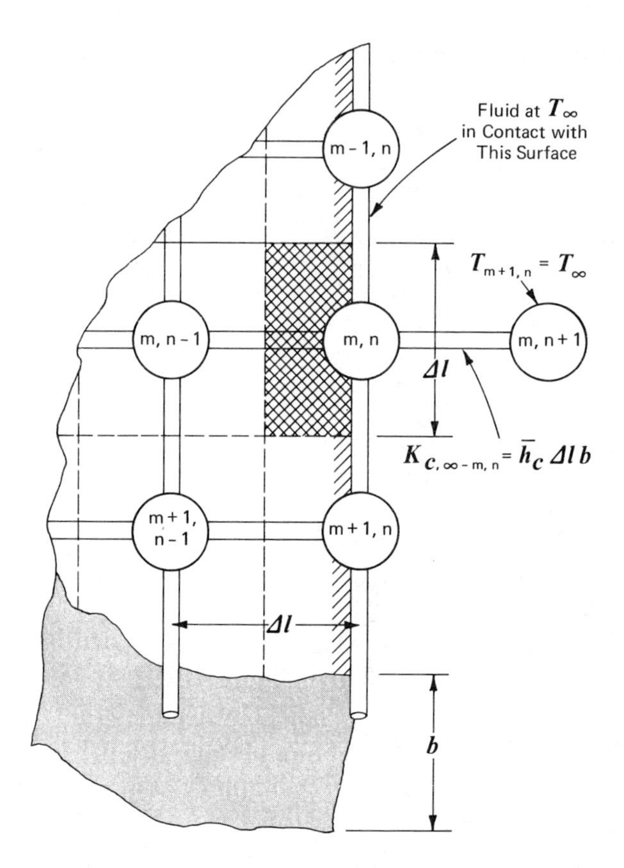

Fig. 3-14. Nodal point at a surface in contact with a fluid at T_∞.

$$+ \bar{h}_c b\Delta 1 (T_\infty - T_{m,n}) + k\,\frac{b\Delta 1}{2}\,\frac{T_{m+1,n} - T_{m,n}}{\Delta 1} = 0 \qquad \text{(3-32)}$$

and the residual equation becomes

$$Q_{m,n}/kb = Q'_{m,n} = \frac{1}{2}(T_{m-1,n} + 2T_{m,n-1} + T_{m+1,n}) + \frac{\bar{h}\Delta l}{k} T_\infty - \left(\frac{\bar{h}\Delta l}{k} + 2\right) T_{m,n} \qquad \text{(3-33)}$$

Other boundary conditions can be treated in a similar fashion. A convenient summary of residual equations is given in Table 3-9 for different geometries and boundary conditions. Attention is directed to the last two conditions which give the residual equations for nodal points in the vicin-

Table 3-9. Summary of heat balances at boundaries of a two-dimensional system.

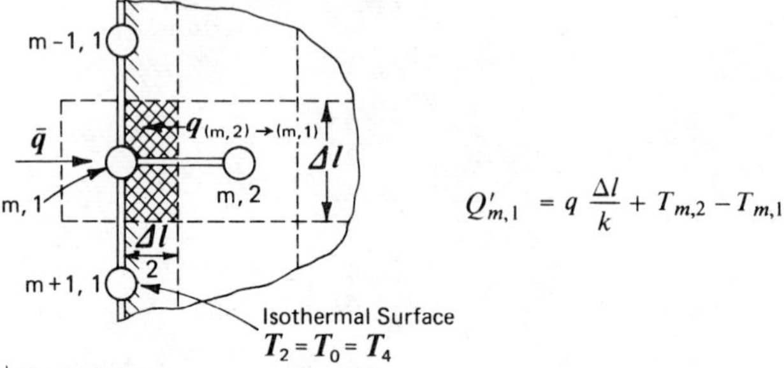

(a) Nodal Point at an Isothermal Boundary

$$Q'_{m,1} = q\frac{\Delta l}{k} + T_{m,2} - T_{m,1}$$

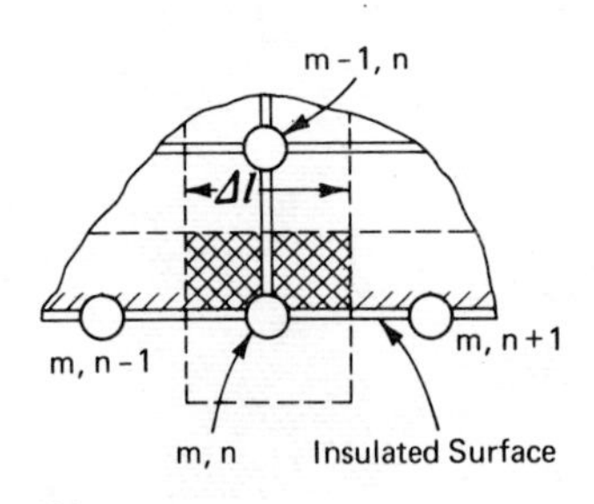

(b) Nodal Point at an Insulated Surface

$$Q'_{m,n} = \tfrac{1}{2}(T_{m,n-1} + T_{m,n+1}) + T_{m-1,n} - 2T_{m,n}$$

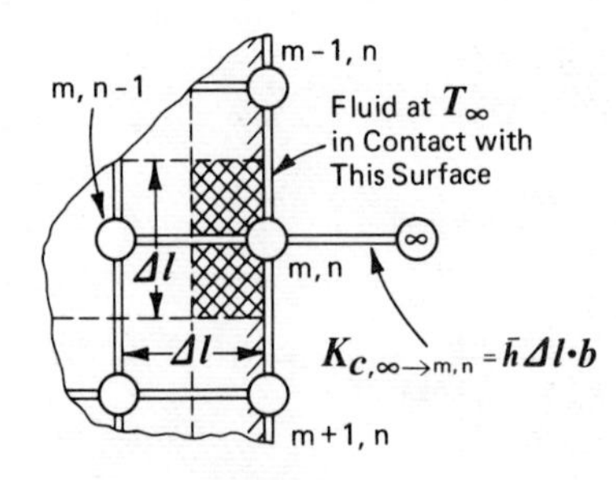

(c) Nodal Point at Surface in Contact with a Fluid

$$Q'_{m,n} = \tfrac{1}{2}(T_{m-1,n} + T_{m+1,n}) + T_{m,n-1} + T_\infty \frac{\bar{h}\Delta l}{k} - T_{m,n}\left(2 + \frac{\bar{h}\Delta l}{k}\right)$$

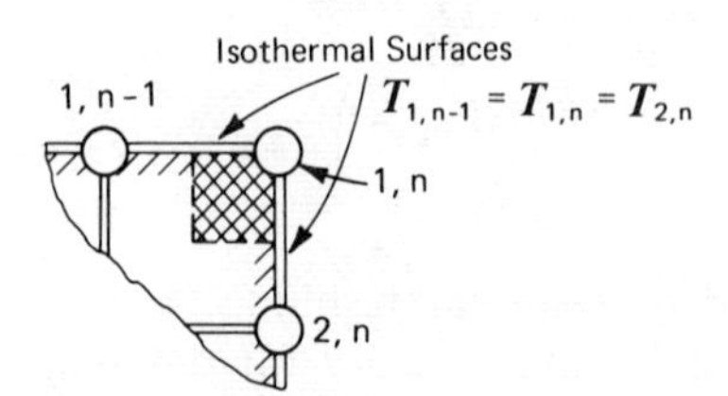

(d) Nodal Point in Exterior Corner Between Isothermal Surfaces

$Q'_{1,n} = 0$ (because no heat can flow to or from it)

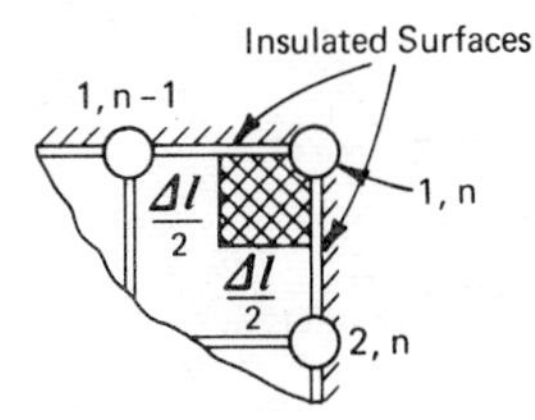

(e) Nodal Point at an Exterior Corner between Insulated Surfaces

$$Q'_{1,n} = \tfrac{1}{2}(T_{1,n-1} + T_{2,n}) - T_{1,n}$$

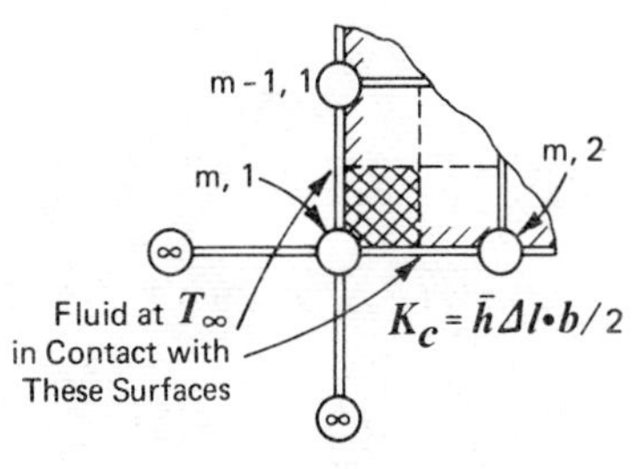

(f) Nodal Point at an Exterior Corner in Contact with a Fluid

$$Q'_{m,1} = \tfrac{1}{2}(T_{m-1,1} + T_{m,2}) + T_\infty \frac{\bar{h}\Delta l}{k} - \left(\frac{\bar{h}\Delta l}{k} + 1\right) T_{m,1}$$

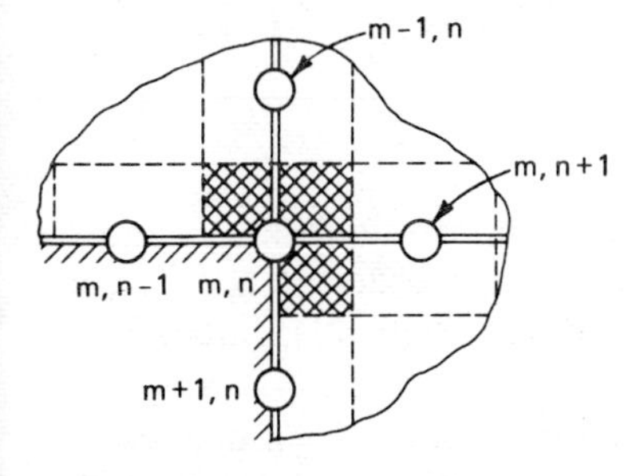

(g) Nodal Point at an Interior Corner between Insulated Surfaces

$$Q'_{m,n} = \tfrac{1}{2}(T_{m,n-1} + T_{m+1,n}) + (T_{m-1,n} + T_{m,1}) - 3T_{m,n}$$

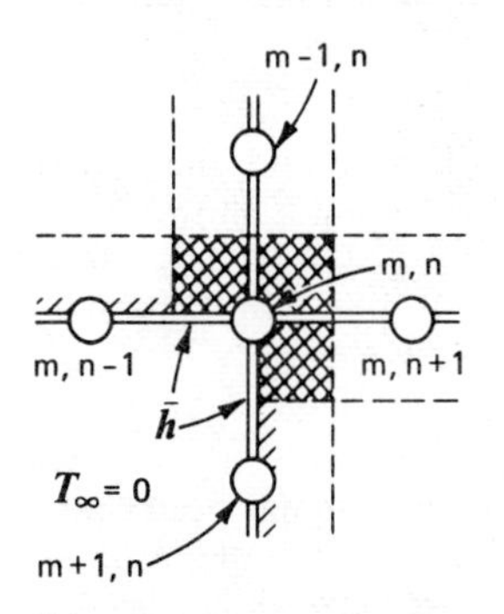

(h) Nodal Point at an Interior Corner whose Surfaces are in Contact, with a Fluid at $T_\infty = 0$

$$Q'_{m,n} = \tfrac{1}{2}(T_{m,n-1} + T_{m+1,n}) + (T_{m-1,n} + T_{m,n}) - \left(3 + \frac{\bar{h}\Delta l}{k}\right) T_{m,n} + \left(\frac{\bar{h}\Delta l}{k}\right) T_\infty$$

Table 3-9. Continued.

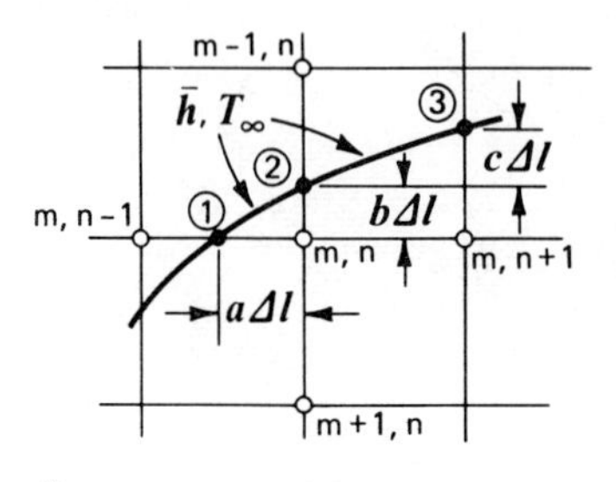

$$Q'_{m,n} = \frac{2}{b(b+1)} T_2 + \frac{2}{a+1} T_{m,n+1} + \frac{2}{b+1} T_{m+1,n} + \frac{2}{a(a+1)} T_1 - 2\left(\frac{1}{a} + \frac{1}{b}\right) T_{m,n}$$

(*i*) Interior Node Near Curved Boundary

$$Q'_2 = \frac{b}{\sqrt{a^2+b^2}} T_1 + \frac{b}{\sqrt{c^2+1}} T_3 + \frac{a+1}{b} T_{m,n} + \frac{h\Delta l}{k}\left(\sqrt{c^2+1} + \sqrt{a^2+b^2}\right) T_\infty - \left[\frac{b}{\sqrt{a^2+b^2}} + \frac{b}{\sqrt{c^2+1}} + \frac{a+1}{b} + \left(\sqrt{c^2+1} + \sqrt{a^2+b^2}\right)\frac{h\Delta l}{k}\right] T_2$$

(*j*) Boundary Node with Convection along Curved Boundary—Node 2 for (*i*) above

ity of curved boundaries of a system subdivided into uniform Δx and Δy increments. This approach neglects terms of the order of Δl^4 as compared to terms of the order of Δl^3 in the interior (19, 21), but it is still sufficiently accurate in practice.

The mechanics of the numerical solution for a two-dimensional problem are identical to those used for the one-dimensional case. The following example illustrates the relaxation method for a two-dimensional system.

EXAMPLE 3-3. Determine the temperature distribution and the rate of heat flow per foot of height for a tall chimney whose cross section is shown in Fig. 3-15. Assume that the inside surface temperature is 500 F, the outside surface temperature is 100 F, and the thermal conductivity of the wall material is 1.0 Btu/hr ft F.

Solution: From the observed symmetry we need consider only the one-eighth section which is shown enlarged in Fig. 3-16. A large mesh has been chosen for the relaxation treatment to reduce the amount of labor. For better accuracy, the mesh should be subdivided once more after the coarser network has been relaxed. The new solution should be started from the results obtained in the previous one.

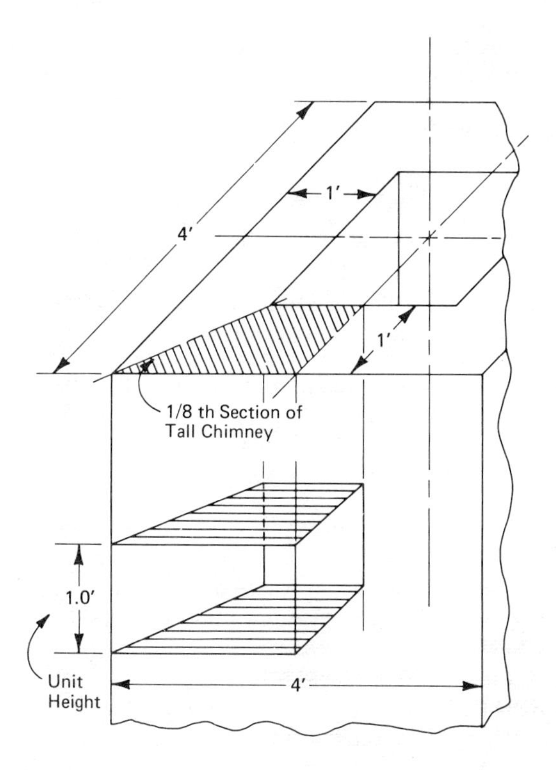

Fig. 3-15. Sketch of chimney for Examples 3-3 and 3-4.

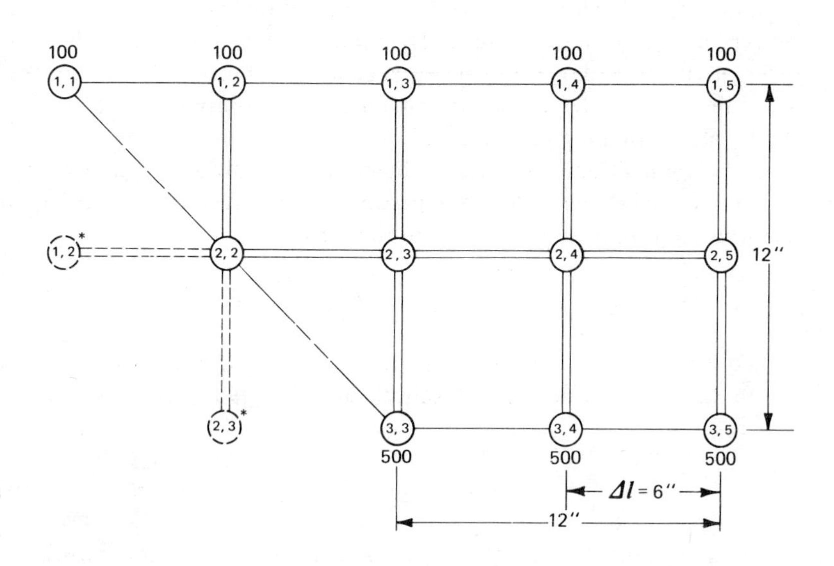

Fig. 3-16.

Since the boundary temperatures are constant, we only need to relax the nodal points (2,2), (2,3), (2,4) and (2,5). Points (2,3) and (2,4) are simple interior points, and Eq. 3-31 applies. For instance, if the initial temperatures at (2,2), (2,3), (2,4) and (2,5) are assumed to be 300 F, the initial residual at (2,4) is

$$Q'_{(2,4)} = T_{(2,3)} + T_{(1,4)} + T_{(2,5)} + T_{(3,4)} - 4\,T_{(2,4)}$$
$$= 300 + 100 + 300 + 500 - 1200 = 0$$

Point (2,2) is located on a line of symmetry, and the influence of points (1,2)* and (2,3)*, the mirror images of points (1,2) and (2,3), is taken into account by doubling the residual due to temperatures at points (1,2) and (2,3). A similar procedure is followed for point (2,5). For instance, the initial residual at point (2,2) is

$$Q'_{(2,2)} = 2(T_{(2,3)} + T_{(1,2)}) - 4T_{(2,2)}$$
$$= 2\,(300 + 100) - 1200 = -400$$

The relaxation process is shown in Table 3-10. The heat-flow rate from the furnace is found by summing the heat-flow rates through the rods terminating at the inner surface, or

$$q = 8k[(500 - 269) + (500 - 292) + 1/2\,(500 - 297)]$$
$$= 4324 \text{ Btu/hr} \qquad \textit{Ans.}$$

The preceding problem could be solved by the relaxation method without a computer because the boundary conditions are simple and only a few nodal points were used. A hand calculation would, however, become very cumbersome if more nodal points were selected or if the heat transfer at the boundaries would be specified by more realistic convection conditions. Under those circumstances the solution should be obtained by means of a digital computer, if available.

In the application of the Gauss-Seidel method to a two-dimensional conduction problem the system of equations to be solved consists of equations for nodal points in the interior of the form

$$T_{m+1,n} + T_{m-1,n} + T_{m,n+1} + T_{m,n-1} - 4T_{m,n} = 0 \tag{3-34}$$

and of equations for nodal points at the boundary. If the temperature at the boundary is prescribed, as in Example 3-3, the equations for boundary nodes are simply

$$T_{m,n} = C_{m,n} \tag{3-35}$$

where the constants $C_{m,n}$ are the prescribed values of the temperature. If there is convection heat transfer at the boundary, the equation for a con-

Table 3-10. Relaxation solution for Example 3-3.

	$T_{2,2}$	$Q'_{2,2}$	$T_{2,3}$	$Q'_{2,3}$	$T_{2,4}$	$Q'_{2,4}$	$T_{2,5}$	$Q'_{2,5}$
Assumed Initial Temperature Distribution	300		300		300		300	
Initial Residuals		−400		0		0		0
Decrease $T_{2,2}$ by 100	200	0		−100				
Decrease $T_{2,3}$ by 25		−50	275	0		−25		
Decrease $T_{2,2}$ by 12	188	−2		−12				
Decrease $T_{2,4}$ by 6				−18	294	−1		−12
Decrease $T_{2,3}$ by 5		−12	270	+2		−6		
Decrease $T_{2,2}$ by 3	185	0		−1				
Decrease $T_{2,5}$ by 3						−9	297	0
Decrease $T_{2,4}$ by 2				−3	292	−1		−2
Decrease $T_{2,3}$ by 1		−2	269	+1		−2		−2
Final Temperatures	185		269		292		297	
Final Residuals		−2		+1		−2		−2

dition corresponding to Fig. 3-14 is

$$\frac{1}{2}\left(T_{m-1,n} + T_{m+1,n} + 2T_{m,n-1}\right) + \frac{h\Delta l}{k}T_{\infty} - \left(\frac{h\Delta l + 2k}{k}\right)T_{m,n} = 0 \qquad \text{(3-36)}$$

Equations corresponding to other boundary conditions can be obtained by means of a heat balance or from Table 3-9 by setting $Q'_{m,n}$ equal to zero.

In the iterative scheme used to solve a set of equations obtained for a two-dimensional system, the iterative equation for interior points is written as

$$T_{m,n}^{t+1} = \frac{1}{4}(T_{m+1,n}^{t} + T_{m-1,n}^{t} + T_{m,n-1}^{t} + T_{m,n+1}^{t}) \tag{3-37}$$

where the superscripts t and $(t + 1)$ refer to the tth and $(t + 1)^{\text{th}}$ approximations in the iterative pattern. Nodal points at a convection boundary are treated similarly, i.e., the entire equation is divided by the coefficient of $T_{m,n}^{t+1}$. For example, the iterative equation for a nodal point at a surface of a solid in contact with a fluid at T_∞ becomes

$$T_{m,n}^{t+1} = \frac{k}{2(h\Delta l + 2k)}(T_{m-1,n}^{t} + T_{m+1,n}^{t} + 2T_{m,n-1}^{t}) + \frac{h\Delta l}{h\Delta l + 2k}T_\infty \tag{3-38}$$

If the temperature at the boundary is prescribed, Eq. 3-34 will apply.

The solution begins with assumed values of $T_{m,n}^{0}$ and known values at the boundary, if available. The iterative equations, e.g. Eqs. 3-37 and 3-38, are used to sweep all the nodal points and to calculate a new set of values $T_{m,n}^{1}$. This sequence is a cyclic, single-step process which is repeated over and over to calculate successively $T_{m,n}^{2}$, $T_{m,n}^{3}$, etc. using in each sweep the known set of temperatures $T_{m,n}^{t}$ to calculate a new set $T_{m,n}^{(t+1)}$. Finally, there obtains

$$T_{m,n}^{t+1} - T_{m,n}^{t} \leq e \tag{3-39}$$

at all points, where e is a prescribed value of the error in the temperature. When this condition has been met, the iteration has converged to the solution of the finite difference approximation of the Laplace's equation for the specific value of Δx used to form the mesh. The following example illustrates this procedure.

EXAMPLE 3-4. For the tall chimney wall whose cross section is shown in Fig. 3-15, determine the temperature distribution and the rate of heat loss per foot height. Assume that the temperature of the gases flowing through the chimney is 100 F above that of the surrounding atmosphere, T_∞, which is taken as the datum and set equal to zero. The unit-surface conductances at the interior ($\bar{h}_i$) and exterior ($\bar{h}_0$) surfaces are 12 and 3 Btu/hr sq ft F respectively. The thermal conductivity of the wall is 1.0 Btu/hr ft F.

Solution: Because of the geometrical symmetry of the system, we need consider only a one-eighth section. We subdivide this section as shown in Fig. 3-17. For the chimney section of Fig. 3-17, the iterative temperature equations are:

$$T_{1,1}{}^{t+1} = \left(\frac{k}{2k + 2\bar{h}_0 \Delta l}\right)\left(2T_{1,2}{}^t + 2\left[\frac{\bar{h}_0 \Delta l}{k}\right] T_0\right)$$

$$T_{1,2}{}^{t+1} = \left(\frac{k}{4k + 2\bar{h}_0 \Delta l}\right)\left(T_{1,1}{}^t + T_{1,3}{}^t + 2T_{2,2}{}^t + 2\left[\frac{\bar{h}_0 \Delta l}{k}\right] T_0\right)$$

$$T_{1,3}{}^{t+1} = \left(\frac{k}{4k + 2\bar{h}_0 \Delta l}\right)\left(T_{1,2}{}^t + T_{1,4}{}^t + 2T_{2,3}{}^t + 2\left[\frac{\bar{h}_0 \Delta l}{k}\right] T_0\right)$$

$$T_{1,4}{}^{t+1} = \left(\frac{k}{4k + 2\bar{h}_0 \Delta l}\right)\left(T_{1,3}{}^t + T_{1,5}{}^t + 2T_{2,4}{}^t + 2\left[\frac{\bar{h}_0 \Delta l}{k}\right] T_0\right)$$

$$T_{1,5}{}^{t+1} + \left(\frac{k}{4k + 2\bar{h}_0 \Delta l}\right)\left(2T_{1,4}{}^t + 2T_{2,5}{}^t + 2\left[\frac{\bar{h}_0 \Delta l}{k}\right] T_0\right)$$

$$T_{2,2}{}^{t+1} = \frac{1}{4}\left(2T_{1,2}{}^t + 2T_{2,3}{}^t\right)$$

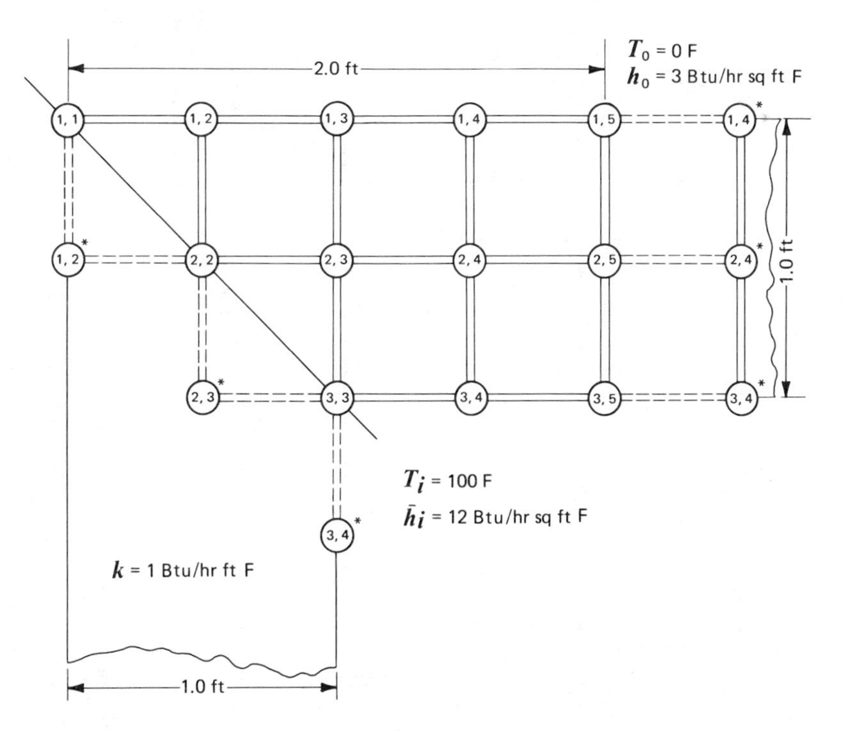

Fig. 3-17.

$$T_{2,3}{}^{t+1} = \frac{1}{4}\left(T_{1,3}{}^{t} + T_{2,4}{}^{t} + T_{3,3}{}^{t} + T_{2,2}{}^{t}\right)$$

$$T_{2,4}{}^{t+1} = \frac{1}{4}\left(T_{1,4}{}^{t} + T_{2,5}{}^{t} + T_{3,4}{}^{t} + T_{2,3}{}^{t}\right)$$

$$T_{2,5}{}^{t+1} = \frac{1}{4}\left(T_{1,5}{}^{t} + 2T_{2,4}{}^{t} + T_{3,5}{}^{t}\right)$$

$$T_{3,3}{}^{t+1} = \left(\frac{k}{3k + \bar{h}_i \Delta l}\right)\left(2T_{2,3}{}^{t} + T_{3,4}{}^{t} + \left[\frac{\bar{h}_i \Delta l}{k}\right] T_i\right)$$

$$T_{3,4}{}^{t+1} + \left(\frac{k}{4k + 2\bar{h}_i \Delta l}\right)\left(T_{3,3}{}^{t} + T_{3,5}{}^{t} + 2T_{2,4}{}^{t} + 2\left[\frac{\bar{h}_i \Delta l}{k}\right] T_i\right)$$

$$T_{3,5}{}^{t+1} = \left(\frac{k}{4k + 2\bar{h}_i \Delta l}\right)\left(2T_{3,4}{}^{t} + 2T_{2,5}{}^{t} + 2\left[\frac{\bar{h}_i \Delta l}{k}\right] T_i\right)$$

The flow diagram showing the computational sequence is shown in Fig. 3-18. The

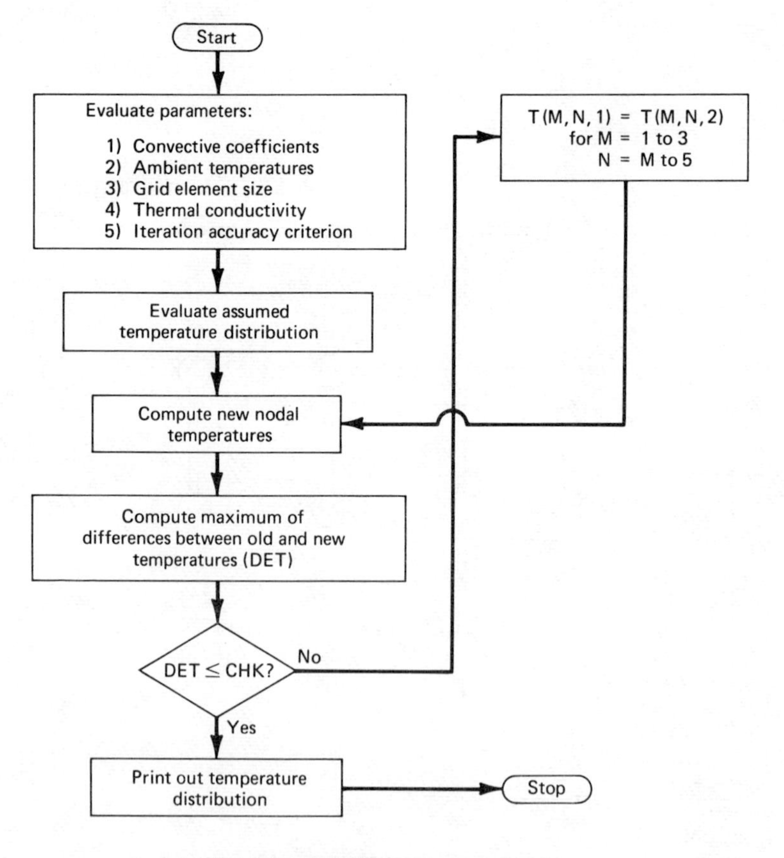

Fig. 3-18. Flow diagram for Example 3-4.

variables used in the program are related to those used in equations 3-34 to 3-39 in Table 3-11. A FORTRAN IV computer program to solve the iterative equations to an accuracy of ± 0.1 degrees F is shown in Table 3-12.

Solution to Example 3-4:

$T_{(1,1)} = 5.3$ F	$T_{(2,3)} = 49.1$ F
$T_{(1,2)} = 13.0$ F	$T_{(2,4)} = 55.2$ F
$T_{(1,3)} = 19.1$ F	$T_{(2,5)} = 56.8$ F
$T_{(1,4)} = 21.9$ F	$T_{(3,3)} = 88.0$ F
$T_{(1,5)} = 22.4$ F	$T_{(3,4)} = 93.3$ F
$T_{(2,2)} = 31.3$ F	$T_{(3,5)} = 93.7$ F

[*Text continues page 122*]

Table 3-11. Correspondence between physical variables and FORTRAN symbols for Example 3-4.

FORTRAN Symbol	Equation Symbol	Description	Units
CHK	e	Termination criterion i.e., if $\mid T(M,N,2) - T(M,N,1) \mid <$ CHK, terminate	F
DL	Δl	Grid element size	ft
HI	$\bar{h}_i$	Inner unit surface conductance	Btu/hr ft^2 F
HO	$\bar{h}_0$	Outer unit surface conductance	Btu/hr ft^2 F
K	k	Thermal conductivity	Btu/hr ft F
SO	$\bar{h}_0 \Delta l/k$	—	
SI	$\bar{h}_i \Delta l/k$	—	
TO	T_0	Outer air gas temperature	F
TI	T_i	Inner flue gas temperature	F
T(M,N,1)	$T^t_{m,n}$	"Old" value of temperature at location (m,n)	F
T(M,N,2)	$T^{t+1}_{m,n}$	"New" or iterated value of temperature at location (m,n)	F
DT	$\mid T^{t+1}_{m,n} - T^t_{m,n} \mid$	—	F
DET	—	Maximum value of DT in the iteration	F

Table 3-12. FORTRAN IV computer program for Example 3-4.

```
      REAL K
      DIMENSION T(3,5,2)
```

Thirty locations are reserved for the array T. The first two subscripts specify the spatial location. The third subscript specifies "old" (subscript 1) or "new" (subscript 2) values of T.

```
      HI=12.
      HO=3.
      TO=0.
      TI=100.
      K=1.
      CHK=.1
      DL=.5
      SO=HO*DL/K
      SI=HI*DL/K
```

The program parameters are evaluated. Refer to Table 3-11 for the definition of each.

```
      DO 1 M=1,3
      DO 1 N=M,5
      T(M,N,1)=(TO+TI)/2.
1     CONTINUE
```

The initial temperature distribution is specified. It is simply the average of the inner and outer ambient temperatures.

```
2     T(1,1,2)=(2.*T(1,2,1)+2.*SO*TO)/(2.+2.*SO)
      T(1,2,2)=(T(1,1,1)+T(1,3,1)+2.*T(2,2,1)+2.*SO*TO)/(4.+2.*SO)
      T(1,3,2)=(T(1,2,1)+T(1,4,1)+2.*T(2,3,1)+2.*SO*TO)/(4.+2.*SO)
      T(1,4,2)=(T(1,3,1)+T(1,5,1)+2.*T(2,4,1)+2.*SO*TO)/(4.+2.*SO)
      T(1,5,2)=(2.*T(1,4,1)+2.*T(2,5,1)+2.*SO*TO)/(4.+2.*SO)
      T(2,2,2)=(2.*T(1,2,1)+2.*T(2,3,1))/4.
      T(2,3,2)=(T(1,3,1)+T(2,4,1)+T(3,3,1)+T(2,2,1))/4.
      T(2,4,2)=(T(1,4,1)+T(2,5,1)+T(3,4,1)+T(2,3,1))/4.
      T(2,5,2)=(T(1,5,1)+2.*T(2,4,1)+T(3,5,1))/4.
      T(3,3,2)=(2.*T(2,3,1)+T(3,4,1)+SI*TI)/(3.+SI)
      T(3,4,2)=(T(3,3,1)+T(3,5,1)+2.*T(2,4,1)+2.*SI*TI)/(4.+2.*SI)
      T(3,5,2)=(2.*T(3,4,1)+2.*T(2,5,1)+2.*SI*TI)/(4.+2.*SI)
```

The nodal temperatures T are computed from the heat balance equations developed earlier.

```
        DET=0.
        DO 3  M=1,3
        DO 3  N=M,5
        DT=ABS(T(M,N,2)-T(M,N,1))
        IF (DT.LE.DET) GO TO 3
        DET=DT
3       CONTINUE
```

The maximum of the absolute values of the differences between the old and new temperature distributions is determined. If it is less than the convergence criterion CHK, the temperature distribution is printed out; if not, the old values of T are replaced by the new ones and the computation is returned to statement 2 for an additional iteration. These latter operations are programmed in the following six statements.

```
        IF (DET.LE.CHK) GO TO 5
        DO 4  M=1,3
        DO 4  N=M,5
        T(M,N,1)=T(M,N,2)
4       CONTINUE
        GO TO 2
```

The remaining cards in the program print out the results as specified in format statements 7 and 8.

```
5       PRINT 55
55      FORMAT (1H1)
        DO 6  M=1,3
        DO 6  N=M,5
        PRINT 7,M,N,T(M,N,2)
6       CONTINUE
        PRINT 8,CHK
7       FORMAT(1H0,*T(*,I1,*,*,I1,*) = *,F5.1,* DEG F *,/)
8       FORMAT(1H0,*ACCURACY OF TEMPERATURES - PLUS OR MINUS *,
       1F3.1, *DEG F*)
        END
```

The process described above is simple to program for digital computation, but convergence of the iterative scheme is relatively slow and two complete sets of interior temperature values, i.e. $2N^2$, as well as the boundary values must be stored in the computer memory. The Liebmann method (21) represents a distinct improvement over these shortcomings because it uses new temperature values in the iteration pattern as soon as they have been calculated instead of after a sweep of the entire field has been completed. For the Liebmann method the difference formulation for an interior node is written

$$T_{m,n}{}^{t+1} = \tfrac{1}{4}(T_{m+1,n}{}^{t} + T_{m-1,n}{}^{t+1} + T_{m,n+1}{}^{t} + T_{m,n-1}{}^{t+1}) \quad \text{(3-40)}$$

If the computation starts in the upper left hand corner ($m = n = 1$), it moves along the first row to the right until new values for the entire row are obtained and then jumps to the second row, etc.; thus $T_{m-1,n}{}^{t+1}$ and $T_{m,n-1}{}^{t+1}$ are known before the need to calculate $T_{m,n}{}^{t+1}$ arises. Consequently, as soon as a new temperature has been calculated, it is immediately substituted for the previous value at the same nodal point. In this manner the storage requirement for internal nodal points is reduced to N^2 and convergence of the iteration process is approximately doubled (see Example 3-5). For additional information on numerical techniques, error estimates, and convergence the reader is referred to References (21) and (22). Finite difference methods for handling inhomogeneous regions are presented in Reference 23.

EXAMPLE 3-5. Repeat Example 3-4, using the Liebmann method. Write a FORTRAN program, using equations of the form of Eq. 3-34 for the iterative scheme. Count the number of iterations required to achieve an accuracy level of ± 0.1 degrees F for the nodal temperatures and compare to the number required with the Gauss-Seidel method in Example 3-4.

Solution: The nodal equations for use in the Liebmann method are:

$$T_{1,1}{}^{t+1} = \left(\frac{k}{2k + 2\bar{h}_0\Delta_l}\right)\left(2T_{1,2}{}^{t} + 2\left[\frac{\bar{h}_0\Delta l}{k}\right]T_0\right)$$

$$T_{1,2}{}^{t+1} = \left(\frac{k}{4k + 2\bar{h}_0\Delta l}\right)\left(T_{1,1}{}^{t+1} + T_{1,3}{}^{t} + 2T_{2,2}{}^{t} + 2\left[\frac{\bar{h}_0\Delta l}{k}\right]T_0\right)$$

$$T_{1,3}{}^{t+1} = \left(\frac{k}{4k + 2\bar{h}_0\Delta l}\right)\left(T_{1,2}{}^{t+1} + T_{1,4}{}^{t} + 2T_{2,3}{}^{t} + 2\left[\frac{\bar{h}_0\Delta l}{k}\right]T_0\right)$$

$$T_{1,4}{}^{t+1} = \left(\frac{k}{4k + 2\bar{h}_0\Delta l}\right)\left(T_{1,3}{}^{t+1} + T_{1,5}{}^{t} + 2T_{2,4}{}^{t} + 2\left[\frac{\bar{h}_0\Delta l}{k}\right]T_0\right)$$

$$T_{1,5}^{t+1} + \left(\frac{k}{4k + 2\bar{h}_0 \Delta l}\right) \left(2T_{1,4}^{t+1} + 2T_{2,5}^{t} + 2\left[\frac{\bar{h}_0 \Delta l}{k}\right] T_0\right)$$

$$T_{2,2}^{t+1} = \frac{1}{4}\,(2T_{1,2}^{t+1} + 2T_{2,3}^{t})$$

$$T_{2,3}^{t+1} = \frac{1}{4}\,(T_{1,3}^{t+1} + T_{2,4}^{t} + T_{3,3}^{t} + T_{2,2}^{t+1})$$

$$T_{2,4}^{t+1} = \frac{1}{4}\,(T_{1,4}^{t+1} + T_{2,5}^{t} + T_{3,4}^{t} + T_{2,3}^{t+1})$$

$$T_{2,5}^{t+1} = \frac{1}{4}\,(T_{1,5}^{t+1} + 2T_{2,4}^{t+1} + T_{3,5}^{t})$$

$$T_{3,3}^{t+1} = \left(\frac{k}{3k + \bar{h}_i \Delta l}\right) \left(2T_{2,3}^{t+1} + T_{3,4}^{t} + \left[\frac{\bar{h}_i \Delta l}{k}\right] T_i\right)$$

$$T_{3,4}^{t+1} + \left(\frac{k}{4k + 2\bar{h}_i \Delta l}\right) \left(T_{3,3}^{t+1} + T_{3,5}^{t} + 2T_{2,4}^{t+1} + 2\left[\frac{\bar{h}_i \Delta l}{k}\right] T_i\right)$$

$$T_{3,5}^{t+1} = \left(\frac{k}{4k + 2\bar{h}_i \Delta l}\right) \left(2T_{3,4}^{t+1} + 2T_{2,5}^{t+1} + 2\left[\frac{\bar{h}_i \Delta l}{k}\right] T_i\right)$$

A FORTRAN IV computer program to solve those iterative equations to an accuracy of ±0.1 degrees F is shown in Table 3-13. The structure and the symbols used in the program are the same as in Example 3-4, with the exception of the variable COUNT. COUNT is an integer variable which is used to count the number of iterations. The statements in the FORTRAN program which differ from Example 3-4 have been underlined in Table 3-13.

Table 3-13. FORTRAN IV computer program for Example 3-5.

```
REAL K
INTEGER COUNT
DIMENSION T(3,5,2)
HI=12.
HO=3.
TO=0.
TI=100.
K=1.
CHK=.1
DL=.5
SO=HO*DL/K
SI=HI*DL/K
```

Table 3-13. Continued.

```
        COUNT=1
        DO 1 M=1,3
        DO 1 N=M,5
        T (M,N,1)=(TO+TI)/2.
1       CONTINUE
2       T(1,1,2)=(2.*T(1,2,1)+2.*SO*TO)/(2.+2.*SO)
        T(1,2,2)=(T(1,1,2)+T(1,3,1)+2.*T(2,2,1)+2.*SO*TO)/(4.+2.*SO)
        T(1,3,2)=(T(1,2,2)+T(1,4,1)+2.*T(2,3,1)+2.*SO*TO)/(4.+2.*SO)-
        T(1,4,2)=(T(1,3,2)+T(1,5,1)+2.*T(2,4,1)+2.*SO*TO)/(4.+2.*SO)
        T(1,5,2)=(2.*T(1,4,2)+2.*T(2,5,1)+2.*SO*TO)/(4.+2.*SO)
        T(2,2,2)=(2.*T(1,2,2)+2.*T(2,3,1))/4.
        T(2,3,2)=(T(1,3,2)+T(2,4,1)+T(3,3,1)+T(2,2,2))/4.
        T(2,4,2)=(T(1,4,2)+T(2,5,1)+T(3,4,1)+T(2,3,2))/4.
        T(2,5,2)=(T(1,5,2)+2.*T(2,4,2)+T(3,5,1))/4.
        T(3,3,2)=(2.*T(2,3,2)+T(3,4,1)+SI*TI)/(3.+SI)
        T(3,4,2)=(T(3,3,2)+T(3,5,1)+2.*T(2,4,2)+2.*SI*TI)/(4.+2.*SI)
        T(3,5,2)=(2.*T(3,4,2)+2.*T(2,5,2)+2.*SI*TI)/(4.+2.*SI)
        DET=0.
        DO 3 M=1,3
        DO 3 N=M,5
        DT=ABS(T(M,N,2)-T(M,N,1))
        IF (DT.LE.DET) GO TO 3
        DET=DT
3       CONTINUE
        IF (DET.LE.CHK) GO TO 5
        COUNT=COUNT+1
        DO 4 M=1,3
        DO 4 N=M,5
        T(M,N,1)=T(M,N,2)
4       CONTINUE
        GO TO 2
5       PRINT 55
55      FORMAT (1H1)
        DO 6 M=1,3
        DO 6 N=M,5
        PRINT 7,M,N,T(M,N,2)
6       CONTINUE
        PRINT 8,CHK
        PRINT 9,COUNT
7       FORMAT(1H0,*T(*,I1,*,*,I1,*) = *,F5.1,* DEG F *,/)
8       FORMAT(1H0,*ACCURACY OF TEMPERATURES - PLUS OR MINUS *,
       1F3.1,* DEG F*)
9       FORMAT(1H0,*NUMBER OF ITERATIONS REQUIRED FOR SPECIFIED
        1ACCURACY*I10)
         END
```

Thirteen iterations are required to achieve the specified accuracy using the simple iteration technique of Example 3-4, whereas only 10 iterations are required for the Liebmann method to converge. This represents a reduction in computational time of about 25%. If a level of accuracy of ± 0.001 degrees F is demanded, the advantage of Liebmann method improves to 35% (see Problem 3-34).

Three-dimensional systems. The application of numerical methods to three-dimensional systems requires no additional concepts. However, an interior nodal point will now have six neighbors and there will be 6 fictitious conducting rods emanating from it. If the body is subdivided into cubes of sides Δl as shown in Fig. 3-19, a heat balance on an interior nodal point, such as point 0 gives

$$Q_o = K_{1-0}(T_1 - T_o) + K_{2-0}(T_2 - T_o) + K_{3-0}(T_3 - T_o) + K_{4-0}(T_4 - T_o) + K_{5-0}(T_5 - T_o) + K_{6-0}(T_6 - T_o) \tag{3-41}$$

where $K_{1-0} = K_{2-0} = K_{3-0} = K_{4-0} = K_{5-0} = K_{6-0} = k\Delta l$

Dividing Eq. 3-33 by $k\Delta l$ yields the residual equation in the form

$$Q'_o = Q_o/k\Delta l = T_1 + T_2 + T_3 + T_4 + T_5 + T_6 - 6T_o \tag{3-42}$$

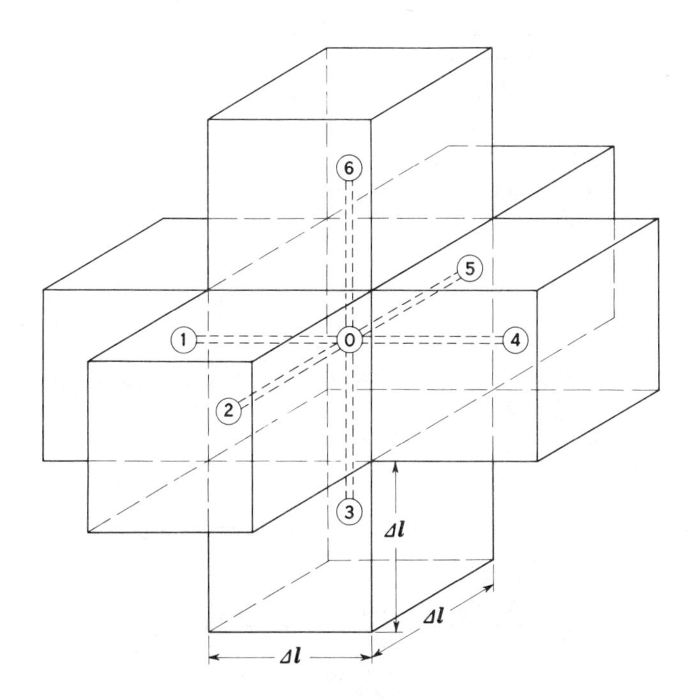

Fig. 3-19. Relaxation network for interior nodal point in a three-dimensional system.

The residual equations for nodal points on system boundaries can be derived from appropriate heat balances, using the approach illustrated previously for one- and two-dimensional systems.

3-7. Closure

In this chapter we have considered solutions to heat-conduction problems in more than one dimension by analytical, graphical, experimental-analogic, and finally by numerical means. Each of these methods has certain advantages which should be recognized before the solution of a problem is initiated. Although it is not possible to give infallible rules for the selection of the most suitable method for a particular problem, the following comments will aid in the selection.

The analytical approach is recommended for problems dealing with systems simple in geometry and boundary conditions. It is accurate and lends itself readily to parameterization. However, when the geometry or the boundary conditions are complex, the analytical approach becomes too involved to be practical.

Systems of complex geometry, but with isothermal and insulated boundaries, are readily amenable to a solution by graphical or analogical methods. These methods become unwieldy, however, when the boundary conditions involve heat transfer through a surface conductance. For such cases the numerical approach is recommended because it can easily be adapted to all kinds of boundary conditions and geometrical shapes. The numerical method is sufficiently flexible to solve also problems involving systems with variable physical properties and nonuniform boundary conditions. Moreover, numerical solutions can be conveniently carried out by a digital computer.

PROBLEMS

The problems below are organized in the following manner: Problems 3-1 through 3-6 deal with analytical methods, 3-7 through 3-16 with flux plots, 3-17 through 3-22 with shape factors, and Problems 3-23 through 3-36 are to be solved by numerical techniques.

3-1. The heat-conduction equation in cylindrical coordinates is

$$\rho c \frac{\partial T}{\partial \theta} = k\left(\frac{\partial^2 T}{\partial r^2} + \frac{1}{r}\frac{\partial T}{\partial r} + \frac{1}{r^2}\frac{\partial^2 T}{\partial \phi^2} + \frac{\partial^2 T}{\partial z^2}\right) + \dot{q}$$

(a) Simplify the above equation by eliminating terms equal to zero for the case of steady-state heat flow without sources or sinks around a right-angle corner such as the one shown in the accompanying sketch. It may be assumed that the corner extends to infinity in the direction perpendicular to the paper. (b) Solve the resulting equation for the temperature distribution by substituting the boundary conditions. (c) Determine the rate of heat flow from T_1 to T_2. Assume $k = 1$ Btu / hr ft F and unit depth perpendicular to the paper.

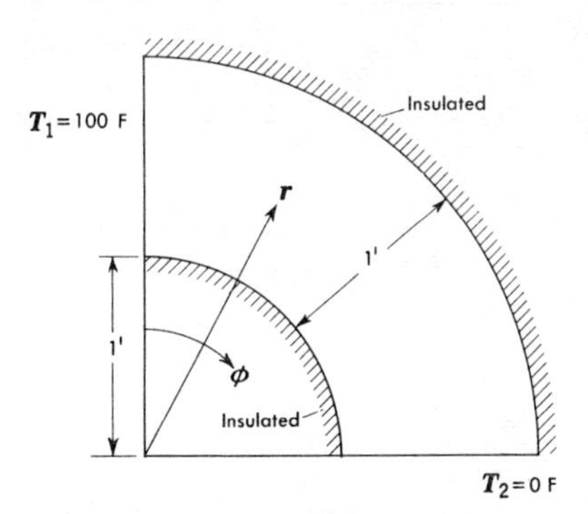

Prob. 3-1

3-2. Derive Eq. 3-7 starting with an energy balance.

3-3. Show that for a semi-infinite plate of width L, having the boundary condition for $T(x, y)$

$$T(0, y) - T_1 = 0$$

$$T(L, y) - T_1 = 0$$

$$T(x, \infty) - T_1 = 0$$

$$T(x, 0) - T_1 = (T_2 - T_1)$$

the temperature distribution is

$$\frac{T - T_1}{T_2 - T_1} = \frac{4}{\pi}\left[e^{-(\pi/L)y} \sin \frac{\pi x}{L} + \frac{1}{3} e^{-(3\pi/L)y} \sin \frac{3\pi x}{L} + \cdots\right]$$

For $T_1 = 0$ and $T_2 = 100$ F, plot isotherms of 25, 50, and 75 F.

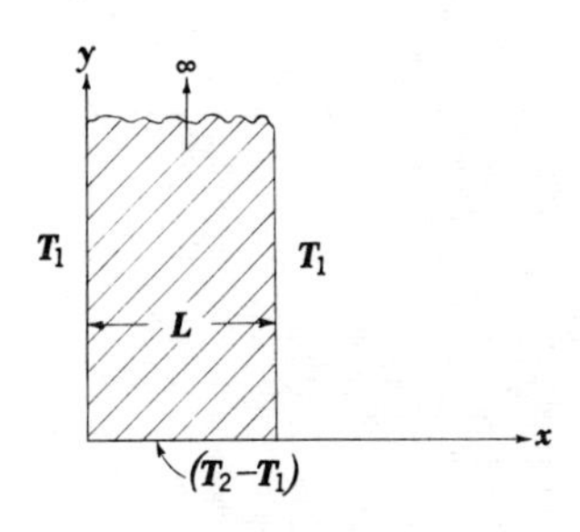

Prob. 3-3

3-4. Show that the temperature distribution in a long rectangular bar, shown in the accompanying sketch, with the bottom side at a uniform temperature of 100 F and the other sides at 0 F is

$$T(x, y) = \frac{400}{\pi} \sum_{n=0}^{\infty} \frac{1}{2n+1} \sin \frac{(2n+1)\pi x}{a} \sinh \frac{(b-y)(2n+1)\pi}{a} \operatorname{cosech} \frac{(2n+1)\pi b}{a}$$

Prob. 3-4

3-5. A rectangular plate 1 ft wide in the x-direction, infinite in the y-direction, has a temperature distribution given by $T(x, 0) = 100 \sin \pi x$ imposed on the $y = 0$ edge. Determine the temperature distribution, $T(x, y)$.

3-6. In a long rectangular bar, $0 \leq x \leq a$, $0 \leq y \leq b$, heat is generated uniformly at the rate $\dot{q}$ Btu/hr ft^3. The boundaries at $x = 0$ and $y = 0$ are insulated while at the boundaries at $x = a$ and $y = b$ heat is convected into a surrounding at zero temperature through a unit surface conductance h_c. Set up the differential equation with the appropriate boundary condition and obtain an expression for the temperature distribution.

3-7. Compare the rate of heat flow from the top to the bottom in the aluminum structure shown in the sketch with the rate of heat flow through a solid

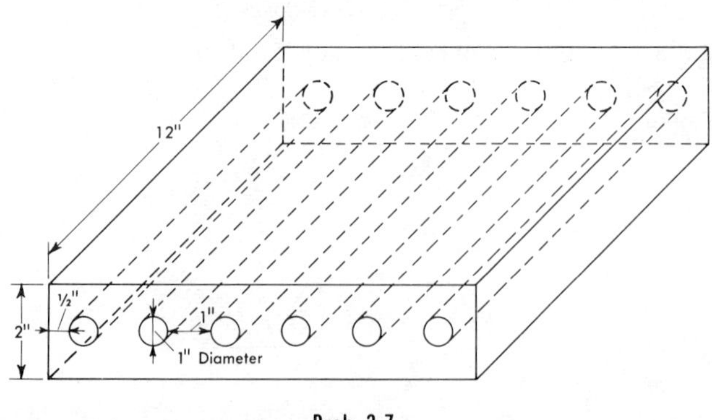

Prob. 3-7

slab. The top is at −34 F, the bottom at −32 F. The holes are filled with insulation which does not conduct heat appreciably.

3-8. Determine by means of the relaxation method the temperatures and heat flow per unit depth in the ribbed insulation shown in the accompanying sketch.

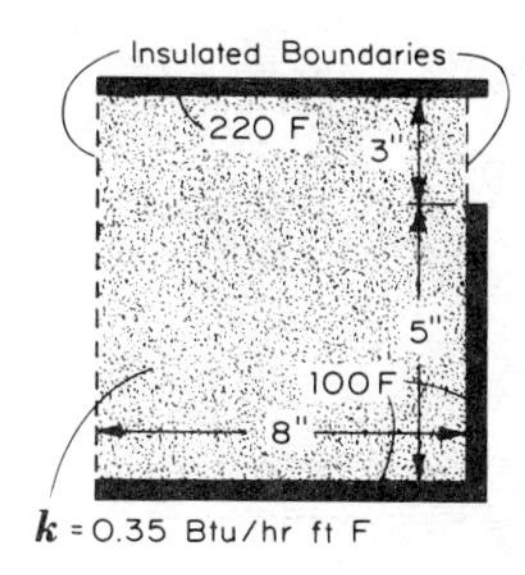

Prob. 3-8

3-9. By means of a flux plot, estimate the rate of heat flow through the object (k = 10 Btu/hr ft F) shown in the sketch. Assume that no heat is lost from the sides.

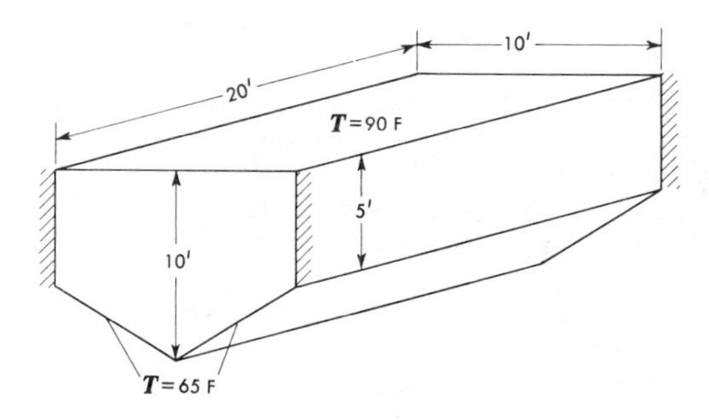

Prob. 3-9

3-10. Determine the rate of heat transfer per foot length from a 2-in.-OD pipe at 300 F placed eccentrically within a larger cylinder of rock wool as shown

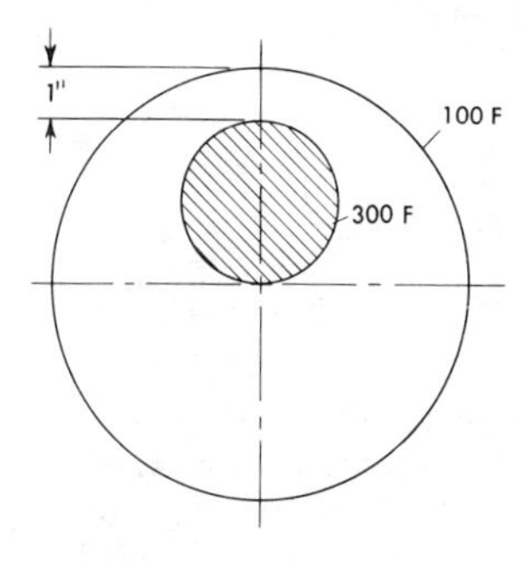

Prob. 3-10

in the sketch. The outside diameter of the larger cylinder is 6 in. and the surface temperature 100 F.

3-11. Determine the rate of heat flow per foot length from the inner to the outer surface of the molded asbestos insulation shown in the accompanying sketch ($k = 0.1$ Btu/hr ft F).

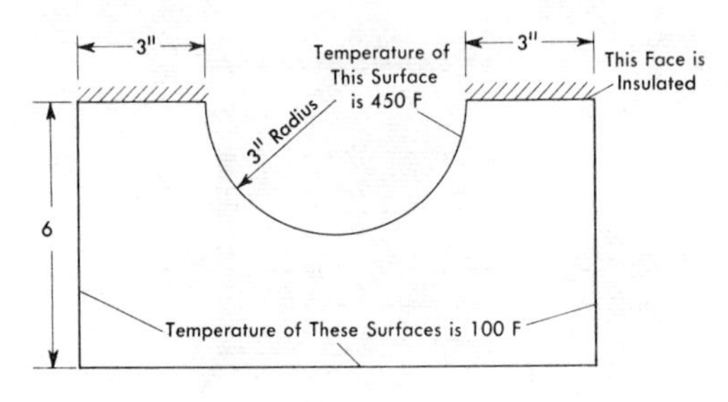

Prob. 3-11

3-12. A long 0.5-in.-diameter electric copper cable is embedded in the center of a 1-ft-square concrete block. If the outside temperature of the concrete is 100 F and the rate of electrical energy dissipation in the cable is 150 w per foot length, determine the temperatures at the outer surface and at the center of the cable.

3-13. A large number of 1.5-in.-OD pipes carrying hot and cold liquids are embedded in concrete stone in an equilateral staggered arrangement with center lines 4.5 in. apart as shown in the sketch. If the pipes in rows A and C are at 60 F while the pipes in rows B and D are at 150 F, determine the rate of heat transfer per foot length from pipe X in row B.

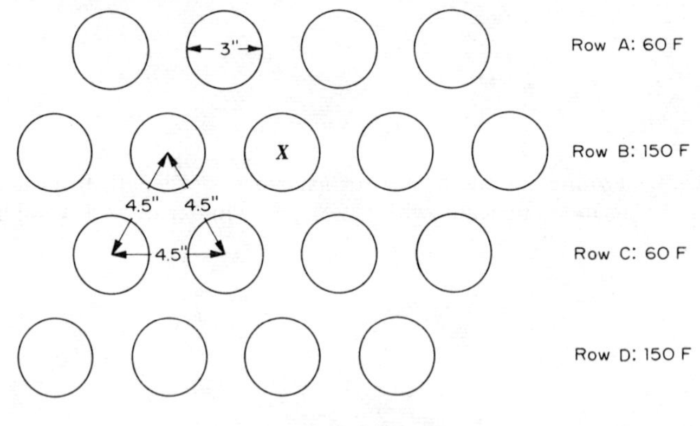

Prob. 3-13

3-14. A plane insulating wall has alternatingly spaced ribs extending from both surfaces as shown in the accompanying sketch. The ribs are made of aluminum so that their temperature is essentially the same as that of the surface to which they are attached. If the upper surface is at 300 F, the lower surface at 110 F, and the space between the surfaces is filled with powdered diatomaceous earth, estimate the rate of heat transfer per square foot of wall. What would be the percentage reduction in the rate of heat transfer if the ribs were removed?

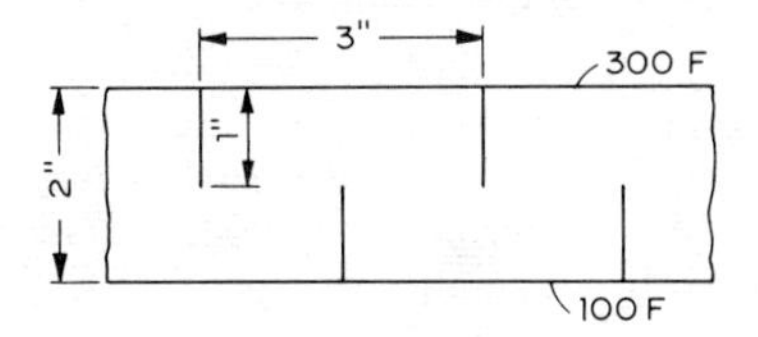

Prob. 3-14

3-15. A long 1-in.-diam electric cable is imbedded in a concrete wall (k = 0.60 Btu/hr ft F) which is 1 by 1 ft, as shown in the sketch below. If the lower surface is insulated, the surface of the cable is 200 F and the exposed surface of the concrete is 100 F. Estimate the rate of energy dissipation per foot of cable.

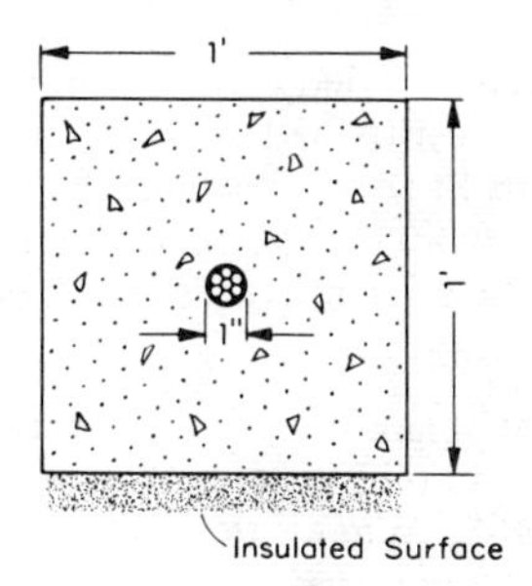

Prob. 3-15

3-16. Determine the temperature distribution and the heat-flow rate per unit depth in the stone concrete block shown. The cross-sectional area of the block is square and the hole is centered.

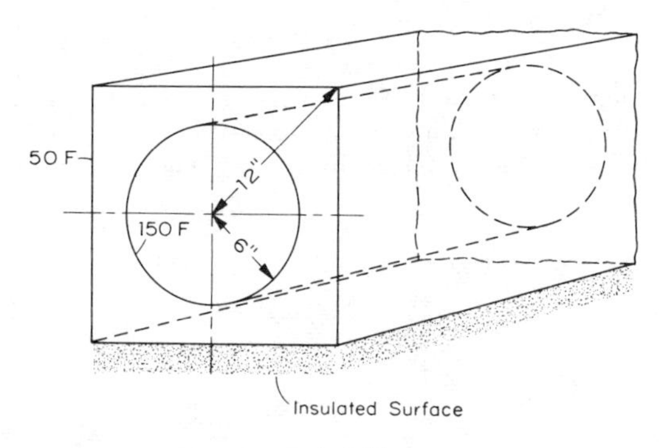

Prob. 3-16

3-17. A 1-ft-OD pipe with a surface temperature of 200 F carries steam over a distance of 300 ft. The pipe is buried with its center line at a depth of 3 ft, the ground surface is 20 F, and the mean thermal conductivity of the soil is 0.4 Btu/hr ft F. Calculate the heat loss per day, and the cost, if steam is worth $0.14

per 10^6 Btu. Also, estimate the thickness of 85 per cent magnesia insulation necessary to achieve the same insulation with a total unit-surface conductance of 4 Btu/hr sq ft F on the outside of the insulation.

3-18. Two long pipes, one having a 10-in. OD and a surface temperature of 300 F, the other having a 5-in. OD and a surface temperature of 100 F, are buried deeply in dry sand with their center lines 15 in. apart. Determine the rate of heat flow from the larger to the smaller pipe per foot length.

3-19. A radioactive sample is to be stored in a protective box with 4-in.-thick walls having interior dimensions of 4 by 4 by 12 in. The radiation emitted by the sample is completely absorbed at the inner surface of the box which is made of concrete. If the outside temperature of the box is 70 F, but the inside temperature is not to exceed 120 F, determine the maximum permissible radiation rate from the sample, in Btu/hr.

3-20. A 6-in.-OD pipe is buried with its center line 50 in. below the surface of the ground (k of soil is 0.20 Btu/hr ft F). An oil having a specific gravity of 0.8 and a specific heat of 0.5 Btu/lb F flows in the pipe at 100 gpm. Assuming a ground-surface temperature of 40 F and a pipe-wall temperature of 200 F, estimate the length of pipe in which the oil temperature decreases by 10 F.

3-21. A 1-in.-OD hot-steam line at 212 F runs parallel to a 2-in.-OD cold-water line at 60 F. The pipes are 2 in. center to center and deeply buried in concrete with $k = 0.50$ Btu/hr ft F. What is the heat transfer per foot of pipe between the two pipes? If the steam were used to heat cold water at 60°F to 100°F, how many gallons per hour could be heated in a 1000-foot length of pipe?

3-22. Calculate the rate of heat transfer between a 6-in.-OD pipe at 250 F and a 4-in.-OD pipe at 100 F. The two pipes are 1000 ft long; they are buried in sand ($k = 0.19$ Btu/hr ft F) 4 ft below the surface ($T_s = 80$ F); they are parallel and separated by 9 in., center to center. *Ans.* 74,700 Btu/hr

3-23. The temperatures at nodal points 1, 2, 3, 4, 5, 6, 7, and 8 in the corner section of the wall shown in plan view in the accompanying sketch are 104, 101, 101, 105, 107, 107, 109, and 110 F respectively. The unit-surface conductance

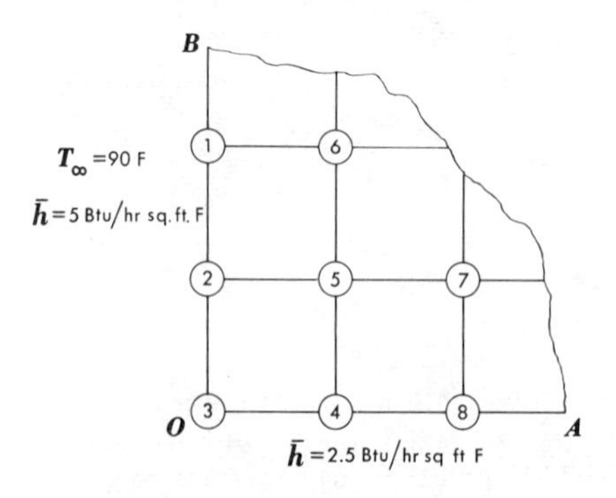

Prob. 3-23

over the surface OA is 2.5, over the surface OB, 5.0 Btu/hr sq ft F. The distance between nodal points is 1 in., and the wall material is masonry brick (k = 0.5 Btu/hr ft F). (a) Compute the residuals at points 2, 3, 4, and 5 per foot thickness. (b) Relax the temperature at the point having the largest residual to eliminate it.

3-24. A thick stainless-steel plate (k = 12 Btu/hr ft F) receives radiant heat at the rate q_r = 1440 Btu/hr sq ft. The temperature distribution at one instant of time is shown in the accompanying sketch. Using the relaxation method, determine the residual Q' at point 2 per foot thickness and determine the temperature change necessary to reduce the residue to zero.

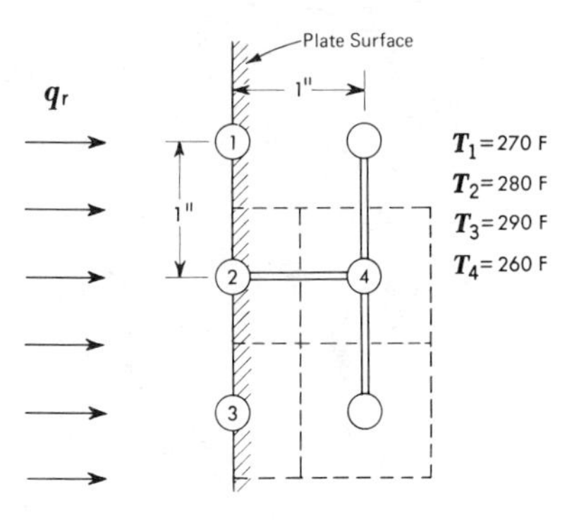

Prob. 3-24

3-25. The temperatures at points 1, 2, 3, 4, 5, 6, and 7 in a corner cross section of the wall shown in plan view are given in the sketch. The lower surface is insulated, the left surface is exposed through a unit-surface conductance of 5 Btu/hr sq ft F to a fluid at T_∞ = 90 F. The thermal conductivity of the wall material is 1 Btu/hr sq ft F/ft and the distance between nodal points is 1 in. Using the relaxation method (a) compute the residual at point 5, (b) compute the residual at point 2.

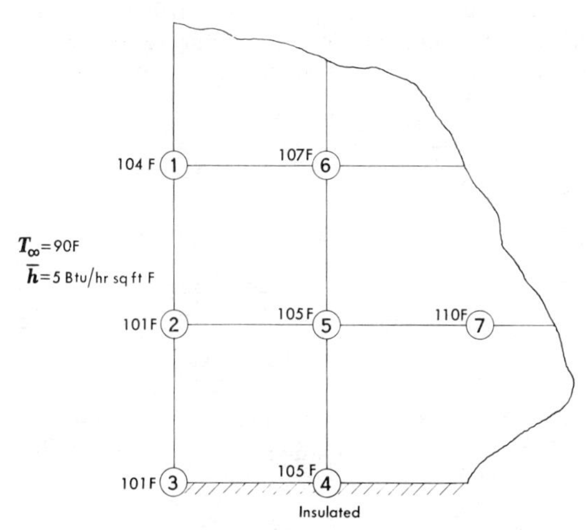

Prob. 3-25

3-26. A section of a quasi two-dimensional wall, exposed through a unit-surface conductance of 6 Btu/hr sq ft F to a fluid at 80 F, is shown in the accompanying sketch. If the distance between nodal points is 1 in. and the temperatures are $T_1 = 100$ F, $T_2 = 105$ F, $T_3 = 110$ F, and $T_4 = 127$ F, calculate the residual at point 2 and indicate by how many degrees it would have to be changed to reduce it to zero. Assume a unit thickness (1 ft) perpendicular to the plane of the paper and a thermal conductivity of 0.5 Btu/hr ft F. *Ans.* 1 F

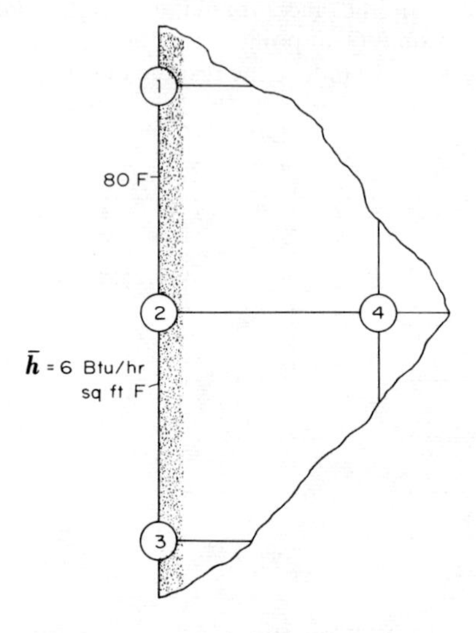

Prob. 3-26

3-27. A turbine blade $2\frac{1}{2}$ in-long, cross-sectional area $A = 0.005$ sq ft, perimeter P of 0.40 ft. is made of stainless steel ($k = 15$ Btu/hr ft F). The temperature of the root, T_w, is 900 F. The blade is exposed to a hot gas at 1600 F, and the unit-surface conductance h is 80 Btu/hr sq ft F. Using the network shown in the accompanying sketch, estimate the temperature distribution and the rate of heat transfer by numerical method and compare your results with those obtained analytically in Sec. 2-5 or Prob. 2-15. Assume that the tip is insulated.

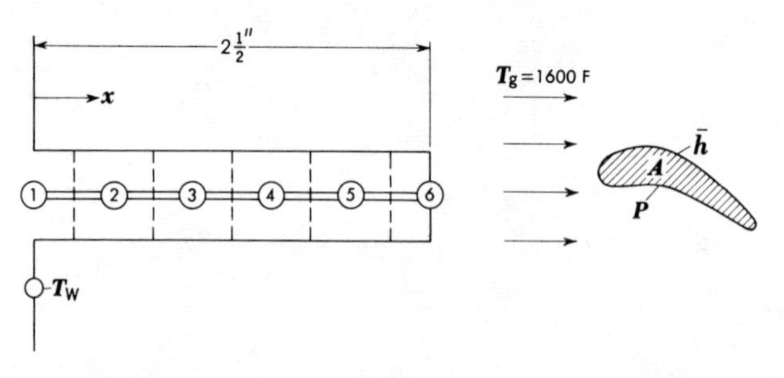

Prob. 3-27

3-28. Determine by a numerical method (a) the temperatures at the 16 equally-spaced points shown in the accompanying sketch to three figure accuracy,

(b) the rate of heat flow per foot thickness. Assume two-dimensional heat flow, $k = 1$ Btu/hr ft F, and make use of the symmetry of the system.

Ans. 20.8, 41.65, 45.8

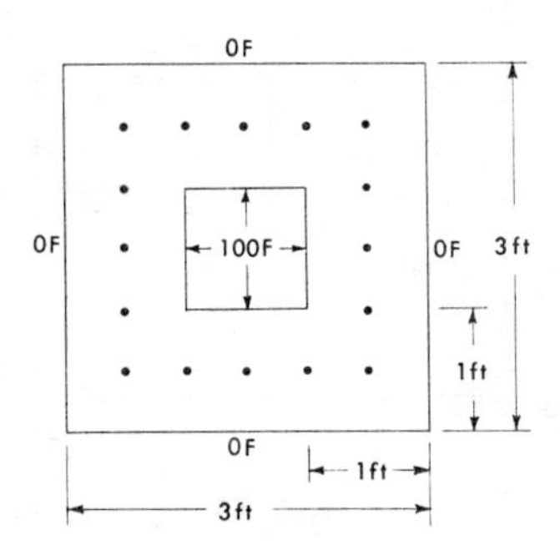

Prob. 3-28

3-29. In a long, 1-ft-square bar shown in the accompanying sketch the left face is maintained at 100 F, the top face at 500 F, and the other two faces are in contact with a fluid at 100 F through a unit-surface conductance of 10 Btu/hr sq ft F. If the thermal conductivity of the bar is 10 Btu/hr ft F, calculate the temperatures at points 1 through 9 by a numerical method.

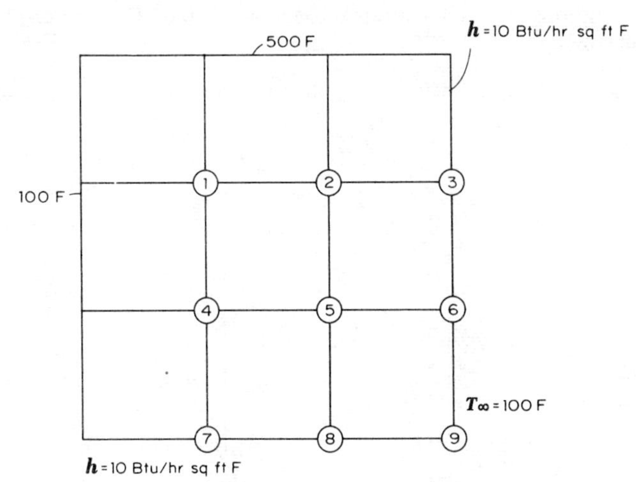

Prob. 3-29

3-30. Repeat Prob. 3-29, but assume that the top face is exposed to thermal radiation and uniformly receives a flux of 10,000 Btu/hr sq ft. Note that three additional nodal points at the upper surface must be considered.

3-31. A tall chimney ($k = 0.4$ Btu/hr ft F) has the cross section shown in the accompanying sketch. If the interior surface temperature is 500 F and the exterior surface temperature is 100 F, calculate the rate of heat loss and the temperature distribution using the Gauss-Seidel numerical method with a 6-in.-square grid and a freehand flux plot.

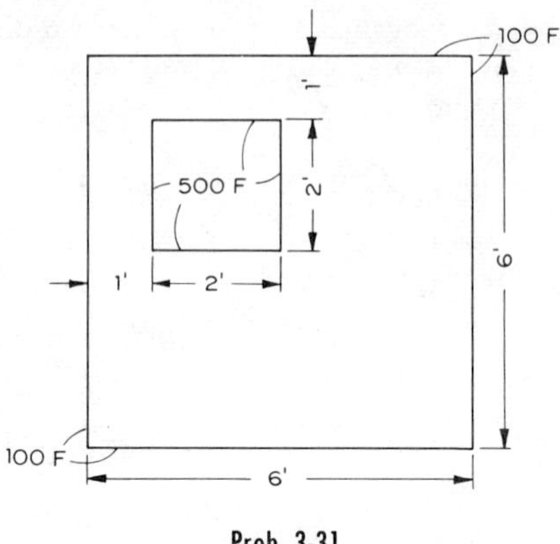

Prob. 3-31

3-32. Repeat Prob. 3-31, but assume that the unit-surface conductance over the interior and exterior surfaces are 5 and 2 Btu/hr sq ft F, respectively, and that the hot gases are at 500 F while the air outside is at 70 F.

3-33. How should the numerical method be altered at an interface between two materials of different thermal conductivities? Illustrate by a simple example.

3-34. Repeat Problem 3-31 using the Liebmann method and compare the number of iterations required to attain an accuracy of 0.01 F in the result with the Gauss-Seidel method.

3-35. Derive Eq. 3-28 in detail, starting from a heat balance similar to Eq. 3-24.

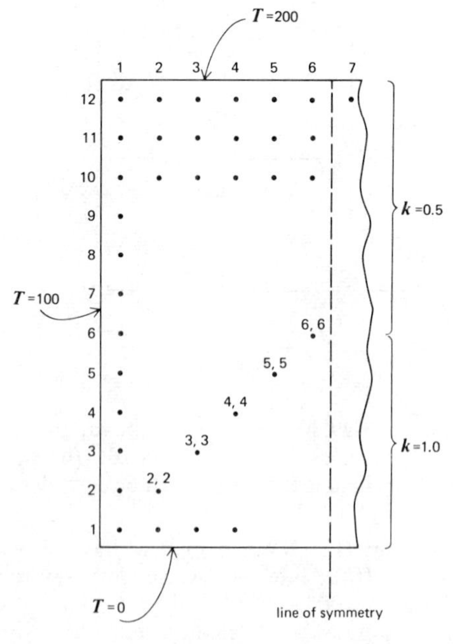

Prob. 3-36

3-36. A two-dimensional heat-flow field is 12 inches square with two opposite boundaries at 100 F and the other two boundaries at 200 F and at 0 F as shown below. Determine the temperature across the lower diagonal, if the thermal conductivity in the lower half is 1 Btu/hr ft F and in the upper half 0.5 Btu/hr ft F, by means of the Gauss-Seidel method, suitably modified as shown in Ref. 23.

Ans. $T_{22} = 51.2$; $T_{33} = 54.8$; $T_{44} = 60.8$; $T_{55} = 68.7$; $T_{66} = 78.3$

REFERENCES

1. P. J. Schneider, *Conduction Heat Transfer.* (Cambridge, Mass.: Addison-Wesley Publishing Company, 1955.)
2. M. S. Carslaw and J. C. Jaeger, *Conduction of Heat in Solids.* (London: Oxford University Press, 1947.)
3. M. Necati Özisik, *Boundary Value Problems of Heat Conduction.* (Scranton, Pennsylvania: International Textbook Company, 1968.)
4. M. Jakob, *Heat Transfer*, Vol. 1. (New York: John Wiley & Sons, Inc., 1949.)
5. O. Lutz, "Graphical Determination of Wall Temperatures for Heat Transfer Through Walls of Arbitrary Shape," (translation from German), *NACA TM* 1280, 1950.
6. L. V. Bewley, *Two Dimensional Fields in Electrical Engineering.* (New York: The Macmillan Company, 1948.)
7. O. Rüdenberg, "Die Ausbreitung der Luft und Erdfelder um Hochspannungsleitungen besonders bei Erd-und Kurzschlüssen," *Electrotech. Z.*, Vol. 46 (1925), pp. 1342–1346.
8. I. Langmuir, E. Q. Adams, and F. S. Meikle, "Flow of Heat through Furnace Walls," *Trans. Am. Electrochem. Soc.*, Vol. 24 (1913), pp. 53–84.
9. L. M. K. Boelter, V. M. Cherry, and H. A. Johnson, *Heat Transfer.* (Berkeley, Calif.: University of California Press, 1942.)
10. A. D. Moore, "Fields from Fluid Flow Mappers," *J. Appl. Physics*, Vol. 20 (1949), pp. 790–804.
11. P. J. Schneider, "The Prandtl Membrane Analogy for Temperature Fields with Permanent Heat Sources or Sinks," *J. Aeronautical Sci.*, Vol. 19 (1952), pp. 644–645.
12. *Instructions for Analog Field Plotter, Catalogues* 112L152G1 and G2, General Electric Company, Schenectady, New York.
13. C. F. Kayan, "An Electrical Geometrical Analogue for Complex Heat Flow," *Trans. ASME*, Vol. 67 (1945), pp. 713–716.
14. C. F. Kayan, "Heat-Transfer Temperature Patterns of a Multicomponent Structure by Comparative Methods," *Trans. ASME*, Vol. 71 (1949), pp. 9–16.
15. H. M. Ellerbrook, Jr., E. F. Schum, and A. J. Nachtigall, "Use of Electric Analogs for Calculation of Temperature Distribution of Cooled Turbine Blades," *NACA TN* 3060, 1953.
16. D. N. de G. Allen, *Relaxation Methods.* (New York: McGraw-Hill Book Company, Inc., 1954.)
17. G. M. Dusinberre, *Numerical Analysis of Heat Flow.* (New York: McGraw-Hill Book Company, Inc., 1949.)
18. R. V. Southwell, *Relaxation Methods in Engineering Science.* (New York: Oxford University Press, 1940.)
19. S. H. Crandall, *Engineering Analysis.* (New York: McGraw-Hill Book Company, Inc., 1956.)

20. H. W. Emmons, "The Numerical Solution of Heat-Conduction Problems," *Trans. ASME*, Vol. 65 (1943), pp. 607–612.

21. L. Lapidus, *Digital Computation for Chemical Engineers*. (New York: McGraw-Hill Book Company, Inc., 1962.)

22. M. L. James, A. M. Smith, and J. C. Wolford, *Applied Numerical Methods for Digital Computation with FORTRAN*. (Scranton, Pennsylvania: International Textbook Company, 1967.)

23. G. J. Trezek and J. G. Witwer, "Finite-Difference Methods for Inhomogeneous Regions," *Trans. ASME, J. Heat Transfer*, Vol. 94 (1972), pp. 321–322.

4
Conduction of heat in the unsteady state

4-1. Transient and periodic heat flow

In the preceding chapters we dealt only with problems involving steady-state heat conduction. However, before steady-state conditions can be reached, some time must elapse after the heat-transfer process is initiated to allow the transient conditions to disappear. For instance, when we determined the rate of heat flow through the furnace wall in Sec. 2-1, we did not consider the period during which the furnace was starting up and the temperatures of the interior, as well as those of the walls, were slowly increasing. We simply assumed that this period of transition had passed and that steady-state conditions had been established. In Sec. 2-3 where we determined the temperature distribution in an electrically heated wire, we also neglected the warming-up period. Yet we know that when we turn on a toaster, it takes some time before the resistance wires attain maximum temperature, although heat generation starts instantaneously when the current begins to flow.

Another type of unsteady-heat-flow problem encountered in engineering involves periodic variations of temperature and heat flow. Periodic heat flow is of importance in internal-combustion engines, air-conditioning, instrumentation, and process control. For example, at the end of a hot day the atmospheric air becomes cooler, and yet the temperatures inside stone buildings remain quite high for several hours after sundown. In the morning, even though the atmosphere has already become warm, the air inside the buildings will remain comfortably cool for several hours. The reason for this phenomenon is the existence of a time lag before temperature equilibrium between the inside of the building and the outdoors can be reached. Another typical example is the periodic heat flow through the walls of engines which are heated only during a portion of their cycle of operation. After the engine has warmed up and operates in the steady state, the temperature at any point in the wall undergoes cyclic variations with time. While the engine is warming up, transient heat-flow phenomena are superimposed on the cyclic variations.

In this chapter we shall consider a number of heat-transfer problems in which either periodic or transient temperature variations are of primary concern. We shall first analyze problems which can be simplified by assuming that the temperature is only a function of time and is uniform throughout the system at any instant. In subsequent sections of this chapter we shall consider analytical and numerical methods for solving problems of unsteady heat flow when the temperature depends not only on the time, but also varies in the interior of the system. Throughout this chapter we shall not be concerned with the mechanisms of heat transfer by convection or radiation. Where these modes of heat transfer affect the boundary conditions of the system, an appropriate value for the unit-surface conductance will simply be specified.

4-2. *Transient heat flow in systems with negligible internal resistance*

Even though there are no materials in nature that possess an infinite thermal conductivity, many transient heat-flow problems can be readily solved with acceptable accuracy by assuming that the internal conductive resistance of the system is so small that the temperature within the system is substantially uniform at any instant. This simplification is justified when the external thermal resistance between the surface of the system and the surrounding medium is so large compared to the internal thermal resistance of the system that it controls the heat-transfer process.

A measure of the relative importance of the thermal resistance within a solid body is the ratio of the internal to the external resistance. This ratio can be written in dimensionless form as $\bar{h}L/k_s$, the *Biot number*, where $\bar{h}$ is the average unit-surface conductance, L is a significant length dimension obtained by dividing the volume of the body by its surface area, and k_s is the thermal conductivity of the solid body. In bodies whose shape resembles a plate, a cylinder, or a sphere, the error introduced by the assumption that the temperature at any instant is uniform will be less than 5 percent when the internal resistance is less than 10 percent of the external surface resistance, i.e., when $\bar{h}L/k_s < 0.1$.

A typical example of this type of transient heat flow is the cooling of a small metal casting or a billet in a quenching bath after its removal from a hot furnace. Suppose that the billet is removed from the furnace at a uniform temperature T_0 and is quenched so suddenly that we can approximate the environmental temperature change by a step. Designate the time at which the cooling begins as $\theta = 0$, assume that the heat-transfer coefficient $\bar{h}$ remains constant during the process, and that the bath temperature T_∞ at a distance far removed from the billet does not vary with time. Then, in accordance with the assumption that the temperature within the body is substantially constant at any instant, an energy balance for the billet over a small time interval $d\theta$ is

The change in internal energy of the billet during $d\theta$	=	the net heat flow from the billet to the bath during $d\theta$

or

$$-c\rho V dT = \bar{h}A_s(T - T_\infty)\, d\theta \tag{4-1}$$

where c = the specific heat of the billet, in Btu/lb F;
ρ = density of the billet, in lb/cu ft;
V = volume of the billet, in cu ft;
T = average temperature of the billet, in F;
$\bar{h}$ = average heat transfer coefficient, in Btu/hr sq ft F;
A_s = surface area of the billet, in sq ft;
dT = temperature change during $d\theta$.

The minus sign in Eq. 4-1 indicates that the internal energy decreases when $T > T_\infty$. The variables T and θ can be readily separated and, for a differential time interval $d\theta$, Eq. 4-1 becomes

$$\frac{dT}{T - T_\infty} = \frac{d(T - T_\infty)}{(T - T_\infty)} = -\frac{\bar{h}A_s}{c\rho V}\, d\theta \tag{4-2}$$

where it is noted that $d(T - T_\infty) = dT$, since T_∞ is constant. With an initial temperature of T_0 and a temperature at time θ of T as limits, integration of Eq. 4-2 yields

$$\ln \frac{T - T_\infty}{T_0 - T_\infty} = -\frac{\bar{h}A_s}{c\rho V}\,\theta$$

or

$$\frac{T - T_\infty}{T_0 - T_\infty} = e^{-(\bar{h}A_s/c\rho V)\theta} \tag{4-3}$$

An electrical network analogous to the thermal network for a lumped single-capacity system is shown in Fig. 4-1b. In this network the capacitor is initially "charged" to the potential T_0 by closing the switch S. When the switch is opened, the energy stored in the capacitance is discharged through the resistance $1/\bar{h}A_s$. The analogy between this thermal system and an electric system is apparent. The thermal resistance is $R = 1/\bar{h}A_s$ and the thermal capacitance is $C = c\rho V$, while R_e and C_e are the electric resistance and capacitance, respectively. To construct an elec-

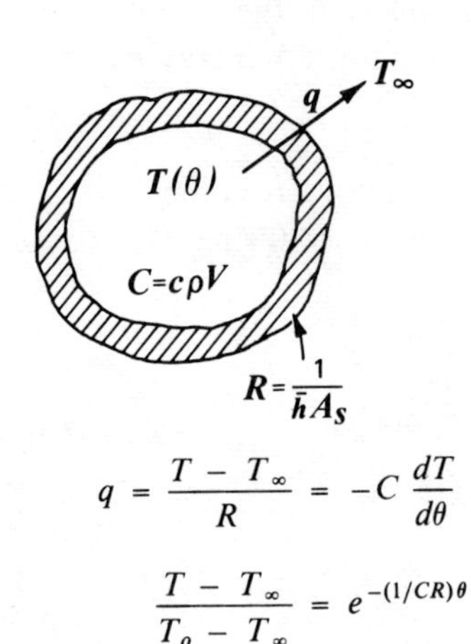

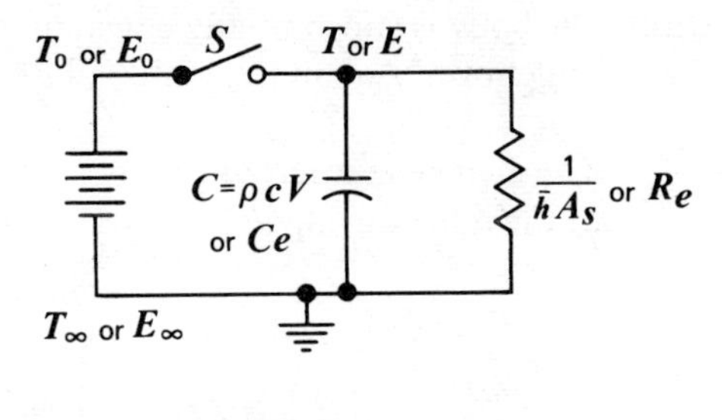

$$q = \frac{T - T_\infty}{R} = -C\frac{dT}{d\theta}$$

$$\frac{T - T_\infty}{T_o - T_\infty} = e^{-(1/CR)\theta}$$

$\theta = 0$ when billet is immersed in fluid and heat begins to flow.

(a)

$$i = \frac{E - E_\infty}{R_e} = -C_e\frac{dE}{d\theta}$$

$$\frac{E - E_\infty}{E_o - E_\infty} = e^{-(1/C_eR_e)\theta}$$

$\theta = 0$ when switch S is opened and the condenser begins to discharge.

(b)

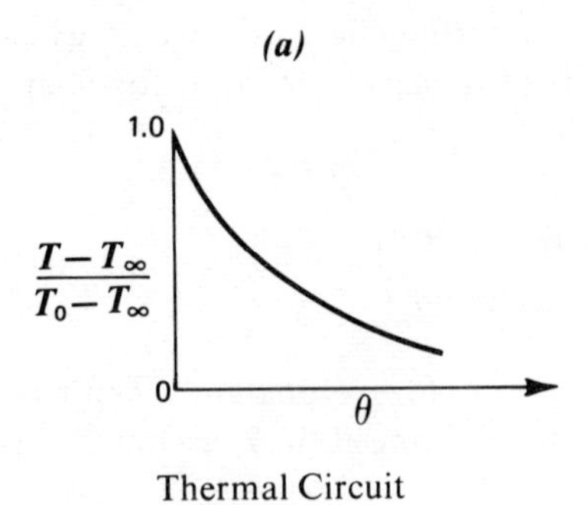

Thermal Circuit

Heat Flow q (Btu/hr)
Thermal Capacity
$C = c\rho V$(Btu/F)
Thermal Resistance
$R = 1/\bar{h}A_s$(F hr/Btu)
Thermal Potential $(T - T_\infty)$ (F)

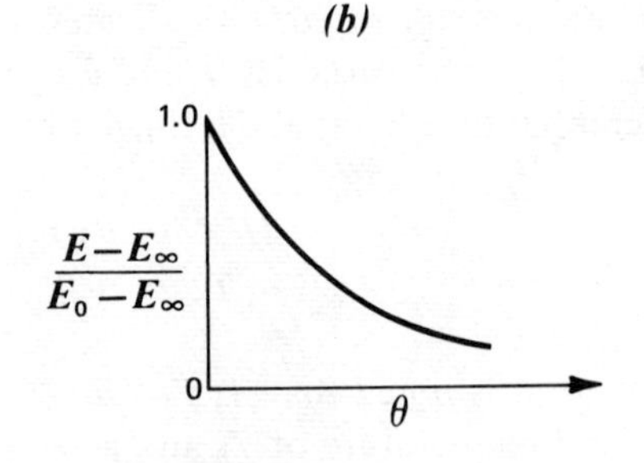

Electrical System

Current Flow i (amps)
Electrical Capacity C_e (farads)
Electrical Resistance R_e (ohms)
Electrical Potential $(E - E_\infty)$ (volts)

Fig. 4-1. Analogy between cooling billet and discharging condenser.

tric system which would behave exactly like the thermal system we would only have to make the ratio $\bar{h}A_s/c\rho V$ equal $1/RC$. In the thermal system we store internal energy, while in the electric system we store electric charge. The flow of energy in the thermal system is called heat, and the flow of charge is called electric current. The quantity $(c\rho V/\bar{h}A)$ is called the time constant of the system since it has the dimensions of time.[1] Observe that when $\theta = (c\rho V/\bar{h}A_s)$ the temperature difference $(T - T_\infty)$ is equal to 36.8 percent of the initial difference $(T_0 - T_\infty)$.

EXAMPLE 4-1. Determine the temperature response of a $\frac{1}{32}$-in.-diam copper wire originally at 300 F when suddenly immersed in (a) water ($\bar{h}$ = 15 Btu/hr sq ft F) at 100 F; (b) air ($\bar{h}$ = 2 Btu/hr sq ft F) at 100 F.

Solution: From Table A-1 in Appendix III we obtain

$$\begin{aligned} k_s &= 216 \text{ Btu/hr ft F} \\ c &= 0.091 \text{ Btu/lb F} \\ \rho &= 558 \text{ lb/cu ft} \end{aligned}$$

The surface area A_s and volume V of the wire are

$$\begin{aligned} A_s \text{ per inch length} &= \pi D = 6.82 \times 10^{-4} \text{ sq ft} \\ V \text{ per inch length} &= \pi D^2/4 = 5.32 \times 10^{-6} \text{ cu ft} \end{aligned}$$

The Biot modulus is

$$\frac{\bar{h}D}{4k_s} = \frac{(15)(1/32)/12}{(4)\,216} \ll 0.1 \text{ for water}$$

Hence, the internal resistance may be neglected and Eq. 4-3 applies. From the data and the properties we have

$$C = c\rho V = 2.72 \times 10^{-4} \text{ Btu/F}$$

$$R = \frac{1}{\bar{h}A_s} = \begin{matrix} 732 \text{ F hr/Btu, for air} \\ 97.6 \text{ F hr/Btu, for water} \end{matrix}$$

The temperature response is given by Eq. 4-3, and we get

$$T(\theta) = 100 + 200e^{-\theta/RC} \quad \text{F}$$

The results are plotted in Fig. 4-2. We note that the time required for the temperature of the wire to reach 200 F is 8.4 min in air, but only 1.1 min in water. *Ans.*

---

[1] Its value is indicative of the rate of response of a single capacity system to a sudden change in the environmental temperature.

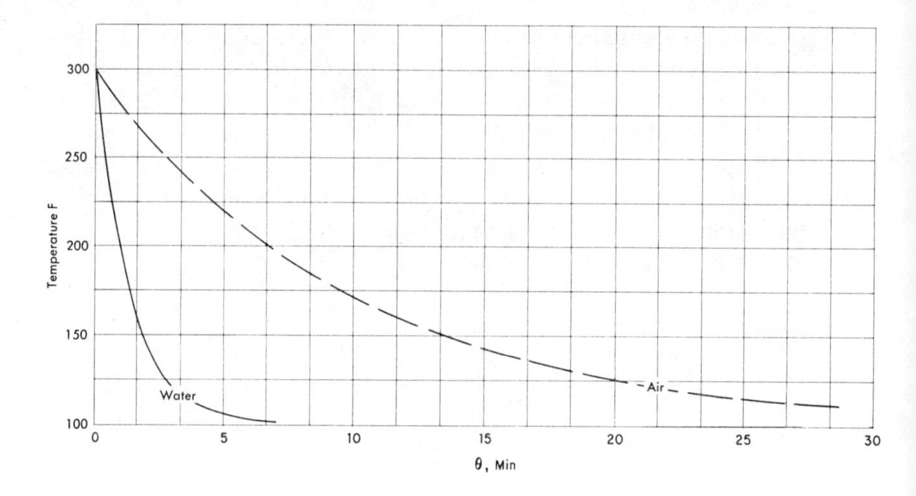

Fig. 4-2. Cooling of wire in air and water.

A thermocouple of $\frac{1}{32}$-in. diameter would therefore lag considerably if it were used to measure rapid changes in air temperature, and it would be advisable to use wire of the smallest available diameter to reduce this lag.

The results of the preceding analysis can be expressed conveniently in terms of dimensionless parameters. Let $V/A_s = L$, where L is a significant length dimension for the body, and multiply the numerator and the denominator of the exponent $\bar{h}\theta/c\rho L$ by Lk_s. Separating the resulting expression into two dimensionless groups gives

$$\frac{\bar{h}\theta L k_s}{c\rho L^2 k_s} = \left(\frac{\bar{h}L}{k_s}\right)\left(\frac{k_s}{\rho c}\frac{\theta}{L^2}\right) = (\text{Bi})\,(\text{Fo})$$

and Eq. 4-3 becomes

$$\frac{T - T_\infty}{T_0 - T_\infty} = e^{-(\text{Bi})(\text{Fo})} \tag{4-4}$$

where Fo is a dimensionless time called the *Fourier modulus*, $a\theta/L^2$, and a is the combination of physical properties $k_s/c\rho$, called the *thermal diffusivity*.

For some problems the change in the internal energy of the system during a given time interval must be determined. From Eq. 4-1 the instantaneous rate of heat flow at any time θ is

$$q = c\rho V \frac{dT}{d\theta} \tag{4-5}$$

From Eq. 4-3 the instantaneous rate of temperature change is

$$\frac{dT}{d\theta} = (T_\infty - T_0)\frac{\bar{h}A_s}{c\rho V} e^{-(\bar{h}A_s/c\rho V)\theta}$$

and we get

$$\frac{q}{\bar{h}A_s(T_\infty - T_0)} = e^{-(\bar{h}A_s/c\rho V)\theta} = e^{-(\mathrm{Bi})(\mathrm{Fo})}$$

Multiplying by $d\theta$ and integrating between $\theta = 0$ and $\theta = \theta$ yields Q, the amount of heat transferred in the time interval θ, which equals the change in internal energy of the system, or

$$\frac{Q}{\bar{h}A_s(T_\infty - T_0)} = \int_0^\theta e^{-(\bar{h}A_s/c\rho V)\theta}\, d\theta$$

$$= [1 - e^{-(\bar{h}A_s/c\rho V)\theta}](c\rho V/\bar{h}A_s) \tag{4-6}$$

The same general method can also be used to estimate the temperature-time history and the internal energy change of a well-stirred fluid in a metal container when the entire system is suddenly immersed in a fluid and heated or cooled by the surrounding medium. If the walls of the container are so thin that their heat capacity is negligible, the temperature-time history of the fluid is

$$\frac{T - T_\infty}{T_0 - T_\infty} = e^{-(UA_s/c\rho V)\theta}$$

where UA_s is the transmittance between the fluid and the surrounding medium and c and ρ are the specific heat and the density of the fluid respectively.

The lumped capacity method of analysis can also be applied to composite systems or bodies. For example, if the walls of the container (shown in Fig. 4-3a) have a substantial thermal capacitance $(c\rho V)_2$, the unit thermal conductance at A_1, the inner surface of the container, is $\bar{h}_1$, the unit thermal conductance at A_2, the outer surface of the container, is $\bar{h}_2$, and the thermal capacitance of the fluid in the container is $(c\rho V)_1$, the temperature-time history of the fluid $T_1(\theta)$ is obtained by solving simultaneously the energy balance equations

Fluid: $$-(c\rho V)_1 \frac{dT_1}{d\theta} = \bar{h}_1 A_1 (T_1 - T_2) \tag{4-7a}$$

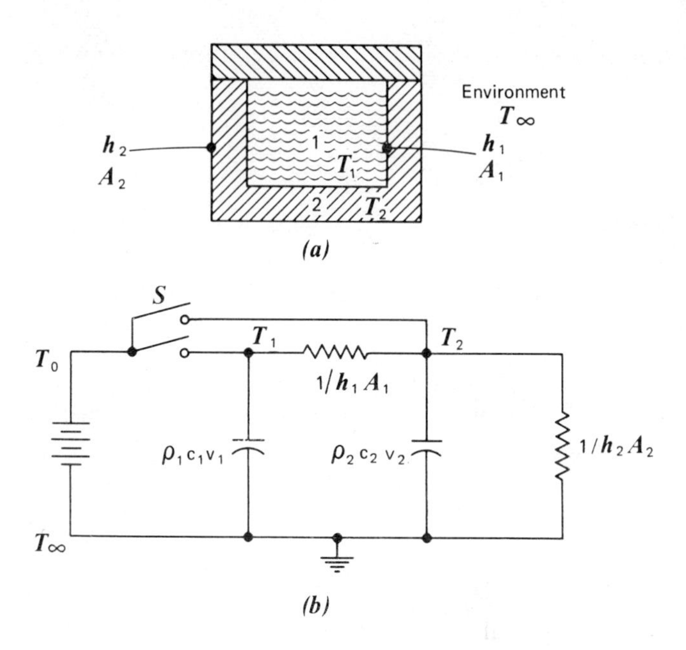

Fig. 4-3. Thermal network for a two-lump best-capacity system.

$$\textit{Container:} \qquad -(c\rho V)_2\frac{dT_2}{d\theta} = \bar{h}_2A_2(T_2 - T_\infty) - \bar{h}_1A_1(T_1 - T_2) \quad \text{(4-7b)}$$

where T_2 is the temperature of the walls of the container.

The above two simultaneous linear differential equations can be solved for the temperature history in each of the bodies. If the fluid and the container are initially at T_0, the boundary conditions for the system are

$$T_1 = T_2 = T_0 \qquad \text{at } \theta = 0$$

which implies that $dT_1/d\theta = 0$ at $\theta = 0$ from Eq. (4-7a).

Equations 4-7a and 4-7b may be rewritten in operator form as

$$\left(D + \frac{h_1A_1}{\rho_1C_1V_1}\right)T_1 - \frac{h_1A_1}{\rho_1C_1V_1}T_2 = 0$$

$$-\frac{h_1A_1}{\rho_2C_2V_2}T_1 + \left(D + \frac{h_1A_1 + h_2A_2}{\rho_2C_2V_2}\right)T_2 = \frac{h_2A_2}{\rho_2C_2V_2}T_\infty$$

where the symbol D denotes differentiation with respect to time. For convenience let

$$K_1 = \frac{h_1 A_1}{\rho_1 c_1 V_1}$$

$$K_2 = \frac{h_1 A_1}{\rho_2 c_2 V_2}$$

$$K_3 = \frac{h_2 A_2}{\rho_2 c_2 V_2}$$

Then

$$(D + K_1)T_1 - K_1 T_2 = 0$$

$$-K_2 T_1 + (D + K_2 + K_3)T_2 = K_3 T_\infty$$

Solving the equations simultaneously, we get for the differential equation involving only T_1

$$[D^2 + (K_1 + K_2 + K_3)D + K_1 K_3]T_1 + K_1 K_3 T_\infty$$

whose general solution is

$$T = T_\infty + Me^{m_1 \tau} + Ne^{m_2 \tau}$$

where m_1 and m_2 are given by

$$m_1 = \frac{-(K_1 + K_2 + K_3) + [(K_1 + K_2 + K_3)^2 - 4K_1 K_3]^{1/2}}{2}$$

$$m_2 = \frac{-(K_1 + K_2 + K_3) - [(K_1 + K_2 + K_3)^2 - 4K_1 K_3]^{1/2}}{2}$$

The arbitrary constants M and N may be obtained by applying the initial conditions

$$T_1 = T_0 \qquad \text{at } \theta = 0$$

and

$$\frac{dT_1}{d\tau} = 0 \qquad \text{at } \theta = 0$$

so that

$$T_0 = T_\infty + M + N$$

$$0 = m_1 M + m_2 N$$

The final solution is

$$\frac{T_1 - T_\infty}{T_0 - T_\infty} = \frac{m_2}{m_2 - m_1} e^{m_1\theta} - \frac{m_1}{m_2 - m_1} e^{m_2\theta} \tag{4-8}$$

The solution for $T_2(\theta)$ is obtained by substituting the relation for T_1 from Eq. 4-8 into Eq. 4-7a.

The network analogy for the two-lump system is shown in Fig. 4-3b. When the switch S is closed, the two thermal capacitances are charged to the potential T_0. At time zero the switch is opened and these capacitances discharge through the thermal resistances shown.

4-3. Periodic heat flow in systems with negligible internal resistance[2]

The preceding analysis has been limited to transient heat flow in systems where the ambient temperature remains constant. There exist, however, many problems in which the temperature of the medium surrounding the system varies with time. For example, there are batch processes in the chemical industry where the temperature of a chemical compound in a container must periodically follow a specified time schedule. The temperature changes of the material in the container are usually induced by heating or cooling the environment. Even if the compound is well stirred, it cannot immediately respond to the variation in the environmental temperature, as a result of its finite thermal capacity. To obtain the desired temperature-time schedule it is therefore necessary to initiate changes in the environmental temperature early enough to allow for the time lag in the system. Other typical examples of periodic-heat-flow problems are encountered in the design of a thermostatic temperature-control unit for a building which is continuously exposed to cyclic temperature variations and the design of the temperature-sensing element of a transducer used in the control and programming of high-vacuum processes such as the purification of vitamins.

The equation describing the temperature-time history of a system exposed to periodic temperature fluctuations is identical to the one derived for a constant ambient temperature. For a differential time interval $d\theta$ we can write Eq. 4-1 in the form

$$c\rho V\, dT = \bar{h}A_s(T_\infty - T)\, d\theta$$

but since T_∞ now varies with time, i.e., $T_\infty = T_\infty(\theta)$, we cannot simply

[2]This section may be omitted without breaking the continuity of the presentation.

separate the variables. To obtain the temperature-time history of a system subjected to a variable environmental temperature we collect the terms containing the system temperature T on the left-hand side. This yields

$$\frac{dT(\theta)}{d\theta} + \frac{\bar{h}A_s}{\rho c V} T(\theta) = \frac{\bar{h}A_s}{\rho c V} T_\infty(\theta) \tag{4-9}$$

a linear nonhomogeneous equation with constant coefficients. The general solution of Eq. 4-9 for a specified variation of $T_\infty(\theta)$ will be composed of the sum of two parts. The first part, called the *particular integral*, satisfies the complete equation and contains no arbitrary constants. Physically, the particular integral is the temperature-time history of the system after the transient phenomena have disappeared. In problems in dynamics and electric-circuit theory, this portion of the solution is often called the steady-state solution. Steady state in periodic phenomena means that the cyclic variations of the system will not change with time. The steady-state response of the system temperature T is caused and sustained by the environmental temperature T_∞, which acts as the driving potential and will generally be of the same form. If the periodic steady state is a cyclic variation of T_∞, then the temperature of the system T will also be cyclic. For example, if T_∞ is sinusoidal, the steady-state response of T will also be sinusoidal.

The second part of the solution, called the *complementary function*, makes the left-hand side of Eq. 4-9 equal to zero. It contains the constants of integration whose values must be obtained from the initial or boundary conditions and represents physically the transient response of the system temperature. The transient response arises because of a lack of initial equilibrium and will decay exponentially as when T_∞ is constant.

In summary, the complete solution to Eq. 4-9 consists of two parts:

$$T = T_{ss} + T_t \tag{4-10}$$

where T_{ss} means the steady-state part and T_t means the transient part of the system temperature T. It should be noted that the boundary conditions and the initial conditions must always be applied to the complete solution, $T = T_{ss} + T_t$, and never to the transient part alone.

EXAMPLE 4-2. Compare the temperature-time response of a bare iron-constant thermocouple with that of a mercury-in-glass thermometer when these instruments are used to measure the temperature-time history of a gas whose temperature is a sinusoidal function of time, θ, i.e., $T_\infty = (100 + 50 \sin 2\pi\theta)$ F. Assume that the overall heat-transfer coefficient for both instruments is equal to 5 Btu/hr sq ft F. The thermocouple is $\frac{1}{32}$ in. in diameter with 2 in. of length immersed. The thermometer is idealized by a mercury cylinder 1 in. long and $\frac{1}{4}$ in. in diameter. The initial temperature of both instruments is 60 F.

Solution: The transient response is obtained by solving Eq. 4-9 for the complementary function. Setting the left-hand side of Eq. 4-9 equal to zero, we obtain the homogeneous equation

$$\frac{dT}{d\theta} + \frac{\bar{h}A_s}{c\rho V} T = 0$$

After separating the variables and integrating, the solution is found to be

$$T_t = C_1 e^{-(\bar{h}A_s/c\rho V)\theta} \tag{4-11}$$

In order to find the steady-state response we must obtain the particular integral. Since the driving potential (i.e., the bath temperature) is sinusoidal, the response also must be sinusoidal. In addition, the driving potential contains a constant term, and therefore the response also must contain a constant. We recall that, when the solutions to linear differential equations are superposed (i.e., simply added), the sum also is a solution. By means of this fact we can construct the type of equation which meets the required conditions, as

$$T_{ss} = C_2 \sin 2\pi\theta + C_3 \cos 2\pi\theta + C_4 \tag{4-12}$$

This expression must satisfy Eq. 4-9 if it is a solution. Hence, we take the derivative of Eq. 4-12

$$\frac{dT_{ss}}{d\theta} = 2\pi C_2 \cos 2\pi\theta - 2\pi C_3 \sin 2\pi\theta$$

and substitute Eqs. 4-12 and the above equation into Eq. 4-9, the original expression for T. If we let $m = \bar{h}A_s/c\rho V$, we obtain, after collecting terms,

$$(2\pi C_2 + mC_3)\cos 2\pi\theta + (mC_2 - 2\pi C_3)\sin 2\pi\theta + mC_4$$
$$= m100 + m50 \sin 2\pi\theta \tag{4-13}$$

This can be an identity for all values of time θ only if the coefficients of like terms on each side of the equation are equal and we have

$$2\pi C_2 + mC_3 = 0 \quad \text{from the cosine terms}$$
$$mC_2 - 2\pi C_3 = 50m \quad \text{from the sine terms}$$
$$C_4 = 100 \quad \text{from the constant term}$$

Solving these equations for C_2 and C_3 simultaneously, we obtain

$$C_2 = \frac{50}{1 + (2\pi/m)^2} \quad \text{and} \quad C_3 = -\frac{(2\pi/m)50}{1 + (2\pi/m)^2}$$

The steady-state temperature response is therefore

$$T_{ss} = 100 + \frac{50}{1 + (2\pi/m)^2} \sin 2\pi\theta - \frac{(2\pi/m)50}{1 + (2\pi/m)^2} \cos 2\pi\theta \qquad \text{(4-14)}$$

Terms such as $C_2 \sin 2\pi\theta - C_3 \cos 2\pi\theta$ can be combined by using the relation

$$C_2 \sin 2\pi\theta - C_3 \cos 2\pi\theta$$

$$= \sqrt{C_2^2 + C_3^2} \left(\frac{C_2}{\sqrt{C_2^2 + C_3^2}} \sin 2\pi\theta - \frac{C_3}{\sqrt{C_2^2 + C_3^2}} \cos 2\pi\theta \right)$$

If we now construct a right triangle with $\sqrt{C_2^2 + C_3^2}$ as the hypotenuse and C_2 and C_3 as the sides, we get $C_2/\sqrt{C_2^2 + C_3^2} = \cos\delta$ and $C_3/\sqrt{C_2^2 + C_3^2} = \sin\delta$. But since $\sin(A - B) = \sin A \cos B - \cos A \sin B$, we get $C_2 \sin 2\pi\theta - C_3 \cos 2\pi\theta = \sqrt{C_2^2 + C_3^2} \sin(2\pi\theta - \delta)$ where δ is equal to the $\tan^{-1}(C_3/C_2)$.

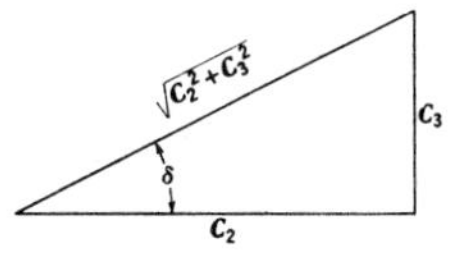

Combining the last two terms of Eq. 4-14 in this manner and adding T_t from Eq. 4-11 gives

$$T = 100 + \frac{50}{1 + (2\pi/m)^2} \sqrt{1 + (2\pi/m)^2} \sin(2\pi\theta - \delta) + C_1 e^{-m\theta} \qquad \text{(4-15)}$$

where $\delta = \tan^{-1}(2\pi/m)$ and represents the time lag in the temperature response of the instruments.

The constant of integration C_1 can now be evaluated from the initial condition, i.e., $T = 60$ F at $\theta = 0$. Substituting this condition into Eq. 4-15 yields

$$T_{\theta=0} = 60 = 100 + \frac{50}{\sqrt{1 + (2\pi/m)^2}} \sin(-\delta) + C_1 \qquad \text{(4-16)}$$

Making use of the trigonometric identity for $\sin\delta$ we obtain

$$C_1 = \frac{100\pi/m}{1 + (2\pi/m)^2} - 40 \qquad \text{(4-17)}$$

Finally, the expression for the temperature-time history of the instrument is

$$T = \left[\frac{100\pi/m}{1 + (2\pi/m)^2} - 40 \right] e^{-m\theta} + \frac{50}{\sqrt{1 + (2\pi/m)^2}} \sin(2\pi\theta - \delta) + 100 \qquad \text{(4-18)}$$

To obtain the time lag δ in units of time we first find the time required for the system to go through one complete cycle. In the problem under consideration, the bath or the steady-state response of the instruments will complete one cycle each hour (i.e., $2\pi\theta$ increases by 2π radians as θ increases by one). The time lag δ in hours is therefore obtained by dividing the lag in radians by the number of radians corresponding to a unit increase of time, which is 2π radians in this case.

In order to plot the results, we determine the numerical value of m for the instruments. By definition we have

$$m = \frac{\bar{h}A_s}{c\rho V} = \frac{\bar{h}}{c\rho} \frac{\pi DL}{(\pi/4) D^2 L} = \frac{4\bar{h}}{c\rho D}$$

For the thermometer, using physical properties of mercury

$$\rho = 849 \text{ lb/cu ft}$$

$$c = 0.0325 \text{ Btu/lb F}$$

$$D = 0.021 \text{ ft} \qquad m = 35.2 \text{ hr}^{-1}$$

For the thermocouple, using properties of iron as an approximation

$$\rho = 475 \text{ lb/cu ft}$$

$$c = 0.12 \text{ Btu/lb F}$$

$$D = 0.0026 \text{ ft} \qquad m = 135 \text{ hr}^{-1}$$

The final equations for the temperature response of the thermometer and thermocouple are respectively within slide-rule accuracy

$$T_{\text{meter}} = 100 + 49.3 \sin (2\pi\theta - 0.178) - 31.35 e^{-35.2\theta}$$

$$T_{\text{couple}} = 100 + 50 \sin (2\pi\theta - 0.0465) - 37.7 e^{-135\theta}$$

where the angles of the sines are in radians.

These results are plotted in Fig. 4-4 showing the steady-state and the transient response separately. We note that the transient response of the thermometer is considerably slower—it takes about 6 min to die out—than that of the thermocouple. This is not unexpected since the time constant of the thermometer is 1.7 min, while the time constant of the thermocouple is less than one half of a minute. The steady-state lag of the thermometer is nearly a minute, while the thermocouple lags less than five seconds behind the temperature of the bath. The reason for this behavior is the large thermal capacitance of the mercury thermometer, which makes this type of instrument unsuitable when high sensitivity and fast response are desired.

The preceding technique can be extended to arbitrary kinds of periodic temperature variations, since nearly any periodic function of time

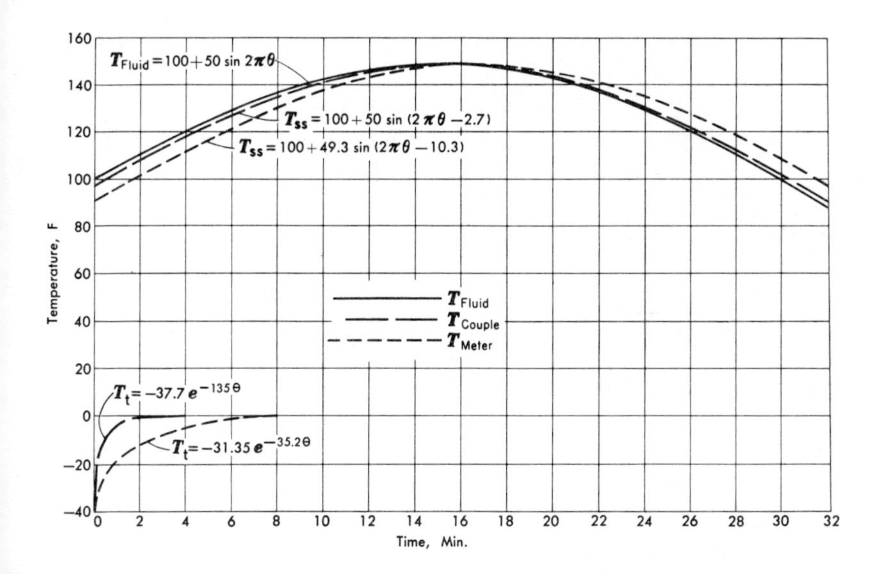

Fig. 4-4. Transient and steady-state response of the thermocouple and the thermometer in Example 4-2.

can be expressed in terms of a series of sine and cosine terms of the form

$$T_\infty(\theta) = \frac{A_0}{2} + A_1 \cos \frac{2\pi}{\theta_0}\,\theta + A_2 \cos \frac{4\pi}{\theta_0}\,\theta + \cdots + B_1 \sin \frac{2\pi}{\theta_0}\,\theta + B_2 \sin \frac{4\pi}{\theta_0}\,\theta + \cdots \tag{4-19}$$

Once the temperature variation in Eq. 4-9 is expressed in this form, a solution can be obtained by algebraic addition of the solutions corresponding to each term of the series. The type of series shown in Eq. 4-19 is called a *Fourier series*. It can be written more compactly as

$$T_\infty(\theta) = \frac{A_0}{2} + \sum_{n=1}^{\infty} \left(A_n \cos \frac{2\pi n}{\theta_0}\,\theta + B_n \sin \frac{2\pi n}{\theta_0}\,\theta \right) \tag{4-20}$$

or in the equivalent form,

$$T_\infty(\theta) = \frac{A_0}{2} + \sum_{n=1}^{\infty} C_n \cos \left(\frac{2\pi n}{\theta_0}\,\theta - \beta_n \right) \tag{4-21}$$

where $\frac{A_0}{2}$ = mean temperature, T_{avg};

$$\theta_0 = \text{period of the first harmonic or fundamental component;}$$
$$C_n = \sqrt{A_n^2 + B_n^2} = \text{the temperature amplitude of the } n\text{th harmonic;}$$
$$\beta_n = \tan^{-1}(B_n/A_n), \text{ the phase angle of the } n\text{th harmonic;}$$
$$n = \text{positive integers 1,2,3,etc.}$$

The solution of Eq. 4-9 can be written compactly as

$$T(\theta) = T_{avg} + \sum_{n=1}^{\infty} \frac{T_{avg}}{\sqrt{1 + \left(\frac{2\pi n}{\theta_0} m\right)^2}} \cos\left(\frac{2\pi n}{\theta_0}\theta - \beta_n - \delta_n\right) + Ce^{-m\theta} \tag{4-22}$$

where C is a constant whose value depends on the initial conditions and δ_n is the lag of the nth harmonic in the response. It should be noted that the coefficient $T_{avg}/\sqrt{1 + [(2\pi n/\theta_0)m]^2}$, often called the *amplitude ratio*, decreases rapidly for higher harmonics, i.e., if n is large. Therefore, the higher harmonics of the ambient-temperature variation have little or no effect on the system temperature, $T(\theta)$. The analogy to problems dealing with electrical or mechanical vibrations is apparent and may assist in the analysis of the thermal problems.

4-4. Transient heat flow in an infinite plate[3]

In Sec. 4-2 we discussed analytic methods for solving a class of transient-heat-flow problems which could be simplified by neglecting the thermal resistance within the system and treating the thermal capacity of the entire system as a lumped parameter. The mathematical description of this type of problem leads to ordinary differential equations. For simple shapes this approach is satisfactory when the Biot modulus is less than 0.1. Systems having a Biot modulus larger than 0.1 can be analyzed by the numerical method to be described in Sec. 4-6. However, for several cases of practical importance, solutions are available in the form of charts which are based on exact solutions and reduce the amount of labor and time required for the analysis. The material in this section serves as an introduction to the mathematical methods for solving the general heat-conduction equation and will also foster an understanding of the technique for using the charts presented in Sec. 4-5.

The equations describing the temperature distribution in a solid having a finite thermal conductivity were derived in Sec. 3-2. These equations are partial-differential equations because the temperature is a function of time as well as location. A detailed treatment of the methods for solving

[3] This section may be omitted without breaking the continuity of the presentation.

the general heat-conduction equation is beyond the scope of this book, and for an extensive treatment reference may be made to the books by Schneider (1), Carslaw and Jaeger (2), and Özisik (3). Only one relatively simple case which can be handled essentially with the tools of ordinary differential equations will be solved here.

To illustrate the analytical method we will determine the temperature-time history in a wall of thickness L, as shown in Fig. 4-5. If edge effects are neglected, the only space coordinate is x and the general conduction equation, Eq. 3-3, reduces to

$$\frac{\partial T}{\partial \theta} = a \frac{\partial^2 T}{\partial x^2} \qquad 0 \leq x \leq L \tag{4-23}$$

if the thermal diffusivity $a = k/c\rho$ is constant. The thermal diffusivity a, which appears in all unsteady-heat-conduction problems is a property of the material, and the time rate of temperature change depends on its numerical value. Qualitatively we observe that, in a material that combines a low thermal conductivity with a large specific heat per unit volume, the rate of temperature change will be slower than in a material that possesses a large thermal diffusivity.

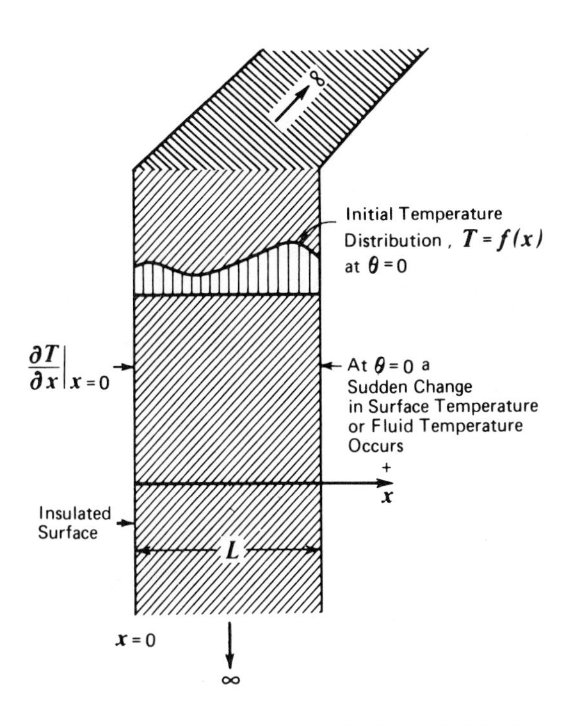

Fig. 4-5. Nomenclature for analytical solution of slab with sudden change in surface temperature.

We shall seek a solution by the same method used in Sec. 3-3. Assume that the solution consists of the product of a function of x and a function of θ, or

$$T(x, \theta) = X(x) \cdot \Theta(\theta) \tag{4-24}$$

Introducing this function in Eq. 4-23 gives

$$X \frac{\partial \Theta}{\partial \theta} = a\Theta \frac{\partial^2 X}{\partial x^2}$$

After separating the variables we obtain

$$\frac{1}{a\Theta} \frac{\partial \Theta}{\partial \theta} = \frac{1}{X} \frac{\partial^2 X}{\partial x^2} = -\lambda^2 \tag{4-25}$$

Since each side in Eq. 4-25 is only a function of one variable, the two sides can only be equal if each is equal to the same constant. If we let this constant be $(-\lambda^2)$, we have two ordinary differential equations with solutions as shown below:

$$\frac{\partial \Theta}{\partial \theta} + a\lambda^2\Theta = 0 : \Theta(\theta) = C_1 e^{-a\lambda^2 D} \tag{4-26}$$

$$\frac{\partial^2 X}{\partial x^2} + \lambda^2 X = 0 : X(x) = C_2 \sin \lambda x + C_3 \cos \lambda x \tag{4-27}$$

where the C's are constants. Observe that the selection of a negative constant for the separation parameter, $(-\lambda^2)$, insures a solution for the temperature which will decay with time. This is a requirement imposed by the second law of thermodynamics. If the constant were a positive number, the temperature would become infinitely high as time increases.

Substituting Eqs. 4-26 and 4-27 in Eq. 4-24 gives

$$T(x, \theta) = e^{-a\lambda^2\theta}[A \cos(\lambda x) + B \sin(\lambda x)] \tag{4-28}$$

where $A = C_1C_2$ and $B = C_1C_3$. The evaluations of the constants A and B depend on the physical boundary conditions.

If the initial temperature distribution is a known function of x, we have as our initial condition

$$\theta = 0 : T(x, 0) = f(x) \tag{i}$$

Suppose that at time $\theta = 0$ the exterior surface temperature at $x = L$ is suddenly changed to T_0 while the surface at $x = 0$ is insulated. The mathematical analysis for this case can be simplified by recognizing first that the temperature datum is arbitrary and that selecting $T_0 = 0$ would simply mean that we are measuring the slab temperature relative to the surface temperature. Another way of achieving this simplification is to choose $[T(x,\theta) - T_0]$ as the variable. The boundary conditions can then be written

$$\theta \geq 0: \text{at } x = 0, \partial(T - T_0)/\partial x = 0 \qquad (ii)$$

$$\text{at } x = L, T - T_0 = 0 \qquad (iii)$$

Introducing the conditions (i) and (ii) into Eq. 4-28 gives

$$\frac{\partial(T(0,0) - T_0)}{\partial x} = 0 = A \sin 0 - B \cos 0.$$

Since cos 0 is unity, $\theta = 0$. Application of conditions (i) and (iii) then gives

$$T(L,0) - T_0 = 0 = A \cos \lambda L$$

This last relation is satisfied for $\lambda = \pi/2L$, $3\pi/2L$, $5\pi/2L$, etc. or in general, when the so-called *characteristic values* of λ are

$$\lambda_n = \frac{(n + 1/2)\pi}{L} \qquad n = 0, 1, 2, 3, \ldots. \qquad (4\text{-}29)$$

Each of these λ_n's gives rise to a separate solution and, since the general solution will be the sum of these individual solutions,

$$[T(x,\theta) - T_0] = \sum_{n=1}^{\infty} e^{-\lambda_n^2 a\theta} A_n \cos(\lambda_n x) \qquad (4\text{-}30)$$

Application of the initial condition (i) gives

$$f(x) = \sum_{n=0}^{\infty} A_n \cos(\lambda_n x) \qquad (4\text{-}31)$$

It can be shown that the characteristic functions, $\cos(\lambda_n x)$ are *orthogonal* between $x = 0$ and $x = L$ and this property implies that

$$\int_0^L \cos \lambda_n x \cdot \cos \lambda_m x \, dx = 0 \quad m \neq n$$

$$\neq 0 \quad m = n \qquad (4\text{-}32)$$

where λ_m is any characteristic value of λ.[4] To obtain a particular value of A_n, multiply both sides of Eq. 4-31 by $\cos \lambda_m x$ and integrate between 0 and L. In accordance with Eq. 4-32 all terms on the right-hand side disappear, except the one involving the square of the characteristic function $\cos \lambda_n x$, and we get

$$\int_0^L f(x) \cos(\lambda_n x)\, dx = A_n \int_0^L \cos^2(\lambda_n x)\, dx$$

From standard integral tables (12) we get

$$\int_0^L \cos^2(\lambda_n x)\, dx = \frac{1}{2\lambda_n}\left(\lambda_n x + \sin(\lambda_n x)\cos(\lambda_n x)\right)\Big|_0^L = \frac{L}{2}$$

and

$$A_n = \frac{2}{L}\int_0^L f(x)\cos(\lambda_n x)\, dx \tag{4-33}$$

The temperature distribution in the slab at any time θ can now be obtained by evaluating the A_n's from Eq. 4-33 and substituting the result in Eq. 4-30. This yields

$$T(x,\theta) - T_0 = \frac{2}{L}\sum_{n=0}^{\infty} e^{-[(n+1/2)\pi/L]^2 a\theta} \cos\left(n + \frac{1}{2}\right)\frac{\pi}{L}x \cdot \int_0^L f(x)\cos\left(n + \frac{1}{2}\right)\frac{\pi x}{L}\, dx \tag{4-34}$$

If the initial temperature in the slab is uniform, i.e., $f(x) = (T_i - T_0)$, where T_i is a constant, Eq. 4-34 for the transient temperature distribution becomes

---

[4]This can be verified by performing the integration which yields

$$\int_0^L \cos\lambda_n x \cos\lambda_m x\, dx = \frac{\sin(\lambda_n - \lambda_m)x}{2(\lambda_n - \lambda_m)}\Bigg|_0^L + \frac{\sin(\lambda_n + \lambda_m)x}{2(\lambda_n + \lambda_m)}\Bigg|_0^L = 0$$

when $m \neq n$ since

$$\lambda_n = \frac{(n + 1/2)\pi}{L} \quad \text{and} \quad \lambda_m = \frac{(m + 1/2)\pi}{L}$$

and m as well as n are integers.

$$T(x,\theta) - T_0 = 2(T_i - T_0) \sum_{n=0}^{\infty} e^{-[(n+1/2)\pi/L]^2 a\theta} \cdot \frac{(-1)^n}{\left(n + \frac{1}{2}\right)\pi} \cdot \cos\left(n + \frac{1}{2}\right)\frac{\pi x}{L} \quad \text{(4-35a)}$$

To present the results graphically it is convenient to write Eq. 4-35a in terms of the following dimensionless parameters:

Dimensionless position: x/L

Dimensionless temperature: $[T(x,\theta) - T_0]/(T_i - T_0)$

Dimensionless time: $a\theta/L^2$ (Fourier modulus)

Introducing these parameters into Eq. 4-35a leads to the dimensionless expression

$$\frac{T(x,\theta) - T_0}{T_i - T_0} = \frac{2}{\pi} \sum_{n=0}^{\infty} e^{-[(n+1/2)\pi]^2 (a\theta/L^2)} \cdot \frac{(-1)^n}{\left(n + \frac{1}{2}\right)} \cdot \cos\left(n + \frac{1}{2}\right)\pi x/L \quad \text{(4-35b)}$$

The rate of heat flow out of the slab per unit area normal to the x-direction is

$$\frac{q}{A} = [-k_s(\partial T/\partial x)_{x=L}]$$

The change in the internal energy per unit area of the slab, Q/A, between time $\theta = 0$ and $\theta = \theta$ is equal to

$$\frac{Q}{A} = \int_0^{\theta} \frac{q}{A}\, d\theta \quad \text{(4-36)}$$

Introducing the derivative of Eq. 4-35b into Eq. 4-36 gives

$$\frac{Q}{Q_i} = \frac{2}{\pi^2} \sum_{n=0}^{\infty} \frac{(-1)^n}{\left(n + \frac{1}{2}\right)^2} [1 - e^{-[(n+1/2)\pi]^2 (a\theta/L^2)}] \quad \text{(4-37)}$$

where $Q_i = AL\rho c(T_i - T_0)$, the internal energy stored initially in the slab, measured relative to the fixed boundary temperature T_0.

It should be noted that exactly the same solution would be obtained for a slab or a wall of thickness $2L$ when *both* surfaces are suddenly changed and maintained at T_0. Since this system is symmetrical about the center plane, no heat can be conducted across it and the temperature gradient is zero as specified by the boundary condition (ii). The only difference between the two results is that in the latter case heat flows across two surfaces, at $x = 0$ and at $x = 2L$. Fig. 4-6 shows the dimensionless temperature distribution and the dimensionless heat-flow rate as a function of the Fourier modulus for a slab of thickness $2L$, initially at temperature T_i, whose surface temperatures are changed at $\theta = 0$ to T_0. With the aid of the curves in Fig. 4-6 the temperature-time history and the rate of heat flow at any time can be readily determined. The following example illustrates the procedure.

EXAMPLE 4-3. A large two-inch-thick sheet of plastic material ($\rho = 50$ lb_m/ft^3, $c = 0.2$ Btu/lb_m F, $k = 0.5$ Btu/hr ft F, $a = 0.05$ ft^2/hr) is initially at 100 F. It is suddenly immersed in boiling water so that its surface temperature rises to 212 F. Determine the temperature at a plane $\frac{1}{2}$ inch from the surface and the increase in internal energy after three minutes.

Solution: The dimensionless time is

$$\frac{a\theta}{(2L)^2} = \frac{0.05\,[\text{ft}^2/\text{hr}] \times 3\,[\text{min}] \times (1/60)\,[\text{hr/min}]}{2^2\,[\text{in}]^2 \times (1/12)^2\,[\text{ft/in}]^2} = 0.09$$

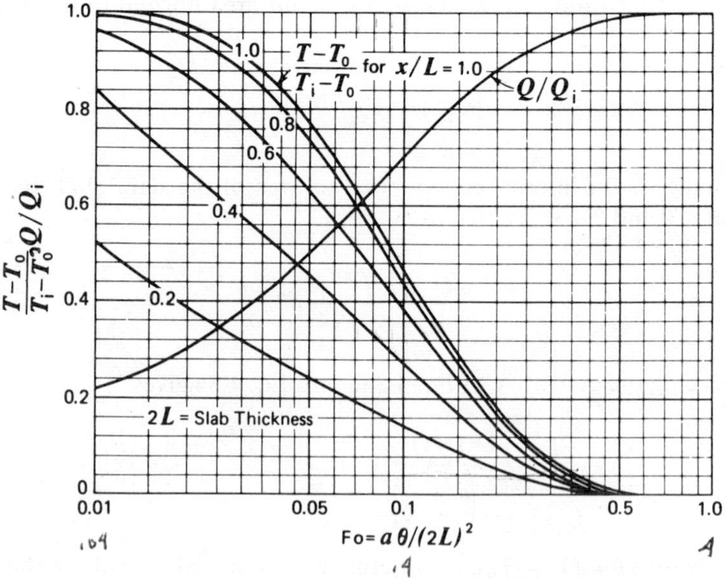

Fig. 4-6. The time variation of the temperature distribution and heat flow in an infinite slab of thickness $2L$ at a temperature T_i which has its surface temperature suddenly changed to T_0. (Note: X is measured from the centerplane.)

From Fig. 4-6, at the location $x/L = 0.5/1$ and at a dimensionless time of 0.09 the temperature ratio $[T(x,\theta) - T_0]/[T_i - T_0]$ is 0.36. The temperature after three minutes is, therefore,

$$T(x,\theta) = 100 + 0.36\,(212 - 100) = 140.4\text{ F} \qquad \textit{Ans.}$$

A boundary condition of more practical significance than the one discussed above occurs when a slab is suddenly exposed to a fluid and heat is transferred by convection at the interfaces. Since this condition gives at any time a temperature profile symmetrical about the center plane where $\partial T/\partial x = 0$, it describes also a slab with one surface insulated and the other surface suddenly exposed to a fluid at a different temperature. An example of such a system is the walls of an uncooled rocket motor which are suddenly exposed to hot gases when the propellants are ignited. The operation of the rocket is critically dependent on the temperature rise and the resulting thermal stresses in the wall. Since heat losses from the outer surface are generally so small that one may assume the surface is insulated, the boundary conditions conform approximately to the problem under consideration if the wall thickness is small compared to the motor diameter.

The constants A and B in the solution to the heat-conduction equation, as given by Eq. 4-28, must now satisfy the convection condition at the surface. Denoting the fluid temperature by T_∞ and the unit surface conductance by $\bar{h}$, the boundary condition (iii) requires that at the interface between the solid and the fluid the rate of heat transfer by conduction from the interior of the slab be equal to the rate of convection heat transfer to the fluid, or

$$\theta \geq 0: x = L, \; - k \left.\frac{\partial T}{\partial x}\right|_{x=L} = \bar{h}(T_{x=L} - T_\infty)$$

or

$$- \partial\, T(\theta, L)/\partial\, x = (\bar{h}/k)[T(\theta, L) - T_\infty]$$

Since, as shown previously, $B = 0$ from conditions (i) and (ii), the solution reduces to

$$T(\theta,x) - T_\infty = e^{-x^2 a\theta} A \cos(\lambda x)$$

Applying condition (i) and (iii) gives

$$-\frac{\partial T(0,L)}{\partial x} = A\lambda \sin \lambda L = (\bar{h}/k)\, A \cos \lambda L$$

or

$$\cot \lambda L = \frac{k_s}{hL} \lambda L = \frac{\lambda L}{\text{Bi}} \tag{4-38}$$

Equation 4-38 is transcendental, and there are an infinite number of characteristic values of λ which will satisfy it. The simplest way to determine the numerical values of λ is to plot cot λL and $\lambda L/\text{Bi}$ against λL. The values of λ at the points of intersection of these curves are the characteristic values and will satisfy the second boundary condition. Figure 4-7 is a plot of these curves, and if $L = 1$ we read off the first few characteristic values as $\lambda_1 = 0.86\text{Bi}$, $\lambda_2 = 3.43\text{Bi}$, $\lambda_3 = 6.44\text{Bi}$, etc. The value $\lambda = 0$ is disregarded because it leads to the trivial solution $T = 0$ (see Eq. 4-30).

A particular solution of Eq. 4-30 corresponds to each value of λ. Of course, the constant A must be evaluated for each value of λ. Therefore, we shall adopt a subscript notation to identify the correspondence between A and λ. For instance, A_1 corresponds to λ_1 or, in general, A_n to λ_n. The complete solution is, as shown previously (Eq. 4-30) the sum of the solutions corresponding to each characteristic value, or

$$T(x,\theta) - T_\infty = \sum_{n=1}^{\infty} e^{-a\lambda_n{}^2\theta} A_n \cos \lambda_n x$$

The constants of this infinite series are evaluated by substituting the initial condition, i.e., the initial temperature distribution, into Eq. 4-30.

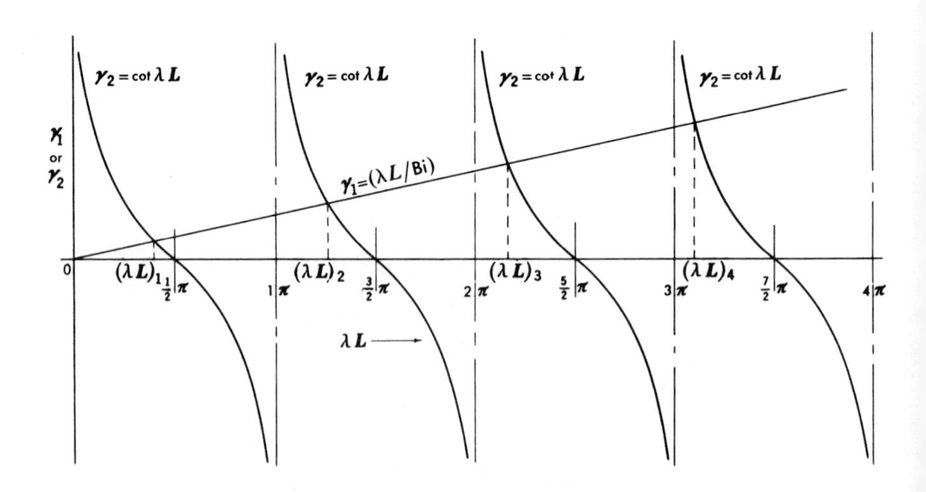

Fig. 4-7. Graphical solution of transcendental Equation 4-38.

For our problem, we have

$$T(x,0) - T_\infty = T_i - T_\infty = \sum_{n=1}^{\infty} A_n \cos \lambda_n x \tag{4-39}$$

Since the characteristic functions, $\cos \lambda_n x$, are orthogonal between $x = 0$ and $x = L$ (Eq. 4-32)

$$\begin{aligned} \int_0^L \cos \lambda_n x \cos \lambda_m x dx &= 0 \qquad \text{if } m \neq n \\ &\neq 0 \qquad \text{if } m = n \end{aligned}$$

where λ_m is any characteristic value of λ. To obtain a particular value of A_n, multiply both sides of Eq. 4-39 by $\cos \lambda_m x$ and integrate between 0 and L. In accordance with Eq. 4-32 all terms on the right-hand side disappear, except the one involving the square of the characteristic function, $\cos \lambda_n x$, and we obtain

$$\int_0^L (T_0 - T_\infty)(\cos \lambda_n x)\, dx = A_n \int_0^L (\cos^2 \lambda_n x)\, dx$$

From standard integral tables (12) we get

$$\begin{aligned} \int_0^L \cos^2 \lambda_n x dx &= \frac{1}{2} x + \frac{1}{2\lambda_n} \sin \lambda_n x \cos \lambda_n x \Big|_0^L \\ &= \frac{L}{2} + \frac{1}{2\lambda_n} \sin \lambda_n L \cos \lambda_n L \end{aligned}$$

and

$$\int_0^L \cos \lambda_n x dx = \frac{1}{\lambda_n} \sin \lambda_n L$$

whence the constant A_n is

$$\begin{aligned} A_n &= \frac{2\lambda_n}{L\lambda_n + \sin \lambda_n L \cos \lambda_n L} \, \frac{\sin \lambda_n L (T_i - T_\infty)}{\lambda_n} \\ &= \frac{2 \sin \lambda_n L (T_i - T_\infty)}{L\lambda_n + \sin \lambda_n L \cos \lambda_n L} \end{aligned} \tag{4-40}$$

As an illustration of the general procedure outlined above, determine A_1 when $\bar{h} = 1$, $k_s = 1$, and $L = 1$. From the graph of Fig. 4-7, the value of λ_1 is 0.86 radians or 49.2 degrees and

$$A_1 = \frac{2 \sin 49.2° (T_i - T_\infty)}{1 \times 0.86 + \sin 49.2° \cos 49.2°} = \frac{2 \times 0.757 (T_i - T_\infty)}{0.86 + 0.757 \times 0.653}$$
$$= 1.12(T_i - T_\infty)$$

Similarly, we obtain

$$A_2 = -0.152(T_i - T_\infty) \qquad \text{and} \qquad A_3 = 0.046(T_i - T_\infty)$$

Note that the series converges rapidly and, for Bi = 1, three terms represent a fairly good approximation for practical purposes.

To express the temperature in the slab in terms of conventional dimensionless moduli, we let $\lambda_n = \delta_n/L$. The final form of the solution, obtained by substituting Eq. 4-40 into Eq. 4-30, is then

$$\frac{T(x,\theta) - T_\infty}{T_i - T_\infty} = \sum_{n=1}^{\infty} e^{-\delta_n^2(\theta a/L^2)} 2 \frac{\sin \delta_n \cos (\delta_n x/L)}{\delta_n + \sin \delta_n \cos \delta_n} \tag{4-41}$$

Note that the time dependence is now contained in the dimensionless Fourier modulus, Fo $= \theta a/L^2$. Furthermore, if we write the second boundary condition in terms of δ_n, we obtain from Eq. 4-38

$$\cot \delta_n = \frac{k_s}{\bar{h}L} \delta_n$$

or

$$\delta_n \tan \delta_n = \frac{\bar{h}L}{k_s} = \text{Bi}$$

Observe that δ_n is a function only of the dimensionless Biot modulus, Bi $= \bar{h}L/k_s$. Hence the temperature can be expressed in terms of three dimensionless quantities: Fo $= \theta a/L^2$, Bi $= \bar{h}L/k_s$, and x/L.

The internal energy change of the slab can be obtained by substituting the derivative of Eq. 4-41 at the surface, i.e., $\partial T/dx\,|_{x=L}$, into Eq. 4-36. This yields, for the internal energy change in the time interval between $\theta = 0$ and $\theta = \theta$,

$$Q = 2(T_i - T_\infty)Lc\rho \sum_{n=1}^{\infty} (1 - e^{-\delta_n^2 \text{Fo}}) \frac{\sin^2 \delta_n}{\delta_n^2 + \delta_n \sin \delta_n \cos \delta_n} \tag{4-42}$$

To make Eq. 4-42 dimensionless, note that $c\rho L(T_i - T_\infty)$ represents the initial internal energy per square foot of the slab relative to T_∞. If we denote $c\rho L(T_i - T_\infty)$ by Q_i, we get

$$\frac{Q}{Q_i} = \sum_{n=1}^{\infty} \frac{2 \sin^2 \delta_n}{\delta_n^2 + \delta_n \sin \delta_n \cos \delta_n} (1 - e^{-\delta_n^2 \text{Fo}}) \tag{4-43}$$

The temperature distribution and the amount of heat transferred at any time may be determined from Eqs. 4-41 and 4-43, respectively. The final expressions are in the form of infinite series. These series have been evaluated, and the results are available in the form of charts. The use of the charts for the problem treated in this section as well as for other cases of practical interest will be taken up in the following section. A complete understanding of the methods by which the mathematical solutions have been obtained, although helpful, is not necessary for using the charts.

4-5. Charts for transient heat conduction

For transient heat conduction in several simple shapes, subject to boundary conditions of practical importance, the temperature distribution and the heat flow have been calculated and the results are available in the form of charts or tables (1,2,3,9,10,11,23). In this section we shall illustrate the application of some of these charts to typical problems of transient heat conduction in solids having Biot moduli larger than 0.1. The charts presented here have been taken from References (10) and (11), and for details of the mathematical solutions, the original references should be consulted.

Flat plate. The first series of charts (Figs. 4-8, 4-9, and 4-10) apply to a large flat plate of thickness $2L$. Initially the temperature of the plate is uniform at T_i. At some instant of time which will be designated as $\theta = 0$, the plate is immersed in a fluid at T_∞. If T_∞ is larger than T_i, heat begins to flow from the fluid to the plate. The rate of heat flow depends on the temperature difference $T_\infty - T_i$, the unit-surface conductance $\bar{h}$ between the plate and the fluid, the physical properties of the plate, and the plate thickness. The temperature distribution and the internal energy in the plate at any instant are functions of the same variables and the functional relationships are given by Eqs. 4-41 and 4-43 respectively, in terms of dimensionless parameters. The results are presented graphically in Figs. 4-8, 4-9, and 4-10. In Fig. 4-8 the dimensionless temperature ratio $[T(0,\theta) - T_\infty] / [T_i - T_\infty]$ at the mid-plane of a large slab is plotted as a function of the dimensionless time modulus $a\theta/L^2$ with the inverse of the Biot modulus, $k_s/\bar{h}L$, as a parameter. Fig. 4-9 gives for any given time θ the ratio of the temperature at an arbitrary position,

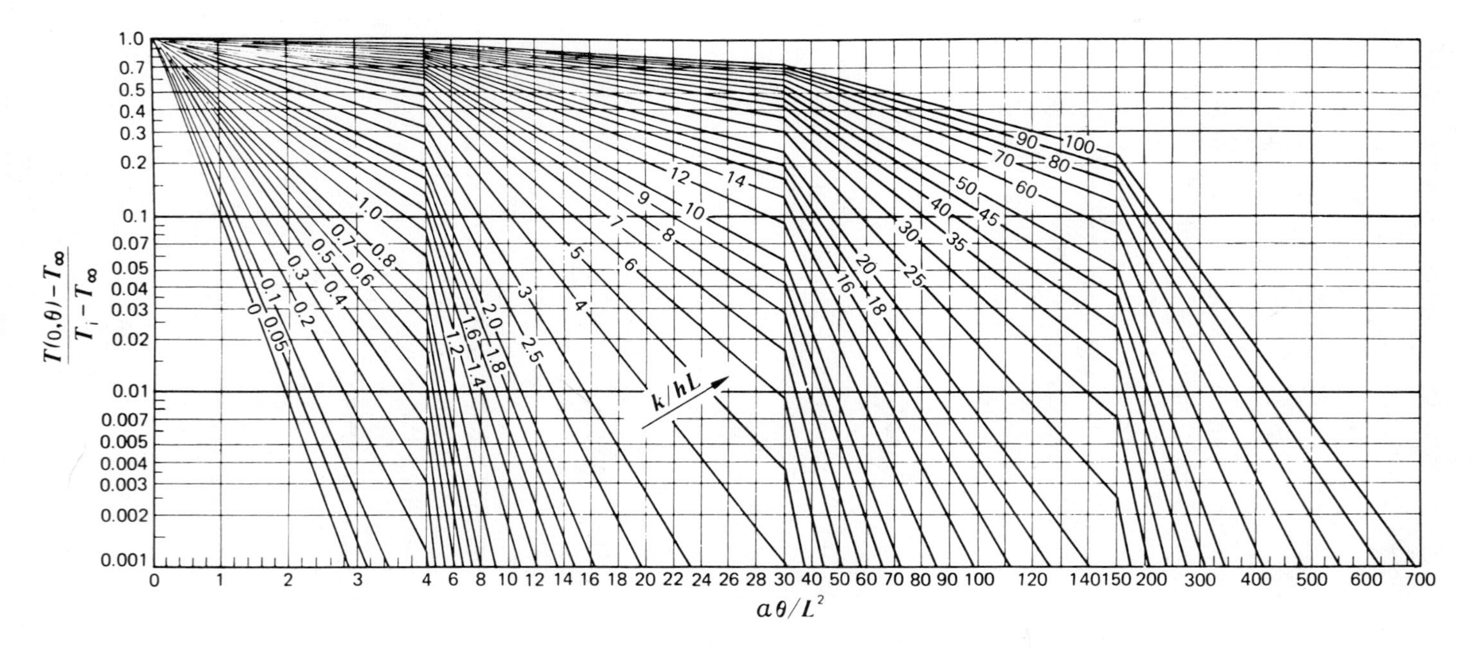

Fig. 4-8. Midplane temperature for an infinite plate of thickness $2L$, from Heisler (11).

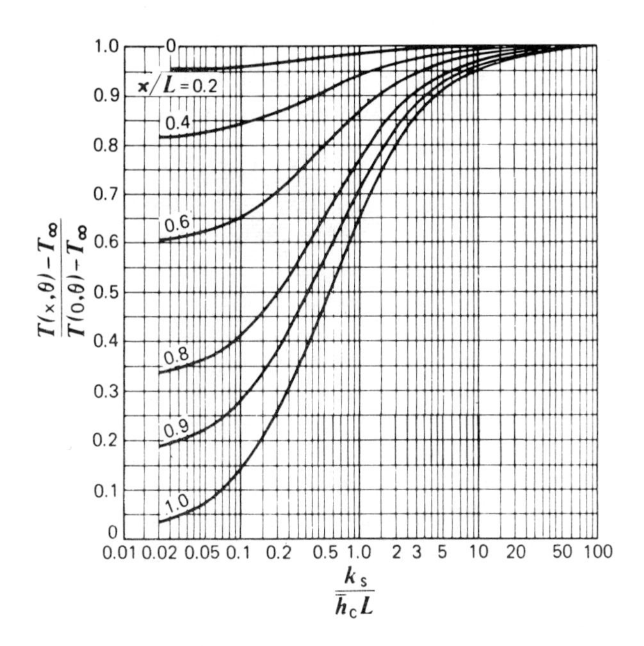

Fig. 4-9. Temperature as a function of center temperature in an infinite plate of thickness $2L$, from Heisler (11).

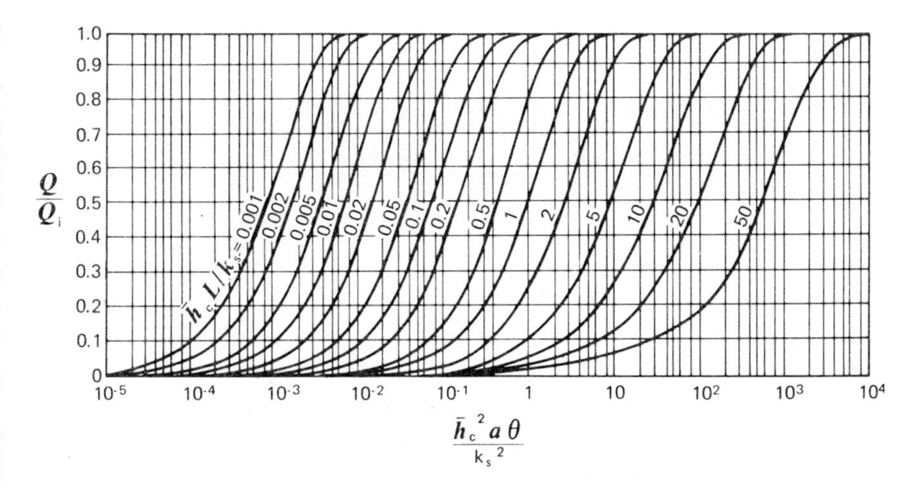

Fig. 4-10. Dimensionless heat loss Q/Q_i of an infinite plate of thickness $2L$ with time, from Gröber (10).

x/L, to the mid-plane temperature, $[T(x,\theta) - T_\infty]/[T(0,\theta) - T_\infty]$, as a function of $k_s/\bar{h}L$ with the dimensionless distance x/L as the parameter.

An inspection of Fig. 4-9 shows that for values of $k_s/\bar{h}L$ larger than 10, i.e. Biot numbers less than 0.1, the center temperature of the slab is nearly equal to the surface temperature. For such cases the assumption that a uniform temperature prevails throughout the body does not introduce a serious error, and the simplified analysis presented in Sec. 4-2 may be used. The justification for this assumption is now substantiated for one system by the results of an exact analysis.

Figure 4-10 is a plot of Q/Q_i vs. $\bar{h}_c^2 a\theta/k_s^2(\text{Fo}\cdot\text{Bi}^2)$ for various values of $(\bar{h}_c L/k_s)$. Here Q represents the total change in internal energy per unit area, i.e., the amount of heat transferred per unit area in the time interval between $\theta = 0$ and $\theta = \theta$ in Btu per square foot; Q_i represents the initial internal energy per unit area relative to the fluid temperature T_∞, i.e., $c\rho L(T_i - T_\infty)$. A positive value of Q indicates, therefore, that heat is transferred from the wall to the fluid, while a negative value of Q shows that the direction of heat flow is into the slab.

EXAMPLE 4-4. A concrete wall, 1 ft thick and originally at 100 F, is suddenly exposed on one side to a hot gas at 1600 F. If the heat-transfer coefficient on the hot side is 5 Btu/hr sq ft F and the other side is insulated, determine (a) the time required to raise the temperature at the insulated face of the slab to 1100 F, (b) the temperature distribution in the wall at that instant, and (c) the heat transferred to the wall per square foot of surface area.

Solution: (a) From Table A-2 of properties we get

$$k_s = 0.54 \text{ Btu/hr ft F}$$

$$c = 0.20 \text{ Btu/lb F}$$

$$\rho = 144 \text{ lb/cu ft}$$

$$a = 0.0187 \text{ sq ft/hr}$$

We note that the insulated face corresponds to the center plane in a slab of thickness $2L$, since $\partial T/\partial x = 0$ for both at $x = 0$. The temperature ratio at the insulated face is

$$\left.\frac{T - T_\infty}{T_i - T_\infty}\right|_{\text{at } x=0} = \frac{1100 \text{ F} - 1600 \text{ F}}{100 \text{ F} - 1600 \text{ F}} = 0.333$$

and the reciprocal of the Biot modulus is

$$\frac{k_s}{\bar{h}L} = \frac{0.54}{(5)(1)} = 0.108$$

From the chart of Fig. 4-8, $a\theta/L^2|_{\text{at } x=0} = 0.70$ under these conditions, and therefore $\theta = 0.70 \times 1/0.0187 = 37.5$ hr. *Ans.*

b) The temperature distribution in the slab at this instant is determined from Fig. 4-9 for the various depth ratios as shown below.

x/L	0.2	0.4	0.6	0.8	1.0
$\dfrac{T(x) - T_\infty}{T_0 - T_\infty}$	0.96	0.84	0.65	0.41	0.14
$T_\infty - T(x)$	480	420	325	205	70

The temperatures at various distances from the insulated face are tabulated below.

x (ft)	0	0.2	0.4	0.6	0.8	1.0
Temp (F)	1100	1120	1180	1275	1395	1530

Ans.

c) The heat transferred to the wall during the process can be obtained from Fig. 4-10. For $\bar{h}L/k_s$ equal to 9.25, Q/Q_i at $h_c^2 a\theta/k_s^2 = 60$ is about 0.70. Thus, we find that

$$Q = c\rho L(T_i - T_\infty)(0.7) = (0.2)(144)(1)(100 - 1600)(0.7) = -30{,}400 \text{ Btu}$$

Ans.

The minus sign indicates that the heat flow is into the wall and that the internal energy of the wall increased during the process.

Long cylinder and sphere. In addition to the plane wall, solutions are also available in chart form for the infinitely long cylinder and the sphere. The mathematical solutions may be obtained by the same method of approach that was used in Sec. 4-4 for the slab, namely by assuming a product solution and separating the variables. The basic differential equations to be solved are

$$\frac{\partial T}{\partial \theta} = a\left(\frac{\partial^2 T}{\partial r^2} + \frac{1}{r}\frac{\partial T}{\partial r}\right) \quad \text{for a long cylinder} \tag{4-44}$$

$$\frac{\partial T}{\partial \theta} = a\left(\frac{\partial^2 T}{\partial r^2} + \frac{2}{r}\frac{\partial T}{\partial r}\right) \quad \text{for a sphere} \tag{4-45}$$

The initial and the boundary conditions for which the solutions to Eqs. 4-44 and 4-45 have been evaluated are described as follows:

1. The initial temperature distribution in the cylinder or the sphere is uniform and equal to T_i, i.e., at $\theta = 0$, $T = T_i$.
2. At time $\theta = 0$, the cylinder or the sphere is exposed to a fluid whose

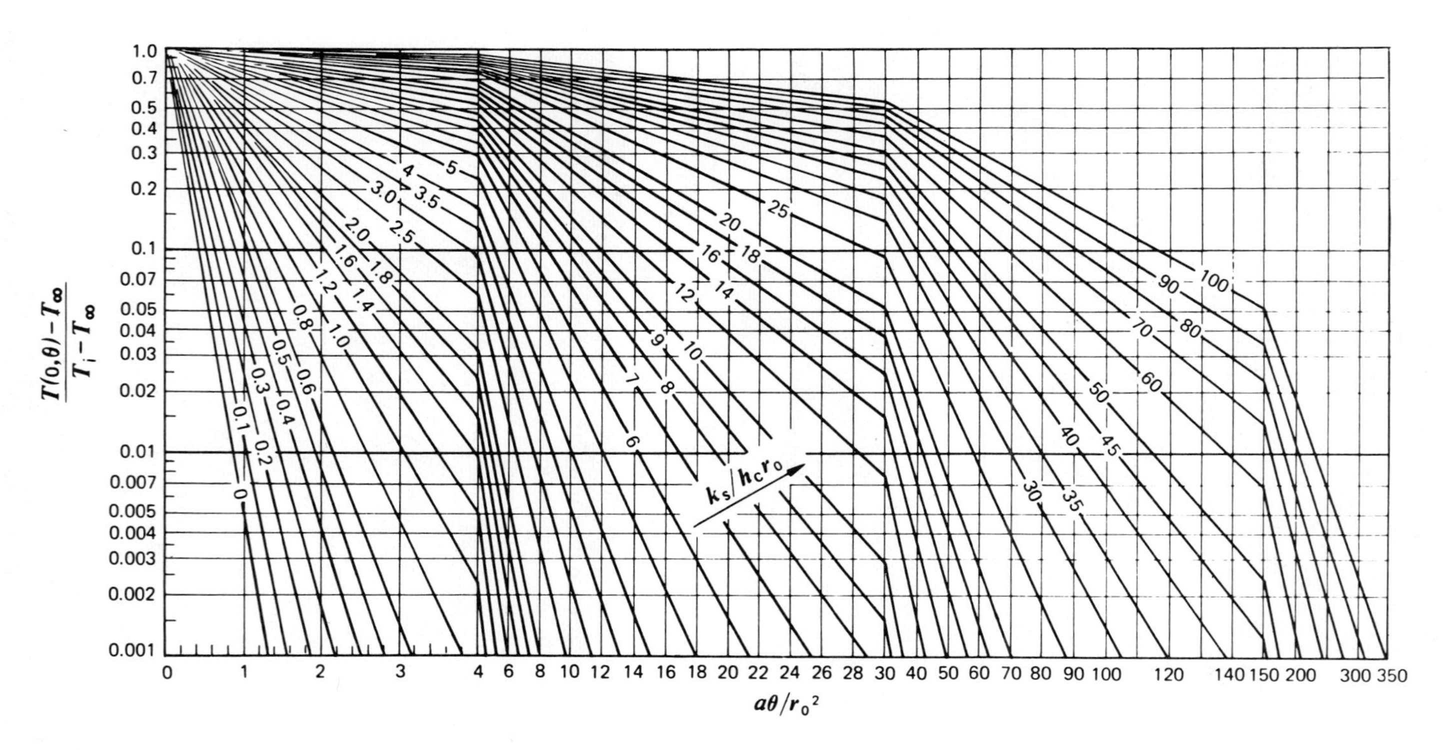

Fig. 4-11. Axis temperature for an infinite cylinder of radius r_0, from Heisler (11).

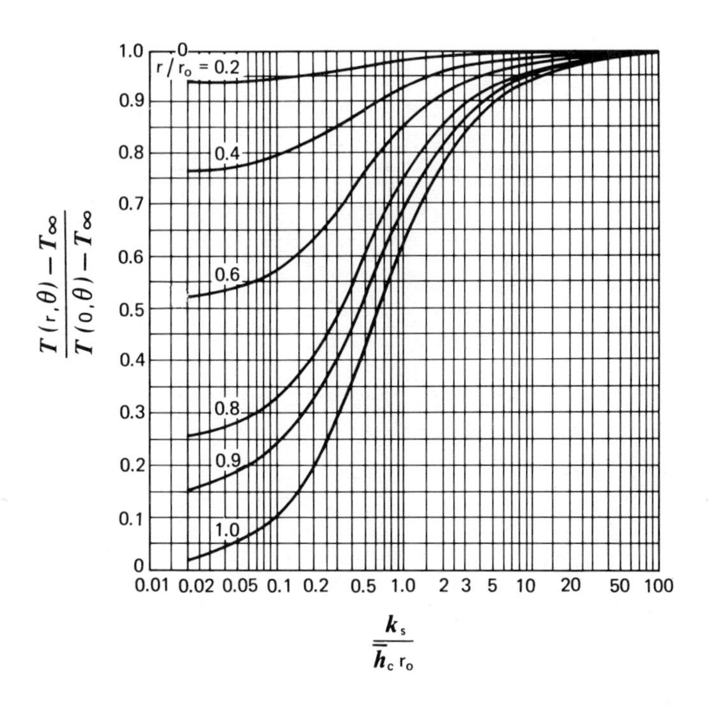

Fig. 4-12. Temperature as a function of axis temperature in an infinite cylinder of radius r_0, from Heisler (11).

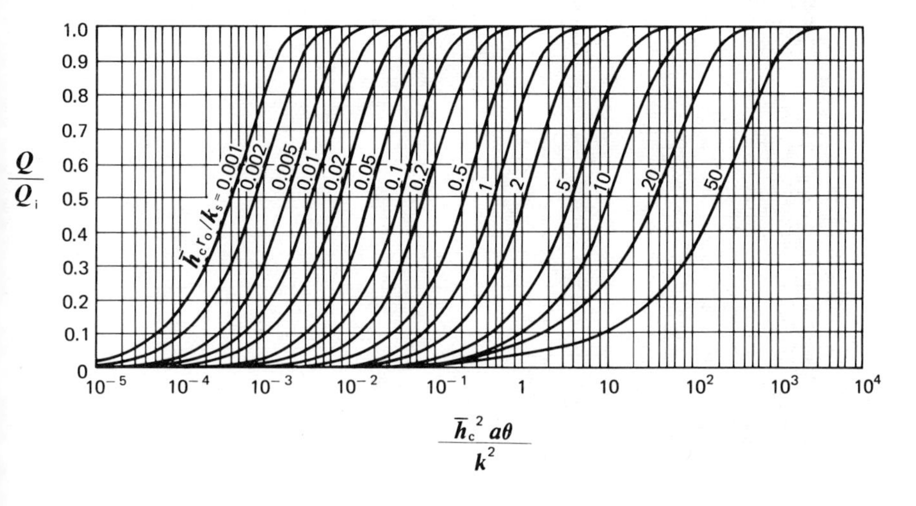

Fig. 4-13. Dimensionless heat loss Q/Q_i of an infinite cylinder of radius r_0 with time, from Gröber (10).

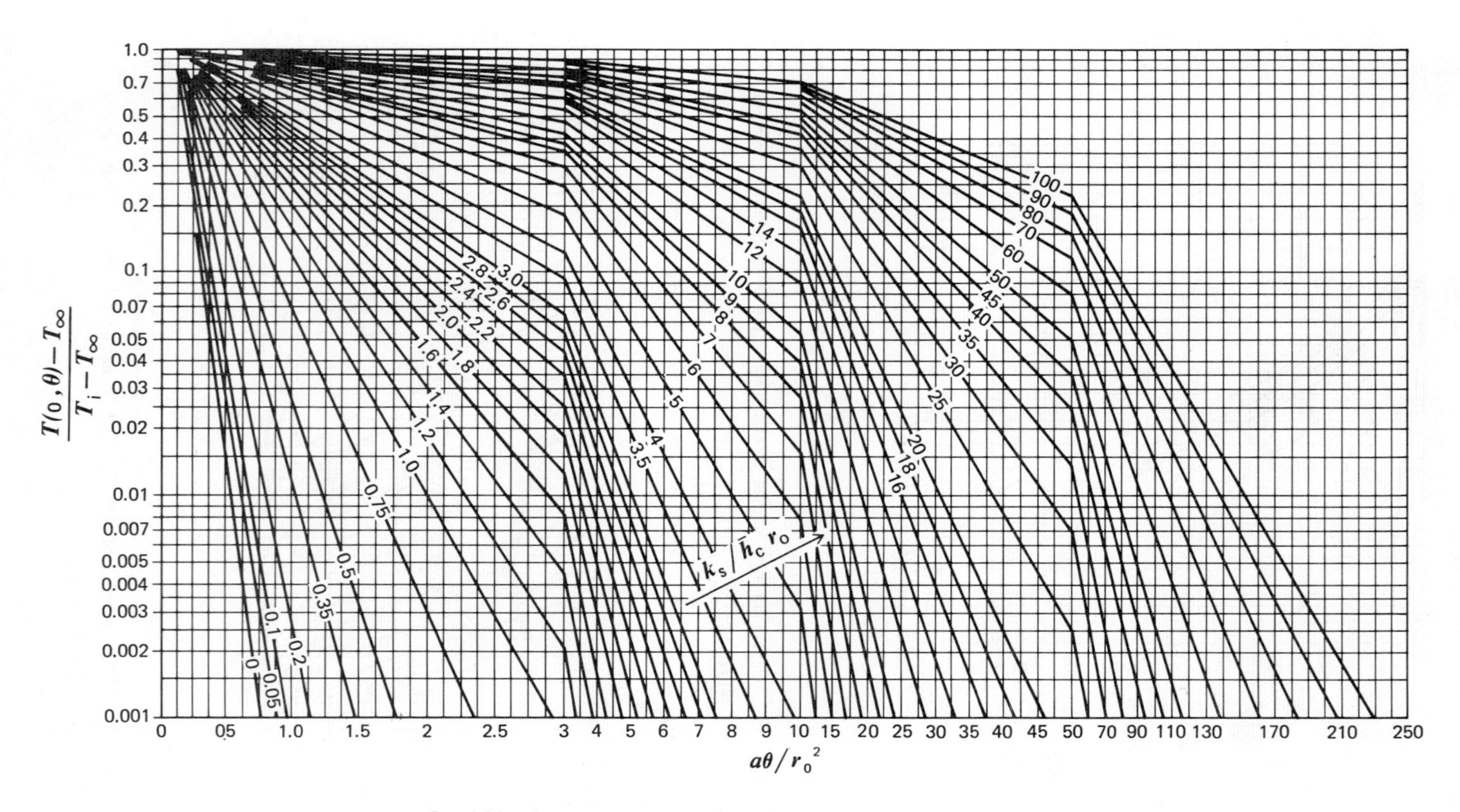

Fig. 4-14. Center temperature for a sphere of radius r_0, from Heisler (11).

temperature is T_∞. This temperature is used as the datum above or below which changes in temperature are measured.

3. The unit-surface conductance, $\bar{h}$, between the surface of the body and the fluid is uniform and does not change with time.

The charts of Figs. 4-11 and 4-14 show the dimensionless center temperature $[T(0,\theta) - T_\infty] / (T_i - T_\infty)$ is a function of the dimensionless time modulus $a\theta/r_0^2$ with $(k_s/h_c r_o)$ as a parameter. Figs. 4-12 and 4-15 give the ratio of the temperature difference at radial distance r, $[T(r,\theta) - T_\infty]$ to the centerline temperature difference $[T(0,\theta) - T_\infty]$ as a function of $k_s/\bar{h}_c r_o$, where r_o is the outside radius, with the ratio r/r_o as a parameter. In Figs. 4-13 and 4-16 the dimensionless heat flow, Q/Q_i, is shown as a function of $\bar{h}_c^2 a\theta/k_s^2$ with the Biot number $\bar{h}_c r_o/k_s$ as a parameter. The initial energy stored is measured with respect to T_∞ and is defined as $Q_i = c\rho\pi r_o^2(T_i - T_\infty)$ per unit length of the cylinder and as $Q_i = c\rho\frac{4}{3}\pi r_o^3(T_i - T_\infty)$ for the sphere.

A change in the environmental temperature may not affect the interior of a body for some time after the temperature change originally occurred. There are numerous practical problems where one is only interested in the temperature distribution and the heat flow during the initial stages of a process or where the body is so large that the tempera-

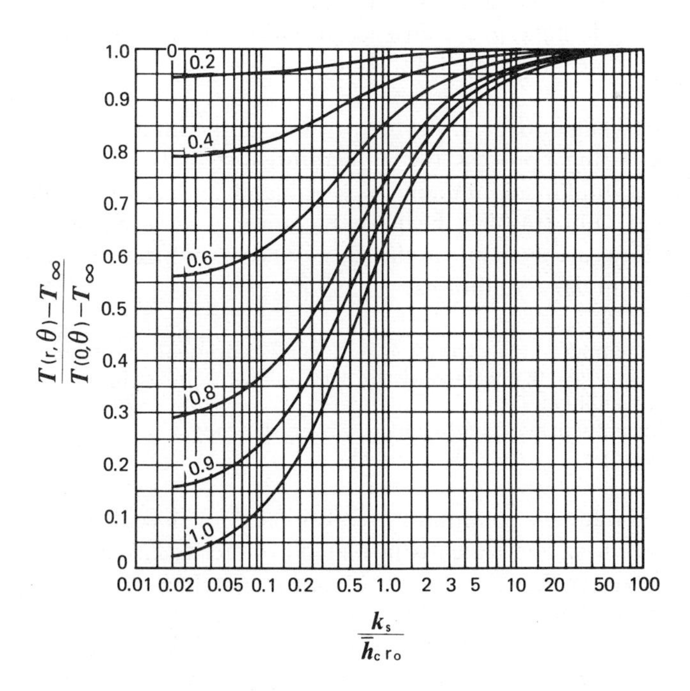

Fig. 4-15. Temperature as a function of center temperature for a sphere of radius r_0, from Heisler (11).

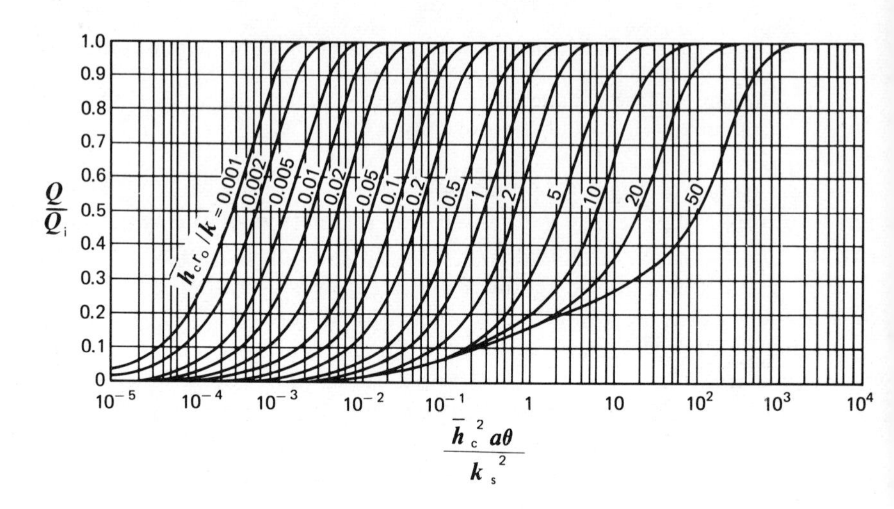

Fig. 4-16. Dimensionless heat loss Q/Q_i of a sphere of radius r_0 with time, from Gröber (10).

ture at the interior is not affected by the temperature changes at the surface. In such cases it is often found that the charts of Figs. 4-8 to 4-16 cannot be read to a sufficient degree of accuracy. However, Boelter (9) and Heisler (11) have prepared special charts for short-time heating or cooling to supplement Figs. 4-8 to 4-16.

Semi-infinite body. If the temperature in the interior of a slab does not change during a process, the temperature distribution near the surface is the same as that in an infinitely thick slab and we have a so-called semi-infinite solid. For transient heat conduction in a semi-infinite solid (Fig. 4-17), solutions are available in the form of charts subject to the following initial and boundary conditions.

1. The temperature distribution in the body is originally uniform at T_o.
2. At time $\theta = 0$, the face of the semi-infinite solid is brought in contact with a fluid at T_∞.
3. The unit-surface conductance $\bar{h}$ over the face $x = 0$ is constant and uniform.

These boundary conditions are also valid for a wall of finite thickness, or for a long rod which is insulated around its circumference when $L/2\sqrt{\theta a}$ is larger than 0.5; they are approximately correct for cylinders and spheres as long as the depth to which the heat conduction has penetrated is small compared with the radius of curvature.

We shall first consider the special case of one-dimensional transient heat conduction in a semi-infinite solid with no thermal resistance at the surface. This assumption simplifies the problem because, at $\theta = 0$, the

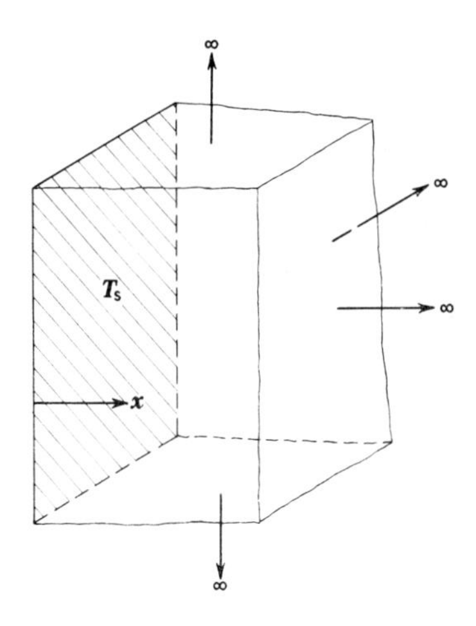

Fig. 4-17. Sketch illustrating nomenclature for semi-infinite slab.

temperature change occurs directly at the surface, i.e., $T_{x=0} = T_\infty$ for $\theta \geq 0$.

For this case[5] the solution to Eq. 4-23 becomes

$$\frac{T - T_\infty}{T_o - T_\infty} = G\left(\frac{x}{2\sqrt{a\theta}}\right) \tag{4-46}$$

where $G(x/2\sqrt{a\theta})$ is the Gaussian error integral defined as

$$G\left(\frac{x}{2\sqrt{a\theta}}\right) = \frac{2}{\sqrt{\pi}} \int_0^{x/2\sqrt{a\theta}} e^{-\beta^2}\, d\beta \tag{4-47}$$

In Fig. 4-18, $G(x/2\sqrt{a\theta})$ is plotted against $x/2\sqrt{a\theta}$, and in Table A-8 the function is tabulated. The curve may be used for convenience in computation. The variable $x/2\sqrt{a\theta}$ is a dimensionless quantity. If a is in sq ft/hr, θ must be expressed in hours and x in feet.

The instantaneous rate of heat flow at the surface can readily be obtained from Eq. 4-46 by evaluating the temperature gradient at the surface, or

$$q = -k_s A \left.\frac{\partial T}{\partial x}\right|_{x=0} = -k_s A \left.\frac{T_o - T_\infty}{\sqrt{\pi a \theta}}\, e^{-x^2/4a\theta}\right|_{x=0}$$

$$= k_s A \frac{T_\infty - T_o}{\sqrt{\pi a \theta}}$$

---

[5]For details of the solution, see References (1), (2), or (3).

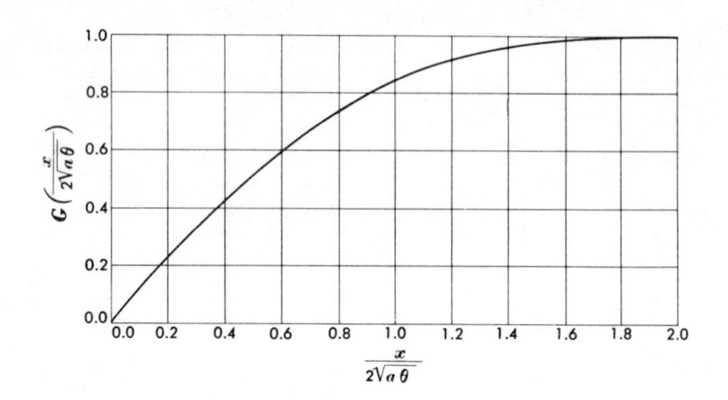

Fig. 4-18. Gaussian error integral, Eq. 4-47 (tabulated in Table A-8).

The total change in internal energy of the slab during the process from $\theta = 0$ to $\theta = \theta$ is

$$Q = \int_0^\theta q d\theta = \int_0^\theta k_s A \frac{T_\infty - T_o}{\sqrt{\pi a}} \theta^{-1/2} d\theta = 2k_s A (T_\infty - T_o) \sqrt{\frac{\theta}{\pi a}} \tag{4-48}$$

EXAMPLE 4-5. In the installation of underground water pipes, it is important to determine the depth to which a change in surface temperature is felt during a 12-hr period. If the original soil temperature is 40 F and the surface temperature suddenly drops to 25 F, determine the depth to which the freezing temperature penetrates. Assume that the soil is dry and $a = 0.012$ sq ft/hr.

Solution: For the dimensionless temperature ratio we have

$$\frac{T - T_\infty}{T_o - T_\infty} = \frac{32\text{ F} - 25\text{ F}}{40\text{ F} - 25\text{ F}} = 0.467$$

From Fig. 4-15 we find that

$$G\left(\frac{x}{2\sqrt{a\theta}}\right) = 0.467$$

when

$$\frac{x}{2\sqrt{a\theta}} = 0.44$$

Solving for x, we get

$$x = 2\sqrt{(0.012)(12)}\,0.44 = 0.334\text{ ft}$$

Thus, freezing will not occur if the pipe is more than 4 in. below the surface.

Ans.

For a finite value of the surface conductance at the face of a semi-infinite slab, the boundary condition at the surface is $hA\,[T - T(0)] = -k_s A\,(\partial T/\partial x)_x = 0$. The solution for this problem is rather involved. The result, as shown in detail by Schneider (1), is

$$\frac{T(x,\theta) - T_i}{T_\infty - T_i} = 1 - \operatorname{erf}\sqrt{x^2/4a\theta} - \left[\exp\left(\frac{\bar{h}x}{k_s} + \frac{\bar{h}^2 a\theta}{k_s^2}\right)\right] \cdot \left[1 - \operatorname{erf}\left(\sqrt{x^2/4a\theta} + \sqrt{\frac{\bar{h}^2 a\theta}{k_s^2}}\right)\right] \quad (4\text{-}49)$$

This solution is presented in graphical form in Fig. 4-19 where the temperature ratio $(T - T_\infty)/(T_o - T_\infty)$ is plotted against a *local Biot modulus* $\bar{h}x/k_s$, where *x is the distance from the face.* The constant parameter for each of the curves is Θ, the dimensionless time parameter $(\bar{h}/k_s)^2 a\theta$, often called the *Boundary Fourier modulus*. The use of this chart is illustrated by the following example.

EXAMPLE 4-6. A cylindrical combustion chamber (10 in. ID) has a refractory lining of 1-in. thickness on the inside to protect the exterior shell. In order to determine the thermal stress, it is necessary to obtain the temperature distribution in the lining 1 min after initiation of combustion. The following data are given:

$$\begin{aligned} T_\infty &= 3000\text{ F} \\ h &= 40\text{ Btu/hr sq ft F} \\ a &= 0.020\text{ sq ft/hr} \\ k &= 0.6\text{ Btu/hr ft F} \\ T_o &= 100\text{ F} \end{aligned}$$

Solution: The time period of interest is short and the radius of curvature of the refractory wall is large compared to the wall thickness. We therefore treat the system as a semi-infinite slab. For $\theta = 1/60$, the Boundary Fourier modulus is

$$\frac{\bar{h}^2 a\theta}{k_s^2} = \frac{(40)^2(0.02)}{(0.6)^2(60)} = 1.48$$

For this value of the time parameter, the temperature at various values of x can be found from Fig. 4-19. The results are tabulated below.

x (in.)	0.0	0.2	0.4	0.6	0.8	1.0
$\bar{h}x/k_s$	0.0	1.11	2.22	3.33	4.44	5.55
$\dfrac{T - T_\infty}{T_o - T_\infty}$	0.38	0.7	0.9	0.97	1.0	1.0
$T - T_\infty$	1100	2040	2610	2810	2900	2900
T (F)	1900	960	390	190	100	100

The temperature distribution permits an analysis of the thermal stress due to the differential expansion of the lining.

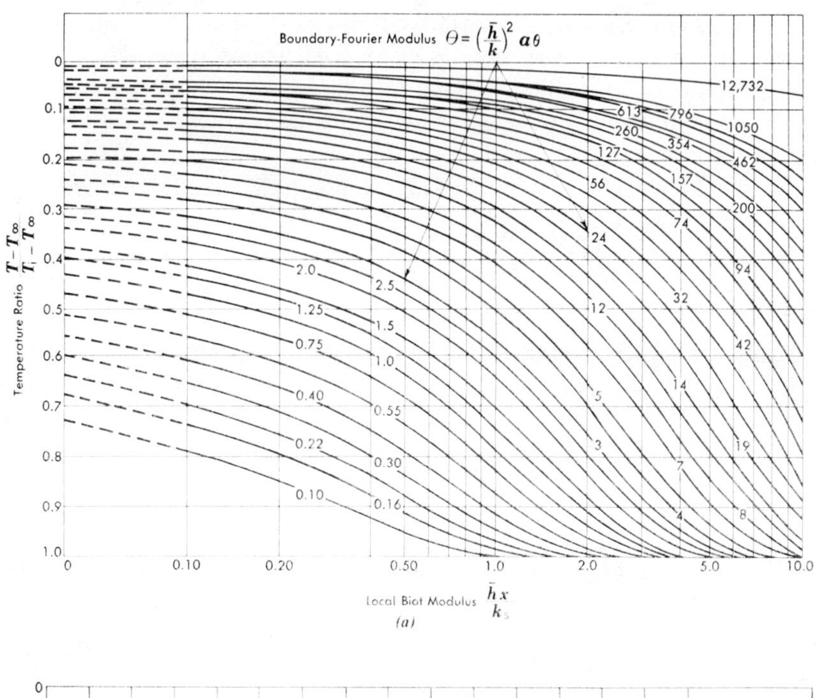

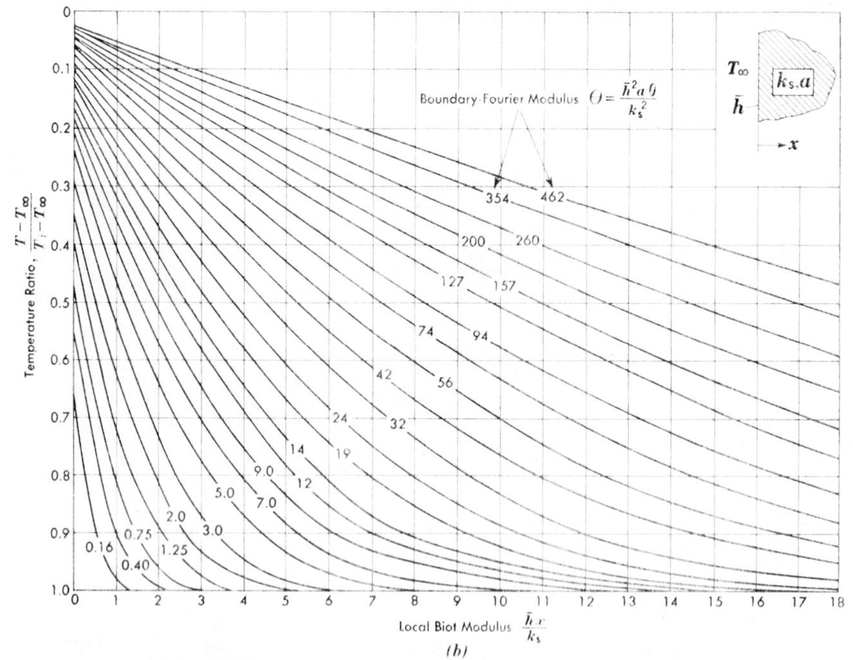

Fig. 4-19. Dimensionless temperature distribution in a semi-infinite slab subjected to a sudden change in environmental temperature. (By permission from L. M. K. Boelter, V. H. Cherry, and H. A. Johnson, *Heat Transfer*, 3d ed., 1942.)

Two- and three-dimensional bodies. The problems considered in this section have so far been limited to one-dimensional heat flow. Since many practical problems involve heat flow in two- and three-dimensional systems, we shall now show how solutions for one-dimensional problems can be combined to yield solutions of certain two- and three-dimensional transient systems. As an example, consider the heating of a long rectangular bar (Fig. 4-20) initially at a uniform temperature T_o. At time $\theta = 0$, the rod is placed into an environment at a temperature T_∞. The unit-surface conductance over both of the longer sides is $\bar{h}_a$, over both of the shorter sides, $\bar{h}_b$.

If we compare this problem with the heating of a large plate of width $2b$, it is physically obvious that the heat flow from the shorter sides accelerates the heating. It can be shown (see, for example, References 1 and 2) that the effect of the shorter sides on the solution for an infinitely long plate of width $2b$ can also be expressed in the form of a product solution. In other words, to obtain the temperature-time history of the rectangular bar we need only multiply the dimensionless temperature ratios for two infinite plates, one $2b$, the other $2a$ wide. Symbolically we write

$$\left(\frac{T - T_\infty}{T_o - T_\infty}\right)_{\text{bar}} = \left(\frac{T - T_\infty}{T_o - T_\infty}\right)_{2a \text{ plate}} \left(\frac{T - T_\infty}{T_o - T_\infty}\right)_{2b \text{ plate}} \tag{4-50}$$

where the temperature ratios at the respective locations for any point in the system may be taken from Figs. 4-8 and 4-9 at *corresponding time parameters.* The Biot moduli are of course different for the two infinite plates from which the solution to the rectangular bar is formed.

EXAMPLE 4-7. In the design of fire-fighting equipment it is necessary to know how long wooden beams can be exposed to fire before they ignite. The beams are long, 2 by 4 in. in cross section, and initially at a uniform temperature of 60 F. The physical properties of the wood are as follows:

$$\rho = 50 \text{ lb/ct ft}$$
$$c = 0.6 \text{ Btu/lb F}$$
$$k = 0.2 \text{ Btu/hr ft F}$$

At the instant the fire breaks out, the beams are exposed to gases at 1200 F and the unit-surface conductance is 3.0 Btu/hr sq ft F over all of the faces. Estimate the time elapsed before the wood reaches the ignition temperature of 800 F.

Solution: The dimensionless temperature ratio when $T = 800$ F is

$$\frac{T - T_\infty}{T_o - T_\infty} = \frac{800 \text{ F} - 1200 \text{ F}}{60 \text{ F} - 1200 \text{ F}} = 0.35$$

According to Eq. 4-50, the temperature ratio for this rectangular beam equals

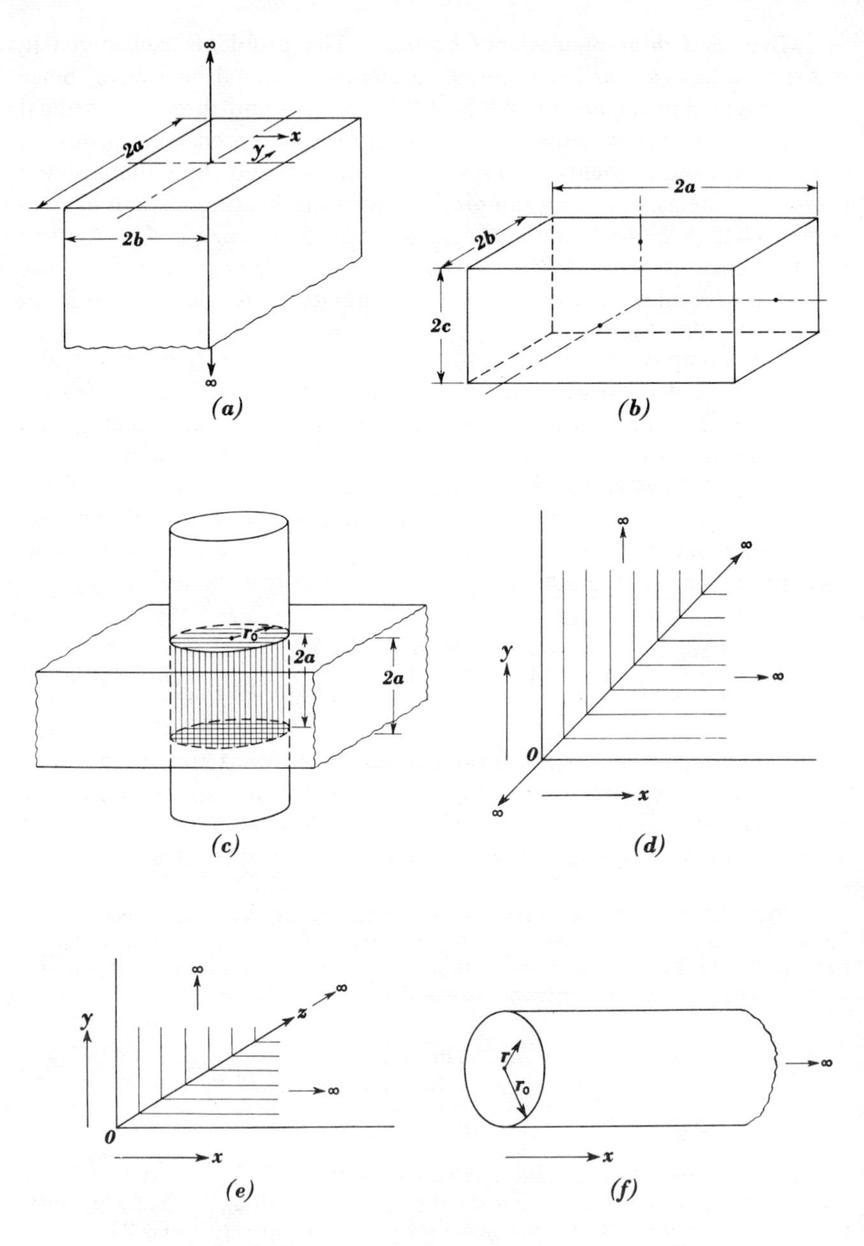

(a) Co-ordinate system for infinitely long rectangular bar.
(b) Co-ordinate system for brick-shaped body.
(c) Co-ordinate system for finite cylinder.
(d) Co-ordinate system for quarter-infinite body.
(e) Co-ordinate system for eighth-infinite body.
(f) Co-ordinate system for semi-infinite cylinder.

Fig. 4-20. Sketches illustrating two- and three-dimensional systems amenable to product solutions.

the product of the temperature ratios for two large plates, one of them 4 in., the other 2 in. thick, or

$$\frac{T - T_\infty}{T_o - T_\infty} = 0.35 = \left(\frac{T - T_\infty}{T_o - T_\infty}\right)_{\text{4-in. plate}} \left(\frac{T - T_\infty}{T_o - T_\infty}\right)_{\text{2-in. plate}}$$

To obtain the time required for the temperature to reach 800 F we use the charts of Figs. 4-8 and 4-9. The surface will reach 800 F first and therefore the graphs for $x/L = 1$ apply. The solution, however, cannot be obtained directly, but requires some trial and error. We assume various values of time. Use Figs. 4-8 and 4-9 to determine the temperature ratios for each of the plates, and multiply these ratios. The value of θ at which the product equals 0.35 in the desired answer.

Assuming $\theta = 0.5$ hr, $a\theta/L^2$ of the wider plate is

$$\frac{a\theta}{L^2} = \frac{k\theta}{c\rho L^2} = \frac{(0.2 \text{ Btu/hr ft F})(0.5 \text{ hr})(12^2 \text{ sq in./sq ft})}{(0.6 \text{ Btu/lb F})(50 \text{ lb/cu ft})(2^2 \text{ sq in.})} = 0.12$$

and since $k/\bar{h}L$ is equal to 0.4, we have

$$\left(\frac{T - T_\infty}{T_o - T_\infty}\right)_{\text{4-in plate}} = 0.46$$

from the graph for $x/L = 1.0$. Similarly, for the other plate ($L = 1$ in.) we find at $\theta = 0.5$

$$\left(\frac{T - T_\infty}{T_o - T_\infty}\right)_{\text{2-in plate}} = 0.48$$

Then the temperature ratio for the wooden beam at $\theta = 0.5$ is

$$\frac{T - T_\infty}{T_o - T_\infty} = (0.46)(0.48) = 0.22$$

Since this is less than 0.35, the assumed value of θ is too large. Repeating the calculations for $\theta = 0.23$, we find that the surface temperature reaches 800 F during this time. *Ans.*

It is suggested that the reader fill in the remaining steps and verify the answer.

The extension of the product solution to a brick-shaped solid (Fig. 4-20b) leads to

$$\left(\frac{T - T_\infty}{T_o - T_\infty}\right)_{\text{brick}} = \left(\frac{T - T_\infty}{T_o - T_\infty}\right)_{2a \text{ plate}} \left(\frac{T - T_\infty}{T_o - T_\infty}\right)_{2b \text{ plate}} \left(\frac{T - T_\infty}{T_o - T_\infty}\right)_{2c \text{ plate}} \tag{4-51}$$

For a cylinder of finite length (Fig. 4-20c) the dimensionless temperature ratio is obtained by forming the product of the temperature ratios for an infinite cylinder and an infinite plate of a width equal to the length of the cylinder, or

$$\left(\frac{T - T_\infty}{T_o - T_\infty}\right)_{\text{cyl } 2a \text{ long}} = \left(\frac{T - T_\infty}{T_o - T_\infty}\right)_{\text{infinite cyl}} \left(\frac{T - T_\infty}{T_o - T_\infty}\right)_{2a \text{ plate}} \tag{4-52}$$

When using the above product solutions, it should be noted that, to satisfy the boundary conditions of the one-dimensional problems to which the charts apply, the coordinates of a two- or three-dimensional system must lie along the axes of symmetry and intersect at the center of the body.

In a similar manner the solution for the semi-infinite body can be used to build up solutions for a quarter-infinite body (Fig. 4-20d), an eighth-infinite body (Fig. 4-20e) and a semi-infinite cylinder (Fig. 4-20f). Using the notation of Fig. 4-20 the respective solutions are

$$\left(\frac{T - T_\infty}{T_o - T_\infty}\right)_{x,y} = \left(\frac{T - T_\infty}{T_o - T_\infty}\right)_x \left(\frac{T - T_\infty}{T_o - T_\infty}\right)_y \tag{4-53}$$

for a quarter-infinite body (an edge bounded by two planes)

$$\left(\frac{T - T_\infty}{T_o - T_\infty}\right)_{x,y,z} = \left(\frac{T - T_\infty}{T_o - T_\infty}\right)_x \left(\frac{T - T_\infty}{T_o - T_\infty}\right)_y \left(\frac{T - T_\infty}{T_o - T_\infty}\right)_z \tag{4-54}$$

for an eighth-infinite body (a corner bounded by three planes), and

$$\left(\frac{T - T_\infty}{T_o - T_\infty}\right)_{r,x} = \left(\frac{T - T_\infty}{T_o - T_\infty}\right)_r \left(\frac{T - T_\infty}{T_o - T_\infty}\right)_x \tag{4-55}$$

for a semi-infinite cylinder of outer radius r_o.

4-6. Numerical method

The charts described in the preceding section are useful in the thermal analysis and design of regular-shaped systems with simple boundary conditions. In many practical situations, however, one encounters systems of more complicated geometrical configurations and with boundary conditions which vary with time. In such cases the thermal analysis is handled by means of a numerical technique, using either hand calculations or a digital computer. The basic steps in preparing a problem for a solution by the numerical methods are the same, regardless of whether the actual solution is carried out by hand or with a computer.

To avoid unnecessary complications the numerical technique will first be applied to the analysis of a one-dimensional system which can be described by the equation

$$k \frac{\partial^2 T}{\partial x^2} = c\rho \frac{\partial T}{\partial \theta} \tag{4-23}$$

To solve Eq. 4-23 numerically we must first express it in the form of a finite difference equation. There are two types of finite difference approximations to this partial differential equation: *forward difference* and *backward difference*. As will be shown later, each of them has certain advantages and disadvantages, but the former is conceptually easier and will be presented first.

To transform Eq. 4-23 into a finite differential equation begin by dividing the system into layers of thickness Δx and label the planes between layers as shown in Fig. 4-21. Then set up a time scale in terms of finite time increments $\Delta\theta$, denoting the number of $\Delta\theta$s which have elapsed by t. At a certain time $\theta = t\Delta\theta$ the actual temperature distribution on

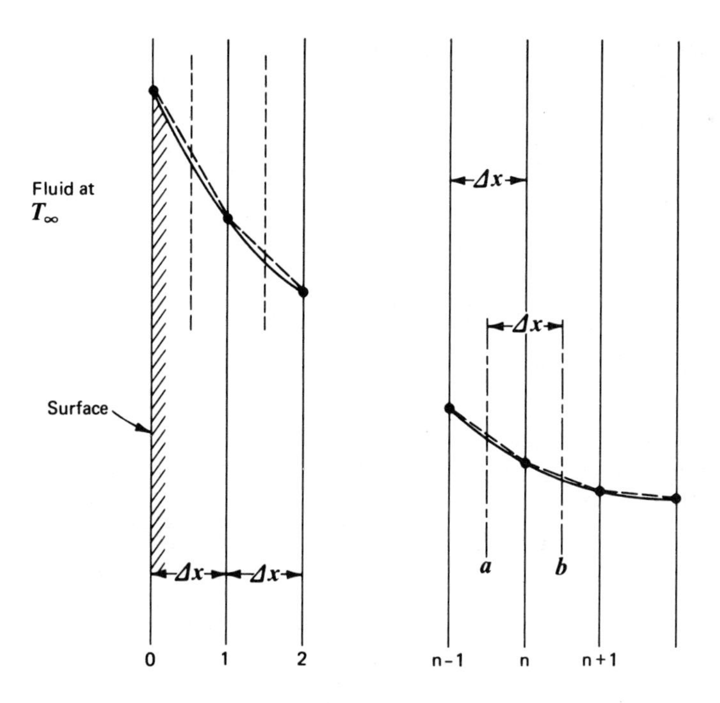

Fig. 4-21. Sketch illustrating finite difference approximation in a one-dimensional system.

both sides of a typical section n in this slab is shown by the solid line in Fig. 4-21. Approximate the temperature distribution by straight line segments between sectional planes and set up a heat balance. It is obvious that the representation of the temperature curve can be made more exact and the accuracy of the temperature gradients for the heat balance can be improved by using smaller subdivisions.

Referring to Fig. 4-22, the average temperature gradient in the section between $(n - 1)$ and n is $(T^t_{n-1} - T^t_n)/\Delta x$, and in the adjacent section between n and $(n + 1)$ the gradient is $(T^t_n - T^t_{n-1})\Delta x$. During a time interval $\Delta\theta$, heat is therefore conducted from plane $(n - 1)$ to cross-sectional plane n and from plane n to plane $(n + 1)$. The difference between the heat flow to and the heat flow from cross-sectional plane n changes the internal energy in a layer *ab* extending $\Delta x/2$ to the left and to the right of plane n. Writing an energy balance for the time interval $\Delta\theta$ can be expressed semantically as

Net heat flow from all neighboring nodes towards node n during $\Delta\theta$	=	Increase in internal energy of material associated with node n

To express this heat balance in symbolic form, denote the number of time increments elapsed by a superscript and the location by a subscript (e.g., T^t_n is the temperature at plane n at a time $t \cdot \Delta\theta$ after the transient has begun. The algebraic expression for the heat balance then becomes

$$\left[\frac{T^t_{n-1} - T^t_n}{R_{n-1,n}} + \frac{T^t_{n+1} - T^t_n}{R_{n,n+1}}\right] \Delta\theta = C_n(T^{t+1}_n - T^t_n) \tag{4-56}$$

where $R_{n-1,n}$ is the thermal resistance between the node $n - 1$ and the node n.

$R_{n,n+1}$ is the thermal resistance between the node n and the node $n + 1$, and

C_n is the thermal capacity of the material associated with node n, the layer *ab*.

The temperature T^{t+1}_n is the "future" value predicted from the known temperatures at time t. Solving Eq. 4-56 for $T_n{}^{t+1}$ gives

$$T^{t+1}_n = T^t_n\left[1 - \Delta\theta\left(\frac{1}{C_nR_{n-1,n}} + \frac{1}{C_nR_{n,n+1}}\right)\right] + \frac{\Delta\theta}{C_nR_{n-1,n}}T^t_{n-1} + \frac{\Delta\theta}{C_nR_{n,n+1}}T^t_{n+1} \tag{4-57}$$

For any interior node $R = \Delta x/kA$ and $C = c\rho\Delta xA$ so that Eq. 4-57 can be written

$$T_n^{t+1} = T_n^t \left[1 - 2\frac{\Delta\theta k}{c\rho\Delta x^2}\right] + \frac{\Delta\theta \cdot k}{c\rho\Delta x^2}(T_{n-1}^t + T_{n+1}^t) \tag{4-58}$$

The numerical procedure becomes particularly simple if the time and distance increments are chosen so that $(2\Delta\theta k/c\rho\Delta x^2) = 1$. Then the "present" temperature at node n, T_n^t, drops out and the "future" temperature at an interior node, T_n^{t+1}, is equal to the arithmetic average of the "present" temperatures at the adjacent nodal points.

It should be noted that the choice of the time and distance intervals determines the coefficient of T_n^t. From the computational viewpoint, the

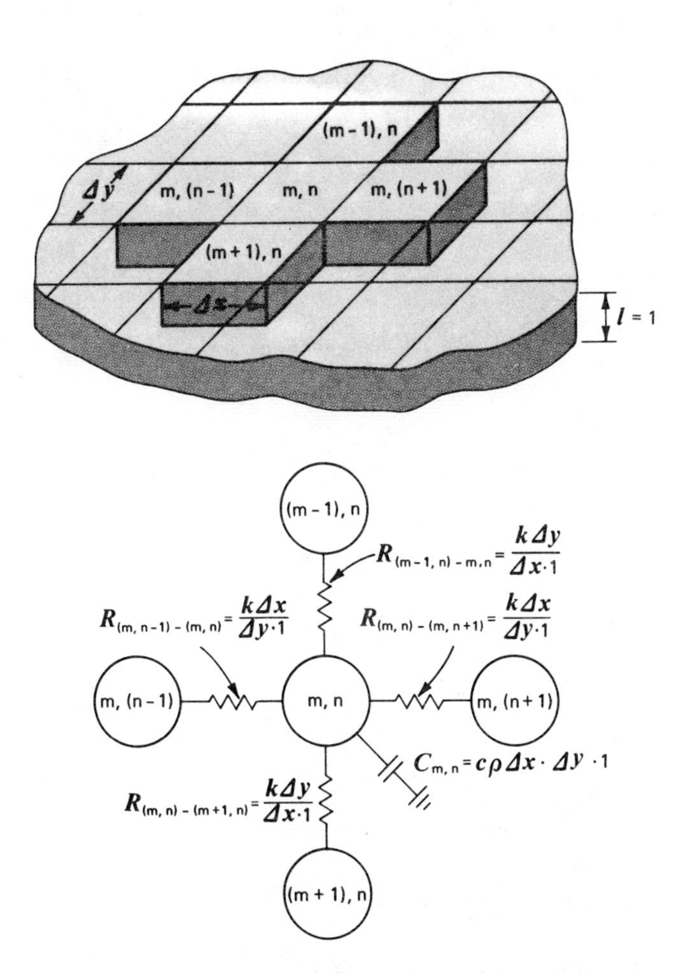

Fig. 4-22. Nodal representation of nonsteady conduction.

larger the values of Δx and $\Delta\theta$, the more rapidly the solution will proceed. On the other hand, the smaller the increments of the independent variables, the more accurate the solution will be. At first glance it may appear that the selection of Δx and $\Delta\theta$ is merely a matter of convenience. This is not the case, however, because if $\Delta\theta$ is too large or Δx is too small, the coefficient of T_n^t in Eq. 4-58, i.e., $[1 - (2\Delta\theta k/c\rho\Delta x^2)]$, can become negative, and a negative coefficient of the "present" temperature of nodal point n would lead to a violation of thermodynamic principles. Suppose, for example, that at the neighboring nodes the temperatures T_{n-1}^t and T_{n+1}^t are equal, but less than T_n^t. If the coefficient of T_n^t were negative, the temperature at nodal point n after time $\Delta\theta$ predicted from Eq. 4-58 would be lower than the adjoining temperatures. This could have happened only if heat had been conducted from a lower to a higher temperature and such a process is thermodynamically impossible. More sophisticated criteria for the stability of numerical solutions of partial differential equations can be deduced on mathematical grounds (e.g. References 13, 14, and 16), but for practical purposes one can simply use the rule that in the explicit numerical solution negative coefficients must be avoided by appropriate choices of $\Delta\theta$ and Δx. This simple rule indicates that a choice of Δx for a network of nodal points places an upper limit on the permissible value of the time interval $\Delta\theta$. For an interior point in a one-dimensional system this limit is $\Delta\theta \leq \frac{1}{2}\frac{\Delta x^2}{a}$.

The expression for the "future" temperature at an interior node in a two-dimensional system with square subdivisions, such as shown in Fig. 4-22, is

$$T_{m,n}^{t+1} = \frac{\Delta\theta}{c\rho\Delta x^2}(T_{m+1,n}^t + T_{m-1,n}^t + T_{m,n+1}^t + T_{m,n-1}^t) + \left[1 - 4\left(\frac{\Delta\theta k}{c\rho\Delta x^2}\right)\right] T_n^t \tag{4-59}$$

and the stability criterion is $4\Delta\theta k/c\rho\Delta x^2 \leq 1$, or $\Delta\theta \leq \frac{1}{4}\frac{\Delta x^2}{a}$.

EXAMPLE 4-8. A 4.8 in. thick plate with a thermal diffusivity $a = 0.25$ sq ft/hr is initially at a uniform temperature of 100 F. At a time $\theta = 0$, it is immersed in a medium at 500 F, and it may be assumed that the thermal resistance at the surface of the plate is so small that the plate surface reaches the temperature of the medium almost instantaneously. Determine the temperature distribution during the initial heating time of 3 min.

Solution: We begin by subdividing the plate into eight equal sections, locating nodal points at both surfaces and at seven equally-spaced intermediate

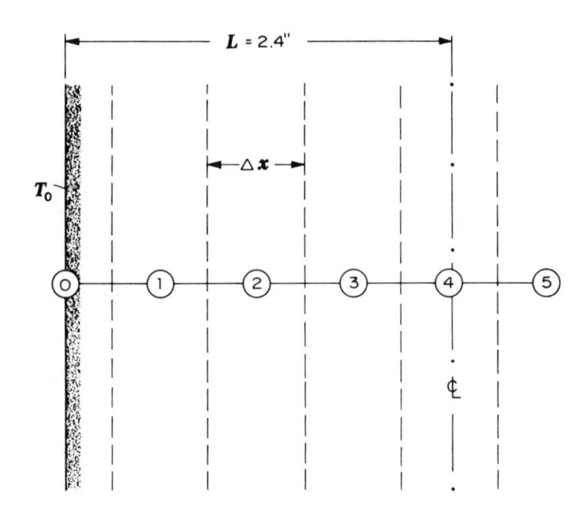

Fig. 4-23. Subdivision of plate for transient-numerical solution.

points as shown in Fig. 4-23. Note that in our method of subdividing the system, the slices associated with the two surface nodal points are only half as thick as those associated with interior nodal points where $\Delta x = 4.8/(8 \times 12) = 0.05$ ft. For our first calculation let $\Delta\theta a/\Delta x^2 = 1/2$ so that $\Delta\theta = (0.05)^2/2 \times 0.25 = 0.005$ hr or 0.30 min. This particular choice reduces Eq. 4-58 to $T_n^{t+1} = (T_{n+1} + T_{n-1})/2$. The results of the numerical calculations are shown in Table 4-1 for one half of the plate. The temperature distribution in the other half will be identical because of the symmetry of the system and nodal point 4 at the mid-plane behaves as though it were located at an insulated surface.

Table 4-1. Plate temperature response with $\Delta x^2/a\Delta\theta = 2$.

Time (min.)	Time ($\Delta\theta$s elapsed)	*Nodal Point Temperature* (F)					
		0	1	2	3	4	5
0	0	500	100	100	100	100	100
0.3	1	500	300	100	100	100	100
0.6	2	500	300	200	100	100	100
0.9	3	500	350	200	150	100	150
1.2	4	500	350	250	150	150	150
1.5	5	500	375	250	200	150	200
1.8	6	500	375	287.5	200	200	200
2.1	7	500	393.7	287.5	243.7	200	243.7
2.4	8	500	393.7	318.7	243.7	243.7	243.7
2.7	9	500	409.4	318.7	281.2	243.7	281.2
3.0	10	500	409.4	345.3	281.2	281.2	281.2

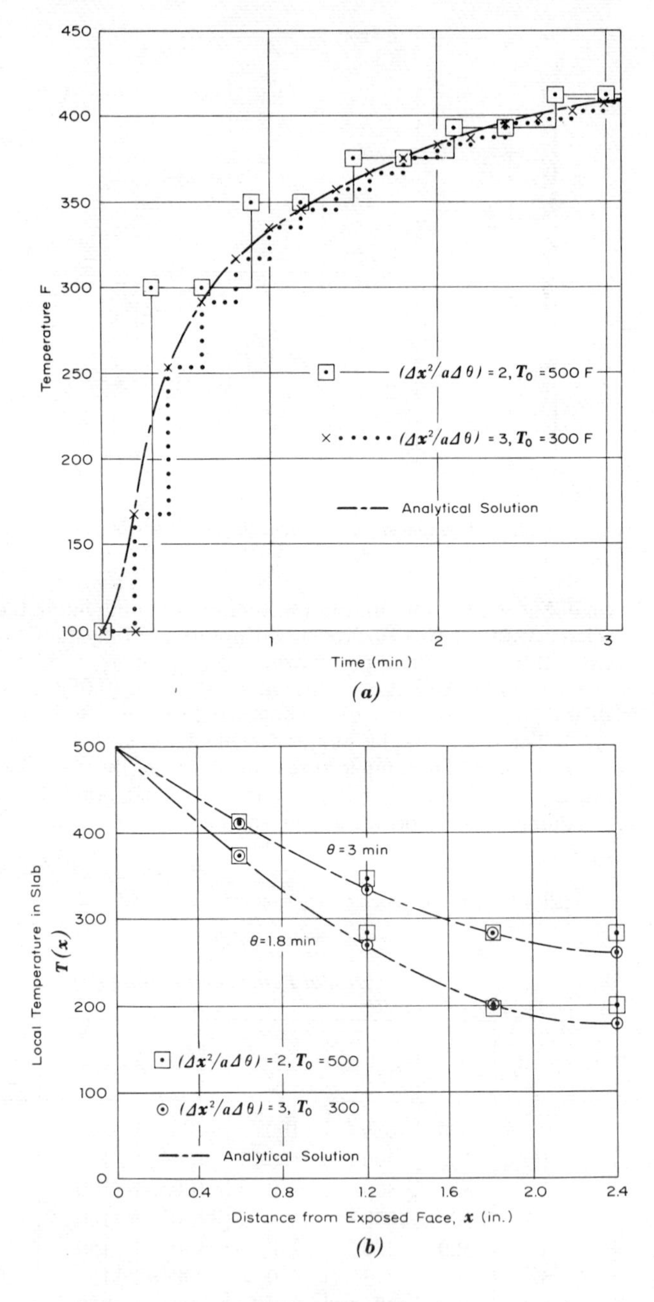

Fig. 4-24. Comparison of analytical solution of results of numerical analysis—transient conduction in a plate: a) temperature response for a nodal point 0.05 ft from the surface of the plate in illustrative example, b) temperature distribution of plate in illustrative example after 1.8 and 3.0 min.

An inspection of Table 4-1 shows that choosing $\Delta\theta = 0.3$ min will give only approximate results unless the slices are very thin. For instance, the temperature at any point increases and then remains constant over two $\Delta\theta$ intervals before increasing again. This is illustrated graphically in Fig. 4-24, where the temperature at a distance of 0.05 ft from the surface of the plate is plotted against time. Moreover, the temperature behavior during the initial 3 or 4 time increments, particularly at the nodal points near the surface, is markedly discontinuous and the derived values are, therefore, only approximate.

Greatly improved results can be obtained without increasing the calculation technique appreciably by taking $\Delta\theta a/\Delta x^2 = 1/3$. Equation 4-58 then shows that $T_n^{t+1} = (T_{n+1} + T_n + T_{n-1})/3$ and the time interval is $\Delta\theta = \Delta x^2/3a = [0.05^2/(3 \times 0.25)]60 = .20$ min. Fifteen steps are now required and the calculations, rounded off to three significant figures, are shown in Table 4-2. To improve the results further, the initial surface temperature has been taken at the mean between the value at $0 > \theta$, i.e., 100 F, and $0 > \theta$, i.e., 500 F, which is 300 F. Afterwards, the surface temperature is put at 500 F, as required by the physical boundary condition. If $\Delta x^2/a\Delta\theta = 4$, the results of the numerical calculation approach the results of the analytical solution so closely that the temperature distributions obtained by the two methods are almost indistinguishable after two $\Delta\theta$ increments.

A FORTRAN IV program to solve this problem by the explicit method is shown in Table 4-3. The symbols for the computer solution are identified in Table 4-4, and a flow chart illustrating the sequence of computer operations is presented in Fig. 4-25.

Values of $\Delta x = 0.6$ in. and $\Delta\theta a/\Delta x^2 = 0.5$ have been used in the computer solution, which corresponds to the hand calculations shown in Table 4-1. Note, however, that the lowest starting value of the spatial and time subscripts, I and J respectively, are 1 instead of 0 as in the hand calculations. This change is necessary because FORTRAN compilers cannot recognize a subscript of 0.

Table 4-2. Plate temperature response with $\Delta x^2/a\Delta\theta = 3$.

TIME (min.)	T_0 (F)	T_1 (F)	T_2 (F)	T_3 (F)	T_4 (F)
0	300	100	100	100	100
0.2	500	167	100	100	100
0.4	500	256	122	100	100
0.6	500	293	159	107	100
0.8	500	318	186	122	105
1.0	500	335	209	138	116
1.2	500	348	227	154	131
1.4	500	358	243	171	146
1.6	500	367	258	187	163
1.8	500	375	271	203	179
2.0	500	382	283	218	195
2.2	500	388	294	232	210
2.4	500	394	304	245	224
2.6	500	399	314	257	238
2.8	500	404	323	269	250
3.0	500	409	332	280	262

Table 4-3. **FORTRAN program for Example 4-8**

```
      REAL L
      DIMENSION T(9,20),TIME(20)
```

The arrays T and TIME are declared with the indicated number of locations reserved for them. The first subscript of T denotes the spatial location, the second the point in time.

```
      ALPHA=.25
      L=4.8/12.
      DX=L/8.
      DTHETA=(DX*DX)/(2.*ALPHA)
      TIME(1)=0.
```

The parameters of the program are evaluated and the initial time TIME(1) is set to zero.

```
      DO 1 1=2,8
      T(I,1)=100.
1     CONTINUE
      T(1,1)=500.
      T(9,1)=500.
```

The preceding statements evaluate the initial temperature distribution within the solid and specify the boundary temperatures.

```
      DO 2 J=2,11
      T(1,J)=500.
      T(9,J)=500.
      DO 12 I=2,8
      T(I,J) = T(I,J-1)*(1.-2.*ALPHA*DTHETA/(DX*DX))+ALPHA*DTHETA*
     1(T(I-1,J-1)+T(I+1,J-1))/(DX*DX)
12    CONTINUE
      TIME(J)=(J-1)*DTHETA
2     CONTINUE
```

The time-varying temperature is computed at each node at each time increment from equations developed above. The values are stored in the array T. The values of time corresponding to each iteration are stored in the array TIME.

```
      PRINT 3
      DO 4 J=1,11
```

```
      PRINT 5,TIME(J),T(1,J),T(2,J),T(3,J),T(4,J),T(5,J),T(6,J),T(7,J),T(8,J),T(9,J)
4     CONTINUE
      CALL EXIT
3     FORMAT (1H1,*TIME-HR*,6X,*T1*,8X,*T2*,8X,*T3*,8X,*T4*,
     18X,*T5*,8X,*T6*,8X,*T7*,8X,*T8*,8X,*T9*,//)
5     FORMAT (1H0,F7.4,9(5X,F5.1),/)
      END
```

The final cards in the program print out the temperature distribution and the value of time as specified in format statements 3 and 5. The results are shown below. The program is terminated by the CALL EXIT statement.

Time-hr	T_1	T_2	T_3	T_4	T_5	T_6	T_7	T_8	T_9
0.0000	500.0	100.0	100.0	100.0	100.0	100.0	100.0	100.0	500.0
.0050	500.0	300.0	100.0	100.0	100.0	100.0	100.0	300.0	500.0
.0100	500.0	300.0	200.0	100.0	100.0	100.0	200.0	300.0	500.0
.0150	500.0	350.0	200.0	150.0	100.0	150.0	200.0	350.0	500.0
.0200	500.0	350.0	250.0	150.0	150.0	150.0	250.0	350.0	500.0
.0250	500.0	375.0	250.0	200.0	150.0	200.0	250.0	375.0	500.0
.0300	500.0	375.0	287.5	200.0	200.0	200.0	287.5	375.0	500.0
.0350	500.0	393.7	287.5	243.7	200.0	243.7	287.5	393.7	500.0
.0400	500.0	393.7	318.7	243.7	243.7	243.7	318.7	393.7	500.0
.0450	500.0	409.4	318.7	281.2	243.7	281.2	318.7	409.4	500.0
.0500	500.0	409.4	345.3	281.2	281.2	281.2	345.3	409.4	500.0

In most practical problems a system exchanges energy with its environment by convection and radiation. For these types of problems the surface conditions are somewhat more complicated than in pure-conduction systems and the boundary conditions must be expressed in finite-difference equations which are different than the equations for the interior of the system. We shall now develop the equations and the associated stability criteria for network points at the surface, the so-called surface point equations.

If the boundary condition involves heat transfer by convection be-

Table 4-4. List of symbols used in FORTRAN programs.

FORTRAN Symbol	Equation Symbol	Description	Units
ALPHA	a	Thermal diffusivity $= k/\rho c$	ft^2/hr
DTHETA	$\Delta\theta$	Time increment	hr
DX	Δx	Distance increment	ft
H, HC	$\bar{h}_c$	Average surface conductance	$Btu/hr\ ft^2\ F$
HMAX	–	Maximum value of $\bar{h}_c$	$Btu/hr\ ft^2\ F$
I	$n + 1$	Distance index	—
J	$t + 1$	Time index	—
K	k	Thermal conductivity	Btu/hr ft F
L	–	Length of solid	ft
T(I,J)	T_{n+1}^{t+1}	Nodal temperature	F
TINF	T_∞	Environment temperature—hot side	F
TIME(J)	θ	Time ($= t \cdot \Delta\theta$)	hr

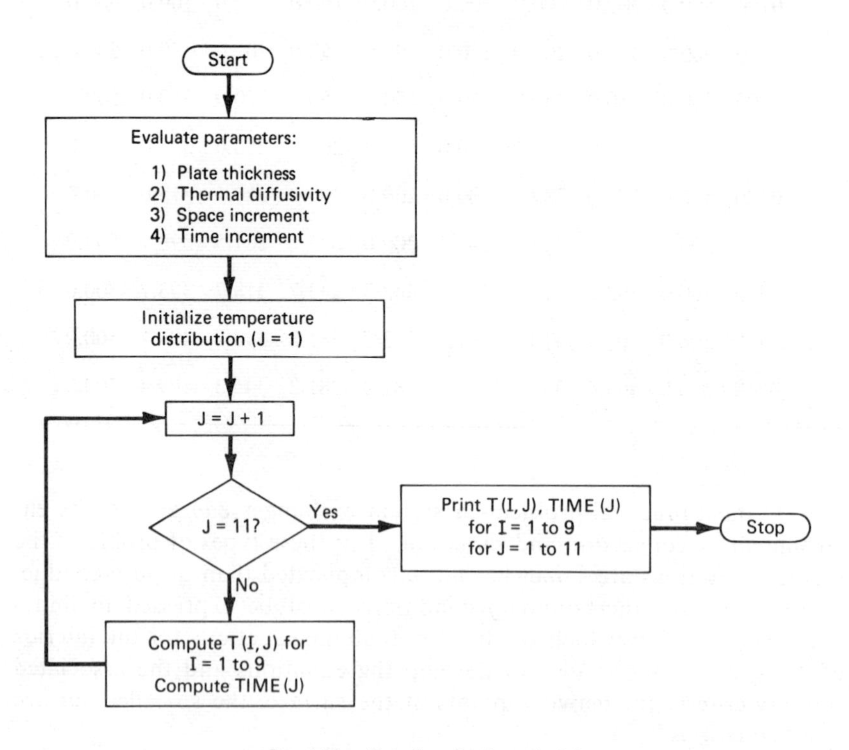

Fig. 4-25. Flow chart for Example 4-8.

tween the surface of the system and its environment at a temperature T_∞ through a unit-surface conductance $\bar{h}$, the finite-difference equation for a nodal point at the surface can be obtained by means of an energy balance. In order to include all types of surface phenomena, we shall assume that the surface is also subject to a heat flux q_0 per unit area taken as positive if it flows into the system. This surface heat flux could be the result of radiation to or from the surface. A frequently encountered source of such an external flux is the sun.

In a one-dimensional system, the volume associated with a surface nodal point is $A\Delta x/2$. In the absence of internal heat generation, the increase in internal energy at the surface node 0 (see Fig. 4-21) during a time interval $\Delta\theta$ is equal to the net heat transfer to the node during that same time interval, or

$$\left(\frac{T'_\infty - T_0{}'}{R_{\infty,0}} + \frac{T_1{}' - T_0{}'}{R_{0,1}} + q_0{}'A\right)\Delta\theta = C_0(T_0{}'^{+1} - T_0{}')$$

Solving for the "future" temperature $T_0{}'^{+1}$ gives

$$T_0{}'^{+1} = T_0{}'\left[1 - \Delta\theta\left(\frac{1}{C_0 R_{\infty,0}} + \frac{1}{C_0 R_{0,1}}\right)\right] + \frac{\Delta\theta}{C_0 R_{\infty,0}} T'_\infty + \frac{\Delta\theta}{C_0 R_{0,1}} T_1{}' + q_0{}' \frac{A\Delta\theta}{C_0}$$

Noting that $C_0 = c\rho A\Delta x/2$, $R_{\infty,0} = 1/\bar{h}A$, and $R_{0,1} = \Delta x/kA$, the future temperature at the surface is

$$T_0{}'^{+1} = T_0{}'\left[1 - \Delta\theta\left(\frac{2k}{c\rho\Delta x^2} + \frac{2\bar{h}}{c\rho\Delta x}\right)\right] + \frac{2\bar{h}\Delta\theta}{c\rho\Delta x} T_\infty{}' + \frac{2k\Delta\theta}{c\rho\Delta x^2} T_1{}' + \frac{2\Delta\theta}{c\rho\Delta x} q_0{}' \qquad \text{(4-60)}$$

and the associated stability criterion is

$$\Delta\theta \leq \frac{1}{\dfrac{2k}{c\rho\Delta x^2} + \dfrac{2\bar{h}}{c\rho\Delta x}} = \frac{1}{2}\frac{\Delta x^2}{a}\left(\frac{1}{1 + (\bar{h}\Delta x/k)}\right) \qquad \text{(4-61)}$$

Since the term in brackets in Eq. 4-61 is always less than one, it is apparent that to satisfy the stability criterion at a surface node with convection requires a smaller time increment than at an interior node. Thus, the surface node becomes the controlling factor for the maximum permissible value for $\Delta\theta$. When the unit surface conductance is large, the permissible value of the time increment may become so small that the computations by the explicit method will require an exorbitant amount of time. In such cases the implicit method described at the end of this chapter should be used.

The finite difference equation for a surface node in two- and three-dimensional systems can be derived similarly. As illustrated schematically in Fig. 4-26, the heat balance at point $(m, 0)$ must also include conduction from nodes $(m - 1, 0)$ and $(m + 1, 0)$ since their temperatures are in general different from $T_{m,0}$. In a form suitable for numerical computations, the equation for a surface node in a two-dimensional system with convection at the surface is

$$T_m^{t+1} = T_m^t \left[1 - \Delta\theta \left(\frac{4k}{c\rho\Delta x^2} + \frac{2\bar{h}}{c\rho\Delta x}\right)\right]$$

$$+ \frac{k\Delta\theta}{c\rho\Delta x^2}(T_{m-1}^t + T_{m+1}^t + 2T_{m,1}^t) + \frac{2\bar{h}}{c\rho\Delta x} T_\infty^{\,t} \tag{4-62}$$

and the associated stability criterion is

$$\Delta\theta \leq \frac{1}{4}\frac{\Delta x^2}{a}\left[\frac{1}{(\bar{h}\Delta x/2k) + 1}\right] \tag{4-63}$$

If the nodal point is in a corner, the first coefficient becomes

$$\left[1 - \Delta\theta \left(\frac{4k}{c\rho\Delta x^2} + \frac{4\bar{h}}{c\rho\Delta x}\right)\right]$$

and the second and third coefficients in Eq. 4-62 double. The corresponding stability criterion for a numerical solution by the explicit method is

$$\Delta\theta \leq \frac{1}{4}\frac{\Delta x^2}{a}\left[\frac{1}{(h\Delta x/k) + 1}\right]. \tag{4-64}$$

When the boundary conditions, e.g., the environmental temperature or the unit-surface conductance, change with time, the value of $\bar{h}$ or T_∞ at time $\theta = t \cdot \Delta\theta$ must be identified by a superscript t just the same as the temperatures at interior nodal points. Such a case is illustrated in the next example.

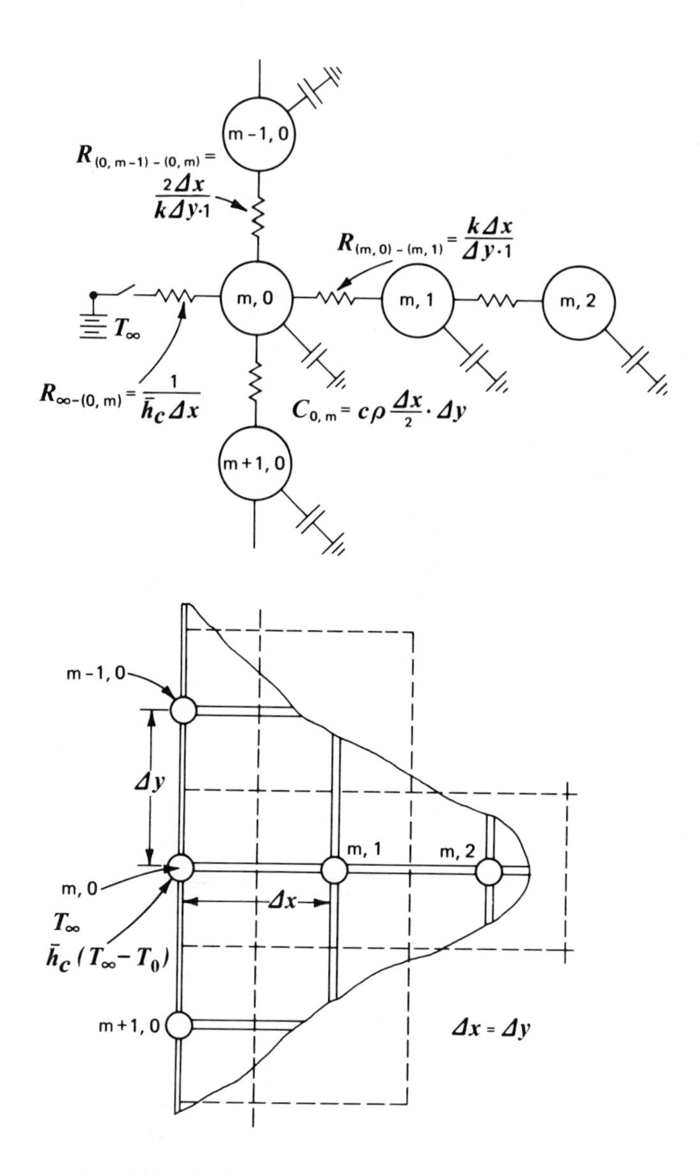

Fig. 4-26. Sketch illustrating heat balance on an exterior nodal point with convection and radiation.

EXAMPLE 4-9. A large plastic plate, 8 in. thick, is exposed to a hot environment on one side and a cold environment on the other. The initial temperature distribution varies linearly from 500 F to 100 F. Suppose the environment temperature on the hot side is suddenly reduced to 100 F and the heat-transfer coefficient at this surface depends upon the temperature difference between the surface and the fluid, as given by the expression

$$\bar{h}_c = 2.0 + 0.02\,(T_0 - T_\infty)$$

Determine the temperature-time history in the plate if k_s = 2.0 Btu/hr ft F and a = 0.003 sq ft/hr.

Solution: At the surface, the equation for the "future" temperature T_0^{t+1} is, from Eq. 4-60

$$T_0^{t+1} = \left[1 - \Delta\theta\left(\frac{2a}{\Delta x^2} + \frac{2\bar{h}_c{}^t a}{k\Delta x}\right)\right] T_0{}^t + \frac{2\bar{h}_c{}^t a\Delta\theta}{k\Delta x} T_\infty + \frac{2a\Delta\theta}{\Delta x^2} T_1{}^t$$

The largest value of $\bar{h}_c{}^t$ occurs at time zero and it is $\bar{h}_c{}^0 = 2 + 0.02\,(500 - 100) = 10$ Btu/hr sq ft F, and

$$\frac{\bar{h}_c \Delta x}{k} = \frac{10 \times 1/12}{2} = 0.416$$

Consequently, a value of $\Delta\theta = \frac{1}{3}\,(\Delta x^2/a)$ yields a positive coefficient for $T_0{}^t$ and will give a stable numerical solution. However, the value of $\bar{h}^t$ depends on the surface temperature which varies with time; the coefficient of $T_0{}^t$ will therefore also vary with time, and it is necessary to construct an auxiliary table to keep track of $T_0{}^t$, $\bar{h}^t$, and the coefficient of the "present" surface temperature $\left[1 - \Delta\theta\left(\frac{2a}{\Delta x^2} + \frac{2\bar{h}_c}{c\rho\Delta x}\right)\right]$. With Δx = 1 in.,

$$\Delta\theta = \frac{(1/12)^2}{0.003 \times 3} = 0.772 \text{ hr}$$

The equation for the "future" temperature at an interior point is

$$T_n^{t+1} = \frac{T_{n+1}^t + T_n^t + T_{n-1}^t}{3}$$

and for the surface temperature it is

$$T_0{}^{t+1} = \frac{1 - 2(h_c{}^t\Delta x/k)}{3} T_0{}^t + \frac{2(\bar{h}^t\Delta x/k)}{3} T_\infty{}^t + \frac{2}{3} T_1{}^t$$

The initial temperature distribution is shown on the first line in Table 4-5 and the subsequent temperature-time history has been calculated and recorded on subsequent lines using the auxiliary table to obtain the coefficients of $T_0{}^t$ at each step. For example, after $1\,\Delta\theta(t = 1)$, the surface temperature $T_0{}^t$ is

$$T_0{}^1 = 0.056 \times 500 + 0.277 \times 100 + 0.667 \times 450 = 356 \text{ F}$$

and after $2\Delta\theta$ the temperature at nodal point 1 is $T_1{}^2 = (356 + 450 + 400)/3 = 402$ F.

Table 4-5. Numerical solution for Example 4-9.

(Temperatures in deg F)

$\Delta\theta$	T_0	T_1	T_2	T_3	T_4	T_5	T_6	T_7	T_8
0	500	450	400	350	300	250	200	150	100
1	356	450	400	350	300	250	200	150	100
2	368	402	400	350	300	250	200	150	100
3	336	390	384	350	300	250	200	150	100
4	327	374	375	345	300	250	200	150	100
5	316	360	365	337	298	250	200	150	100
6	307	347	354	333	295	249	200	150	100
7	298	336	345	328	292	248	200	150	100
8	291	326	337	322	289	247	199	150	100
9	283	318	328	316	286	245	198	150	100

Auxiliary table for Example 4-9.

$\Delta\theta$	θ (hr)	$T_0^t - T_\infty$ (F)	$\bar{h}_c^t$	$\bar{h}_t^t \Delta x/k_s$
0	0	400	10.0	0.416
1	0.772	256	7.2	0.30
2	1.544	268	7.4	0.31
3	2.316	236	6.7	0.28
4	3.088	227	6.5	0.27
5	3.860	216	6.3	0.26
6	4.632	207	6.1	0.25
7	5.404	198	6.0	0.25
8	6.176	191	5.8	0.24

The reader should independently verify the subsequent steps in the solution shown in Table 4-5. No comparison with an analytical solution can be made because the complexity of the boundary condition makes a mathematical analysis prohibitively difficult.

A flow chart illustrating the sequence of operations for a computer solution is shown in Fig. 4-27 and a FORTRAN IV program to solve this problem is shown in Table 4-6, using the symbols identified in Table 4-4. As in the hand calculations a value of $\Delta x = 1$ inch is used, but the maximum permissible value of $\Delta\theta$ is selected according to the stability criterion

$$\Delta\theta = \frac{1}{2}\frac{\Delta x^2}{a}\left[\frac{1}{1 + (\bar{h}_{c,\max}\Delta x/k)}\right]$$

An inspection of the stability limits for Eq. 4-61 or Eq. 4-64 shows that for large values of $\bar{h}\Delta x/k$, i.e., for systems with a large ratio of surface conductance to thermal conductivity, the value of $\Delta x^2/a\Delta\theta$ required for stability can become very small. This means that either very many subdivisions or a very small time increment must be used. In either case,

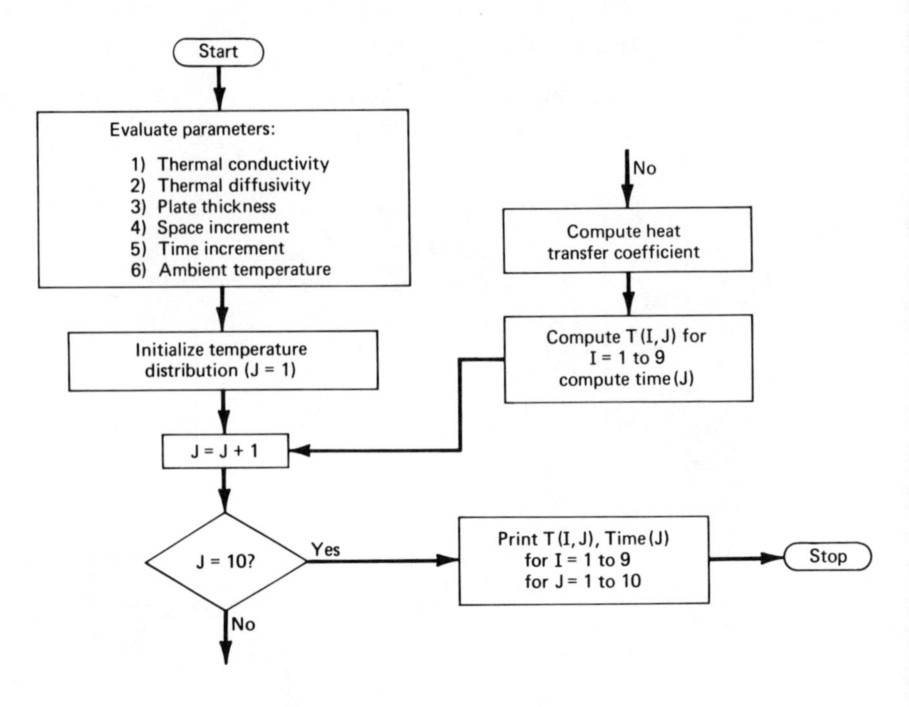

Fig. 4-27. Flow chart for Example 4-9.

Table 4-6. FORTRAN program for Example 4-9.

```
H(Z,ZINF)=2.+.02*(Z-ZINF)
REAL K,L
DIMENSION T(9,20),TIME(20)
```

The function H is defined with the general arguments Z and ZINF. When the function is to be evaluated later in the program these arguments will be replaced by T(I, J), the time-varying surface temperature, and TINF, respectively. The arrays T and TIME are declared with appropriate space reserved for the problem.

```
K=2.
ALPHA=.003
L=8./12.
DX=L/8.
HMAX=H(500.,100.)
DTHETA=1./(2.*ALPHA/(DX*DX)+2.*HMAX*ALPHA/(K*DX))
TIME(1)=0.
TINF=100.
```

Table 4-6. Continued.

The physical and computational parameters are specified and the value TIME(1) is initialized at zero.

```
      T(1,1)=500.
      T(9,1)=100.
      DO 2 I=2,8
      T(I,1)=T(1,1)+(I-1)*(T(9,1)-T(1,1))*DX/L
2     CONTINUE
```

The initial linear temperature distribution is specified.

```
      DO 5 J=2,10
      HC=H(T(1,J-1),TINF)
      T(1,J)=(1.-DTHETA*(2.*ALPHA/(DX*DX)+2.*HC*ALPHA/(K*DX)))*T(1,J-1)
     12.*HC*ALPHA*DTHETA*TINF/(K*DX)+2.*ALPHA*DTHETA*T(2,J-1)/(DX*DX)
      DO 4 1=2,8
      T(I,J)=T(I,J-1)*(1.-2.*ALPHA*DTHETA/(DX*DX))+ALPHA*DTHETA*(T(I-1,
     1J-1)+T(I+1,J-1))/(DX*DX)
4     CONTINUE
      T(9,J)=100.
      TIME(J)=(J-1)*DTHETA
5     CONTINUE
```

The convective coefficient is evaluated and the temperature distribution computed at each time point. These values are stored in the array T. The values of TIME are computed.

```
      PRINT 8
      DO 6 J=1,10
      PRINT 7,TIME(J),T(1,J),T(2,J),T(3,J),T(4,J),T(5,J),T(6,J),T(7,J)1(8,J),T(9,J)
6     CONTINUE
      CALL EXIT
7     FORMAT (1H0,F7.1,5X,F5.1,8(5X.F5.1),/)
8     FORMAT 1H1,*TIME-HR*,6X,        *T1*,8X,*T2*,8X,*T3*,8X,*T4*,8X,
     1*T5*,8X,*T6*,8X,*T7*,8X,*T8*,8X,*T9*,//)
      END
```

The final statements print out TIME and T at each time interval as specified by formats 7 and 8. The results are shown below. The program is then terminated.

Table 4-6. Continued.

Time-hr	T_1	T_2	T_3	T_4	T_5	T_6	T_7	T_8	T_9
0.0	500.0	450.0	400.0	350.0	300.0	250.0	200.0	150.0	100.0
.8	347.1	450.0	400.0	350.0	300.0	250.0	200.0	150.0	100.0
1.6	369.3	396.0	400.0	350.0	300.0	250.0	200.0	150.0	100.0
2.5	329.7	388.0	380.9	350.0	300.0	250.0	200.0	150.0	100.0
3.3	326.3	364.9	372.5	343.3	300.0	250.0	200.0	150.0	100.0
4.1	310.1	354.0	359.5	338.3	297.6	250.0	200.0	150.0	100.0
4.9	302.7	340.5	350.1	331.4	295.2	249.2	200.0	150.0	100.0
5.7	293.3	330.5	340.1	325.2	291.7	248.1	199.7	150.0	100.0
6.5	286.2	320.8	331.5	318.7	288.1	246.4	199.2	149.9	100.0
7.4	279.2	312.4	323.2	312.4	284.2	244.5	198.5	149.7	100.0

the amount of labor may be excessive if the problem is to be solved by hand calculation. If a digital computer is available, small subdivisions will not offer any difficulties unless the memory storage in the machine is not adequate.

When it is desirable or necessary in a numerical solution to use a larger value of $\Delta\theta$ than a system with a convection boundary condition allows, stability can be obtained if the heat capacity associated with the surface nodal point is neglected and a steady-state heat balance is used to obtain the surface temperature $T_0{}'$. In the absence of any internal heat generation $\dot{q}$ and external heat flux q_0, a steady-state heat balance at time t on the nodal point 0, i.e., at the surface (see Fig. 4-21), gives in a one-dimensional system

$$\bar{h}_c(T_\infty{}' - T_0{}') + \frac{k}{\Delta x}(T_1{}' - T_0{}') = 0 \tag{4-65}$$

Solving for $T_0{}'$ we get

$$T_0{}'\left(\frac{k}{\Delta x} + \bar{h}_c\right) = T_0{}'\bar{h}_c\left(\frac{k}{\bar{h}_c\Delta x} + 1\right) = \bar{h}_c T_\infty{}' + \frac{k}{\Delta x}T_1{}'$$

$$= \bar{h}_c\left(T_\infty{}' + \frac{k}{\bar{h}_c\Delta x}T_1{}'\right) \tag{4-66}$$

or

$$T_0{}^t = \frac{1}{1 + (k/\bar{h}_c\Delta x)} T_\infty{}^t + \frac{1}{1 + (\bar{h}_c\Delta x/k)} T_1{}^t \tag{4-67}$$

If this value of $T_0{}^t$ is now used to compute the future temperature $T_1{}^{t+1}$ at nodal point 1, the finite-difference form of the transient heat balance at nodal point 1 becomes

$$\frac{c\rho\Delta x}{\Delta\theta}(T_1{}^{t+1} - T_1{}^t) = \frac{k}{\Delta x}(T_0{}^t - T_1{}^t) + \frac{k}{\Delta x}(T_4{}^t - T_1{}^t) \tag{4-68}$$

or

$$T_1{}^{t+1} = T_1{}^t\left(1 - \frac{2a\Delta\theta}{\Delta x^2}\right) + \frac{a\Delta\theta}{\Delta x^2} T_0{}^t + T_4{}^t \tag{4-69}$$

Substituting the present value for $T_0{}^t$ from Eq. 4-67 in Eq. 4-69, we get

$$T_1{}^{t+1} = T_1{}^t\left[1 - \frac{a\Delta\theta}{\Delta x^2}\left(2 - \frac{1}{1 + (\bar{h}_c\Delta x/k)}\right)\right] + \frac{a\Delta\theta}{\Delta x^2}\left[\frac{1}{1 + (k/\bar{h}_c\Delta x)} T_\infty{}^t + T_2{}^t\right] \tag{4-70}$$

For stability in the interior, $(\Delta x^2/a\Delta\theta)$ must not be less than 2. Since the coefficient of $T_1{}^t$ in Eq. 4-70 will be positive for any value of $(\Delta x^2/a\Delta\theta)$ larger than two, the solution will be stable. It should be noted, however, that this stability has been achieved by neglecting the heat capacity of the half-slice associated with nodal point 0, and as a result the surface temperature will not be known with a high degree of accuracy when large subdivisions are used.

EXAMPLE 4-10. Repeat Example 4-9 using $\Delta x^2/a\Delta\theta = 2$, but neglect the heat capacity of the half-slice associated with the surface nodal point.

Solution: The time interval $\Delta\theta = \Delta x^2/2a = 1.16$ hr and $\bar{h}_c{}'\Delta x/k = \bar{h}_c{}'/24$. The numerical solution is shown in Table 4-7, and the reader is urged to verify the numerical calculations independently. A comparison of the temperature distribution after about 7 hours in Tables 4-5 and 4-7 shows that the numerical solution which neglects the heat capacity associated with the half-slice at the surface yields lower temperatures in the vicinity of the surface exposed to the cooling medium. However, despite the rather crude subdivisions of the system and the large $\Delta\theta$ steps, the temperature profiles agree within 6 percent, and an

Table 4-7. Numerical solution for Example 4-10.

(Temperatures in deg F)

$\Delta\theta$	θ (hrs)	T_0	T_1	T_2	T_3	T_4	T_5	T_6	T_7	T_8
0	0	500	450	400	350	300	250	200	150	100
1	1.16	347	450	400	350	300	250	200	150	100
2	2.32	310	372	400	350	300	250	200	150	100
3	3.48	303	356	361	350	300	250	200	150	100
4	4.64	276	332	353	330	300	250	200	150	100
5	5.80	275	316	331	326	290	250	200	150	100
6	6.96	267	305	321	311	288	245	200	150	100

Auxiliary table for Example 4-10.

$\Delta\theta$	$T_0 - T_\infty$	$\bar{h}_c{}^t$	$\frac{\bar{h}_c{}^t \Delta x}{k}$	$\left(\frac{1}{1 + \bar{h}_c \Delta x/k}\right)^t$	$\left(\frac{1}{1 + k/\bar{h}_c \Delta x}\right)^t$	$\frac{2\bar{h}_c{}^t \Delta x}{k} + 1$	$\left(2 - \frac{1}{1 + \bar{h}_c \Delta x/k}\right)^t$	$\frac{1}{2}\left(\frac{1}{1 + k/\bar{h}_c \Delta x}\right)^t$
0	400	10	0.416	0.707	0.294	1.832	0.650	0.147
1	247	6.94	0.288	0.775	0.224	1.576	0.610	0.112
2	210	6.20	0.258	0.795	0.205	1.516	0.602	0.102
3	203	6.06	0.252	0.80	0.201	1.504	0.601	0.100
4	176	5.52	0.230	0.813	0.187	1.460	0.593	0.093
5	175	5.5	0.229	0.815	0.186	1.458	0.592	0.093

The stability of numerical methods in systems involving radiation boundary conditions is discussed in Reference 22.

even closer agreement can be expected after a few additional time increments. The stability of numerical methods in systems involving radiation boundary conditions is discussed in Reference 22.

In the implicit formulation of the finite difference equation for a transient heat conduction system, we set up the heat balance for a nodal point from the point of view of an observer at time $(t+1)\Delta\theta$. To satisfy an energy balance for a one-dimensional transient system from that viewpoint, the internal energy storage which has occurred at nodal point n (see Fig. 4-21) during the last time interval between $(t+1)\Delta\theta$ and $t\Delta\theta$ must equal the net heat flow at time $(t+1)\Delta\theta$. Symbolically, this yields for a system without heat sources

$$(T_n^{t+1} - T_n^t)\,C_n = \Delta\theta\left[\frac{T_{n-1}^{t+1} - T_n^{t+1}}{R_{n-1,n}} + \frac{T_{n-1}^{t+1} - T_n^{t+1}}{R_{n,n+1}}\right] \tag{4-71}$$

For a one-dimensional system with heat sources an energy balance yields the relation

$$(T_n^{t+1} - T_n^t)\,C_n = \Delta\theta\left[\frac{T_{n-1}^{t+1} - T_n^{t+1}}{R_{n-1,n}} + \frac{T_{n+1}^{t+1} - T_n^{t+1}}{R_{n,n+1}} + \dot{q}_n\,\Delta V_n\right] \tag{4-72}$$

where $\dot{q}_n$ is the rate of heat generation per unit volume at the nodal point n and ΔV_n is the volume associated with nodal point n.

Solving Eq. 4-71 for T_n^{t+1} at an interior node where $R_{n-1,n} = R_{n,n+1} = \Delta x/kA$ and $C_n = c\rho A\Delta x$ gives

$$T_n^{t+1} = \frac{T_n^t + (\Delta\theta k/c\rho\Delta x^2)(T_{n-1}^{t+1} + T_{n+1}^{t+1})}{1 + (2\Delta\theta k/c\rho\Delta x^2)} \tag{4-73}$$

If in a finite difference system all the resistances between nodes, the capacitances associated with each node, and an initial temperature distribution are known, equations of the type of Eq. 4-73 can be written for each nodal point. Special care must of course be taken to satisfy the boundary conditions. The final result is a set of equations for the future temperature of each nodal point in terms of the present temperature at that point and the future temperatures at neighboring points. There will be as many equations as there are unknown future temperatures, and once this set of equations is solved the resulting future temperatures become the initial temperatures for the next time increment $\Delta\theta$. The coefficients are always positive and the implicit technique will therefore be stable with any time increment step. In essence, the implicit technique reduces to the solution of a set of simultaneous algebraic equations at each time increment, and the same methods discussed previously in connection with the solution of steady-state systems (relaxation, iteration, or matrix inversion) can therefore be used; the latter two are recommended for solution by means of a digital computer.

EXAMPLE 4-11. In a heat-treating operation a large 2-ft-thick slab of steel is suddenly placed into water at 200 F. If the initial temperature of the slab is 1600 F and the heat-transfer coefficient between the surface of the bar and the water is 1000 Btu/hr sq ft F, calculate the time required for the center of the slab to reach 500 F by the implicit, finite difference method. Use five nodal points.

Solution: Making a heat balance similar to that of Eq. 4-9 for a one-dimensional system (as shown in Figure 4-22) yields the following equation for the temperature change at an interior point j during the time interval $(t + 1)\Delta\theta$ and $t\Delta\theta$.

$$\frac{(\rho c A \Delta x)(T_j^{t+1} - T_j^t)}{\Delta\theta} = \frac{kA}{\Delta x}\left[(T_{j-1}^{t+1} - T_j^{t+1}) + (T_{j+1}^{t+1} - T_j^{t+1})\right]$$

Solving for the nodal temperature T_j^{t+1}:

$$T_j^{t+1}\left(\frac{1}{a}\frac{\Delta x^2}{\Delta\theta} + 2\right) = \frac{1}{a}\frac{\Delta x^2}{\Delta\theta}T_j^t + T_{j-1}^{t+1} + T_{j+1}^{t+1}$$

The expression for T_j^{t+1} involves other unknowns at the $t + 1$ time level; hence the value of T_j^{t+1} cannot be determined from one equation as in the explicit method. Instead, all nodal temperatures are determined at once by solving all heat balances simultaneously by matrix inversion.

The heat balance at a surface point (node 1) gives:

$$T_1^{t+1}\left(\frac{1}{2a}\frac{\Delta x^2}{\Delta\theta} + \frac{h\Delta x}{k} + 1\right) = \frac{h\Delta x}{k}T_\infty + T_2^{t+1} + \frac{1}{2}\frac{\Delta x^2}{a\Delta\theta}T_1^t$$

The heat balances for all 5 nodal points may be written in concise matrix form

$$\begin{bmatrix} \left(\frac{P}{2} + \frac{h\Delta x}{k}\right) & -1 & 0 & 0 & 0 \\ -1 & P & -1 & 0 & 0 \\ 0 & -1 & P & -1 & 0 \\ 0 & 0 & -1 & P & -1 \\ 0 & 0 & 0 & -1 & \left(\frac{P}{2} + \frac{h\Delta x}{k}\right) \end{bmatrix} \cdot \begin{bmatrix} T_1^{t+1} \\ T_2^{t+1} \\ T_3^{t+1} \\ T_4^{t+1} \\ T_5^{t+1} \end{bmatrix} = \begin{bmatrix} \frac{Q}{2}T_1^t + \frac{h\Delta x}{k}T_\infty \\ QT_2^t \\ QT_3^t \\ QT_4^t \\ \frac{Q}{2}T_5^t + \frac{h\Delta x}{k}T_\infty \end{bmatrix}$$

where $P = \left(\dfrac{\Delta x^2}{a\Delta\theta} + 2\right)$

$$Q = \frac{\Delta x^2}{a\Delta\theta}$$

This is a matrix equation of the form $A \cdot X = B$ which may be solved for X by finding the inverse of matrix A, A^{-1}, or $X = A^{-1} \cdot B$.

The transient temperature distribution is computed by executing such a matrix inversion at each time t. A computer program which can perform these computations is shown in Table 4-8 and the relation between physical and computer symbols is shown in Table 4-9.

Table 4-8. Computer program for Example 4-11.

```
      DIMENSION A(10,10),B(10,1),X(10,1)
      DIMENSION DIGITS(1),SCRA(10,13)
      REAL K,L
```

The first dimension statement reserves the appropriate number of spaces for the coefficient matrix A, the unknown vector X, and the right hand side vector B. The second dimension statement reserves the appropriate spaces in the computational arrays required by the matrix inversion routine LINQZ.

```
      L=2.
      TINF=200.
      TO=1600.
      H=1000.
      K=26.
      RHO=490.
      C=.11
      T=0.
      ALPHA=K/(RHO*C)
      READ 1,N,DT
1     FORMAT (I2,F7.4)
      DX=L/N
      Q=DX*DX/(ALPHA*DT)
      P=Q+2.
      NN=N/2.+1
      NP1=N+1
      DT=.005
```

The preceding statements specify the physical and computational parameters.

```
      DO 3 I=1,NP1
      X(I)=TO
3     CONTINUE
```

The initial temperature distribution is specified above.

```
        DO 2 I=2,N
        A(I,I)=P
        A(I,I-1) = - 1.
        A(I,I + 1) = - 1.
2       CONTINUE
        A(1,1)=P/2.+H*DX/K
        A(NP1,NP1)=P/2.+H*DX/K
        A(1,2)=-1.
        A(NP1,N)=-1.
```

The non-zero elements of the coefficient matrix are specified next.

```
6       DO 4 I=2,N
        B(I)=Q*X(I)
4       CONTINUE
        B(1)=(Q*X(1)/2.)+TINF*H*DX/K
        B(NP1)=(Q*X(NP1)/2.)+TINF*H*DX/K
```

Then the values of the right-hand side vector elements are specified.

```
        CALL LINQZ(NP1,10,1,A,B,X,D,DIGITS,SCRA,KERR)
        T=T+DT
        PRINT 5,T,X(NN)
5       FORMAT (1H0,*TIME (HR) = *,F7.4,* CENTER TEMP (DEG F) =
        *,F7.2,//)
```

The matrix inversion routine LINQZ is called to compute A^{-1} and to perform the multiplication $A^{-1} \cdot B$. (LINQZ is only one of many such routines. The reader should use his own scheme or one available among the software of the computer). The value of T (time) is incremented by DT and the temperature and time are printed out in order to observe the time-temperature history of the slab.

```
        IF (X(NN).LT.500.) CALL EXIT
        GO TO 6
        END
```

When the value of T_3 reaches 500 F the program is terminated by the CALL EXIT command. Until $T_3 < 500$ F, the IF statement returns the computation to statement 6.

Table 4-9. Correspondence between physical symbols and FORTRAN symbols in Example 4-11.

FORTRAN Symbol	Text Symbol	Description	Units
A	A	Coefficient matrix	—
B	B	Right-hand side vector	F
C	c	Heat capacity	Btu/lb F
DT	$\Delta\theta$	Time increment	hr
DX	Δx	Spatial increment	ft
H	h	Convective coefficient	Btu/hr ft^2 F
I	j	Nodal location	—
K	k	Thermal conductivity	Btu/hr ft F
L	—	Slab half-thickness	ft
N	—	Number of spatial increments	—
NN	—	$(\frac{1}{2}N + 1)$ Center node location	—
NPI	—	$(N + 1)$ Number of nodes	—
P	—	$\Delta x^2/a\Delta\theta + 2$	—
Q	—	$\Delta x^2/a\Delta\theta$	—
RHO	ρ	Density	16/ft^3
T	t	Time	hr
TINF	T_∞	Ambient temperature	F
TO	T_j^1	Initial temperature	F
X(I)	T_j^{t+1}	Unknown temperature vector	F

The time required for T_3 to reach 500 F is 1.61 hr if a value of $\Delta\theta = .005$ hr is used. For a more precise computation more nodal points and a smaller $\Delta\theta$ may be used. The reader can verify that the answer obtained with the relatively coarse grid used is, very close to the solution found using the transient response chart in Figure 4-8.

The implicit method has the advantage that *any* time increment can be used. In fact, the time increment can be varied during the calculation. It has the disadvantage of requiring a complete set of calculations (i.e. iteration or matrix inversion) at each $\Delta\theta$ step. This may lead to excessive storage requirements in a computer when the number of nodes becomes large. The implicit method is, therefore, used in practice when the physical or boundary conditions impose excessively small time increments for the convergence of the solution by the explicit formulation. References 24 to 29 give more details on the mathematical foundations of these computational procedures and techniques.

PROBLEMS

The problems below have been organized in the following manner: Problems 4-1 through 4-12 deal with conduction in systems which can be treated by the lumped parameter method (Bi < 0.1), 4-13 through 4-17 with analytical solutions of periodic conduction problems, 4-18 through 4-22 are conduction problems with analytical solutions, 4-23 through 4-40 are problems which can be solved by means of the charts in the book, and finally, 4-41 through 4-51 are problems which are to be treated by numerical means.

4-1. A $\frac{1}{4}$-in.-diameter mild-steel rod at 100 F is suddenly immersed in a liquid at 200 F with $\bar{h}_c$ = 20 Btu/hr sq ft F. Determine the time required for the rod to warm to 190 F.

4-2. A spherical shell satellite (1 ft OD, $\frac{1}{2}$-in. wall thickness, made of stainless steel) re-enters the atmosphere from outer space. If its original temperature is 100 F, the effective average temperature of the atmosphere is 2000 F, and the effective heat-transfer coefficient is 20 Btu/hr sq ft F, estimate the temperature of the shell after re-entry, assuming the time of re-entry is 10 min and the interior of the shell is evacuated.

4-3. A thin-wall cylindrical vessel (3 ft in diam) is filled to a depth of 4 ft with water at an initial temperature of 60 F. The water is well stirred by a mechanical agitator. Estimate the time required to heat the water to 120 F if the tank is suddenly immersed into oil at 220 F. The overall heat-transfer coefficient between the oil and the water is 50 Btu/hr sq ft F, and the effective heat-transfer surface area is 45 sq ft.

4-4. A thin-wall jacketed tank, heated by condensing steam at 14.7 psia, contains 200 lb of agitated water (assume uniform water temperature). The heat-transfer area of the jacket is 10 sq ft and the overall conductance U = 40 Btu/hr sq ft F based on that area. Determine the heating time required for an increase in temperature from 60 to 140 F. *Ans.* 18 min

4-5. The heat-transfer coefficients for the flow of 80 F air over a $\frac{1}{2}$-in. diameter sphere are measured by observing the temperature-time history of a copper ball of the same dimension. The temperature of the copper ball (c = 0.09 Btu/ lb F, ρ = 558 lb/cu ft) was measured by two thermocouples, one located in the center, the other near the surface. Both of the thermocouples registered, within the accuracy of the recording instruments, the same temperature at a given instant. In one test run the initial temperature of the ball was 150 F and in 1.15 min the temperature decreased by 20 F. Calculate the heat-transfer coefficient for this case.

4-6. A spherical stainless steel vessel at 200 F contains 100 lb_m of water at the same temperature. If the entire system is suddenly immersed in ice water, determine (a) the time required for the water in the vessel to cool to 60 F, and (b) the temperature of the walls of the vessel at that time. Assume that:

1. The unit-surface conductance at the inner surface is 3 Btu/hr sq ft F.
2. The unit-surface conductance at the outer surface is 4 Btu/hr sq ft F.
3. The wall of the vessel is 1 in. thick.

4-7. A copper wire, $\frac{1}{32}$ in. OD, 2 in. long, is placed in an air stream whose temperature rises as $T_{air} = (50 + 25\theta)$ F, where θ is the time in seconds. If the initial temperature of the wire is 50 F, determine its temperature after 2 sec, 10 sec, and 1 min. The unit-surface conductance between the air and the wire is 7 Btu/hr sq ft F.

4-8. Ball bearings are to be hardened by quenching them in a water bath at a temperature of 100 F. Suppose you are asked to devise a continuous process in which the balls could roll from a soaking oven at a uniform temperature of 1600 F into the water where they are carried away by a rubber conveyer belt. The rubber conveyor belt would, however, not be satisfactory if the surface temperature of the balls leaving the water is above 200 F. If the surface coefficient of heat transfer between the balls and the water may be assumed to be equal to 104 Btu/hr sq ft F, (a) find an approximate relation giving the minimum allowable cooling time in the water as a function of the ball radius for balls up to $\frac{1}{2}$ in. in diam. (b) Calculate the cooling time, in seconds, required for a ball having a 1-in. diam. (c) Calculate the total amount of heat in Btu/hr which would have to be removed from the water bath in order to maintain its temperature uniform if 100,000 balls of 1-in. diam are to be quenched per hour.

4-9. The thermometer well, described in Example 2-11, is subjected to a sudden temperature rise of 100 F. Derive the equation describing the response of the thermometer, lumping the heat capacity of the well and the thermometer separately. For a unit-surface conductance at the outer surface of the well of 12 Btu/hr sq ft F and a thermal resistance between the inner surface of the well and the mercury thermometer of 0.01 hr sq ft F/Btu, plot the response of the thermometer as a function of time.

4-10. Estimate the time required to heat the center of a 5-lb roast in a 400 F oven to 300 F. State your assumptions carefully and compare your results with cooking instructions in a standard cookbook.

4-11. A large 1-in.-thick copper plate is placed between two air streams. The unit-surface conductance on the one side is 5 Btu/hr sq ft F and on the other side is 10 Btu/hr sq ft F. If the temperature of both streams is suddenly changed from 100 to 200 F, determine how long it will take for the copper plate to reach a temperature of 180 F.

4-12. A 3-lb aluminum household iron has a 50 watt heating element. The surface area is 0.5 sq ft. The ambient temperature is 70 F and the surface heat-transfer coefficient is 2.0 Btu/hr sq ft F (assumed constant). How long after the iron is plugged in will its temperature reach 220 F?

4-13. A slab of material having a thermal diffusivity of 0.05 sq ft/hr is 2 in. thick and of relatively large breadth and width. The slab is being held at a mean temperature of 1000 F in a gas stream having a mean temperature of 1000 F. The gas temperature is controlled by an on-off controller which produces an essentially triangular gas-temperature variation of 25 F amplitude and 10-min period. Presuming the film conductance to be 20 Btu/hr sq ft F and the heat transfer to be convective only, comment on the adequacy of the control system if the slab temperature should not depart from the mean value of 1000 F by more than 5.0 F at any point in the slab.

4-14. Estimate the depth in moist soil at which the annual temperature variation will be 10 percent of that at the surface.

4-15. Two gas streams are passed alternately for a duration of 5 min each over the surface of a steel plate. The one stream is hot (1000 F), the other cold (100 F), but for both streams $\bar{h}$ = 5 Btu/hr sq ft F. Determine the temperature variation with time of the plate surface if the imposed free stream temperature variation is approximated by the Fourier series (θ is in min).

$$T_\infty = 100 + \frac{900}{\pi}\left[(\pi/2) + 2\left(\sin\frac{\pi\theta}{5} + \frac{1}{3}\sin\frac{3\pi\theta}{5} + \frac{1}{5}\sin\frac{5\pi\theta}{5} + \cdots\right)\right]$$

4-16. A small aluminum sphere of diameter D, initially at a uniform temperature T_0, is immersed in a liquid whose temperature varies sinusoidally according to

$$T_\infty - T_m = A \sin \omega\theta$$

where T_m = time-average temperature of the liquid
A = amplitude of the temperature fluctuation
ω = frequency of the fluctuations

If the heat transfer coefficient between the fluid in the sphere, $\bar{h}_c$, is constant and the system may be treated as a "lumped capacity," derive an expression for the sphere temperature as a function of time.

4-17. You are asked to heat and cool a glass bar 1 by 1 by 3 in. The bar is first to be placed in a deep freeze where the temperature is −100 F, and the heat-transfer coefficient is 2 Btu/hr sq ft F. Then the bar is to be placed in a hot box where the temperature is 165 F and the heat-transfer coefficient is 2 Btu/hr sq ft F. The bar is to be removed whenever the temperature at the center reaches −65 F or 120 F respectively. If the cycling is to be repeated 100 times, how long will the test take?

4-18. A large plate of thickness L initially has a linear temperature distribution varying from $T = 0$ at $x = 0$ to $T = T_0$ at $x = L$. The temperature at both surfaces is suddenly changed to 2 T_0 and held constant thereafter. Derive a relation for the temperature as a function of time, distance, and thermal properties.

4-19. Water, while flowing through a pipe, is heated by steam condensing on the outside of the pipe. (a) Assuming the overall conductance is constant along the pipe, derive an expression for the temperature as a function of pipe length. (b) For a unit conductance of 100 Btu/hr sq ft F based on the inside diam of 2 in., steam temperature of 220 F, and a water flow rate of 500 lb_m/min, what length will be required to raise the temperature of the water from 60 F to 150 F? (c) What will be the final water temperature if the pipe length is made twice that calculated in (b)? *Ans.* (b) L = 7.91 ft, (c) 189.4 F

4-20. Derive the conduction equation for an infinitely long cylinder in the unsteady state without heat generation in cylindrical coordinates starting with an energy balance.

4-21. A wire of perimeter P and cross-sectional area A emerges from a die at a temperature T above ambient and with a velocity U. Specify the temperature distribution along the wire in the steady state if the exposed length downstream from the die is quite long. State clearly and try to justify all assumptions.

4-22. A long, slender metal rod of length L is attached at its base to a wall at 0 F. The curved surface of the rod is insulated. The end of the rod is in contact with a fluid at 0 F, where the unit-surface conductance at the interface $\bar{h}$ is constant and uniform. If the initial temperature in the rod is given by $T(x,0) = f(x)$, show that the temperature distribution after time θ is

$$T(x,\theta) = \sum_{n=1}^{\infty} C_n e^{-a\lambda_n{}^2\theta} \sin \lambda_n x$$

where $$C_n = \frac{2\lambda_n L}{(\lambda_n L - \sin \lambda_n L \cos \lambda_n L) L} \int_0^L f(x) \sin \lambda_n x dx$$

Calculate the temperature at the end ($x = L$) of a 0.1-in. diam, 2-in.-long stainless-steel rod as a function of time, if the initial temperature distribution is linear, with 100 F at the end, and $\bar{h} = 10$ Btu/hr sq ft F at the end.

4-23. A stainless-steel cylindrical billet ($k = 10$ Btu/hr ft F, $a = 0.15$ sq ft/hr) is heated to 1100 F preparatory to a forming process. If the minimum temperature permissible for forming is 900 F, how long may the billet be exposed to air at 100 F if the average unit-surface conductance is 15 Btu/hr sq ft F? The shape of the billet is shown in the accompanying sketch.

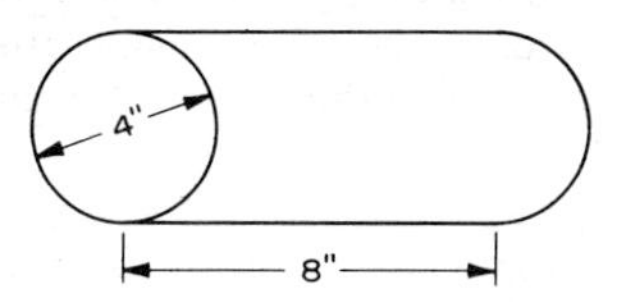

Prob. 4-23

4-24. In the vulcanization of tires, the carcass is placed into a jig, and steam at 300 F is admitted suddenly to both sides. If the tire thickness is 1 in. and the initial temperature 70 F, estimate the time required for the central layer to reach 270 F.

4-25. A long copper cylinder 2 ft in diam and initially at a uniform temperature of 100 F is placed in a water bath at 200 F. Assuming that the heat-transfer coefficient between the copper and the water is 220 Btu/hr sq ft F, calculate the time required to heat the center of the cylinder to 150 F. As a first approximation neglect the temperature gradient within the cylinder; then repeat your calculation without this simplifying assumption and compare your results.

Ans. 4.8 min; 8.1 min

4-26. A steel sphere with a diameter of 3 in. is to be hardened by first heating it to a uniform temperature of 1600 F and then quenching it in a large bath of water which is at a temperature of 100 F. The following data apply:

Surface coefficient $\bar{h}_c$ = 104 Btu / hr sq ft F.
Thermal conductivity of steel = 26 Btu / hr ft F.
Specific heat of steel = 0.15 Btu / lb F.
Density of steel = 490 lb / cu ft.

Calculate: (a) time elapsed in cooling the surface of the sphere to 400 F and (b) time elapsed in cooling the center of the sphere to 400 F.

4-27. A fireproof safe is to be constructed. Its walls consist of two $\frac{1}{16}$-in. steel sheets with a layer of asbestos board between them. Using the chart for a slab, estimate the thickness of asbestos required to give 1 hr of fire protection on the basis that, for an outside temperature of 1500 F, the inside temperature is not to rise above 250 F during this period. The heat-transfer coefficient at the exterior surface is 5 Btu / hr sq ft F.

4-28. A 1-in.-thick sheet of plastic initially at 70 F is placed between two heated steel plates which are maintained at 280 F. The plastic is to be heated just long enough for its mid-plane temperature to reach 270 F. If the thermal conductivity of the plastic is 0.092 Btu / hr ft F, the thermal diffusivity is 0.0029 sq ft / hr, and the thermal resistance at the interface between the plates and the plastic is negligible, calculate: (a) the required heating time, (b) the temperature at a plane $\frac{1}{4}$ in. from the steel plate at the moment the heating is discontinued, and (c) the time required for the plastic to reach a temperature of 270 F $\frac{1}{4}$ in. from the steel plate. *Ans.* (a) 49 min; (b) 274 F; (c) 43 min.

4-29. A turnip (assume spherical) weighing 1 lb is dropped into water boiling at atmospheric pressure. If the initial temperature of the turnip is 62 F, how long does it take to reach 197 F at the center? Assume that:

$\bar{h}_c$ = 300 Btu / hr sq ft F c_p = 0.95 Btu / lb F
k = 0.3 Btu / hr ft F ρ = 65 lb / cu ft

4-30. An egg, which for the purposes of this problem can be assumed to be a 2-in.-diameter sphere having the thermal properties of water, is initially at a temperature of 40 F. It is immersed in boiling water at 212 F for 15 min. The heat-transfer coefficient from the water to the egg may be assumed to be 1000 Btu / hr sq ft F. What is the temperature of the egg center at the end of the cooking period?

4-31. A long wooden rod 1 in. OD is placed at 100 F into an airstream at 1500 F. The unit-surface conductance between the rod and air is 5 Btu / hr sq ft F. If the ignition temperature of the wood is 800 F, ρ = 50 lb/cu ft, k = 0.1 Btu / hr ft F, and c = 0.6 Btu / lb F, determine the time between initial exposure and ignition of the wood.

4-32. In the inspection of a sample of meat intended for human consumption, it was found that certain undesirable organisms were present. In order to make the meat safe for consumption, it is ordered that the meat be kept at a temperature of at least 250 F for a period of at least 20 min during the preparation.

Assume that a slab of this meat, 1 in. thick, is originally at a uniform temperature of 80 F; that it is to be heated from both sides in a constant temperature oven; and that the maximum temperature which meat can withstand is 310 F. Assume furthermore that the surface coefficient of heat transfer remains constant and is 4 Btu/hr ft F. The following data may be taken for the sample of meat:

Specific heat = 1.0 Btu/lb F.
Density = 80 lb/cu ft.
Thermal conductivity = 0.4 Btu/hr ft F.

Calculate the *minimum total time* of heating required to fulfill the safety regulation.

4-33. A frozen-food company freezes its spinach by first compressing it into large slabs and then exposing the slab of spinach to a low-temperature cooling medium. The large slab of compressed spinach is initially at a uniform temperature of 70 F; it must be reduced to an average temperature over the entire slab of −30 F. The temperature at any part of the slab, however, must never drop below −60 F. The cooling medium which passes across both sides of the slab is at a constant temperature of −130 F. The following data may be used for the spinach:

Density = 5 lb/cu ft.
Thermal conductivity = 0.5 Btu/hr ft F.
Specific heat = 0.5 Btu/lb F.

Present a detailed analysis outlining a method to estimate the maximum thickness of the slab of spinach which may be safely cooled in 60 min. *Ans.* 12 in.

4-34. To determine the heat-transfer coefficient between a heated steel ball and cooler ground or crushed mineral solids experimentally, a series of SAE 1040 steel balls were heated to a temperature of 700 C and the center temperature-time history of each was measured with a thermocouple while it was cooling in a bed of crushed iron ore which was placed in a steel drum, rotating horizontally at about 30 rpm. For a 2-in.-diameter ball, the time required for the temperature difference between the ball center and the surrounding ore to decrease from 500 to 250 C was found to be 64, 67, and 72 sec respectively in three different test runs. Determine the average unit-surface conductance between the ball and the ore. Compare the results obtained by assuming the thermal conductivity to be infinite with those obtained by taking the internal thermal resistance of the ball into account. *Ans.* ~54 Btu/hr sq ft F

4-35. A cylindrical mild-steel billet, 1 in. OD, 3 in. long, initially at 1000 F is cooled in a large vessel filled with oil at 200 F. The average unit-surface conductance between the oil and the steel during cooling is 100 Btu/hr sq ft F. Determine the time required to cool (a) the center of the billet and (b) the surface of the billet to 500 F.

4-36. A mild-steel cylindrical billet, 10 in. in diameter, is to be raised to a minimum temperature of 1400 F by passing it through a 20-ft-long strip type furnace. If the furnace gases are at 2800 F and the over-all unit-surface conductance on the outside of the billet is 12 Btu/hr sq ft F, determine the maximum speed at which a continuous billet entering at 400 F can travel through the furnace.

4-37. A solid lead cylinder 2 ft in diam and 2 ft long, initially at a uniform temperature of 250 F, is dropped into a 70 F liquid bath in which the unit-surface conductance $\bar{h}_c$ is 200 Btu/hr sq ft F. Plot the temperature-time history of the center of this cylinder and compare it with the time histories of a 2-ft in diameter, infinitely long lead cylinder and a lead slab of 2-ft thickness.

4-38. A long 2 ft OD solid steel (k = 12 Btu/hr ft F) cylindrical billet at 60 F room temperature is placed in an oven where the temperature is 500 F. If the average unit-surface conductance is 30 Btu/hr sq ft F, estimate the time required for the center temperature to increase to 450 F by (a) using the appropriate chart, (b) dividing the solid into *two* equal lumped thermal capacities with appropriate thermal resistances between them. Also (c) determine the instantaneous surface heat fluxes when the center temperature is 450 F.

4-39. Repeat Prob. 4-38a but assume that the billet is only 4 ft long with the average unit-surface conductance at both ends equal to 24 Btu/hr sq ft F.

4-40. A large billet of steel originally at a temperature of 500 F is placed in a radiant furnace where the surface temperature is held at 2200 F. Assuming the billet infinite in extent, compute the temperature at point P shown in the accompanying sketch after 25 min have elapsed. The average steel properties are: k = 23 Btu/hr ft F, ρ = 460 lb_m/cu ft, and c = 0.12 Btu/lb_m F. *Ans.* 1904 F

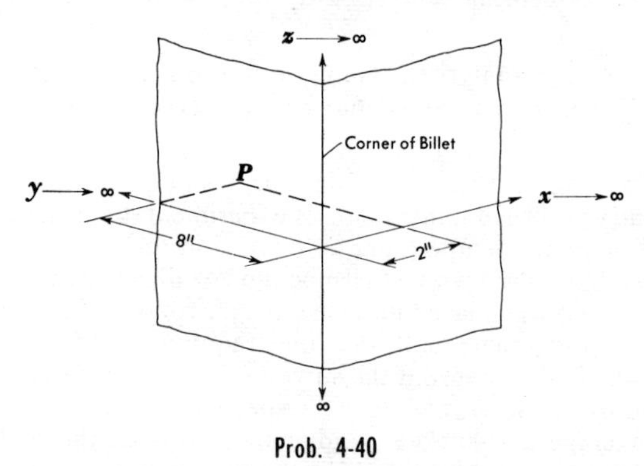

Prob. 4-40

4-41. A long 1-in.-diameter stainless-steel rod (k = 10 Btu/hr sq ft F, a = 0.48 sq ft/hr) originally at 0 F throughout is suddenly exposed on one end to a fluid at 1000 F. If the circumference of the rod is insulated and the unit-surface conductance between the fluid and the exposed end of the rod is 60 Btu/hr sq ft F, perform the steps indicated below to determine the time required for the exposed end of the rod to reach 500 F by numerical means, subdividing the rod into 1-in.-long sections. (a) Show the finite-difference equation for interior points. (b) Show the finite-difference equation for the end point. (c) Verify the stability of solution and select an appropriate value of M. (d) Determine the $\Delta\theta$ increment. (e) Carry out the numerical solution.

4-42. Repeat Example 4-41 using the numerical method with M = 2 and M = 3.

4-43. A long steel rod (k = 10 Btu/hr ft F, a = 0.48 sq ft/hr) originally at 0 F is suddenly exposed on one end to a heat source at 3000 F. If the circumference of the rod is insulated and the unit-surface conductance between the heat source and the rod is 50 Btu/hr sq ft F, determine the time required for the exposed end of the rod to reach 1500 F by numerical means. Compare your answer with that obtained from an appropriate chart in the book.

4-44. Determine the temperature distribution in a 4.8-in.-thick plate initially at a uniform temperature of 100 F if its thermal diffusivity is 0.25 sq ft/hr and it is suddenly immersed in a medium at 500 F with negligible thermal resistances at both surfaces. Use $M = 4$.

4-45. A large steel plate (thermal diffusivity of 0.5 sq ft/hr and thermal conductivity of 25 Btu/hr ft F) is 1 in. thick. At zero time it begins to receive heat on one side at the rate of 30,000 Btu/hr sq ft while the other side is exposed to a fluid at 0 F through a unit-surface conductance of 1000 Btu/hr sq ft F. If the initial temperature of the entire plate is 100 F, determine by the numerical method the temperature at the mid-plane after 5 sec.

4-46. A large slab of 1-ft-thick steel armor plate (k = 10 Btu/hr ft F, a = 0.12 sq ft/hr) is initially at a uniform temperature of 1300 F. One surface is maintained at 1300 F while air is blown over the other surface at a velocity which gives rise to an average heat-transfer coefficient of 20 Btu/hr sq ft F. The temperature of the air varies with time as $T_\infty = (600 - 10\theta)$ F, where θ is in minutes. Determine the surface temperature and the distribution after 1 hr has elapsed.

4-47. A 2-ft-thick concrete wall has initially a sinusoidal temperature distribution, 50 F at both surfaces and 150 F in the center. The left and right surfaces are suddenly changed to 75 and 100 F, respectively. Find the time necessary for the maximum temperature in the wall to drop to 120 F by numerical means.

4-48. A large slab of ice, having a uniform temperature of −10 F, is suddenly subjected to a surface temperature of 32 F. Compute by numerical means the time required to raise the temperature of the ice 1 ft below the surface to 0 F.

4-49. A steel plate (a = 0.5 sq ft/hr, k = 25 Btu/hr ft F) receives heat on one side at the rate of 3000 Btu/hr sq ft and is exposed on the other side to a fluid at 0 F through a unit-surface conductance of 1000 Btu/hr sq ft F. If the plate is 1 in. thick and the initial temperature of the plate is 100 F, determine by numerical means the temperature distribution after 5 sec have elapsed.

4-50. The temperature distribution in a 6-in.-thick magnesite wall (k = 2.2 Btu/hr ft F, a = 0.06 sq ft/hr) is linear, 500 F on the hot side and 100 F on the cold side. At time zero the air temperature on the hot side is suddenly reduced to 100 F. The heat-transfer coefficient between the slab and the air is a function of the temperature difference given by

$$\bar{h} = 2.0 + 0.01\,(T_{surface} - T_{air}) \qquad \text{Btu/hr sq ft F}$$

Determine by means of the numerical method the time required to cool the hot side to 200 F. *Ans.* $\cong$1 hr

4-51. A long concrete bar of triangular cross section having the dimensions shown in the accompanying sketch is initially at a temperature of 60 F. It is suddenly placed into an environment at 160 F and is being heated through a unit-surface conductance of 5 Btu/hr sq ft F over its two upper surfaces, while the 10-ft lower surface on which the bar rests is insulated. Using a 1 by 1 ft grid, estimate numerically the temperature distribution after 1 hr of heating and also the time required for the minimum temperature in the slab to reach 80 F.

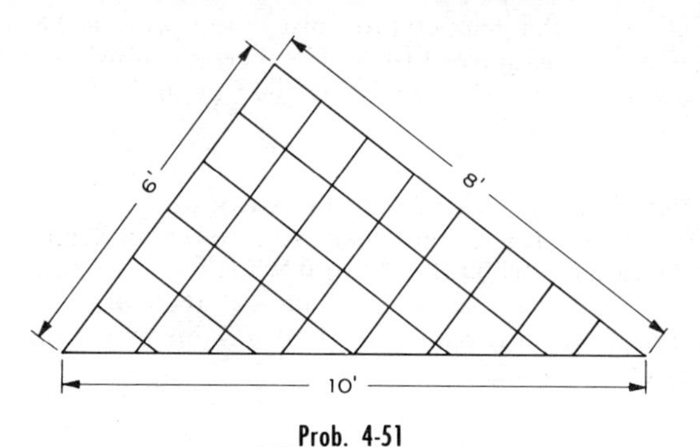

Prob. 4-51

REFERENCES

1. P. J. Schneider, *Conduction Heat Transfer.* (Cambridge, Mass.: Addison-Wesley Publishing Company, 1955.)

2. H. S. Carslaw and J. C. Jaeger, *Conduction of Heat in Solids.* (Oxford: Clarendon Press, 1947.)

3. M. N. Özisik, *Boundary Value Problems of Heat Conduction.* (Scranton, Pennsylvania: International Textbook Company, 1968.)

4. V. Paschkis and M. D. Baker, "A Method for Determining Unsteady State Heat Transfer by Means of an Electrical Analogy," *Trans. ASME*, Vol. 64 (1942), pp. 105–112.

5. C. B. Neel, Jr., "An Investigation Utilizing an Electrical Analogue of Cyclic De-Icing of a Hollow Steel Propeller with an External Blade Shoe," *NACA TN* 2852, 1952.

6. D. I. Lawson and J. H. McGuire, "The Solution of Transient Heat Flow Problems by Analogous Electrical Networks," *Proc. (A) Inst. Mech. Engrs.*, Vol. 167, No. 3 (1953), pp. 275–287.

7. G. A. Hawkins and J. T. Agnew, "The Solution of Transient Heat Conduction Problems by Finite Differences," *Eng. Bull. Res. Series* 98, Purdue Univ., 1947.

8. G. M. Dusinberre, *Numerical Analysis of Heat Flow.* (New York: McGraw-Hill Book Company, Inc., 1949.)

9. L. M. K. Boelter, V. H. Cherry, and H. A. Johnson, *Heat Transfer*, 3d ed. (Berkeley: University of California Press, 1942.)

10. H. Gröber, S. Erk, and U. Grigull, *Fundamentals of Heat Transfer.* (New York, N.Y.: McGraw-Hill Book Co., 1961.)

11. M. P. Heisler, "Temperature Charts for Induction and Constant Temperature Heating," *Trans. ASME*, Vol. 69 (1947), pp. 227–236.

12. B. O. Peirce, *A Short Table of Integrals.* (Boston: Ginn & Company, 1929.)

13. G. Leppert, "A Stable Numerical Solution for Transient Heat Flow," *J. Am. Soc. Naval Engrs.*, Vol. 65 (1953), pp. 741–752.

14. C. M. Fowler, "Analysis of Numerical Solutions of Transient Heat Flow Problems," *Quart. Appl. Math.*, Vol. 3 (1945), pp. 361–376.

15. M. Jakob and G. A. Hawkins, *Elements of Heat Transfer*, 3d ed. (New York: John Wiley & Sons, Inc., 1957.)

16. G. R. Gaumer, "The Stability of Three Finite-Difference Methods of Solving for Transient Temperatures," *Proc. Fifth U.S. Navy Symp. on Aeroballistics*, Vol. 1, Paper No. 32, U.S. Naval Ord. Lab., White Oak, Md., October, 1961.

17. H. G. Elrod, Jr., "New Finite Difference Technique for Solution of the Heat Conduction Equation, Especially Near Surfaces with Convective Heat Transfer," *Trans. ASME*, Vol. 79 (1957), pp. 1519–1526.

18. H. G. Elrod, Jr., "Improved Lumped Parameter Method for Transient Heat Conduction Calculations with Flat-Slab and Cylindrical Elements," *Trans. ASME*, Ser. C, Vol. 82 (1960), pp. 181–188.

19. G. Liebmann, "A New Electrical Analog Method for the Solution of Transient Heat-Conduction Problems," *Trans. ASME*, Vol. 78 No. 3 (1956), pp. 655–666.

20. H. Schenck, *Fortran Methods in Heat Flow*. (New York, N.Y.: The Ronald Press Comp., 1963.)

21. G. M. Dusinberre, *Heat Transfer Calculations By Finite Differences*. (Scranton, Pa.: International Textbook Co., 1961.)

22. G. R. Gaumer, "Stability of Three Finite Difference Methods of Solving for Transient Temperatures," *ARS Journal*, Vol. 32 No. 10 (1962), pp. 1595–1596.

23. P. J. Schneider, *Temperature Response Charts*. (New York, N.Y.: John Wiley & Sons, Inc., 1963.)

24. Gordon D. Smith, *Numerical Solution of Partial Differential Equations with Exercises and Worked Solutions*. (London: Oxford University Press, 1965.)

25. Robert D. Richtmeyer, *Difference Methods for Initial Value Problems*. (New York: Interscience Publishers, Inc. 1957.)

26. J. Cranck and P. Nicolson, "A Practical Method for Numerical Evaluation of Solutions of Partial Differential Equations of the Heat Conduction Type," *Proc. Camb. Phil. Soc.*, Vol. 43 (1947), pp. 50–67.

27. D. W. Peaceman and H. H. Rachford, "The Numerical Solution of Parabolic and Elliptic Differential Equations," *J. Soc. Indust. Appl. Math.*, Vol. 3 (1955), pp. 28–41.

28. H. Z. Barakat and J. A. Clark, "On the Solution of Diffusion Equations by Numerical Methods," *Trans. ASME, J. Heat Transfer*, Vol. 88 (1966), pp. 421–427.

29. R. F. Thomas and M. D. J. MacRoberts, "RATH Thermal Analysis Programs," LA-3264-MS, UC-32, *Mathematics and Computers* (March 1965), Los Alamos Scientific Laboratory, Los Alamos, N.M.

5
Heat transfer by radiation

5-1. Thermal radiation

When a body is placed in an enclosure whose walls are at a temperature below that of the body, the temperature of the body will decrease even if the enclosure is evacuated. The process by which heat is transferred from a body by virtue of its temperature, without the aid of any intervening medium, is called thermal radiation. This chapter deals with the characteristics of thermal radiation and radiation exchange, i.e., heat transfer by radiation.

The physical mechanism of radiation is not yet completely understood. Radiant energy is sometimes envisioned to be transported by electromagnetic waves, at other times by photons. Neither viewpoint completely describes the nature of all observed phenomena. It is known, however, that radiation travels with the speed of light, c, equal to about 3×10^8 m/sec in a vacuum. From the viewpoint of electromagnetic theory, the waves travel at that speed. Alternatively, from a quantum point of view, energy is transported by photons which travel at the speed of light. Although all the photons have the same velocity, there exists always a distribution of energy among them. The energy associated with each photon is $h\nu$, where h is a constant and ν is the frequency of the radiation. The energy spectrum can also be described in terms of the wavelength of radiation, λ, which is related to the propagation velocity and the frequency by

$$\lambda = c/\nu$$

The unit of wavelength used in this chapter is the micron, μ, equal to 10^{-6} meters or 3.94×10^{-5} inches.

Radiation phenomena are usually classified by their characteristic wavelength (Fig. 5-1). Electromagnetic phenomena encompass many types of radiation, from short wavelength γ- and x-rays to long wavelength radiowaves. The wavelength of radiation depends on how it is produced. For example, a metal bombarded by high frequency electrons emits x-rays while certain crystals can be excited to emit long wavelength radiowaves. Thermal radiation is defined as *radiant energy emitted by a medium by virtue of its temperature.* In other words, the emission of thermal radiation is governed by the temperature of the emitting body. The wavelength range encompassed by thermal radiation falls approximately between 0.1 and 100μ. This range is usually subdivided into the ultraviolet, the visible, and the infrared as shown in Fig. 5-1. The sun, with an effective surface temperature of about 10,000 F, emits most of its energy below 3μ, whereas a lamp filament at 2,000 F emits over 90 percent of its radiation between 1 and 10μ. A greenhouse is warm inside even when the outside air is cool: glass permits radiation at the wavelength of the sun to pass, but it is opaque to radiation in the wavelength range emitted by the interior of the greenhouse. Thus solar radiation may enter, but once it has been absorbed, it cannot leave the interior of the greenhouse.

5-2. Blackbody radiation

A blackbody, or ideal radiator, is a body which emits and absorbs at any temperature the maximum possible amount of radiation at any given wavelength. The ideal radiator is a theoretical concept which sets

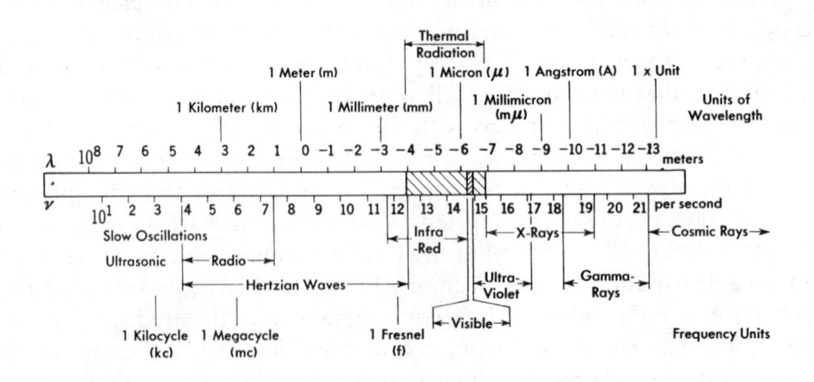

Fig. 5-1. Electromagnetic wave spectrum.

an upper limit to the emission of radiation in accordance with the second law of thermodynamics. It is a standard with which the radiation characteristics of other media are compared.

The total energy emission per unit time and per unit area from a blackbody at a wavelength λ in the wavelength range $d\lambda$ will be denoted by $E_{b\lambda} d\lambda$. This quantity $E_{b\lambda}$ is usually called the *spectral (or monochromatic) blackbody emissive power.* The adjective *spectral* is used to denote the dependance of the radiation upon wavelength, while *monochromatic* refers to a radiation quantity at a given wavelength λ. A relationship which shows how the emissive power of a blackbody is distributed among the different wavelengths was derived by Max Planck, in 1900, by means of his quantum theory. According to Planck's law

$$E_{b\lambda}(T) = \frac{C_1}{\lambda^5(e^{C_2/\lambda T} - 1)} \tag{5-1}$$

where $E_{b\lambda}$ = monochromatic emissive power of a blackbody at temperature T in Btu/hr ft^2 micron,

λ = wavelength in microns (μ),

T = absolute temperature of the body in degrees R (deg F abs.),

C_1 = 1.1870×10^8 Btu/micron^4/ft^2 hr, and

C_2 = 2.5896×10^4 micron R.

The monochromatic emissive power for a blackbody at various temperatures is plotted in Fig. 5-2 as a function of wavelength. One observes that at temperatures below 10,000 R the emission of radiation energy is appreciable between 0.2 and about 50μ. The wavelength at which the monochromatic emissive power is a maximum, $E_{\lambda b}(\lambda_{max}, T)$, shifts with increasing temperature to shorter wavelengths. The relationship between the wavelength λ_{max} at which $E_{b\lambda}$ is a maximum and the absolute temperature is given by Wien's displacement law (1) as

$$\lambda_{max} T = 5215.6 \, \mu \, \text{R} \tag{5-2}$$

The visible range of wavelengths extends only over a narrow region from about 0.4 to 0.7μ, shown as a shaded band in Fig. 5-2. Only a very small amount of the total energy falls into this range of wavelengths at temperatures below 1200 F. At higher temperatures, the amount of radiant energy within the visible range increases and the human eye begins to detect the radiation. The sensation produced on the retina and transmitted to the optical nerve depends on the temperature, a phenomenon which is still used to estimate the temperatures of metals during heat treatment. At about 1300 F an amount of radiant energy sufficient to be observed is emitted at wavelengths between 0.6 to 0.7μ, and an object at that temperature glows with a dull-red color. As the temperature is further increased, the color changes to bright red and yellow, becoming nearly

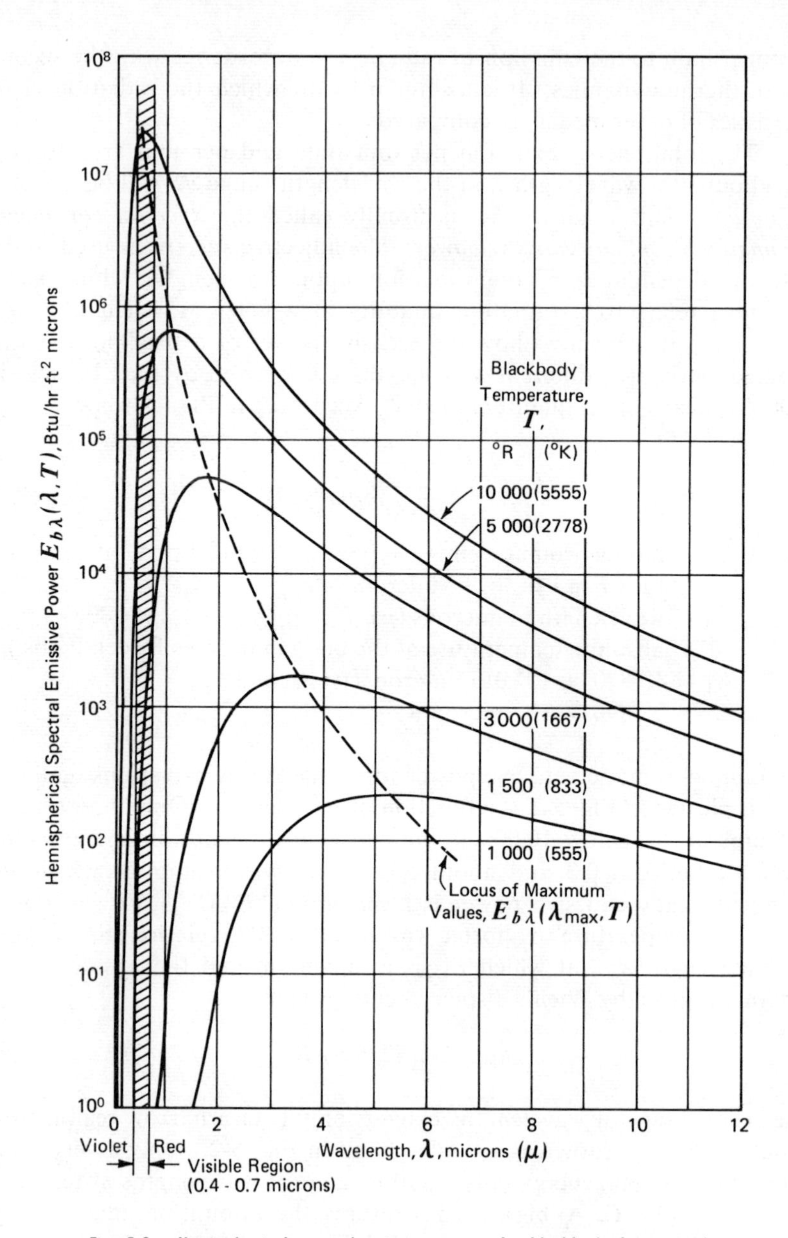

Fig. 5-2. Hemispherical spectral emissive power of a blackbody for several different temperatures.

white at about 2400 F. At the same time also, the brightness increases because more and more of the total radiation falls within the visible range.

It will be recalled from Chapter 1 that the total emission of radiation per unit surface area per unit time from a blackbody is related to the fourth power of the absolute temperature according to the Stefan-Boltzman law

$$E_b(T) = q_r/A = \sigma T^4 \tag{5-3}$$

where A = area of the blackbody emitting the radiation, and
T = absolute temperature of the area A in degrees R.

The total emissive power given by Eq. 5-3 represents the total thermal radiation emitted over the entire wavelength spectrum. At a given temperature T the area under a curve such as shown in Fig. 5-2 is E_b. The total emissive power and the monochromatic emissive power are related by

$$\int_0^\infty E_{b\lambda} d\lambda = \sigma T^4 \tag{5-4}$$

Substituting Eq. 5-1 for $E_{b\lambda}$ and performing the integration indicated above (see Problem 5-8) will show that the Stefan-Boltzmann constant σ and the constants C_1 and C_2 in Planck's law are related by

$$\sigma = \left(\frac{\pi}{C_2}\right)^4 \frac{C_1}{15} = 1.714 \times 10^{-9} \text{ Btu/hr ft}^2 \text{ R}^4 \tag{5-5}$$

For engineering calculations involving real surfaces it is often important to know the energy radiated at a specified wavelength or in a finite band between specific wavelengths λ_1 and λ_2, i.e., $\int_{\lambda_1}^{\lambda_2} E_{b\lambda}(T)\, d\lambda$. Numerical calculations for such cases are facilitated by the use of the auxiliary curves shown in Figs. 5-3, 5-4, and 5-5. The construction of these curves and their application is illustrated below.

At any given temperature the monochromatic emissive power is a maximum at the wavelength $\lambda_{max} = 5{,}215.6/T$ according to Eq. 5-2. Substituting λ_{max} in Eq. 5-1 gives $E_{b\lambda max}(T)$, or

$$E_{b\lambda max}(T) = \frac{C_1 T^5}{5215.6^5(e^{C_2/5215.6} - 1)}$$

$$= 2.161 \times 10^{-13}\, T^5 \text{ Btu/hr sq ft } \mu \tag{5-6}$$

This relationship is shown graphically in Fig. 5-3 where $E_{b\lambda max}(T)$ is plotted as a function of the absolute temperature T in degrees R.

If we divide the monochromatic emissive power of a blackbody $E_{b\lambda}(T)$ by its maximum emissive power at the same temperature

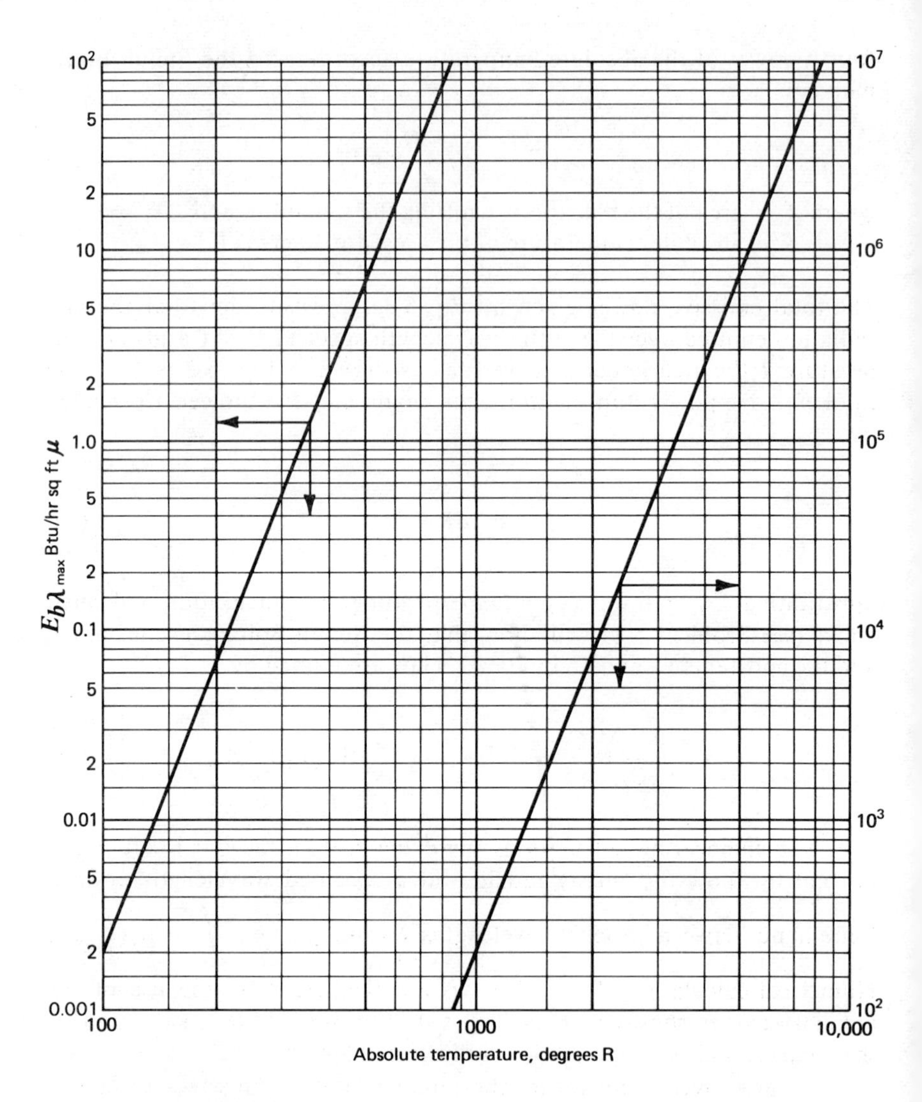

Fig. 5-3. Maximum monochromatic emissive power as a function of absolute temperature.

$E_{b\lambda\max}(T)$, we obtain the dimensionless ratio

$$\frac{E_{b\lambda}(T)}{E_{b\lambda\max}(T)} = \frac{C_1}{\lambda^5(e^{C_2/\lambda T} - 1)} \frac{\lambda_{\max}{}^5(e^{C_2/\lambda_{\max} T} - 1)}{C_1}$$

$$= \left(\frac{5215.6}{\lambda T}\right)^5 \left(\frac{e^{4.965} - 1}{e^{25896/\lambda T} - 1}\right) \qquad (5\text{-}7)$$

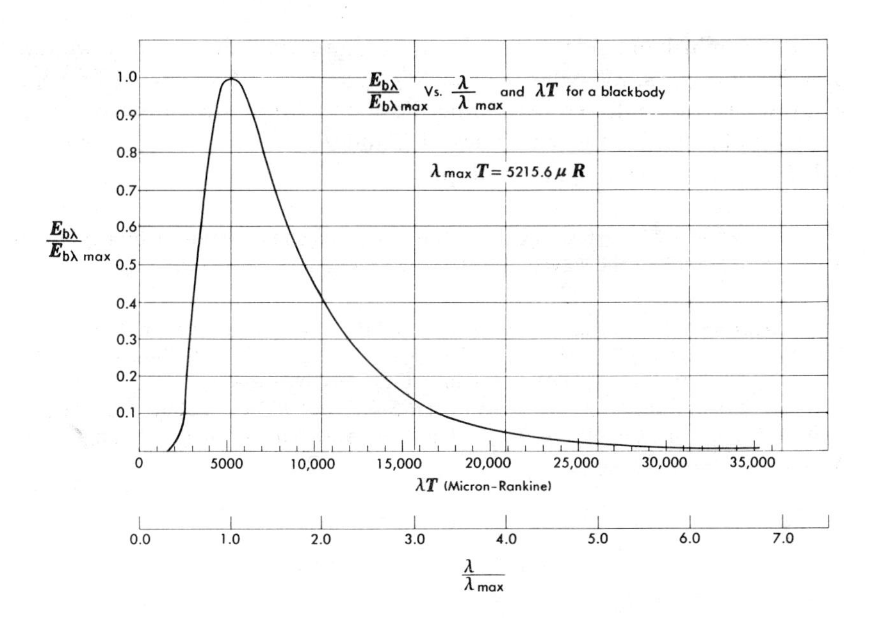

Fig. 5-4. Ratio of monochromatic emissive power to maximum monochromatic emissive power at λ/λ_{max} as a function of λT.

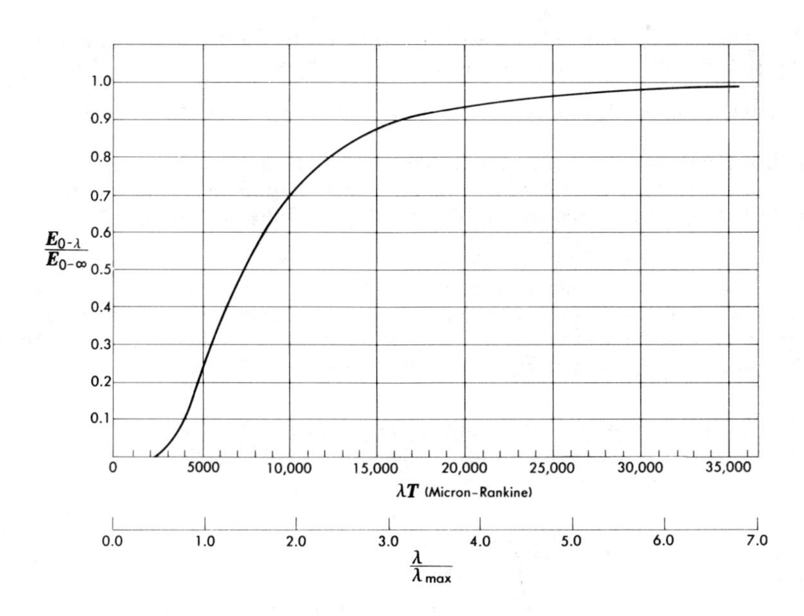

Fig. 5-5. Fraction of total emissive power in spectral region between $\lambda = 0$ and λ as a function of λT.

Since the right-hand side of Eq. 5-7 is a unique function of λT, we can plot $E_{b\lambda}/E_{b\lambda \max}$ vs. λT or vs $\lambda/\lambda_{\max}$ ($\lambda/\lambda_{\max} = 5{,}215.6\lambda/T$ from Eq. 5-2) as shown in Fig. 5-4. To determine the monochromatic emissive power $E_{b\lambda}$ for a blackbody at given values of λ and T, read $E_{b\lambda}/E_{b\lambda \max}$ from Fig. 5-4, $E_{b\lambda \max}$ from Fig. 5-3, and multiply.

EXAMPLE 5-1. Determine the monochromatic emissive power of a tungsten filament at 2,000 F at the wavelength at which it is a maximum and at 5μ.

Solution: From Fig. 5-3, at $T = 2{,}460$ R, $E_{b\lambda \max}$ is approximately 20,000 Btu/hr ft$^2\mu$; from Eq. 5-6, $E_{b\lambda \max} = 2.16 \times 10^{-13} \times 2{,}460^5 = 19{,}400$ Btu/hr ft$^2\mu$. From Eq. 5-2 the wavelength at which the emissive power is a maximum is $\lambda_{\max} = 5{,}215.6/2{,}460 = 2.12\mu$. From Fig. 5-4 $E_{b\lambda}$ at $\lambda T = 12{,}300\mu$ R is 28 percent of $E_{b\lambda \max}$, or 5430 Btu/hr ft$^2\mu$. The blackbody emission in the range from $\lambda = 0$ to λ is given by

$$E_{b,0-\lambda}(T) = \int_0^{\lambda} E_{b\lambda}(T)\, d\lambda \tag{5-8}$$

The foregoing expression may be recast in dimensionless form as

$$\frac{E_b(0 - \lambda T)}{\sigma T^4} = \int_0^{\lambda T} \frac{E_{b\lambda}}{\sigma T^5}\, d(\lambda T) \tag{5-9}$$

The quantity $E_b(0 - \lambda T)/\sigma T^4$ is the fraction of the total blackbody emissive power at temperature T contained in the range between 0 and λT. The results are shown graphically in Fig. 5-5 where $\int_0^{\lambda} E_{b\lambda}(T)\, d\lambda/\sigma T^4$ is plotted. A numerical tabulation is presented in Table 5-1. With the aid of Fig. 5-5 or Table 5-1, one can easily determine the quantity of radiation in a spectral band between λ_1 and λ_2. The procedure is illustrated in the following example.

EXAMPLE 5-2. Silica glass transmits 92 percent of the incident radiation in the wavelength range between 0.35 and 2.7μ and is opaque at longer and shorter wavelengths. Estimate the percent of solar radiation which the glass will transmit. The sun may be assumed to radiate as a blackbody at 10,000 R.

Solution: For the wavelength range within which the glass is transparent, $\lambda T = 3500$ at the lower limit and 27,000 at the upper limit. From Fig. 5-5 or Table 5-1 we find

$$\frac{\int_0^{0.35} E_{b\lambda} d\lambda}{\int_0^{\infty} E_{b\lambda} d\lambda} = 6 \text{ percent}$$

Table 5-1. Radiation functions from R. V. Dunkle, Trans. ASME, 76, 549 (1954).

λT	$\frac{E_{\lambda b} \times 10^5}{\sigma T^5}$	$\frac{E_{b(0-\lambda T)}}{\sigma T^4}$	λT	$\frac{E_{\lambda b} \times 10^5}{\sigma T^5}$	$\frac{E_{b(0-\lambda T)}}{\sigma T^4}$	λT	$\frac{E_{\lambda b} \times 10^5}{\sigma T^5}$	$\frac{E_{b(0-\lambda T)}}{\sigma T^4}$
1000	.0000394	0	7200	10.089	.4809	13400	2.714	.8317
1200	.001184	0	7400	9.723	.5007	13600	2.605	.8370
1400	.01194	0	7600	9.357	.5199	13800	2.502	.8421
1600	.0618	.0001	7800	8.997	.5381	14000	2.416	.8470
1800	.2070	.0003	8000	8.642	.5558	14200	2.309	.8517
2000	.5151	.0009	8200	8.293	.5727	14400	2.219	.8563
2200	1.0384	.0025	8400	7.954	.5890	14600	2.134	.8606
2400	1.791	.0053	8600	7.624	.6045	14800	2.052	.8648
2600	2.753	.0098	8800	7.304	.6195	15000	1.972	.8688
2800	3.872	.0164	9000	6.995	.6337	16000	1.633	.8868
3000	5.081	.0254	9200	6.697	.6474	17000	1.360	.9017
3200	6.312	.0368	9400	6.411	.6606	18000	1.140	.9142
3400	7.506	.0506	9600	6.136	.6731	19000	.962	.9247
3600	8.613	.0667	9800	5.872	.6851	20000	.817	.9335
3800	9.601	.0850	10000	5.619	.6966	21000	.702	.9411
4000	10.450	.1051	10200	5.378	.7076	22000	.599	.9475
4200	11.151	.1267	10400	5.146	.7181	23000	.516	.9531
4400	11.704	.1496	10600	4.925	.7282	24000	.448	.9589
4600	12.114	.1734	10800	4.714	.7378	25000	.390	.9621
4800	12.392	.1979	11000	4.512	.7474	26000	.341	.9657
5000	12.556	.2229	11200	4.320	.7559	27000	.300	.9689
5200	12.607	.2481	11400	4.137	.7643	28000	.265	.9718
5400	12.571	.2733	11600	3.962	.7724	29000	.234	.9742
5600	12.458	.2983	11800	3.795	.7802	30000	.208	.9765
5800	12.282	.3230	12000	3.637	.7876	40000	.0741	.9881
6000	12.053	.3474	12200	3.485	.7947	50000	.0326	.9941
6200	11.783	.3712	12400	3.341	.8015	60000	.0165	.9963
6400	11.480	.3945	12600	3.203	.8081	70000	.0092	.9981
6600	11.152	.4171	12800	3.071	.8144	80000	.0055	.9987
6800	10.808	.4391	13000	2.947	.8204	90000	.0035	.9990
7000	10.451	.4604	13200	2.827	.8262	100000	.0023	.9992
						∞	0	1.0000

and

$$\frac{\int_0^{2.7} E_{b\lambda}\,d\lambda}{\int_0^{\infty} E_{b\lambda}\,d\lambda} = 96.9 \text{ percent}$$

Thus 90.9 percent of the total radiant energy incident upon the glass from the sun is in the wavelength range between 0.35 and 2.7μ and 83.6 percent of the solar radiation is transmitted through the glass. *Ans.*

As mentioned before, the concept of a blackbody is an idealization because all surfaces reflect some of the incident radiation. For laboratory purposes, however, a blackbody can be approximated by a cavity, such as a hollow sphere, whose interior walls are maintained at a uniform temperature T. If a small hole is provided in the wall, any radiation entering through it is partly absorbed and partly reflected at the interior surfaces. The reflected radiation, as shown schematically in Fig. 5-6, will not immediately escape from the cavity but will first strike repeatedly the interior surface. Each time it strikes, a part of it is absorbed; when the original radiation beam finally reaches the hole again and escapes, it has been so weakened by repeated reflection that the energy leaving the cavity is negligible. This is true regardless of the surface and composition of the wall of the cavity. Thus, a small hole in the walls surrounding a large cavity acts like a blackbody because practically all the radiation incident upon it is absorbed.

In a similar manner, the radiation emitted by the interior surface of a cavity is absorbed and reflected many times and eventually fills the cavity uniformly. If a blackbody at the same temperature as the interior surface is placed into the cavity, it receives radiation uniformly, i.e., it is irradiated isotropically. The blackbody absorbs all of the incident radiation and, since the system consisting of the blackbody and the cavity is at a uniform temperature, the rate of emission of radiation by the body must equal its rate of irradiation. Otherwise there would be a net transfer of energy as heat between two bodies at the same temperature in an isolated system, an obvious violation of the second law of thermodynamics. Denoting the rate at which radiant energy from the walls of the cavity is incident on the blackbody, i.e., the *blackbody irradiation*, by G_b and the rate at which the blackbody emits energy by E_b, we thus obtain $G_b = E_b$. This means that the irradiation in a cavity whose walls are at a temperature T is equal to the emissive power of a blackbody at the same temperature. A small hole in the wall of a cavity will not disturb this condition appreciably, and the radiation escaping from it will therefore have blackbody characteris-

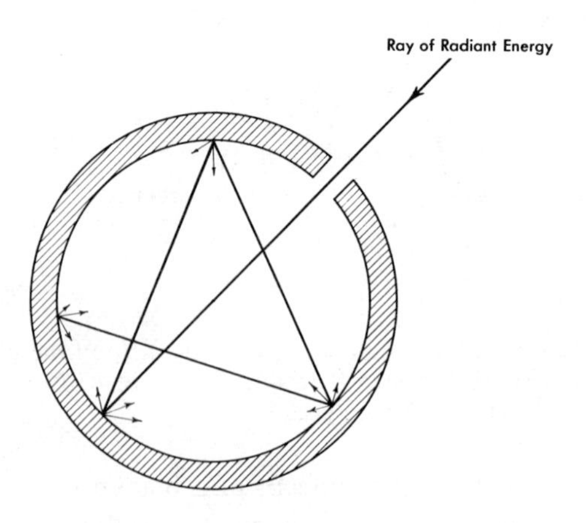

Fig. 5-6. Reflection of radiation in a cavity.

tics. Since this radiation is independent of the nature of the surface, it follows that the emissive power of a blackbody depends only on its temperature.

Intensity of radiation. In our discussion so far we have only considered the total amount of radiation leaving a surface, i.e., the emissive power. This concept, however, is inadequate for a heat-transfer analysis when the amount of radiation passing in a given direction and intercepted by some other body is sought. The amount of radiation passing in a given direction is described in terms of the *intensity of radiation*, I. Before defining the intensity of radiation, we must have measures of the direction and the space into which a body radiates. The rate of radiation heat flow per unit surface area of a body which passes in a given direction can be measured by determining the radiation through an element on the surface of a hemisphere constructed around the radiating surface. If the radius of this hemisphere equals unity, the hemisphere has a surface area of 2π and subtends a solid angle of 2π steradians about a point at the center of its base. The surface area on such a unit sphere has the same numerical value as the so-called solid angle ω measured from the radiating surface element and can be used to define simultaneously the direction and the space into which radiation from a body propagates.

Referring to Fig. 5-7, the intensity of radiation I is defined as the energy emitted per unit area per unit time into a unit solid angle centered around the direction θ of the pencil of rays per unit emitting surface area *projected normal to the direction* θ. To illustrate the concept of intensity,

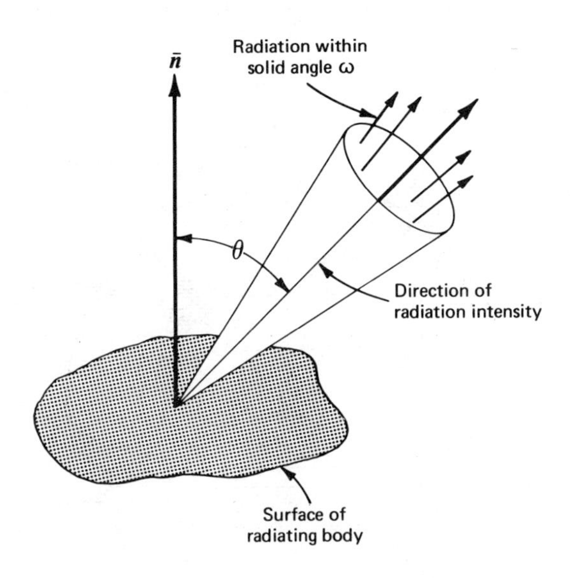

Fig. 5-7. Intensity of radiation.

let $d(q/A)_r$ represent the radiant energy per unit time and unit surface area that passes within a differential solid angle denoted by $d\omega$ inclined at an angle θ with respect to the normal of the emitting surface. The intensity is then given by

$$I = \frac{d(q/A)_r}{d\omega \cos\theta} \tag{5-10}$$

Observe that to conform to the definition of intensity, which is based on the area normal to the direction θ, the surface area of the radiating body is multiplied by $\cos\theta$. To relate the intensity of radiation to the emissive power, one simply determines the energy from a surface radiating into a hemispherical enclosure placed above it, as shown in Fig. 5-8. Since the hemisphere will intercept all the radiant rays emanating from the surface, the total amount of radiation passing through the hemispherical surface equals the emissive power.

The differential solid angle $d\omega$ subtended by a surface receiving radiation is equal to the area of that surface projected in the direction of the incident radiation divided by the square of the distance between the emitting and receiving surfaces. As illustrated in Fig. 5-8, the differential solid angle $d\omega$ can be expressed in terms of the radius r and the angles θ and ψ of a spherical co-ordinate system. It is equal to a surface element on the hemisphere, dA_s, divided by the square of the hemisphere radius, or

$$d\omega = \frac{dA_s}{r^2} = \frac{r d\theta \cdot r \sin\theta \, d\psi}{r^2} \tag{5-11}$$

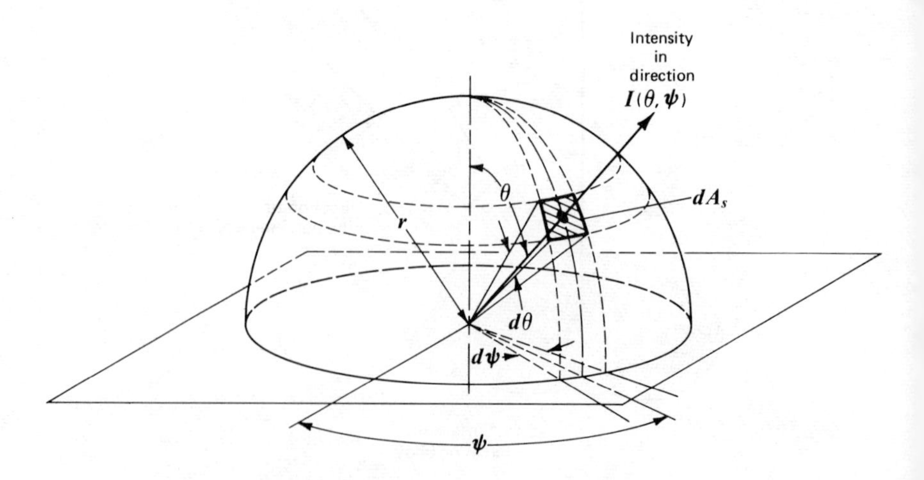

Fig. 5-8. Nomenclature for intensity of radiation in a spherical co-ordinate system.

Integrating the intensity defined by Eq. 5-10 over the entire hemisphere yields the total rate of radiation emission

$$(q/A)_r = \int_0^{2\pi} \int_0^{\pi/2} I(\psi,\theta) \cos\theta \sin\theta \, d\theta \, d\psi \tag{5-12}$$

In order to integrate Eq. 5-12, the variation of the intensity with θ and ψ must be known. As discussed more fully in Sec. 5-3, the intensity of real surfaces exhibits no appreciable variation with ψ, but does vary with θ. Although this variation can be taken into account, for most engineering calculations it may be assumed that the surface is *diffuse*, i.e., that the intensity of radiation leaving a surface is uniform in all angular directions. Blackbody radiation is perfectly diffuse and radiation from industrially rough surfaces approaches diffuse characteristics. If the intensity from a surface is independent of direction, it is said to conform to Lambert's cosine law. For such a surface, integration of Eq. 5-12 yields

$$(q/A)_r = E_b = \pi I_b \tag{5-13}$$

Thus, for a diffuse surface the emissive power equals π times the intensity.

The concept of intensity can be applied to the total radiation over the entire wavelength spectrum as well as to radiation at a given wavelength. The relation between the total and the monochromatic intensity I_λ is simply

$$I(\psi, \theta) = \int_0^\infty I_\lambda(\psi, \theta) \, d\lambda \tag{5-14}$$

If the surface radiates diffusely, also $E_\lambda = \pi I_\lambda$ since I_λ is uniform in all directions.

5-3. Radiation properties

Most surfaces encountered in engineering applications do not behave like blackbodies. To characterize the radiation properties of nonblack surfaces, dimensionless quantities such as the emittance and absorptance are used to relate the emitting and absorbing capabilities of a surface to those of a blackbody. Designating E_λ as the monochromatic emissive power of a real (non-black) surface, the monochromatic hemispherical emittance of the surface ϵ_λ is defined by the relation

$$E_\lambda = \epsilon_\lambda E_{b\lambda} \tag{5-15}$$

In other words, ϵ_λ is the percentage of blackbody radiation emitted by the surface at the wavelength λ. The hemispherical monochromatic absorptance of a surface, α_λ, is defined as the percentage of the total radiation incident on the surface which is absorbed.

An important relation between ϵ_λ and α_λ can be obtained from Kirchhoff's radiation law which states in essence that the monochromatic emittance is equal to the monochromatic absorptance for any surface. A rigorous derivation of this law has been presented by Planck (Ref. 33), but the essential features can be illustrated more simply from the following consideration. Suppose we placed a small body inside a black enclosure whose walls are fixed at temperature T. After thermal equilibrium is established also, the body must attain the temperature of the walls. In accordance with the second law of thermodynamics the body must, under these conditions, emit at every wavelength as much radiation as it absorbs. If the monochromatic radiation per unit time per unit area incident on the body is $G_{b\lambda}$, the equilibrium condition is expressed by

$$E_\lambda = \alpha_\lambda G_{b\lambda} \tag{5-16}$$

or

$$E_\lambda / \alpha_\lambda = G_{b\lambda} \tag{5-17}$$

But since the incident radiation depends only on the temperature of the enclosure, it would be the same on any other body in thermal equilibrium with the enclosure, irrespective of the absorptance of the body's surface. One can therefore conclude that the ratio of the monochromatic emissive power to the absorptance at any given wavelength is the same for all bodies at thermal equilibrium. Since the absorptance must always be less than unity and can be equal to one only for a perfect absorber, i.e., a blackbody, Eq. 5-17 shows also that at any given temperature the emissive power is a maximum for a blackbody. Thus, when $\alpha_\lambda = 1$, $E_\lambda = E_{b\lambda}$, and $G_{b\lambda} = E_{b\lambda}$ in Eq. 5-17. Replacing E_λ by $\epsilon_\lambda E_{b\lambda}$ in Eq. 5-16 gives

$$\epsilon_\lambda E_{b\lambda} = \alpha_\lambda G_{b\lambda} = \alpha_\lambda E_{b\lambda}$$

which shows that at any wavelength λ at temperature T,

$$\epsilon_\lambda(\lambda, T) = \alpha_\lambda(\lambda, T) \tag{5-18}$$

as stated at the outset.

Although the above relation was derived under the condition that the body is in equilibrium with its surroundings, it is actually a general relation which applies under any conditions. The reason for this is that both α_λ and ϵ_λ are surface properties which depend solely on the condition of the surface and its temperature. We can, therefore, conclude that unless changes in temperature cause physical alteration in the surface characteristics, the hemispherical monochromatic absorptance equals the monochromatic emittance of a surface.

The total hemispherical emittance for a non-black surface is obtained from Eqs. 5-8 and 5-15. Combining these two relations we find that at a given temperature T, the total hemispherical emittance is

$$\epsilon(T) = \frac{E(T)}{E_b(T)} = \frac{\int_0^\infty \epsilon_\lambda(\lambda) E_{b\lambda}(\lambda, T)\, d\lambda}{\int_0^\infty E_{b\lambda}(\lambda, T)\, d\lambda} \tag{5-19}$$

This relation shows that when the monochromatic emittance of a surface is a function of wavelength, it will vary with the temperature of the surface, even though the monochromatic emittance is solely a surface property. The reason for this variation is that the percentage of the total radiation that falls within a given wavelength band depends on the temperature of the emitting surface.

EXAMPLE 5-3. The hemispherical emittance of an aluminum paint is approximately 0.2 below $\lambda = 3\mu$ and 0.8 at longer wavelengths. Determine the total emittance of a surface covered with this paint at 80 F and 940 F.

Solution: From Table 5-1 we determine that $(E_{b,0-\lambda T}/\sigma T^4)$ for $\lambda T = 3 \times (460 + 80) = 1{,}600$ is 0.0001, while for $\lambda T = 3 \times (460 + 940) = 4{,}200$ the fraction of the total radiation below 3μ is 0.1267. Thus, the emittance at 80 F is 0.8, while at 940 F it is $0.2 \times 0.1267 + 0.8 \times 0.8733 = 0.763$. *Ans.*

The reason for the difference in the total emittance is that at the higher temperature the percentage of the total emissive power in the low emittance region of the point is appreciable, while at the lower temperature practically all the radiation is emitted at wavelengths above 3 microns.

Similarly, the total absorptance of a surface can also be obtained from basic definitions. Consider a surface at temperature T subject to incident radiation given by

$$G = \int_0^\infty G_\lambda(\lambda^*, T^*)\, d\lambda \tag{5-20}$$

where the superscript* is used to denote the conditions of the source.

If the variation of absorptance in the wavelength of the receiving surface is given by $\alpha_\lambda(\lambda)$ which, according to Eq. 5-18, is the same function as $\epsilon_\lambda(\lambda)$, the total absorptance is

$$\alpha(\lambda^*, T^*) = \frac{\int_0^\infty \alpha_\lambda(\lambda)\, G_\lambda(\lambda^*, T^*)\, d\lambda}{\int_0^\infty G_\lambda(\lambda^*, T^*)\, d\lambda} \tag{5-21}$$

We observe that the total absorptance of a surface depends on the temperature and on the spectral characteristics of the incident radiation. Therefore, although the relation $\epsilon_\lambda = \alpha_\lambda$ is always valid, the total values of absorptance and emittance are in general different. They are only equal when either of the following two conditions prevails:

(a) If $G_\lambda(\lambda^*, T^*) = E_b(\lambda, T)$, i.e., when the incident hemispherical radiation has the spectral distribution of a black source at the temperature of the surface. Under this condition the receiver is in thermal equilibrium with its surroundings and no net heat transfer occurs.
(b) If ϵ_λ and α_λ are uniform over the entire wavelength spectrum. Such surfaces are called *gray bodies*. Even though most real materials do not meet this specification exactly, it is often possible to choose suitable average values for the emittance and absorptance to make the gray body assumption acceptable for engineering analysis.

EXAMPLE 5-4. The aluminum paint in Example 5-3 is used to cover a surface which is maintained at 80 F. In one installation the surface is subjected to solar radiation, in another it receives radiation from a black source at 940 F. Determine the effective total absorptance of the surface for both conditions.

Solution: From Table 5-1, 98 percent of the total radiation from the sun (T = 10,460 R) falls below 3 μ (λT = 31,280). Thus, the effective absorptance for solar radiation is, from Eq. 5-21

$$\alpha(\lambda^*, T^*) = \int_0^3 \alpha(\lambda) G_\lambda(\lambda^*, T^*)\, d\lambda + \int_3^\infty \alpha(\lambda)\, G_\lambda(\lambda^* T^*)\, d\lambda$$

$$= 0.2 \times 0.98 + 0.8 \times 0.02 = 0.212$$

For a source at 940 F the absorptance may be calculated in a similar manner. However, it can also be obtained from Example 5-3 since condition 1 for $\alpha = \epsilon$ is satisfied and, therefore, $\alpha = 0.763$.

Radiation from real surfaces. Radiation from real surfaces differs in several aspects from blackbody radiation. According to Kirchhoff's law, a real surface always radiates less than a blackbody at the same temperature. If at a given temperature the ratio of the monochromatic emissive power of a body to the monochromatic emissive power of a blackbody at the same wavelength is constant over the entire wavelength spectrum, the body is said to be *gray* and its emissive power E_g is given by

$$E_g = \epsilon_g \sigma T^4 \qquad \text{Btu/hr sq ft}$$

The shape of a spectroradiometric curve for a gray surface is similar to that of a black surface at the same temperature, but the height is reduced by the numerical value of the emittance. For the purpose of heat-transfer calculations, surfaces are usually regarded as gray even though the characteristics of most surfaces deviate from gray-body specifications.

Fig. 5-9 shows the spectral characteristics of a blackbody at 3,000 F and the hemispherical monochromatic emissive power spectrum of an industrial metal surface at the same temperature. Superimposed on the spectral curve for the real surface is the monochromatic emissive power of a gray body with $\epsilon_\lambda = \epsilon = 0.6$. This value for the emittance was chosen so that the total emissive power of the real surface should equal that of a gray body. Although the detailed spectral characteristics of the real surface differ from those of the gray body, for purposes of analysis the fit would be considered sufficiently close to approximate the real surface characteristics by those of a gray body with an emittance of 0.6 at all wavelengths. It would also be reasonable to assume for a heat-transfer analysis that the absorptance of the surface is 0.6, unless the incident

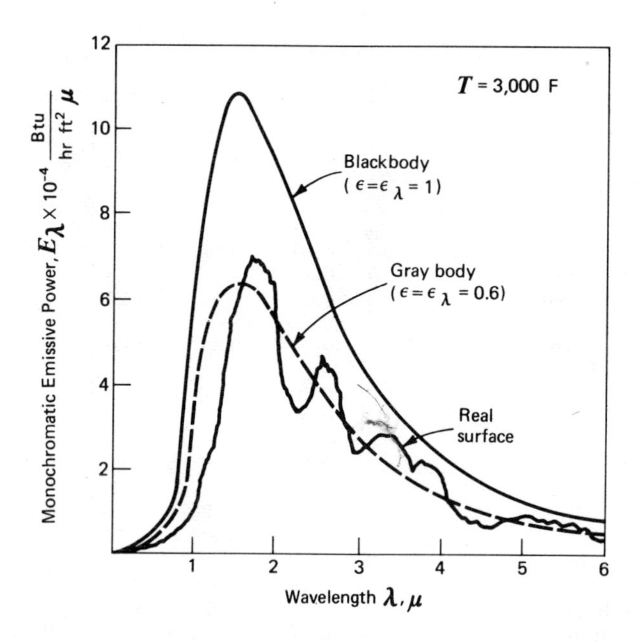

Fig. 5-9. Comparison of emissive power of black and gray bodies with that of a real surface.

radiation is in the solar spectrum, i.e., below $2\,\mu$ where the spectral emittance of the real surface is considerably less. The effective solar absorptance for this surface is approximately 0.3; this value, instead of 0.6, should be used if a gray-body assumption is made to calculate the portion of incident solar energy which is absorbed by the surface.

For convenience, the total hemispherical emittances of a selected group of industrially important surfaces at different temperatures are presented in Table 5-2. A more extensive tabulation of experimentally measured radiation properties for many surfaces have been prepared by Gubareff, et al. (36) and some general features and trends of the results are discussed below.

Fig. 5-10 shows the measured monochromatic emittance (or absorptance) of some electrical conductors as a function of wavelength (3). Polished surfaces of metals have low emittances, but, as shown in Fig. 5-11, the presence of an oxide layer may increase the emittance value appreciably. The monochromatic emittance of an electrical conductor (e.g., see the curves for Al or Cu in Fig. 5-10) increases with decreasing wavelength. Consequently in accordance with Eq. 5-19 the total emittance of electrical conductors increase with increasing temperature, as illustrated in Fig. 5-12 for several metals and one dielectric.

As a group, electrical nonconductors exhibit the opposite trend and generally possess high values of infrared emittances. Fig. 5-13 illustrates

Table 5-2. Emittance of various surfaces.

Material	Wavelength and Average Temperature				
	9.3μ 100 F	5.4μ 500 F	3.6μ 1000 F	1.8μ 2500 F	0.6μ Solar
Metals					
Aluminum					
Polished	0.04	0.05	0.08	0.19	~0.3
Oxidized	0.11	0.12	0.18		
24-ST weathered	0.4	0.32	0.27		
Surface roofing	0.22				
Anodized (at 1000 F)	0.94	0.42	0.60	0.34	
Brass					
Polished	0.10	0.10			
Oxidized	0.61				
Chromium					
Polished	0.08	0.17	0.26	0.40	0.49
Copper					
Polished	0.04	0.05	0.18	0.17	
Oxidized	0.87	0.83	0.77		
Iron					
Polished	0.06	0.08	0.13	0.25	0.45
Cast, oxidized	0.63	0.66	0.76		
Galvanized, new	0.23	....	....	0.42	0.66
Galvanized, dirty	0.28	....	....	0.90	0.89
Steel plate, rough	0.94	0.97	0.98		

Table 5-2. Continued.

Material	Wavelength and Average Temperature				
	9.3μ 100 F	5.4μ 500 F	3.6μ 1000 F	1.8μ 2500 F	0.6μ Solar
Oxide	0.96	....	0.85	....	0.74
Molten	....	....	....	0.3–0.4	
Magnesium	0.07	0.13	0.18	0.24	0.30
Molybdenum filament	....	....	~0.09	~0.15	~0.2*
Silver					
Polished	0.01	0.02	0.03	....	0.11
Stainless steel					
18–8, polished	0.15	0.18	0.22		
18–8, weathered	0.85	0.85	0.85		
Steel tube					
Oxidized	....	0.80			
Tungsten filament	0.03	....	....	~0.18	0.35†
Zinc					
Polished	0.02	0.03	0.04	0.06	0.46
Galvanized sheet	~0.25				
Building and Insulating Materials					
Asbestos paper	0.93	0.93			
Asphalt	0.93	....	0.9	...	0.93
Brick					
Red	0.93	....	....	....	0.7
Fire clay	0.9	....	~0.7	~0.75	
Silica	0.9	....	~0.75	0.84	
Magnesite refractory	0.9	....	....	~0.4	
Enamel, white	0.9				
Marble, white	0.95	....	0.93	....	0.47
Paper, white	0.95	....	0.82	0.25	0.28
Plaster	0.91				
Roofing board	0.93				
Enameled steel, white	....	....	....	0.65	0.47
Asbestos cement, red	....	....	....	0.67	0.66
Paints					
Aluminized lacquer	0.65	0.65			
Cream paints	0.95	0.88	0.70	0.42	0.35
Lacquer, black	0.96	0.98			
Lampblack paint	0.96	0.97	....	0.97	0.97
Red paint	0.96	....	....	....	0.74
Yellow paint	0.95	....	0.5	....	0.30
Oil paints (all colors)	~0.94	~0.9			
White (ZnO)	0.95	....	0.91	....	0.18
Miscellaneous					
Ice	~0.97‡				
Water	~0.96				
Carbon					
T-carbon, 0.9 per cent ash	0.82	0.80	0.79		
Filament	~0.72	....	....	0.53	
Wood	~0.93				
Glass	0.90	....	....	....	(Low)

* At 5000 F.
† At 6000 F.
‡ At 32 F.
SOURCE: Refs. 11, 15–18.

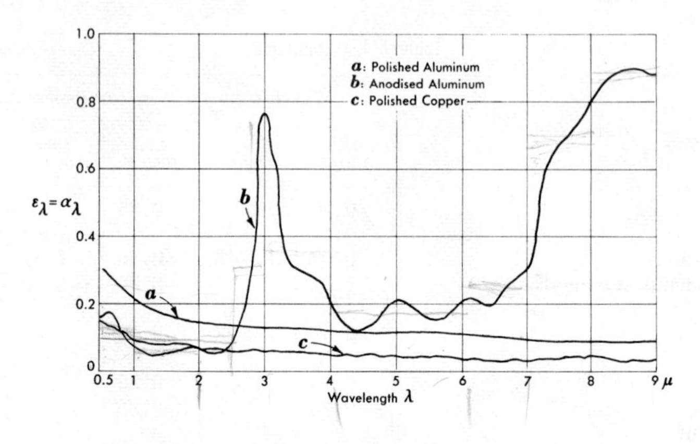

Fig. 5-10. Variation of monochromatic absorptance or emittance with wavelength for electrical conductors at room temperatures: (a) polished aluminum, (b) anodised aluminum, (c) polished copper.

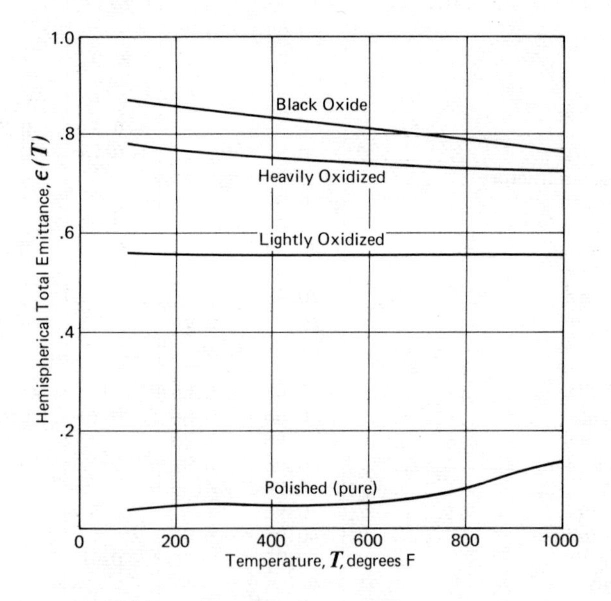

Fig. 5-11. Effect of oxide coating on hemispherical total emittance of copper. (Data from Gubareff et al., Ref. 36.)

the variation of the monochromatic emittance of several electrical nonconductors with wavelength.

For heat-transfer calculations an average emittance or absorptance for the wavelength band in which the bulk of the radiation is emitted or absorbed is desired. The wavelength band of interest depends on the

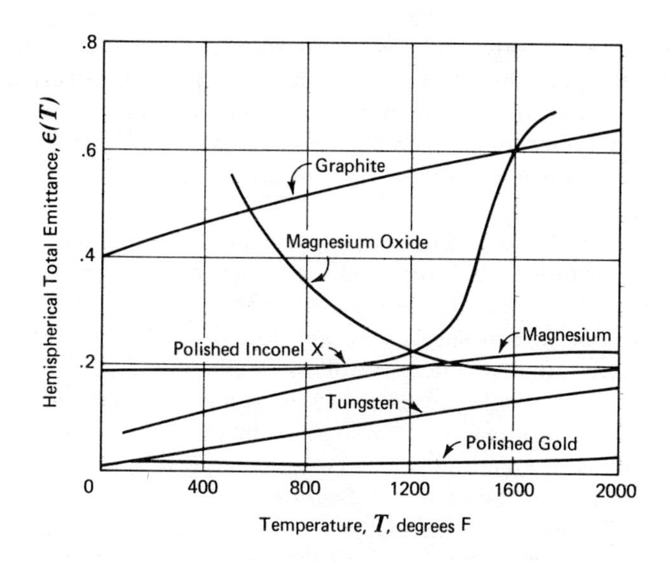

Fig. 5-12. Effect of temperature on hemispherical total emittance of several metals and one dielectric. (Data from Gubareff et al., Ref. 36.)

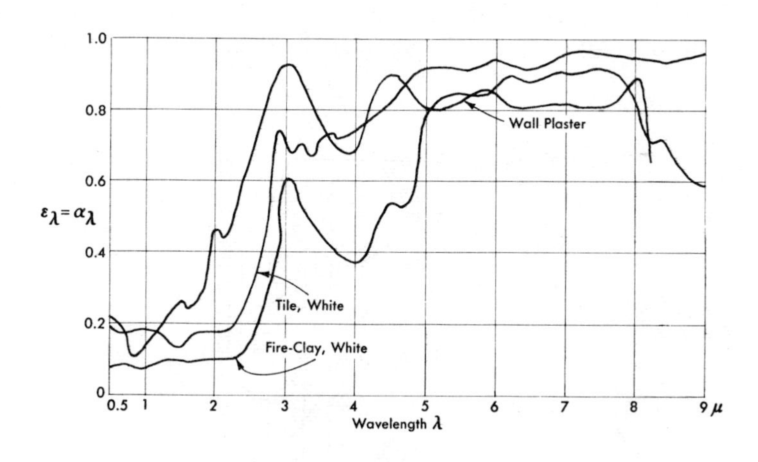

Fig. 5-13. Variation of monochromatic absorptance or emittance with wavelength for electrical nonconductors. (According to W. Sieber, Ref. 3.)

temperature of the body from which the radiation originates, as pointed out in Sec. 5-1. If the distribution of the monochromatic emittance is known, the total emittance can be evaluated from Eq. 5-19 and the total absorptance can be calculated from Eq. 5-21, if also the temperature and the spectral characteristics of the source are specified. Sieber (3) evaluated the total absorptance of the surfaces of several materials as a function of source temperature, with the receiving surfaces at room temperature and the emitter a blackbody. His results are shown in Fig. 5-14 in which the ordinate is the total absorptance for radiation normal to the surface, while the abscissa is the source temperature. We observe that the absorptance of aluminum, typical of good conductors, increases with increasing source temperature, whereas the absorptance of nonconductors exhibits the opposite trend.

In addition to its variation with wavelength, the emittance of many bodies also has directional properties which do not conform to Lambert's cosine law. This is illustrated in Figs. 5-15 and 5-16, where the directional emittances ϵ_θ of several substances are plotted in polar diagrams. For surfaces whose radiation intensity follows Lambert's cosine law and depends only on the projected area, the emittance curves would be semicircles. Fig. 5-15 shows that, for nonconductors such as wood, paper, and oxide films, the emittance decreases at large values of the emission angle θ, whereas for polished metals, the opposite trend is observed (see Fig. 5-16). For example, the emittance of polished chromium, which is widely used as a radiation shield, is as low as 0.06 in the normal direction but increases to 0.14 when viewed from an angle θ of 80 deg. Experimental data on the directional variation of emittance are scant, and until more information becomes available a satisfactory approximation for engineering calculations is to assume for polished metallic surfaces a mean value of $\epsilon/\epsilon_n = 1.2$ and for nonmetallic surfaces $\epsilon/\epsilon_n = 0.96$, where

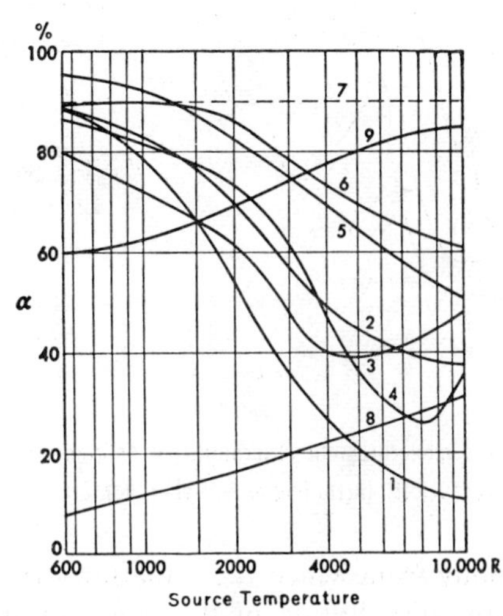

(1) White fire clay
(2) Asbestos
(3) Cork
(4) Wood
(5) Porcelain
(6) Concrete
(7) Roof shingles
(8) Aluminum
(9) Graphite

Fig. 5-14. Variation of total absorptance with source temperature for several materials at room temperature. (According to W. Sieber, Ref. 3.)

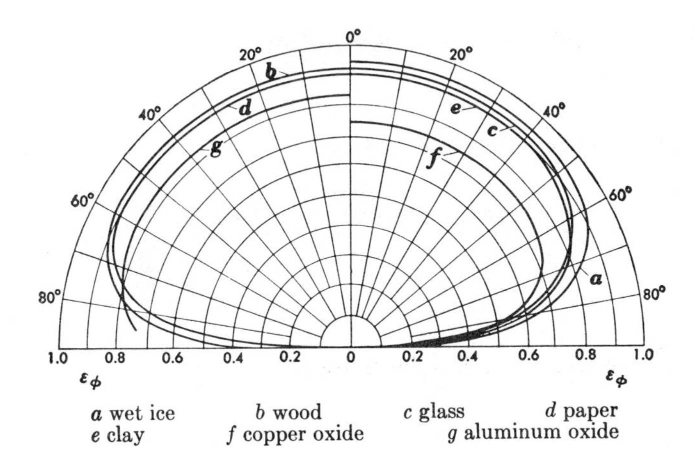

Fig. 5-15. Directional variation of emittance for several electrical nonconductors. (By permission from E. Schmidt and E. Eckert, "Uber die Richtungsverteilung der Warmestrahlung," *Forsch. Gebeite Ingenieurwesen*, Vol. 6, 1935.)

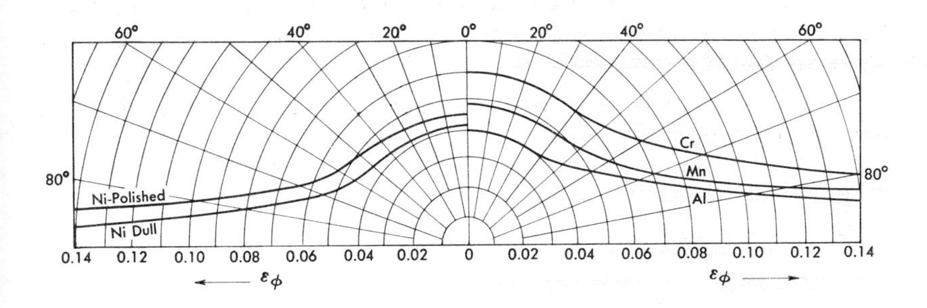

Fig. 5-16. Directional variation of emittance for several metals. (By permission from E. Schmidt and E. Eckert, "Uber die Richtungsverteilung der Warmestrahlung," *Forsch. Gebeite Ingenieurwesen*, Vol. 6, 1935.)

ϵ is the average emittance through a hemispherical solid angle of 2π steradians and ϵ_n is the emittance in the direction of the normal to the surface.

Reflectance and transmittance. When a surface does not absorb all of the incident radiation, the portion not absorbed will either be transmitted or reflected. Most solids are opaque and do not transmit radiation. The portion of the irradiation which is not absorbed is, therefore, reflected back into hemispherical space. This may be characterized by the monochromatic hemispherical reflectance, ρ_λ, defined as

$$\rho_\lambda = \frac{\text{radiant energy reflected/time-area-wavelength}}{G_\lambda} \tag{5-22}$$

or by the total reflectance, ρ, defined as

$$\rho = \frac{\text{radiant energy reflected/time-area}}{\int_0^\infty G_\lambda d_\lambda} \tag{5-23}$$

For nontransmitting materials the relations

$$\rho_\lambda = 1 - \alpha_\lambda \tag{5-24}$$

and

$$\rho = 1 - \alpha$$

must obviously hold at every wavelength and over the entire spectrum, respectively.

For the most general case of a material which partly absorbs, partly reflects, and partly transmits radiation incident on its surface, we define τ_λ as the fraction transmitted at wavelength λ and τ as the fraction of the total incident radiation which is transmitted. Referring to Fig. 5-17 the monochromatic relation

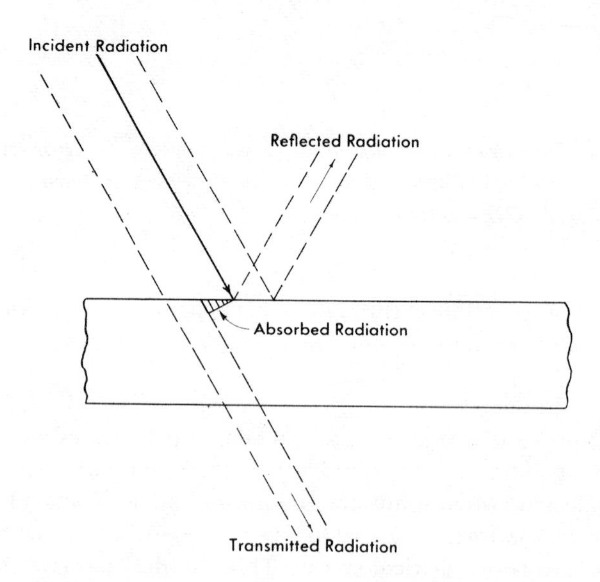

Fig. 5-17. Reflection, absorption, and transmission of radiation.

$$\rho_\lambda + \alpha_\lambda + \tau_\lambda = 1 \qquad (5\text{-}25)$$

and the total relation between reflectance, absorptance and transmittance

$$\rho + \alpha + \tau = 1$$

apply. Glass, rock salt, and other inorganic crystals are examples of the few exceptions among the solids which, unless very thick, are to a certain degree transparent to radiation of certain wavelengths. Many liquids and all gases are also transparent.

There are two basic types of radiation reflections: *specular* and *diffuse*. If the angle of reflection is equal to the angle of incidence, the reflection is called specular. On the other hand, when an incident beam is reflected uniformly in all directions, the reflection is called diffuse. No real surface is either specular or diffuse. In general, reflection from highly polished and smooth surfaces approaches specular characteristics, while reflection from industrially "rough" surfaces approaches diffuse characteristics. An ordinary mirror reflects specularly in the visible wavelength range, but not necessarily over the longer-wavelength range of thermal radiation.

The influence of surface characteristics on thermal-radiation properties of materials is not yet fully understood and is, therefore, still a topic of research. (Methods of analyses which consider specular as well as diffuse reflections are presented in Refs. 24 and 27.) But for engineering calculations industrially-plated, machined, or painted surfaces may be treated as though they were completely diffuse. This was verified experimentally by Schornhorst and Viskranta (37) who made measurements over temperature ranges from 500 to 1300 R with sandblasted and electroplated stainless steel, rough and smooth electroplated gold, and a surface covered with white paint. The data were compared with predictions based on diffuse, specular, and diffuse-specular models and were found to agree with the diffuse model within the accuracy of the measurements for all the surfaces tested. Similar conclusions were also drawn by Hering and Smith (38) in a study of surface roughness effects on radiation heat transfer.

5-4. *The radiation shape factor*

In most practical problems involving radiation, the intensity of thermal radiation passing between surfaces is not appreciably affected by the presence of intervening media because, unless the temperature is so high as to cause ionization or dissociation, monatomic and most diatomic gases as well as air are transparent. Moreover, since most industrial surfaces can be treated as diffuse emitters and reflectors of radiation in a heat-transfer analysis, a key problem in calculating radiation heat transfer between surfaces is to determine the fraction of the total diffuse radiation leaving one surface which is intercepted by another surface and vice versa. The fraction of diffusely distributed radiation leaving a surface A_i that reaches surface A_j is called the radiation shape factor F_{i-j}. The first subscript appended to the radiation shape factor denotes the surface from which the radiation emanates while the second subscript denotes the surface receiving the radiation.

Consider two black surfaces A_1 and A_2, as shown in Fig. 5-18. The radiation leaving A_1 and arriving at A_2 is

$$q_{1\to 2} = E_{b1} A_1 F_{12} \tag{5-26a}$$

and the radiation leaving A_2 and arriving at A_1 is

$$q_{2\to 1} = E_{b2} A_2 F_{21} \tag{5-26b}$$

Since both surfaces are black, all the incident radiation will be absorbed and the net rate of energy exchange, $q_{1\rightleftharpoons 2}$, is

$$q_{1\rightleftharpoons 2} = E_{b1} A_1 F_{12} - E_{b2} A_2 F_{21} \tag{5-27}$$

If both surfaces are at the same temperature $E_{b1} = E_{b2}$ and there can be no net heat flow between them. Therefore, $q_{1\rightleftharpoons 2} = 0$ and since neither areas nor shape factors are functions of temperature

$$A_1 F_{12} = A_2 F_{21} \tag{5-28}$$

Eq. 5-28 is known as the reciprocity theorem. The net rate of transfer between any two black surfaces, A_1 and A_2, can thus be written in two

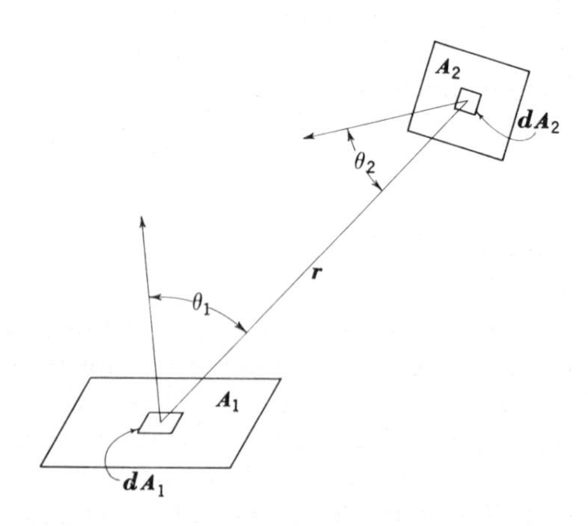

Fig. 5-18. Geometrical shape-factor notation.

forms

$$q_{1\rightleftharpoons 2} = A_1 F_{12}(E_{b1} - E_{b2}) = A_2 F_{21}(E_{b1} - E_{b2}) \tag{5-29}$$

Inspection of Eq. 5-29 reveals that the net rate of heat flow between two black bodies can be determined by evaluating the radiation from either one of the surfaces to the other surface and replacing its emissive power by the difference of the emissive powers of the two surfaces. Since the end result is independent of the choice of the emitting surface, one selects that surface whose shape factor can be determined more easily. For example, the shape factor F_{1-2} for any surface A_1 completely enclosed by another surface is unity. In general, however, the determination of a shape factor for any but the most simple geometric configuration is rather complex.

To determine the fraction of the energy leaving surface A_1 that strikes surface A_2, consider first the two differential surfaces dA_1 and dA_2. If the distance between them is r, then $dq_{1\rightarrow 2}$, the rate at which radiation from dA_1 is received by dA_2, is, from Eq. 5-10, given by

$$dq_{1\rightarrow 2} = I_1 \cos\theta_1 \, dA_1 \, d\omega_{1-2} \tag{5-30}$$

where I_1 = intensity of radiation from dA_1;
$dA_1 \cos\theta_1$ = projection of area element dA_1 as seen from dA_2;
$d\omega_{1-2}$ = solid angle subtended by receiving area dA_2 with respect to center point of dA_1.

The subtended angle $d\omega_{1-2}$ is equal to the projected area of the receiving surface in the direction of the incident radiation divided by the square of the distance between dA_1 and dA_2, or, using the nomenclature of Fig. 5-18,

$$d\omega_{1-2} = \cos\theta_2 \frac{dA_2}{r^2} \tag{5-31}$$

Substituting Eqs. 5-31 and 5-13 for $d\omega_{1-2}$ and I_1 respectively in Eq. 5-30 yields

$$dq_{1\rightarrow 2} = E_{b1}\, dA_1 \left(\frac{\cos\theta_1 \cos\theta_2 \, dA_2}{\pi r^2}\right) \tag{5-32}$$

where the term in parentheses is equal to the fraction of the total radiation emitted from dA_1 that is intercepted by dA_2. By analogy, the fraction of the total radiation emitted from dA_2 that strikes dA_1 is

$$dq_{2\rightarrow 1} = E_{b2}\, dA_2 \left(\frac{\cos\theta_2 \cos\theta_1 \, dA_1}{\pi r^2}\right) \tag{5-33}$$

so that the net rate of radiant heat transfer between dA_1 and dA_2 is

$$dq_{1\rightleftharpoons 2} = (E_{b1} - E_{b2}) \frac{\cos\theta_1 \cos\theta_2 \, dA_1 \, dA_2}{\pi r^2} \tag{5-34}$$

To determine $q_{1\rightleftharpoons 2}$, the net rate of radiation between the entire surfaces A_1 and A_2, we simply integrate the fraction in the preceding equation over both surfaces and obtain

$$q_{1\rightleftharpoons 2} = (E_{b1} - E_{b2}) \int_{A_1} \int_{A_2} \frac{\cos\theta_1 \cos\theta_2 \, dA_1 \, dA_2}{\pi r^2} \tag{5-35}$$

The double integral is conveniently written in shorthand notation either as $A_1 F_{1-2}$ or $A_2 F_{2-1}$, where F_{1-2} is called the shape factor evaluated on the basis of area A_1 and F_{2-1} is called the shape factor evaluated on the basis of A_2. The method of evaluation of the double integral is illustrated below.

EXAMPLE 5-5. Determine the geometric shape factor for a very small disk A_1 and a large parallel disk A_2 located a distance L directly above the smaller one, as shown in Fig. 5-19

Solution: From Eq. 5-19 the geometric shape factor is

$$A_1 F_{1-2} = \int_{A_1} \int_{A_2} \frac{\cos\theta_1 \cos\theta_2}{\pi r^2} \, dA_1 \, dA_2$$

but since A_1 is very small the shape factor is given by

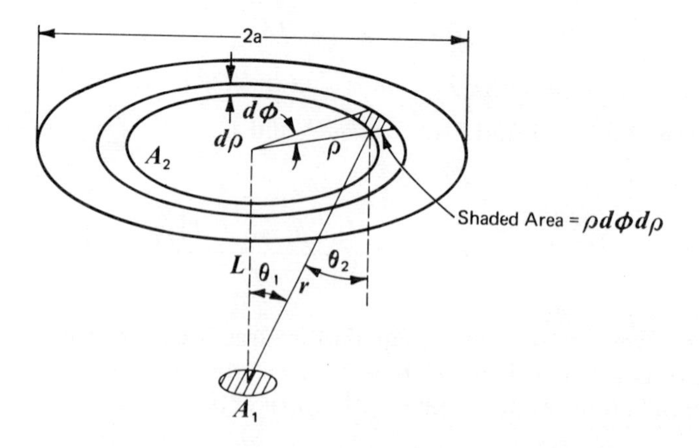

Fig. 5-19. Nomenclature for the evaluation of the shape factor between a small disk and a large disk located parallel directly above.

$$A_1 F_{1-2} = \frac{A_1}{\pi} \int_{A_2} \frac{\cos\theta_1 \cos\theta_2}{r^2} dA_2$$

From Fig. 5-19, $\cos\theta_1 = \cos\theta_2 = L/r$, $r = \sqrt{\rho^2 + L^2}$, and $dA_2 = \rho d\phi d\rho$. Substituting these relations, we obtain

$$A_1 F_{1-2} = \frac{A_1}{\pi} \int_0^a \int_0^{2\pi} \frac{L^2}{(\rho^2 + L^2)^2} \rho d\rho d\phi$$

which can be integrated directly to yield

$$A_1 F_{1-2} = \frac{A_1 a^2}{a^2 + L^2} = A_2 F_{2-1} \qquad \textit{Ans.}$$

The preceding example shows that the determination of a shape factor by evaluating the double integral of Eq. 5-35 is generally very tedious.

Table 5-3. Geometric shape factors for use in Eqs. 5-29 and 5-36.

Surfaces Between Which Radiation Is Being Interchanged	Shape Factor, F_{1-2}
1. Infinite parallel planes.	1
2. Body A_1 completely enclosed by another body, A_2. Body A_1 cannot see any part of itself.	1
3. Surface element dA (A_1) and rectangular surface (A_2) above and parallel to it, with one corner of rectangle contained in normal to dA.	See Fig. 5-20
4. Element dA (A_1) and parallel circular disk (A_2) with its center directly above dA. (See Example 5-3.)	$a^2/(a^2 + L^2)$
5. Two parallel and equal squares, rectangles or disks of width or diameter D, a distance L apart.	See Fig. 5-22
6. Two parallel disks of unequal diameter, distance L apart with centers on same normal to their planes, smaller disk A_1 of radius a, larger disk of radius b.	$\frac{1}{2a^2}\left[L^2 + a^2 + b^2 - \sqrt{(L^2 + a^2 + b^2)^2 - 4a^2b^2}\right]$
7. Two rectangles in perpendicular planes with a common side.	See Fig. 5-21
8. Radiation between an infinite plane A_1 and one or two rows of infinite parallel tubes in a parallel plane A_2 if the only other surface is a refractory surface behind the tubes.	See Fig. 5-23

Fortunately the shape factors for a large number of geometrical arrangements have been evaluated and a majority of them can be found in Refs. 5, 6, 7, 8, and 39. A selected group of practical interest is summarized in Table 5-3 and Figs. 5-20, 5-21, 5-22, and 5-23. The data from the graphi-

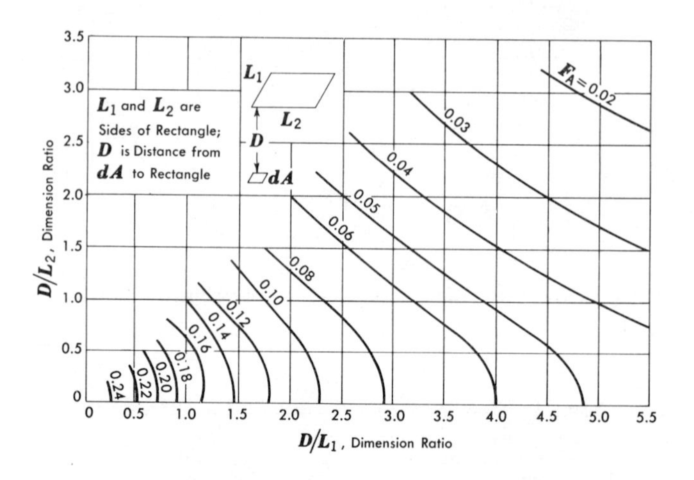

Fig. 5-20. Shape factor for a surface element and a rectangular surface parallel to it. (By permission from H. C. Hottel, "Radiant Heat Transmission," *Mechanical Engineering*, Vol. 52, 1930.)

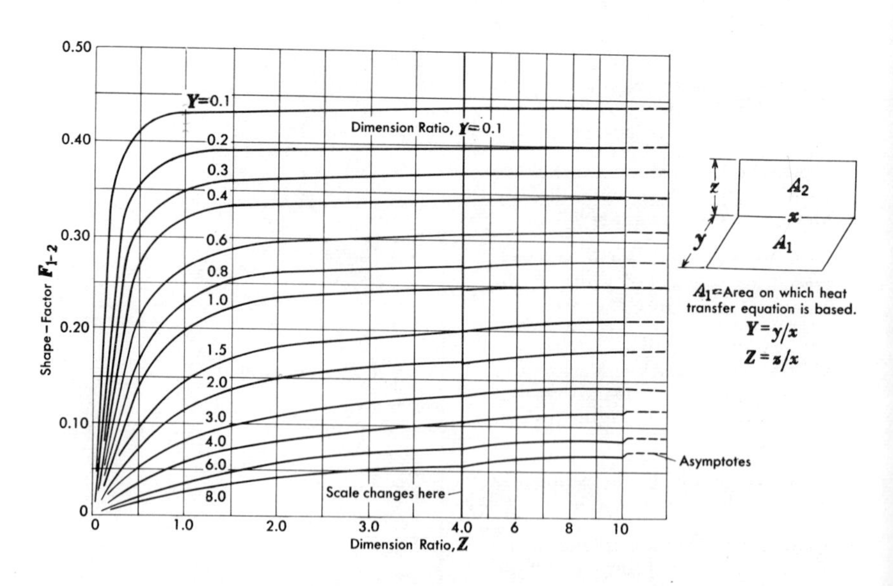

Fig. 5-21. Shape factor for adjacent rectangles in perpendicular planes. (By permission from H. C. Hottel, "Radiant Heat Transmission," *Mechanical Engineering*, Vol. 52, 1930.)

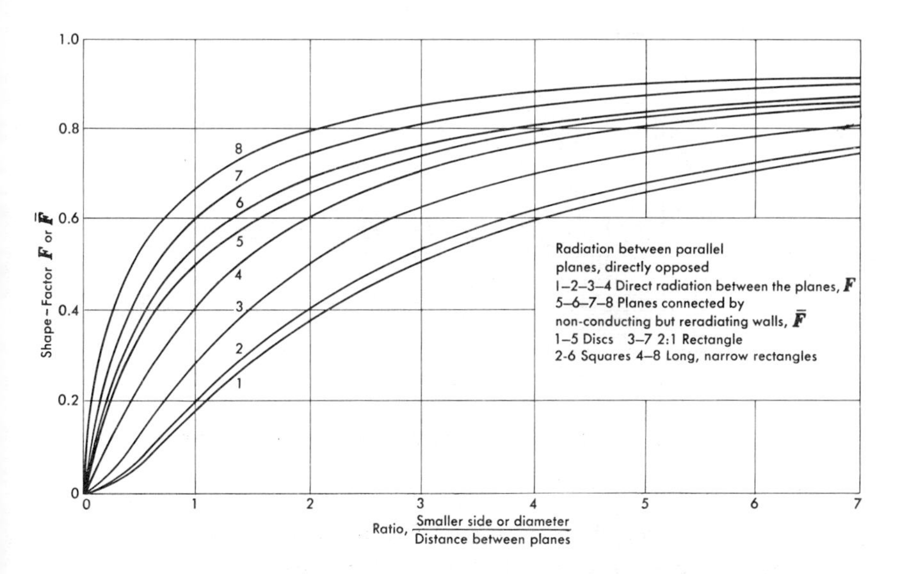

Fig. 5-22. Shape factors for equal and parallel squares, rectangles, and disks. The curves labeled 5,6,7 and 8 allow for continuous variation in the side-wall temperatures from top to bottom. (By permission from H. C. Hottel, "Radiant Heat Transmission," *Mechanical Engineering*, Vol. 52, 1930.)

cal solutions of Figs. 5-20 and 5-21 can be extended by simple arithmetical addition and subtraction of shape factors to permit the evaluation of a shape factor for geometrical arrangements which can be built up from these elementary cases.

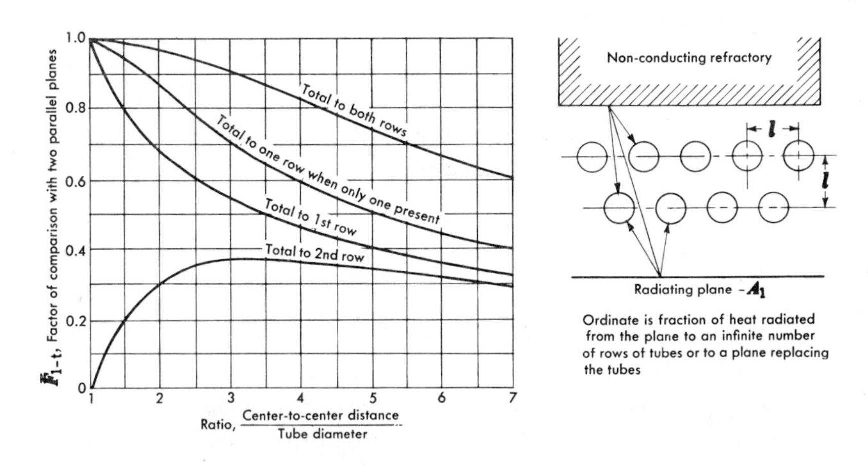

Fig. 5-23. Shape factor for a plane and one or two rows of tubes above and parallel to it. (By permission from H. C. Hottel, "Radiant Heat Transmission," *Mechanical Engineering*, Vol. 52, 1930.)

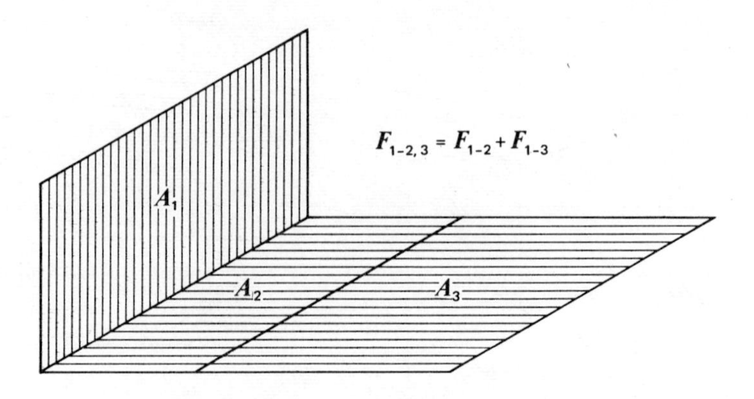

Fig. 5-24. Notation for evaluation shape factors for nonadjacent rectangles.

For example, the shape-factor for radiation from surface A_1 to the combined surfaces of A_2 and A_3 in Fig. 5-24 is simply

$$F_{1-2,3} = F_{1-2} + F_{1-3} \tag{5-36}$$

i.e., the total shape factor is equal to the sum of its parts. Rewriting Eq. 5-36 as

$$A_1 F_{1-2,3} = A_1 F_{1-2} + A_1 F_{1-3}$$

and using the reciprocity relations

$$A_1 F_{1-2,3} = (A_2 + A_3) F_{2,3-1}$$
$$A_1 F_{1-2} = A_2 F_{2-1}$$
$$A_1 F_{1-3} = A_3 F_{3-1}$$

yields

$$(A_2 + A_3) F_{2,3-1} = A_2 F_{2-1} + A_3 F_{3-1} \tag{5-37}$$

Thus, the total radiation received by A_1 is the sum of the radiant energy fractions from A_2 and A_3. This simple relation can be used to evaluate the shape factor F_{1-3} in terms of the shape factors for perpendicular rectangles with a common edge given in Fig. 5-21. Other combinations can be obtained in a similar manner. The following example illustrates the numerical evaluation procedure.

EXAMPLE 5-6. A room 12 ft on one side by 24 ft on the other has a ceiling height of 12 ft. Determine the shape factor of the floor with respect to a small window of area A_1 located in the ceiling 6 ft from two walls.

Solution: If the floor is divided into four rectangles, two 6 by 6 ft each and two 6 by 18 ft each, each rectangle meets the condition of the graphical solution presented in Fig. 5-24. The shape factor of the entire floor area will be the sum of the shape factors for each of the rectangles. The dimensionless ratios D/L_1 and D/L_2 for each of the smaller rectangles are $12/6 = 2.00$. From Fig. 5-22 the shape factor for one section is about 0.06. For each of the larger rectangles $D/L_1 = 0.66$, $D/L_2 = 2.0$, and, from Fig. 5-22, the shape factor between the window and one of the larger rectangles is 0.10. The shape factor for the entire floor is therefore 0.32. Thus, 32 percent of the total emissive power from the window will strike the floor, and $A_1 F_{1-2} = A_2 F_{2-1} = 0.32\, A_1$. *Ans.*

5-5. Radiation in enclosures with black surfaces

To determine the net radiation heat transfer to or from a surface, it is necessary to account for radiation coming from all directions. This procedure is facilitated by figuratively constructing an enclosure around the surface and specifying the radiation characteristics of each surface. The surfaces comprising the enclosure for a given surface i are all the surfaces that can be seen by an observer standing on surface i in the surrounding space. The enclosure need not necessarily consist only of solid surfaces, but may include open spaces denoted as "windows." Each such open window may be assigned an equivalent blackbody temperature corresponding to the entering radiation. If no radiation enters, a window acts like a blackbody at zero temperature which absorbs all outgoing radiation and emits and reflects none.

The net rate of radiation loss from a typical surface A_i in an enclosure (see Fig. 5-25) consisting of N black surfaces is equal to the difference between the emitted radiation and the absorbed radiation, or

$$q_{i \rightleftharpoons \text{enclosure}} = A_i(E_{bi} - G_i) \tag{5-38}$$

where G_i is the radiation incident on surface i per unit time and unit area, called the irradiation.

The radiation incident on A_i comes from the other N surfaces in the enclosure. From a typical surface j, the radiation incident on i is $E_{bj} A_j F_{j-i}$. Summing the contributions from all N surfaces gives

$$A_i G_i = E_{b1} A_1 F_{1-i} + E_{b2} A_2 F_{2-i} + \cdots + E_{bN} A_N F_{N-i}$$

which can be written compactly in the form

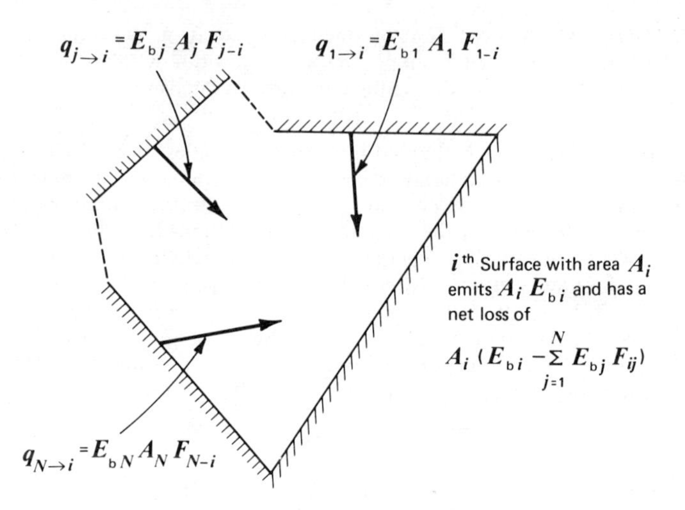

Fig. 5-25. Schematic diagram of enclosure of N black surfaces with energy quantities incident upon and leaving surface i.

$$A_i G_i = \sum_{j=1}^{N} E_{bj} A_j F_{j-i} \tag{5-39}$$

Using the reciprocity law $A_i F_{i-j} = A_j F_{j-i}$ and substituting Eq. 5-39 for G_i in Eq. 5-38 yields for the net rate of radiation heat loss from *any* surface in an enclosure of black surfaces

$$q_{i \rightleftharpoons \text{enclosure}} = A_i \left(E_{bi} - \sum_{j=1}^{N} E_{bj} F_{ij} \right) \tag{5-40}$$

An alternate approach to the problem is by extension of Eq. 5-26. Since the radiant energy leaving any surface i must impinge on the N surfaces forming the enclosure,

$$\sum_{j=1}^{N} F_{i-j} = 1.0 \tag{5-41}$$

Equation 5-41 includes a term F_{i-i} which is not zero when a surface is concave so that some radiation leaving surface i will be directly incident on it. The total emissive power of A_i is therefore distributed between the N surfaces according to

$$A_i E_{bi} = \sum_{j=1}^{N} E_{bi} A_i F_{i-j} \tag{5-42}$$

Introducing Eq. 5-42 for $A_i E_{bi}$ in Eq. 5-40 gives the net rate of heat loss from surface i in the form

$$q_{i \rightleftharpoons \text{enclosure}} = \sum_{j=1}^{N} (E_{bi} - E_{bj}) A_i F_{i-j} \tag{5-43}$$

Thus, the net heat loss may be calculated by summing the differences in emissive power and weighting each by the appropriate area-shape factor.

An inspection of Eq. 5-43 shows that there is also an analogy between heat flow by radiation and the flow of electric current. If the blackbody emissive power E_b is considered to act as a potential and the area-shape factor $A_i F_{ij}$ as the conductance between two nodes at potentials E_{bi} and E_{bj}, then the resulting net flow of heat is analogous to the flow of electric current in an analogous network. Examples of networks for blackbody enclosures consisting of three and four heat-transfer surfaces at given temperatures are shown in Figs. 5-26a and 5-26b, respectively.

In engineering problems there are situations when, for one or more surfaces in an enclosure, not the temperature but the heat flux is prescribed. In such cases the temperatures of these surfaces are unknown. For the case when the net heat-transfer rate q_r from one surface A_k is prescribed, while for all the other surfaces of the enclosure the temperature is specified, Eq. 5-40 can be re-arranged to solve for T_k. Since $E_{bk} = \sigma T_k^4$ one obtains

$$T_k = \left[\frac{\sum_{j=1}^{N} \sigma T_j^4 F_{k-j} + (q_r/A)_k}{\sigma(1 - F_{k-k})} \right]^{1/4} \tag{5-44}$$

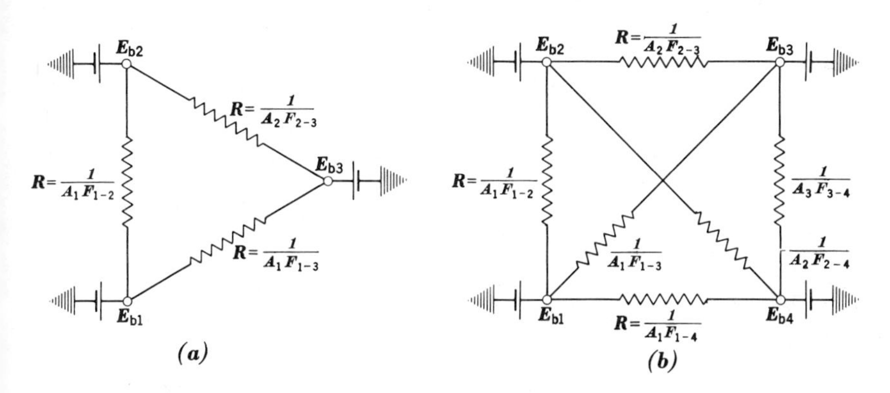

Fig. 5-26. Equivalent networks for radiation in blackbody enclosures consisting of three and four surfaces.

where $j = k$ is specifically excluded from the summation. Once T_k is known, the heat-transfer rates at all other surfaces can be obtained from Eq. 5-43.

Of special interest is the case of a no-flux or adiabatic surface which diffusely reflects and emits radiation at the same rate at which it receives it. Under steady-state conditions the interior surfaces of refractory walls in industrial furnaces can be treated as adiabatic surfaces. The interior walls of these surfaces receive heat by convection as well as radiation and lose heat to the outside by conduction. In practice, however, the heat flow by radiation is so much larger than the difference between the heat flow by convection to and the heat flow by conduction from the surface that the walls act essentially as reradiators, i.e. no-flux surfaces.

A simplified sketch of a pulverized-fuel furnace is shown in Fig. 5-27a. The floor is assumed to be at a uniform temperature T_1 radiating to a nest of oxidized-steel tubes at T_2 which fill the ceiling of the furnace. The side walls are assumed to act as reradiators at *a uniform temperature* T_R. If we neglect radiation between the tubes and the ceiling and assume that the floor and the tubes are black, the equivalent network representing the radiation exchange between the floor and the tubes in the presence of the reradiating walls is that shown in Fig. 5-27b. A part of the radiation emitted from A_1 goes directly to A_2, while the rest strikes A_R and is reflected from there. Of the reflected radiation, a part is returned to A_1, a part to A_2, and the rest to A_R for further reflection. However, since the refractory wall must get rid of all the incident radiation either by reflection or radiation, its emissive power will act in the steady state like a floating potential whose actual value, i.e., its emissive power and temperature, depends only on the relative values of the conductances between E_R and E_{b1} and E_R and E_{b2}. Thus, the net effect of this rather complicated radiation pattern can be represented in the equivalent network by two parallel heat-flow paths between A_1 and A_2, one having an effective conductance of $A_1 F_{1-2}$, the other having an effective conductance equal to

$$1/(1/A_1 F_{1-R} + 1/A_2 F_{2-R})$$

The net heat flow by radiation between a black heat source and a black heat sink in such a simple furnace is then equal to

$$q_{1\rightleftharpoons 2} = A_1(E_{b1} - E_{b2})\left(F_{1-2} + \frac{1}{1/F_{1-R} + A_1/A_2 F_{2-R}}\right) \tag{5-45}$$

If neither of the surfaces can see any part of itself, F_{1-R} and F_{2-R} can be eliminated by using Eqs. 5-28 and 5-41. This yields after some simplification

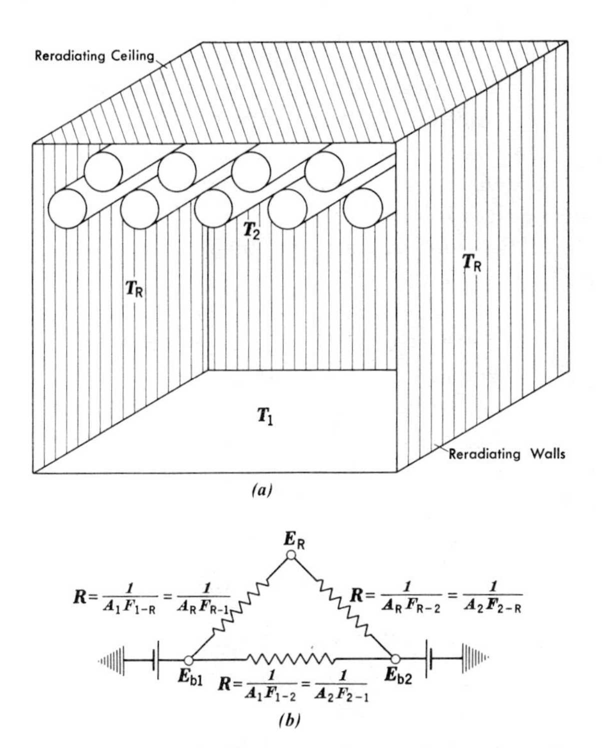

Fig. 5-27. Simplified sketch of a furnace and equivalent network for radiation in an enclosure consisting of two black surfaces and an adiabatic surface.

$$q_{1\rightleftharpoons 2} = A_1\sigma(T_1^4 - T_2^4)\frac{A_2 - A_1F_{1-2}^2}{A_1 + A_2 - 2A_1F_{1-2}} = A_1\bar{F}_{1-2}(E_{b1} - E_{b2}) \tag{5-46}$$

where $\bar{F}_{1-2}$ is the effective shape factor for the configuration shown in Fig. 5-27b. The same result would of course be obtained from Eqs. 5-43 and 5-44. The details of this derivation are left as an exercise (see Prob. 5-16).

5-6. Radiation in enclosures with gray surfaces

In the preceding section, radiation between black surfaces was considered. The assumption that a surface is black simplifies heat-transfer calculations because all of the incident radiation is absorbed. In practice one may generally neglect reflections without introducing serious errors if the absorptivity of the radiating surfaces is larger than 0.9. There are, however, numerous problems involving surfaces of low absorptance and emittance, especially in installations where radiation is undesirable. For example, the inner walls of a thermos bottle are silvered in order to reduce the heat flow by radiation. Also, thermocouples for high-temperature work are frequently surrounded by radiation shields to reduce the difference between the indicated temperature and the temperature of the medium to be measured.

If the radiating surfaces are not black, the analysis becomes exceedingly difficult unless the surfaces are considered to be gray. The analysis in this section is limited to gray surfaces which follow Lambert's cosine law and also reflect diffusely. The radiation from such surfaces can be treated conveniently in terms of the *radiosity* J, which is defined as the rate at which radiation leaves a given surface per unit area. The radiosity is the sum of radiation emitted, reflected, and transmitted. For opaque bodies which transmit no radiation, the radiosity from a typical surface i can be defined

$$J_i = \rho_i G_i + \epsilon_i E_{bi} \tag{5-47}$$

where J_i = radiosity, in Btu/hr sq ft;
G_i = irradiation or radiation per unit time incident on unit surface area, in Btu/hr sq ft;
E_{bi} = blackbody emissive power, in Btu/hr sq ft;
ρ_i = reflectance;
ϵ_i = emittance.

Consider the ith surface having area A_i in an enclosure consisting of N surfaces as shown in Fig. 5-25. To maintain surface i at temperature T_i a certain amount of heat, q_i, must be supplied from some external source to make up for the net radiative loss in a steady-state condition. The net rate of heat transfer from a surface i by radiation is equal to the difference between the outgoing and the incoming radiation. Using the terminology of Eq. 5-47, the net rate of heat loss is the difference between the radiosity and the irradiation, or

$$q_i = A_i(J_i - G_i) \tag{5-48}$$

It should be noted that Eq. 5-48 is strictly valid only when the temperature as well as the irradiation over A_i are uniform. To satisfy both of these conditions simultaneously, it is sometimes necessary to subdivide a physical surface into smaller sections for the purpose of analysis.

If the surfaces exchanging radiation are gray, $\epsilon_i = \alpha_i$ and $\rho_i = (1 - \epsilon_i)$ for each of them. The irradiation G_i can then be eliminated from Eq. 5-48 by combining it with Eq. 5-47. This yields

$$q_i = \frac{A_i\epsilon_i}{\rho_i}(E_{bi} - J_i) = \frac{A_i\epsilon_i}{1 - \epsilon_i}(E_{bi} - J_i) \tag{5-49}$$

Another relation for the net rate of heat loss by radiation from A_i can be obtained by evaluating the irradiation in terms of the radiosity of all the other surfaces which can be seen from it. The incident radiation G_i can be evaluated by the same approach used previously in a blackbody enclosure. The incident radiation consists of the portions of radiation from the other N surfaces which impinge on A_i. If the surface A_i can see a part of itself, also a portion of the radiation emitted by A_i will contribute to the irradiation. The shape factors for diffusely reflecting gray surfaces are obviously the same as for black surfaces since they depend only on geometrical relations defined by Eq. 5-35. We can, therefore, write in symbolic form

$$A_iG_i = J_1A_1F_{1-i} + J_2A_2F_{2-i} + \cdots + J_iA_iF_{i-i} + \cdots + J_jA_jF_{j-i} + \cdots + J_NA_NF_{N-i} \tag{5-50}$$

Using the reciprocity relations

$$A_1F_{1-i} = A_iF_{i-1}$$
$$A_2F_{2-i} = A_iF_{i-2}$$
$$A_NF_{N-i} = A_iF_{i-N}$$

Eq. 5-50 can be written so that the only area appearing is A_i:

$$A_iG_i = J_1A_iF_{i-1} + J_2A_iF_{i-2} + \cdots + J_iA_iF_{i-i} + \cdots + J_jA_jF_{j-i} + \cdots + J_NA_NF_{N-i}$$

This can be written compactly as

$$G_i = \sum_{j=1}^{N} J_jF_{i-j} \tag{5-51}$$

Eq. 5-51 is identical to Eq. 5-42 for a black enclosure, except that the blackbody emissive power has been replaced by the radiosity. Substituting the summation of Eq. 5-51 for G_i in Eq. 5-48 yields

$$q_i = A_i\left(J_i - \sum_{j=1}^{N} J_jF_{i-j}\right) \tag{5-52}$$

Eqs. 5-49 and 5-52 can be written for each of the N surfaces of the enclosure giving 2 N equations for 2 N unknowns. There will always be N

unknown J's while the remaining unknowns will consist of q's or T's, depending on what boundary conditions are specified. The J's can always be eliminated giving N equations relating the N unknown temperatures and net rates of radiation transfer.

In terms of an equivalent electrical circuit we could write Eq. 5-49 in the form

$$q_i = \frac{E_{bi} - J_i}{(1 - \epsilon_i)/A_i\epsilon_i} \tag{5-53}$$

and consider the rate of radiation heat transfer q_i as the current in a network between potentials E_{bi} and J_i with a resistance of $(1 - \epsilon_i)/A_i\epsilon_i$ between them. Since the effect of the system geometry on the net radiation between any two gray surfaces A_i and A_k emitting radiation at the rate J_i and J_k respectively is the same as for geometrically similar black surfaces, it can be expressed in terms of the geometric shape factor defined by Eq. 5-35. The direct radiation exchange between any two opaque and diffuse surfaces A_i and A_j is given by the equation

$$q_{i \rightleftharpoons j} = (J_i - J_j)A_iF_{i-j} = (J_i - J_j)A_jF_{j-i} \tag{5-54}$$

Equations 5-49 and 5-54 provide the basis for determining the net rate of radiant heat transfer between gray bodies in a gray enclosure by means of an equivalent network. The effect of the reflectance and emittance can be taken into account by connecting a *blackbody potential node* E_b, to each of the nodal points in the network by means of a *finite resistance* $(1 - \epsilon)/A\epsilon$. In the case of a blackbody this resistance is zero since $\epsilon = 1$. In Fig. 5-28 the equivalent networks for radiation in an enclosure consisting of two and four gray bodies are shown. For two-component gray enclosures, such as two parallel and infinite plates, concentric cylinders of infinite height, and concentric spheres, the network reduces to a single line of resistances in series as shown in Fig. 5-28a.

To illustrate the procedure for calculating radiation heat transfer between gray surfaces we will derive an expression for the rate of radiation heat transfer between two long concentric cylinders of area A_1 and A_2 at temperatures T_1 and T_2 respectively, and compare the result with the network in Fig. 5-28a.

Referring to Fig. 5-29, the shape factor for the smaller cylinder of area A_1 relative to the larger cylinder which encloses it, F_{1-2}, is 1.0. From Eq. 5-28, $A_1F_{1-2} = A_2F_{2-1}$ and $F_{2-1} = A_1/A_2$. Since surface 2 can partly view itself, from Eq. 5-41 we have also $F_{2-2} = 1 - (A_1/A_2)$. From Eqs. 5-49 and 5-52 the net rate of heat losses from A_1 and A_2 are

$$q_1 = \frac{A_1\epsilon_1}{1 - \epsilon_1}(E_{b1} - J_1) = A_1(J_1 - J_2)$$

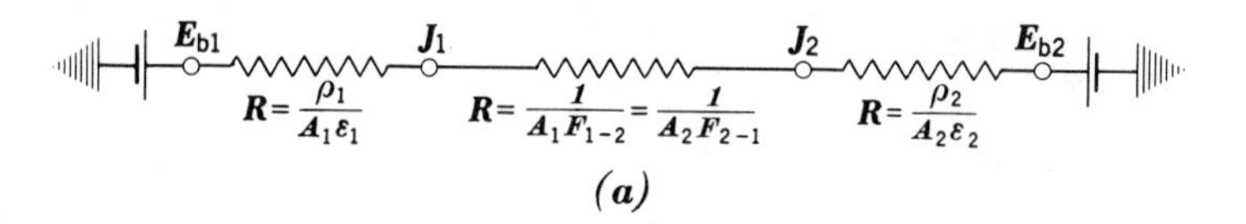

(a)

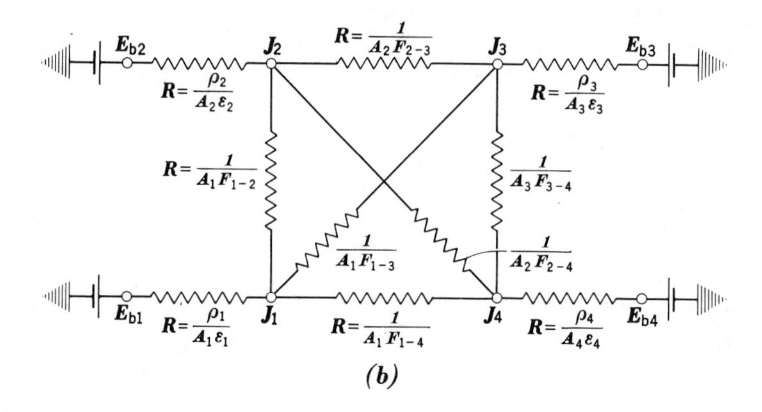

(b)

Fig. 5-28. Equivalent networks for radiation in gray enclosures consisting of two and four surfaces: a) two gray-body surfaces, b) four gray-body surfaces.

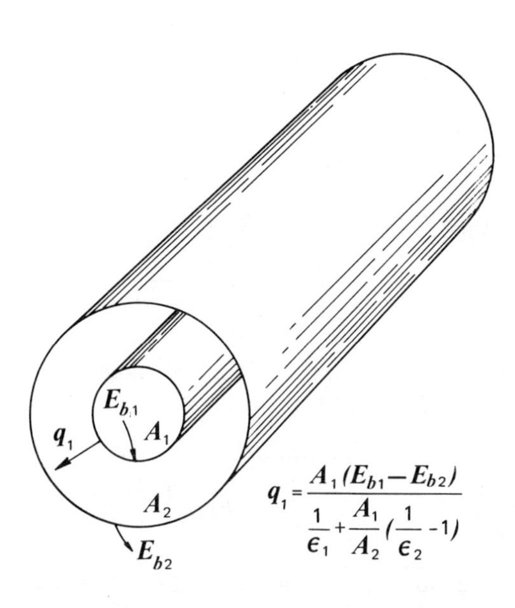

Fig. 5-29. Radiation exchange between two gray cylindrical surfaces.

and

$$q_2 = \frac{A_2\epsilon_2}{1 - \epsilon_2}(E_{b2} - J_2) = A_2(J_2 - J_1F_{2-1} - J_2F_{2-2})$$

Substituting the appropriate expressions for F_{2-1} and F_{2-2} yields the relation $q_2 = A_1(-J_1 + J_2) = -q_1$, as expected from an overall heat balance. Eliminating J_2 and substituting for J_1 in the heat loss equation for A_1 gives

$$q_1 = \frac{A_1(E_{b1} - E_{b2})}{\dfrac{1}{\epsilon_1} + \dfrac{A_1}{A_2}\left(\dfrac{1 - \epsilon_2}{\epsilon_2}\right)} \tag{5-55}$$

From the equivalent network in Fig. 5-28a the sum of the three resistances is

$$\frac{1 - \epsilon_1}{\epsilon_1 A_1} + \frac{1}{A_1F_{1-2}} + \frac{1 - \epsilon_2}{\epsilon_2 A_2} = \frac{1}{A_1}\left[\frac{1}{\epsilon_1} + \frac{A_1}{A_2}\left(\frac{1 - \epsilon_2}{\epsilon_2}\right)\right]$$

which gives the identical result for the net rate of heat loss from A_1, as expected.

The net rate of heat transfer in simple systems where radiation is transferred only between two gray surfaces can also be written in terms of an equivalent conductance $A_1\mathfrak{F}_{1-2}$ in the form

$$q_{1\rightleftharpoons 2} = A_1\mathfrak{F}_{1-2}(E_{b1} - E_{b2}) \tag{5-56}$$

where A_1 is the smaller of the two surfaces, and $\mathfrak{F}_{1-2}$ is given below. For two concentric cylinders or spheres,

$$\mathfrak{F}_{1-2} = \frac{1}{\dfrac{1 - \epsilon_1}{\epsilon_1} + 1 + \dfrac{A_1(1 - \epsilon_2)}{A_2\epsilon_2}}$$

For two equal parallel plates spaced a finite distance apart,

$$\mathfrak{F}_{1-2} = \frac{1}{\dfrac{1 - \epsilon_1}{\epsilon_1} + \dfrac{1}{F_{1-2}} + \dfrac{1 - \epsilon_2}{\epsilon_2}}$$

where F_{1-2} can be obtained from Fig. 5-22. For two infinitely large parallel plates,

$$\mathcal{F}_{1-2} = \frac{1}{\dfrac{1}{\epsilon_1} + \dfrac{1}{\epsilon_2} - 1}$$

For a small gray body of area A_1 inside a large enclosure of area A_2 ($A_1 \ll A_2$),

$$\mathcal{F}_{1-2} = \epsilon_1$$

In many real problems radiation heat transfer will cause the internal energy and the temperature of a body to change. The heat transfer rate should then be interpreted as a quasi-steady-state result. Under those circumstances the solution will require a transient analysis similar to that presented in Chapter 4, with the surface temperature of the body a function of time.

EXAMPLE 5-7. Liquefied oxygen (boiling temperature, −297 F) is to be stored in a spherical container of 1-ft-diameter. The system is insulated by an evacuated space between the inner sphere and a surrounding 1.5-ft-ID concentric sphere. Both spheres are made of aluminum ($\epsilon = 0.03$), and the temperature of the outer sphere is 30 F. Estimate the rate of heat flow by radiation to the oxygen in the container.

Solution: Although the internal energy of the oxygen will change, its temperature will remain constant since it is undergoing a change in phase. The absolute temperatures of the surfaces are

$$T_1 = 460 - 297 = 163\text{R}$$

$$T_2 = 460 + 30 = 490\text{R}$$

From Eq. 5-56 the rate of heat loss from the inner sphere is

$$q_1 = \frac{A_1 \sigma (T_1^4 - T_2^4)}{\dfrac{1}{\epsilon_1} + \dfrac{A_1}{A_2}\left(\dfrac{1 - \epsilon_2}{\epsilon_2}\right)} = \frac{\pi \times 0.1714 (1.63^4 - 4.9^4)}{\dfrac{1}{0.03} + \dfrac{1}{2.25}\left(\dfrac{0.97}{0.03}\right)}$$

or $q_1 = -6.5$ Btu/hr. Since the radiation loss from A_1 is negative, the heat is actually transferred to the oxygen, as expected.

The radiant heat flow in an enclosure consisting of two gray surfaces connected by re-radiating surfaces can also be solved without difficulty by means of the equivalent circuit. According to Eqs. 5-53 and 5-54, it is only necessary to replace E_{b1} and E_{b2}, the potentials used in Sec. 5-7 for black surfaces, by J_1 and J_2 and connect the new potentials with the resistances $\rho_1/\epsilon_1 A_1$ and $\rho_2/\epsilon_2 A_2$ to their respective blackbody potentials E_{b1} and E_{b2}. The resulting network is shown in Fig. 5-30, and from it

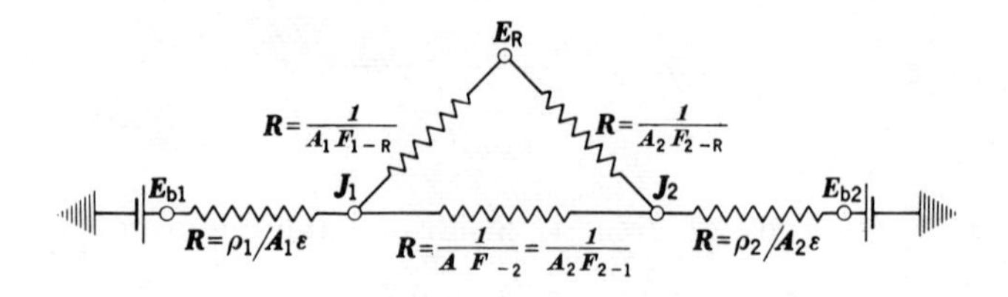

Fig. 5-30. Equivalent circuit for radiation in an enclosure consisting of two gray surfaces connected by a re-radiating surface.

we see that the total conductance between E_1 and E_2 is now

$$A_1\mathfrak{F}_{1-2} = \frac{1}{\dfrac{\rho_1}{\epsilon_1 A_1} + \dfrac{\rho_2}{\epsilon_2 A_2} + \dfrac{1}{A_1[F_{1-2} + 1/(1/F_{1-R} + A_1/A_2F_{2-R})]}}$$

where the denominator of the last term is the conductance for the blackbody network given by Eq. 5-46. The expression for the conductance can be recast into the more convenient form

$$A_1\mathfrak{F}_{1-2} = \frac{1}{\dfrac{1}{A_1}\left(\dfrac{1}{\epsilon_1} - 1\right) + \dfrac{1}{A_2}\left(\dfrac{1}{\epsilon_2} - 1\right) + \dfrac{1}{A_1\bar{F}_{1-2}}}$$

where $A_1\bar{F}_{1-2}$ is the effective conductance for the blackbody network, equal to the last term in the denominator of the original expression. The equation for the net radiant heat transfer per unit time between two gray surfaces at uniform temperatures in the presence of re-radiating surfaces can then be written

$$q_{1\rightleftharpoons 2} = A_1\mathfrak{F}_{1-2}\sigma(T_1^4 - T_2^4)$$

For enclosures consisting of several surfaces the radiation heat transfer from any one of them can be calculated by drawing the equivalent circuit and performing a circuit analysis. This analysis can be made by applying Kirchhoff's current law which states that the sum of the current entering a given node is zero. When a computer is available, the same result can be obtained more simply by a matrix method outlined below.

The problem at hand is solving N equations in N unknowns. The equations are obtained by evaluating the emittances of the surfaces and the shape factors between them and writing Eqs. 5-49 and 5-52 for each nodal point.

For a gray enclosure consisting of three surfaces at specified temperatures this procedure yields

$$q_1/A_1 = \epsilon_1/(1 - \epsilon_1)(E_{b1} - J_1) = J_1 - J_1F_{11} - J_2F_{12} - J_3F_{13}$$

$$q_2/A_2 = \epsilon_2/(1 - \epsilon_2)(E_{b2} - J_2) = J_2 - J_1F_{21} - J_2F_{22} - J_3F_{23}$$

$$q_3/A_3 = \epsilon_3/(1 - \epsilon_3)(E_{b3} - J_3) = J_3 - J_1F_{31} - J_2F_{32} - J_3F_{33}$$

In the above set of 3 linear equations J_1, J_2, and J_3 are unknowns. The equations can be recast in the more convenient form

$$a_{11}J_1 + a_{12}J_2 + a_{13}J_3 = C_1$$

$$a_{21}J_1 + a_{22}J_2 + a_{23}J_3 = C_2$$

$$a_{31}J_1 + a_{32}J_2 + a_{33}J_3 = C_3$$

where the coefficients in the first row are $a_{11} = 1 - F_{11} + \epsilon_1/(1 - \epsilon_1)$, $a_{12} = -F_{12}$, $a_{13} = -F_{13}$, and $C_1 = E_{b1}\epsilon_1/(1 - \epsilon_1)$. The coefficients in rows 2 and 3 can be obtained similarly, e.g., $a_{21} = -F_{21}$.

If, for example, the heat flux at surface A_1 is specified instead of the temperature, the coefficients a_{12} and a_{13} remain unchanged, but $C_1 = q_1/A_1$ and $a_{11} = (1 - F_{11})$. Observe also that for any black surface at temperature T, the radiosity must equal σT^4 and is no longer an unknown.

To simplify the calculation procedure for N surfaces, define the following matrix representations:

$$[A] = \begin{bmatrix} a_{11} & a_{12} & \cdots & a_{1N} \\ a_{21} & a_{22} & \cdots & \\ a_{31} & & & \\ \vdots & & & \\ a_{N1} & A_{N2} & \cdots & a_{NN} \end{bmatrix}, \quad [C] = \begin{bmatrix} C_1 \\ C_2 \\ \vdots \\ C_N \end{bmatrix}, \quad [J] = \begin{bmatrix} J_1 \\ J_2 \\ \vdots \\ J_N \end{bmatrix}$$

The set of equations to be solved can then be written compactly

$$[A][J] = [C] \tag{5-57}$$

If $[A]^{-1}$ represents the inverse of matrix $[A]$, the solution for the radiosities is given by

$$[J] = [A]^{-1}[C] \tag{5-58}$$

The inverse of matrix $[A]$ is a matrix having the elements

$$[A]^{-1} = \begin{bmatrix} b_{11} & b_{12} & \cdots & b_{1N} \\ b_{21} & \cdots & & \\ \vdots & & & \\ b_{N1} & n_{N2} & \cdots & b_{NN} \end{bmatrix}$$

and the solution for the radiosities can be written

$$J_1 = b_{11}C_1 + b_{12}C_2 + b_{13}C_3 + \cdots + b_{1N}C_N,$$
$$J_2 = b_{21}C_1 + \cdots$$
$$\vdots$$
$$J_N = b_{N1}C_1 + b_{N2}C_2 + \cdots \qquad\qquad + b_{NN}C_N.$$

For a mathematical treatment of matrix algebra and inversion techniques the reader is referred to Ref. 31.

In practical terms the problem of solving the simultaneous linear algebraic equations for the radiosities reduces to the inversion of a matrix. The process of matrix inversion is usually laborious when done by hand, but all modern digital computer facilities have matrix inversion and matrix multiplication routines available in the software library. With such a routine, the inversion operation can be carried out simply and the radiosities evaluated. Once the radiosities are known, the unknowns, i.e., either the rate of heat flow or the temperature, can be obtained from Eq. 5-49 for each surface. The following examples illustrate this procedure.

EXAMPLE 5-8. The temperatures of the top and bottom surfaces of the frustum of the cone shown in Figure 5-31a are maintained at 1000 R and 2000 R respectively, while the side (A_2) is perfectly insulated ($q_2 = 0$). If all surfaces are assumed to be gray and diffuse, determine the net radiative exchange between the top and bottom surfaces, i.e. A_3 and A_1.

Solution: From Table 5-3 we find that $F_{31} = 0.333$ and from Eq. 5-41 we obtain $F_{32} = 1 - F_{31} = .667$.

According to the reciprocity theorem $A_1F_{13} = A_3F_{31}$ and $A_2F_{23} = A_3F_{32}$. Therefore, $F_{13} = 0.147$ and $F_{23} = 0.130$. From Eq. 5-41 we get $F_{12} = 1 - F_{13} = 0.853$ and by reciprocity, $F_{21} = F_{12}A_1/A_2 = 0.372$. Finally, $F_{22} = 1 - F_{21} - F_{23} = 0.498$.

According to the general relations given by equations 5-49 and 5-52 the system of equations to be solved for this problem may be written:

$$E_{b1} \cdot \frac{\epsilon_1}{1 - \epsilon_1} = J_1\left(1 - F_{11} + \frac{\epsilon_1}{1 - \epsilon_1}\right) + J_2(-F_{12}) + J_3(-F_{13})$$

$$0 = J_1(-F_{21}) + J_2(1 - F_{22}) + J_3(-F_{23})$$

$$E_{b3} \cdot \frac{\epsilon_3}{1 - \epsilon_3} = J_1(-F_{31}) + J_2(-F_{32}) + J_3\left(1 - F_{33} + \frac{\epsilon_3}{1 - \epsilon_3}\right)$$

or in matrix notation $[A]\cdot[J] = [C]$.

A FORTRAN subroutine called LINEQZ which solves linear algebraic systems of equations of the form $[A] \cdot [X] = [B]$ will be used to evaluate all the J's. The net rate of heat transfer between top and bottom, i.e., the value of

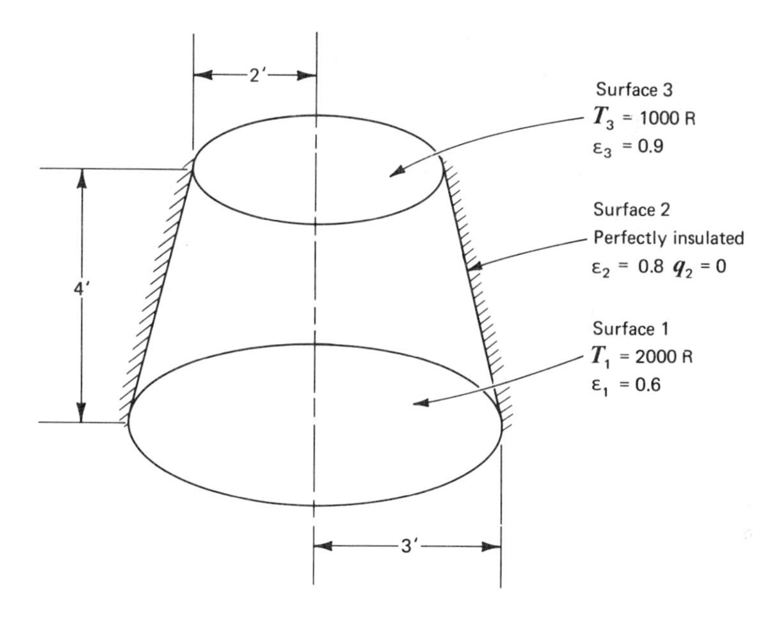

Fig. 5-31a. Schematic sketch of cone for Example 5-8.

$q_{3\rightleftharpoons 1}$, can then be determined from Equation 5-54. Fig. 5-31b gives the flow diagram for the computer operations to solve this problem. The FORTRAN program and the solution are presented in Table 5-4. The symbols used in the FORTRAN program are defined in Table 5-5. The symbols which appear in the program but are not listed in Table 5-5 are associated only with the subroutine LINEQZ and do not have significance in the solution of the heat transfer problem.

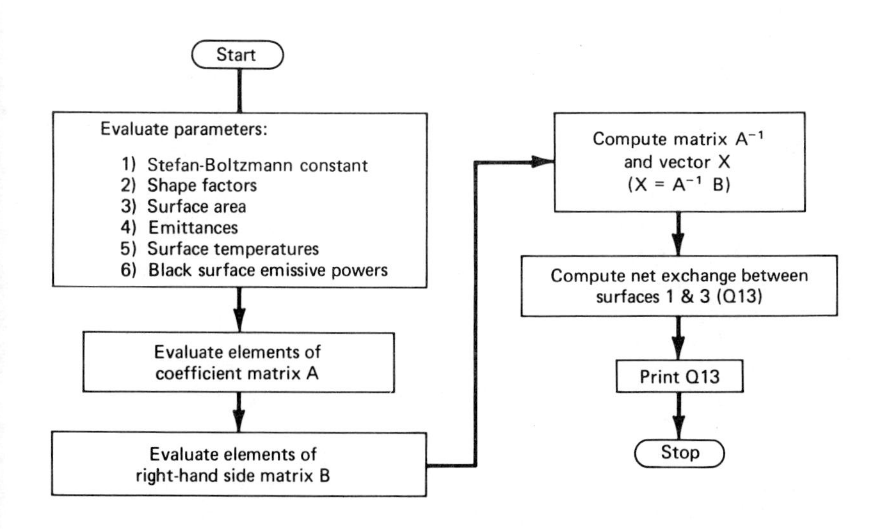

Fig. 5-31b. Flow chart for Example 5-8.

Table 5-4. FORTRAN program for Example 5-8.

```
DIMENSION A(3,3),B(3,1),X(3,1),DIGITS(1)
DIMENSION SCRA1(3,3),SCRA2(3),SCRA3(3),SCRA4(3)
DIMENSION F(3,3),AR(3),EPS(3),T(3),EB(3)
```

The arrays required for the program are declared. See Table 5-5 for the precise definition of each. The arrays SCRA1, SCRA2, etc. are peculiar to the matrix inversion subroutine LINEQZ.

```
PI=4.*ATAN(1.)
SIGMA=.1714E-08
F(1,1)=0.
F(1,2)=.853
F(1,3)=.147
F(2,1)=.372
F(2,2)=.498
F(2,3)=.130
F(3,1)=.333
F(3,2)=.667
F(3,3)=0.
AR(1)=9.*PI
EPS(1)=.6
EPS(3)=.9
T(1)=2000.
T(3)=1000.
EB(1)=SIGMA*T(1)**4.
EB(3)=SIGMA*T(3)**4.
```

The physical parameters, e.g. shape factors, emittance, etc., are evaluated.

```
A(1,1)=1.-F(1,1)+EPS(1)/(1.-EPS(1))
A(1,2)=-F(1,2)
A(1,3)=-F(1,3)
A(2,1)=-F(2,1)
A(2,2)=1.-F(2,2)
A(2,3)=-F(2,3)
A(3,1)=-F(3,1)
A(3,2)=-F(3,2)
A(3,3)=1.-F(3,3)+EPS(3)/(1.-EPS(3))
```

The values of the elements of the coefficient matrix A in the equation $[A]\cdot[X] = [B]$ are specified.

Table 5-4. Continued.

```
      B(1)=EB(1)*EPS(1)/(1.-EPS(1))
      B(2)=0.
      B(3)=EB(3)*EPS(3)/(1.-EPS(3))
```

The values of the right hand side vector B are specified.

```
      CALL LINEQZ(3,3,1,A,B,X,D,DIGITS,SCRA1,SCRA2,SCRA3,SCRA4)
```

The matrix inversion subroutine LINEQZ is called to compute the solution vector $X = A^{-1} \cdot B$. This is one of many matrix inversion routines. The reader may use any routine available to him and specify the arguments accordingly.

```
      Q13=(X(1)-X(3))*AR(1)*F(1,3)
      PRINT 1,Q13
1     FORMAT(1H1,* NET EXCHANGE BETWEEN SURFACES 1 AND 3,Q13,= *,
     1F12.1,*BTU/HR*)
      END
```

The net exchange $q_{1\rightleftharpoons 3}$ is computed from Eq. 5-54. This value is printed out and the program is terminated.

```
      NET EXCHANGE BETWEEN SURFACES 1 AND 3, Q13, = 79913.6 BTU/HR
```

EXAMPLE 5-9. Determine the temperature of surface 1 for the cone shown in Fig. 5-31a if $q_1 = 10^5$ Btu/hr ft^2 and $\epsilon_3 = 1$. Assume that all other parameters are the same as in Example 5-8.

Solution: From equations 5-49 and 5-52 the following system of equations must be solved for J_1, J_2 and J_3:

$$q_1/A_1 = J_1(1 - F_{11}) + J_2(-F_{12}) + J_3(-F_{13})$$

$$0 = J_1(-F_{21}) + J_2(1 - F_{22}) + J_3(-F_{23})$$

$$E_{b3} = J_3$$

Once the J_i's are known, Equation 5-49 gives T_1. The FORTRAN program for the solution of this problem is shown in Table 5-6. Since it is very similar to the preceding program the flow diagram is essentially the same as that used in Example 5-8.

Table 5-5. Symbols used in FORTRAN program for Example 5-8.

FORTRAN Symbol	Heat Balance Equation Notation	Description	Units
A(I,J)	a_{ij}	Coefficient matrix elements	—
AR(1), AR(3)	A_1, A_3	Lower and upper surface areas	ft^2
B(I)	C_i	Right hand side matrix elements	$Btu/hr\ ft^2$
EB(1), EB(3)	E_{b1}, E_{b3}	Blackbody emissive powers	$Btu/hr\ ft^2$
EPS(1), etc.	ϵ_1, etc.	Total hemispheric emittance	—
F(1,1), F(1,2), etc.	F_{11}, F_{12}, etc.	Shape factors	—
PI	π	3.14159...	—
Q31	$q_{3 \rightleftharpoons 1}$	Net exchange between surfaces 3 and 1	Btu/hr
SIGMA	σ	Stefan-Boltzmann constant (0.1714×10^{-8})	$Btu/hr\ ft^2\ R^4$
T(1),T(3)	T_1, T_3	Surface temperatures	R
X(I)	J_i	Radiosities (elements of solution vector)	$Btu/hr\ ft^2$

Table 5-6. FORTRAN program for Example 5-9.

```
      DIMENSION A(3,3),B(3,1),X(3,1),DIGITS(1)
      DIMENSION SCRA1(3,3),SCRA2(3),SCRA3(3),SCRA4(3)
      DIMENSION F(3,3),AR(3),EPS(3),T(3),EB(3)
COMMENT EVALUATE CONSTANTS AND PARAMETERS
      PI=4.*ATAN(1.)
      SIGMA=.1714E-08
      F(1,1)=0.
      F(1,2)=.853
      F(1,3)=.147
      F(2,1)=.372
      F(2,2)=.498
      F(2,3)=.130
      F(3,1)=.333
      F(3,2)=.667
      F(3,3)=0.
      AR(1)=9.*PI
      EPS(1)=0.6
      Q1=100000.
```

Table 5-6. Continued.

```
        T(3) = 1000.
        EB(3) = SIGMA * T(3) * * 4.
COMMENT EVALUATE ELEMENTS OF COEFFICIENT MATRIX
        A(1,1) = 1.-F(1,1)
        A(1,2) = -F(1,2)
        A(1,3) = -F(1,3)
        A(2,1) = -F(2,1)
        A(2,2) = 1.-F(2,2)
        A(2,3) = -F(2,3)
        A(3,1) = 0.
        A(3,2) = 0.
        A(3,3) = 1.
COMMENT EVALUATE ELEMENTS OF RIGHT HAND SIDE MATRIX
        B(1) = Q1/AR(1)
        B(2) = 0.
        B(3) = EB(3)
COMMENT CALL SUBROUTINE TO SOLVE SYSTEM OF EQUATIONS
        CALL LINEQZ(3,3,1,A,B,X,D,DIGITS,SCRA1,SCRA2,SCRA3,SCRA4)
COMMENT COMPUTE T(1) FROM EQUATION 5-49
        T(1)=((X(1)*Q1+(1.-EPS(1))/(AR(1)*EPS(1)))/SIGMA)**.25
        PRINT 2,T(1)
2       FORMAT (1H1,*TEMPERATURE OF SURFACE 1, T(1), = *,F7.1,*DEG R*)
        END

        TEMPERATURE OF SURFACE 1, T(1), = 1681.0 DEG R
```

The method of approach used to calculate heat transfer in gray surface enclosures can easily be adapted to non-gray surfaces. If the surface properties are functions of wavelength, they can be approximated by gray "bands" within which an average value of emittance and absorptance is used. Then the same calculation method used previously for gray enclosures can be used to determine the radiation heat transfer within each band. The following example illustrates the procedure.

EXAMPLE 5-10. Determine the rate of heat transfer between two 1-ft by 1-ft parallel flat plates placed two inches apart, if one plate (A) is at 2040 F and the other (B) at 540 F. Plate A has an emittance of 0.1 between 0 and 2.5μ and an emittance of 0.9 for wavelengths longer than 2.5μ. The emittance of plate B is 0.9 between 0 and 4.0μ and 0.1 at longer wavelengths.

Solution: The shape factor F_{A-B} for two parallel rectangular plates ($L/D = 6$) is found from Fig. 5-22 to be 0.725. The radiosity of A is given by

$$\int_0^\infty J_{\lambda A} d\lambda = \int_0^\infty \epsilon_{\lambda A} E_{b\lambda A}\, d\lambda + \int_0^\infty \rho_{\lambda A} G_{\lambda A} d\lambda$$

while the radiosity of B is given by

$$\int_0^\infty J_{\lambda B} d\lambda = \int_0^\infty \epsilon_{\lambda B} E_{b\lambda B} d\lambda + \int_0^\infty \rho_{\lambda B} G_{\lambda B} d\lambda$$

However, using spectral bands between 0 and 2.5μ, 2.5 and 4.0μ, and 4.0μ or larger, the system obeys gray surface radiation laws within each band and the rate of heat transfer can be calculated from Eq. 5-56 in three steps as shown below:

Band 1:

$$q_{A \rightleftharpoons B}\Big|_0^{2.5\mu} = \mathfrak{F}_{A-B}(\epsilon_A = 0.1, \epsilon_B = 0.9) \left[\frac{E_{b,0-2.5}(T_A)}{E_{b,0-\infty}(T_A)} \sigma T_A^4 - \frac{E_{b,0-2.5}(T_B)}{E_{b,0-\infty}(T_B)} \sigma T_B^4\right]$$

Band 2:

$$q_{A \rightleftharpoons B}\Big|_{2.5\mu}^{4.0\mu} = \mathfrak{F}_{A-B}(\epsilon_A = 0.9, \epsilon_B = 0.9) \left[\frac{E_{b,2.5-4.0}(T_A)}{E_{b,0-\infty}(T_A)} \sigma T_A^4 - \frac{E_{b,2.5-4.0}(T_B)}{E_{b,0-\infty}(T_B)} \sigma T_B^4\right]$$

Band 3:

$$q_{A \rightleftharpoons B}\Big|_{4.0\mu}^{\infty} = \mathfrak{F}_{A-B}(\epsilon_A = 0.9, \epsilon_B = 0.1) \left[\frac{E_{b,4.0-\infty}(T_A)}{E_{b,0-\infty}(T_A)} \sigma T_A^4 - \frac{E_{b,4.0-\infty}(T_B)}{E_{b,0-\infty}(T_B)} \sigma T_B^4\right]$$

where $\mathfrak{F}_{A-B} = \dfrac{1}{\dfrac{1}{\epsilon_A} + \dfrac{1}{\epsilon_B} - 0.72}$ from Figs. 5-28a and 5-22.

The percentage of the total radiation within a given band is obtained from Table 5.1. For example, $(E_{b,0-2.5}/E_{b,0-\infty})$ for a temperature of $T_A = 2{,}500$ R is 0.375 and for a temperature of $T_B = 1{,}000$ R it is about 0.004. Thus, for the first band,

$$q_{A \rightleftharpoons B}^{1}\Big|_0^{2.5\mu} = 0.094 \times 0.1714(0.375 \times 25^4 - 0.004 \times 10^4)$$

$$= 2{,}380 \text{ Btu/hr}$$

Similarly, for the second band,

$$q_{A \rightleftharpoons B}^{2} \Big|_{2.5\mu}^{4.0\mu} = 18{,}700 \text{ Btu/hr}$$

and for the third band,

$$q_{A \rightleftharpoons B}^{3} \Big|_{4.0\mu}^{\infty} = 1{,}170 \text{ Btu/hr}$$

Finally, summing over all three bands, the total rate of radiation heat transfer is

$$q_{A \rightleftharpoons B} \Big|_{0}^{\infty} = \sum_{N=1}^{N=3} q_{A \rightleftharpoons B}^{N} = 2{,}380 + 18{,}700 + 1{,}770 = 22{,}850 \text{ Btu/hr}$$

It should be noted that most of the radiation is transferred within the second band where both surfaces are nearly black.

Enclosures consisting of several non-gray surfaces can be treated in a similar manner by dividing the radiation spectrum into finite bands within which the radiation properties can be approximated by constant values. This procedure can become particularly useful when the enclosure is filled with a gas which only absorbs and emits radiation at certain wavelengths.

5-7. Radiation in gas-filled enclosures

The method of analysis outlined in the preceding sections can be extended to solve problems in which heat is transferred by radiation in an enclosure containing a medium which is both absorbing and transmitting. Various glasses and many gases are examples of such media. To illustrate the method of approach, consider first radiation between two plates when the space between them is filled with a "gray" gas, as shown in Fig. 5-32a. Designate the radiation properties of the gas by the subscript m and assume that the gas does not contain any particles which could reflect and scatter radiation. Kirchhoff's law applies under these conditions and

$$\alpha_m + \tau_m = 1 \quad \text{or} \quad \epsilon_m + \tau_m = 1 \tag{5-59}$$

where τ_m is the portion of radiation transmitted by the gas, called the transmittance.

The portion of the total radiation leaving surface 1 which arrives at

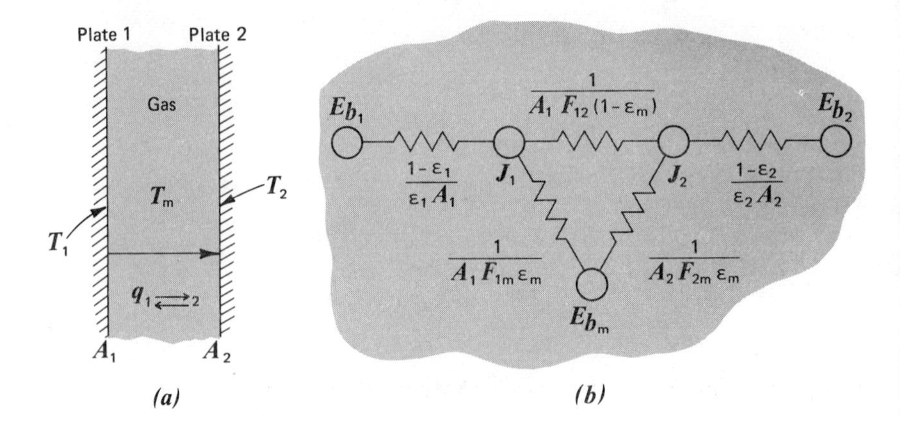

Fig. 5-32. Radiation between two parallel plates when the space between them is filled by a gas: a) schematic sketch of system, b) equivalent circuit.

surface 2 is

$$J_1 A_1 F_{12} \tau_m$$

and that from surface 2 which reaches 1 is

$$J_2 A_2 F_{21} \tau_m$$

The net rate of heat transfer between the two surfaces is therefore

$$q_{1\rightleftarrows 2} = A_1 F_{12} \tau_m (J_1 - J_2) = \frac{J_1 - J_2}{1/A_1 F_{12}(1 - \epsilon_m)} \tag{5-60}$$

Thus, the equivalent resistance between nodal points J_1 and J_2 for this case in a network will be $1/A_1 F_{12}(1 - \epsilon_m)$.

Radiation heat transfer occurs also between each of the surfaces and the gas. If the gas is at temperature T_m it will emit radiation at a rate

$$J_m = \epsilon_m E_{bm}$$

The fraction of the energy emitted by the gaseous medium which reaches surface 1 is

$$A_m F_{m-1} J_m = A_m F_{m-1} \epsilon_m E_{bm}$$

Similarly, the fraction of the radiation leaving A_1 which is absorbed by the transparent medium is

$$J_1 A_1 F_{1m} \alpha_m = J_1 A_1 F_{1m} \epsilon_m$$

The net rate of heat transfer by radiation between the gas and surface 1 is the difference between the radiation emitted by the gas towards A_1 and the radiation emanating from A_1 which is absorbed by the gas. Thus

$$q_{m \rightleftharpoons 1} = A_m F_{m1} \epsilon_m E_{bm} - J_1 A_1 F_{1m} \epsilon_m$$

Using the reciprocity theorem, $A_1 F_{1m} = A_m F_{m1}$, and the net exchange can be written in the form

$$q_{m \rightleftharpoons 1} = \frac{E_{bm} - J_1}{1/A_1 F_{1m} \epsilon_m} \tag{5-61a}$$

Similarly, the net exchange between the gas and A_2 is

$$q_{m \rightleftharpoons 2} = \frac{E_{bm} - J_2}{1/A_2 F_{2m} \epsilon_m} \tag{5-61b}$$

Using the above relations to construct an equivalent circuit, radiation between two surfaces at T_1 and T_2 respectively, separated by an absorbing medium at T_m, can be represented as shown in Fig. 5-32b. On the other hand, if the gas is not maintained at a specified temperature but has reached an equilibrium temperature at which it emits radiation at the same rate at which it absorbs it, E_{bm} becomes a floating node in the network. For this case the net rate of heat transfer between A_1 and A_2 is

$$q_{1 \rightleftharpoons 2} = \frac{\sigma A_1 (T_1^4 - T_2^4)}{\dfrac{1-\epsilon_1}{\epsilon_1 A_1} + \dfrac{1-\epsilon_2}{\epsilon_2 A_2} + \dfrac{1}{A_1[F_{1-2} + 1/(1/F_{1-m} + A_1/A_2 F_{2-m})]}} \tag{5-62}$$

More complicated situations involving several surfaces can be treated by an extension of the matrix method as shown in References (24) and (27).

5-8. Radiation properties of gases and vapors

In this section we shall consider some basic concepts of gaseous radiation. A comprehensive treatment of this subject is beyond the scope of this text, and the reader should consult References (9), (24), and (27) for details of the theoretical background and complete derivations of the calculation techniques.

Elementary gases such as O_2, N_2, H_2 and dry air have a symmetrical molecular structure and neither emit nor absorb radiation unless they are heated to extremely high temperatures at which they become ionized plasmas and at which electronic energy transformations occur. On the other hand, gases which have polar molecular forms with an electronic moment such as a dipole or quadrupole absorb and emit radiation in limited spectral ranges, called bands. In practice, the most important of these gases are H_2O, CO_2, CO, SO_2, NH_3, and the hydrocarbons. These gases are asymmetric in one or more of their modes of vibration. During molecular collisions, rotation and vibrations of individual atoms in a molecule can be excited so that atoms which possess free electrical charges can emit electromagnetic waves. Similarly, when radiation of the appropriate wavelength impinges on such a gas, it can be absorbed in the process. We shall restrict our consideration here to the evaluation of the radiation properties of H_2O and CO_2. They are the most important gases in thermal radiation calculations and also illustrate the basic principles of gaseous radiation.

Typical changes in energy level due to changes in vibrational frequency or rotation manifest themselves in a strong peak at the wavelength corresponding to the vibrational transformation, with multiple rotational energy changes slightly above or below the peak. This process results in absorption or emission bands. The shape and width of these bands depend on the temperature and on the pressure of the gas while the magnitude of the monochromatic absorptance is primarily a function of the thickness of the gas layer. The absorption spectra of steam and carbon dioxide shown in Figs. 5-33 and 5-34 illustrate the complexity of the process. The most important absorption bands for steam lie between 1.7 and 2.0μ, 2.2 and 3.0μ, 4.8 and 8.5μ, and 11 and 25μ; those for CO_2 lie between 2.4 and 3μ, 4 and 4.8μ, and 12.5 and 17μ.

Experimental measurements generally yield the absorptance of a gas layer over a band width corresponding to the width of the spectrometer slit used. Thus, experimental data are usually presented in terms of the monochromatic absorptance, as shown in Figs. 5-33 and 5-34. For many engineering calculations, however, the quantity of primary interest is the effective total absorptance or emittance. This quantity assumes that the gas is gray and, as shown below, its value depends not only on the pressure, temperature, and composition, but also on the geometry of the radiating gas.

Whereas for opaque solids the emission and absorption of radiation are surface phenomena, in calculating the radiation emitted or absorbed by a gas layer its thickness, pressure, and shape as well as its surface area must be taken into account. When monochromatic radiation at an intensity $I_{\lambda 0}$ passes through a gas layer of thickness L, the radiant-energy absorption in a differential distance dx is governed by the relation

$$dI_{\lambda x} = -k'_{\lambda} I_{\lambda x} dx \tag{5-63}$$

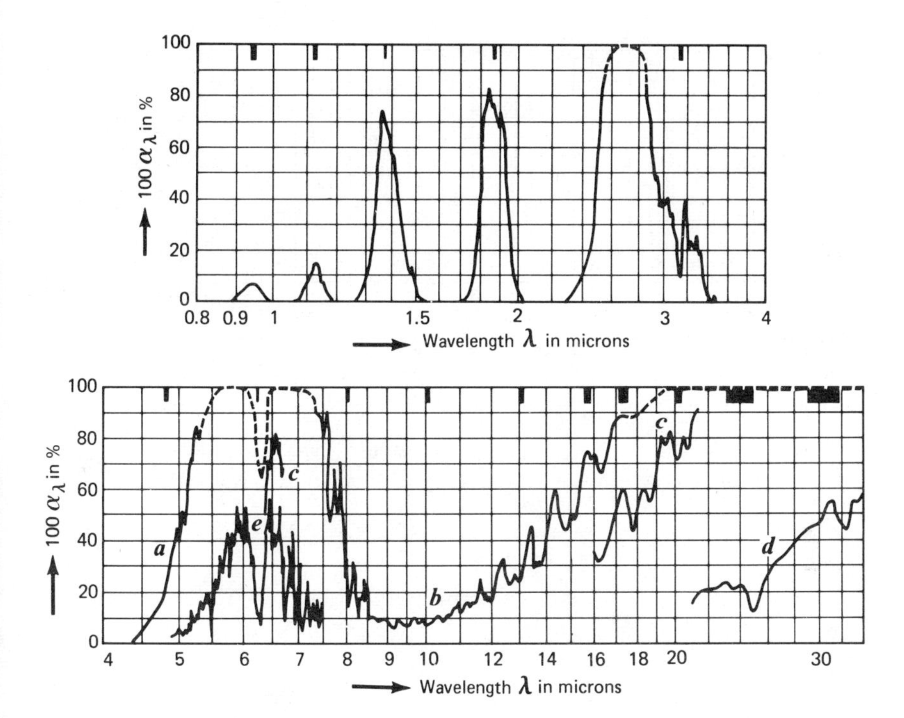

Fig. 5-33. Monochromatic absorption coefficient of steam, etc. Shaded rectangles near upper edges of the diagrams represent width of the spectrometer slit, measured in wavelength units. Upper diagram, wavelength from 0.8 to 4, steam temperature 260.6 F, thickness of layer 109 cm; lower diagram, wavelength from 4 to 34. (a) Temperature 260.6 F, thickness of layer 109 cm. (b) Temperature 260.6 F, thickness of layer 104 cm. (c) Temperature 260.6 F, thickness of layer 32.4 cm. (d) Temperature 177.8 F, thickness of layer 32.4 cm, air-steam mixture corresponding to a steam layer approximately 4 cm thick. (e) Room temperature, layer of moist air 220 cm thick, corresponding to a layer of steam at atmospheric pressure approximately 7 cm thick.

where $I_{\lambda x}$ = intensity at a distance x;
k'_λ = monochromatic absorption coefficient, a proportionality constant whose value depends on the pressure and temperature of the gas.

Integration between the limits $x = 0$ and $x = L$ yields

$$I_{\lambda L} = I_{\lambda 0} e^{-k'_\lambda L} \tag{5-64}$$

where $I_{\lambda L}$ is the intensity of radiation at L. The difference between the intensity of radiation entering the gas at $x = 0$ and the intensity of radia-

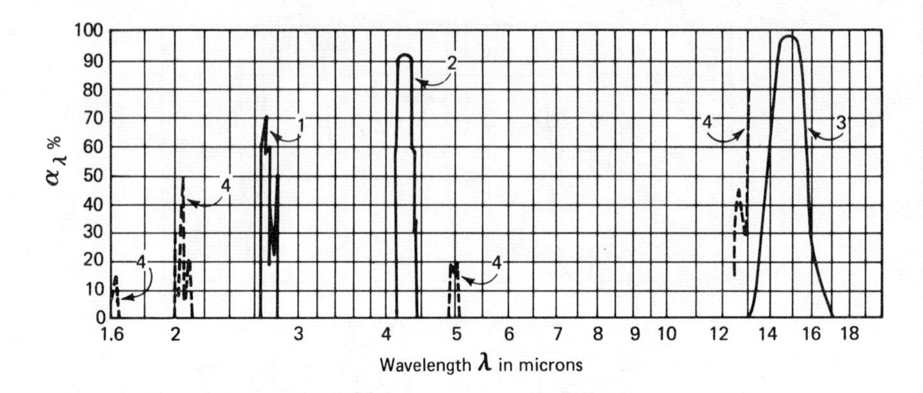

Fig. 5-34. Coefficient of absorption of carbon dioxide. Curve 1, thickness 5 cm. Curve 2, thickness 3 cm. Curve 3, thickness 6.3 cm. Curve 4, thickness 100 cm.

tion leaving the gas layer at $x = L$ is the amount of energy absorbed by the gas

$$I_{\lambda 0} - I_{\lambda L} = I_{\lambda 0}(1 - e^{-k'_\lambda L}) \tag{5-65}$$

The quantity in the parentheses represents the monochromatic absorptance of the gas $\alpha_{G\lambda}$, or, according to Kirchhoff's law, also the emittance at the wavelength λ, $\epsilon_{G\lambda}$. To obtain effective values of the emittance or absorptance, a summation over all of the radiation bands is necessary. We observe that, for large values of L, i.e., for thick layers, gas radiation approaches blackbody conditions within the wavelengths of its absorption bands.

For gas bodies of finite dimensions, however, the effective absorptance or emittance depends on the shape and the size of the gas body, since radiation is not confined to one direction. The precise method of calculating the effective absorptance or emittance is quite complex (12, 13, 14), but for engineering calculations an approximate method developed by Hottel (9) yields results of satisfactory accuracy. Hottel evaluated the effective total emittances of a number of gases at various temperatures and pressures and presented the results of his calculations in graphs similar to those shown in Figs. 5-35 and 5-36. The graphs apply strictly only to a system in which a hemispherical gas mass of radius L radiates to an element of surface located at the center of the base of a hemisphere. However, for shapes other than hemispheres, an effective beam length can be calculated. Table 5-7 lists the constants by which the characteristic dimensions of several simple shapes are to be multiplied to obtain an equivalent mean hemispherical beam length L for use in Figs. 5-35 and

Table 5-7. Average lengths of radiant beams in various gas shapes.

Shape	L
1. Sphere..................................	$\frac{2}{3} \times$ diameter
2. Infinite cylinder.........................	$1 \times$ diameter
3. Space between infinite parallel planes......	$1.8 \times$ distance between planes
4. Cube....................................	$\frac{2}{3} \times$ side
5. Space outside infinite bank of tubes with centers on equilateral triangles; tube diameter equals clearance.....................	$2.8 \times$ clearance
6. Same as (5) except tube diameter equals one-half clearance...........................	$3.8 \times$ clearance

SOURCE: Ref. 14.

5-36. For rough calculations, L can be taken as 3.4 × volume/surface area.

The curves in Figs. 5-35 and 5-36 give the total effective emittances of water vapor and carbon dioxide at a total pressure p_T of 1 atmosphere

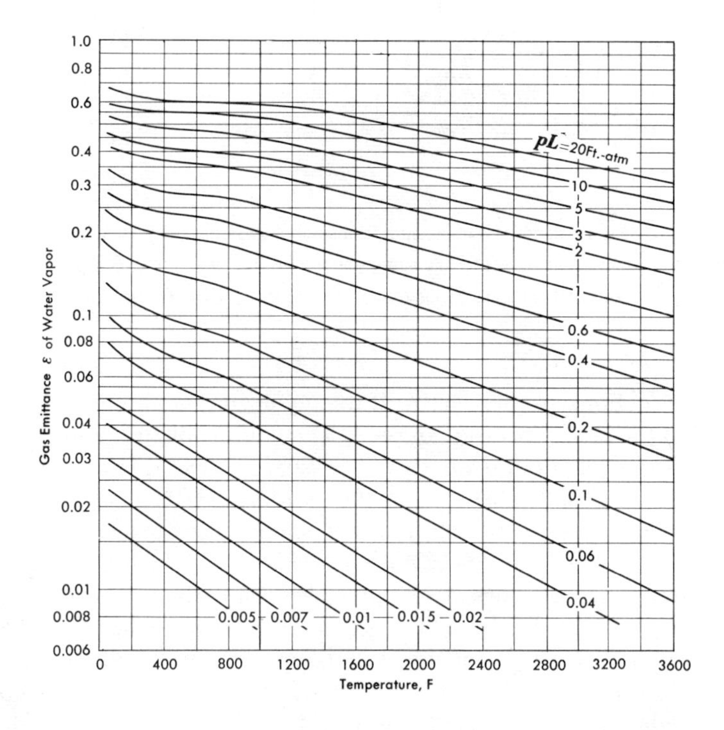

Fig. 5-35. Emittance of water vapor for a hypothetical system at 1 atmosphere total pressure and 0 partial pressure. (By permission from H. C. Hottel and R. S. Egbert, "Radiant Heat Transmission from Water Vapor," *AIChE Trans.*, Vol. 38, 1942)

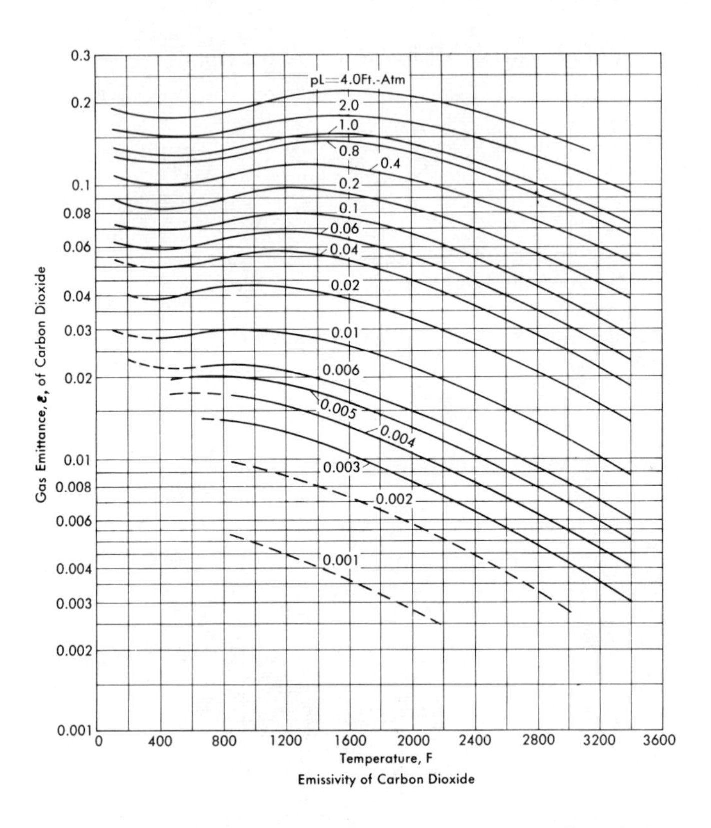

Fig. 5-36. Emittance of carbon dioxide measured experimentally at 1 atmosphere total pressure. (By permission from H. C. Hottel and R. B. Egbert, "Radiant Heat Transmission from Water Vapor," *AIChE Trans.*, Vol. 38, 1942)

(atm) as functions of the temperature, in degrees Fahrenheit, and the product of the partial pressure of the gas p, in atmospheres, and the hemispherical beam length of the body of the gas L, in feet. The effects of the actual values of total pressures on the emittance of CO_2 and both total and partial pressures on the emittance of water vapor are accounted for by means of the auxiliary charts in Figs. 5-37 and 5-38 which give correction factors to the curves in Figs. 5-35 and 5-36.[1] For carbon dioxide the emittance obtained from Fig. 5-36 at one atmosphere must be multiplied by the correction factor C_p from Fig. 5-38 to compensate for the broadening of the absorption bands with total pressure. For water vapor the correction procedure is illustrated in Example 5-11.

[1]Charts similar to those shown for CO_2 and H_2O are available in Ref. 9 for a number of other gases.

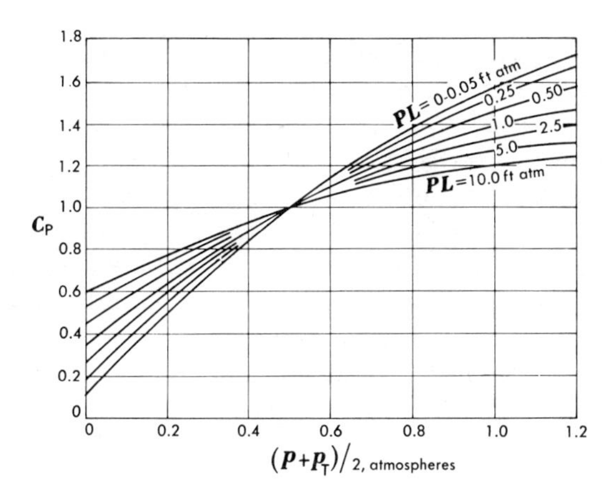

Fig. 5-37. Factor C_p for correcting emittance of water vapor to values of p and p_T other than 0 and 1 atmosphere. (By permission from H. C. Hottel and R. B. Egbert, "Radiant Heat Transmission from Water Vapor," *AIChE Trans.*, Vol. 38, 1942)

EXAMPLE 5-11. Determine the total effective emittance of water vapor at 2000 F at a partial pressure p of 0.1 atm when the total pressure p_T is 2 atm and the equivalent hemispherical beam length of the gas body is 5 ft.

Solution: From Fig. 5-35 we obtain the emittance reduced to $p = 0$. For pL equal to 0.5, $\epsilon_G = 0.125$ at a temperature of 2000 F. From the curves of Fig. 5-37, $C_p = 1.45$ at $(p + p_T)/2 = 1.05$. The effective emittance is therefore 0.181. *Ans.*

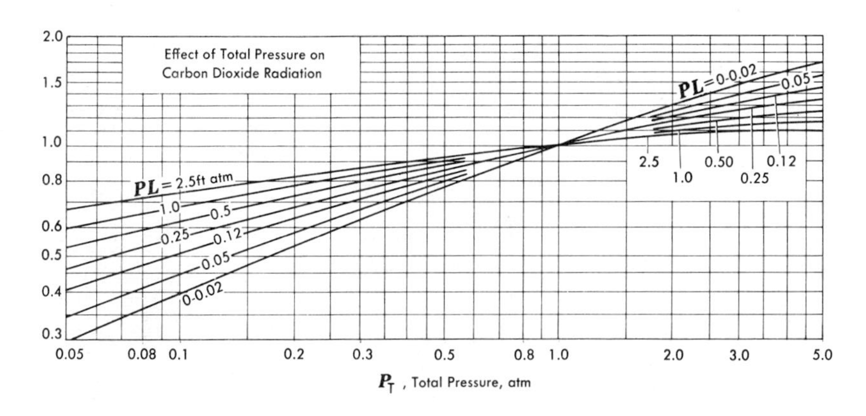

Fig. 5-38. Factor C_p for correcting emittance of CO_2 at 1 atmosphere total pressure to emittance at p_T atmosphere.

When both CO_2 and H_2O are present, the emittance can be estimated by adding the emittances of the two constituents. The value obtained by a simple addition is slightly too large because some of the absorption bands of these two gases overlap. A correction may be applied as shown by Hottel (9), but the error incurred by simple addition of emittances is not appreciable.

To calculate the rate of heat flow by radiation between a nonluminous gas at T_G and the walls of a blackbody container at T_w the absorptance α_G of the gas should be evaluated at the temperature T_w and the emittance ϵ_G at the temperature T_G. The net rate of radiant heat flow is then equal to the difference between the emitted and absorbed radiation, or

$$q_r = \sigma A_G(\epsilon_G T_G^4 - \alpha_G T_w^4) \tag{5-66}$$

EXAMPLE 5-12. Flue gas at 2000 F containing 5 percent water vapor flows at atmospheric pressure through a 2-ft-square flue made of refractory brick. Estimate the rate of heat flow per foot length from the gas to the wall if the inner-wall surface temperature is 1850 F and the average unit-surface convective conductance is 2 Btu/hr sq ft F.

Solution: The rate of heat flow from the gas to the wall by convection is

$$\begin{aligned} q_c &= \bar{h}_c A (T_{\text{gas}} - T_{\text{wall}}) \\ &= (2)(4)(2 \times 1)(150) = 2400 \text{ Btu/hr ft length} \end{aligned}$$

To determine the rate of heat flow by radiation, we calculate first the effective beam length, or

$$L = \frac{3.4 \times \text{volume}}{\text{surface area}} = \frac{(3.4)(4)}{8} = 1.7 \text{ ft}$$

The product of partial pressure and L is

$$pL = (0.05)(1.7) = 0.085 \text{ ft-atm}$$

From Fig. 5-35, for $pL = 0.085$ and $T_G = 2000$ F, we find $\epsilon_G = 0.035$. Similarly, we find $\alpha_G = 0.039$ at $T_w = 1850$ F. The pressure correction is negligible since $\bar{C}_p \cong 1$ according to Fig. 5-37. Assuming that the brick surface is black, the net rate of heat flow from the gas to the wall by radiation is, according to Eq. 5-66,

$$q_r = 0.171 \times 8[0.035(24.6)^4 - 0.039(23.1)^4] = 2340 \text{ Btu/hr}$$

The total heat flow from the gas to the duct is therefore 4740 Btu/hr. It is interesting to note that the small amount of moisture in the gas contributes about one-half of the total heat flow. *Ans.*

A recent review of radiation properties of gases (26) showed that when the radiation properties of H_2O and CO_2 are evaluated from the graphs in Figs. 5-33 to 5-35, they can be used for industrial heat transfer calculations with satisfactory accuracy as long as the enclosure surface is not highly reflecting.

The calculation of the radiant heat transfer in a gas-filled enclosure becomes considerably more complicated when the enclosure surfaces are not black and reflect a part of the incident radiation. When the emittance of the enclosure is larger than 0.7, an approximate answer may be obtained by multiplying the rate of heat flow calculated from Eq. 5-66 by $(\epsilon_s + 1)/2$, where ϵ_s is the emittance of the enclosure surface. When the enclosure walls have smaller emittances, the procedure outlined in Sec. 5-7 can be used provided the assumption that all surfaces as well as the gas are "gray" is acceptable. If one or more of the surfaces are not gray or if the gas can not be treated as a "gray" body, a band approximation procedure similar to that used in Example 5-10 must be used. Details for such refinement in the calculation procedures are presented in References (24), (25), (26), (27), and (28).

5-9. *Radiation combined with convection and conduction*

In the preceding sections of this chapter we have considered radiation as an isolated phenomenon. Energy exchange by radiation is the predominant heat-flow mechanism at high temperatures because the rate of heat flow depends on the fourth power of the absolute temperature. In many practical problems, however, convection and conduction can not be neglected, and in this section we shall consider problems which involve two or all three modes of heat flow simultaneously.

To include radiation in a thermal network involving convection and conduction it is often convenient to define a unit thermal radiative conductance, or radiant-heat-transfer coefficient, $\bar{h}_r$, as

$$\bar{h}_r = \frac{q_r}{A_1(T_1 - T_2')} = \mathfrak{F}_{1\text{-}2}\left[\frac{\sigma(T_1^4 - T_2^4)}{T_1 - T_2'}\right] \tag{5-67}$$

where A_1 = area upon which $\mathfrak{F}_{1\text{-}2}$ is based, in sq ft;

$T_1 - T_2'$ = a reference temperature difference, in F, in which T_2' may be chosen equal to T_2 or any other convenient temperature in the system;

$\bar{h}_r$ = radiant-heat-transfer coefficient, in Btu/hr sq ft F.

Once a radiant-heat-transfer coefficient has been calculated, it can be treated similarly to the convective-heat-transfer coefficient, because the rate of heat flow becomes linearly dependent on the temperature difference and radiation can be incorporated directly in a thermal network for which the temperature is the driving potential. A knowledge of the value of $\bar{h}_r$ is also essential in determining the overall conductance $\bar{h}$ for a surface to or from which heat flows by convection and radiation, since according to Eq. 1–25

$$\bar{h} = \bar{h}_c + \bar{h}_r \tag{1-25}$$

If $T_2 = T_2'$, the bracket in Eq. 5–67 is called the *temperature factor* F_T, and

$$\bar{h}_r = \mathfrak{F}_{1-2} F_T \tag{5-68}$$

Values of F_T for ordinary Fahrenheit temperatures are given in Fig. 5-39. The use of these curves is illustrated in the following example.

EXAMPLE 5-13. A hot-air duct having an outside diameter of 9 in. and a surface temperature of 200 F is located in a large room whose walls are at 70 F. The air in the room is at 80 F and the heat-transfer coefficient for free convection between the duct and the air is 1 Btu/hr sq ft F. Estimate the rate of heat transfer per foot of duct if (a) the duct is bare tin ($\epsilon = 0.1$) and (b) the duct is painted with white lacquer ($\epsilon = 0.9$).

Solution: (a) The duct may be considered as a small gray body in black surroundings and, from Eq. 5-56, $A_1 \ \mathfrak{F}_{1-2} = A_1 \epsilon_1$. From Fig. 5-39 we have $F_T = 1.5$, and therefore $\bar{h}_r = 1.5 \ \epsilon_1$. The thermal network is shown in Fig. 5-40. Note that there are two heat-flow paths in parallel but the lower temperature potentials are not equal. The total heat-flow rate is given by

$$q_{\text{total}} = q_r + q_c = A_1 \bar{h}_r (T_1 - 70) + A_1 \bar{h}_c (T_1 - 80)$$

For the bare duct the total heat-flow rate is found to be 326 Btu/hr ft of which 14 percent is due to radiation. *Ans.*

b) If the duct were painted, the total rate of heat flow would increase to 698 Btu/hr ft, of which the contribution of radiation represents 60 percent. *Ans.*

EXAMPLE 5-14. A butt-welded thermocouple (Fig. 5-41) having an emittance of 0.8 is used to measure the temperature of a transparent gas flowing in a large duct whose walls are at a temperature of 440 F. The temperature indicated by the thermocouple is 940 F. If the convective-heat-transfer coefficient between the surface of the couple and the gas $\bar{h}_c$ is 25 Btu/hr sq ft F, estimate the *true* gas temperature.

Solution: The temperature of the thermocouple is below the gas temperature because the couple loses heat by radiation to the wall. Under steady-state conditions the rate of heat flow by radiation from the thermocouple junction to the wall equals the rate of heat flow by convection from the gas to the couple. We

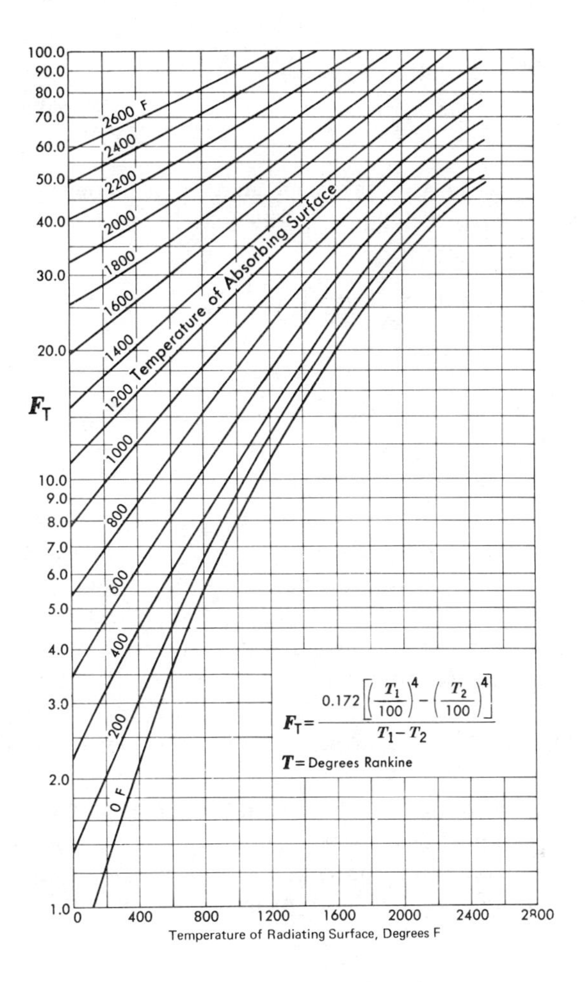

Fig. 5-39. Temperature factor, F_T, as a function of temperature in degrees Fahrenheit.

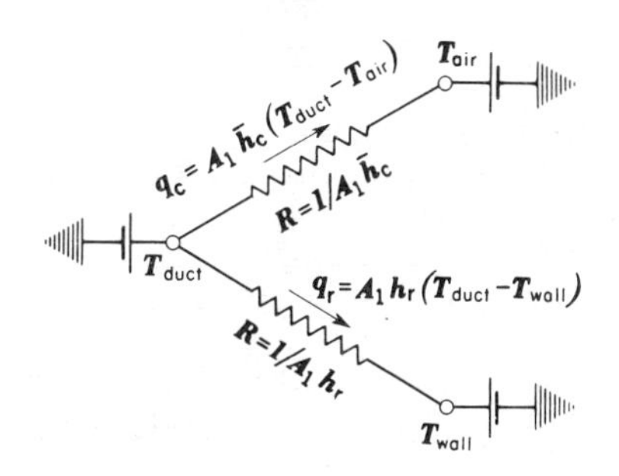

Fig. 5-40. Thermal network for Example 5-13.

can write this heat balance as

$$q = \bar{h}_c A_T (T_G - T_T) = A_T \epsilon \sigma (T_T^4 - T_{wall}^4)$$

where A_T is the surface area, T_T the temperature of the thermocouple, and T_G the temperature of the gas. Substituting the data of the problem we obtain

$$\frac{q}{A_T} = 0.8 \times 0.1714 \left[\left(\frac{1400}{100}\right)^4 - \left(\frac{900}{100}\right)^4\right] = 4410 \text{ Btu/hr sq ft}$$

and the true gas temperature is

$$T_G = \frac{q}{\bar{h}_c A_T} + T_T = \frac{4410}{25} + 940 = 1116 \text{ F} \qquad \textit{Ans.}$$

In systems where heat is transferred simultaneously by convection and radiation, it is frequently not possible to determine the radiant-heat-transfer coefficient directly. Since the temperature factor F_T contains the temperatures of the radiation emitter and the receiver, it can be evaluated only when both of these temperatures are known. If one of the temperatures depends on the rate of heat flow, that is, if one of the potentials in the network is "floating," one must assume a value for the floating potential and then determine if that value will satisfy continuity of heat flow in the steady state. If the rate of heat flow to the potential node is not equal to the rate of heat flow from the node, another temperature must be assumed. The trial-and-error process is continued until the energy balance is satisfied. The general technique is illustrated in the next example.

EXAMPLE 5-15. Determine the correct gas temperature in Example 5-14 if the thermocouple had been shielded by a thin cylindrical radiation shield having an inside diameter four times as large as the outer diameter of the thermocouple. Assume that the convective-heat-transfer coefficient of the shield is 20 Btu/hr sq ft F on both sides and that the emittance of the shield, made of stainless steel is 0.3 at 1000 F.

Solution: A sketch of the physical system is shown in Fig. 5-41a. Heat flows by convection from the gas to the thermocouple and its shield. At the same time, heat flows by radiation from the thermocouple to the inside surface of the shield, is conducted through the shield, and flows by radiation from the outer surface of the shield to the walls of the duct. If we assume that the temperature of the shield is uniform (that is, if we neglect the thermal resistance of the conduction path because the shield is very thin), the thermal network is as shown in Fig. 5-41b. The temperature of the duct wall T_w and the temperature of the thermocouple T_T are known, while the temperatures of the shield T_s and of the gas T_G must be determined. The latter two temperatures are floating potentials. A heat balance on the shield can be written as

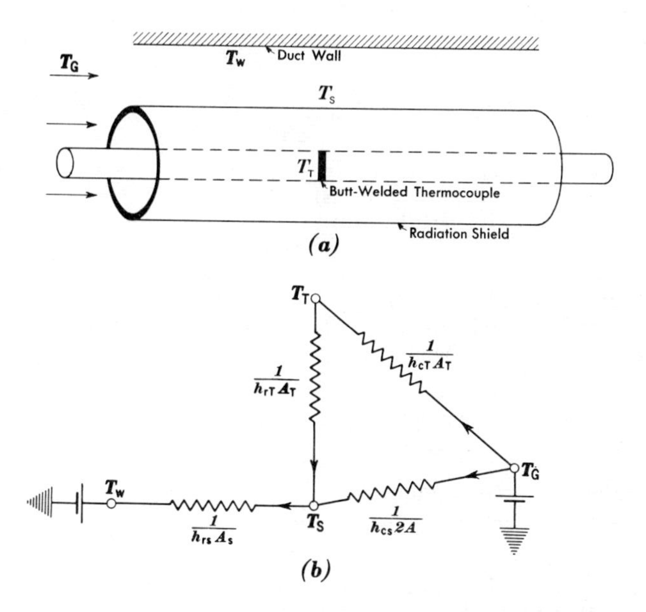

Fig. 5-41. Physical system and thermal network for Example 5-15.

$$\begin{matrix}\text{Rate of heat flow from} \\ T_G \text{ and } T_T \text{ to } T_s\end{matrix} = \begin{matrix}\text{rate of heat flow} \\ \text{from } T_s \text{ to } T_w\end{matrix}$$

or

$$\bar{h}_{cs} 2A_s(T_G - T_s) + h_{rT}A_T(T_T - T_s) = \bar{h}_{rs}A_s(T_s - T_w)$$

A heat balance on the thermocouple yields

$$\bar{h}_{cT}A_T(T_G - T_T) = \bar{h}_{rT}A_T(T_T - T_s)$$

where the nomenclature is given in the sketch. Taking A_T as unity, A_s equals 4 and we obtain from Eq. 5-56

$$A_T\,\mathfrak{F}_{T-s} = \frac{1}{\dfrac{1-\epsilon_T}{A_T\epsilon_T} + \dfrac{1}{A_T} + \dfrac{1-\epsilon_s}{A_s\epsilon_s}} = \frac{1}{\dfrac{0.2}{0.8} + 1 + \dfrac{0.7}{4 \times 0.3}} = 0.547$$

and

$$A_s\,\mathfrak{F}_{s-w} = A_s\epsilon_s = (4)(0.3) = 1.2$$

Assuming a shield temperature of 900 F, we have, according to Eq. 5-67,

$$\bar{h}_{rT}A_T = A_T\,\mathfrak{F}_{T-s}F_T = (0.547)(18.1) = 9.85$$

and

$$\bar{h}_{rs}A_s = A_s\,\mathfrak{F}_{s-w}F_T = (1.2)(11.4) = 12.5$$

Substituting these values into the first heat balance permits the evaluation of the gas temperature and we get

$$T_G = \frac{\bar{h}_{rs}A_s(T_s - T_w) - \bar{h}_{rT}A_T(T_T - T_s)}{\bar{h}_{cs} \times 2A_s} + T_s$$

$$= \frac{5750 - 581}{(20)(2)(4)} + 900 = 932 \text{ F}$$

Since the temperature of the gas can not be less than that of the thermocouple, the assumed shield temperature was too low. Repeating the calculations with a new shield temperature of 930 F yields $T_G = 970$ F. We now substitute this value to see if it satisfies the second heat balance and get:

Heat flow rate by convection *to* thermocouple $= 25\, A_T(970 - 940) = 750$ Btu/hr

Net heat flow rate by radiation *from* thermocouple $= \bar{h}_{rT}A_T(T_T - T_s) = 203$ Btu/hr

Since the rate of heat flow to the thermocouple exceeds the rate of heat flow from the thermocouple, our assumed shield temperature was too high. Repeating the calculations with an assumed shield temperature of 923 F yields a gas temperature of 966 F, which satisfies the heat balance on the thermocouple. *Ans.*

The details of this calculation are left as an exercise (see Prob. 5-47).

A comparison of the results in Examples 5-14 and 5-15 shows that the indicated temperature of the unshielded thermocouple differs from the true gas temperature by 176 F, while the shielded couple reads only 26 F less than the true gas temperature. A double shield would reduce the temperature error to less than 10 F for the conditions specified in the example.

5-10. *Solar radiation*

Solar radiation plays an important role in many environmental processes. All sources of energy used by man derive from the sun, and plants depend on solar energy for photosynthesis and growth. By virtue of its interaction with nitrogen oxide in the atmosphere, solar energy also affects the density of smog and air pollution. Although solar energy is at present not used for industrial purposes, there exists growing interest in the direct utilization of solar energy for heating homes and distilling fresh water from sea water. Solar radiation is also an important factor in the design of spacecraft.

Calculations of solar radiation. The rate at which solar energy impinges on a surface of unit area placed normal to the sun at the outer fringes of the earth's atmosphere, the so-called *solar constant*, is about 442 Btu/hr sq ft (19). The rate at which solar radiation reaches the earth is, however, substantially less than 442 Btu/hr sq ft because part of the radiation is absorbed and scattered as it passes through the 90-mile-thick layer of air, water vapor, carbon dioxide, and dust which envelops the earth. The amount of solar radiation received by a surface on the earth depends on the location, the time of day, the time of year, the weather, and the tilt of the surface (28).

The diminution of the solar radiation by the earth's atmosphere depends on the length of the path, which in turn depends on the position of the sun. The radiant energy incident upon a surface on the earth placed normal to the rays of the sun G_n can be estimated (20) from the equation

$$G_n = G_o \tau_a^{m} \tag{5-69}$$

where G_o = solar constant;
m = *relative air mass*, defined as the ratio of the actual path length to the shortest possible path;
τ_a = transmission coefficient for unit air mass.

The value of τ_a is slightly less in the summer than in the winter because the atmosphere contains more water vapor during the summer. It also varies with the condition of the sky, ranging from 0.81 on a clear day to 0.62 on a cloudy one. A mean value of 0.7 is generally considered acceptable for most purposes.

The value of m depends on the position of the sun given by the *zenith distance* z, the angle between the zenith and the direction of the sun. Assuming that the thickness of the atmosphere is negligible compared to the radius of the earth, the relative air mass is equal to secant z. This relation is sufficiently accurate for z between 0 and 80 deg, and beyond this angle solar radiation is almost negligible.

If the receiving surface is not normal to the direction of the sun, the incident radiation per unit area G_i will be reduced by the cosine of i, the angle between the sun direction and the surface normal, or

$$G_i = G_n \cos i \tag{5-70}$$

If the receiving surface is horizontal, as in a solar evaporator, then $\cos i = \cos z$.

The determination of the angle between the sun direction and the surface normal requires a knowledge of the sun's position in the sky relative to an observer on the surface. The sun's position[2] depends on at least

---

[2]For more detailed information about celestial and terrestrial coordinates as well as conventional methods of measuring the azimuth and the latitude with a transit, see Chapter 17 of *Elementary Surveying* by R. C. Brinker and W. C. Taylor, International Textbook Company, 3d ed., 1955.

two simultaneous motions because the earth revolves in the ecliptic plane once every 365.25 days around the sun's ecliptic axis and spins at the same time like a gyroscope around its own celestial axis, which is tilted 23.5 deg with respect to the ecliptic axis, at the rate of $(\pi/12)$ radians/hr.

When the sun is viewed from the earth (see Fig. 5-42), the zenith angle varies with the latitude of the location, the time of day, and the solar declination. The latitude of a location, ϕ, can be obtained from an atlas or a globe. The time of day is expressed in terms of the *hour angle*, h, which indicates the apparent rotation of the celestial sphere about the earth's axis. In other words, it is the angle through which the earth must turn to bring the meridian of a particular location directly under the sun. The hour angle is measured in degrees westward from local noon (i.e., from the south meridian). As a result of the earth's rotation, h varies from zero at local noon to a maximum at sunrise or sunset. The maximum value of h depends on the latitude and the solar declination, δ_s. The latter can be obtained directly for any day of the year from an ephemeris.[3] It can be shown (e.g., Reference 21) that the equation

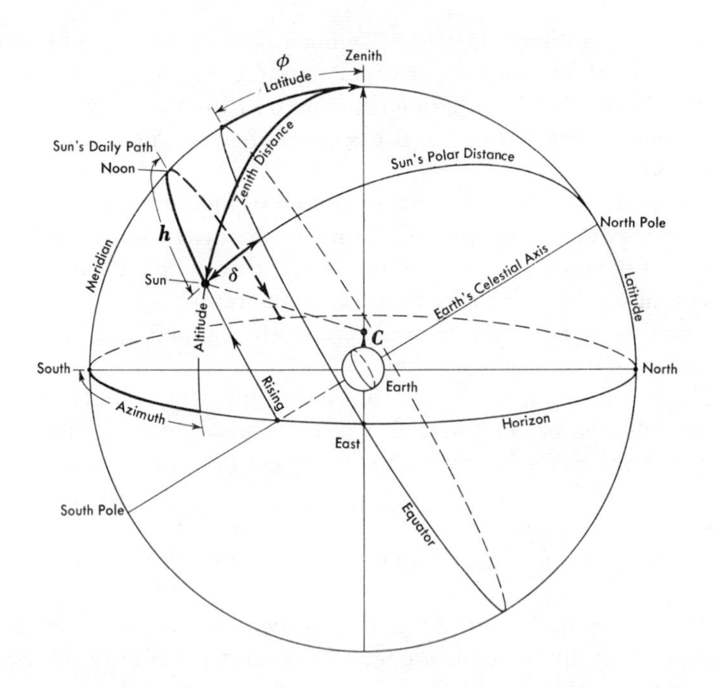

Fig. 5-42. Celestial sphere and sun's coordinates relative to observer on earth at point C.

---

[3]The *American Ephemeris and Nautical Almanac* is published yearly and may be obtained from the Superintendent of Documents, Washington, D. C.

$$\cos z = \sin \phi \sin \delta_s + \cos \phi \cos \delta_s \cos h \tag{5-71}$$

relates the zenith angle to the terrestrial coordinates. Combining Eqs. 5-70 and 5-71 gives the rate at which radiant energy is received at a horizontal surface, the *local insolation*, as

$$G_i = G_n (\sin \phi \sin \delta_s + \cos \phi \cos \delta_s \cos h) \tag{5-72}$$

The amount of solar radiation received during a 24-hr period, obtained by integration of the equation $dQ = G_i \, d\theta$ between sunrise and sunset, is

$$Q = \frac{24}{\pi} G_o \sin \phi \sin \delta_s (H - \tan H) \tag{5-73}$$

where H is the total hour angle traversed by the sun between noon (zero) and sunrise or sunset. Its value can also be obtained from the ephemeris.

For a surface which is tilted (Fig. 5-43) at an angle ψ degrees to the horizontal and whose normal faces α degrees westward (measured along the horizon from the south meridian), the normal solar irradiation G_n can be divided into two components respectively perpendicular to and parallel to the tilted surface. Only the perpendicular component G_i impinges on the surface. The ratio of the effective radiation component to the normal intensity is given by

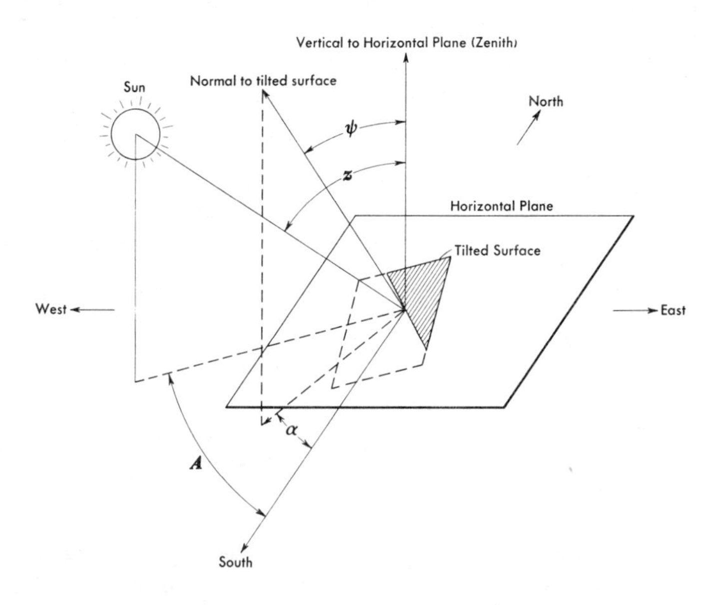

Fig. 5-43. Definition of solar and surface angles for Eq. 5-74.

$$\frac{G_i}{G_n} = \cos i = \cos |z - \psi| - \sin z \sin \psi + \sin z \sin \psi \cos |A - \alpha| \tag{5-74}$$

where A, the azimuth of the sun, is $\sin^{-1} [\cos \delta_s \sin h/\cos (90\text{-}z)]$. Brown and Marco have prepared the graphs shown in Fig. 5-44 from which the values of the pertinent angles in Eq. 5-74 can be obtained for the hours from 6 A.M. to 6 P.M. for various northern latitudes. They recommend that the curves for a solar azimuth A of 30 deg be used for latitudes from 25 to 35 deg and the curve for A of 45 deg be used for 40 to 50 deg lati-

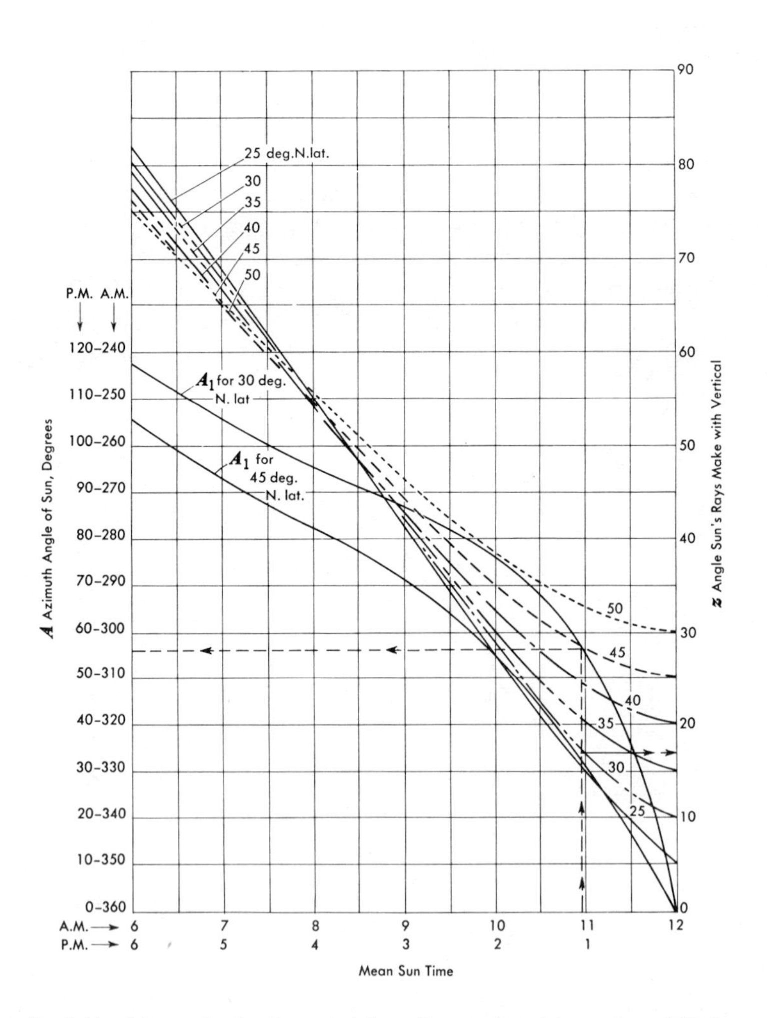

Fig. 5-44. Solar angles for the period from May to August in northern latitudes. (By permission from A. I. Brown and S. M. Marco, *Introduction to Heat Transfer*, 2d. ed. New York: McGraw-Hill Book Company, Inc. 1951).

tude. When local and sun times do not coincide, a correction of 1 hr for every 15 deg of longitude by which the location departs from the standard meridian should be applied.

Equilibrium temperature. The total amount of radiant energy received by a body on the earth is the sum of the direct radiation from the sun and the diffuse radiation scattered from the atmosphere. The latter may amount to only 10 percent of the total direct radiation reaching a horizontal surface on a bright sunny day; in partly cloudy weather it may amount to 50 percent, while on completely overcast days it comprises the total radiation. The diffuse radiation is relatively independent of the geometry of the receiver.

If the body is up in the sky, as for example an airplane, or is tilted so that it can "see" the earth's surface, it also receives terrestrial radiation.

The net rate of heat flow to or from a body, q_{net}, can be calculated from the equation

$$q_{net} = q_s + q_a + q_t + q_c + q_k - q_r \tag{5-75}$$

where q_s = portion of the direct solar radiation absorbed;
q_a = portion of the atmospheric radiation absorbed;
q_t = portion of the terrestrial radiation absorbed;
q_r = radiation emitted;
q_k = net conduction to the body;
q_c = net convection to the body.

The equilibrium temperature attained by a surface in the open can be calculated from Eq. 5-75 by setting the right-hand side equal to zero.

The fraction of the incident radiation absorbed by a surface depends on the value of its absorption coefficient, which is in turn a function of the

Table 5-8. Spectral distribution of solar energy, normal solar irradiation, and transmission coefficient.

	m (See Eq. 5–38)					
	0	1	2	3	4	5
Ultraviolet 0.29–0.40 μ	7.2*	4.3*	2.7*	1.5*	1.1*	0.6*
Visible 0.4–0.7 μ	40.8*	45.2*	44.3*	47.6*	40.4*	38.1*
Above 0.8 μ	52.0*	50.5*	53.0*	55.8*	58.5*	61.3*
Normal solar irradiation, G_n Btu/sq ft hr	442	310	248	203	170	143
Transmission coefficient	....	0.702	0.748	0.771	0.788	0.799

*In percent of the total radiation within the wavelength range shown in the first column

SOURCE: Parry Moon, "Solar Radiation Curves for Engineering Use," *J. Franklin Inst.*, Vol. 230 (1940), pp. 583–618, with correction for more recent value of the solar constant from Ref. 15.

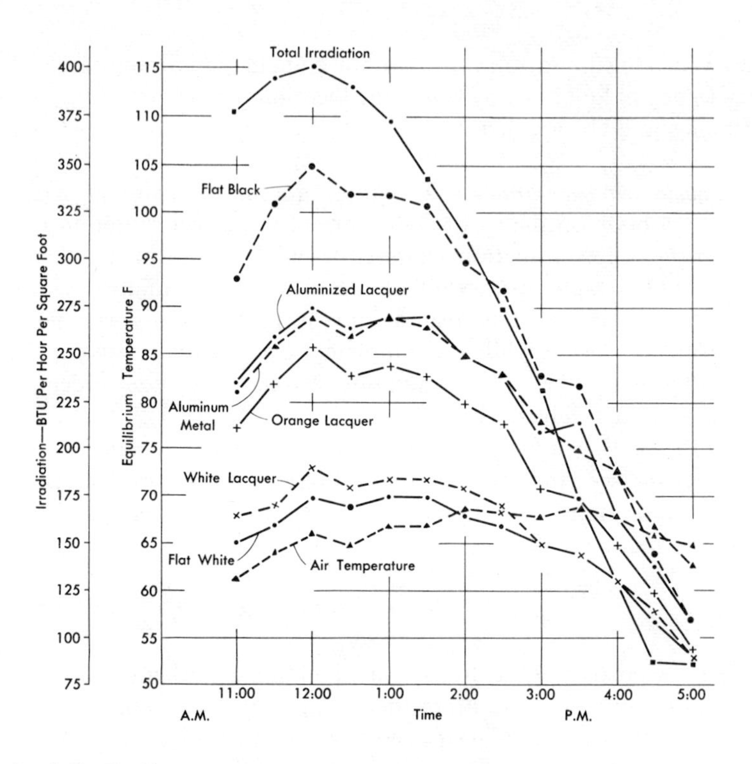

Fig. 5-45. Equilibrium temperatures of various surfaces. (Courtesy of Professors R. V. Dunkle and J. T. Gier, Ref. 23).

spectral distribution of the incoming energy. The spectral distribution of solar radiation is shown in Table 5-8. While 99 percent of the solar energy is contained between 0.25 and 3.0μ, the terrestrial and emitted radiation, on the other hand, fall largely in the long-wavelength portion of the spectrum.

Figure 5-45 shows the results of some measurements made by Gier and Dunkle (23). We observe that surfaces having large absorptances in the solar-wavelength range reach substantially higher equilibrium temperatures than surfaces with low absorptances. The lowest equilibrium temperatures are obtained by white paints which are *selective emitters.* They are poor absorbers (i.e., good reflectors) for the spectral range of solar radiation, but good absorbers, and consequently also good emitters, in the spectral range associated with relatively low temperatures. Since these paints emit more radiation than, for example, aluminum, an equally good reflector of solar energy, their equilibrium temperatures are lower than those attained by the metal surface. This property can be used to advantage on roofs of houses in a sunny climate, where it is desirable to keep the roof temperature as low as possible for the comfort of the occupants.

EXAMPLE 5-16. Calculate the equilibrium temperature of a polished-aluminum surface at 2 P.M. on a clear day. The surface faces southwest, is inclined 15 deg with the horizontal, is located at 30 deg north latitude at a longitude where the local time is 1 hr ahead of sun time. The atmosphere is at 50 F and the unit convective surface conductance is 2.0 Btu/hr sq ft F. Conduction effects may be neglected.

Solution: The direct solar radiation incident on the surface is calculated from Eqs. 5-69 and 5-74, where, for the specified conditions,

$$\begin{aligned} \alpha &= 45 \text{ deg (southwest direction)} \\ \psi &= 15 \text{ deg} \\ A &= 56 \text{ deg (from Fig. 5-35 at 1 P.M.)} \\ z &= 17 \text{ deg (from Fig. 5-35 at 1 P.M. sun time)} \end{aligned}$$

The irradiation on a square foot of surface placed normal to the sun is, from Eq. 5-69,

$$G_n = G_o \tau_o^{\,m} = (442)(0.7^{\sec 17}) = 305 \text{ Btu/hr sq ft}$$

From Eq. 5-72 we have

$$\begin{aligned} \cos i &= \cos|17 - 15| - \sin 17 \sin 15 + \sin 17 \sin 15 \cos|56 - 45| \\ &= 0.999 - (0.297 \times 0.259) + (0.297 \times 0.259 \times 0.97) \\ &= 0.997 \end{aligned}$$

The irradiation is found from Eq. 5-69:

$$G_i = G_n \cos i = (305)(0.997) = 304 \text{ Btu/hr sq ft}$$

The absorptance of polished aluminum for solar radiation is about 0.3 (Table 5-2). The rate of heat absorption is thus

$$q_s = \alpha_{\text{solar}} G_i = (.3)(304) = 91.2 \text{ Btu/hr sq ft}$$

On a sunny day the atmospheric radiation is about 10 percent of the direct solar radiation or

$$q_a = (0.1)(91.2) = 9.12 \text{ Btu/hr sq ft}$$

Since the surface can not see any part of the earth, we have

$$q_t = 0$$

The rate at which the surface emits radiation is given by

$$q_r = \epsilon \sigma T_s^{\,4} = (0.04)(0.172)(T_s/100)^4$$

where ϵ is taken at a temperature of 100 F from Table 5-2. The rate of heat flow by convection to the surrounding air is

$$q_c = \bar{h}_c(T_s - T_\infty) = 2.0[T_s - (460 + 50)]$$

Equilibrium is established when the surface temperature has reached a value at which the rate of heat flow to the surface equals the rate of heat flow from the surface, that is, when $q_{\text{net}} = 0$ in Eq. 5-73. Then we have

$$q_s = q_r + q_c$$

or

$$103.3 = 6.88 \times 10^{-11} T_s^4 + 2T_s - 1020$$

This equation is solved by trial and error for the equilibrium surface temperature and we get $T_s = 97$ F. *Ans.*

Nocturnal radiation. Whereas in the daytime atmospheric radiation consists largely of reflected solar energy, measurements of nocturnal radiation taken on cold clear nights indicate that the effective sky temperature for an object on the earth at that time is of the order of 410 Rankine (−50 F) (11). This accounts for the freezing of water during the night even when the atmospheric temperature is above 32 F.

PROBLEMS

The problems below are organized in the following manner: Problems 5-1 through 5-8 deal with spectral characteristics of radiation, 5-9 through 5-15 with the evaluation of geometrical shape factors, 16-23 with radiation in blackbody enclosures, 24–35 with radiation in gray body enclosures, 36–41 with gaseous radiation, 42–50 with radiation combined with convection and conduction, 51–55 with solar and atmospheric radiation, and 56–66 are problems requiring analysis and design.

5-1. For an ideal radiator (hohlraum) with a 4-in.-diam opening, located in black surroundings at 60 F, calculate for hohlraum temperatures of 212 F and 1040 F, (a) the net radiant-heat-transfer rate, in Btu/hr; (b) the wavelength at which the emission is a maximum, in microns; (c) the monochromatic emission at λ_{max}, in Btu/hr sq ft μ; (d) the wavelengths at which the monochromatic emission is 1 percent of the maximum value.
Ans. (a) 19.5, 747; (b) 7.72, 3.46; (c) 30, 1652; and (d) 2.54 and 50.8, 1.14 and 22.8

5-2. A tungsten filament is heated to 5000 R. At what wavelength is the maximum amount of radiation emitted? What fraction of the total energy is in the visible range (0.4 to 0.75 microns)? Assume that the filament radiates as a gray body.

5-3. Determine the total average hemispherical emittance and the emissive power of a surface which has a spectral hemispherical emittance of 0.8 at wavelengths less than 1.5 μ, 0.6 from 1.5 to 2.5 μ, and 0.4 at wavelengths longer than 2.5 μ. The surface temperature is 1540 F. *Ans.* 0.446

5-4. Show that (a) $(E_{b\lambda})_1/(E_{b\lambda})_2 = T_2^5/T_1^5$, and (b) $E_{b\lambda}/T^5 = f(\lambda T)$. Also, for λT = 10,000 R − μ, (c) calculate $E_{b\lambda}/T^5$ and check your result with Table A-4.

5-5. Compute the average emittance of anodized aluminum at 200 F and 1200 F from the spectral curve in Fig. 5-10. Assume $\epsilon_\lambda = 0.8$ for $\lambda > 9\mu$.

5-6. A large body of nonluminous gas at a temperature of 2000 F has emission bands between 2.5 and 3.5 μ and between 5 and 8 μ. At 2000 F the effective emittance in the first band is 0.8 and in the second 0.6. Determine the emissive power of this gas in Btu/hr sq ft.

5-7. A flat plate is in a solar orbit 93,000,000 miles from the sun. It is always oriented normal to the rays of the sun and both sides of the plate have a finish which has a spectral absorptance of 0.95 at wavelengths shorter than 3 μ and a spectral absorptance of 0.06 at wavelengths longer than 3μ. Assuming that both surfaces are diffuse and that the sun is a 10,000 R blackbody source, determine the equilibrium temperature of the plate. *Ans.* 1060 R

5-8. By substituting Eq. 5-1 for $E_{b\lambda}(T)$ in Eq. 5-4 and performing the integration over the entire spectrum, derive a relationship between σ and the constants C_1 and C_2 in Eq. 5-1.

5-9. Derive an expression for the geometric shape factor F_{1-2} for a rectangular surface A_1, 1 by 20 ft, placed parallel to and centered 5 ft above a 20-ft-square surface A_2.

5-10. Determine the shape factor F_{1-4} for the geometrical configuration shown below.

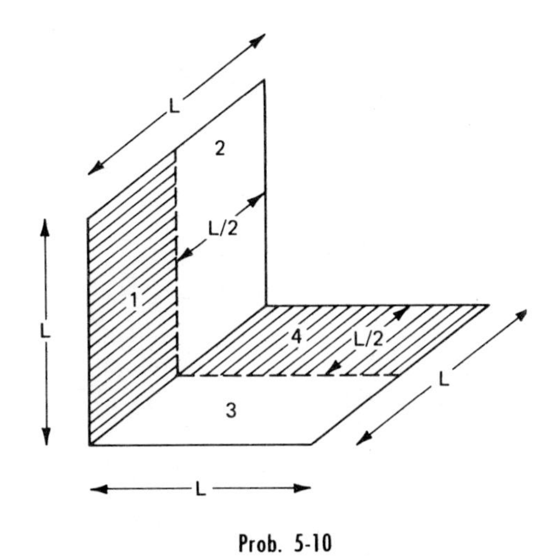

Prob. 5-10

5-11. Determine the shape factor F_{2-1} for the geometrical configuration shown below.

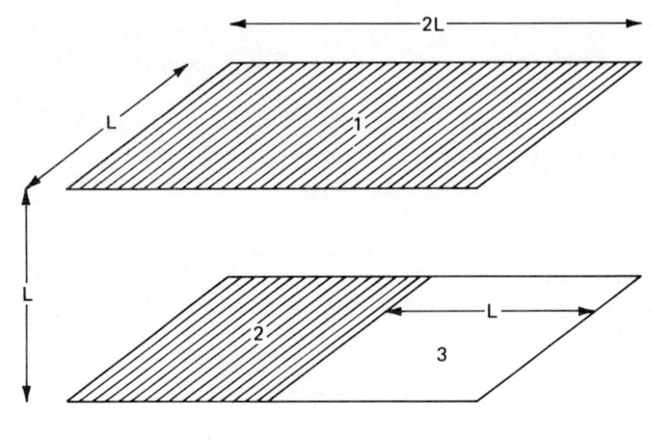

Prob. 5-11

5-12. Determine the ratio of the total hemispherical emittance to the normal emittance for a nondiffuse surface if the intensity of emission varies as the cosine of the angle measured from the normal. *Ans.* $\frac{2}{3}$

5-13. Using basic shape-factor definitions, estimate the equilibrium temperature of the planet Mars which has a diameter of 4150 miles and revolves around the sun at a distance of 141×10^6 miles. Assume that both the planet Mars and the sun act as blackbodies with the sun having an equivalent blackbody temperature of 10,000 R. Then repeat your calculations assuming that the albedo of Mars (the fraction of the incoming radiation returned to space) is 0.15.

5-14. Show that the moon would appear as a disk if its surface were perfectly diffuse.

5-15. A 4-in.-diam cylindrical enclosure of black surfaces, as shown in the

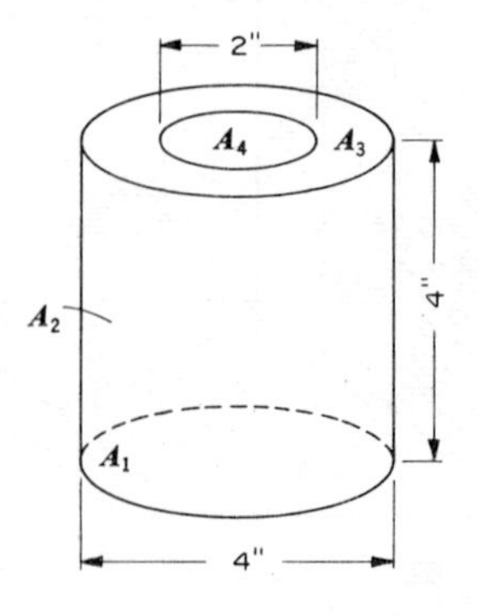

Prob. 5-15

accompanying sketch, has a 2-in. hole in the top cover. Assuming the walls of the enclosure are at the same temperature, determine the percentage of the total radiation emitted from the walls which will escape through the hole in the cover.

5-16. Show that the temperature of the re-radiating surface T_R in Fig. 5-27a is

$$T_R = \left(\frac{A_1 F_{1R} T_1^4 + A_2 F_{2R} T_2^4}{A_1 F_{1R} + A_2 F_{2R}} \right)^{1/4}$$

5-17. A radiation source is to be built, as shown in the diagram, for an experimental study of radiation. The base of the hemisphere is to be covered by a circular plate having a centered hole of radius $R/2$. The underside of the plate is to be held at 540 F by heaters embedded in its surface. The heater surface is black. The hemispherical surface is well-insulated on the outside. Assume gray diffuse processes and uniform distribution of radiation. (a) Find the ratio of the radiant intensity at the opening to the intensity of emission at the surface of the heated plate. (b) Find the radiant energy loss through the opening in Btu/hr for R = 1.0 ft. (c) Find the temperature of the hemispherical surface.

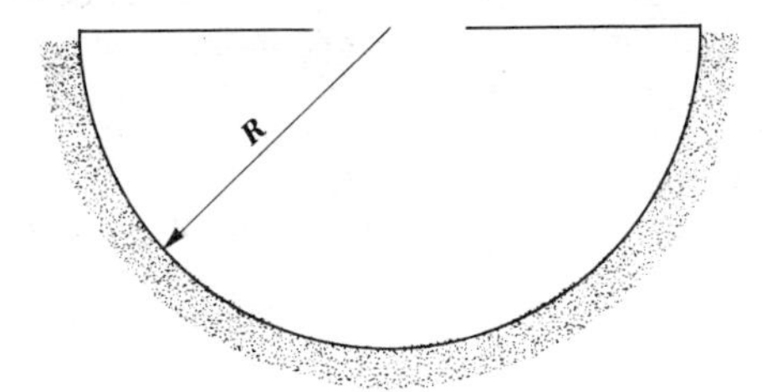

Prob. 5-17

5-18. The radiant-heating ceiling of a 12- by 20-ft room is 8 ft from the floor and is maintained at 110 F while the room air is 50 F. Assuming both surfaces are black, estimate the net rate of heat transfer per square foot of floor surface at 80 F, located (a) in the center of the room, (b) in the corner of the room.

Ans. 18.5, 7.1 Btu/hr sq ft

5-19. A large slab of steel 4 in. thick has in it a 4-in.-diam hole, with axis normal to the surface. Considering the sides of the hole to be black, specify the rate of heat loss from the hole. The plate is at 1000 F, the surroundings at 80 F.

Ans. 1110 Btu/hr

5-20. A 0.5-ft-diam black disk is placed halfway between two black 10-ft-diam disks which are 20 ft apart with all disk surfaces parallel to each other. If the surroundings are at 0 R, determine the temperature of the two larger disks required to maintain the smaller disk at 1000 F. What would be the effect of replacing the small disk by a black sphere of the same diameter?

Ans. 2180 R, 2350 R

5-21. Show that $A_1\bar{F}_{1-2}$ for two black parallel planes of equal area connected by re-radiating walls at a constant temperature is:

$$A_1\bar{F}_{1-2} = A_1\left(\frac{1 + F_{1-2}}{2}\right)$$

Compare the results from this problem with the curves in Fig. 5-22, where allowance is made for a continuous variation in the temperature of the re-radiating walls.

5-22. Calculate the net radiant-heat-transfer rate if the two surfaces in Prob. 5-9 are black and connected by a refractory surface of 500-sq-ft area. A_1 is at 540 F and A_2 is at 40 F. What is the refractory surface temperature?

5-23. A black sphere (1 in. diam) is placed in a large infrared heating oven whose walls are maintained at 700 F. The temperature of the air in the oven is 200 F and the heat-transfer coefficient for convection between the surface of the sphere and the air is 5 Btu/hr sq ft F. Estimate the *net rate of heat flow* to the sphere when its surface temperature is 100 F.

5-24. A hollow cylindrical container is shown in the accompanying sketch. Designating the top, bottom, and walls, respectively, by subscripts 1, 2, and 3 and assuming that all surfaces are gray and that areas A_1 and A_2 are at the same temperature T while A_3 is at temperature T_3, which is lower than T, derive an equation for the rate of radiant heat transfer from A_3 to the top and bottom surfaces of the container.

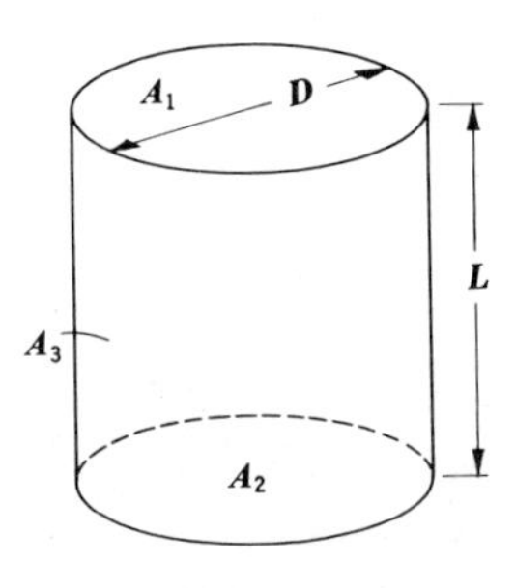

Prob. 5-24

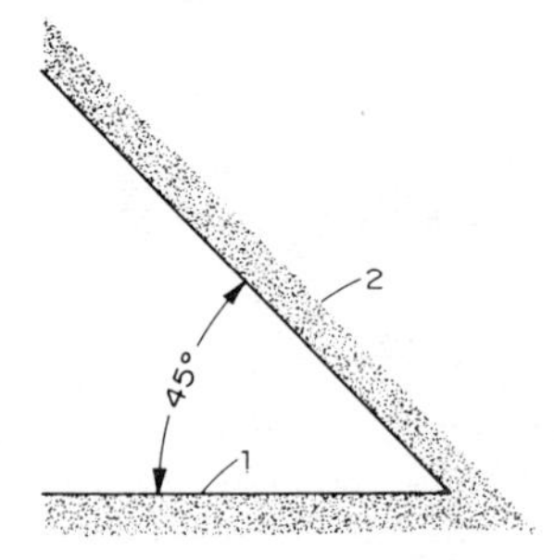

Prob. 5-25

5-25. The wedge-shaped cavity shown in the accompanying sketch consists of two long strips joined along one edge. Surface 1 is 1 ft wide and has an emittance of 0.4 and a temperature of 1540 F. The other wall has a temperature of 1000 F and is black. The enclosure may be assumed to be at a temperature of 0 R. Assuming gray diffuse processes and uniform flux distribution, calculate the rate of energy loss from surfaces 1 and 2 per foot length.

5-26. Derive an equation for the net rate of radiant heat transfer from surface 1 in the system shown in the accompanying sketch. Assume that each surface is at a uniform temperature and that the geometrical shape factor F_{1-2} is 0.1.

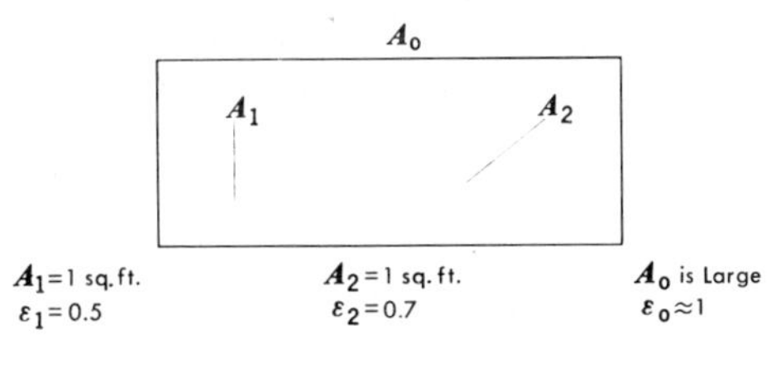

Prob. 5-26

5-27. Two 5-ft-square and parallel flat plates are 1 ft apart. Plate A_1 is maintained at a temperature of 1540 F and A_2 at 460 F. The emittances are 0.5 and 0.8. Considering the surroundings black at 0 R and including multiple inter-reflections, determine (a) the net radiant exchange and (b) the heat input required by surface A_1 to maintain its temperature.

5-28. Two concentric spheres, 8 and 12 in. in diam, with the space between them evacuated, are to be used to store liquid air (−220 F) in a room at 68 F. If the surfaces of the spheres have been flashed with aluminum and the liquid air has a latent heat of vaporization of 90 Btu/lb, determine the number of pounds of liquid air evaporated per hour.

5-29. Determine the steady-state temperatures of two radiation shields placed in the evacuated space between two infinite planes at temperatures of 540 F and 40 F. The emittances of all surfaces are 0.8.

5-30. Derive an expression for the net rate of heat transfer between a small gray sphere of area A_1, emittance ϵ_1, and temperature T_1, and a small flat disk radiating from both sides (area A_2, emittance ϵ_2, and temperature T_2). These two bodies are separated by a distance, large compared to the area of either, and are enclosed in the center of a very large, well-insulated chamber with black walls.

5-31. If in Prob. 5-30 the emittance of A_1 is 0.5, A_2 is a zero flux surface, and A_3 is black, draw the thermal circuit and evaluate all thermal resistances numerically. Then determine the rate of heat transfer from the interior of the enclosure to the surroundings of the system, assuming that the interior surfaces A_1 and A_3 of the enclosure are at 2000 R and the surroundings are at 0 R.

5-32. Three thin sheets of polished aluminum are placed parallel to each other so that the distance between them is very small compared to the size of the sheets. If one of the outer sheets is at 540 F, whereas the other outer sheet is at 140 F, calculate the net rate of heat flow by radiation and the temperature of the intermediate sheet. Convection may be ignored.

Ans. Approx. 15 Btu/hr sq ft, 410 F

5-33. Determine the rate of heat transfer between two 1 by 1 ft parallel flat plates placed 2 in. apart and connected by re-radiating walls. Assume that plate A is maintained at 2040 F and plate B at 540 F. (a) Plate A has an emittance of 0.9 over the entire spectrum and plate B has an emissivity of 0.1. (b) Plate A has an emittance of 0.1 between 0 and 2.5 μ and an emittance of 0.9 at wavelengths longer than 2.5 μ, while plate B has an emittance of 0.1 over the entire spectrum. (c) The emittance of plate A is the same as in part (b), and plate B has an emittance of 0.9 between 0 and 4.0 μ and an emittance of 0.1 at wavelengths larger than 4.0 μ.

5-34. Determine the rate of heat transfer per unit area by radiation from the surface of a dimpled flat plate at 1000 R in space where the environmental temperature is practically zero. The plate has an emittance of 0.2 and has hemispherical indentations 1 in. in diam spaced regularly in a square array on 2-in. centers. Compare the rate of heat transfer from the dimpled plate with that of a flat plate of the same cross-sectional area.

5-35. Show that for two parallel gray plates of emittance ϵ_1 and ϵ_2 respectively, if surface A_1 is heated at a uniform rate q_1 while surface A_2 is at T_2, the surface temperature of A_1 is

$$T_1 = \left[\frac{q_1}{\sigma}\left(\frac{1}{\epsilon_1} + \frac{1}{\epsilon_2} - 1\right) + T_2^4\right]^{1/4}$$

5-36. A small sphere (1 in. diam) is placed in a heating oven whose cavity is a 1-ft cube filled with air at 14.7 psia contains 3 percent water vapor at 1000 F, and whose walls are at 2000 F. The emittance of the sphere is equal to 0.4 − 0.0001 T, where T is the surface temperature in F. When the surface temperature of the sphere is 1000 F, determine (a) the total irradiation received *by* the walls of the oven *from* the sphere, (b) the net heat transfer by radiation between the sphere and the walls of the oven, and (c) the radiant-heat-transfer coefficient.

5-37. A gas leaving a lime kiln contains 20 percent CO_2 and 80 percent N_2 and O_2. This gas flows through a square duct, 6 by 6 in. at the rate of 0.4 lb/sq ft sec. The average temperature of the inside surface of the duct is 800 F, and the emissivity is 0.9. The gas enters the duct at 2000 F, leaves at 1000 F, and has an average specific heat of 0.28 Btu/lb F. The heat-transfer coefficient by convection is 1.5 Btu/hr sq ft F. (a) How long must the duct be to cool the gas to 1000 F? (b) What percentage of heat lost is transferred by radiation? (c) If the velocity of the gas were doubled and the average temperature and length of duct kept the same, what would be the temperature of the gas leaving? (When the velocity is doubled, $\bar{h}_c$ will increase to 2.6 Btu/hr sq ft F.) (*Courtesy of the American Institute of Chemical Engineers.*)

5-38. The radiant section of a small thermal-cracking combustion chamber has a volume of 3200 cu ft and a total surface of 2000 sq ft. A single-row tube-curtain of 4 in. schedule 40 pipes on 7-in. centers covers 760 sq ft of the wall. When the furnace is fired with gas of composition $(CH_2)_x$ at a rate of 2000 lb/hr, an Orsat analysis shows that 13 percent CO_2 are present. Estimate the rate of heat transfer to the tubes under these conditions, using the simplest furnace model that can be justified.

Additional data:

Tube emittance = 0.9
Fuel heating value = 250,000 Btu/lb-mole of C.

Mean molar heat capacity of combustion products = 8.2 Btu/lb-mole F.

Mean radiating gas temperature is 200 F above the bridge-wall gas-temperature.

Air and fuel enter at 60 F.

5-39. A 2-ft-radius hemisphere (1000 F surface temperature) is filled with a gas mixture at 500 F and 2-atm pressure containing 6.67 percent CO_2 and water vapor at 0.5 percent relative humidity. Determine the emittance and absorptance of the gas, and the net rate of radiant heat flow to the gas.

5-40. Two infinitely large black plane surfaces are 1 ft apart and the space between them is filled by an isothermal gas mixture at 1,500 R and atmospheric pressure consisting of 25% CO_2, 25% H_2O, and 50% N_2 by volume. If one of the surfaces is maintained at 500 R and the other at 2,500 F respectively, calculate

a) the effective emittance of the gas at its temperature (*ans.* 0.27)

b) the effective absorptance of the gas to radiation from the 2,500 R surface (*ans.* 0.17)

c) the effective absorptance of the gas to radiation from the 500 R surface (*ans.* 0.9)

d) the net rate of heat transfer to the gas per square foot of surface area (*ans.* 7,000 Btu/hr-ft^2)

e) the net rate of heat transfer from the hotter surface (*ans.* 6.5×10^4 Btu/hr ft^2)

5-41. Calculate the rate of heat transfer by radiation from the gas in Example 5-40 assuming the absorptance of the surface of the flue is 0.8 instead of 1.0 and the gas pressure is two instead of one atmosphere.

5-42. A manned spacecraft capsule has a shape of a cylinder 8 ft in diameter and 30 ft long. The air inside the capsule is maintained at 55 F and the convection-heat-transfer coefficient on the interior surface is 3 Btu/hr sq ft F. Between the outer skin and the inner surface is a 6-in. layer of glass-wool insulation having a thermal conductivity of 0.01 Btu/hr ft F. If the emittance of the skin is 0.05 and there is no aerodynamic heating or irradiation from astronomical bodies, calculate the total heat-transfer rate into space at 0 R.

5-43. A package of electronic equipment is enclosed in a sheet-metal box which has a 1-sq-ft base and is 6 in. high. The equipment uses 1200 w of electrical power and is placed on the floor of a large room. The emittance of the walls of the box is 0.80 and the room air and the surrounding temperature is 70 F. Assuming that the average temperature of the container wall is uniform, estimate that temperature.

5-44. An 8-in. oxidized steel pipe at a surface temperature of 900 F passes through a large room in which the air and the walls are at 100 F. If the heat-transfer coefficient by convection from the surface of the pipe to the air in the room is 5 Btu/hr sq ft F, estimate the total heat loss per foot length of pipe.

5-45. A $\frac{1}{4}$-in.-thick sheet of polished stainless steel is suspended in a comparatively large vacuum-drying oven with black walls. The dimensions of the sheet are 12 by 12 in. and its specific heat is 0.135 Btu/lb F. If the walls of the oven are uniformly at 300 F and the metal is to be heated from 50 to 250 F, estimate how long the sheet should be left in the oven if (a) heat transfer by convection may be neglected and (b) the unit-surface conductance for convection is 0.5 Btu/hr sq ft F. *Ans.* (a) 2.8 hr; (b) 0.4 hr

5-46. Two parallel gray surfaces having an emittance of 0.9 are at temperatures of 65 F and 20 F respectively, with an air space between them. Experiments have shown that the heat-transfer coefficient for natural convection from each surface is given by $\bar{h}_c = 0.20\Delta T^{1/4}$, where ΔT is the temperature difference between any one of the surfaces and the main body of air contained between the two surfaces in deg F and $\bar{h}_c$ is the average heat-transfer coefficient for free convection in Btu/hr sq ft F. Calculate (a) the total rate of heat transfer per square foot between the surfaces and (b) the per cent reduction in the rate of heat transfer in part (a) when a thin aluminum sheet with an emissivity of 0.04 is placed halfway between the two surfaces. *Ans.* 42 Btu/hr sq ft, 88%

5-47a. Calculate the equilibrium temperature of a thermocouple in a large air duct if the air temperature is 2000 F, the duct-wall temperature 500 F, the emittance of the couple 0.5, and the convective-heat-transfer coefficient, $\bar{h}_c$, is 20 Btu/hr sq ft F.

5-47b. Repeat Prob. 5-47a with the addition of a radiation shield ($\epsilon = 0.1$, $\bar{h}_c = 20$ Btu/hr sq ft F) using a FORTRAN computer program.

5-48. A rectangular flat water tank is placed on the roof of a house with its lower portion perfectly insulated. A sheet of glass whose transmission characteristics are tabulated below is placed $\frac{1}{4}$ in. above the water surface. Assuming that the average incident solar radiation is 200 Btu/hr sq ft, calculate the equilibrium water temperature for a water depth of 6 in. if the unit-convective conductance at the top of the glass is 1.5 Btu/hr sq ft F and the surrounding air temperature is 70 F. Disregard interreflections.

$$\begin{aligned} \tau_\lambda \text{ of glass} &= 0 \quad \text{for wavelength from 0 to } 0.35\ \mu \\ &= 0.92 \text{ for wavelength from 0.35 to } 2.7\ \mu \\ &= 0 \quad \text{for wavelength larger than } 2.7\ \mu \\ \tau_\lambda \text{ of glass} &= 0.08 \text{ for all wavelengths} \end{aligned}$$

5-49. A thermocouple is used to measure the temperature of a flame in a combustion chamber. If the thermocouple temperature is 1400 F and the walls of the chamber are at 800 F, what is the error in the thermocouple reading due to radiation to the walls? Assume all surfaces are black and the convection coefficient is 100 Btu/hr sq ft F on the thermocouple. *Ans.* 168 F

5-50. If the thermocouple of Prob. 5-28 is enclosed by a thin cylindrical shield $\frac{1}{4}$ in. in diam placed with its axis in the direction of the flow, what will be the error in the thermocouple reading? Assume the shield is black and long enough to allow neglect of end effects. Also, the thermocouple surface is negligible compared to the shield surface. *Ans*. 32 F

5-51. The overhanging eaves on the south side of a house are designed to shade its entire glass wall. (a) If the eaves are 10 ft from the ground, how far should they overhang to shade the wall on May 10 at 12:00 noon? What will be the effect of these eaves at 4:00 o'clock on the same day? Latitude is 34 N. (b) If the solar transmittance of the glass wall is 0.85, how much solar energy is supplied to the room per hour at 12:00 noon on January 10? Assume the glass wall area is 120 sq ft.

5-52. A series of surfaces, insulated to prevent heat loss from the underside, are placed in the sunlight. Estimate the equilibrium temperature for each of several surfaces with the sun at zenith for a clear atmosphere, an air temperature of 70 F, a relative humidity of 100 percent, and a convective-heat-transfer coefficient of 3 Btu/hr sq ft F. The materials are (a) polished aluminum, (b) polished silver, (c) white (ZnO) painted surface, and (d) lamp-black painted surface.

5-53. A metal plate is placed in the sunlight. The incident radiant energy G is 250 Btu/hr sq ft. The air and the surroundings are at 50 F. The heat-transfer coefficient by free convection from the upper surface of the plate is 3 Btu/hr sq ft F. The plate has an average emittance of 0.9 at solar wavelengths and 0.1 at long wavelengths. Neglecting conduction losses on the lower surface, determine the equilibrium temperature of the plate. *Ans*. $\simeq$ 121 F

5-54. The irradiation received by a stratospheric balloon was found to be 394 Btu/hr sq ft. If the transmittance (i.e., the percent of the emitted radiation which reaches the receiver) of the earth's atmosphere is about 82 percent, the distance from the earth to the sun is 93,000 miles, and the radius of the sun is 433,000 miles, estimate the temperature of the sun. The emittance of the sun may be taken as unity.

5-55. Calculate the equilibrium temperature of the earth if the concentration of CO_2 in the atmosphere were twice as large as it is today.

5-56. One hundred pounds of carbon dioxide are stored in a high-pressure cylinder 10 in. in diam (OD), 4 ft long and $\frac{1}{2}$ in. thick. The cylinder is fitted with a safety rupture diaphragm designed to fail at 2000 psia (with the specified charge, this pressure will be reached when the temperature increases to 120 F). During a fire, the cylinder is completely exposed to the irradiation from flames at 2000 F (ϵ = 1.0). For the specified conditions, c_v = 0.60 Btu/lb F for CO_2. Neglecting the convective heat transfer, determine the time the cylinder may be exposed to this irradiation before the diaphragm will fail if the initial temperature is 70 F and (a) the cylinder is bare oxidized steel (ϵ = 0.79), (b) the cylinder is painted with aluminum paint (ϵ = 0.30). *Ans*. (a) 0.47 min, (b) 1.24 min

5-57. An inside room of a house, 24 ft square with a 12-ft ceiling, is to be heated by means of a panel-heating installation located in the ceiling. The ceiling is painted with an oil paint and heated uniformly by hot-water pipes imbedded in it. The floor is made of oak and is to be maintained at 80 F. To maintain comfort in the room it is necessary to transfer radiant heat between the ceiling and the floor at the rate of 6000 Btu/hr. (a) Determine the required ceiling temperature, (1) neglecting the side walls, and (2) assuming that the side walls are nonconducting, but re-radiating. (b) Determine the temperature of the side walls for case (2). (c) Estimate the rate of heat flow by convection to the room for case (2) if the air temperature is 86 F and the unit-surface conductance for free convection is 1.0 Btu/hr sq ft F. (d) Estimate the total heat loss from the heater per hour.

5-58. A spacecraft on an interplanetary flight uses a radiator condenser which may be idealized by a stack of thin discs of 5 ft diam spaced 1 ft apart as shown in the accompanying sketch. Assuming that the surface temperature of each disk is 1000 R, that the surface is gray and has an emittance of 0.85, and that the surrounding space is at 0 R, determine (a) the shape factor of one disk surface relative to space and (b) the rate of heat rejection per disk into space.

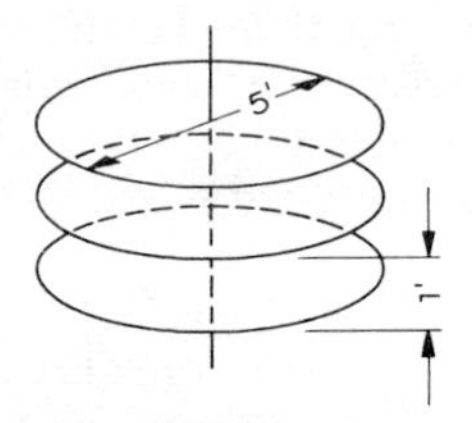

Prob. 5-58

5-59. A hydrogen bomb may be approximated by a fireball at a temperature of 7200 K according to a report published in 1950 by the Atomic Energy Commission. (a) Calculate the total rate of radiant-energy emission in Btu/hr, assuming that the gas radiates as a blackbody and has a diameter of 1 mile. (b) If the surrounding atmosphere absorbs radiation below 0.3 μ, determine the percent of the total radiation emitted by the bomb which is absorbed by the atmosphere. (c) Calculate the rate of irradiation on a 1-sq-ft area of the wall of a house 25 miles from the center of the blast if the blast occurs at an altitude of 10 miles and the wall faces in the direction of the blast. (d) Estimate the total amount of radiation absorbed, assuming that the blast lasts approximately 10 sec and that the wall is covered by a coat of red paint. (e) If the wall were made of oak whose inflammability limit is estimated to be 700 F and which had a thickness of 1 in., determine whether or not the wood would catch on fire. Justify your answer by an engineering analysis stating carefully all assumptions.

5-60. An electric furnace is to be used for batch heating a certain material (specific heat of 0.16 Btu/lb F) from 70 to 1400 F. The material is placed on the furnace floor which is 6 by 12 ft in area as shown in the accompanying sketch. The side walls of the furnace are made of a refractory material. Parallel to the plane of the roof but several inches below it, a grid of round resistor rods is installed. The resistors are $\frac{1}{2}$ in. in diameter and are spaced 2 in. center to center. The resistor temperature is to be maintained at 2000 F, under which conditions the emissivity of the resistor surface is 0.6. If the top surface of the stock may be assumed to have an emissivity of 0.9, estimate the time required for heating a 6-ton batch. External heat losses from the furnace may be neglected, the temperature gradient through the stock may be considered negligibly small, and steady-state conditions may be assumed. *Ans.* Approx. 2 hr

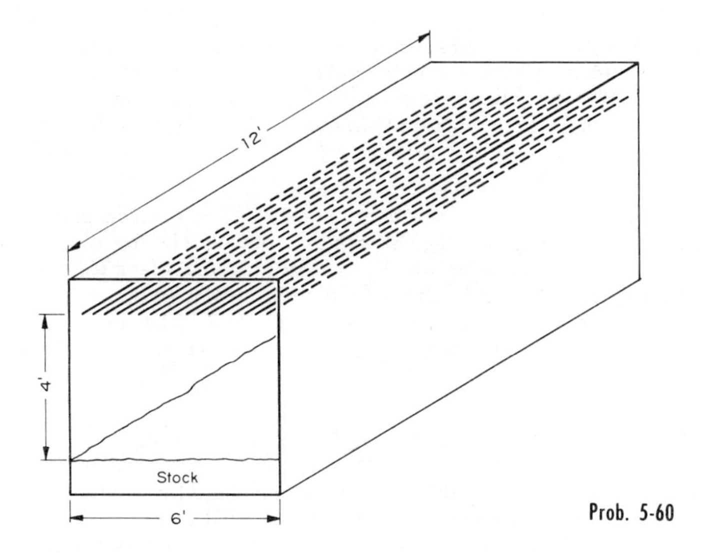

Prob. 5-60

5-61. A solid copper sphere 6 in. in diam is placed in an evacuated 1- by 1- by 1-ft cubical furnace. The furnace walls are maintained at a uniform temperature of 1540 F. The copper sphere is initially at 60 F and is to be removed from the furnace when it reaches a temperature of 600 F. The emittance of the surface of the copper sphere is 0.07 while the emittance of the furnace walls is 0.15. Neglecting temperature gradients in the copper, calculate (a) the rate of heat transfer to the copper when its temperature is 300 F and (b) the time required to heat the copper to 600 F. *Ans.* (a) 1400 Btu/hr; (b) 1.4 hr

5-62. Mercury is to be evaporated at 605 F in a furnace. The mercury flows through a 1-in. BWG No. 18 gauge stainless-steel tube, which is placed in the center of the furnace whose cross section, perpendicular to the tube axis, is a square 8 by 8 in. The furnace is made of brick having an emissivity of 0.85, with the walls maintained uniformly at 1800 F. If the surface heat-transfer coefficient on the inside of the tube is 500 Btu/hr sq ft F and the emittance of the outer surface of the tube is 0.60, calculate the rate of heat transfer per foot of tube, neglecting convection within the furnace. *Ans.* 5550 Btu/hr ft

5-63. A 2-ft-square section of panel heater is installed in the corner of the ceiling of a room having a 9- by 12-ft floor area with an 8-ft ceiling. If the surface of the heater, made from oxidized iron, is at 300 F and the walls and the air of the room are at 68 F in the steady state, determine (a) the rate of heat transfer to the room by radiation; (b) the rate of heat transfer to the room by convection ($\bar{h}_c \simeq$ 2 Btu/hr sq ft F); (c) the cost of heating the room at 1 cent per kwhr in cents per hour.

5-64. A 1-in.-diam cylindrical refractory crucible for melting lead is to be built for thermocouple calibration. An electrical heater immersed in the metal is shut off at some temperature above the melting point. The fusion-cooling curve is obtained by observing the thermocouple emf as a function of time. Neglecting heat losses through the wall of the crucible, estimate the cooling rate (Btu/hr) for the molten lead surface (melting point 621.2 F, surface emissivity 0.8) if the crucible depth above the lead surface is (a) 1 in., (b) 5 in. Assume that the emittance of the refractory surface is unity and the surroundings are at 70 F. (c) Noting that the crucible would hold about 0.2 lb of lead for which the heat of fusion is 10 Btu/lb, comment on the suitability of the crucible for the purpose intended.

5-65. A spherical satellite circling the sun is to be maintained at a temperature of 60 F. The satellite rotates continuously and is covered partly with solar cells having a gray surface with an absorptance of 0.1. The rest of the sphere is to be covered by a special coating which has an absorptance of 0.8 for solar radiation and an emittance of 0.2 for the emitted radiation. Estimate the portion of the surface of the sphere which can be covered by solar cells. The solar irradiation may be assumed to be 450 Btu/hr per square foot of surface perpendicular to the rays of the sun.

5-66. The thermal conductivity of a gas is measured in an apparatus described below. The gas is placed in the annulus formed between two cylinders. The inner cylinder is heated to a uniform temperature by means of an electrical heating wire, while the outer cylinder is maintained at a constant temperature by placing the entire assembly in a water bath. Convection is reduced to a minimum by maintaining a low temperature difference across the annulus. The inner cylinder is 1 ft long and $\frac{1}{2}$ in. in diam while the outer cylinder has a 0.70 in. ID. Under steady-state conditions with a heat input to the inner cylinder of 4 Btu/hr, the temperature of the inner cylinder was found to be 200 F while the temperature of the outer cylinder was maintained uniform at 190 F. The emittance of the two surfaces was found by independent measurements to be 0.20. Neglecting end effects and convection, calculate the thermal conductivity of the gas at an average temperature of 195 F. Assume that the gas does not absorb any radiation.

Ans. 0.20 Btu/hr ft F

REFERENCES

1. N. W. Snyder, "A Review of Thermal Radiation Constants," *Trans. ASME*, Vol. 65 (1954), pp. 537–540.

2. R. V. Dunkle, "Thermal-Radiation Tables and Applications," *Trans. ASME*, Vol. 65 (1954), p. 549–552.

3. W. Sieber, "Zusammensetzung der von Werk-und Baustoffen Zurückgeworfenen Wärmestrahlung," *Z. Tech. Physik*, Vol. 22 (1941), pp. 130–135.

4. E. Schmidt and E. Eckert, "Über die Richtungsverteilung der Wärmestrahlung von Oberflächen," *Forsch. Gebiete Ingenieurw.*, Vol. 6 (1935), pp. 175–183.

5. H. C. Hottel, "Radiant Heat Transmission Between Surfaces Separated by Non-Absorbing Media," *Trans. ASME*, FSP-53-19b, Vol. 53 (1931), pp. 265–271.

6. H. C. Hottel, "Radiant Heat Transmission," *Mech. Eng.*, Vol. 52 (1930), pp. 699–704.

7. D. C. Hamilton and W. R. Morgan, "Radiant-Interchange Configuration Factors," *NACA TN* 2836, December, 1952.

8. F. W. Hutchinson, *Industrial Heat Transfer*. (New York: The Industrial Press, 1952.)

9. H. C. Hottel, Chapter 2 of *Heat Transmission* (by W. C. McAdams), 3d ed. (New York: McGraw-Hill Book Company, Inc., 1954.)

10. A. K. Oppenheim, "The Network Method of Radiation Analysis," *ASME Paper* 54-A75, 1954.
11. M. Fischenden and O. A. Saunders, *The Calculation of Heat Transmission.* (London: His Majesty's Stationery Office, 1932.)
12. S. Chandrasekhar, *Radiative Transfer.* (Oxford: Clarendon Press, 1950.)
13. H. C. Hottel, "Heat Transmission by Radiation from Non-Luminous Gases," *Trans. Am. Inst. Chem. Engrs.*, Vol. 19 (1927), pp. 173–205.
14. H. C. Hottel and R. B. Egbert, "Radiant Heat Transmission from Water Vapor," *Trans. Am. Inst. Chem. Engrs.*, Vol. 38 (1942), pp. 531–565.
15. F. A. Brooks, "Solar Energy and Its Use for Heating Water in California," *Bull.* 602, Col. of Agric., Univ. of Calif., 1936.
16. N. W. Snyder, J. T. Gier, and R. V. Dunkle, "Total Normal Emissivity Measurements on Aircraft Materials Between 100 and 800 F," *ASME Paper* 54-A-189, 1954.
17. H. Schmidt and E. Furthman, "Üeber die Gesamtstrahlung fester Köerper," *Mitt. K. W. Inst. Eisenforsch*, Abh. 109, Dusseldorf, 1928.
18. W. H. McAdams, *Heat Transmission*, 3d ed. (New York: McGraw-Hill Book Company, Inc., 1954.)
19. F. S. Johnson, "The Solar Constant," *J. of Meteorology*, Vol. 11 (1954), pp. 431–439.
20. H. Heywood, "Solar Energy: Past, Present and Future Applications," *Engineering*, Vol. 176 (1956), pp. 377–380.
21. W. J. Humphreys, *Physics of the Air.* (New York: McGraw-Hill Book Company, Inc., 1940.)
22. F. Kreith, "Thermal Design of High Altitude Balloons and Instrument Packages," *ASME*, Trans. ser. C, J. Heat Transfer, vol. 92, (1970) pp. 307–332.
23. J. T. Gier and R. V. Dunkle, "Selective Spectral Characteristics of Solar Collectors," *Trans. Tuscon Conference on Applied Solar Energy*, Vol. 2, 1957.
24. E. M. Sparrow and R. D. Cess, *Radiation Heat Transfer*, (Belmont, California: Wadsworth Publishing Co., Inc., 1966.)
25. J. A. Wibelt, *Engineering Radiation Heat Transfer.* (New York: Holt, Rinehart, and Winston, 1966.)
26. C. L. Tien, "Thermal Radiation Properties of Gases," *Advances in Heat Transfer*, vol. 5, pp. 254–321. (New York: Academic Press, 1968.)
27. R. Siegel and J. R. Howell, *Thermal Radiation Heat Transfer*, vols. I and II. (Washington, D.C.: NASA SP-164, 1968.)
28. B. Gebhart, *Heat Transfer, 2nd ed.* (New York, N.Y.: McGraw-Hill Book Company, Inc., 1971.)
29. J. T. Gier, R. V. Dunkle, and J. T. Bevans, "Measurement of Absolute Spectral Reflectivity from 1.0 to 15.0 Microns," *J. Opt. Soc.*, Vol. 44 (1954), pp. 558–562.
30. H. Gröber, S. Erk and U. Grigull, Translated by J. R.Moszynski, *Fundamentals of Heat Transfer.* (New York: McGraw-Hill Book Company, Inc., 1961.)
31. F. B. Hildebrandt, *Methods of Applied Mathematics.* (Englewood Cliffs, New Jersey: Prentice-Hall, Inc., 1952.)
32. R. Goldstein, "Measurements of infrared absorption by water vapor at temperatures to 1000 K," *J. Quant. Spectr. Radiative Transfer*, Vol. 4 (1964), pp. 343–352.
33. M. Planck, *The Theory of Heat Radiation.* (New York: Dover Publications, Inc., 1959.)
34. D. K. Edwards and W. Sun, "Correlations for absorption by the 9.4- and 10.4-micron CO_2 bands." *Applied Optics*, Vol. 3 (1964), p. 1501.

35. D. K. Edwards, B. J. Flornes, L. K. Glassen, and W. Sun, "Correlations of absorption by water vapor at temperatures from 300 to 1100 K." *Applied Optics*, Vol. 4 (1965), pp. 715–722.

36. G. G. Gubareff, J. E. Janssen, and R. H. Torborg, *Thermal Radiation Properties Survey*. (Minneapolis, Minnesota: Honeywell Research Center, 1960.)

37. J. R. Schornhorst and R. Viskranta, "An Experimental Examination of the Validity of the Commonly Used Methods of Radiant-Heat Transfer Analysis", *ASME Trans.*, ser. C, J. Heat Transfer, vol. 90, (1968) pp.429–436.

38. R. G. Hering and T. F. Smith,"Surface Roughness Effects on Radiant Energy Interchange," *ASME* Trans., ser. C, J. *Heat Transfer*, vol. 93 (1971), pp. 88–96.

39. B. T. F. Chung and P. S. Sumitra, "Radiation Shape Factors from Plane Point Sources," *ASME*, Trans., ser. C, J. Heat Transfer, vol. 94, (1972) pp. 328–330.

6
Fundamentals of convection

6-1. The convective-heat-transfer coefficient

In the preceding chapters, attention has been focused on heat transfer by conduction and radiation. In an effort to simplify the work and to emphasize the methods for calculating heat transfer by conduction and radiation, an effort has been made to eliminate, as much as possible, problems related to heat transfer by convection. However, from the illustrative examples it has probably become apparent already that there are hardly any practical problems which can be solved without a knowledge of the mechanisms by which heat is transferred between the surface of a solid conductor and the surrounding medium. In our work so far we simply specified the unit convective surface conductance at the solid fluid interface and did not investigate the details of the transfer mechanism. We evaluated the rate of heat transfer by convection between a solid boundary and a fluid by means of the equation

$$q_{\text{surface to fluid}} = A\bar{h}_c(T_s - T_\infty) \tag{1-13}$$

The convection equation in this form seems quite simple. The simplicity is misleading, however, because Eq. 1-13 is a definition of the average unit thermal convective conductance $\bar{h}_c$ rather than a law of heat transfer by convection. The convective-heat-transfer coefficient is actually a complicated function of the fluid flow, the thermal properties of the fluid medium, and the geometry of the system. Its numerical value is in general not uniform over a surface, and depends also on the location where the fluid temperature T_∞ is measured.

Although Eq. 1-13 is generally used to determine the rate of heat flow by convection between a surface and the fluid in contact with it, this relation is inadequate to explain the convective heat-flow mechanism. A meaningful analysis which will eventually lead to a quantitative evaluation of the convective heat-transfer coefficient must start with a study of the dynamics of the fluid flow. In this and the following chapters we shall follow this line of approach and investigate the influence of flow conditions, fluid properties, and boundary shapes on the convective-heat-transfer coefficient.

6-2. Energy transport mechanism and fluid flow

The transfer of heat between a solid boundary and a fluid takes place by a combination of conduction and mass transport. If the boundary is at a higher temperature than the fluid, heat flows first by conduction from the solid to fluid particles in the neighborhood of the wall. The energy thus transmitted increases the internal energy of the fluid and is carried away by the motion of the fluid. When the heated fluid particles reach a region at a lower temperature, heat is again transferred by conduction from warmer to cooler fluid.

Since the convective mode of energy transfer is so closely linked to the fluid motion, it is necessary to know something about the mechanism of fluid flow before the mechanism of heat flow can be investigated. One of the most important aspects of the hydrodynamic analysis is to establish whether the motion of the fluid is *laminar* or *turbulent*.

In laminar, or streamline, flow, the fluid moves in layers, each fluid particle following a smooth and continuous path. The fluid particles in each layer remain in an orderly sequence without passing one another. Soldiers on parade provide a somewhat crude analogy to laminar flow. They march along well-defined lines, one behind the other, and maintain their order even when they turn a corner or pass an obstacle.

In contrast to the orderly motion of laminar flow, the motion of fluid particles in turbulent flow rather resembles a crowd of commuters in a railroad station during the rush hour. The general trend of the motion is from the gate toward the train, but superimposed upon this motion are the deviations of individuals according to their instantaneous direction and their ability to pass the less agile members of the crowd. Yet if one could obtain a statistical average of the motion of a large number of individuals, it would be steady and regular. The same applies to fluid particles in turbulent flow. The path of any individual particle is zigzag and irregular, but on a statistical basis the overall motion of the aggregate of fluid particles is regular and predictable.

When a fluid flows in laminar motion along a surface at a temperature different from that of the fluid, heat is transferred only by molecular conduction within the fluid as well as at the interface between the fluid and the surface. There exist no turbulent mixing currents or eddies by which energy stored in fluid particles is transported across streamlines. Heat is transferred between fluid layers by molecular motion on a submicroscopic scale.

In turbulent flow, on the other hand, the conduction mechanism is modified and aided by innumerable eddies which carry lumps of fluid across the streamlines. These fluid particles act as carriers of energy and transfer energy by mixing with other particles of the fluid. An increase in the rate of mixing (or turbulence) will therefore also increase the rate of heat flow by convection.

The fluid motion can be induced by two processes. The fluid may be set in motion as a result of density differences due to a temperature variation in the fluid. This mechanism is called *free*, or *natural*, *convection*. The motion observed when a pot of water is heated on a stove or the motion of air in the desert on a calm day after sunset are examples of free convection. When the motion is caused by some external energy, such as a pump or a blower, we speak of *forced convection*. The cooling of an automobile radiator by the air blown over it by the fan is an example of forced convection. (The term *radiator* is obviously poorly chosen because the heat flow is *not* primarily by radiation; *convector* would be a more appropriate term.)

6-3. Boundary-layer fundamentals

When a fluid flows along a surface, irrespective of whether the flow is laminar or turbulent, the particles in the vicinity of the surface are slowed down by virtue of viscous forces. The fluid particles adjacent to the surface stick to it and have zero velocity relative to the boundary.[1] Other fluid particles attempting to slide over them are retarded as a result of an interaction between faster- and slower-moving fluid, a phenomenon which gives rise to shearing forces. In laminar flow the interaction, called viscous shear, takes place between molecules on a submicroscopic scale. In turbulent flow an interaction between lumps of fluid on a macroscopic scale, called turbulent shear, is superimposed on the viscous shear.

---

[1]This is strictly true only when the mean free path of the molecules is small compared to the boundary-layer thickness. In rarefied gases the molecules may slide or slip along a surface.

The effects of the viscous forces originating at the boundary extend into the body of the fluid, but a short distance from the surface the velocity of the fluid particles approaches that of the undisturbed free stream. The fluid contained in the region of substantial velocity change is called the *hydrodynamic boundary layer.* The thickness of the boundary layer has been defined as the distance from the surface at which the local velocity reaches 99 percent of the external velocity u_∞.

The concept of a boundary layer was introduced by the German scientist, Prandtl, in 1904. The boundary layer essentially divides the flow field around a body into two domains: a thin layer covering the surface of the body where the velocity gradient is great and the viscous forces are large, and a region outside this layer where the velocity is nearly equal to the free-stream value and the effects of viscosity are negligible. By means of the boundary-layer concept, the equations of motion, usually called the Navier-Stokes equations, can be reduced to a form in which they can be solved; the effects of viscosity on the flow can be determined; and the frictional drag along a surface can be calculated. The boundary-layer concept is also of great importance, as we shall see, to an understanding of convective heat transfer.

The shape of the velocity profile within the boundary layer depends on the nature of the flow. Consider, for example, the flow of air over a flat plate, placed with its surface parallel to the stream. At the leading edge of the plate ($x = 0$ in Fig. 6-1), only the fluid particles in immediate contact with the surface are slowed down, while the remaining fluid continues at the velocity of the undisturbed free stream in front of the plate. As the fluid proceeds along the plate, the shearing forces cause more and more of the fluid to be retarded, and the thickness of the boundary layer increases. The growth of the boundary layer and typical velocity profiles at various stations along the plate are shown in Fig. 6-1.

The velocity profiles near the leading edge are representative of laminar boundary layers. However, the flow within the boundary layer remains laminar only for a certain distance from the leading edge and then

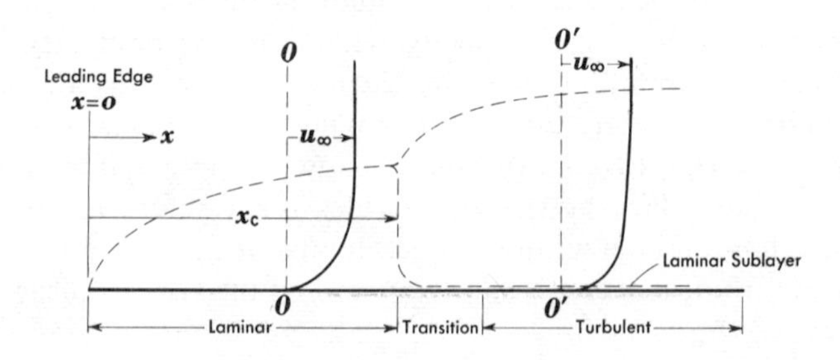

Fig. 6-1. Velocity profiles for laminar and turbulent boundary layers in flow over a field plate. (Vertical scale enlarged for clarity.)

becomes turbulent. We do not know enough about the mechanism of transition to predict precisely when the transition will occur, but the phenomenon leading to the growth of disturbances in a laminar boundary layer can be described (see Reference 1 for details). There are always small disturbances and waves in a flowing fluid, but as long as the viscous forces are large they will prevent disturbances from growing. As the laminar boundary layer thickens, the ratio of viscous forces to inertia forces decreases, and eventually a point is reached at which disturbances will no longer decay, but will grow with time. Then the boundary layer becomes unstable and the transition from laminar to turbulent flow begins. Eddies and vortexes form and destroy the laminar regularity of the boundary-layer motion. Quasi-laminar motion persists only in a thin layer in the immediate vicinity of the surface. This portion of a generally turbulent boundary layer is called the *laminar sublayer.* The region between the laminar sublayer and the completely turbulent portion of the boundary layer is called the *buffer layer.* The structure of the flow in a turbulent boundary layer is shown schematically on an enlarged scale in Fig. 6-2.

The distance from the leading edge at which the boundary layer becomes turbulent is called the *critical length* x_c (Fig. 6-1). This distance is usually specified in terms of a dimensionless quantity called the local critical Reynolds number $u_\infty \rho x_c/\mu$, which is an indication of the ratio of inertial to viscous forces at which disturbances begin to grow. Experimental results have shown that the point of transition depends on the surface contour, the surface roughness, the disturbance level, and even on the heat transfer. When the flow is calm and no disturbances occur, laminar flow can persist in the boundary layer at Reynolds numbers as high as 5×10^6. If the surface is rough, or disturbances are intentionally introduced into the flow, as for example by means of a grid, the flow may become turbulent at Reynolds numbers as low as 8×10^4. Under average conditions, the flow over a flat plate becomes turbulent at a distance from the leading edge x_c where the local Reynolds number $u\rho x_c/\mu$ is approximately equal to 5×10^5.

In view of the difference in the flow characteristics, the frictional

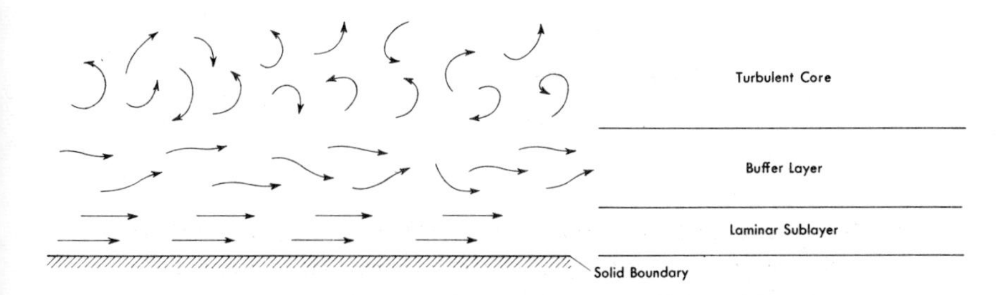

Fig. 6-2. Structure of a turbulent flow field near a solid boundary.

forces as well as the heat transfer are governed by different relations in laminar and turbulent boundary layers. Also the limiting conditions under which the flow will follow a given contour, and the boundary-layer theory can be applied, depends on whether the flow is laminar or turbulent.

Even when the contour of the surface over which the fluid flows is curved, the flow in the boundary layer is, at least qualitatively, similar to the flow in the boundary layer on a flat plate. The contour of the body becomes very important, however, in the determination of the point at which the boundary layer separates from the surface. The separation of flow occurs mainly because the kinetic energy of the fluid in the boundary layer is dissipated by viscosity within the layer. As long as the main stream is accelerating, the external pressure is decreasing along the direction of flow and the forces at the edge of the boundary layer oppose the retardation of the fluid by the wall shear. On the other hand, when the flow is decelerating, as for example in a low-speed diffuser, the external pressure as well as the shearing forces tend to decelerate the fluid. A local reversal of the flow in the boundary layer will then occur when the kinetic energy of the fluid in the boundary layer can no longer overcome the adverse pressure gradient. Near this point the boundary layer separates as shown in Fig. 6-3. Beyond the point of separation, the flow near

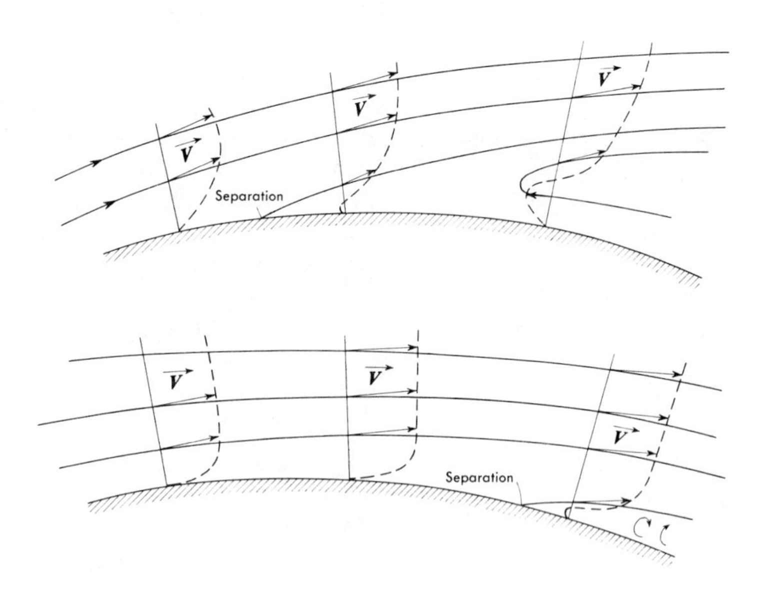

Fig. 6-3. Sketches illustrating separation of boundary layer. Top: streamlines and flow pattern near separation point of laminar boundary layer. Bottom: streamlines and flow pattern near separation point of turbulent boundary layer.

the surface consists of highly irregular eddies and vortexes and cannot be treated by boundary-layer theory.

A more advanced boundary-layer theory allows us to calculate the point at which the flow separates from the surface (1). Generally speaking, a turbulent boundary layer will not separate as easily as a laminar boundary layer because the kinetic energy of the fluid particles is larger in a turbulent layer. In flow over a streamlined object, separation takes place near the rear, if it occurs at all. In flow over bluff objects, on the other hand, separation occurs nearer to the front. The problem of separation is too complicated to be taken up in detail here, and the reader interested in additional information on this subject should consult References 1, 27, and 28 of the bibliography at the end of this chapter.

6-4. The Nusselt modulus

From the description of the mechanism of convective energy transport, we recall that both conduction and mass transport play a role. Since the thermal conductivity of fluids, except for liquid metals, is relatively small, the rapidity of the energy transfer depends largely on the mixing motion of the fluid particles.

When the fluid velocity and the turbulence are small, the transport of energy is not aided materially by mixing currents on a macroscopic scale. On the other hand, when the velocity is large and the mixing between warmer and colder fluid contributes substantially to the energy transfer, the conduction mechanism becomes less important. Consequently, to transfer heat by convection through a fluid at a given rate, a larger temperature gradient is required in a region of low velocity than in a region of high velocity.

Applying these qualitative observations to heat transfer from a solid wall to a fluid in turbulent flow, we can roughly sketch the temperature profile. In the immediate vicinity of the wall, heat can only flow by conduction because the fluid particles are stationary relative to the boundary. We naturally expect a large temperature drop in this layer. As we move further away from the wall, the movement of the fluid aids in the energy transport and the temperature gradient will be less steep, eventually, leveling out in the main stream. For air flowing turbulently over a flat plate, the temperature distribution shown in Fig. 6-4 illustrates these ideas qualitatively.

The foregoing discussion suggests a method for evaluating the rate of heat transfer between a solid wall and a fluid. Since at the interface (i.e., at $y = 0$) heat flows only by conduction, the rate of heat flow can be calculated from the equation

$$q_{\text{surface}\to\text{fluid}} = -k_f A \left.\frac{\partial T}{\partial y}\right|_{y=0} \tag{6-1}$$

This approach has indeed been used, but for engineering purposes the concept of the convective-heat-transfer coefficient is much more convenient. In order not to lose sight of the physical picture, we shall relate the heat-transfer coefficient defined by Eq. 1-13 to the temperature gradient at the wall. Equating Eqs. 6-1 and 1-13 we obtain

$$q_{\text{surface}\to\text{fluid}} = -k_f A \left.\frac{\partial T}{\partial y}\right|_{y=0} = \bar{h}_c A (T_s - T_\infty) \tag{6-2}$$

Since the magnitude of the temperature gradient in the fluid will be the same regardless of the reference temperature, we can write $\partial T = \partial (T - T_s)$. Introducing a significant length dimension of the system L to specify the geometry of the object from which heat flows, we can write Eq. 6-2 in dimensionless form as

$$\frac{\bar{h}_c L}{k_f} = \frac{-\left.\dfrac{\partial T}{\partial y}\right|_{y=0}}{\dfrac{T_s - T_\infty}{L}} = \left.\frac{\partial \left(\dfrac{T_s - T}{T_s - T_\infty}\right)}{\partial \left(\dfrac{y}{L}\right)}\right|_{y=0} \tag{6-3}$$

The combination of the convective heat-transfer coefficient $\bar{h}_c$, the significant length L, and the thermal conductivity of the fluid k_f in the form $\bar{h}_c L/k_f$ is called the Nusselt modulus, or *Nusselt number*, $\overline{\text{Nu}}$. The Nusselt number is a dimensionless quantity.

Inspection of Eq. 6-3 shows that the Nusselt number could be in-

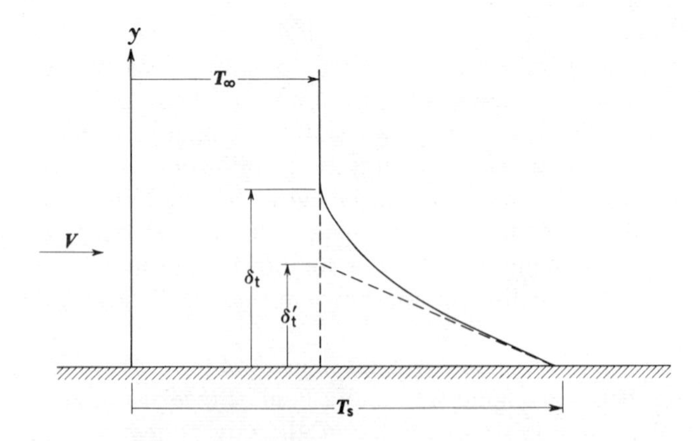

Fig. 6-4. Temperature distribution in a turbulent boundary layer for a fluid flowing over a heated plate.

terpreted physically as the ratio of the temperature gradient in the fluid immediately in contact with the surface to a reference temperature gradient $(T_s - T_\infty)/L$. In practice the Nusselt number is a convenient measure of the convective heat-transfer coefficient because, once its value is known, the convective heat-transfer coefficient can be calculated from the relation

$$\bar{h}_c = \overline{\text{Nu}}\,\frac{k_f}{L} \tag{6-4}$$

We observe that, for a given value of the Nusselt number, the convective heat-transfer coefficient is directly proportional to the thermal conductivity of the fluid but inversely proportional to the significant length dimension describing the system.

The temperature distribution for a fluid flowing past a hot wall, as sketched by the solid line in Fig. 6-4, shows that the temperature gradient in the fluid is confined to a relatively thin layer, δ_t, in the vicinity of the surface. We shall now simplify the true picture by replacing the actual temperature distribution by the dashed straight line shown in Fig. 6-4. The dashed line is tangent to the actual temperature curve at the wall and physically represents the temperature distribution in a hypothetical layer of fluid of thickness δ_t' which, if completely stagnant, offers the same thermal resistance to the flow of heat as the actual boundary layer. In this stagnant layer, heat can flow only by conduction and the rate of heat transfer per unit area is

$$\frac{q}{A} = k_f\,\frac{T_s - T_\infty}{\delta_t'} = \bar{h}_c(T_s - T_\infty) \tag{6-5}$$

An inspection of Eq. 6-5 shows that $\bar{h}_c$ may be expressed as

$$\bar{h}_c = \frac{k_f}{\delta_t'} \tag{6-6}$$

and the Nusselt number as

$$\overline{\text{Nu}} = \bar{h}_c\,\frac{L}{k_f} = \frac{L}{\delta_t'} \tag{6-7}$$

While this picture is considerably oversimplified, it does illustrate the fact that the thinner the boundary layer δ_t', the larger will be the convective conductance. To transfer large quantities of heat rapidly, one attempts to reduce the boundary-layer thickness as much as possible. This can be accomplished by increasing the velocity and/or the turbulence of the fluid. If insulation of the surface is the desired aim, a thick stagnant layer is beneficial. In fact, most commercial insulating materials simply trap air in small spaces to eliminate its mixing motion while at the same time taking advantage of its low thermal conductivity to reduce the transfer of heat.

6-5. Evaluation of convection heat-transfer coefficients

There are four general methods available for the evaluation of convection heat-transfer coefficients:

1. Dimensional analysis combined with experiments.
2. Exact mathematical solutions of the boundary-layer equations.
3. Approximate analyses of the boundary layer by integral methods.
4. The analogy between heat, mass, and momentum transfer.

All four of these techniques have contributed to our understanding of convective heat-transfer. Yet, no single method can solve all the problems because each one has limitations which restrict its scope of application.

Dimensional analysis is mathematically simple and has found a wide range of application. The chief limitation of this method is that results obtained by it are incomplete and quite useless without experimental data. It contributes little to our understanding of the transfer process, but facilitates the interpretation and extends the range of application of experimental data by correlating them in terms of dimensionless groups.

There are two different methods for determining dimensionless groups suitable to correlate experimental data. The first of these methods, discussed in the following section, requires only the listing of the variables pertinent to a phenomenon. This technique is simple to use, but if a pertinent variable is omitted, erroneous results ensue. In the second method the dimensionless groups and similarity conditions are deduced from the differential equations describing the phenomenon. This method is preferable when the phenomena can be described mathematically, but the solution of the resulting equations is too involved to be practical. An illustration of this technique is presented in Sec. 7-2.

Exact mathematical analyses require the simultaneous solution of the equations describing the fluid motion and the transfer of energy in the moving fluid. The method presupposes that the physical mechanisms are sufficiently well understood to be described in mathematical language. This preliminary requirement limits the scope of exact solutions because complete mathematical equations describing the fluid flow and the heat-transfer mechanisms can be written only for laminar flow. Even for laminar flow the equations are quite complicated, but solutions have been obtained for a number of simple systems such as flow over a flat plate or a circular cylinder.

Exact solutions are important because the assumptions made in the course of the analysis can be specified accurately and their validity can be checked by experiment. They also serve as a basis of comparison and as a check on simpler, but approximate methods. Furthermore, the development of electronic computers has increased the range of problems amenable to mathematical solution, and results of computations for different systems are continually being published in the literature.

The details of the mathematical solution are quite complicated. They are, however, not essential to a correct application of the results. We shall here only derive the boundary-layer equations to introduce the fundamental concepts, indicate how they can be solved, and finally illustrate the application of the results for the simple case of flow over a flat plate. For details regarding the methods of solution of the boundary-layer equations in geometrically more complex systems, the reader is referred to the translation of Schlichting's treatise on boundary-layer theory (1).

The approximate analysis of the boundary layer avoids the detailed mathematical description of the flow in the boundary layer. Instead, a plausible but simple equation is used to describe the velocity and temperature distributions in the boundary layer. The problem is then analyzed on a macroscopic basis by applying the equation of motion and the energy equation to the aggregate of the fluid particles contained within the boundary layer. This method is relatively simple; moreover, it yields solutions to problems which can not be treated by an exact mathematical analysis. In those instances where other solutions are available, they agree within engineering accuracy with the solutions obtained by this approximate method. The technique is not limited to laminar flow, but can also be applied to turbulent flow.

The analogy between heat, mass, and momentum transfer is a useful tool for analyzing turbulent transfer processes. Our knowledge of turbulent-exchange mechanisms is insufficient to write mathematical equations describing the temperature distribution directly, but the transfer mechanism can be described in terms of a simplified model. According to one such model which has been widely accepted, a mixing motion in a direction perpendicular to the mean flow accounts for the transfer of momentum as well as energy. The mixing motion can be described on a statistical basis by a method similar to that used to picture the motion of gas molecules in the kinetic theory. There is by no means general agreement that this model corresponds to conditions actually existing in nature, but for practical purposes its use can be justified by the fact that experimental results are substantially in agreement with analytical predictions based on the hypothetical model.

6-6. Dimensional analysis

Dimensional analysis differs from other methods of approach in that it does not yield equations which can be solved. Instead, it combines several variables into dimensionless groups, such as the Nusselt number, which facilitate the interpretation and extend the range of application of experimental data. In practice, convective heat-transfer coefficients are generally calculated from empirical equations obtained by correlating experimental data with the aid of dimensional analysis.

The most serious limitation of dimensional analysis is that it gives no information about the nature of a phenomenon. In fact, to apply dimensional analysis it is necessary to know beforehand what variables influence the phenomenon, and the success or failure of the method depends on the proper selection of these variables. It is therefore important to have at least a preliminary theory or a thorough physical understanding of a phenomenon before a dimensional analysis can be performed. However, once the pertinent variables are known, dimensional analysis can be applied to most problems by a routine procedure which is outlined below.[2]

Primary dimensions and dimensional formulas. The first step is to select a system of primary dimensions. The choice of the primary dimensions is arbitrary, but the dimensional formulas of all pertinent variables must be expressible in terms of them. We shall use the primary dimensions of length L, time θ, temperature T, and mass M.

The dimensional formula of a physical quantity follows from definitions or physical laws. For instance, the dimensional formula for the length of a bar is $[L]$ by definition.[3] The average velocity of a fluid particle is equal to a distance divided by the time interval taken to traverse it. The dimensional formula of velocity is therefore $[L/\theta]$, or $[L\theta^{-1}]$, i.e., a distance or length divided by a time. The units of velocity could be expressed in feet per second, miles per hour, or knots, since they all are a length divided by a time.

The dimensional formulas and the symbols of physical quantities occurring frequently in heat-transfer problems are given in Table 6-1. The primary dimensions in the $ML\theta T$ column of Table 6-1 avoid the use of dimensional constants such as g_c or J. This standardizes the method, but conversion factors may have to be inserted in the final results (i.e., the dimensionless quantities) to comply with the system of units used (see Appendix II for conversion factors). For convenience the dimensional

---

[2]The algebraic theory of dimensional analysis will not be developed here. For a rigorous and comprehensive treatment of the mathematical background, Chapters 3 and 4 of Reference 2 are recommended.

[3]A square bracket [ ] denotes that the quantity has the dimensional formula stated within the bracket.

Table 6-1. Some physical quantities with associated symbols, dimensions, and units.

Quantity	Symbol	Dimensions		Units in the Engineering System
		$ML\theta T$ System	$ML\theta TFQ$ System	
Length	L, x	L	L	ft
Time	θ	θ	θ	sec or hr
Mass	M	M	M	lb_m
Force	F	ML/θ^2	F	lb_f
Temperature	T	T	T	F
Heat	Q	ML^2/θ^2	Q	Btu
Velocity	V	L/θ	L/θ	ft/sec
Acceleration	a, g	L/θ^2	L/θ^2	ft/sec^2
Dimensional conversion factor	g_c	None	$ML/\theta^2 F$	32.2 lb_m ft/sec² lb_f
Energy conversion factor	J	None	FL/Q	778 ft-lb_f/Btu
Work	W	ML^2/θ^2	FL	ft-lb_f
Pressure	p	$M/\theta^2 L$	F/L^2	lb_f/sq ft
Density	ρ	M/L^3	M/L^3	lb_m/cu ft
Internal energy	u	L^2/θ^2	Q/M	Btu/lb_m
Enthalpy	h	L^2/θ^2	Q/M	Btu/lb_m
Specific heat	c	$L^2/\theta^2 T$	Q/MT	Btu/lb_m F
Dynamic viscosity	μ_f	$M/L\theta$	$F\theta/L^2$	lb_f-sec/sq ft
Absolute viscosity	μ	$M/L\theta$	$M/L\theta$	lb_m/ft-sec
Kinematic viscosity	$\nu = \mu/\rho$	L^2/θ	L^2/θ	sq ft/sec
Thermal conductivity	k	$ML/\theta^3 T$	$Q/LT\theta$	Btu/hr ft F
Thermal diffusivity	a	L^2/θ	L^2/θ	sq ft/hr
Thermal resistance	R	$T\theta^3/ML^2$	$T\theta/Q$	F hr/Btu
Coefficient of expansion	β	$1/T$	$1/T$	$1/F$
Surface tension	σ	M/θ^2	F/L	lb_f/ft
Shear per unit area	τ	$M/L\theta^2$	F/L^2	lb_f/sq ft
Unit surface conductance	h	$M/\theta^3 T$	$Q/\theta L^2 T$	Btu/hr sq ft F
Mass flow rate	m	M/θ	M/θ	lb_m/sec

formulas are also listed in the $ML\theta TFQ$ system. In this system, sometimes called the engineering system, there are six primary dimensions.[4]

Buckingham π-theorem. To determine the number of independent dimensionless groups required to obtain a relation describing a physical phenomenon, the Buckingham π (pi) theorem may be used.[5] According to this rule, the required number of independent dimensionless groups which can be formed by combining the physical variables pertinent to a

[4] Since the number of primary quantities is increased by two, the conversion constants g_c and J, whose dimensional formulas can be derived from the primary dimensions, must be included among the physical quantities.

[5] A more rigorous rule, proposed by van Driest (3), shows that the π-theorem holds as long as the set of simultaneous equations formed by equating the exponents of each primary dimension to zero is linearly independent. If one equation in the set is a linear combination of one or more of the other equations, i.e., if the equations are linearly dependent, then the number of dimensionless groups is equal to the total number of variables n minus the number of independent equations.

problem is equal to the total number of these physical quantities n (for example, density, viscosity, heat-transfer coefficient, etc.) minus the number of primary dimensions m required to express the dimensional formulas of the n physical quantities. If we call these groups π_1, π_2, etc., the equation expressing the relationship among the variables has a solution of the form

$$F(\pi_1, \pi_2, \pi_3, \cdots) = 0 \tag{6-8}$$

In a problem involving five physical quantities and three primary dimensions, $n - m$ is equal to two and the solution either has the form

$$F(\pi_1, \pi_2) = 0 \tag{6-9}$$

or the form

$$\pi_1 = f(\pi_2)$$

Experimental data for such a case can be presented conveniently by plotting π_1 against π_2. The resulting empirical curve reveals the functional relationship between π_1 and π_2 which can not be deduced from dimensional analysis.

For a phenomenon which can be described in terms of three dimensionless groups (i.e., if $n - m = 3$), Eq. 6-8 has the form

$$F(\pi_1, \pi_2, \pi_3) = 0 \tag{6-10}$$

but can also be written as

$$\pi_1 = f(\pi_2, \pi_3)$$

For such a case, experimental data can be correlated by plotting π_1 against π_2 for various values of π_3. Sometimes it is possible to combine two of the π's in some manner and to plot this parameter against the remaining π on a single curve.

Determination of dimensionless groups. A simple method for determining dimensionless groups will now be illustrated by applying it to the problem of correlating experimental convection heat-transfer data for a fluid flowing across a heated tube. Exactly the same approach could be used for heat transfer in flow through a heated tube.

From the description of the convective heat-transfer process, it is reasonable to expect that the physical quantities listed in Table 6-2 are pertinent to the problem.

There are seven physical quantities and four primary dimensions.

Table 6-2

Variable	Symbol	Dimensional Equation
Tube diameter	D	$[L]$
Thermal conductivity of the fluid	k	$[ML/\theta^3 T]$
Velocity of the fluid	V	$[L/\theta]$
Density of the fluid	ρ	$[M/L^3]$
Viscosity of the fluid	μ	$[M/L\theta]$
Specific heat at constant pressure	c_p	$[L^2/\theta^2 T]$
Heat-transfer coefficient	$\bar{h}_c$	$[M/\theta^3 T]$

We therefore expect that three dimensionless groups will be required to correlate the data. To find these dimensionless groups, we write π as a product of the variables, each raised to an unknown power

$$\pi = D^a k^b V^c \rho^d \mu^e c_p^f \bar{h}_c^g \tag{6-11}$$

and substitute the dimensional formulas

$$\pi = [L]^a[ML/\theta^3 T]^b[L/\theta]^c[M/L^3]^d\,[M/L\theta]^e[L^2/\theta^2 T]^f[M/\theta^3 T]^g \tag{6-12}$$

For π to be dimensionless, the exponents of each primary dimension must separately add up to zero. Equating the sum of the exponents of each primary dimension to zero, we obtain the set of equations

$$\begin{aligned} b + d + e + g &= 0 \qquad \text{for } M \\ a + b + c - 3d - e + 2f &= 0 \qquad \text{for } L \\ -3b - c - e - 2f - 3g &= 0 \qquad \text{for } \theta \\ -b - f - g &= 0 \qquad \text{for } T \end{aligned}$$

Evidently any set of values of a, b, c, d, and e that simultaneously satisfies this set of equations will make π dimensionless. There are seven unknowns, but only four equations. We can therefore choose values for three of the exponents in each of the dimensionless groups. The only restriction on the choice of the exponents is that each of the selected exponents be independent of the others. An exponent is independent if the determinant formed with the coefficients of the remaining terms does not vanish (i.e., is not equal to zero).

Since $\bar{h}_c$, the convective heat-transfer coefficient, is the variable we eventually want to evaluate, it is convenient to set its exponent g equal to unity. At the same time we let $c = d = 0$ to simplify the algebraic manipulations. Solving the equations simultaneously, we obtain $a = 1$, $b = -1$, $e = f = 0$, and the first dimensionless group is

$$\pi_1 = \frac{\bar{h}_c D}{k}$$

which we recognize as the *Nusselt number*, $\overline{\mathrm{Nu}}$.

For π_2 we select g equal to zero, so that $\bar{h}_c$ will not appear again, and let $a = 1$ and $f = 0$. Simultaneous solution of the equations with these choices yields $b = 0$, $c = d = 1$, $e = -1$, and

$$\pi_2 = \frac{VD\rho}{\mu}$$

This dimensionless group is a *Reynolds number*, Re_D, with the tube diameter as the length parameter.

If we let $e = 1$ and $c = g = 0$, we obtain the third dimensionless group

$$\pi_3 = \frac{c_p \mu}{k}$$

which is known as the *Prandtl number*, Pr.

We observe that, although the heat-transfer coefficient is a function of six variables, with the aid of dimensional analysis, the seven original variables have been combined into three dimensionless groups. According to Eq. 6-10, the functional relationship can be written

$$\overline{\mathrm{Nu}} = f(\mathrm{Re}_D, \mathrm{Pr})$$

and experimental data can now be correlated in terms of three variables instead of the original seven. The importance of this reduction in the variables becomes apparent when we attempt to correlate experimental data.

Correlation of experimental data. Suppose that, in a series of tests with air flowing over a 1-in.-OD pipe, the heat-transfer coefficient has been measured experimentally at velocities ranging from 0.1 to 100 fps. This range of velocities corresponds to Reynolds numbers based on the diameter, $VD\rho/\mu$, ranging from 50 to 50,000. Since the velocity was the only variable in these tests, the results are correlated in Fig. 6-5 by plotting the heat-transfer coefficient $\bar{h}_c$ against the velocity V. The resulting curve permits a direct determination of $\bar{h}_c$ at any velocity for the system used in the tests, but it cannot be used to determine the heat-transfer coefficients for cylinders which are larger or smaller than the one used in the tests. Neither could the heat-transfer coefficient be evaluated if the air were under pressure and its density were different from that used in the

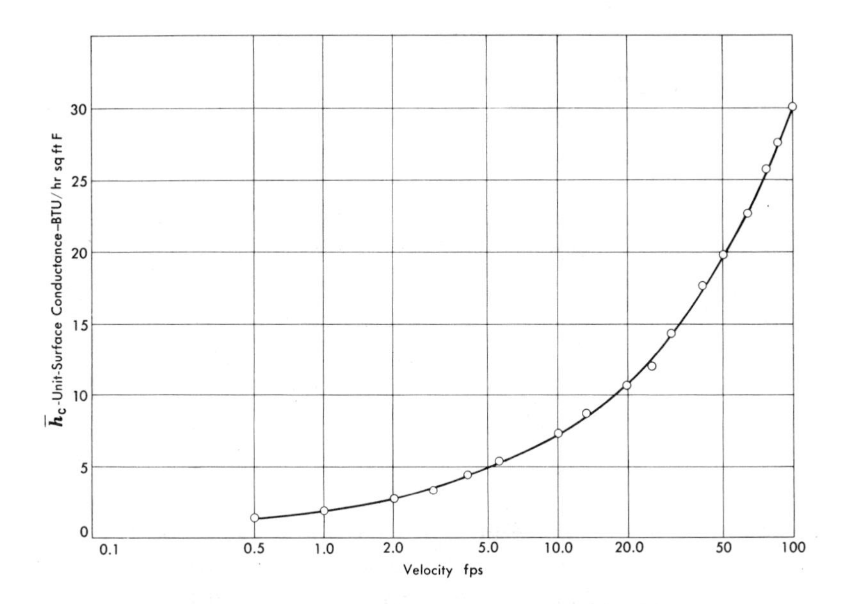

Fig. 6-5. Variation of heat-transfer coefficient with velocity for flow of air over a 1-in.-OD pipe.

tests. Unless experimental data could be correlated more effectively, it would be necessary to perform separate experiments for every cylinder diameter, every density, etc. The amount of labor would obviously be enormous.

With the aid of dimensional analysis, however, the results of one series of tests can be applied to a variety of other problems. This is illustrated by Fig. 6-6, where the data of Fig. 6-5 are replotted in terms of pertinent dimensionless groups. The abscissa in Fig. 6-6 is the Reynolds number $VD\rho/\mu$, and the ordinate is the Nusselt number $\bar{h}_c D/k$. This correlation of the data permits the evaluation of the heat-transfer coefficient for air flowing over any size of pipe or wire as long as the Reynolds number of the system falls within the range covered in the experiment.

Experimental data obtained with air alone do not reveal the dependence of the Nusselt number on the Prandtl number since the Prandtl number is a combination of physical properties whose value does not vary appreciably for gases. To determine the influence of the Prandtl number it is necessary to use different fluids. According to the preceding analysis, experimental data with several fluids whose physical properties yield a wide range of Prandtl numbers are necessary to complete the correlation (see Prob. 6-17).

In Fig. 6-7 the experimental results of several independent investigations for heat transfer between air, water, and oils in cross-flow over a tube or a wire are plotted for a wide range of temperatures, cylinder sizes,

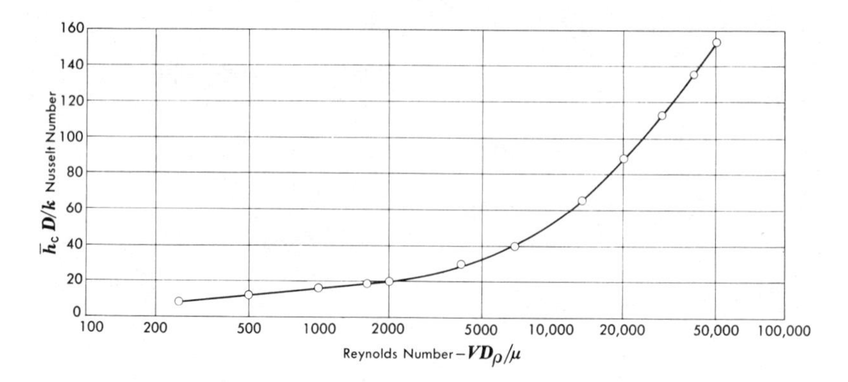

Fig. 6-6. Variation of a Nusselt number with a Reynolds number for flow of air over a 1-in.-OD pipe.

and velocities. The ordinate in Fig. 6-7 is the dimensionless quantity[6] $\overline{\text{Nu}}/\text{Pr}^{0.3}$ and the abscissa is Re_D. An inspection of the results shows that all of the data follow a single line reasonably well, so that they can be correlated empirically. For example, in the range of Reynolds numbers between 3 and 100 a straight line on the log-log plot is a satisfactory approximation to the best correlation, shown by a heavy dotted line in Fig. 6-7. The slope of this straight line is approximately 0.4 and its ordinate value at Re_D of unity is 0.82. The empirical correlation equation within the range of Reynolds numbers between 3 and 100 is therefore

$$\overline{\text{Nu}}/\text{Pr}^{0.3} = 0.82\ \text{Re}_D{}^{0.4}$$

Principle of similarity. The remarkable result of Fig. 6-7 can be explained by the principle of similarity. According to this principle, often called the model law, the behavior of two systems will be similar if the ratios of their linear dimensions, forces, velocities, etc., are the same. Under conditions of forced convection in geometrically similar systems, the velocity fields will be similar provided the ratio of inertia forces to viscous forces is the same in both fluids. The Reynolds number is the ratio of these forces, and consequently we expect similar flow conditions in forced convection for a given value of the Reynolds number. The Prandtl number is the ratio of two molecular-transport properties, the kinematic viscosity $\nu = \mu/\rho$, which affects the velocity distribution, and

[6]Combining the Nusselt number with the Prandtl number for plotting the data is simply a matter of convenience. As mentioned previously, any combination of dimensionless parameters is satisfactory. The selection of the most convenient parameter is usually made on the basis of experience by trial and error with the aid of experimental results, although sometimes the characteristic groups are suggested by the results of analytical analyses.

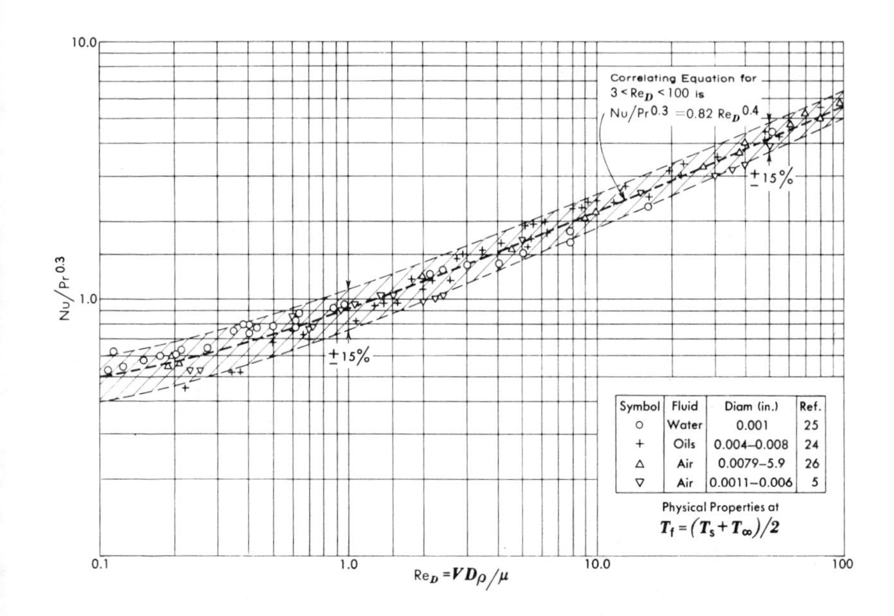

Fig. 6-7. Correlation of experimental heat-transfer data for various fluids in cross flow over cylinders of various diameters.

the thermal diffusivity $k/\rho c_p$, which affects the temperature profile. In other words, it is a dimensionless group which relates the temperature distribution to the velocity distribution. Hence, in geometrically similar systems having the same Prandtl and Reynolds numbers, the temperature distribution will be similar. According to its definition (see Eq. 6-3), the Nusselt number is numerically equal to the ratio of the temperature gradient at a fluid-to-surface interface to a reference-temperature gradient. We expect therefore that, in systems having similar geometries and similar temperature fields, the numerical values of the Nusselt numbers will be equal. This fact is borne out by the experimental results in Fig. 6-7.

6-7. Laminar boundary layer on a flat plate[7]

In the preceding section we determined dimensionless groups for correlating experimental data of heat transfer by forced convection. We found that the Nusselt number depends on the Reynolds number and the Prandtl number, i.e.,

---

[7] In the remainder of this chapter the mathematical details may be omitted in an introductory course without breaking the continuity of the presentation.

$$\mathrm{Nu} = \phi(\mathrm{Re})\psi(\mathrm{Pr}) \tag{6-13}$$

To determine the functional relationship in this equation it is necessary to resort either to experiments or to analytical methods.

In this and the following sections of the chapter we shall consider analytical methods of approach and apply them to the problem of heat transfer between a flat plate and an incompressible fluid flowing parallel to its surface. This system has been selected primarily because it is the simplest to analyze. However, the results obtained from this analysis have many practical applications. They are good approximations to forced convection in flow over the surfaces of streamlined bodies or in the inlet regions of pipes and ducts. In some cases appropriate transformations can reduce the equations for the flow of a compressible fluid, or the equations for the flow over wedges and cones, to the same form as those of the boundary-layer equations for the flat plate. The results for this case are therefore of considerable value; for more advanced boundary-layer problems and digital computer calculation methods, the reader should consult References 1 and 30.

In view of the difference in the flow characteristics, the frictional forces as well as the heat transfer are governed by different relations for laminar and turbulent types of boundary layers. We will first consider the laminar boundary layer, which is amenable to both an exact mathematical treatment and an approximate boundary-layer analysis. The turbulent boundary layer is taken up in Sec. 6-9.

Continuity and momentum equations. The equations of motion for boundary layer flow can be obtained by means of mass and force-and-momentum balances. As any other dynamic process, the flow of a fluid is governed by Newton's second law of motion which states that the summation of forces acting on a body in a given direction is equal to the time rate of change of its momentum in that direction, or

$$\Sigma F = \frac{1}{g_c} \frac{d(mV)}{d\theta}$$

In this form Newton's second law applies to a system of constant mass. In fluid dynamics, however, it is usually not convenient to deal with elements of mass; instead, one defines an elemental control volume, such as that shown in Fig. 6-8. Mass can flow in and out of this volume which is fixed in space, and a force-and-momentum balance in the x-direction for this system can be written

$$\Sigma F_x = \text{increase in momentum flux in } x\text{-direction}$$

where ΣF_x are the external forces acting on the control volume and the momentum flux in the x-direction is the product of the mass flow rate through the control volume and the component of the velocity in the x-direction there.

To derive the equations governing the flow in the boundary layer, consider an elemental control volume having the shape of a parallelepiped with dimensions $dx \cdot dy \cdot 1$ (see Fig. 6-8) and assume

1. The flow is two-dimensional, i.e., the velocity distribution is the same in any plane perpendicular to the z axis (i.e., parallel to the surface of the paper).
2. The fluid is incompressible.
3. The flow is steady with respect to time.
4. The pressure changes in the direction perpendicular to the surface are negligible.
5. The physical properties are constant.
6. Viscous shear forces in the y-direction are negligible.

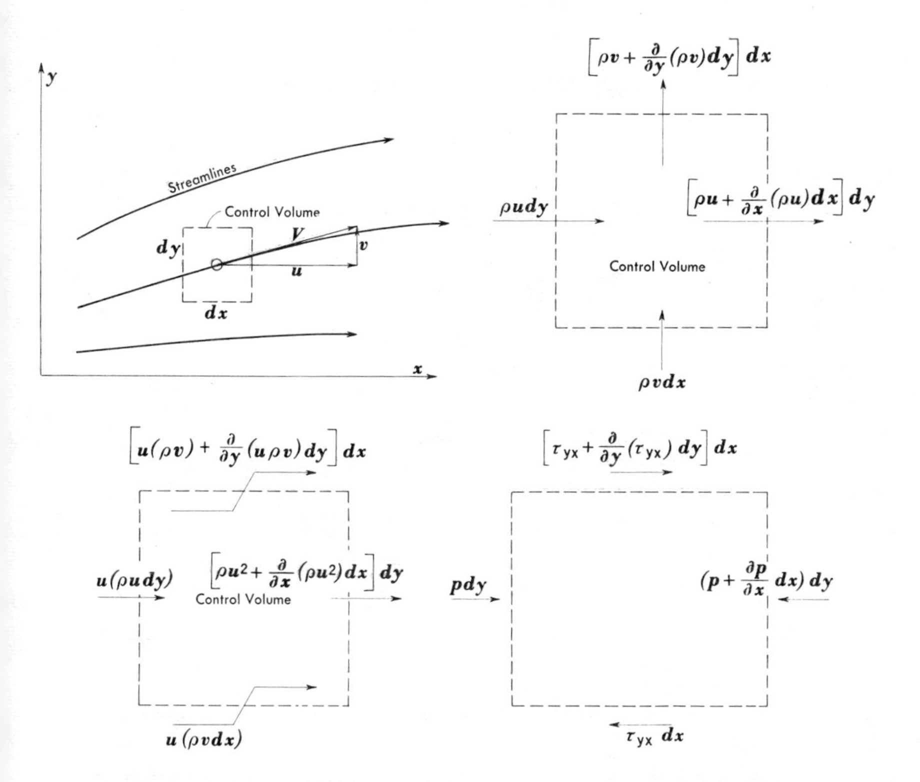

Fig. 6-8. Notation for continuity and momentum equations. Upper left: control volume in boundary layer. Upper right: mass flow through surface of control volume. Lower left: momentum fluxes in x-direction through surface of control volume. Lower right: forces acting on surface of control volume.

The mass flow rate entering through the left face is

$$\rho u dy$$

while the corresponding momentum flux is

$$(\rho u dy)u = \rho u^2 dy.$$

The mass flow rate leaving through the right face is

$$\rho \left(u + \frac{\partial u}{\partial x} dx \right) dy$$

and the corresponding momentum flux is

$$\rho \left(u + \frac{\partial u}{\partial x} dx \right)^2 dy.$$

The mass flow rate entering through the bottom face is

$$\rho v dx$$

and the mass flow rate leaving through the top is

$$\rho \left(v + \frac{\partial v}{\partial y} dy \right) dx.$$

A mass balance on the control volume gives

$$\rho u dy + \rho v dx = \rho \left(u + \frac{\partial u}{\partial x} dx \right) dy + \rho \left(v + \frac{\partial v}{\partial y} dy \right) dx$$

or

$$\frac{\partial u}{\partial x} + \frac{\partial v}{\partial y} = 0 \qquad \text{(6-14)}$$

Equation 6-14 is the *equation of continuity* for an incompressible two-dimensional flow.

To complete the momentum analysis, note that the momentum in the x-direction entering through the bottom face is

$$(\rho v dx) \cdot u$$

and the momentum in the x-direction leaving through the top face is

$$\rho\left(v + \frac{\partial v}{\partial y}\,dy\right)\left(u + \frac{\partial u}{\partial y}\,dy\right)dx$$

The forces in the x-direction acting on the elemental volume are due to pressure and viscous shear. The pressure force on the left face is $p\,dy$ and that on the right face is $\left[p + \left(\frac{\partial p}{\partial x}\right)dx\right]dy$; thus, the net pressure force in the x-direction is

$$-\frac{\partial p}{\partial x}\,dxdy$$

Viscous shear is the result of molecular interaction between faster- and slower-moving layers of fluid. It gives rise to a frictional stress τ, which is proportional to the velocity gradient normal to the direction of flow. The factor of proportionality is a property of the fluid and is called the *dynamic viscosity* μ_f. For flow over a flat plate (Fig. 6-9) when the change of velocity occurs only in the y direction perpendicular to the surface, the shearing stress in a plane parallel to the plate is

$$\tau_{yx} = \mu_f\frac{du}{dy} = \frac{\mu}{g_c}\frac{du}{dy} \tag{6-15}$$

where τ_{yx} = shearing stress in lb_f/sq ft;
u = velocity, in ft/sec;
y = distance, in ft;
μ_f = dynamic viscosity, in lb_f sec/sq ft;
μ = absolute viscosity, in lb_m/sec ft.

The dynamic viscosity is physically the same property of the fluid as the

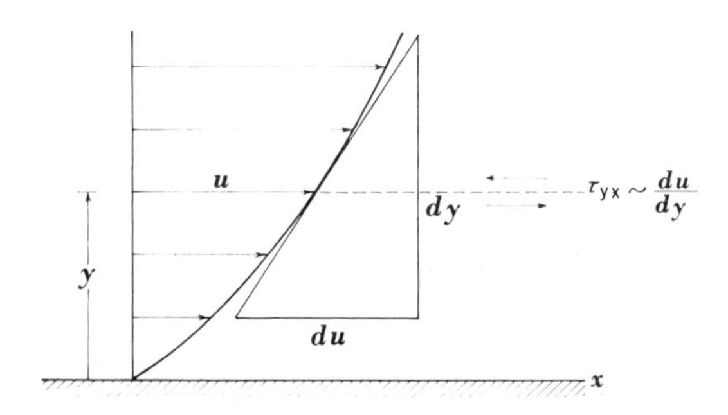

Fig. 6-9. Shearing stress in a laminar boundary layer.

absolute viscosity. The difference between them is merely the system of units used in their evaluation and the reason for distinguishing between μ_f and μ is to avoid errors in numerical computations by calling attention to the fact that some references list numerical values of viscosity in lb_f sec/sq ft, whereas this book uses the units lb_m/sec ft throughout. The subscript notation for the shearing stress τ indicates the axis to which the shear-affected area is perpendicular by the first letter and the direction of the stress by the second letter, e.g., τ_{yx} is the shear in the x direction on a plane perpendicular to the y axis.

At the lower face the shear acting on the fluid within the control volume is

$$(\tau_{yx})\, dx = \left(\frac{\mu}{g_c}\frac{\partial u}{\partial y}\right) dx$$

and at the upper face the shear is

$$\left[\tau_{yx} + \frac{\partial}{\partial y}(\tau_{yx})\, dy\right] dx = \left[\frac{\mu}{g_c}\frac{\partial u}{\partial y} + \frac{\partial}{\partial y}\left(\frac{\mu}{g_c}\frac{\partial u}{\partial y}\right) dy\right] dx$$

Since the wall is stationary, the shear on the fluid at the lower face of the control volume (Fig. 6-8) acts in a direction opposite to that of the flow (i.e., in the negative direction), while the shear on the upper face is caused by fluid tending to pull in the direction of motion. The net positive shear force is thus

$$\left[\left(\tau_{yx} + \frac{\partial \tau_{yx}}{\partial y} dy\right) - \tau_{yx}\right] dx = \frac{\partial}{\partial y}\left(\frac{\mu}{g_c}\frac{\partial u}{\partial y}\right) dxdy$$

Equating the sum of the viscous shear and pressure forces to the net momentum transfer in the x-direction gives

$$\mu \frac{\partial^2 u}{\partial y^2} dxdy - \frac{\partial p}{\partial x} dxdy = \rho\left(u + \frac{\partial u}{\partial x} dx\right)^2$$
$$- \rho u^2 dy + \rho\left(v + \frac{\partial v}{\partial y} dy\right)\left(u + \frac{\partial u}{\partial y} dy\right) dx - \rho v u dx$$

Neglecting higher order terms and simplifying the above relation with the aid of Eq. 6-14 yields

$$\rho\left(u \frac{\partial u}{\partial x} + v \frac{\partial u}{\partial y}\right) = \mu \frac{\partial^2 u}{\partial y^2} - \frac{\partial p}{\partial x} \tag{6-16}$$

For flow over a flat plate the pressure gradient is zero.

Boundary-layer thickness and skin friction. Equation 6-16 must be solved simultaneously with the continuity equation (Eq. 6-14) in order to determine the velocity distribution, the boundary-layer thickness, and the friction force at the wall. These equations are solved by first defining a stream function, $\psi(x,y)$, which automatically satisfies the continuity equation, or

$$u = \frac{\partial \psi}{\partial y} \qquad \text{and} \qquad v = -\frac{\partial \psi}{\partial x}$$

Introducing the new variable

$$\eta = y\sqrt{u_\infty / \nu x}$$

we can let

$$\psi = \sqrt{\nu x u_\infty}\, f(\eta)$$

where $f(\eta)$ denotes a dimensionless stream function. In terms of $f(\eta)$, the velocity components are

$$u = \frac{\partial \psi}{\partial y} = \frac{\partial \psi}{\partial \eta}\frac{\partial \eta}{\partial y} = u_\infty \frac{d[f(\eta)]}{d\eta}$$

and

$$v = -\frac{\partial \psi}{\partial x} = \frac{1}{2}\sqrt{\frac{\nu u_\infty}{x}}\left\{\frac{d[f(\eta)]}{d\eta}\eta - f(\eta)\right\}$$

Expressing $\partial u/\partial x$, $\partial u/\partial y$, and $\partial^2 u/\partial y^2$ in terms of η and inserting the resulting expressions in the momentum equation yields the ordinary, nonlinear, third-order differential equation

$$f(\eta)\frac{d^2[f(\eta)]}{d\eta^2} + 2\frac{d^3[f(\eta)]}{d\eta^3} = 0$$

which can be solved subject to the three boundary conditions that

$$\text{at } \eta = 0,\ f(\eta) = 0,\ \frac{d[f(\eta)]}{d\eta} = 0$$

and

$$\text{at } \eta = \infty,\ \frac{d[f(\eta)]}{d\eta} = 1$$

The solution to this differential equation was obtained numerically by Blasius, in 1908 (6). The significant results are shown in Figs. 6-10 and 6-11.

In Fig. 6-10 the Blasius velocity profiles in the laminar boundary on a flat plate are plotted in dimensionless form together with experimental data obtained by Hansen (13). The ordinate in Fig. 6-10 is the local velocity in the x direction u divided by the free stream velocity u_∞, and the abscissa is a dimensionless distance parameter, $(y/x)\sqrt{(\rho u_\infty x)/\mu}$. We note that a single curve is sufficient to correlate the velocity distributions at all stations along the plate. The velocity u reaches 99 percent of the free-stream value u_∞ at $(y/x)\sqrt{(\rho u_\infty x)/\mu} = 5.0$. If we define the hydrodynamic boundary-layer thickness as that distance from the surface at which local velocity u reaches 99 percent of the free-stream value u_∞, the boundary-layer thickness δ becomes

$$\delta = \frac{5x}{\sqrt{\mathrm{Re}_x}} \tag{6-17}$$

where $\mathrm{Re}_x = (\rho u_\infty x)/\mu$, the local Reynolds number. Equation 6-17

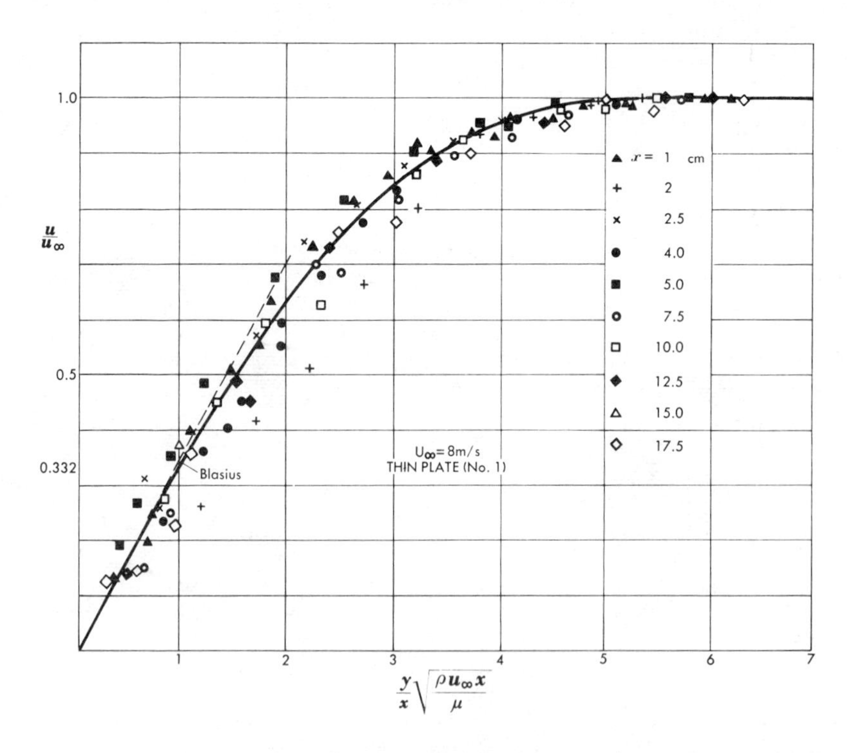

Fig. 6-10. Velocity profile in the laminar boundary layer according to Blasius with experimental data of Hansen (13). (Courtesy of National Advisory Committee for Aeronautics, NACA TM 585)

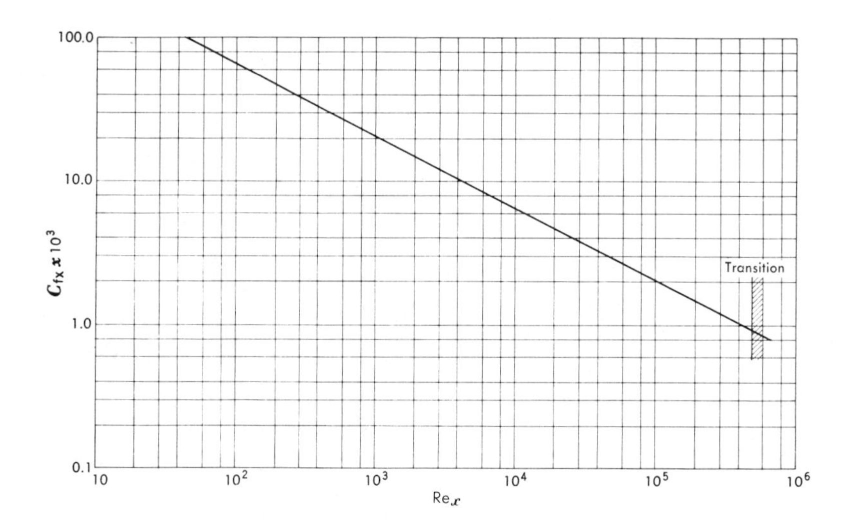

Fig. 6-11. Variation of local friction coefficient with dimensionless distance from leading edge for laminar flow over a flat plate.

satisfies the qualitative description of the boundary-layer growth, δ being zero at the leading edge ($x = 0$) and increasing with x along the plate. At any station, i.e., a given value of x, the thickness of the boundary layer is inversely proportional to the square root of the local Reynolds number. Hence, an increase in velocity will result in a decrease of the boundary-layer thickness.

The shear force at the wall can be obtained by substituting the velocity gradient at $y = 0$ into Eq. 6-15. From Fig. 6-10 we see that

$$\left.\frac{\partial(u/u_\infty)}{\partial(y/x)\sqrt{\mathrm{Re}_x}}\right|_{y=0} = 0.332$$

and thus at any specified value of x the velocity gradient at the surface is

$$\left.\frac{\partial u}{\partial y}\right|_{y=0} = 0.332\,\frac{u_\infty}{x}\sqrt{\mathrm{Re}_x}$$

Substituting this velocity gradient in Eq. 6-15, the wall shear per unit area τ_s becomes

$$\tau_s = \left.\frac{\mu}{g_c}\frac{\partial u}{\partial y}\right|_{y=0} = 0.332\,\frac{\mu}{g_c}\frac{u_\infty}{x}\sqrt{\mathrm{Re}_x} \qquad \text{(6-18)}$$

We note that the wall shear near the leading edge is very large and decreases with increasing distance from the leading edge.

For a graphical presentation it is more convenient to use dimensionless coordinates. Dividing both sides of Eq. 6-18 by the velocity pressure of the free stream $\rho u_\infty^2/2g_c$, we obtain

$$C_{fx} = \frac{\tau_s}{\rho u_\infty^2/2g_c} = 0.664/\sqrt{\text{Re}_x} \tag{6-19}$$

where C_{fx} is a dimensionless number called the *local drag or friction coefficient.* Figure 6-11 is a plot of C_{fx} against Re_x and shows the variation of the local friction coefficient graphically.

In many practical cases the average friction coefficient for a plate of finite length L is more important than the local friction coefficient. The average friction coefficient is obtained by integrating Eq. 6-19 between the leading edge, $x = 0$, and $x = L$. For laminar flow over the flat plate we get

$$\bar{C}_f = \frac{1}{L}\int_0^L C_{fx}\,dx = 1.33/\sqrt{\frac{u_\infty \rho L}{\mu}} \tag{6-20}$$

Thus, the average friction coefficient $\bar{C}_f$ is equal to twice the value of the local friction coefficient at $x = L$.

Energy equation. To evaluate the rate of heat transfer by convection we must determine the temperature gradient at the surface. The equation governing the temperature distribution in the boundary layer is obtained with the aid of the first law of thermodynamics, the principle of

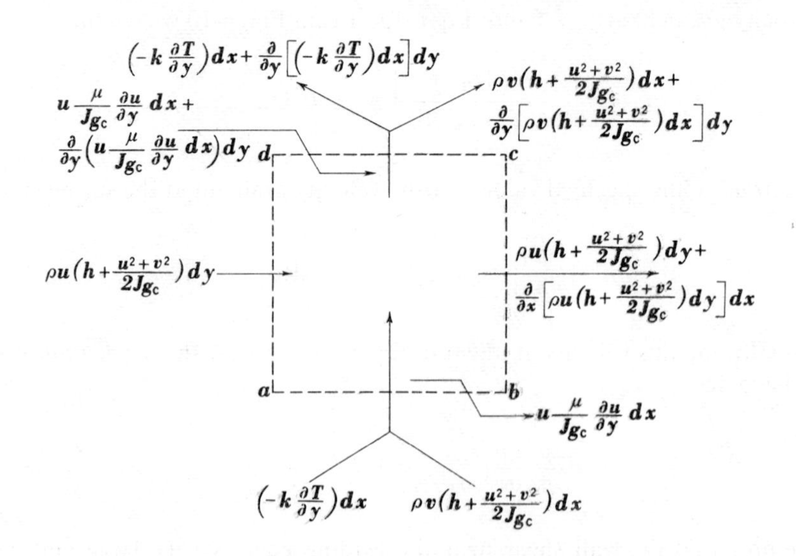

Fig. 6-12. Control volume in the boundary layer illustrating energy balance.

conservation of energy. Since we are dealing with a moving fluid, energy stored in fluid particles is transported by their motion. The rate of transport depends on the velocity of the fluid particles, and it is therefore always necessary to solve the hydrodynamic problem before the temperature distribution can be obtained.

To derive the equation governing the temperature distribution, consider the elementary control volume in the boundary layer shown in Fig. 6-12. Let the surfaces *ab*, *bc*, *cd*, and *da* define the boundaries of the system, and make an energy balance under the same assumptions used previously for the hydrodynamic equations. The energy equation for the system can be expressed semantically as

Influx of enthalpy and kinetic energy + rate of heat inflow by conduction + rate at which work is done by frictional shear *on* the fluid in control volume = efflux of enthalpy and kinetic energy + rate at heat outflow by conduction + rate at which work is done as a result of frictional shear *by* the fluid in control volume

or in symbolic form as

$$\rho u\left(h + \frac{u^2 + v^2}{2g_c J}\right)dy + \rho v\left(h + \frac{u^2 + v^2}{2g_c J}\right)dx$$

$$-k\left(\frac{\partial T}{\partial y}\right)dx + \frac{1}{J}\left[u\frac{\mu}{g_c}\frac{\partial u}{\partial y}dx + \frac{\partial}{\partial y}\left(u\frac{\mu}{g_c}\frac{\partial u}{\partial y}dx\right)dy\right]$$

$$= \rho u\left(h + \frac{u^2 + v^2}{2g_c J}\right)dy + \frac{\partial}{\partial x}\left[\rho u\left(h + \frac{u^2 + v^2}{2g_c J}\right)dy\right]dx$$

$$+ \rho v\left(h + \frac{u^2 + v^2}{2g_c J}\right)dx + \frac{\partial}{\partial y}\left[\rho v\left(h + \frac{u^2 + v^2}{2g_c J}\right)dx\right]dy$$

$$- k\left(\frac{\partial T}{\partial y}\right)dx + \frac{\partial}{\partial y}\left[-k\left(\frac{\partial T}{\partial y}\right)dx\right]dy + \frac{1}{J}\left(u\frac{\mu}{g_c}\frac{\partial u}{\partial y}\right)dx \tag{6-21}$$

The frictional work terms represent the work done by shearing forces on the surface of the control volume as faster fluid particles slide over slower ones. At the lower surface, the fluid inside the control volume exerts a force on the fluid outside because the former moves faster. The force times distance per unit time (i.e., velocity) $u\ (\mu/g_c)\ (\partial u/\partial y)$ represents the rate at which work is done *by* the fluid in the control volume. Similarly, the last term in square brackets on the left-hand side of Eq. 6-21 represents the rate at which work is done *on* the fluid in the control volume.

Conduction along the x direction has been omitted because, in the boundary layer, the term $-k(\partial T/\partial x$ is negligible compared to $-k(\partial T/\partial y)$ and the convection terms.

The term $h + (u^2 + v^2)/2g_cJ$ can be written c_pT_o for fluids having a constant specific heat. T_o is the stagnation temperature, i.e., the temperature reached by the fluid when it is isentropically slowed down to zero velocity. For low-speed flow, $T \simeq T_o$ because the kinetic energy of the flow is negligible. For high-speed flow, on the other hand, especially at supersonic velocities, this simplification is not permissible (see Sec. 6-12).

Adding up the terms of Eq. 6-21 and dropping those of higher order (i.e., terms involving triple products of d quantities) we obtain after simplifying[8]

$$\rho c_p u \frac{\partial T_o}{\partial x} + \rho c_p v \frac{\partial T_o}{\partial y} = k \frac{\partial^2 T}{\partial y^2} + \frac{\partial}{\partial y}\left(u \frac{\mu}{g_c} \frac{\partial u}{\partial y}\right) \qquad (6\text{-}22)$$

The last term of Eq. 6-22 represents the net rate at which shearing forces perform work *on* the fluid in the control volume. The mechanical energy or frictional power increases the internal energy of the fluid in the control volume appreciably only at high velocities, but for low subsonic flow in the main stream the frictional power term is small compared to the other terms and can be neglected. With these simplifications, Eq. 6-22 becomes

$$u \frac{\partial T}{\partial x} + v \frac{\partial T}{\partial y} = a \frac{\partial^2 T}{\partial y^2} \qquad (6\text{-}23)$$

where $a = k/\rho c_p$.

The velocities in the energy equation, u and v, have the same values at any point (x, y) as in the dynamic equation. For the case of the flat plate, Pohlhausen (7) used the velocities calculated previously by Blasius to obtain the solution of the heat-transfer problem. Without considering the details of this mathematical solution, we can obtain significant results by comparing Eq. 6-23, the heat-transfer equation for the boundary layer, with Eq. 6-16, the momentum equation for the boundary layer. The two equations are similar; in fact, a solution for the velocity distribution

---

[8] In this and in subsequent equations the work-to-heat conversion constant J has been ommitted.

$u(x, y)$ is also a solution for the temperature distribution $T(x, y)$ if $\nu = a$ and if the temperature of the plate T_s is constant. We can easily verify this by replacing the symbol T in Eq. 6-23 by the symbol u and noting that the boundary conditions for both T and u are identical. If we use the surface temperature as our datum and let the variable in Eq. 6-23 be $(T - T_s)/(T_\infty - T_s)$, then the boundary conditions are:

$$\text{at } y = 0: \quad \frac{T - T_s}{T_\infty - T_s} = 0 \quad \text{and} \quad \frac{u}{u_\infty} = 0$$

$$\text{at } y \rightarrow \infty: \quad \frac{T - T_s}{T_\infty - T_s} = 1 \quad \text{and} \quad \frac{u}{u_\infty} = 1$$

where T_∞ is the free-stream temperature.

The condition that $\nu = a$ corresponds to a Prandtl number of unity since

$$\text{Pr} = \frac{c_p \mu}{k} = \frac{\nu}{a}$$

For Pr = 1 the velocity distribution is therefore identical to the temperature distribution. An interpretation in terms of physical processes is that the transfer of momentum is analogous to the transfer of heat when Pr = 1. The physical properties of most gases are such that they have Prandtl numbers ranging from 0.65 to 1.0, and the analogy is therefore satisfactory. Liquids, on the other hand, have Prandtl numbers considerably different from unity, and the preceding analysis cannot be applied directly (36).

Using the analytical results of Pohlhausen's work, the temperature distribution in the laminar boundary layer for Pr = 1 can be modified empirically to include fluids having Prandtl numbers different from unity. In Fig. 6-13 theoretically calculated temperature profiles in the boundary layer are shown for values of Pr of 0.6, 0.8, 1.0, 3.0, 7.0, 15, and 50. We now define a thermal boundary-layer thickness δ_{th} as the distance from the surface at which the temperature difference between the wall and the fluid reaches 99 percent of the free-stream value. Inspection of the temperature profiles shows that the thermal boundary layer is larger than the hydrodynamic boundary layer for fluids having Pr less than unity, but smaller when Pr is larger than one. According to Pohlhausen's calculations, the relationship between the thermal and hydrodynamic boundary layer is approximately

$$\delta_{th} = \delta / \text{Pr}^{1/3} \tag{6-24}$$

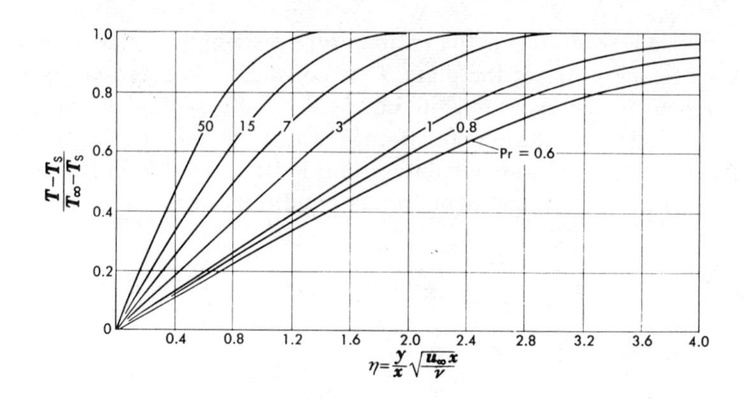

Fig. 6-13. Temperature distribution in a fluid flowing over a heated plate for various Prandtl numbers.

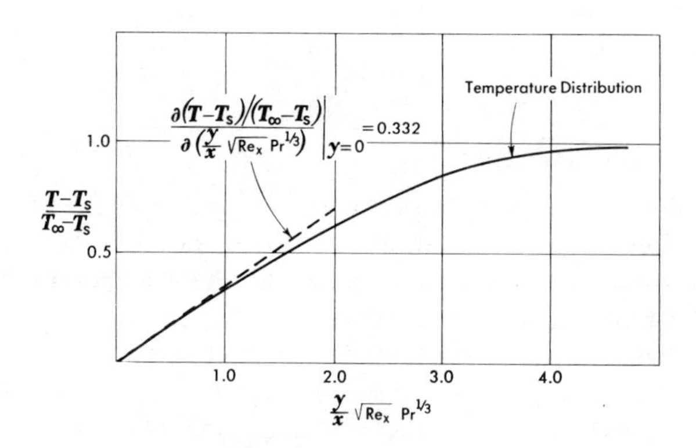

Fig. 6-14. Dimensionless correlation of temperature profiles for flow over a flat plate at constant temperature.

Using the same correction factor, i.e., $\mathrm{Pr}^{1/3}$, at any distance from the surface, the curves of Fig. 6-13 are replotted in Fig. 6-14. The new abscissa is $\mathrm{Pr}^{1/3}(y/x)\sqrt{\mathrm{Re}_x}$ and the ordinate is the dimensionless temperature $(T - T_s)/(T_\infty - T_s)$, where T is the local fluid temperature of the fluid, T_s the surface temperature of the plate, and T_∞ the free-stream temperature. This modification of the ordinate brings the temperature profiles for a wide range of Prandtl numbers together on the curve for $\mathrm{Pr} = 1$.

Evaluation of convective-heat-transfer coefficient. The rate of heat transfer by convection and the convective-heat-transfer coefficient can now be determined. The dimensionless temperature gradient at the surface (i.e., at $y = 0$) is

$$\left.\frac{\partial\left(\dfrac{T - T_s}{T_\infty - T_s}\right)}{\partial\left(\dfrac{y}{x}\sqrt{\mathrm{Re}_x}\,\mathrm{Pr}^{1/3}\right)}\right|_{y=0} = 0.332$$

Therefore, at any specified value of x

$$\left.\frac{\partial T}{\partial y}\right|_{y=0} = 0.332\,\frac{\mathrm{Re}_x^{1/2}\,\mathrm{Pr}^{1/3}}{x}\,(T_\infty - T_s) \tag{6-25}$$

and the local rate of heat transfer by convection per unit area becomes on substituting $\partial T/\partial y$ from Eq. 6-25 in Eq. 6-1

$$\frac{q}{A} = -k\left.\frac{\partial T}{\partial y}\right|_{y=0} = -0.332k\,\frac{\mathrm{Re}_x^{1/2}\,\mathrm{Pr}^{1/3}}{x}\,(T_\infty - T_s) \tag{6-26}$$

The total rate of heat transfer from a plate of width b and length L, obtained by integrating q from Eq. 6-26 between $x = 0$ and $x = L$, is

$$q = 0.664k\mathrm{Re}_L^{1/2}\,\mathrm{Pr}^{1/3}b\,(T_s - T_\infty) \tag{6-27}$$

where $\mathrm{Re}_L = u_\infty L/\nu$.

The local convective-heat-transfer coefficient is

$$h_{cx} = \frac{q}{A\,(T_s - T_\infty)} = 0.332\,\frac{k}{x}\,\mathrm{Re}_x^{1/2}\,\mathrm{Pr}^{1/3} \tag{6-28}$$

and the corresponding local Nusselt number is

$$\mathrm{Nu}_x = \frac{h_{cx}\,x}{k} = 0.332\,\mathrm{Re}_x^{1/2}\,\mathrm{Pr}^{1/3} \tag{6-29}$$

The average Nusselt number, $\bar{h}_c L/k$, is obtained by integrating the right-hand side of Eq. 6-28 between $x = 0$ and $x = L$ and dividing the result by L to obtain $\bar{h}_c$, the average value of h_{cx}; multiplying $\bar{h}_c$ by L/k gives

$$\overline{\mathrm{Nu}}_L = 0.664\,\mathrm{Re}_L^{1/2}\,\mathrm{Pr}^{1/3} \tag{6-30}$$

The average value of the Nusselt number over a length L of the plate is therefore twice the local value of Nu_x at $x = L$. It can easily be verified that the same relation between the average and local value holds also for the heat-transfer coefficient, that is,

$$\bar{h}_c = 2h_{c(x=L)} \tag{6-31}$$

In practice, the physical properties in Eqs. 6-24 to 6-30 vary with temperature, while for the purpose of analysis it was assumed that the physical properties are constant. Experimental data have been found to agree satisfactorily with the results predicted analytically if the properties are evaluated at a mean temperature halfway between that of the wall and the free-stream temperature.

EXAMPLE 6-1. Air at 60 F and at a pressure of 1 atm is flowing over a plate at a velocity of 10 fps. If the plate is 1 ft wide and at 140 F, calculate the following quantities at $x = 1$ ft and $x = x_c$.

a) Boundary-layer thickness.
b) Local friction coefficient.
c) Average friction coefficient.
d) Local drag or shearing stress due to friction.
e) Thickness of thermal boundary layer.
f) Local convective heat-transfer coefficient.
g) Average convective heat-transfer coefficient.
h) Rate of heat transfer by convection.

Solution: Properties of air at 100 F from Table A-3 are:

$$\begin{aligned} \rho &= 0.071 \text{ lb}_m/\text{cu ft} \\ c_p &= 0.240 \text{ Btu/lb}_m \text{ F} \\ \mu &= 1.285 \times 10^{-5} \text{ lb}_m/\text{ft sec} \\ k &= 0.0154 \text{ Btu/hr ft F} \\ \text{Pr} &= 0.72 \end{aligned}$$

The local Reynolds number at $x = 1$ ft is

$$\text{Re}_{x=1} = \frac{u_\infty \rho x}{\mu} = \frac{(10 \text{ ft/sec})(0.071 \text{ lb}_m/\text{cu ft})(1 \text{ ft})}{1.285 \times 10^{-5} \text{ lb}_m/\text{ft sec}} = 55{,}200$$

Assuming that the critical Reynolds number is 5×10^5, the critical distance is

$$x_c = \frac{5 \times 10^5 \mu}{u_\infty \rho} = \frac{(5 \times 10^5)(1.285 \times 10^{-5} \text{ lb}_m/\text{ft sec})}{(10 \text{ ft/sec})(0.071 \text{ lb}_m/\text{cu ft})} = 9 \text{ ft}$$

The desired quantities are determined by substituting appropriate values of the variable into the pertinent equations. The results of the calculations are shown in Table 6-3, and it is suggested that the reader verify them.

A useful relation between the local Nusselt number Nu_x and the corresponding friction coefficient C_{fx} is obtained by dividing Eq. 6-29 by $\text{Re}_x \text{Pr}^{1/3}$, or

Table 6-3

Part	Symbol	Unit	Eq. Used	Result (x = 1 ft)	Result (x = 9 ft)
a	δ	ft	6–17	0.0212	0.064
b	C_{fx}	......	6–19	0.00282	0.00094
c	$\bar{C}_f$	......	6–20	0.00564	0.00188
d	τ_s	lb_f/sq ft	6–18	3.12×10^{-4}	1.04×10^{-4}
e	δ_{th}	ft	6–24	0.0236	0.0715
f	h_{cx}	Btu/hr sq ft F	6–28	1.03	0.36
g	$\bar{h}_c$	Btu/hr sq ft F	6–31	2.06	0.72
h	q_c	Btu/hr	6–27	206	648

$$\left(\frac{\mathrm{Nu}_x}{\mathrm{Re}_x \mathrm{Pr}}\right)\mathrm{Pr}^{2/3} = \frac{0.322}{\mathrm{Re}_x^{1/2}} = \frac{C_{fx}}{2} \tag{6-32}$$

The dimensionless ratio $\mathrm{Nu}_x/\mathrm{Re}_x \mathrm{Pr}$ is known as the *Stanton number*, St_x. According to Eq. 6-32 the Stanton number times the Prandtl raised to the two-thirds power is equal to one-half the value of the friction coefficient. This relation between heat transfer and fluid friction was proposed by Colburn (4) and illustrates the interrelationship of the two processes.

6-8. Approximate boundary-layer analysis

The mathematical difficulties of an exact solution of the boundary layer equations in Sec. 6-7 can be circumvented by an approximate analysis which simplifies the mathematical manipulations and whose results agree with satisfactory accuracy with exact solutions. Instead of writing the equations of motion and heat transfer for a differential control volume, in the approximate method we write these equations for the aggregate of particles in the boundary layer. For this purpose we choose a control volume (Fig. 6-15) bounded by the two planes *ab* and *cd* which are perpendicular to the wall and a distance dx apart, the surface of the plate, and a parallel plane in the free stream at a distance l from the surface. Under steady-state conditions the mass flow rate into the control volume through the face *ab* is

$$\int_0^l \rho u dy$$

and the associated momentum flux is

$$\int_0^l \rho u^2 dy$$

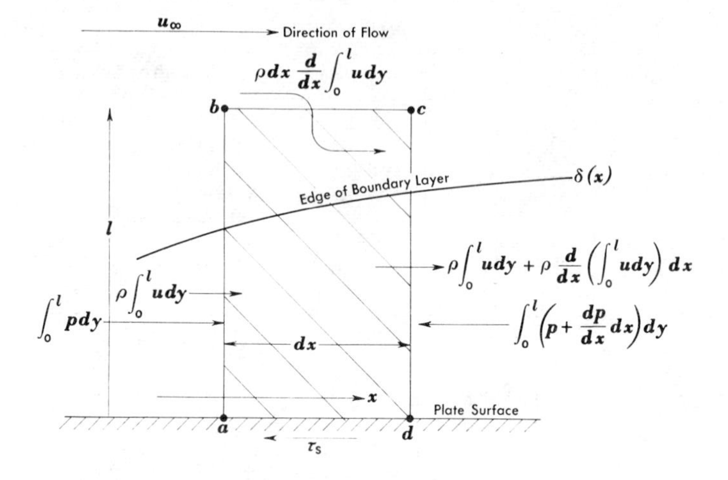

Fig. 6-15. Control volume for approximate momentum analysis of a boundary layer.

The mass flow rate out of the control volume through the face cd is

$$\int_0^l \rho u dy + \frac{d}{dx}\left(\int_0^l \rho u dy\right) dx$$

and the momentum flux out through cd is

$$\int_0^l \rho u^2 dy + \frac{d}{dx}\left(\int_0^l \rho u^2 dy\right) dx$$

No mass can enter the control volume through the plate and the difference in the mass flow out through cd and the mass flow in through ab must, therefore, enter through the upper face bd. Since the velocity at $y = 1$ is approximately equal to the free stream velocity u_∞, the x-momentum flux associated with the fluid entering through bd is

$$u_\infty \frac{d}{dx}\left(\int_0^l \rho u dy\right) dx$$

The net x-momentum flux out of the control volume is therefore

$$\frac{d}{dx}\left(\int_0^l \rho u^2 dy\right) dx - u_\infty \frac{d}{dx}\left(\int_0^l \rho u dy\right) dx$$

We can recast the second term in this expression into a more useful form by using the so-called "product formula" of integral calculus

$$d(r \cdot q) = rdq + qdr$$

or

$$rdq = d(r \cdot q) - q \cdot dr$$

If we treat the integral $\int_0^l \rho u dy$ as the q function and the free-stream velocity $u_\infty(x)$ as the r function

$$u_\infty \frac{d}{dx}\left(\int_0^l \rho u dy\right) dx = \frac{d}{dx}\left(u_\infty \int_0^l \rho u dy\right) dx - \frac{du_\infty}{dx}\left(\int_0^l \rho u dy\right) dx$$

But since $u_\infty(x)$ is not a function of y it can be placed under the integral sign and this gives

$$u_\infty \frac{d}{dx}\left(\int_0^l \rho u dy\right) dx = \frac{d}{dx}\left(\int_0^l \rho u u_\infty dy\right) dx - \frac{du_\infty}{dx}\left(\int_0^l \rho u dy\right) dx$$

The increase in x-momentum flux is equal to the summation of the forces in the x direction acting on the surface of the control volume. These forces, considered positive in the direction of flow, are

1. The shearing stress at the bottom surface, $-\tau_s dx = -\dfrac{\mu}{g_c} dx \left.\dfrac{\partial u}{\partial y}\right|_{y=0}$
2. The pressure on face ab, $\int_0^l pdy$.
3. The pressure on face cd, $-\left[\int_0^l pdy + \dfrac{d}{dx}\left(\int_0^l pdy\right) dx\right]$.

Since the velocities on both sides of the face bc are equal, no shearing stress exists there.

Equating the forces to the rate of momentum increase yields finally

$$\tau_s + \frac{d}{dx}\left(\int_0^l pdy\right) = \frac{d}{dx}\left(\int_0^l \frac{\rho}{g_c} u(u_\infty - u) dy\right) - \frac{du_\infty}{dx}\left(\int_0^l \frac{\rho}{g_c} udy\right) dx \tag{6-33}$$

Equation 6-33 is the *von Karman momentum integral equation of the boundary layer for incompressible flow over a flat plate.* It applies also to flow over slightly curved boundaries if x is measured along the surface and y normal to it; but if the surface over which the fluid flows is curved, the free-stream velocity u_∞ and the pressure p_∞ vary with x. For incompressible flow a relationship between u_∞ and p_∞ can be obtained from Bernouilli's equation

$$p_\infty + \frac{\rho u_\infty^2}{2g_c} = \text{constant}$$

or

$$\frac{dp_\infty}{dx} = -\frac{\rho u_\infty}{g_c}\frac{du_\infty}{dx}$$

Since the boundary layer is very thin, it may be assumed (1) that the pressure at any x location is constant throughout the boundary layer, i.e., $p(x) = p_\infty(x)$, and

$$\int_0^l \frac{dp}{dx}\,dy = \int_0^l \frac{dp_\infty}{dx}\,dy = -\frac{\rho}{g_c}\frac{du_\infty}{dx}\int_0^l u_\infty\,dy \qquad \text{(6-34)}$$

Substituting Eq. 6-34 for the pressure term in Eq. 6-33 gives

$$\tau_s = \frac{d}{dx}\int_0^l \frac{\rho u}{g_c}(u_\infty - u)\,dy + \frac{\rho}{g_c}\frac{du_\infty}{dx}\int_0^l (u_\infty - u)\,dy$$

Since the integrals in both terms on the right-hand side are zero for $y > \delta$, their upper limits can be replaced by δ. For flow over a flat plate u_∞ does not vary with x and

$$\frac{du_\infty}{dx} = 0 = \int_0^l \frac{dp_\infty}{dx}\,dy$$

For the constant pressure condition, Eq. 6-33 becomes therefore simply

$$g_c\frac{\tau_s}{\rho} = \frac{d}{dx}\int_0^l u(u_\infty - u)\,dy$$

If one assumes a physically reasonable velocity distribution in the boundary layer, then the momentum integral equation can be used to determine the boundary-layer thickness and the wall friction for specified geometries and flow conditions. The results naturally become more accurate the more closely the assumed velocity distribution resembles actual conditions. It has been found, however, that even a very rough assumption for the velocity distribution will yield satisfactory results. For this reason the approximate method is a powerful tool in engineering analysis. Example 6-2 illustrates the method.

EXAMPLE 6-2. Determine the hydrodynamic boundary-layer thickness for laminar flow over a flat plate by means of the von Karman momentum equation of the boundary layer. Assume a straight-line velocity distribution in the boundary layer.

Solution: The equation describing the velocity distribution for a linear increase in velocity from $u = 0$ at $y = 0$ to $u = u_\infty$ at $y = \delta$ is $u = u_\infty y/\delta$. The shearing stress at the wall τ_s is then

$$g_c \tau_s = \mu \left.\frac{\partial u}{\partial y}\right|_{y=0} = \frac{\mu u_\infty}{\delta}$$

Substituting for u and τ_s in Eq. 6-33 yields

$$\rho u_\infty^2 \frac{d}{dx} \int_0^\delta \left(1 - \frac{y}{\delta}\right) \frac{y}{\delta}\, dy = \frac{\mu u_\infty}{\delta}$$

Evaluating the integral above yields

$$\int_0^\delta \frac{y}{\delta}\, dy - \int_0^\delta \frac{y^2}{\delta^2}\, dy = \frac{1}{\delta} \left.\frac{y^2}{2}\right|_0^\delta - \frac{1}{\delta^2} \left.\frac{y^3}{3}\right|_0^\delta = \frac{\delta}{6}$$

Then, we get

$$\frac{\rho u_\infty^2}{6} \frac{d\delta}{dx} = \frac{\mu u_\infty}{\delta}$$

which yields

$$\delta d\delta = d\left(\frac{\delta^2}{2}\right) = \frac{6\mu}{\rho u_\infty}\, dx$$

Integrating the above equation gives the boundary-layer thickness δ as

$$\delta = \sqrt{(12\mu x)/(\rho u_\infty)} = 3.46x/\sqrt{\mathrm{Re}_x} \qquad \textit{Ans.}$$

The boundary-layer thickness calculated by means of a linear approximation to the velocity distribution is about 30 percent less than the value obtained by Blasius (see Eq. 6-17). However, the approximate method can be considerably improved by taking a velocity distribution which resembles the true conditions more closely. Eckert (9) used a cubic parabola of the form

$$\frac{u}{U_\infty} = C_1 \frac{y}{\delta} - C_2 \left(\frac{y}{\delta}^3\right) \quad \text{(6-35)}$$

and obtained, by substituting the above relation for u in Eq. 6-33,

$$\delta = 4.64x / \sqrt{\text{Re}_x} \quad \text{(6-36)}$$

a value only 8 percent less than that of the exact analysis. Since most of the experimental measurements are only accurate to within 10 percent, the results of the approximate analysis are satisfactory in practice.

To determine the rate of convective heat transfer to or from a surface we make an energy balance for the aggregate of fluid particles within the control volume of Fig. 6-16. To simplify the problem we shall neglect the shear work due to the frictional forces along the wall and assume also that the physical properties are independent of the temperature.

Energy is convected into and out of the control volume as a result of the fluid motion, and there is also heat flow by conduction across the interface. The energy flow rates across the individual faces of the control volume are listed in Table 6-4. To satisfy the principle of conservation of energy in the steady state, the rate of energy influx must equal the rate of energy efflux. Equating the net rate of convective energy outflow to the net rate of heat inflow by conduction we obtain

$$\frac{\partial}{\partial x} \int_0^l (T_{o\infty} - T_o)\, u dy = \frac{k}{\rho c_p} \frac{\partial T}{\partial y}\bigg|_{y=0}$$

Since the total temperature T_o equals the free-stream total temperature

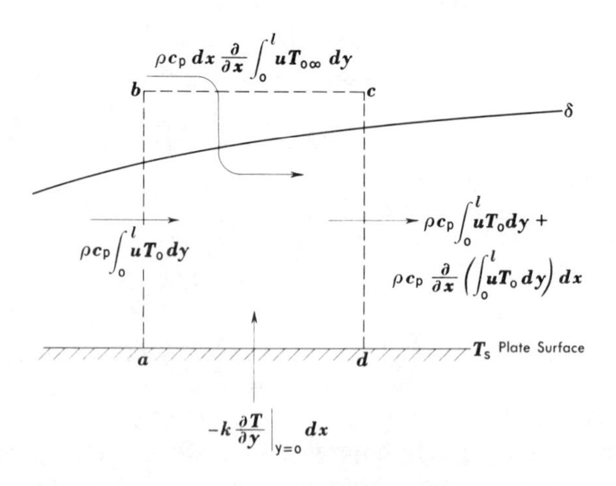

Fig. 6-16. Control volume for approximate energy balance in a boundary layer.

Table 6-4

Face	Mass-Flow Rate	Heat-Flow Rate
ab	$\rho \int_0^l u dy$	$\rho c_p \int_0^l u T_o dy$
bc	$\rho dx \frac{\partial}{\partial x} \int_0^l u dy$	$\rho c_p dx \frac{\partial}{\partial x} \int_0^l u T_{o\infty} dy$
cd	$\rho\left[\int_0^l u dy + \frac{\partial}{\partial x}\left(\int_0^l u dy\right) dx\right]$	$\rho c_p \int_0^l u T_o dy + \rho c_p \frac{\partial}{\partial x}\left(\int_0^l u T_o dy\right) dx$
da	0	$-k \left.\frac{\partial T}{\partial y}\right\vert_{y=0} dx$

$T_{o\infty}$ outside the thermal boundary layer (i.e., $y > \delta_{th}$), the integrand becomes zero for values of y larger than δ_{th}. We therefore replace l, the upper limit of the integral, by δ_{th}, and the heat-transfer equation of the boundary layer becomes

$$\frac{\partial}{\partial x} \int_0^{\delta_{th}} (T_{o\infty} - T_o) u dy = \frac{k}{\rho c_p} \left.\frac{\partial T}{\partial y}\right|_{y=0} \tag{6-37}$$

If we restrict our analysis to low-speed flow in which the kinetic energy is negligible compared with the enthalpy, the total temperatures in Eq. 6-37 are equal to the static temperatures for all practical purposes, i.e., $T_0 \cong T$ and $T_{o\infty} \cong T_\infty$.

To determine the convective heat-transfer coefficient we now select a suitable shape for the temperature distribution which meets the physical boundary conditions. Near the surface, where heat flows by conduction, the temperature gradient should be linear, and at $y = 0$, the fluid temperature should be equal to the plate temperature. At the edge of the thermal boundary layer (i.e., at $y = \delta_{th}$) the temperature should smoothly approach the free-stream temperature. Mathematically these boundary conditions are

$$\frac{\partial T}{\partial y} = C \quad \text{and} \quad (T - T_s) = 0 \quad \text{at } y = 0$$

$$(T - T_s) = (T_\infty - T_s) \quad \text{and} \quad \frac{\partial (T - T_s)}{\partial y} = 0 \quad \text{at } y = \delta_{th}$$

A cubic parabola of the form

$$T - T_s = C_1 y + C_2 y^3 \tag{6-38}$$

satisfies these boundary conditions if the constants C_1 and C_2 are selected appropriately. The conditions at $y = 0$ are automatically satisfied for any value of C_1 and C_2. At $y = \delta_{th}$ we have

$$T_\infty - T_s = C_1\delta_{th} + C_2\delta_{th}^3$$

and

$$\left.\frac{\partial(T - T_s)}{\partial y}\right|_{y=\delta} = C_1 + 3C_2\delta_{th^2} = 0$$

Solving for C_1 and C_2, and substituting these expressions in Eq. 6-38 yields

$$\frac{T - T_s}{T_\infty - T_s} = \frac{3}{2}\left(\frac{y}{\delta_{th}}\right) - \frac{1}{2}\left(\frac{y}{\delta_{th}}\right)^3 \tag{6-39}$$

Using Eqs. 6-39 and 6-35 for $(T - T_s)$ and u respectively, the integral in Eq. 6-37 can be written as

$$\int_0^{\delta_{th}} (T_\infty - T)\,u\,dy = \int_0^{\delta_{th}} [(T_\infty - T_s) - (T - T_s)]\,u\,dy$$

$$= (T_\infty - T_s)u_\infty \int_0^{\delta_{th}} \left[1 - \frac{3}{2}\frac{y}{\delta_{th}} + \frac{1}{2}\left(\frac{y}{\delta_{th}}\right)^3\right]\left[\frac{3}{2}\left(\frac{y}{\delta}\right) - \frac{1}{2}\left(\frac{y}{\delta}\right)^3\right] dy$$

Performing the multiplication under the integral sign we obtain

$$(T_\infty - T_s)u_\infty \int_0^{\delta_{th}}\left[\left(\frac{3}{2\delta}\right)y - \left(\frac{9}{4\delta\delta_{th}}\right)y^2 + \left(\frac{3}{4\delta\delta_{th}^3}\right)y^4 \right.$$
$$\left. -\left(\frac{1}{2\delta^3}\right)y^3 + \left(\frac{3}{4\delta_{\text{th}}\delta^3}\right)y^4 - \left(\frac{1}{4\delta_{\text{th}}^3\delta^3}\right)y^6 \; dy\right]$$

which yields after integrating

$$(T_\infty - T_s)u_\infty\left[\frac{3}{4}\frac{\delta_{th}^2}{\delta} - \frac{3}{4}\frac{\delta_{th}^2}{\delta} + \frac{3}{20}\frac{\delta_{th}^2}{\delta}\right.$$
$$\left. - \frac{1}{8}\frac{\delta_{th}^4}{\delta^3} + \frac{3}{20}\frac{\delta_{th}^4}{\delta^3} - \frac{1}{28}\frac{\delta_{th}^4}{\delta^3}\right]$$

If we let $\zeta = \delta_{th}/\delta$, the above expression can be written

$$(T_\infty - T_s)u_\infty\delta\left(\frac{3}{20}\zeta^2 - \frac{3}{280}\zeta^4\right)$$

For fluids having a Prandtl number equal to or larger than unity, ζ is equal to or less than unity and the second term in the bracket can be neglected compared to the first.[9] Substituting this approximate form for the integral in Eq. 6-37, we obtain

$$\frac{3}{20}u_\infty(T_s - T_\infty)\zeta^2\frac{\partial\delta}{\partial x} = -a\left.\frac{\partial T}{\partial y}\right|_{y=0} = \frac{3}{2}a\frac{T_s - T_\infty}{\delta\zeta}$$

or

$$\frac{1}{10}u_\infty\zeta^3\delta\frac{\partial\delta}{\partial x} = a$$

From Eq. 6-36 we obtain

$$\delta\frac{\partial\delta}{\partial x} = 10.75\frac{\nu}{u_\infty}$$

and with this expression we get

$$\zeta^3 = \frac{10}{10.75}\frac{a}{\nu}$$

or

$$\delta^{th} = 0.9\delta\mathrm{Pr}^{-1/3} \tag{6-40}$$

Except for the numerical constant (0.9 compared with 1.0) the foregoing result is in agreement with the exact calculations of Pohlhausen (Eq. 6-24).

The rate of heat flow by convection from the plate per unit area is, from Eqs. 1-1 and 6-39,

$$\frac{q}{A} = -k\left.\frac{\partial T}{\partial y}\right|_{y=0} = -\frac{3}{2}\frac{k}{\delta_{th}}(T_\infty - T_s)$$

Substituting Eqs. 6-36 and 6-40 for δ_{th} yields

[9]This assumption is not valid for liquid metals, which have $\mathrm{Pr} \ll 1$.

$$\frac{q}{A} = -\frac{3}{2}\,\frac{k}{x}\,\frac{\mathrm{Pr}^{1/3}\mathrm{Re}_x^{\ 1/2}}{(0.9)(4.64)}(T_\infty - T_s) = 0.36\,\frac{k}{x}\,\mathrm{Re}_x^{\ 1/2}\mathrm{Pr}^{1/3}(T_s - T_\infty) \tag{6-41}$$

and

$$\mathrm{Nu}_x = \frac{q}{A(T_s - T_\infty)}\,\frac{x}{k} = 0.36\,\mathrm{Re}_x^{\ 1/2}\,\mathrm{Pr}^{1/3} \tag{6-42}$$

This result is in agreement with the exact analysis (Eq. 6-29) except for the numerical constant, which is about 9 percent larger.

The foregoing example illustrates the usefulness of the approximate boundary-layer analysis. Guided by a little physical insight and intuition, this technique yields satisfactory results without the mathematical complications inherent in the exact boundary-layer equations. The approximate method has been applied to many other problems, and the results are available in the literature.

6-9. Analogy between heat and momentum transfer in turbulent flow

In a majority of practical applications the flow in the boundary layer is turbulent rather than laminar. It is therefore not surprising that many famous scientists, such as Osborn Reynolds, G. I. Taylor, Ludwig Prandtl, and T. von Karman, have studied problems dealing with turbulent-exchange mechanisms. Although these men as well as many others have contributed considerably to our understanding of turbulent flow, so far no one has succeeded in predicting friction and heat-transfer coefficients by a direct analysis. The reason for this lack of success is the extreme complexity of turbulent motion. In turbulent flow, irregular velocity fluctuations are always superimposed upon the motion of the main stream, and the fluctuating components can not be described by simple equations. Yet, it is precisely these fluctuations which are primarily responsible for the transfer of heat as well as momentum in turbulent flow.

Qualitatively the exchange mechanism in turbulent flow can be pictured as a magnification of the molecular exchange in laminar flow. In steady laminar flow, physical properties such as temperature and pressure remain constant at any point and fluid particles follow well-defined streamlines. Heat and momentum are transferred across streamlines only by molecular diffusion. The amount of cross flow is so small that, when a colored dye is injected at some point into the fluid, it follows a streamline without appreciable diffusion. In turbulent flow, on the other hand, the color will be distributed over a wide area a short distance downstream from the point of injection. The mixing mechanism consists of rapidly fluctuating eddies which transport blobs of fluid in an irregular manner. Groups of particles collide with each other at random, establish cross flow on a macroscopic scale, and effectively mix the fluid. Since the mixing in turbulent flow is on a macroscopic scale with groups of particles transported in a zigzag path through the fluid, the exchange mechanism is many times more effective than in laminar flow. As a result, the rates of heat and momentum transfer in turbulent flow and the associated friction and heat-transfer coefficients are many times larger than in laminar flow.

Instantaneous streamlines in turbulent flow are highly jagged, and it would be a hopelessly difficult task to trace the path of individual fluid elements. However, if the flow at a point is averaged over a period of time, long as compared with the period of a single fluctuation, the time-mean properties and the velocity of the fluid are constant if the average flow remains steady. It is therefore general practice to describe each fluid property and the velocity in turbulent flow in terms of a *mean value* which does not vary with time and a *fluctuating component* which is a function of time. To simplify the problem, consider a two-dimensional flow (Fig. 6-17) in which the mean value of velocity is parallel to the x direction. The instantaneous velocity components u and v can then be expressed in the form

$$
\begin{aligned}
u &= \bar{u} + u' \\
v &= v'
\end{aligned}
\tag{6-43}
$$

where the bar over a symbol denotes the temporal mean value, and the prime denotes the instantaneous deviation from the mean value. According to the model used to describe the flow,

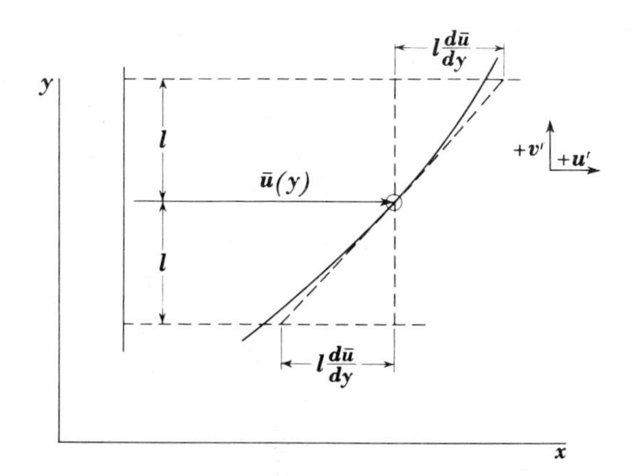

Fig. 6-17. Sketch illustrating mixing length for momentum transfer.

$$\bar{u} = \frac{1}{\theta^*} \int_0^{\theta^*} u d\theta \tag{6-44}$$

where θ^* is large compared with the period of the fluctuations. Figure 6-18 shows qualitatively the time variation of u and u'. From Eq. 6-44 or from an inspection of the graph it is apparent that the time average of u' is zero, i.e. $\bar{u}' = 0$. A similar argument shows that $\bar{v}'$ and $(\overline{\rho v})'$ are also zero.

The fluctuating velocity components continuously transport mass, and consequently momentum, across a plane normal to the y direction. The instantaneous rate of transfer in the y direction of x-momentum per unit area at any point is

$$-(\rho v)'(\bar{u} + u')$$

where the minus sign, as will be shown later, takes account of the statistical correlation between u' and v'.

The time average of the x-momentum transfer gives rise to an *apparent turbulent shear or Reynolds stress* τ_t defined by

$$g_c \tau_t = -\frac{1}{\theta^*} \int_0^{\theta^*} (\rho v)'(\bar{u} + u')\, d\theta \tag{6-45}$$

Breaking this term up into two parts, the time average of the first is zero, or

$$\frac{1}{\theta^*} \int_0^{\theta^*} (\rho v)' \bar{u} d\theta = 0$$

since $\bar{u}$ is a constant and the time average of $(\rho v)'$ is zero. Integrating the second term, Eq. 6-45 becomes

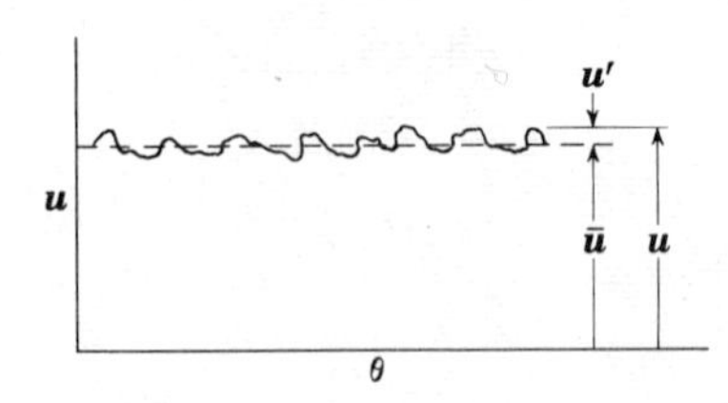

Fig. 6-18. Sketch illustrating time variation of instantaneous velocity.

$$g_c\tau_t = -\frac{1}{\theta^*}\int_0^{\theta^*}(\rho v)'u'\,d\theta = -\overline{(\rho v)'u'} \qquad \text{(6-45a)}$$

or if ρ is constant

$$g_c\tau_t = -\rho(\overline{v'u'}) \qquad \text{(6-46)}$$

It is not difficult to visualize that the time averages of the mixed products of velocity fluctuations, such as for example $\overline{v'u'}$, differ from zero. From Fig. 6-17 we can see that the particles which travel upward ($v' > 0$) arrive at a layer in the fluid in which the mean velocity $\bar{u}$ is larger than in the layer from which they come. Assuming that the fluid particles preserve on the average their original velocity $\bar{u}$ during their migration, they will tend to slow down other fluid particles after they have reached their destination and thereby give rise to a negative component u'. Conversely, if v' is negative, the observed value of u' at the new destination will be positive. On the average, therefore, a positive v' is associated with a negative u', and vice versa. The time average of $u'v'$ is therefore on the average not zero but a negative quantity. The turbulent shearing stress defined by Eq. 6-46 is thus positive and has the same sign as the corresponding laminar shearing stress (Eq. 6-15),

$$\tau_{yx} = \mu_f\frac{d\bar{u}}{dy} = \frac{\rho}{g_c}\nu\frac{d\bar{u}}{dy}$$

It should be noted, however, that the laminar shearing stress is a true stress, whereas the apparent turbulent shearing stress is simply a concept introduced to account for the effects of the momentum transfer by turbulent fluctuations. This concept allows us to express the total shear stress in turbulent flow as

$$\tau = \frac{\text{viscous force}}{\text{unit area}} + \frac{1}{g_c}\text{(turbulent momentum flux)} \qquad \text{(6-47)}$$

To relate the turbulent momentum flux to the time-average velocity gradient, $d\bar{u}/dy$, Prandtl (10) postulated that fluctuations of macroscopic blobs of fluid in turbulent flow are, on the average, similar to the motion of molecules in a gas, i.e., they travel on the average a distance l perpendicular to $\bar{u}$ (Fig. 6-17) before coming to rest in another y plane. This distance l is known as Prandtl's mixing length and corresponds qualitatively to the mean free path of a gas molecule. Prandtl further argued that the fluid particles retain their identity and physical properties during the cross motion and that the turbulent fluctuation arises chiefly from the difference in the time-mean properties between y planes spaced a distance

l apart. According to this argument, if a fluid particle travels from the layer y to the layer $y + l$,

$$u' \simeq l \frac{d\bar{u}}{dy} \tag{6-48}$$

With this model we can write the turbulent shearing stress in a form analogous to the laminar shearing stress as

$$g_c \tau_t = -\rho \overline{v'u'} = \rho \epsilon_M \frac{d\bar{u}}{dy} \tag{6-49}$$

where the symbol ϵ_M is called the eddy viscosity or the turbulent exchange coefficient for momentum. The eddy viscosity ϵ_M is formally analogous to the kinematic viscosity ν, but whereas ν is a physical property, ϵ_M depends on the dynamics of the flow. Combining Eqs. 6-48 and 6-49 shows that $\epsilon_M = -\overline{v'l}$. Substituting Eqs. 6-15 and 6-49 in Eq. 6-47 gives the total shearing stress in the form

$$\tau = \frac{\rho}{g_c} (\nu + \epsilon_M) \frac{d\bar{u}}{dy} \tag{6-50}$$

In turbulent flow ϵ_M is much larger than ν and the viscous term may therefore be neglected.

The transfer of energy as heat in a turbulent flow can be pictured in an analogous fashion. Consider a two-dimensional time-mean temperature distribution as shown in Fig. 6-19. The fluctuating velocity components continuously transport fluid particles and the energy stored in

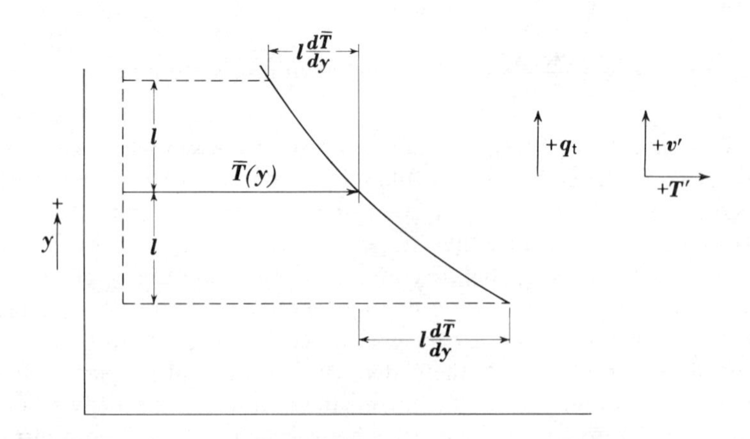

Fig. 6-19. Sketch illustrating mixing length for energy transfer.

them across a plane normal to the y direction. The instantaneous rate of energy transfer per unit area at any point in the y direction is

$$(\rho v')(c_p T) \tag{6-51}$$

where $T = \bar{T} + T'$. Following the same line of reasoning which led to Eq. 6-46, the time average of energy transfer due to the fluctuations, called the turbulent rate of heat transfer q_t, is

$$q_t = A\rho c_p \overline{v'T'} \tag{6-52}$$

Using Prandtl's concept of mixing length, we can relate the temperature fluctuation to the time-mean temperature gradient by the equation

$$T' \simeq l \frac{d\bar{T}}{dy} \tag{6-53}$$

This means physically that, when a fluid particle migrates from the layer y to another layer a distance l above or below, the resulting temperature fluctuation is caused chiefly by the difference between the time-mean temperatures in the layers. Assuming that the transport mechanisms of temperature (or energy) and velocity are similar, the mixing lengths in Eqs. 6-48 and 6-53 are equal. The product $\overline{v'T'}$, however, is positive on the average because a positive v' is accompanied by a positive T', and vice versa.

Combining Eqs. 6-52 and 6-53, the turbulent rate of heat transfer per unit area becomes

$$\frac{q_t}{A} = c_p \rho \overline{v'T'} = -c_p \rho \overline{v'l} \frac{d\bar{T}}{dy} \tag{6-54}$$

where the minus sign is a consequence of the second law of thermodynamics (see Sec. 1-3). To express the turbulent heat flux in a form analogous to the Fourier conduction equation we define ϵ_H, a quantity called the turbulent exchange coefficient for temperature, eddy diffusivity of heat, or eddy heat conductivity, by the equation $\epsilon_H = \overline{v'l}$. Substituting ϵ_H for $\overline{v'l}$ in Eq. 6-54, gives

$$\frac{q_t}{A} = -c_p \rho \epsilon_H \frac{d\bar{T}}{dy} \tag{6-55}$$

The total rate of heat transfer per unit area normal to the mean stream velocity can then be written as

$$\frac{q}{A} = \frac{\text{molecular conduction}}{\text{unit area}} + \frac{\text{turbulent transfer}}{\text{unit area}}$$

or in symbolic form as

$$\frac{q}{A} = -c_p \rho (a + \epsilon_H) \frac{d\overline{T}}{dy} \tag{6-56}$$

where $a = k/c_p\rho$, the molecular diffusivity of heat. The contribution to the heat transfer by molecular conduction is proportional to a, and the turbulent contribution is proportional to ϵ_H. For all fluids except liquid metals, ϵ_H is much larger than a in turbulent flow. The ratio of the molecular kinematic viscosity to the molecular diffusivity of heat ν/a has previously been named the Prandtl number. Similarly, the ratio of the turbulent eddy viscosity to the eddy diffusivity ϵ_M/ϵ_H could be considered a turbulent Prandtl number Pr_t. According to the Prandtl mixing-length theory, the turbulent Prandtl number is unity, since $\epsilon_M = \epsilon_H = \overline{v'l}$.

Although the model postulated by Prandtl in his treatment of turbulent flow is certainly grossly oversimplified, experimental results indicate it is at least qualitatively correct. Isakoff and Drew (11) found that Pr_t for the heating of mercury in turbulent flow inside a tube may vary from 1.0 to 1.6, and Forstall and Shapiro (12) found that Pr_t is about 0.7 for gases. The latter investigators also showed that Pr_t is substantially independent of the value of the laminar Prandtl number as well as of the type of experiment. For practical calculations it is usually satisfactory to assume that Pr_t is unity. With this simplification we can relate the turbulent heat flux to the turbulent shear stress by combining Eqs. 6-49 and 6-55 and obtain

$$\frac{q_t}{A} = -g_c \tau_t c_p \frac{d\overline{T}}{d\overline{u}} \tag{6-57}$$

This relation was originally drived in 1874 by the British scientist Osborn Reynolds and is called the *Reynolds analogy* in his honor. It is a good approximation whenever the flow is turbulent, and can be applied to turbulent boundary layers as well as to turbulent flow in pipes or ducts. However, the Reynolds analogy does not hold in the laminar sublayer. Since this layer offers a large thermal resistance to the flow of heat, Eq. 6-57 does in general not suffice for a quantitative solution. Only for fluids having a Prandtl number of unity can it be used directly to calculate the rate of heat transfer. This special case will now be considered.

6-10. Reynolds analogy for turbulent flow over a flat plate

In this section we shall derive for flow over a plane surface a relation between the heat transfer and the skin friction for a Prandtl number of unity. In the following section we shall show how to calculate the skin friction and consider some improvements over the simple analogy.

In two-dimensional flow the shearing stress in the laminar sublayer τ_{yx} is (Eq. 6-15)

$$g_c \tau_{yx} = \mu \frac{du}{dy}$$

and the rate of heat flow per unit area across any plane perpendicular to the y direction is (Eq. 1-1)

$$\frac{q}{A} = -k \frac{dT}{dy}$$

Combining Eqs. 1-1 and 6-15 yields

$$\frac{q}{A} = -g_c \tau_{yx} \frac{k}{\mu} \frac{dT}{du} \tag{6-58}$$

An inspection of Eqs. 6-57 and 6-58 shows that if $c_p = k/\mu$ (i.e., for $\Pr = 1$), the same equation of heat flow applies in the laminar and turbulent layers.

To determine the rate of heat transfer from a flat plate to a fluid with $\Pr = 1$ flowing over it in turbulent flow, we replace k/μ by c_p and separate the variables in Eq. 6-58. Assuming that q and τ are constant, we get the equation

$$\frac{q_s}{A \tau_s c_p g_c} du = -dT \tag{6-59}$$

where the subscript s is used to indicate that both q and τ are taken at the surface of the plate. Integrating Eq. 6-59 between the limits $u = 0$ when $T = T_s$, and $u = u_\infty$ when $T = T_\infty$, yields

$$\frac{q_s}{A \tau_s c_p g_c} u_\infty = (T_s - T_\infty) \tag{6-60}$$

But since by definition

$$h_{cx} = \frac{q_s}{A(T_s - T_\infty)} \qquad \text{and} \qquad \tau_{sx} = C_{fx} \frac{\rho u_\infty^2}{2g_c}$$

Eq. 6-60 can be written as

$$\frac{h_{cx}}{c_p \rho u_\infty} = \frac{\text{Nu}}{\text{Re}_x \text{Pr}} = \frac{C_{fx}}{2} \tag{6-61}$$

Equation 6-61 is satisfactory for gases in which Pr is approximately unity. Colburn (4) has shown that Eq. 6-61 can also be used for fluids having Prandtl numbers ranging from 0.6 to about 50 if it is modified in accordance with experimental results to read

$$\frac{\mathrm{Nu}_x}{\mathrm{Re}_x \mathrm{Pr}} \mathrm{Pr}^{2/3} = \mathrm{St}_x \mathrm{Pr}^{2/3} = \frac{C_{fx}}{2} \tag{6-62}$$

where the subscript x denotes the distance from the leading edge of the plate.

6-11. *Turbulent flow over plane surfaces*

To apply the analogy between heat transfer and momentum transfer in practice it is necessary to know the skin-friction coefficient C_{fx}. For turbulent flow over a plane surface the empirical equation for the local friction coefficient

$$C_{fx} = 0.0576 \left(\frac{u_\infty x}{\nu}\right)^{-1/5} \tag{6-63}$$

is in good agreement with experimental results (1) in the Reynolds number range between 5×10^5 and 10^7 as long as no separation occurs. Assuming that the turbulent boundary layer starts at the leading edge, the average friction coefficient over a plane surface of length L can be obtained by integrating Eq. 6-63, or

$$\bar{C}_f = \frac{1}{L} \int_0^L C_{fx}\, dx = 0.072 \left(\frac{u_\infty L}{\nu}\right)^{-1/5} \tag{6-64}$$

In reality, however, a laminar boundary layer precedes the turbulent boundary layer between $x = 0$ and $x = x_c$. Since the local frictional drag of a laminar boundary layer is less than the local frictional drag of a turbulent boundary layer at the same Reynolds number, the average drag calculated from Eq. 6-64 without correcting for the laminar portion of the boundary layer is too large. The actual drag can be closely estimated, however, by assuming that, behind the point of transition, the turbulent boundary layer behaves as though it had started at the leading edge.

Adding the laminar friction drag between $x = 0$ and $x = x_c$ to the turbulent drag between $x = x_c$ and $x = L$ gives, per unit width,

$$\bar{C}_f = [0.072\, \mathrm{Re}_L^{-1/5} L - 0.072\, \mathrm{Re}_{x_c}^{-1/5} x_c + 1.33\, \mathrm{Re}_{x_c}^{-1/2} x_c]/L$$

For a critical Reynolds number of 5×10^5 this yields

$$\bar{C}_f = 0.072(\text{Re}_L^{-1/5} - 0.0464\, x_c/L) \tag{6-65}$$

Substituting Eq. 6-63 for C_{fx} in Eq. 6-62 yields the local Nusselt number at any value of x larger than x_c, or

$$\text{Nu}_x = \frac{h_{cx}x}{k} = 0.0288\,\text{Pr}^{1/3}\left(\frac{u_\infty x}{\nu}\right)^{0.8} \tag{6-66}$$

We observe that the local heat-transfer coefficient h_{cx} for heat transfer by convection through a turbulent boundary layer decreases with the distance x as $h_{cx} \propto 1/x^{0.2}$. Equation 6-66 shows that, in comparison with laminar flow where $h_{cx} \propto 1/x^{1/2}$, the heat-transfer coefficient in turbulent flow decreases less rapidly with x and that the turbulent-heat-transfer coefficient is much larger than the laminar heat-transfer coefficient at a given value of the Reynolds number.

The average conductance in turbulent flow over a plane surface of length L can be calculated to a first approximation by integrating Eq. 6-66 between $x = 0$ and $x = L$, or

$$\bar{h}_c = \frac{1}{L}\int_0^L h_{cx}dx$$

In dimensionless form we get

$$\overline{\text{Nu}}_L = \frac{\bar{h}_c L}{k} = 0.036\,\text{Pr}^{1/3}\,\text{Re}_L^{0.8} \tag{6-67}$$

Equation 6-67 neglects the existence of the laminar boundary layer and is therefore valid only when $L \gg x_c$. The laminar boundary layer can be included in the analysis if Eq. 6-28 is used between $x = 0$ and $x = x_c$, and Eq. 6-66 between $x = x_c$ and $x = L$ for the integration of h_{cx}. This yields for $\text{Re}_c = 5 \times 10^5$

$$\overline{\text{Nu}}_L = 0.036\,\text{Pr}^{1/3}(\text{Re}_L^{0.8} - 23{,}200) \tag{6-68}$$

EXAMPLE 6-3. The crankcase of an automobile is approximately 30 in. long, 12 in. wide, and 4 in. deep. Assuming that the surface temperature of the crankcase is 160 F, estimate the rate of heat flow from the crankcase to atmospheric air at 40 F at a road speed of 60 mph. Assume that the vibration of the engine and the chassis induce the transition from laminar to turbulent flow so near to the leading edge that, for practical purposes, the boundary layer is turbulent over the entire surface. Neglect radiation and use for the front and rear surfaces the same average convective heat-transfer coefficient as for the bottom and sides.

Solution: Using physical properties of air at 100 F from Table A-3 in Appendix III, the Reynolds number is

$$\mathrm{Re}_L = \frac{u_\infty \rho L}{\mu} = \frac{(88 \text{ ft/sec})(0.071 \text{ lb}_m/\text{cu ft})(30/12 \text{ ft})}{1.285 \times 10^{-5} \text{lb}_m/\text{ft sec}}$$

$$= 1.21 \times 10^6$$

From Eq. 6-67 the average Nusselt number is

$$\overline{\mathrm{Nu}}_L = 0.036 \,\mathrm{Pr}^{1/3}\, \mathrm{Re}_L{}^{0.8}$$

$$= (0.036)\,(0.896)\,(73{,}480) = 2370$$

and the average convective heat-transfer coefficient becomes

$$\bar{h}_c = \overline{\mathrm{Nu}}_L \frac{k}{L} = \frac{(2370)(0.0154 \text{ Btu/hr ft F})}{30/12 \text{ ft}}$$

$$= 14.55 \text{ Btu/hr sq ft F}$$

The overall area is 4.84 sq ft and the rate of heat loss is therefore

$$q = \bar{h}_c A\,(T_s - T_\infty) = (14.55)\,(4.84)\,(160\text{–}40) = 8430 \text{ Btu/hr} \qquad \textit{Ans.}$$

The thickness of a turbulent boundary layer in flow over a plane surface can be calculated by means of the Karman integral relations. To improve the accuracy of the calculations, we shall use a velocity distribution determined by experiment. Figure 6-20 shows several velocity profiles measured by Van der Hegge-Zynen (14). Near the wall the velocity increases linearly with the distance from the surface. This is the region called the laminar sublayer, although some recent measurements suggest that it is not completely devoid of turbulence. In the fully turbulent portion of the boundary layer the velocity increases with the one-seventh power of distance and can be represented by the equation

$$\frac{u}{u_\infty} = \left(\frac{y}{\delta}\right)^{1/7} \tag{6-69}$$

Between the laminar sublayer and the turbulent portion of the boundary layer is a transition region where the turbulence level is variable. Because the laminar sublayer as well as the transition layer are very thin we shall, as a first approximation, neglect both of them and use Eq. 6-69 to evaluate the momentum change in the integral equations. This approximation cannot be used, however, to determine the shearing stress because, according to Eq. 6-69, the velocity gradient is

$$\frac{du}{dy} = \frac{1}{7}\,\frac{u_\infty}{\delta^{1/7} y^{6/7}}$$

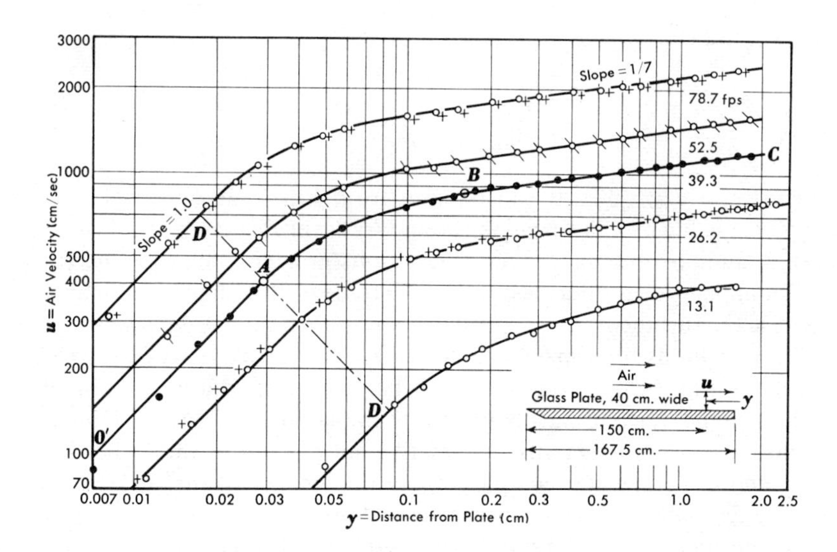

Fig. 6-20. Velocity distribution in turbulent boundary layers over plane surfaces after Van der Hegge-Zynen (14).

which would lead to infinitely large shearing stress at the wall (i.e., at $y = 0$). To overcome this difficulty we shall use an experimentally determined relation for the shearing stress.

In the Reynolds-number range between 10^5 and 10^7 the relation

$$g_c\tau_s = 0.0228\,\rho u_\infty^{\,2}\left(\frac{\nu}{u_\infty\delta}\right)^{1/4} \qquad (6\text{-}70)$$

is in good agreement with experimental results obtained by Schultz-Grunow (15). Substituting Eq. 6-70 for the shearing stress and Eq. 6-69 for the velocity distribution in the integral of Eq. 6-33 gives

$$\frac{7}{72}\,\rho u_\infty^{\,2}\,\frac{d\delta}{dx} = 0.0228\,\rho u_\infty^{\,2}\left(\frac{\nu}{u_\infty\delta}\right)^{1/4}$$

Separation of the variables yields

$$\delta^{1/4}d\delta = 0.235\left(\frac{\nu}{u_\infty}\right)^{1/4} dx$$

from which we obtain the boundary-layer thickness in the form

$$\delta = 0.376\left(\frac{\nu}{u_\infty}\right)^{1/5} x^{4/5}$$

or

$$\frac{\delta}{x} = 0.376 \, \mathrm{Re}_x^{-1/5} \tag{6-71}$$

where $\mathrm{Re}_x = (u_\infty x/\nu)$. It can be seen from Eqs. 6-36 and 6-71 that, at any given value of x, a turbulent boundary layer increases at a faster rate than a laminar boundary layer. Despite its greater thickness, the turbulent boundary layer offers less resistance to heat flow than a laminar layer because the turbulent eddies produce continuous mixing between warmer and cooler fluids on a macroscopic scale. An inspection of the velocity profiles in Fig. 6-20 shows, however, that the eddies diminish in intensity in the buffer layer and hardly penetrate the laminar sublayer. Unless the Prandtl number equals unity, the relative magnitudes of the eddy conductivity and the molecular conductivity in the vicinity of the surface have a pronounced effect on the heat-transfer coefficient.

The effect of the diminution of the turbulent mixing near the surface on the heat-transfer coefficient for fluids having Prandtl numbers larger than unity was considered by Prandtl (16,17), von Karman (18), and most recently also by Deissler (19) in their respective improvements of the Reynolds analogy. Martinelli (20) also treated the problem of heat transfer to liquid metals, which have very small Prandtl numbers.

Prandtl divided the flow field into a laminar and a turbulent layer but neglected the buffer layer in his analysis. The relation for flow over plane surfaces, derived in detail in Ref. 21, is

$$\frac{\mathrm{Nu}_x}{\mathrm{Re}_x \mathrm{Pr}} = \frac{C_{fx}/2}{1 + 2.1 \, \mathrm{Re}_x^{-0.1}(\mathrm{Pr} - 1)} \tag{6-72}$$

We observe that, for $\mathrm{Pr} = 1$, Eq. 6-72 reduces to the simple Reynolds analogy. The second term in the denominator is a measure of the thermal resistance in the laminar sublayer. We see that this portion of the total thermal resistance increases as the Prandtl number becomes larger and accounts for most of the thermal resistance when the Prandtl number is very large.

Prandtl's analysis was later refined by von Karman (18), who divided the flow field into three zones: a laminar sublayer adjacent to the surface in which the eddy diffusivity is zero and heat flows only by conduction; next to it a buffer layer in which both conduction and convection contribute the heat-transfer mechanism (i.e., $k/c\rho$ and ϵ_H are of the same order of magnitude); and, finally, a turbulent region in which conduction is negligible compared to convection, and the Reynolds analogy applies. He used experimental data for the velocity distribution and the shear stress to evaluate ϵ_M from Eq. 6-50 and assumed $\epsilon_M = \epsilon_H$ in his analysis. He also postulated that the physical properties of the fluid are independent of the temperature. With these simplifications he determined the thermal resistances in each of the three zones. The results of von Karman's analysis are given below for flow over a flat plate:

Thermal resistance of laminar sublayer $\qquad \dfrac{5\,\mathrm{Pr}}{c_p\sqrt{\rho g_c \tau_s}}$

Thermal resistance of buffer layer $\qquad \dfrac{5\ln(5\,\mathrm{Pr}+1)}{c_p\sqrt{\rho g_c \tau_s}}$

Thermal resistance of the turbulent region $\qquad \dfrac{5(1+\ln 6)+u_\infty/\sqrt{\tau_s g_c/\rho}}{c_p\sqrt{\rho g_c \tau_s}}$

Adding the thermal resistances and introducing the definitions for the Stanton number St and the local drag-friction coefficient C_{fx} yields, after some rearrangement, the expression

$$\mathrm{St}_x = \frac{\mathrm{Nu}_x}{\mathrm{Re}_x\mathrm{Pr}} = \frac{C_{fx}/2}{1+5\sqrt{C_{fx}/2}\left[(\mathrm{Pr}-1)+\ln\dfrac{5\,\mathrm{Pr}+1}{6}\right]} \tag{6-73}$$

for the local value of the Stanton number for flow over a plane surface at a given value of x. The average value of St or $\bar{h}_c$ over a surface of length L can be obtained by numerical or graphical integration.

To apply any of the equations relating the Stanton number and the friction coefficient in practice, the physical properties must be evaluated at some appropriate mean temperature. It is general practice to evaluate the physical properties at the *mean film temperature* T_f defined as $T_f = (T_s + T_\infty)/2$. This procedure is purely empirical, but has been found satisfactory for moderate-temperature ranges.

The three-distinct-layer concept is somewhat of an oversimplification of the real situation but is satisfactory for Prandtl numbers less than 25 or 30. For larger Prandtl numbers it is preferable to assume turbulent eddy generation near the outer edges of turbulent boundary layers and continuous damping of these eddies as they approach the wall. Some progress has been made recently with this approach (19,22), and the reader is referred to the original papers for details. An extensive review of the analogies is presented in Ref. 23.

6-12. *Heat transfer in high-speed flow*

Convection heat transfer in high-speed flow is important for systems such as aircraft and missiles when the velocity approaches or exceeds the velocity of sound. For a perfect gas the accoustical velocity, a, can be

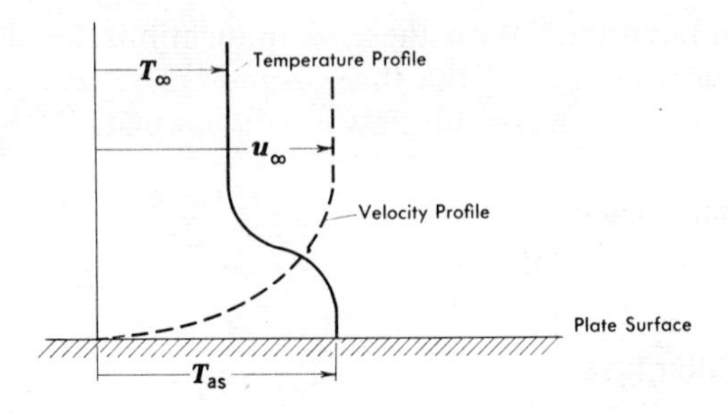

Fig. 6-21. Velocity and temperature distribution in high-speed flow over an insulated plate.

obtained from the relation

$$a = \sqrt{\gamma RT/\mathfrak{M}} \tag{6-74}$$

where γ = specific heat ratio, c_p/c_u (about 1.4 for air)
R = universal gas constant
T = absolute temperature, and
$\mathfrak{M}$ = molecular weight of the gas.

When the velocity of a gas flowing over a heated or cooled surface is of the order of the accoustical velocity or larger, the flow field can no longer be described solely in terms of the Reynolds number, but also the ratio of the gas flow-velocity to the accoustical velocity, i.e., the Mach number $M = u_\infty/a_\infty$, must be considered. When the gas velocity in a flow system reaches a value of about one-half of the speed of sound, the effects of viscous dissipation in the boundary layer become important. Under such conditions the temperature of a surface over which a gas is flowing can actually exceed the free-stream temperature. For flow over an adiabatic surface, e.g., a perfectly insulated wall, Fig. 6-21 shows the velocity and temperature distributions schematically. The high temperature at the surface is the combined result of the heating due to viscous dissipation and the temperature rise of the fluid as the kinetic energy of the flow is converted to internal energy while the flow decelerates through the boundary layer. The actual shape of the temperature profile depends on the relation between the rate at which viscous shear work increases the internal energy of the fluid and the rate at which heat is conducted towards the free stream.

Although the processes in a high-speed boundary layer are not adiabatic, it is general practice to relate them to adiabatic processes. The conversion of kinetic energy in a gas being slowed down adiabatically to zero velocity is described by the relation

$$i_0 = i_\infty + u_\infty^2/2Jg_c \tag{6-75}$$

where i_0 is the stagnation enthalpy and i_∞ is the enthalpy of the gas in the free stream. For an ideal gas Eq. 6-75 becomes

$$T_0 = T_\infty + u_\infty^2/2Jg_c c_p$$

or in terms of the Mach number

$$\frac{T_0}{T_\infty} = 1 + \frac{\gamma - 1}{2} M_\infty^2 \tag{6-76}$$

where T_0 is the stagnation temperature and T_∞ is the free-stream temperature.

In a real boundary layer the fluid is not brought to rest reversibly because the viscous shearing process is thermodynamically irreversible. To account for the irreversibility in a boundary-layer flow we define a recovery factor r as

$$r = \frac{T_{as} - T_\infty}{T_0 - T_\infty} \tag{6-77}$$

where T_{as} is the adiabatic surface temperature.

Experiments (31) have shown that in laminar flow,

$$r = \mathrm{Pr}^{1/2} \tag{6-78}$$

whereas in turbulent flow

$$r = \mathrm{Pr}^{1/3} \tag{6-79}$$

When a surface is not insulated, the rate of heat transfer by convection between a high-speed gas and that surface is governed by the relation

$$q_c/A = -k\frac{\partial T}{\partial y}\bigg|_{y=0}$$

The influence of heat transfer to and from the surface on the temperature distribution is illustrated in Fig. 6-22. We observe that in high-speed flow heat can be transferred to the surface even when the surface temperature is above the free-stream temperature. This phenomenon is the result of viscous shear, often called aerodynamic heating. The heat transfer in high-speed flow over a flat surface can be predicted (1) from the boundary-layer energy equation

$$u\frac{\partial T}{\partial x} + v\frac{\partial T}{\partial y} = a\frac{\partial^2 T}{\partial y^2} + \frac{\mu}{\rho c_p}\left(\frac{\partial u}{\partial x}\right)^2$$

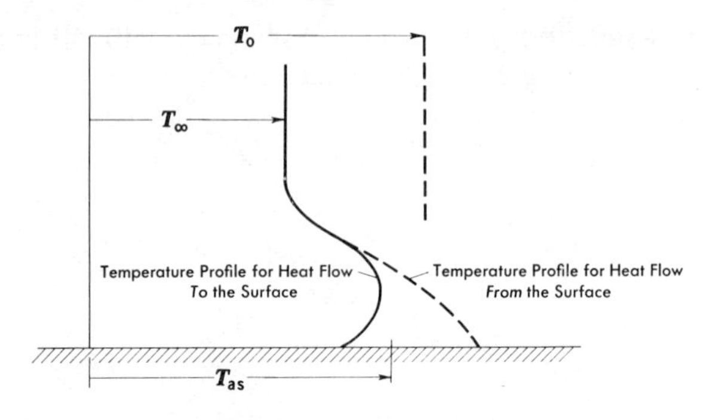

Fig. 6-22. Temperature profiles in high-speed boundary layer for heating and cooling.

where the last term accounts for the viscous dissipation. However, for most practical purposes the rate of heat transfer can be calculated with the same relations used for low speed flow, if the average convection heat-transfer coefficient is redefined by the relation

$$q_c/A = \bar{h}_c(T_s - T_{as}) \tag{6-80}$$

which will yield a zero heat flow when the surface temperature T_s equals the adiabatic surface temperature.

Since in high-speed flow the temperature gradients in a boundary layer are large, also variations in the physical properties of the fluid will be substantial. Eckert (32) has shown, however, that the constant property heat-transfer equations can still be used if all the properties are evaluated at a reference temperature T^* given by the relation

$$T^* = T_\infty + 0.5(T_s - T_\infty) + 0.22(T_{as} - T_\infty) \tag{6-81}$$

The local values of the heat-transfer coefficient, defined by the relation

$$h_{cx} = (q/A)/(T_s - T_{as})$$

can be obtained from the following equations:

Laminar Boundary Layer ($\mathrm{Re}_x^* < 10^5$):

$$\mathrm{St}_x^* = \left(\frac{h_{cx}}{c_p \rho u_\infty}\right)^* = 0.332(\mathrm{Re}_x^*)^{-1/2}(\mathrm{Pr}^*)^{-2/3} \tag{6-82}$$

Turbulent Boundary Layer ($10^5 < \mathrm{Re}_x^* < 10^7$):

$$\mathrm{St}_x^* = \left(\frac{h_{cx}}{c_p \rho u_\infty}\right)^* = 0.0288(\mathrm{Re}_x^*)^{-1/5}(\mathrm{Pr}^*)^{-2/3} \tag{6-83}$$

Turbulent Boundary Layer ($10^7 < \mathrm{Re}_x^* < 10^9$):

$$\mathrm{St}_x^* = \left(\frac{h_{cx}}{c\rho u_\infty}\right)^* = \frac{2.46}{(ln\mathrm{Re}_x^*)^{2.584}}(\mathrm{Pr}^*)^{-2/3} \tag{6-84}$$

because experimental data for local friction coefficients in high-speed gas flow (32) in the Reynolds number range between 10^7 and 10^9 are correlated by the relation

$$C_{fx} = \frac{4.92}{(ln\mathrm{Re}_x^*)^{2.584}} \tag{6-85}$$

If the average value of the heat-transfer coefficients is to be determined, the above expressions must be integrated between $x = 0$ and $x = L$ as shown in Sec. 6-7 for low-speed flow. However, the integration may have to be done numerically in most practical cases because the reference temperature T^* is not the same for the laminar and turbulent portions of the boundary layer, as shown by Eqs. 6-78 and 6-79.

When the speed of a gas is exceedingly high the boundary layer may become so hot that the gas begins to dissociate. In such situations Eckert (32) recommends that the heat-transfer coefficient be based on the enthalpy difference and be defined by the relation

$$q_c/A = h_{ci}(i_s - i_{as}) \tag{6-86}$$

If an enthalpy recovery factor is defined by

$$r_i = \frac{i_{as} - i_\infty}{i_0 - i_\infty} \tag{6-87}$$

the same relation used previously to calculate the reference temperature can be used to calculate a reference enthalpy, or

$$i^* = i_\infty + 0.5(i_s - i_\infty) + 0.22(i_{as} - i_\infty) \tag{6-88}$$

The local Stanton number is then redefined as

$$\mathrm{St}_{x,i}^* = \frac{h_{c,i}}{\rho^* u_\infty} \tag{6-89}$$

and used in Eqs. 6-83, 6-84, and 6-85 to calculate the heat-transfer coefficient. It should be noted that the enthalpies in the above relations are the total values, which include the chemical energy of dissociation as well as the internal energy. As shown in Ref. 32, this method of calculation is in excellent agreement with experimental data.

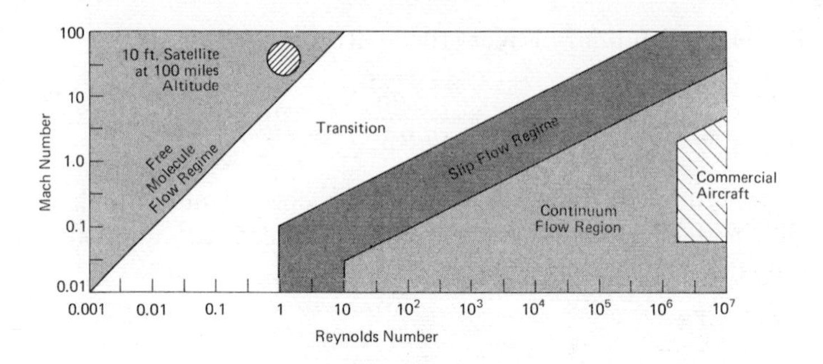

Fig. 6-23. Flow regimes.

In some situations, e.g., extremely high altitudes, the fluid density may be so small that the distance between gas molecules becomes of the same order of magnitude as the boundary layer. In such cases the fluid cannot be treated as a continuum and it is necessary to subdivide the flow processes into regimes. These flow regimes are characterized by the ratio of the molecular free path to a significant physical scale of the system, called the Knudsen number, Kn. Continuum flow corresponds to small values of Kn, while at larger values of Kn, molecular collisions occur primarily at the surface and in the main stream. Since energy transport is by free motion of molecules between the surface and the main stream, this regime is called "free-molecule." Between the free-molecule and the continuum regime is a transition range, called the "slip-flow" regime because it is treated by assuming temperature and velocity "slip" at fluid-solid interfaces. Fig. 6-23 shows a map of these flow regimes. For a treatment of heat transfer and friction in these specialized flow systems the reader is referred to Refs. 33, 34, and 35.

6-13. *Closure*

In this chapter we have studied the principles of heat transfer by forced convection. We have seen that the transfer of heat by convection is intimately related to the mechanics of the fluid flow, particularly to the flow in the vicinity of the heat-transfer surface. We have also observed that the nature of heat transfer as well as flow phenomena depend greatly on whether the fluid far away from the surface is in laminar or in turbulent flow.

To become familiar with the basic principles of boundary layer theory and forced-convection heat transfer, we have considered the problem of convection in flow over a flat plate in some detail. This system is geometrically very simple, but it illustrates the most important features of forced convection. In subsequent chapters we shall treat heat transfer by

Table 6-5. Summary of useful equations for calculating friction and heat-transfer coefficients in flow over flat surfaces.*

LAMINAR FLOW		
Local friction coefficient	$C_{fx} = 0.664\, \mathrm{Re}_x^{-0.5}$	$\mathrm{Re}_x < 5 \times 10^5$
Local Nusselt number at distance x from leading edge	$\mathrm{Nu}_x = 0.332\, \mathrm{Re}_x^{0.5}\, \mathrm{Pr}^{0.33}$	$\mathrm{Pr} > 0.1, \mathrm{Re}_x > 5 \times 10^5$
	$\mathrm{Nu}_x = 0.565(\mathrm{Re}_x\, \mathrm{Pr})^{0.5}$	$\mathrm{Pr} < 0.1, \mathrm{Re}_x < 5 \times 10^5$
Average friction coefficient	$\bar{C}_f = 1.33\, \mathrm{Re}_L^{-0.5}$	$\mathrm{Re}_L < 5 \times 10^5$
Average Nusselt number between $x = 0$ and $x = L$	$\overline{\mathrm{Nu}}_L = 0.664\, \mathrm{Re}_L^{0.5}\, \mathrm{Pr}^{0.33}$	$\mathrm{Pr} > 0.1, \mathrm{Re}_L < 5 \times 10^5$
	$\overline{\mathrm{Nu}}_L = 1.13(\mathrm{Re}_L\, \mathrm{Pr})^{0.5}$	$\mathrm{Pr} < 0.1, \mathrm{Re}_L < 5 \times 10^5$
TURBULENT FLOW		
Local friction coefficient	$C_{fx} = 0.0576\, \mathrm{Re}_x^{-0.2}$	$\mathrm{Re}_x > 5 \times 10^5, \mathrm{Pr} > 0.5$ (applies to both local equations)
Local Nusselt number at distance x from leading edge	$\mathrm{Nu}_x = 0.0288\, \mathrm{Re}_x^{0.8}\, \mathrm{Pr}^{0.3}$ (For $\mathrm{Pr} < 0.1$ use Eq. 6-73 or see Ref. 20)	
Average friction coefficient	$\bar{C}_f = 0.072[\mathrm{Re}_L^{-0.2} - 0.0464(x_{\mathrm{crit}}/L)]$	$\mathrm{Re}_L > 5 \times 10^5, \mathrm{Pr} > 0.5$ (applies to both average equations)
Average Nusselt number between $x = 0$ and $x = L$ with transition at $\mathrm{Re}_{x,\mathrm{crit}} = 5 \times 10^5$	$\overline{\mathrm{Nu}}_L = 0.036\, \mathrm{Pr}^{0.33}[\mathrm{Re}_L^{0.8} - 23{,}200]$	

*Applicable to low-speed flow (Mach number < 0.5) of gases and liquids if all physical properties at the mean film temperature, $T_f = (T_s + T_\infty)/2$.

$$C_{fx} = \tau_s/(\rho u_\infty^2/2g_c), \qquad \bar{C}_f = (1/L)\int_0^L C_{fx}\, dx$$

$$\mathrm{Nu}_x = h_c x/k, \qquad \overline{\mathrm{Nu}} = \bar{h}_c L/k, \qquad \bar{h}_c = (1/L)\int_0^L h_c(x)\, dx$$

$$\mathrm{Re}_x = \rho u_\infty x/\mu, \qquad \mathrm{Re}_L = \rho u_\infty L/\mu, \qquad \mathrm{Pr} = c_p \mu/k$$

convection in geometrically more complicated systems. In the next chapter we shall examine free-convection phenomena. In Chapter 8, heat transfer by convection to and from fluids flowing inside of pipes and ducts will be taken up. In Chapter 9, forced convection in flow over the exterior surfaces of bodies such as cylinders, spheres, tubes, and tube bundles will be considered. The application of the principles of forced-convection heat transfer to the selection and design of heat-transfer equipment will be taken up in Chapter 11.

For the convenience of the reader, a summary of the equations used to calculate the heat-transfer and the friction coefficients in low-speed flow of gases and liquids over flat, or only slightly curved, plane surfaces is presented in Table 6-5.

PROBLEMS

The problems below have been organized in the following manner: Problems 6-1 through 6-5 deal with the evaluation of dimensionless parameters, 6-6 through 6-17 with dimensional analysis, 6-18 through 6-33 with the evaluation of heat transfer and friction coefficients in flow over a flat plate, 6-34 through 6-40 with boundary layers, 6-41 through 6-48 with thermal design, and 6-49 through 6-54 with high-speed flow.

6-1. Evaluate the Reynolds number from the following data:

$$\begin{aligned} D &= 6 \text{ in.} \\ u_\infty &= 1.0 \text{ ft/sec.} \\ \rho &= 30 \text{ slug/ft}^3 \\ \mu &= 90 \text{ lb}_\text{m}/\text{ft hr} \end{aligned}$$

Ans. 19,333

6-2. Evaluate the Prandtl number from the data below:

$$\begin{aligned} c_p &= 0.5 \text{ Btu/lb}_\text{m} \text{ F} \\ k &= 2 \text{ Btu/ft hr F} \\ \mu &= 5 \text{ dyne sec/cm}^2 \end{aligned}$$

6-3. Evaluate the Nusselt number for the following condition:

$$\begin{aligned} D &= 6 \text{ in.} \\ k &= 2 \text{ Kcal/m hr C} \\ h &= 18 \text{ Btu/ft}^2 \text{ hr F} \end{aligned}$$

Ans. 7.7

6-4. Evaluate the Stanton number for the data below:

$$\begin{aligned} D &= 4 \text{ in.} \\ V &= 12 \text{ ft/sec.} \end{aligned}$$

$$\rho = 1280\ \text{lb}_m/\text{ft}^3$$
$$\mu = 30\ \text{slug}/\text{ft hr}$$
$$c_p = 0.95\ \text{Btu}/\text{lb}_m$$
$$\bar{h}_c = 3.0\ \text{Btu}/\text{hr ft}^2\ \text{F}$$

6-5. Evaluate the dimensionless groups $\bar{h}_c D/k$, $u_\infty D\rho/\mu$, $c_p\mu/k$, and $\bar{h}_c/c_p G$ for water, ethyl alcohol, mercury, hydrogen, air, and saturated steam over as wide a temperature range as possible and plot the results vs. temperature. For the purpose of these calculations let $D = 1$ ft, $V = 1$ ft/sec, and $\bar{h}_c = 1$ Btu/hr sq ft F.

6-6. Replot the data points of Fig. 6-6 on a log-log paper and find an equation approximating the best correlation line. Compare your results with Fig. 6-7. Then, suppose steam at 1 atm and 212 F is flowing across a 2-in.-OD pipe at a velocity of 20 fps. Using the data in Fig. 6-7 estimate the Nusselt number, the heat-transfer coefficient, and the rate of heat transfer per ft length of pipe if the pipe is at 400 F.

6-7. The average Reynolds number for air passing in turbulent flow over a 60-in.-long flat plate is 2.4×10^6. Under these conditions the average Nusselt number was found to be equal to 4150. Determine the average heat-transfer coefficient for an oil having a thermal conductivity of 0.11 Btu/hr sq ft F, a specific heat of 0.8 Btu/lb, and a viscosity of 22 lb/ft hr at the same Reynolds number in flow over the same plate.

6-8. A long solid cylinder of radius r_o, initially at a uniform temperature T_o, is suddenly immersed in a fluid at temperature T_∞ with a unit-surface conductance $\bar{h}$. Show by means of dimensional analysis that the temperature distribution can be expressed in terms of the following four parameters:

$$[T(r) - T_\infty]/(T_o - T_\infty), \quad r/r_o, \quad a\theta/r_o^2, \quad \bar{h}r_o/k.$$

6-9. The dimensionless ratio $V/\sqrt{Lg}$, called Froude number, is a measure of similarity between the shapes of the waves produced by a ship model and by its prototype. A 500 ft long cargo ship is designed to run at 20 knots, and a 5 ft. geometrically similar model is towed in a water channel to study the wave resistance. What should be the towing speed?

6-10. The torque due to the frictional resistance of the oil film between a rotating shaft and its bearing is found to be dependent on the force F normal to the shaft, the speed of rotation N of the shaft, the dynamic viscosity μ of the oil, and the shaft diameter D. Establish a correlation between the variables by using dimensional analysis. *Ans.* $[T/(F^3/N\mu)]^{1/2} = \phi(N\mu D^2/F)$

6-11. When a sphere falls freely through a homogeneous fluid, it reaches a terminal velocity at which the weight of the sphere is balanced by the buoyant force and the frictional resistance of the fluid. Make a dimensional analysis of this problem and indicate how experimental data for this problem could be correlated. Neglect compressibility effects and the influence of surface roughness.

6-12. Experiments have been performed on the temperature distribution in a homogeneous long cylinder (0.40 ft diameter, thermal conductivity of 0.12 Btu/hr ft F) with uniform internal heat generation. By dimensional analysis determine the relation between the steady-state temperature at the center of the cylinder T_c, the diameter, the thermal conductivity, and the rate of heat generation. Take the temperature at the surface as your datum. What is the equation for the center temperature if the difference between center and surface temperature is 80 F when the heat generation rate is 960 Btu/hr cu ft?

6-13. Experimental data for the transient cooling of a thick slab are to be correlated by dimensional analysis. The temperature of the slab is originally uniform at T_o. At time $\theta = 0$, the temperature at face $x = 0$ is suddenly lowered to T_s. Thermocouples are imbedded at various depths. Determine dimensionless groups relating T_x, the temperature at x, to the cooling time θ.

6-14. The convection equations relating the Nusselt, Reynolds, and Prandtl numbers can be rearranged to show that for gases the heat-transfer coefficient $\bar{h}_x$ depends on the absolute temperature T and the group $\sqrt{u_\infty/x}$. This formulation is of the form $\bar{h}_x = CT^n \sqrt{u_\infty/x}$, where n and C are constants. Indicate clearly how such a relationship could be obtained for the laminar flow case from $\mathrm{Nu}_x = 0.332\ \mathrm{Re}_x^{0.5}\ \mathrm{Pr}^{0.333}$ for the condition $0.5 < \mathrm{Pr} < 5.0$. State restrictions on method if any such restrictions are necessary.

6-15. Experimental pressure-drop data obtained in a series of tests in which water was heated while flowing through an electrically heated tube of 0.527 in. ID, 38.6 in. long, are tabulated below.

Mass Flow Rate m (lb/sec)	Fluid Bulk Temperature T_b (F)	Surface Temperature T_s (F)	Pressure Drop with Heat Transfer Δp_{ht} (psi)
3.04	90	126	9.56
2.16	114	202	4.74
1.82	97	219	3.22
3.06	99	248	8.34
2.15	107	283	4.45

Isothermal pressure-drop date for the same tube are given in terms of the dimensionless friction coefficient $f = (\Delta p/\rho V^2)(D/2L)g_c$ and the Reynolds number based on the pipe diameter, $\mathrm{Re}_D = VD/\nu$ below.

Re_D	1.71×10^5	1.05×10^5	1.9×10^5	2.41×10^5
f	0.00472	0.00513	0.00463	0.00445

By comparing the isothermal with the nonisothermal friction coefficients at similar bulk Reynolds numbers, derive a dimensionless equation for the nonisothermal friction coefficients in the form

$$f = \text{constant} \times \mathrm{Re}_D{}^n(\mu_s/\mu_b)^m$$

where μ_s = viscosity at surface temperature;
μ_b = viscosity at bulk temperature;
n and m = empirical constants.

6-16. Tabulated below are some experimental data obtained by heating n-butyl alcohol at a bulk temperature of 60 F while flowing over a heated flat plate (1 ft long, 3 ft wide, surface temperature of 140 F). Correlate the experimental data by appropriate dimensionless numbers and compare the line which best fits the data with Eq. 6-30.

Velocity (fps)	0.26	1.0	1.6	3.74
Unit-Surface Conductance (Btu/hr sq ft F)	11.4	23	34.6	69

6-17. Tabulated below are reduced test data from measurements made to determine the heat-transfer coefficient inside tubes at Reynolds numbers only slightly above transition and at relatively high Prandtl numbers (as associated with oils). Tests were made in a double-tube exchanger with a counterflow of water to provide the cooling. The pipe used to carry the oils was $\frac{5}{8}$-in. OD, 18 BWG, 121 in. long. Correlate the data in terms of appropriate dimensionless parameters.

Test No.	Fluid	$\bar{h}_c$	ρV	c_p	k	μ_b	μ_f
11	10C oil	87.0	1,072,000	0.471	0.0779	13.7	19.5
19	10C oil	128.2	1,504,000	0.472	0.0779	13.3	19.1
21	10C oil	264.8	2,460,000	0.486	0.0776	9.60	14.0
23	10C oil	143.8	1,071,000	0.495	0.0773	7.42	9.95
24	10C oil	166.5	2,950,000	0.453	0.0784	23.9	27.3
25	10C oil	136.3	1,037,000	0.496	0.0773	7.27	11.7
36	1488 pyranol	140.7	1,795,000	0.260	0.0736	12.1	16.9
39	1488 pyranol	133.8	2,840,000	0.260	0.0740	23.0	29.2
45	1488 pyranol	181.4	1,985,000	0.260	0.0735	10.3	12.9
48	1488 pyranol	126.4	3,835,000	0.260	0.0743	40.2	53.5
49	1488 pyranol	105.8	3,235,000	0.260	0.0743	39.7	45.7

where $\bar{h}_c$ = mean surface heat-transfer coefficient, based on the mean temperature difference, Btu/hr sq ft F;
ρV = mass velocity, lb_m/hr sq ft;
c_p = specific heat, Btu/lb_m F;
k = thermal conductivity, Btu/hr ft F (based on average bulk temperature);
μ_b = viscosity, based on average bulk (mixed mean) temperature, lb_m/hr ft;
μ_f = viscosity, based on average film temperature, lb_m/hr ft.

Hint: Start by correlating $\overline{Nu}$ and Re_D irrespective of the Prandtl numbers, since the influence of the Prandtl number on the Nusselt number is expected to be relatively small. By plotting $\overline{Nu}$ vs. Re on log-log paper, one can guess the nature of the correlation equation, $\overline{Nu} = f_1$ (Re). A plot of $\overline{Nu}/f_1$(Re) vs. Pr will then reveal the dependence upon Pr. For the final equation, the influence of the viscosity variation should also be considered.

One possible answer: $$\overline{Nu}_D = 0.0067 \frac{\rho V D}{\mu_b} \left(\frac{c_p \mu_b}{k_b}\right)^{0.2} \left(\frac{\mu_b}{\mu_f}\right)^{0.3}$$

6-18. The average friction coefficient for flow over a 2-ft-long plate is 0.01. What is the value of the drag force in lb_f per foot width of the plate for the following fluids: (a) air at 60 F, (b) steam at 212F and 15 psia, (c) water at 100 F, (d) mercury at 200 F, and (e) ethyl alcohol at 212F?

6-19. The average Nusselt number for flow over a 2-ft-long plate is 100. What is the value of the average surface conductance for the following fluids: (a) air at 60 F, (b) steam at 212 F and 15 psia, (c) water at 100 F, and (d) mercury at 200 F, and (e) ethyl alcohol at 212 F.

6-20. Plot the velocity and temperature distributions in the laminar boundary layer for air at 60 F flowing over a flat plate at $\mathrm{Re}_x = 10^4$ if the free-stream velocity is 1.0 fps and the surface temperature is 160 F using (a) the Blasius solution, (b) an assumed straight line, and (c) a cubic parabola.

6-21. Hydrogen at 60 F and at a pressure of 1 atm is flowing along a flat plate at a velocity of 10 fps. If the plate is 1 ft wide and at 160 F, calculate the following quantities at $x = 1$ ft and at the distance corresponding to the transition point, i.e., $\mathrm{Re}_x = 5 \times 10^5$. (Take properties at 110 F.)

a) Hydrodynamic boundary layer thickness, in inches.
b) Local friction coefficient, dimensionless.
c) Average friction coefficient, dimensionless.
d) Drag force, in $\mathrm{lb_f}$.
e) Thickness of thermal boundary layer, in inches.
f) Local convective-heat-transfer coefficient, in Btu/hr sq ft. F.
g) Average convective-heat-transfer coefficient, in Btu/hr sq ft F.
h) Rate of heat transfer, in Btu/hr.

6-22. Repeat Prob. 6-21 for $x = 10$ ft and $u_\infty = 200$ fps, (a) taking the laminar boundary layer into account and (b) assuming that the turbulent boundary layer starts at the leading edge.

6-23. Determine the rate of heat loss in Btu/hr from the wall of a building in a 10-mph wind blowing parallel to its surface. The wall is 80 ft long, 20 ft high, its surface temperature is 80 F, and the temperature of the ambient air is 40 F.

6-24. Plot the local heat-transfer coefficient as a function of length for air at 1000 F flowing over a 5-ft-long flat plate at 3000 F with a velocity of 100 fps.

6-25. A spacecraft heat exchanger is to operate in a nitrogen atmosphere at a pressure of about 1.5 psia and 100 F. For a flat-plate heat exchanger designed to operate on earth in a standard atmosphere of 14.7 psi and 100 F in turbulent flow, estimate the ratio of heat-transfer coefficients on the earth to that in nitrogen assuming forced circulation cooling of the plate surface at the same velocity in both cases.

6-26. A thin flat plate 6 in. square is suspended from a balance into a uniformly flowing stream of glycerin in such a way that the glycerin flows parallel to and along the top and bottom surfaces of the plate. The total drag on the plate is measured and found to be 13 $\mathrm{lb_f}$. If the glycerin flows at the rate of 50 fps and is at a temperature of 112 F, what is the heat-transfer coefficient $\bar{h}_c$ in Btu/hr sq ft F?

6-27. Mercury at 60 F flows over and parallel to a flat surface at a velocity of 10 fps. Calculate the thickness of the hydrodynamic boundary layer at a distance 12 in. from the leading edge of the surface.

6-28. A thin flat plate 6 in. square is tested for drag in a wind tunnel with air at 100 fps, 14.7 psia, and 60 F flowing across and parallel to the top and bottom surfaces. The observed total drag force is 0.0150 lb. Calculate the rate of heat transfer from this plate when the surface temperature is maintained at 250 F. Neglect radiation. *Ans.* 370 Btu/hr

6-29. Mercury at 60 F flows parallel to the short side of a thin flat smooth plate with a velocity of 1 ft/sec. The plate is 6 in. wide and 1 ft long and its surface temperature is 160 F. Find:

a) the local friction coefficient at the middle point of the plate, and the total drag force on the plate
b) the temperature of the mercury at a point 4 in. from the leading edge and 0.05 in. from the surface of the plate
c) the Nusselt number at the end of the plate.

6-30. Water at a velocity of 8 ft/sec flows parallel to the surface of a 3-ft-long horizontal, smooth and thin flat plate. Determine the local thermal and hydrodynamic boundary-layer thicknesses, and the local friction coefficient, at the midpoint of the plate. What is the rate of heat transfer from the plate to the water per ft width of the plate, if the surface temperature is kept uniformly at 300 F, and the temperature of the main water stream is 60 F?

6-31. A thin flat plate is placed in an atmospheric pressure air stream flowing parallel to it at a velocity of 15 ft/sec. The temperature at the surface of the plate is maintained uniformly at 400 F, and that of the main air stream is 70 F. Calculate the temperature and horizontal velocity at a point 1 ft from the leading edge and 0.03 in. above the surface of the plate. *Ans.* 340.5 F

6-32. Find the horizontal component of the force required to hold in position in the air stream the plate of Problem 31. Consider only 1 ft length of the plate measured along the leading edge, and assume the plate to be 2 ft wide.

6-33. The surface temperature of a thin flat plate located parallel to an air stream is 196 F. The free stream velocity is 200 ft/sec and the temperature of the air is 32 F. The plate is 24 in. wide and 18 in. long in the direction of the air stream. Neglecting the end effect of the plate and assuming that the flow in the boundary layer changes abruptly from laminar to turbulent at a transition Reynolds number of $N_{\mathrm{Re}_{tr}} = 4.10^5$, find:

a) the average heat transfer coefficient in the laminar and turbulent regions
b) the rate of heat transfer for the entire plate, considering both sides
c) the average friction coefficient in the laminar and turbulent regions
d) the total drag force.

Also, plot the heat transfer coefficient and local friction coefficient as a function of the distance from the leading edge of the plate.
[*Ans.* a) 15.9, 50.4 Btu/ft^2 hr F; b) 28,340 Btu/hr; c) 0.0021, 0.0041; d) 0.934 lb$_f$]

6-34. The thickness of the laminar sublayer has been estimated to be given by $y\sqrt{\tau_s/\rho}/\nu = 5.0$. Compare this estimate with the experimental data shown in Fig. 6-20.

6-35. The boundary-layer-displacement thickness δ^* is defined as the distance by which a plane surface, past which a fluid is flowing, would have to be shifted into the stream to obtain the same flow rate with an inviscid fluid as with the real fluid. Mathematically δ^* is defined by the equation

$$\delta^* = \int_0^\infty \left(1 - \frac{u}{u_\infty}\right) dy$$

Show that $\delta^* \simeq \delta/3$ for laminar flow past a flat plate.

6-36. Assuming a linear velocity distribution and a linear temperature distribution in the boundary layer over a flat plate, derive a relation between the thermal and hydrodynamic boundary-layer thicknesses and the Prandtl number.

6-37. Derive the integral momentum boundary-layer equation for steady incompressible two-dimensional flow over a flat porous wall through which fluid is injected with a velocity v_o normal to the surface.

6-38. Assuming a velocity distribution of the type $u = a + by + cy^2$, derive by means of the integral method for flow over a flat plate a relation between the boundary-layer thickness and the Reynolds number.

6-39. Show that the energy equation (Eq. 6-21) can be expressed in the form

$$\rho u c_p \frac{\partial T}{\partial x} + \rho v c_p \frac{\partial T}{\partial y} = \frac{u}{g_c}\frac{\partial p}{\partial x} + k\frac{\partial^2 T}{\partial y^2} + \frac{\mu}{g_c}\left(\frac{\partial u}{\partial y}\right)^2$$

Hint: Multiply Eq. 6-16 by u and subtract the resulting expression from Eq. 6-21

6-40. A fluid at temperature T_∞ is flowing at a velocity u_∞ over a flat plate which is at the same temperature as the fluid for a distance x_o from the leading edge, but at a temperature T_s beyond this point. Show by means of the integral boundary-layer equations that ζ, the ratio of the thermal boundary-layer thickness to the hydrodynamic boundary-layer thickness, over the heated portion of the plate is approximately

$$\zeta \simeq \mathrm{Pr}^{-1/3}\left[1 - \left(\frac{x_o}{x}\right)^{3/4}\right]^{1/3}$$

if the flow is laminar.

Hint: Assume that the temperature distribution is a cubic parabola and use T_s as your datum to simplify the boundary conditions, i.e., let

$$T - T_s = ay + cy^3$$

Also, inspect each equation and drop those terms which are small in comparison with others. Show also that, for the partially heated plate, the Nusselt number at x, if $x > x_o$, is approximately

$$\mathrm{Nu}_x \simeq 0.33 \left(\frac{\mathrm{Pr}}{1 - (x_o/x)^{3/4}} \right)^{1/3} \mathrm{Re}_x^{\,1/2}$$

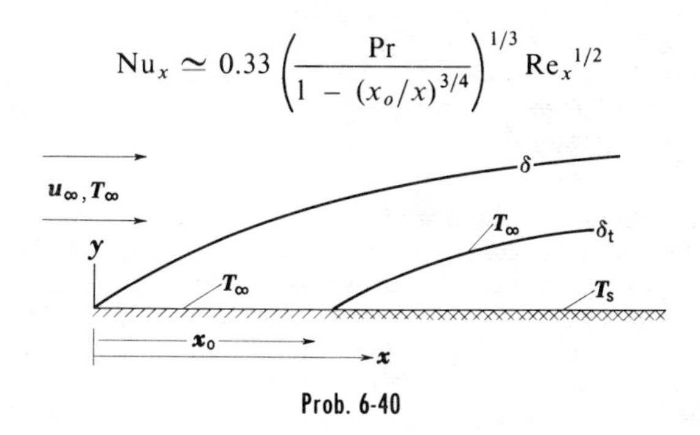

Prob. 6-40

6-41. A refrigeration truck is traveling at 50 mph on a desert highway where the air temperature is 140 F. The body of the truck may be idealized as a rectangular box, 10 ft wide, 7 ft high, and 20 ft long, at a surface temperature of 50 F. Assume that the heat transfer from the front and back of the truck may be neglected, that the stream does not separate from the surface, and that the boundary layer is turbulent over the whole surface. If, for every 12,000 Btu/hr of heat loss one ton capacity of the refrigerating unit is necessary, calculate the required tonnage of the refrigeration unit.

6-42. Wind at 50 mph is blowing in a direction parallel to the short side of the flat roof of a ranch house. The temperature at the surface of the roof is 35 F and the air temperature is −5 F. The roof measures 30 ft by 60 ft. Neglecting the end effect and assuming no separation of the air stream from the roof, calculate the heat loss through the roof.

6-43. The wing of an airplane has a polished chromium skin. At a 10,000-ft altitude it receives 200 Btu/hr sq ft by solar radiation. Assuming that the interior surface of the wing's skin is well insulated and the wing has a chord of 20-ft length, i.e., $L \simeq 20$ ft, estimate the equilibrium temperature of the wing at a flight speed of 500 fps.

6-44. A cooling fin for a heat exchanger, situated parallel to an atmospheric pressure air stream, measures 3 in. along the leading edge and 18 in. in the flow direction. Its base temperature is 190 F, and the air is at 50 F. The velocity of the air is 90 ft/sec. Determine the total drag force and the total rate of heat transfer from the fin to the air.

6-45. A 1-in.-diam, 6-in.-long transite rod (k = 0.56 Btu/hr ft F, ρ = 100 lb/cu ft, c = 0.20 Btu/lb F) on the end of a 1-in.-diam wood rod at a uniform temperature of 212 F is suddenly placed into a 60 F, 100 ft/sec air stream flowing parallel to the axis of the rod. Estimate the center line temperature of the transite rod 8 min after cooling starts. Assume radial heat conduction, but include radiation losses, based on an emissivity of 0.90, to black surroundings at air temperature.

6-46. Air at 100 fps flows between two parallel flat plates spaced 2 in. apart. Estimate the distance from the entrance where the boundary layers meet.

6-47. For Prob. 6-13 estimate the frictional pressure drop in the entrance section, taking into account the pressure drop due to the frictional drag as well as the pressure drop due to the momentum change. Assume that the velocity at the inlet is uniform and that the velocity profiles of both boundary layers can be approximated by cubic parabolas.

6-48. A preliminary design study for a nuclear moon-rocket reactor is to be made. The reactor under consideration consists of a stack of parallel flat plates which are 2 ft square, 2 in. apart, and heated to 3000 F. Gaseous hydrogen at 20-atm pressure and a temperature of 0 F enters at one end at a velocity of 200 fps and is heated as it passes between the plates. (a) Calculate the average heat-transfer coefficient assuming that transition occurs at a Reynolds number of 5×10^5. (b) Estimate the average heat-transfer coefficient assuming that a turbulence screen is placed at the entrance so that the entire boundary layer is turbulent. (c) For the turbulent flow conditions, determine the temperature of the hydrogen at the exit assuming that the surface temperature of the plate is uniformly at 3000 F. State all your assumptions clearly. (d) If the plates are $\frac{1}{10}$-in. thick and have a thermal conductivity of 10 Btu/hr ft F, determine the required rate of heat generation within the plate and the maximum temperature at the center of the plate assuming that uniform heat generation by nuclear fission occurs within the plates. (e) How many times would the hydrogen have to pass between the two plates in order to heat it to an average temperature of 2000 F?

6-49. A long plate at a uniform temperature T_1 moves at a velocity V_1 relative to a stationary insulated plate placed parallel to the moving plate at a distance h. (a) Show that the temperature of the insulated surface will exceed the temperature of the moving plate by Pr $(V_1^2/2c_p)$ if the space between the plates is filled by an ideal gas and the velocity profile of the gas is linear, and (b) derive an equation for the rate of heat transfer to each of the plates.

6-50. A flat plate is exposed to air with a temperature of 0 F, a pressure of 5 psia, and a velocity parallel to the plate of 2000 fps. How long is the laminar boundary layer, and what is the adiabatic wall temperature in the laminar region?

6-51. Air at a static temperature of 70 F and a static pressure of 0.1 psia flows at zero angle of attack over a thin electrically heated flat plate at a velocity of 800 fps. If the plate is 4-in. long in the direction of flow and 24 in. in the direction normal to the flow, determine the rate of electrical heat dissipation necessary to maintain the plate at an average temperature of 130 F.

6-52. Air at 50 F and 14.7 psia flows at 800 fps over a thermally nonconducting flat plate. What is the plate temperature 10 ft downstream from the leading edge? How much does this temperature differ from that which exists 1 in. from the leading edge?

6-53. Air at 65 F and 0.1 psia flows over a thin flat strip of metal, 1 in. long in the direction of flow, at a velocity of 900 fps. Determine (a) the surface temperature of the plate at equilibrium and (b) the rate of heat removal required per foot length if the surface temperature is to be maintained at 100 F.

6-54. An airplane model wing can be idealized as a flat plate, 2 ft long in the direction of flow and 3 ft wide. The wing is placed in an air flow at $M_\infty = 3.0$, $p_\infty = 0.05$ atm, and $T_\infty = 420$ R. Determine the temperature of this wing at a distance of 0.5 and 1.5 ft from the leading edge, if no cooling were provided, and estimate the rate at which heat must be removed from the surface of the wing to maintain its temperature at 100 F. *Ans.* 1056 and 1093 R, 5×10^4 Btu/hr

REFERENCES

1. H. Schlichting, *Boundary Layer Theory,* 6th ed. (translated by J. Kestin). (New York: McGraw-Hill Book Company, Inc., 1968.)
2. H. L. Langhaar, *Dimensional Analysis and Theory of Models.* (New York: John Wiley & Sons, Inc., 1951.)
3. E. R. Van Driest, "On Dimensional Analysis and the Presentation of Data in Fluid Flow Problems," *J. Appl. Mech.*, Vol. 13 (1940), p. A-34.
4. A. P. Colburn, "A Method of Correlating Forced Convection Heat Transfer Data and a Comparison with Fluid Friction," *Trans. Am. Inst. Chem. Engrs.*, Vol. 29 (1933), pp. 174–210.
5. W. J. King, "The Basic Laws and Data of Heat Transmission," *Mech. Eng.*, Vol. 54 (1932), pp. 410–415.
6. M. Blasius, "Grenzschichten in Flüssigkeiten mit Kleiner Reibung," *Z. Math. u. Phys.*, Vol. 56, No. 1 (1908).
7. E. Pohlhausen, "Der Wärmeaustausch zwischen festen Körpern und Flüssigkeiten mit kleiner Reibung und kleiner Wärmeleitung," *ZAMM*, Vol. 1 (1921), p. 115.
8. T. von Karman, "Über laminare und turbulente Reibung," (translation) *NACA TM* 1092, 1946.
9. E. R. G. Eckert and R. M. Drake, *Heat and Mass Transfer*, 2nd ed. (New York: McGraw-Hill Book Company, Inc., 1959.)
10. L. Prandtl, "Über die ausgebildete Turbulenz," *ZAMM*, Vol. 5 (1925), p. 136; *Proc. 2nd Int. Cong. of Appl. Mech.*, Zurich (1926).
11. S. E. Isakoff and T. B. Drew, "Heat and Momentum Transfer in Turbulent Flow of Mercury," Inst. Mech. Eng. and ASME, *Proc. General Discussion on Heat Transfer* (1951), pp. 405–409.
12. W. Forstall, Jr. and A. H. Shapiro, "Momentum and Mass Transfer in Co-axial Gas Jets," *J. Appl. Mech.*, Vol. 17 (1950), p. 399.
13. M. Hansen, "Velocity Distribution in the Boundary Layer of a Submerged Plate," *NACA TM* 585, 1930.
14. Van der Hegge-Zynen, "Measurements of the Velocity Distribution in the Boundary Layer along a Plane Surface," *Thesis*, Delft, 1924. (Delft: I. Waltman, 1924.)
15. F. Schultz-Grunow, "A New Resistance Law for Smooth Plates," *Luftfahrt Forsch.*, Vol. 17 (1940), pp. 239–246: (translation) *NACA TM* 986, 1941.
16. L. Prandtl, "Bemerkungen über den Wärmeübergang im Rohr," *Phys. Zeit.*, Vol. 29 (1928), p. 487.
17. L. Prandtl, "Eine Beziehung zwischen Wärmeaustauch and Ströhmungswiederstand der Flüssigkeiten," *Phys. Zeit.*, Vol. 10 (1910), p. 1072.
18. T. von Karman, "The Analogy between Fluid Friction and Heat Transfer," *Trans. ASME*, Vol. 61 (1939), pp. 705–711.
19. R. G. Deissler, "Investigation of Turbulent Flow and Heat Transfer in Smooth Tubes Including the Effects of Variable Properties," *Trans. ASME*, Vol. 73 (1951), pp. 101–107.

20. R. C. Martinelli, "Heat Transfer to Molten Metals," *Trans. ASME*, Vol. 69 (1947), pp. 947–959.

21. J. M. Coulson and J. V. Richardson, *Chemical Engineering*, Vol. I. (New York: McGraw-Hill Book Company, Inc., 1954).

22. K. Goldmann, "Heat Transfer to Supercritical Water and Other Fluids with Temperature Dependent Properties," *Chem. Eng. Prog. Symp. Series Nuclear Eng.*, Part 1, Vol. 50, No. 11 (1954), pp. 105–110.

23. J. G. Knudsen and D. L. Katz, "Fluid Dynamics and Heat Transfer," *Eng. Res. Bull.* 37, (Ann Arbor: Univ. of Michigan, 1953).

24. A. H. Davis, "Convective Cooling of Wires in Streams of Viscous Liquids," *Phil. Mag.*, Vol. 47 (1924), pp. 1057–1091.

25. E. L. Diret, W. James, and M. Stracy, "Heat Transmission from Fine Wires to Water," *Ind. Eng. Chem.*, Vol. 39 (1947), pp. 1098–1103.

26. R. Hilpert, "Wärmeabgabe von geheizten Drähten und Rohren," *Forsch. Gebiete Ingenieurw.*, Vol. 4 (1933), pp. 215–224.

27. D. Coles, "The Law of the Wake in the Turbulent Boundary Layer," *J. Fluid Mech.*, Vol. 1, Part 2 (1956), pp. 191–225.

28. E. R. Van Driest, "Calculation of the Stability of the Laminar Boundary Layer in a Compressible Fluid on a Flat Plate with Heat Transfer," *J. Aero. Sci.*, Vol. 19 (1952), pp. 801–813.

29. A. H. Shapiro, *The Dynamics and Thermodynamics of Compressible Fluid Flow*, Vol. 1. (New York: The Ronald Press Co., 1954).

30. S. V. Pantakar and D. B. Spalding, *Heat and Mass Transfer in Boundary Layers*, 2nd ed. London: International Textbook Co., 1970).

31. J. Kaye, "Survey of Friction Coefficients, Recovery Factors, and Heat Transfer Coefficients for Supersonic Flow," *J. Aeronautical Sci.*, Vol. 21, No. 2 (1954), pp. 117–129.

32. E. R. A. Eckert, "Engineering Relations for Heat Transfer and Friction in High-Velocity Laminar and Turbulent Boundary Layer Flow over Surface with Constant Pressure and Temperature," *Trans. ASME*, Vol. 78 (1956), pp. 1273–1284.

33. E. R. Van Driest, "Turbulent Boundary Layer in Compressible Fluids," *J. Aeronautical Sci.*, Vol. 18, No. 3 (1951), pp. 145–161.

34. A. K. Oppenheim, "Generalized Theory of Convective Heat Transfer in a Free-Molecule Flow," *J. of the Aero. Sci.*, Vol. 20 (1953), pp. 49–57.

35. W. D. Hayes and R. F. Probstein, *Hypersonic Flow Theory*. (New York, N.Y.: Academic Press, Inc., 1959.)

36. B. Gebhart, Heat Transfer (New York, N.Y.: McGraw-Hill Book Company, Inc., 2d ed., 1971.)

7
Free convection

7-1. Introduction

Free-convection heat transfer occurs whenever a body is placed in a fluid at a higher or a lower temperature than that of the body. As a result of the temperature difference, heat flows between the fluid and the body and causes a change in the density of the fluid layers in the vicinity of the surface. The difference in density leads to downward flow of the heavier fluid and upward flow of the lighter. If the motion of the fluid is caused solely by differences in density resulting from temperature gradients, without the aid of a pump or a fan, the associated heat-transfer mechanism is called *natural* or *free convection.* Free-convection currents transfer internal energy stored in the fluid in essentially the same manner as forced-convection current. However, the intensity of the mixing motion is generally less in free convection, and consequently the heat-transfer coefficients are lower than in forced convection.

Although free-convection heat-transfer coefficients are relatively low, many devices depend largely on this mode of heat transfer for cooling. In the electrical-engineering field, transmission lines, transformers, rectifiers, and electrically heated wires such as the filament of an incandescent lamp or the heating elements of an electric furnace are cooled by free convection. As a result of the heat generated internally, the temperature of these bodies rises above that of the surroundings. As the temperature difference increases, the rate of heat flow also increases until a state of equilibrium is reached where the rate of heat generation is equal to the rate of heat dissipation.

Free convection is the dominant heat-flow mechanism from steam radiators, walls of a building, or the stationary human body in a quiescent atmosphere. The determination of the heat load on air-conditioning or refrigeration equipment requires, therefore, a knowledge of free-convection heat-transfer coefficients. Free convection is also responsible for heat losses from pipes carrying steam or other heated fluids. Recently natural convection has been proposed in nuclear-power applications to cool the surfaces of bodies in which heat is generated by fission (1).

In all of the aforementioned examples the body force responsible for the convection currents is the gravitational attraction. Gravity, however, is not the only body force which can produce free convection. In certain aircraft applications there are components such as the blades of gas turbines and helicopter ramjets which rotate at high speeds. Associated with these rotative speeds are large centrifugal forces whose magnitudes, like the gravitational force, are also proportional to the fluid density and hence can generate strong free-convection currents. Cooling of rotating components by free convection is therefore feasible even at high heat fluxes.

The fluid velocities in free-convection currents, especially those generated by gravity, are generally low, but the characteristics of the flow in the vicinity of the heat-transfer surface are similar to those in forced convection. A boundary layer forms near the surface and the fluid velocity at the interface is zero. Figure 7-1 shows the velocity and temperature distributions near a heated flat plate placed in a vertical position in air (3). At a given distance from the bottom of the plate, the local upward velocity increases with increasing distance from the surface to reach a maximum value at a distance between 0.1 and 0.2 in., then decreases and approaches zero again about 1 to 2 in. from the surface. Although the velocity profile is different from that observed in forced convection over a flat plate where the velocity approaches the free-stream velocity asymptotically, in the vicinity of the surface the characteristics of both types of boundary layer are similar. In free convection, as in forced convection, the flow may be laminar or turbulent, depending on the distance from the leading edge, the fluid properties, the body force, and the temperature difference between the surface and the fluid.

The temperature field in free convection (Fig. 7-1) is similar to that observed in forced convection. Hence, the physical interpretation of the Nusselt number presented in Sec. 6-4 applies. For practical application however, Newton's equation (Eq. 1-13)

$$dq = h_c dA\,(T_s - T_\infty)$$

is generally used. The reason for writing the equation for a differential area dA is that, in free convection, the heat-transfer coefficient h_c is not

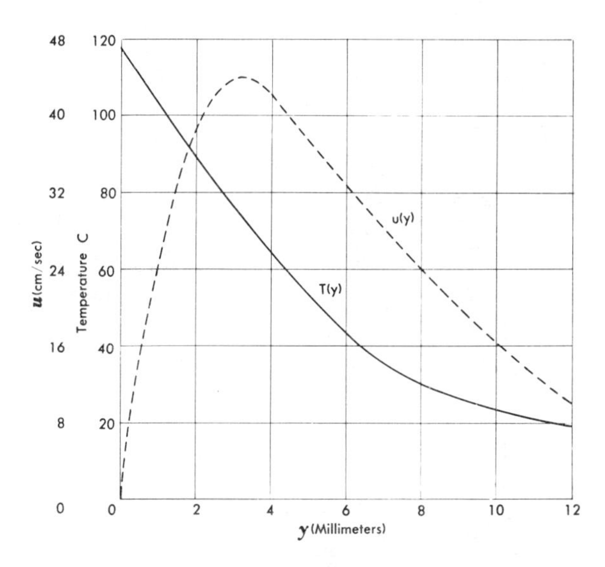

Fig. 7-1. Temperature and velocity distributions in the vicinity of a heated flat plate placed vertically in still air. (After E. Schmidt and W. Beckman, Ref. 3).

uniform over a surface. As in forced convection over a flat plate, we shall therefore distinguish between a local value of h_c and an average value $\bar{h}_c$ obtained by averaging h_c over the entire surface. The temperature T_∞ refers to a point in the fluid sufficiently removed from the body that the temperature of the fluid is not affected by the presence of a heating (or cooling) source.

An exact evaluation of the heat-transfer coefficient for free convection from the boundary layer is very difficult. The problem has only been solved for simple geometries, such as a vertical flat plate and a horizontal cylinder (3,4,19). We shall not discuss these specialized solutions here. Instead, we shall set up the differential equations for free convection from a vertical flat plate using only fundamental physical principles. From these equations, without actually solving them, we shall determine the similarity conditions and associated dimensionless moduli which correlate experimental data. In Sec. 7-3 pertinent experimental data for various shapes of practical interest will be presented in terms of these dimensionless moduli, and their physical significance will be discussed.

7-2. Similarity parameters for free convection

In the analysis of free convection we shall make use of a phenomenon observed by the Greeks over 2000 years ago and phrased by Archi-

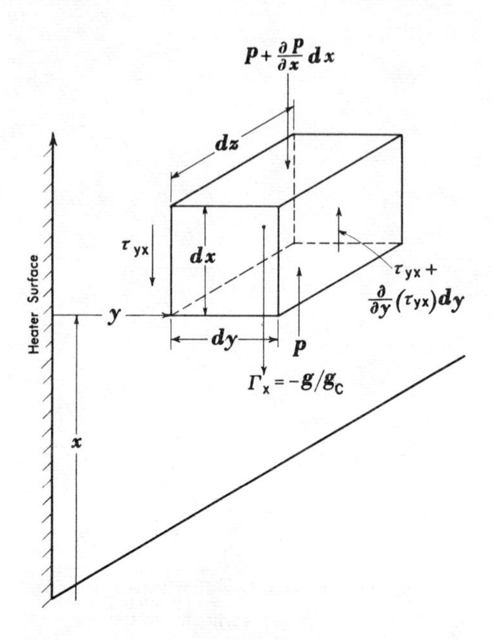

Fig. 7-2. Sketch illustrating forces acting on a fluid element in free-convection flow.

medes somewhat as follows: A body immersed in a fluid experiences a buoyant or lifting force equal to the mass of the displaced fluid. Hence, a submerged body rises when its density is less than that of the surrounding fluid and sinks when its density is greater. The buoyant effect is the driving force in free convection.

For the purpose of analysis, consider a domestic heating panel which can be idealized by a vertical flat plate, very long and wide in the plane perpendicular to the floor so that the flow is two-dimensional (Fig. 7-2). When the heater is turned off, the panel is at the same temperature as the surrounding air. The gravitational or body force acting on each fluid element is in equilibrium with the hydrostatic pressure gradient, and the air is motionless. When the heater is turned on, the fluid in the vicinity of the panel will be heated and its density will decrease. Hence, the body force (defined as the force per unit mass) on a unit volume in the heated portion of the fluid is less than in the unheated fluid. This imbalance causes the heated fluid to rise, a phenomenon which is well-known from experience. In addition to the buoyant force, there are pressure forces and also frictional forces acting when the air is in motion. Once steady-state conditions have been established, the total force on a volume element, $dxdydz$ in the positive x direction perpendicular to the floor consists of the following:

1. The force due to the pressure gradient

$$pdydz - \left(p + \frac{\partial p}{\partial x}\, dx\right) dydz = -\frac{\partial p}{\partial x}\,(dxdydz)$$

2. The body force $\Gamma_x \rho\,(dxdydz)$, where $\Gamma_x = -g/g_c$, since gravity alone is active.
3. The frictional shearing forces due to the velocity gradient

$$(-\tau_{yx})\, dxdz + \left(\tau_{yx} + \frac{\partial \tau_{yx}}{\partial y}\, dy\right) dxdz$$

Since $\tau_{yx} = \mu\,(\partial u/\partial y)/g_c$ in laminar flow, the net frictional force is

$$\left(\frac{\mu}{g_c}\,\frac{\partial^2 u}{\partial y^2}\right) dxdydz$$

Forces due to the deformation of the fluid element will be neglected in view of the low velocity.[1]

The rate of change of momentum of the fluid element is $\rho dxdydz\,[u\,(\partial u/\partial x) + v\,(\partial u/\partial y)]$ as shown in Sec. 6-7. Applying Newton's second law to the elemental volume yields

$$\rho\left(u\,\frac{\partial u}{\partial x} + v\,\frac{\partial u}{\partial y}\right) = -g_c\,\frac{\partial p}{\partial x} - \rho g + \mu\,\frac{\partial^2 u}{\partial y^2} \tag{7-1}$$

after canceling $dxdydz$. The unheated fluid far removed from the plate is in hydrostatic equilibrium, or $g_c(\partial p_e/\partial x) = -\rho_e g$ where the subscript e denotes equilibrium conditions. At any elevation the pressure is uniform and therefore $\partial p/\partial x = \partial p_e/\partial x$. Substituting $\rho_e g$ for $-(\partial p/\partial x)$ in Eq. 7-1 gives

$$\rho\left(u\,\frac{\partial u}{\partial x} + v\,\frac{\partial u}{\partial y}\right) = (\rho_e - \rho)g + \mu\,\frac{\partial^2 u}{\partial y^2} \tag{7-2}$$

A further simplification can be made by assuming that the density ρ depends only on the temperature, and not on the pressure. For an incompressible fluid this is self-evident, but for a gas it implies that the vertical dimension of the body is small enough that the hydrostatic density ρ_e is

---

[1]The effects of the compression work and frictional heating are discussed in Reference 1.

constant. With these assumptions the buoyant term can be written

$$g(\rho_e - \rho) = g(\rho_\infty - \rho) = -g\rho\beta(T_\infty - T) \tag{7-3}$$

where β is the coefficient of thermal expansion, defined as

$$\beta = \frac{\rho_\infty - \rho}{\rho(T - T_\infty)} \tag{7-4}$$

For an *ideal gas* (i.e., $\rho = p/\mathcal{R}T$) the coefficient of expansion is

$$\beta = \frac{\rho_\infty/\rho - 1}{T - T_\infty} = \frac{T/T_\infty - 1}{T - T_\infty} = \frac{1}{T_\infty} \tag{7-5}$$

and

$$g\rho\beta(T_\infty - T) = -g\rho\left(\frac{T}{T_\infty} - 1\right)$$

The equation of motion for free convection is obtained finally by substituting the buoyant term as expressed by Eq. 7-3 into Eq. 7-2 and we get

$$\rho\left(u\frac{\partial u}{\partial x} + v\frac{\partial u}{\partial y}\right) = g\rho\beta(T - T_\infty) + \mu\frac{\partial^2 u}{\partial y^2} \tag{7-6}$$

Equation 7-6 is identical to the boundary-layer equation for forced convection over a flat plate except for the term $g\rho\beta(T - T_\infty)$, which appears as a result of the body force which was ignored in forced convection.

The problem now is to determine the conditions for which the velocity field in one free-convection system is similar to the velocity field in another. The boundary conditions are the same for all free-convection systems, that is the velocity is zero both at the surface and a distance far removed from the surface. Hence, dynamic similarity for different systems exists if Eq. 7-6 applies.

Let us first write Eq. 7-6 for system A as

$$\rho_A\left(u_A\frac{\partial u_A}{\partial x_A} + v_A\frac{\partial u_A}{\partial y_A}\right) = \rho_A g_A \beta_A(T - T_\infty)_A + \mu_A\frac{\partial^2 u_A}{\partial {y_A}^2} \tag{7-7}$$

Now consider another system, B, related to system A by the equations

$$\begin{aligned} u_B &= C_V u_A & \beta_B &= C_\beta \beta_A \\ v_B &= C_V v_A & (T - T_\infty)_B &= C_T(T - T_\infty)_A \\ x_B &= C_L x_A & \mu_B &= C_\mu \mu_A \\ y_B &= C_L y_A & \rho_B &= C_\rho \rho_A \\ g_B &= C_g g_A & & \end{aligned}$$

These equations state that a velocity in system B is equal to C_V, a velocity constant or reference quantity, times the velocity in system A; the viscosity in system B is equal to a constant C_μ times the viscosity in system A; etc. Equation 7-6 applies also to system B, or

$$\rho_B\left(u_B\frac{\partial u_B}{\partial x_B} + v_B\frac{\partial u_B}{\partial y_B}\right) = \rho_B g_B \beta_B (T - T_\infty)_B + \mu_B \frac{\partial^2 u_B}{\partial y_B^2} \tag{7-7a}$$

We can express the equation of motion for system B in terms of the quantities pertaining to system A by inserting the relations previously listed. Then Eq. 7-7a becomes

$$\frac{C_\rho C_V^2}{C_L}\left[\rho_A\left(u_A\frac{\partial u_A}{\partial x_A} + v_A\frac{\partial u_A}{\partial y_A}\right)\right]$$

$$= C_T C_\rho C_g C_\beta[\rho_A g_A \beta_A (T - T_\infty)_A] + \frac{C_\mu C_V}{C_L^2}\left[\mu_A\frac{\partial^2 u_A}{\partial y_A^2}\right] \tag{7-8}$$

The next step is crucial in this type of analysis and should be noted carefully. Equation 7-8, the equation of motion for system B, is identical to the equation of motion of system A if the coefficients of each of the terms in square brackets are identical. Then, the solutions of the equations of motion for both systems (the boundary conditions being similar) will be the same and the systems are said to be dynamically similar. Therefore, the dynamic similarity requirements are that

$$\frac{C_\rho C_V^2}{C_L} = C_T C_\rho C_g C_\beta = \frac{C_\mu C_V}{C_L^2} \tag{7-9}$$

To see the physical significance of Eq. 7-9, we substitute for the reference quantities (i.e., the C's), the equalities relating systems A and B in the tabulation (for instance $C_\beta = \beta_B/\beta_A$, $C_\mu = \mu_B/\mu_A$, etc.). To simplify the relationships we shall use the symbol V for the significant velocity and L for the significant length. Then we have

$$\frac{\rho_B V_B^2/L_B}{\rho_A V_A^2/L_A} = \frac{\rho_B g_B \beta_B (T - T_\infty)_B}{\rho_A g_A \beta_A (T - T_\infty)_A} = \frac{\mu_B V_B/L_B^2}{\mu_A V_A/L_A^2} \tag{7-10}$$

Any combination of terms in the above similarity equation is permissible, but only those combinations which have some physical significance are of practical use. However, it is not always obvious which of the many possibilities is most convenient and significant. Often a trial-and-error approach, with some experimental data as a guide, is required to find the right combination.

If we combine the first and the last term of Eq. 7-10 we get

$$\frac{\rho_B V_B L_B}{\mu_B} = \frac{\rho_A V_A L_A}{\mu_A} \tag{7-11}$$

which are equivalent expressions of the Reynolds number. The equality of the Reynolds numbers means that the ratios of inertia forces to frictional forces are identical at corresponding points.

Combining the second and the third term of Eq. 7-10 we obtain

$$\frac{\rho_B g_B \beta_B (T - T_\infty)_B L_B^2}{\mu_B V_B} = \frac{\rho_A g_A \beta_A (T - T_\infty)_A L_A^2}{\mu_A V_A} \tag{7-12}$$

that is, the ratios of buoyant to frictional forces are equal.

From the physical aspects of the problem we recall that the velocity of the fluid is not an independent quantity, but depends upon the buoyant driving force. Hence, we can eliminate V from Eq. 7-12 by substituting its value from the Reynolds number. We then obtain

$$\frac{\rho_B^2 g_B \beta_B (T - T_\infty)_B L_B^3}{\mu_B^2} = \frac{\rho_A^2 g_A \beta_A (T - T_\infty)_A L_A^3}{\mu_A^2} \tag{7-13}$$

The dimensionless modulus $\rho^2 g\beta(T - T_\infty)L^3/\mu^2$ is called the Grashof number, Gr, *and represents the ratio of buoyant to viscous forces.*[2] Consistent units are:

Engineering System				SI System			
ρ	lb_m/cu ft	L	ft	ρ	kg/cu m	L	m
μ	lb_m/sec ft	$(T - T_\infty)$	F	μ	kg/m sec	$(T - T_\infty)$	K
β	1/R	g	ft/sec^2	β	1/K	g	m/sec^2

When the buoyancy is the only driving force, the fluid velocity is determined entirely by the quantities contained in the Grashof modulus. Therefore, the Reynolds number is superfluous for free convection, and *equality of the Grashof numbers establishes dynamic similarity.*

The equation (6-23) describing the temperature field in free convection is

$$\rho c_p \left(u \frac{\partial T}{\partial x} + v \frac{\partial T}{\partial y} \right) = k \frac{\partial^2 T}{\partial y^2}$$

This equation is identical to the heat-transfer equation for forced convec-

---

[2]In Table A-3 the combination $\rho^2 g\beta/\mu^2$ is listed to facilitate numerical computations.

tion over a flat plate, and its derivation has been presented previously (Sec. 6-7). For similarity of temperature fields in forced convection, we found that the Prandtl numbers, $c_p\mu/k$, must be equal. This applies also to free convection. Therefore, when geometrically similar bodies are cooled or heated by free convection, both the velocity and temperature fields are similar provided Gr and Pr are equal at corresponding points. It follows also from the same arguments used in the case of forced convection that, when the Grashof and Prandtl numbers are equal, the Nusselt numbers for the bodies are the same. Hence, experimental results for free-convection heat transfer can be correlated by an equation of the type

$$\text{Nu} = \phi(\text{Gr})\psi(\text{Pr}) \tag{7-14}$$

where ϕ and ψ denote functional relationships.

The Prandtl number of gases having the same number of atoms per molecule is nearly constant. For a group of gases having the same number of atoms, Eq. 7-14 can therefore be reduced to

$$\text{Nu} = \phi(\text{Gr}) \tag{7-15}$$

As a first approximation, data for different fluids can also be correlated on a single curve. If the velocities are sufficiently small that inertia forces can be neglected in comparison with the forces of friction and buoyancy, the left-hand side of Eq. 7-9, which represents the inertia forces, can be discarded (see Prob. 7-15). Then, the similarity condition is

$$C_\rho C_g C_\beta C_T = \frac{C_\mu C_V}{C_L^2} \tag{7-16}$$

By substituting the equality relations for systems A and B it can easily be verified that the dimensional similarity parameter is (Gr Pr). The product of Gr and Pr is called the Rayleigh number, Ra. Hence, when the inertia forces are negligible, the Nusselt number becomes a function of a single variable and we have

$$\text{Nu} = \phi(\text{Gr Pr}) \tag{7-17}$$

Using an equation of this type, experimental data from various sources for free convection from horizontal wires and tubes are correlated in Fig. 7-3 by plotting $\bar{h}_c D/k$, the average Nusselt number, against $c_p\rho^2 g\beta\Delta T D^3/\mu k$, the product of the Grashof and Prandtl numbers. The physical properties are evaluated at the arithmetic mean temperature. We observe that data for fluids as different as air, glycerine, and water are well correlated over a range of Grashof numbers from 10^{-5} to 10^7 for

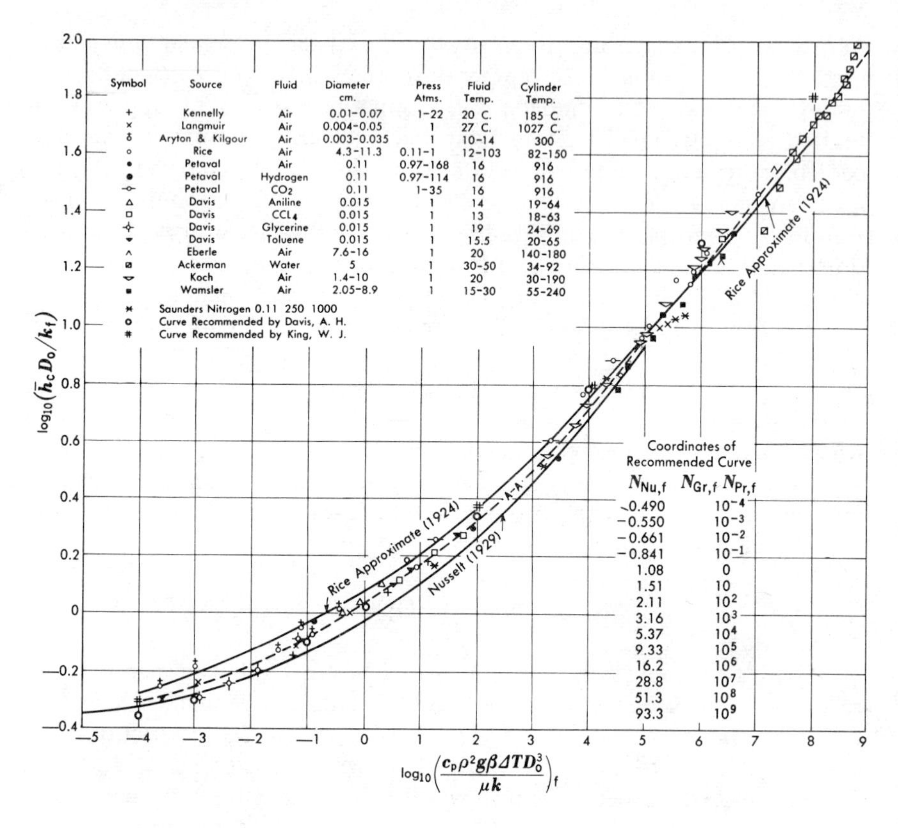

Fig. 7-3. Correlation of data for free-convection heat transfer from horizontal cylinders in gases and liquids. (By permission from W. H. McAdams, *Heat Transmission*, 3d ed., New York: McGraw-Hill Book Company Inc., 1954).

cylinders ranging from small wires to large pipes. King (9) has shown that the correlation in Fig. 7-3 gives approximate results also for three-dimensional shapes such as short cylinders and blocks if the characteristic length dimension is determined by the equation

$$\frac{1}{L} = \frac{1}{L_{\text{hor}}} + \frac{1}{L_{\text{vert}}}$$

where L_{vert} is the height and L_{hor} the average horizontal dimension of the body.

A similar correlation for free convection from vertical plates and vertical cylinders is shown in Fig. 7-4.[3] The ordinate is $\bar{h}_c L/k$, the average

[3] According to Ref. 31, a vertical cylinder of diameter D may be treated as a flat plate of height L when $D/L > 35\,\mathrm{Gr}_L^{-1/4}$.

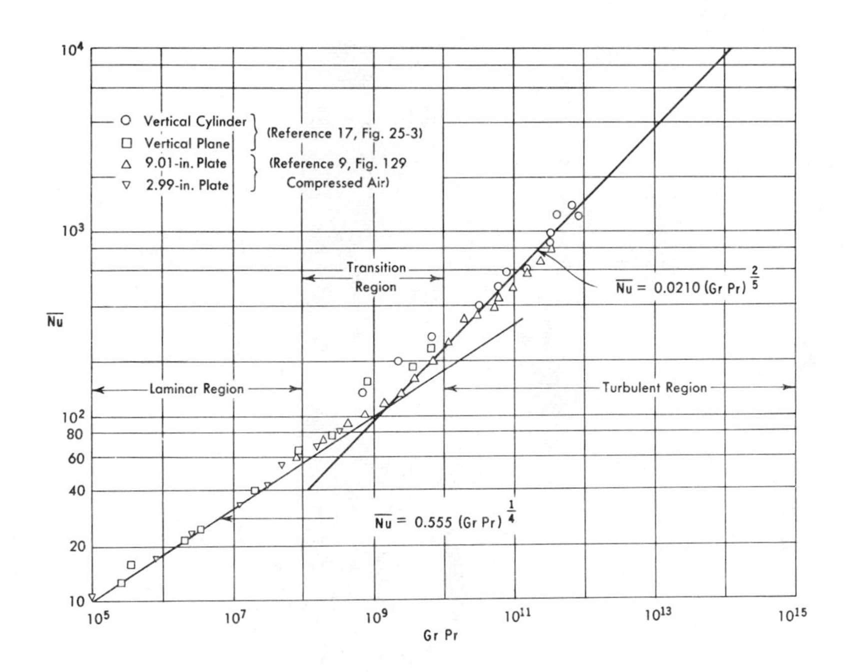

Fig. 7-4. Correlation of data for free-convection heat transfer from vertical plates and cylinders.

Nusselt number based on the height of the body, and the abscissa is the product of Gr and Pr, i.e., $c_p\rho^2\beta g\Delta TL^3/\mu k$. We note that there is a change in the slope of the line correlating the experimental data at a Grashof Number of 10^9. The reason for the change in slope is that the flow is laminar up to a Grashof number of about 10^8, passes through a transition regime between 10^8 and 10^{10}, and becomes fully turbulent at Grashof numbers above 10^{10}. This behavior of the flow is illustrated in the photographs of Fig. 7-5. These pictures show lines of constant density in free convection from a vertical flat plate to air at atmospheric pressure obtained with a *Mach-Zehnder* (6, 7) optical interferometer. This instrument produces interference fringes which are recorded by a camera. The fringes are the result of density gradients caused by temperature gradients in gases. The spacing of the fringes is a direct measure of the density distribution, which is related to the temperature distribution. Figure 7-5 shows the fringe pattern observed near a heated vertical flat plate in air, 3 ft. high and 1.5 ft. wide. We observe that the flow is laminar for about 20 in. from the bottom of the plate. Transition to turbulent flow begins at 21 in., corresponding to a critical Grashof number of about 4×10^8. Near the top of the plate, turbulent flow is approached. This type of behavior is typical of free convection on vertical surfaces, and under normal conditions the critical value of Grashof number is usually taken at 10^9.

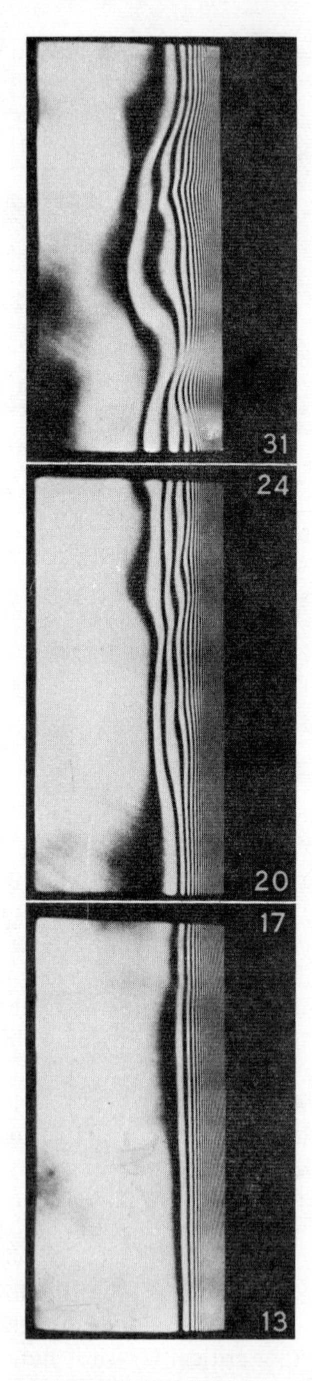

Fig. 7-5. Interference photograph illustrating laminar and turbulent free-convection flow of air along a vertical flat plate. (Courtesy of Professor E. R. G. Eckert).

When the physical properties of the fluid vary considerably with temperature and the temperature difference between the body surface T_s and the surrounding medium T_∞ is large, satisfactory results can be obtained by evaluating the physical properties in Eq. 7-14 at the mean temperature $(T_s + T_\infty)/2$. However, when the surface temperature is not known, a value must be assumed initially. It can then be used to calculate the unit-surface conductance to a first approximation. The surface temperature is then recalculated with this value of the surface conductance and if there is a discrepancy between the assumed and the calculated value of T_s, the latter is used to recalculate the heat-transfer coefficient for the second approximation.

7-3. Evaluation of unit-surface conductance

After experimental data have been correlated by dimensional analysis, it is general practice to write an equation for the line faired through the data and to compare the experimental results with those obtained by analytic means. In this section the results of some experimental studies on free convection for a number of geometric shapes of practical interest are presented. Each shape is identified by a characteristic dimension, such as its distance from the leading edge x, length L, diameter D, etc. The characteristic dimension is attached as a suffix to the dimensionless parameters Nu and Gr. Average values of the Nusselt number for a given surface are identified by a bar, i.e., $\overline{\text{Nu}}$; local values are without a bar. All physical properties are to be evaluated at the arithmetic mean between the surface temperature T_s and the temperature of the undisturbed fluid T_∞. The temperature difference in the Grashof number ΔT represents the absolute value of the difference between the temperatures T_s and T_∞. All of the equations to be discussed apply strictly to bodies immersed in an effectively infinite medium in which the flow pattern is influenced only by the body transferring the heat. The accuracy with which in practice the unit-surface conductance can be predicted from any of the equations is generally no better than 20 percent, because most experimental data scatter by as much as ± 15 percent or more and in a majority of engineering applications stray currents due to some interaction with surfaces other than the one transferring the heat are unavoidable.

Vertical planes and cylinders. The local value of the heat-transfer coefficient for laminar free convection from an isothermal vertical plate or cylinder at a distance x from the leading edge is

$$h_{cx} = 0.41 \frac{k}{x} (\text{Pr} \cdot \text{Gr}_x)^{1/4} \tag{7-18}$$

Eq. 7-18 shows that the heat transfer coefficient decreases with the distance from the leading edge to the 1/4 power. The leading edge is the lower edge for a heated surface and the upper edge for a surface cooler than the surrounding fluid. The average value of the heat-transfer coefficient for a height L is obtained by integrating Eq. 7-18 and dividing by L, or

$$\bar{h}_{cL} = \frac{1}{L} \int_0^L h_{cx}\,dx = 0.555 \frac{k}{L} (\mathrm{Gr}_L \cdot \mathrm{Pr})^{1/4} \qquad (7\text{-}19a)$$

In dimensionless form, the *average Nusselt number is*

$$\overline{\mathrm{Nu}}_L = \frac{\bar{h}_c L}{k} = 0.555 (\mathrm{Gr}_L \cdot \mathrm{Pr})^{1/4} \qquad (7\text{-}19b)$$

in the range $10 < \mathrm{Gr}_L \mathrm{Pr} < 10^9$ (30). For a vertical plane submerged in a liquid metal ($\mathrm{Pr} < 0.03$), the average Nusselt number in laminar flow is (28)

$$\overline{\mathrm{Nu}}_L = \frac{\bar{h}_{cL} L}{k} = 0.68 (\mathrm{Gr}_L \cdot \mathrm{Pr}^2)^{1/4} \qquad (7\text{-}19c)$$

In the turbulent region, the value of h_{cx}, the local heat-transfer coefficient, is nearly constant over the surface. In fact, McAdams (9) recommends for $\mathrm{Gr} > 10^9$ the equation

$$\overline{\mathrm{Nu}}_L = \frac{\bar{h}_c L}{k} = 0.10 (\mathrm{Gr}_L \mathrm{Pr})^{1/3} \qquad (7\text{-}20)$$

according to which the heat-transfer coefficient is independent of the length L.

A theoretical analysis by Sparrow and Gregg (10), supported by experimental data by Dotson (11), indicates that the equations for laminar free convection from a vertical flat plate apply to a constant surface temperature as well as to a uniform heat flux over the surface. In the latter case the surface temperature T_s is to be taken at one-half of the total height of the plate. Other types of correlations for constant heat flux are presented in Refs. (33) and (34).

If a heated plane is inclined somewhat from the vertical, as shown in Fig. 7-6, the body force along the x-axis is $[g\beta(T - T_\infty)\cos\alpha]$ and the average Nusselt number for the upper surface can be obtained by using an effective Grashof number $(\rho^2 g\beta\Delta T \cos\alpha L^3/\mu^2)$ in Eqs. 7-19 or 7-20.

EXAMPLE 7-1. The maximum allowable surface temperature at the center of an electrically heated vertical plate, 6 in. high and 4 in. wide, is 270 F. Esti-

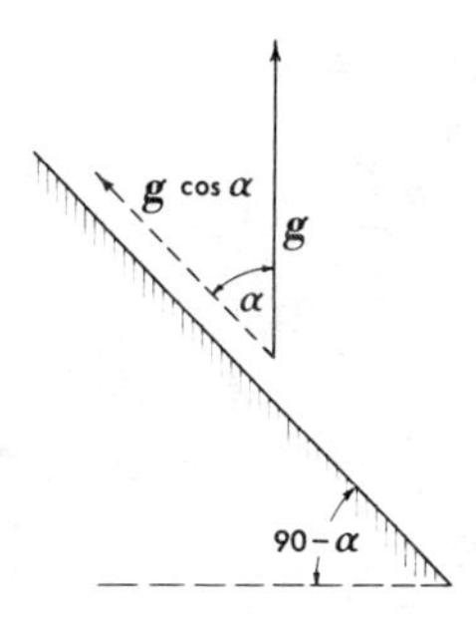

Fig. 7-6. Sketch illustrating body force acting on a fluid on or near an inclined surface.

mate the maximum rate of heat dissipation from both sides of the plate in 70 F atmospheric air if the unit-surface conductance for radiation $\bar{h}_r$ is 1.5 Btu/hr sq ft F for the specified maximum surface temperature.

Solution: The arithmetic mean temperature is 170 F and the corresponding value of Gr_L is found to be $1.2 \times 10^6 L^3 (T_s - T_\infty)$, from the last column in Table A-3 by interpolation. For the specified conditions we get

$$\mathrm{Gr}_L = (1.2 \times 10^6)(6/12)^3(200) = 3 \times 10^7$$

Since the Grashof number is less than 10^9, the flow is laminar. For air at 170 F the Prandtl number is 0.71 and GrPr is therefore 2.1×10^7. From Fig. 7-4 the average Nusselt number is 38 at GrPr of 2.1×10^7 and therefore

$$\bar{h}_c = 38 \times k_f/L = 38 \frac{0.0172}{0.5} \frac{\text{Btu/hr ft F}}{\text{ft}} = 1.31 \text{ Btu/hr sq ft F}$$

The maximum total heat-dissipation rate is therefore

$$q = A(\bar{h}_c + \bar{h}_r)(T_s - T_\infty)$$

$$= \left[\frac{(2)(6)(4)}{144} \text{ sq ft}\right] [(1.31 + 1.5) \text{ Btu/hr sq ft F}](200 \text{ F})$$

$$= 187 \text{ Btu/hr}$$ *Ans.*

Note that more than half of the heat is transferred by radiation.

Horizontal plates. For square plates with a surface warmer than the surrounding medium facing upward or a cooler surface facing downward, McAdams (9) recommends the equation

$$\overline{\mathrm{Nu}}_L = \frac{\bar{h}_c L}{k} = 0.14(\mathrm{Gr}_L \mathrm{Pr})^{1/3} \tag{7-21}$$

in the turbulent range, Gr_L from 2×10^7 to 3×10^{10}, and

$$\overline{Nu}_L = \frac{\bar{h}_c L}{k} = 0.54 (Gr_L Pr)^{1/4} \tag{7-22}$$

in the laminar range, Gr_L from 10^5 to 2×10^7, where L is length of the side of the square. For heated plates facing downward and cooled plates facing upward, the equation

$$\overline{Nu}_L = \frac{\bar{h}_c L}{k} = 0.27 (Gr_L Pr)^{1/4} \tag{7-23a}$$

is recommended (9) in the laminar range, (i.e., Gr from 3×10^5 to 3×10^{10}). Data in the turbulent range are lacking. As a first approximation, the foregoing three equations can be applied to horizontal circular disks if L is replaced by 0.9 D, where D is the diameter of the disk, and to rectangular surfaces if L is taken as the mean between the two sides.

Experimental data for a cooled circular horizontal plate facing down in a liquid metal are correlated by the relation (29)

$$\overline{Nu}_D = \frac{\bar{h}_c D}{k} = 0.26 (Gr_D Pr^2)^{0.35} \tag{7-23b}$$

Horizontal cylinders, spheres, and cones. The temperature field around a horizontal cylinder heated in air is illustrated in Fig. 7-7, which shows interference fringes photographed by Eckert and Soehnghen (7). The flow is laminar over the entire surface. The closer spacing of the interference fringes over the lower portion of the cylinder indicates a steeper temperature gradient and consequently a larger local unit-surface conductance than over the top portion. The variation of the surface conductance with angular position α is shown in Fig. 7-8 for two Grashof numbers. The experimental results do not differ appreciably from the theoretical calculations of Herman (4) who derived the equation

$$Nu_{D\alpha} = 0.604\, Gr_D^{1/4} \phi(\alpha) \tag{7-24}$$

for air, i.e., Pr = 0.74. The angle α is measured from the horizontal position and numerical values of the function $\phi(\alpha)$ are as follows:

α	−90	−60	−30	0	30	60	75	90
$\phi(\alpha)$	0.76	0.75	0.72	0.66	0.58	0.46	0.36	0
	Bottom half			Top half				

An equation for the average heat-transfer coefficient from single horizontal wires or pipes in free convection, recommended by McAdams (9)

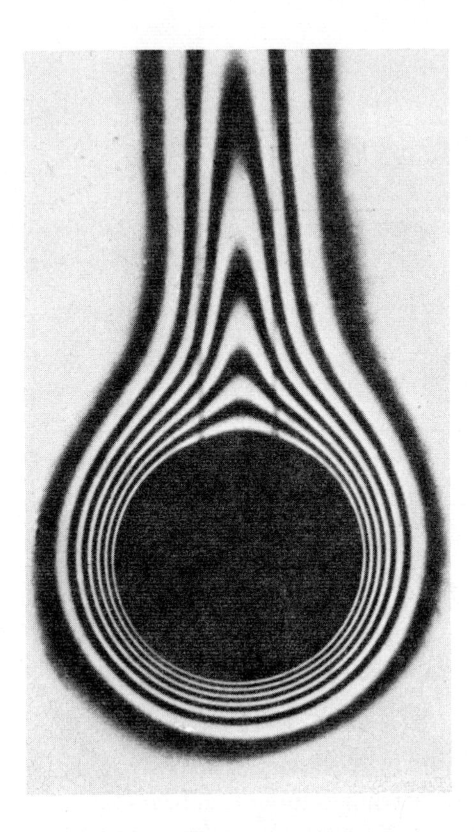

Fig. 7-7. Interference photograph illustrating temperature field around a horizontal cylinder in laminar flow. (Courtesy of Professor E. R. G. Eckert).

on the basis of the experimental data in Fig. 7-3, is

$$\overline{\mathrm{Nu}}_D = 0.53\,(\mathrm{Gr}_D\,\mathrm{Pr})^{1/4} \tag{7-25}$$

This equation is valid for Prandtl numbers larger than 0.5 and Grashof numbers ranging from 10^3 to 10^9. For very small diameters, Langmuir has shown that the rate of heat dissipation per unit length is nearly independent of the wire diameter, a phenomenon he applied in his invention of the coiled filaments in gas-filled incandescent lamps. The average unit-surface conductance for Gr_D less than 10^3 is most conveniently evaluated from the curve A–A drawn through the experimental points in Fig. 7-3 in the low Grashof-number range.

The onset of turbulence in free-convection flow over horizontal cylinders for fluids other than liquid metals occurs at a value of $\mathrm{Gr}_D\,\mathrm{Pr}/D^3$ of about 10^{11} (12). In turbulent flow it has been observed (12) that the heat flux can be increased substantially without a corresponding

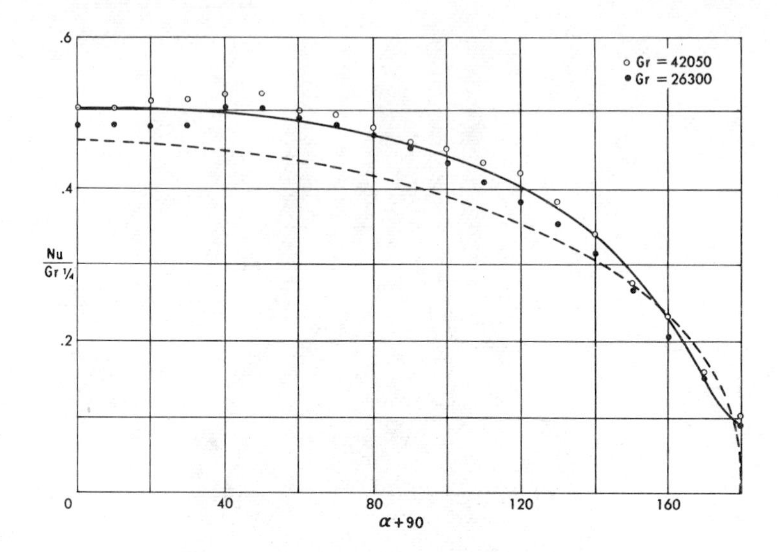

Fig. 7-8. Local dimensionless unit-surface conductance along the circumference of a horizontal cylinder in laminar free convection (dashed line according to Ref. 4). (Courtesy of U.S. Air Force, from "Studies on Heat Transfer in Laminar Free Convection with the Zehnder-Mach Interferometer," by E. R. G. Eckert and E. E. Soehngen, Ref. 7).

increase in the surface temperature. It appears that in free convection the turbulent-exchange mechanism increases in intensity as the rate of heat flow is increased and thereby reduces the thermal resistance.

Simplified equations for horizontal cylinders at moderate surface temperatures are given in Reference 9 *for air* at atmospheric pressure. They are

$$\bar{h}_c = 0.18\,\Delta T^{1/3} \tag{7-26}$$

for Gr_D from 10^9 to 10^{12}, and

$$\bar{h}_c = 0.27\left(\frac{\Delta T}{D}\right)^{1/4} \tag{7-27}$$

for Gr_D from 10^3 to 10^9, where $\bar{h}_c$ is in Btu/hr sq ft F and ΔT is in F.

For *liquid metals* in laminar flow the equation

$$\overline{\mathrm{Nu}}_D = 0.53(\mathrm{Gr}_D\,\mathrm{Pr}^2)^{1/4} \tag{7-28}$$

correlates the available data (12) for horizontal cylinders.

For *free convection to or from spheres* of diameter D the empirical equation

$$\overline{\mathrm{Nu}}_D = 2 + 0.45(\mathrm{Gr}_D \cdot \mathrm{Pr})^{1/4} \tag{7-29}$$

is recommended (35). For very small spheres, as the Grashof number approaches zero, the Nusselt number approaches a value of 2, i.e. $(\bar{h}_c D/k) \rightarrow 2$. This condition corresponds to pure conduction through a stagnant layer of fluid surrounding the sphere.

EXAMPLE 7-2. A $1\frac{1}{2}$-in. OD pipe carrying slightly wet steam at 15 psig is installed in a location where it is covered by water after a heavy rain but is exposed to air under normal conditions. Compare the rate of heat transfer to air with the rate of heat transfer to water, assuming that both fluids are at 50 F.

Solution: From steam tables we find that the temperature of the steam at 15 psig is 250 F. Assuming that the pipe temperature equals the steam temperature because the heat-transfer coefficient inside the pipe is large (Table 1-2), the mean film temperature is 150 F. From Table A-3, the product of the Grashof and Prandtl numbers is

$$\mathrm{Gr}_D \mathrm{Pr} = (1.2 \times 10^9)(200)\left(\frac{1.50}{12}\right)^3 = 4.7 \times 10^8 \qquad \text{for water}$$

$$\mathrm{Gr}_D \mathrm{Pr} = (0.85 \times 10^6)(200)\left(\frac{1.5}{12}\right)^3 = 3.3 \times 10^5 \qquad \text{for air}$$

From Fig. 7-3 the respective Nusselt numbers are therefore

$$\overline{\mathrm{Nu}}_D = 77 \qquad \text{for water}$$

$$\overline{\mathrm{Nu}}_D = 12.6 \qquad \text{for air}$$

The respective heat-transfer coefficients are therefore

$$\bar{h}_c = \mathrm{Nu}\,\frac{k}{D} = 77\,\frac{0.384\ \mathrm{Btu/hr\ ft\ F}}{1.5/12\ \mathrm{ft}} = 23.6\ \mathrm{Btu/hr\ sq\ ft\ F} \qquad \text{for water}$$

$$\bar{h}_c = 12.6\,\frac{0.0164}{1.5/12} = 1.66\ \mathrm{Btu/hr\ sq\ ft\ F} \qquad \text{for air}$$

The rate of heat loss by convection per foot length of pipe is therefore

$$q_c = \bar{h}_c A(T_s - T_\infty) = (236)\,\pi(1.5/12)(200) = 18{,}500\ \mathrm{Btu/hr\ ft} \qquad \text{in water}$$

$$q_c = (1.66)\,\pi(1.5/12)(200) = 130\ \mathrm{Btu/hr\ ft} \qquad \text{in air}$$

There is no appreciable radiation in water, but in air the heat transfer by radiation is from Eq. 1-7.

$$q_r = (0.171)(0.9)\,\pi(1.5/12)\left[(7.10)^4 - (5.10)^4\right] = 115\ \mathrm{Btu/hr\ ft}$$

if the emissivity of the pipe is 0.9. The total heat-transfer rate in air is therefore 245 Btu/hr ft, which is only a small fraction of the heat-transfer rate when the pipe is covered by water. *Ans.*

Experimental data for free convection from vertical cones with vertex angles between 3 and 12 degrees have been correlated (36) by the equation

$$\overline{\mathrm{Nu}}_L = 0.63\,(1 + 0.72\,\epsilon)\,\mathrm{Gr}_L{}^{1/4}$$

where $\epsilon = \dfrac{2}{\mathrm{Gr}_L{}^{1/4}\tan(\phi/2)}$

ϕ = vertex angle

L = slant height of the cone.

An extensive treatment of transition and stability in natural convection systems is presented in Ref. 31.

Free convection in enclosed spaces. Empirical formulas for calculation of free-convection heat transfer in an enclosed air space between vertical walls can be presented in the following form (14):

$$\overline{\mathrm{Nu}}_\delta = \begin{cases} 0.18\,\mathrm{Gr}_\delta{}^{1/4}\left(\dfrac{L}{\delta}\right)^{-1/9} & \text{for } 2000 < \mathrm{Gr}_\delta < 20{,}000 \quad (7\text{-}30) \\[2ex] 0.065\,\mathrm{Gr}_\delta{}^{1/3}\left(\dfrac{L}{\delta}\right)^{-1/9} & \text{for } 20{,}000 < \mathrm{Gr}_\delta < 11 \times 10^6 \quad (7\text{-}31) \end{cases}$$

where Gr_δ is defined by

$$\mathrm{Gr}_\delta = \frac{\rho^2 g\beta(T_1 - T_2)\delta^3}{\mu^2} \tag{7-32}$$

where T_1 and T_2 are the temperatures of the walls on either side of the enclosed space and δ is the thickness of the air space as shown in Fig. 7-9. The height of the air space is designated by L. It should be noted that the length in the Grashof number is the thickness of the air space. For Grashof numbers below 2000, the heat transfer is essentially all conduction, so that $\overline{\mathrm{Nu}}_\delta = 1.0$ for this region.

For free convection in horizontal air spaces the following relations are recommended (14):

$$\overline{\mathrm{Nu}}_\delta = \begin{cases} 0.195\,\mathrm{Gr}_\delta{}^{1/4} & \text{for } 10^4 < \mathrm{Gr}_\delta < 4 \times 10^5 \quad (7\text{-}33) \\ 0.068\,\mathrm{Gr}_\delta{}^{1/3} & \text{for } 4 \times 10^5 < \mathrm{Gr}_\delta \quad (7\text{-}34) \end{cases}$$

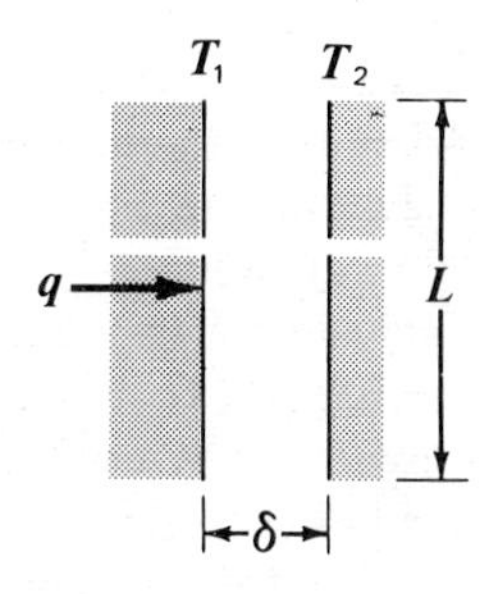

Fig. 7-9. Nomenclature for free convection in enclosed vertical spaces.

The experiments of Globe and Dropkin (27) are useful for calculation of free convection through liquids in horizontal spaces. The results of their work indicate that

$$\overline{\mathrm{Nu}}_\delta = 0.069\,\mathrm{Gr}_\delta^{1/3}\,\mathrm{Pr}^{0.407} \qquad \text{for } 3 \times 10^5 < \mathrm{Gr}_\delta \mathrm{Pr} < 7 \times 10^9 \tag{7-35}$$

For free convection inside spherical cavities of diameter D the relation (13)

$$\frac{D\bar{h}_c}{k} = C(\mathrm{Gr}_D \cdot \mathrm{Pr})^n \tag{7-36}$$

is recommended with the constants C and n selected from the tabulation below

$\mathrm{Gr}_D \cdot \mathrm{Pr}$	C	n
$10^4 - 10^9$	0.59	1/4
$10^9 - 10^{12}$	0.13	1/3

7-4. Convection from rotating cylinders, disks, and spheres

Heat transfer by convection between a rotating body and a surrounding fluid is of importance in the thermal analysis of shafting, flywheels, turbine rotors, and other rotating components of various machines. Convection from a heated rotating horizontal cylinder to ambient air has been studied by Anderson and Saunders (15). Turbulence begins to appear at a critical *peripheral-speed Reynolds number*, $\mathrm{Re}_\omega = \omega\pi D^2/\nu$, of about 50, where ω is the rotational speed in rad/sec. With heat transfer the critical speed is reached when the circumferential speed of the cylinder surface becomes approximately equal to upward free-convection velocity at the side of a heated stationary cylinder.

Below the critical velocity simple free convection, characterized by the conventional Grashof number $\beta g(T_s - T_\infty)D^3/\nu^2$, controls the rate of heat transfer. At speeds greater than critical ($\mathrm{Re}_\omega > 8000$ in air) the peripheral speed Reynolds number $\pi D^2\omega/\nu$ becomes the controlling parameter. The combined effects of the Reynolds, Prandtl, and Grashof numbers on the average Nusselt number for a horizontal cylinder rotating in air above the critical velocity can be expressed by the empirical equation (16)

$$\overline{\mathrm{Nu}}_D = \frac{\bar{h}D}{k} = 0.11\,[0.5\,\mathrm{Re}_\omega^{\,2} + \mathrm{Gr}_D)\,\mathrm{Pr}]^{0.35} \qquad (7\text{-}37)$$

Heat transfer from a rotating disk has been investigated experimentally by Cobb and Saunders (17) and theoretically, among others, by Millsaps and Pohlhausen (18) and Kreith and Taylor (19). The boundary layer on the disk is laminar and of uniform thickness at rotational Reynolds numbers $\omega D^2/\nu$ below about 10^6. At higher Reynolds numbers the flow becomes turbulent and the boundary layer thickens with increasing radius (see Fig. 7-10). The average Nusselt number for a disk rotating in air is (17, 20)

$$\overline{\mathrm{Nu}}_D = \frac{\bar{h}_c D}{k} = 0.35\left(\frac{\omega r_o^{\,2}}{\nu}\right)^{1/2} \qquad (7\text{-}38)$$

for $\mathrm{Re}_D < 5 \times 10^5$.

In the turbulent flow regime of a disk rotating in air (17), the local value of the Nusselt number at a radius r is approximately given by

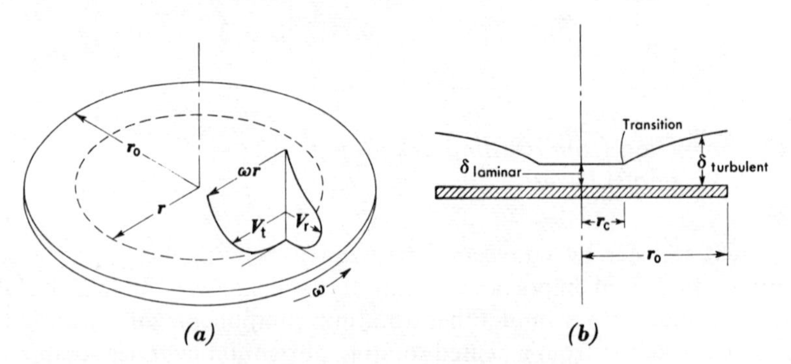

Fig. 7-10. Velocity and boundary layer profiles for a disk rotating in an infinite environment.

$$\text{Nu}_r = \frac{h_c r}{k} = 0.0195\,(\omega r^2/\nu)^{0.8} \tag{7-39}$$

and the average value of the Nusselt number for laminar flow between $r = 0$ and r_c, and turbulent flow in the outer ring between $r = r_c$ and r_o is approximately

$$\overline{\text{Nu}}_{r_o} = \frac{\bar{h}_c r_o}{k} = 0.015\left(\frac{\omega r_o^2}{\nu}\right)^{0.8} - 100\left(\frac{r_c}{r_o}\right)^2 \tag{7-40}$$

For a disk rotating in a fluid having a Prandtl number larger than unity, the local Nusselt number can be obtained according to Reference 21 from the equation

$$\overline{\text{Nu}}_r = \frac{\text{Re}_r \text{Pr}\,(\sqrt{C_{Dr}/2})}{5\,\text{Pr} + 5\ln(5\,\text{Pr} + 1) + (\sqrt{2/C_{Dr}}) - 14} \tag{7-41}$$

where C_{Dr} is the local drag coefficient at radius r which, according to Reference 22, is given by the relation

$$\frac{1}{\sqrt{C_{Dr}}} = -2.05 + 4.07\log_{10}\text{Re}_r\sqrt{C_{Dr}} \tag{7-42}$$

For a sphere of diameter D rotating in an infinite environment with $\text{Pr} > 0.7$ in the laminar flow regime ($\text{Re}_D = \omega D^2/\nu < 5 \times 10^4$) the average Nusselt number ($\bar{h}_c D/k$) can be obtained from the equation

$$\overline{\text{Nu}}_D = 0.43\,\text{Re}_D^{0.5}\,\text{Pr}^{0.4} \tag{7-43}$$

while in the Reynolds-number range between 5×10^4 and 7×10^5 the equation

$$\overline{\text{Nu}}_D = 0.066\,\text{Re}^{0.67}\,\text{Pr}^{0.4} \tag{7-44}$$

correlates the available experimental data.

7-5. Combined forced and free convection

In Chapter 6 forced convection in flow over a flat surface was treated, and the preceding sections of this chapter dealt with heat transfer in natural-convection systems. In this section the interaction between free- and forced-convection processes will be considered.

In any heat-transfer process density gradients occur and in the presence of a force field natural-convection currents arise. If the forced-convection effects are very large, the influence of natural-convection currents may be negligible and, similarly, when the natural-convection forces are very strong, the forced-convection effects may be negligible. The questions we wish to consider now are under what circumstances can either forced or free convection be neglected and what are the conditions when both effects are of the same order of magnitude.

To obtain an indication of the relative magnitudes of free- and forced-convection effects, we consider the differential equation describing the uniform flow over a vertical flat plate with the buoyancy effect and the free-stream velocity U_∞ going in the same direction. This would be the case when the plate is heated and the flow is upward, or when the plate is cooled and the flow is downward. Taking the flow direction as x and assuming that the physical properties are uniform except for the temperature effect on the density, the Navier-Stokes boundary-layer equation for uni-dimensional flow with free convection is

$$u \frac{\partial u}{\partial x} + v \frac{\partial u}{\partial y} = -\frac{1}{\rho}\frac{\partial p}{\partial x} + \frac{\mu}{\rho}\frac{\partial u}{\partial y^2} + g\beta(T - T_\infty) \tag{7-45}$$

This equation can be generalized by a method similar to that outlined in Sec. 7-2. Substituting X for x/L, Y for y/L, Θ for $(T - T_\infty)/(T_o - T_\infty)$, P for $(p - p_\infty)/\rho U_\infty{}^2/2g_c)$, U for u/U_∞ and V for v/U_∞ in Eq. 7-45 gives

$$U\frac{\partial U}{\partial X} + V\frac{\partial U}{\partial Y} = -\frac{1}{2}\frac{\partial P}{\partial X} + \left(\frac{\mu}{\rho U_\infty L}\right)\frac{\partial^2 U}{\partial Y^2} + \left(\frac{g\beta L^3(T_o - T_\infty)}{\nu^2}\right)\frac{\nu^2}{U_\infty{}^2 L^2}\Theta \tag{7-46}$$

In the region near the surface, i.e., in the boundary layer, $\partial U/\partial X$ and U are of the order of unity. Since U changes from 1 at $x = 0$ to a very small value at $x = 1$, and u is of the same order of magnitude as U_∞, the left-hand side of Eq. 7-46 is of the order of unity. Similar reasoning indicates that the first two terms on the right-hand side as well as Θ are of the order of unity. Consequently, the buoyancy effect will influence the velocity distribution, on which, in turn, the temperature distribution depends, if the coefficient of Θ is of the order of 1 or larger, i.e., if

$$\frac{[g\beta L^3(T_o - T_\infty)]/\nu^2}{(U_\infty L/\nu)^2} = \frac{\mathrm{Gr}_L}{\mathrm{Re}_L{}^2} \cong 1 \tag{7-47}$$

In other words, the ratio of $\mathrm{Gr}/\mathrm{Re}^2$ gives a qualitative indication of the influence of buoyancy on forced convection, and when the Grashof number is of the same order of magnitude or larger than the square of the Reynolds number, free convection effects cannot be ignored, compared to forced convection. Similarly, in a natural-convection process the influence of forced convection becomes significant when the square of the Reynolds number is of the same order of magnitude as the Grashof number.

Several special cases have been treated in the literature (24, 25, 26). For example, for laminar forced convection over a vertical flat plate Sparrow and Gregg (24) showed that for Prandtl numbers between 0.01 and 10 the effect of buoyancy on the average heat-transfer coefficient for pure forced convection will be less than 5 per cent if $\mathrm{Gr}_L \leq 0.225\,\mathrm{Re}_L^{\,2}$.

Acrivos (26) showed that for Prandtl numbers between 0.07 and 10, forced convection has negligible effects on natural convection from a vertical flat plate if the Grashof number is at least ten times larger than the square of the Reynolds number. In the region where both free- and forced-convection effects are of the same order of magnitude, heat transfer is increased by buoyancy effects acting in the direction of flow and decreased when acting in the opposite direction.

PROBLEMS

7-1. An empirical equation proposed by Heilman (*Trans. ASME*, Vol. 51, 1929, p. 287) for the unit-surface conductance in free convection from long horizontal cylinders to air is

$$\bar{h}_c = \frac{1.016\,(T_s - T_\infty)^{0.266}}{D^{0.2}T_f^{\,0.181}}$$

The corresponding equation in dimensionless form is

$$\frac{h_c D}{k_f} = C\mathrm{Gr}_f^{\,m}\mathrm{Pr}_f^{\,n}$$

By comparing the two equations, determine those values of the constants C, m, and n in the latter equation which will give the same results as the first equation.

7-2. Consider a design for a nuclear reactor using free-convection heating of liquid bismuth. The reactor core is to be constructed of parallel vertical plates, 6 ft tall and 4 ft wide, in which heat is generated uniformly. Estimate the maximum possible heat-dissipation rate from each plate if the surface temperature of the plate is not to exceed 1600 F and the lowest allowable bismuth temperature is 600 F.

7-3. A 10-gal tank full of water at 60 F is to be heated to 120 F by means of a $\frac{3}{8}$-in. OD copper steam coil having 10 turns of 12 in. diameter. The steam is at atmospheric pressure, and its thermal resistance is negligibly small. Neglecting heat losses from the tank, estimate the heating time required.

7-4. An 8-in. diameter sphere containing liquid air (−220 F) is covered with 2-in.-thick glass wool. Estimate the rate of heat transfer to the liquid air from the surrounding air at 70 F by convection and radiation. How would you reduce the heat transfer?

7-5. A horizontal $2\frac{3}{8}$-in. OD, $2\frac{1}{16}$-in. ID steam pipe carrying saturated steam at 50 psia is covered by 1-in.-thick molded-asbestos insulation. Estimate the rate of heat loss to surrounding air at 70 F for a 100-ft length. What would be the quality of the steam at the outlet if it is saturated at the inlet? The unit-surface conductance at the steam side is 2000 Btu/hr sq ft F and the average velocity is 10 fps.

7-6. Calculate the rate of heat transfer from a 10-in. diameter sphere suspended from a fine wire in air at 70 F if the sphere is rotating at 2000 rpm and has a surface temperature of 300 F.

7-7. Estimate the rate of heat transfer by free convection and radiation across a $\frac{1}{2}$-in. air space formed between two horizontal 24-ST aluminum sheets, the upper one of which is maintained at 300 F while the lower one remains at 70 F.

7-8. Repeat Prob. 7-7 for the case in which the air space is divided in half by a very thin sheet of bright aluminum foil, placed parallel to the surface.

7-9. A vertical isothermal plate 1 ft high is suspended in an atmospheric air stream flowing at 6 fps in a vertical direction. If the air is at 60 F, estimate the plate temperature for which the free-convection effect on the heat-transfer coefficient will be less than 5 percent. *Ans.* 250 F

7-10. If the plate in Example 7-9 is at a surface temperature of 100 F, determine the maximum air velocity for which forced-convection effects on the heat-transfer coefficient are negligible. *Ans.* 0.5 fps

7-11. Starting with the equation

$$Nu_D = 0.53\left(\frac{Pr^2}{0.452 + Pr}\,Gr\right)^{1/4}$$

show that, if Pr is much larger than unity,

$$Nu \simeq (Gr\,Pr)^{1/4}$$

and when Pr is much less than unity (e.g., liquid metals)

$$Nu \simeq (Gr\,Pr^2)^{1/4}$$

7-12. Consider a thin vertical flat plate L feet high and 1 ft wide at a temperature difference between surrounding medium ($Pr = 1$) and plate surface of ΔT. If heat exchange is taking place by free convection in laminar flow, derive an expression for the lifting force acting on the plate as a result of the temperature difference ΔT.

7-13. A light oil is maintained at 150 F in a 2-ft-square sump tank by ten 2-ft-long, $\frac{1}{2}$-in. OD tubes which are widely spaced and arranged horizontally in the lower third of the 6 ft tank depth. The tube surface temperature is maintained at 50 F by cooling water circulated at a high rate through the tubes. Estimate the oil cooling rate in Btu/hr if the heat-transfer area is 2.62 sq ft.

Ans. ~6500 Btu/hr

7-14. A thermocouple ($\frac{1}{32}$-in. OD) is located horizontally in a large enclosure whose walls are at 100 F. The enclosure is filled with a transparent quiescent gas which has the same properties as air. The electromotive force (emf) of the thermocouple indicates a temperature of 450 F. Estimate the true gas temperature if the emissivity of the thermocouple is 0.8.

7-15. Starting with Eqs. 7-2 and 7-6 verify the validity of Eq. 7-16 under the assumption that inertia forces are negligible.

7-16. Show from Eq. 7-39 that if $V_r = 0.162\omega r(y/\delta)^{1/7}[1 - (y/\delta)]$, $\delta = 0.526r(r^2\omega/\nu)^{-1/5}$ (see *ZAMM*, Vol. 1, 1921, p. 231), and $(T - T_\infty) = (T_s - T_\infty)[1 - (y/\delta)^{\frac{1}{7}}]$, the average Stanton number for turbulent flow of a fluid with Pr = 1 on a rotating disk of radius r_o is given by

$$\overline{\mathrm{St}} = \frac{\bar{h}_c}{c_p \rho \omega r_o} = 0.0116\,(\nu/r_o{}^2\omega)^{1/5}$$

7-17. A mild steel, 1-in. OD shaft, rotating in 70 F air at 20,000 rpm, is attached to two bearings, 2 ft apart. If the temperature at the bearings is 200 F, determine the temperature distribution along the shaft. HINT: Show that for high rotational speeds Eq. 7-38 approaches $\mathrm{Nu}_D = 0.076\,(\pi D^2\omega/\nu)^{0.7}$.

7-18. Estimate the rate of heat transfer from one side of a 6-ft-diameter disk rotating at 600 rpm in 70 F air, if its surface temperature is 120 F.

7-19. A 4 ft by 4 ft flat, chromeplated plate, supported horizontally on 6-ft legs, is exposed to the sun at 12 o'clock noon on May 1. If the air temperature is 80 F, (a) determine the equilibrium temperature on an average day; (b) determine the equilibrium temperature for an irradiation of 350 Btu/sq ft hr.

7-20. Estimate the equilibrium temperature of a polished aluminum plate mounted on an insulating pad when exposed on a clear day to the noon sun. The irradiation is 255 Btu/sq ft hr and the ambient temperature is 80 F. Assume that the effective sky temperature is also 80 F.

7-21. A cubical furnace having external dimensions of 20 ft on a side rests on a concrete floor. If the sides and the top are at 200 F and the surrounding air is 70 F, find the total rate of heat loss from the furnace neglecting any losses from the base to the concrete. Assume that the emissivity of the surface is 0.9.

7-22. A 1-in. OD electrical transmission line carrying 5000 amp and having a resistance of 1×10^{-6} ohms per foot of length is placed horizontally in still air at 95 F. Determine the surface temperature of the line in the steady state (a) if radiation is neglected, and (b) if radiation is taken into account. *Ans.* (a) 290 F

7-23. A flat metal plate 4 in. square is mounted in a vertical position in a milling machine. While a rotating cutter is shaping the plate, a coolant coil at 90 F ($Pr = 450$, $\mu = 700 \times 10^{-7}$ lb/ft sec, $\rho = 50$ lb/cu ft, $\beta = 0.38 \times 10^{-3} F^{-1}$, $k = 0.7$ Btu hr ft F) is flowing down over both sides of the plate at a velocity of 0.2 fps. If under these conditions the plate temperature is 140 F, determine the rate of heat generation due to friction by the cutting tool.

7-24. A thin sheet of galvanized iron 2 ft square is placed vertically in air at 40 F on a clear day. If the solar irradiation is 100 Btu/hr sq ft, estimate the equilibrium temperature of the plate.

7-25. The needles of conifers may be idealized as small horizontal cylinders. If a conifer needle has an average diameter of $\frac{1}{32}$ in. and is at a temperature of 90 F, estimate its rate of heat loss by free convection to air at 60 F per inch length of needle. Express your answer in calories per second.

7-26. Assuming a temperature distribution given by $(T - T_\infty)/(T_w - T_\infty) = (1 - y/\delta)^2$ and a velocity distribution given by $u/u_\infty = (y/\delta)(1 - y/\delta)^2$, show by means of the integral momentum equation that for a vertical wall at a uniform temperature of T_w, the ratio of the boundary-layer thickness to the distance from the leading edge can be expressed in the form

$$\delta/x = 3.93\, Pr^{-1/2} (0.952 + Pr)^{1/4} Gr_x^{-1/4}$$

and that the Nusselt number at a given distance x from the leading edge is given by the equation

$$Nu_x = 0.508\, Pr^{1/2} (0.952 + Pr)^{-1/4} Gr_x^{1/4}$$

7-27. A 1000-watt tungsten heating element is mounted inside a 1-ft-diameter polished copper container having a spherical shape. The container is filled with argon under 2 atm pressure and is suspended in a large room filled with atmospheric air at 70 F. Estimate the temperature of the surface of the container for steady-state conditions.

7-28. An 8- by 8-ft steel sheet $\frac{1}{16}$ in. thick is removed from an annealing oven at a uniform temperature of 800 F and placed into a large room at 70 F in a horizontal position. (a) Calculate the rate of heat transfer from the steel sheet immediately after its removal from the furnace, considering both radiation and convection. (b) Determine the time required for the steel sheet to cool to a temperature of 100 F. HINT: This will require numerical integration.

Ans. (a) About 625,000 Btu/hr; (b) About 17 min

7-29. A so-called swimming-pool nuclear reactor consists of 20 parallel vertical plates 1 ft wide and 2 ft high, spaced a distance of 2 in. apart. Calculate the power level at which the reactor may operate safely in 80 F water if the plate surface is not to exceed a temperature of 200 F. State all your assumptions.

7-30. Find the temperature at the center of a horizontal brass rod 4 in. long and $\frac{1}{4}$ in. in diameter. One end of the rod is at 100 F and the other end protrudes into air at 60 F. The total unit-surface conductance by free convection and by radiation may be assumed to be uniform over the entire surface of the rod, and the emissivity of the surface may be taken as 0.9. *Ans.* 89 F

7-31. The maximum allowable surface temperature at the center of an electrically heated vertical plate, 6 in. high and 4 in. wide, is 270 F. Estimate maximum rate of heat dissipation from both sides of the plate in 70 F atmospheric air if the unit-surface conductance for radiation heat transfer is 1.5 Btu/hr sq ft F at the maximum specified temperature.

REFERENCES

1. S. Ostrach, "New Aspects of Natural-Convection Heat Transfer," *Trans. ASME*, Vol. 75 (1953), pp. 1287–1290.
2. E. Griffith and A. H. Davis, "The Transmission of Heat by Radiation and Convection," *Special Report* 9, Food Investigation Board, British Dept. of Sci. and Ind. Res., 1922.
3. E. Schmidt and W. Beckman, "Das Temperatur und Geschwindigkeitsfeld vor einer wärmeabgebenden senkrechten Platte bei natürlicher Konvection," *Tech. Mech. u. Thermodynamic*, Bd. 1, No. 10 (October, 1930), pp. 341–349; cont. Bd. 1, No. 11 (November, 1930), pp. 391–406.
4. R. Herman, "Wärmeübergang bei freier Ströhmung am wagrechten Zylinder in zwei-atomic Gasen," *VDI—Forschungsheft*, No. 379 (1936); translated in *NACA TM* 1366, November, 1954.
5. E. R. G. Eckert and T. W. Jackson, "Analysis of Turbulent Free Convection Boundary Layer on Flat Plate," *NACA Report* 1015, July, 1950.
6. E. R. G. Eckert and E. Soehnghen, "Interferometric Studies on the Stability and Transition to Turbulence of a Free-Convection Boundary Layer," *Proc. of the General Discussion on Heat Transfer* (London: ASME-IME, 1951), pp. 321–323.
7. E. R. G. Eckert and E. Soehnghen, "Studies on Heat Transfer in Laminar Free Convection with the Zehnder-Mach Interferometer," *USAF Tech. Report* 5747, December, 1948.
8. E. R. G. Eckert, *Introduction to the Transfer of Heat and Mass.* (New York: McGraw-Hill Book Company, Inc., 1951.)
9. W. H. McAdams, *Heat Transmission*, 3d ed. (New York: McGraw-Hill Book Company, Inc., 1954.)
10. E. M. Sparrow and J. L. Gregg, "Laminar Free Convection from a Vertical Flat Plate," *Trans. ASME*, Vol. 78 (1956), pp. 435–440.
11. J. P. Dotson, *Heat Transfer from a Vertical Flat Plate by Free Convection*, M. S. Thesis, Purdue University, May, 1954.
12. S. C. Hyman, C. F. Bonilla, and S. W. Ehrlich, "Heat Transfer to Liquid Metals and Non-metals at Horizontal Cylinders," *AIChE Symposium on Heat Transfer*, Atlantic City, 1953, pp. 21–33.

13. F. Kreith, "Thermal Design of High Altitude Balloons and Instrument Packages", *J. Heat Trans.*, Vol. 92 (1970), pp. 307–332.

14. M. Jacob, *Heat Transfer*, Vol. I (New York: John Wiley & Sons, Inc., 1949).

15. J. T. Anderson and O. A. Saunders, "Convection from an Isolated Heated Horizontal Cylinder Rotating About its Axis," Proc. Roy. Soc., A., Vol. 217 (1953), pp. 555–562.

16. W. M. Kays and I. S. Bjorklund, "Heat Transfer from a Rotating Cylinder with and without Cross Flow," *Trans. ASME*, ser. C, Vol. 80 (1958), pp. 70–78.

17. E. C. Cobb and O. A. Saunders, "Heat Transfer from a Rotating Disk," Proc. Roy. Soc., A., Vol. 220 (1956), pp. 343–351.

18. K. Millsap and K. Pohlhausen, "Heat Transfer by Laminar Flow from a Rotating Plate," *J. of the Aero. Sci.*, Vol. 19 (1952), pp. 120–126.

19. F. Kreith and J. H. Taylor, Jr., "Heat Transfer from a Rotating Disk in Turbulent Flow," *ASME Paper* No. 56-A-146, 1956.

20. C. Wagner, "Heat Transfer from a Rotating Disk to Ambient Air," *J. of Appl. Phys.*, Vol. 19 (1948), pp. 837–841.

21. F. Kreith, J. H. Taylor, and J. P. Chang, "Heat and Mass Transfer from a Rotating Disk," *Trans. ASME*, ser. C, Vol. 81 (1959), pp. 95–105.

22. T. Theodorsen and A. Regier, "Experiments on Drag of Revolving Disks, Cylinders, and Streamlined Rods at High Speeds," *NACA Rep.* No. 793, Washington, D.C., 1944.

23. F. Kreith, L. G. Roberts, J. A. Sullivan, and S. N. Sinha, "Convection Heat Transfer and Flow Phenomena of Rotating Spheres," *International Journal of Heat and Mass Transfer*, Vol. 6 (1963), pp. 881–895.

24. E. M. Sparrow and J. L. Gregg, "Buoyancy Effects in Forced Convection Flow and Heat Transfer," *Trans. ASME*, *Journal of Applied Mechanics*, Sec. E, Vol. 81 (1959), pp. 133–135.

25. Y. Mori, "Buoyancy Effects in Forced Laminar Convection Flow Over a Horizontal Flat Plate," *Trans. ASME*, *Journal of Heat Transfer*, Sec. C, Vol. 83 (1961), pp. 479–482.

26. A. Acrivos, "Combined Laminar Free- and Forced-Convection Heat Transfer in External Flows," *AIChE Journal*, Vol. 4 (1958), pp. 285–289.

27. S. Globe and D. Dropkin, "Natural Convection Heat Transfer in Liquids Confined by Two Horizontal Plates and Heated From Below." *Trans. ASME*, ser. C, Vol. 81 (1959), pp. 24–28.

28. O. E. Dwyer, "Liquid-Metal Heat Transfer," Chapt. 5, Sodium and NaK Supplement to the *Liquid Metals Handbook*, 1970 ed., (Washington D.C.: Atomic Energy Commission.)

29. J. S. McDonald and T. J. Connally, "Investigation of Natural Convection Heat Transfer in Liquid Sodium," *Nucl. Sci. Eng.* Vol. 8 (1960), pp. 369–377.

30. J. Gryzagoridis, "Natural Convection from a Vertical Flat Plate in the Low Grashof Number Range," *Int. J. Heat Mass Transfer*, Vol. 14 (1971), pp. 162–164.

31. B. Gebhart, *Heat Transfer*, *2nd ed.*, Chapt. 8. (New York: McGraw Hill Book Co., 1970).

32. C. Y. Warner and V. S. Arpaci, "An Experimental Investigation of Turbulent Natural Convection in Air at Low Pressure along a Vertical Heated Flat Plate," *Int. J. Heat and Mass Trans.*, Vol. 11 (1968), p. 397.

33. G. C. Vliet, "Natural Convection Local Heat Transfer on Constant

Heat Flux Inclined Surfaces," *ASME Trans.*, ser. *C, J. Heat Transfer*, vol. 91 (1969), pp. 511–516.

34. G. C. Vliet and C. K. Lin, "An Experimental Study of Turbulent Natural Convection Boundary Layers,' *ASME Trans.*, ser. *C, J. Heat Transfer*, Vol. 91 (1969), pp. 517–531.

35. T. Yuge, "Experiments on Heat Transfer from Spheres Including Combined Natural and Forced Convection," *ASME Trans.*, ser. *C, J. Heat Transfer*, Vol. 82 (1960), pp. 214–220.

36. P. H. Oosthuizen and E. Donaldson, "Free Convection Heat Transfer from Vertical Cones," *Trans. ASME*, ser. *C, J. Heat Transfer*, Vol. 94 (1972), pp. 330–331.

8

Forced convection inside tubes and ducts

8-1. Introduction

The heating and cooling of fluids flowing inside conduits are among the most important heat-transfer processes in engineering. The design and analysis of all types of heat exchangers requires a knowledge of the heat-transfer coefficient between the wall of the conduit and the fluid flowing inside it. The sizes of boilers, economizers, superheaters, and preheaters depend largely on the unit-convective conductance between the inner surface of the tubes and the fluid. Also, in the design of air-conditioning and refrigeration equipment, it is necessary to evaluate heat-transfer coefficients for fluids flowing inside ducts. Once the heat-transfer coefficient for a given geometry and specified flow conditions is known, the rate of heat transfer at the prevailing temperature difference can be calculated from the equation (Eq. 1-13)

$$q_c = \bar{h}_c A\,(T_{\text{surface}} - T_{\text{fluid}})$$

The same relation can also be used to determine the area required to transfer heat at a specified rate for a given temperature potential.

The heat-transfer coefficient $\bar{h}_c$ can be calculated from the Nusselt number $\bar{h}_c D_H/k$, as shown in Sec. 6-4. For flow in long tubes or conduits

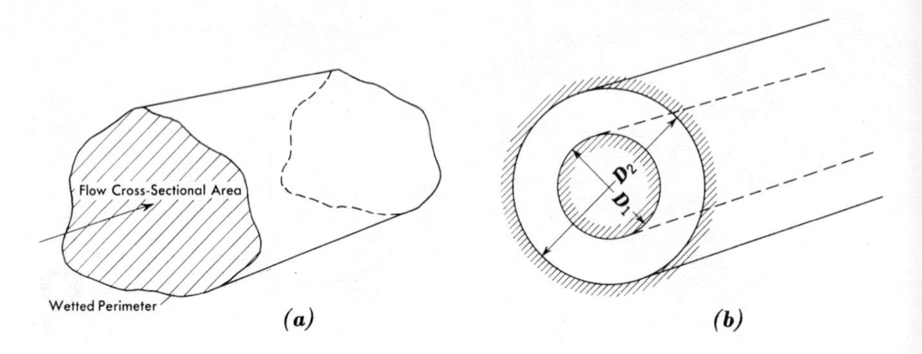

Fig. 8-1. Hydraulic diameter.

(Fig. 8-1a) the significant length in the Nusselt number is the hydraulic diameter D_H, defined as

$$D_H = 4 \frac{\text{flow cross-sectional area}}{\text{wetted perimeter}} \tag{8-1}$$

For a tube or a pipe the flow cross-sectional area is $\pi D^2/4$, the wetted perimeter is πD, and therefore the inside diameter of the tube equals the hydraulic diameter. For an annulus formed between two concentric tubes (Fig. 8-1b) we have

$$D_H = 4 \frac{(\pi/4)(D_2^2 - D_1^2)}{\pi(D_1 + D_2)} = D_2 - D_1 \tag{8-1a}$$

In engineering practice the Nusselt number for flow in conduits is usually evaluated from empirical equations based on experimental results, although in recent years semi-analytic methods of approach have made considerable strides toward an understanding of the basic principles of forced convection in tubes and annuli. From a dimensional analysis, as shown in Sec. 6-6, the experimental results obtained in forced-convection heat-transfer experiments can be correlated by an equation of the form

$$\text{Nu} = \phi(\text{Re})\psi(\text{Pr}) \tag{8-2}$$

where the symbols ϕ and ψ denote functions of the Reynolds number and Prandtl number respectively.

Selection of reference fluid temperature. The convective heat-transfer coefficient used to build the Nusselt number for heat transfer to a fluid flowing in a conduit is defined by Eq. 1-13. The numerical value of

$\bar{h}_c$, as mentioned previously, depends on the choice of the reference temperature in the fluid. For flow over a plane surface the temperature of the fluid far away from the heat source is generally constant, and its value is a natural choice for the fluid temperature in Eq. 1-13. In heat transfer to or from a fluid flowing in a conduit, the temperature of the fluid does not level out but varies both along the direction of mass flow and in the direction of heat flow. At a given cross section of the conduit, the temperature of the fluid at the center could be selected as the reference temperature in Eq. 1-13. However, the center temperature is difficult to measure in practice; furthermore, it is not a measure of the change in internal energy of all the fluid flowing in the conduit. It is therefore a common practice, and one we shall follow here, to use the average bulk temperature T_b as the reference fluid temperature in Eq. 1-13. The average bulk temperature at a station of the conduit is often called the cup mixing temperature because it is the temperature which the fluid passing a cross-sectional area of the conduit during a given time interval would assume if the fluid were collected and mixed in a cup.

The use of the fluid bulk temperature as the reference temperature in Eq. 1-13 allows us to make heat balances readily because, in the steady state, the difference in the average bulk temperature between two sections of a conduit is a direct measure of the rate of heat transfer, or

$$q = mc_p \Delta T_b$$

where q = rate of heat transfer to fluid, in Btu/hr;
m = flow rate, in lb_m/hr;
c_p = specific heat at constant pressure, in Btu/lb_mF;
ΔT_b = difference in bulk temperature between cross sections in question.

The problems associated with variations of the bulk temperature in the direction of flow will be considered in detail in Chapter 11, where the analysis of heat exchangers is taken up. For preliminary calculations, it is common practice to use the bulk temperature halfway between the inlet and the outlet section of a duct as the reference temperature in Eq. 1-13. This procedure is satisfactory when the wall temperature of the duct is constant, but requires some modification when the heat is transferred between two fluids separated by a wall as, for example, in a heat exchanger where one fluid flows inside a pipe while another passes over the outside of the pipe. Although this type of problem is of considerable practical importance, it will not concern us in this chapter, where the emphasis is placed on the evaluation of convective heat-transfer coefficients, which can be determined in a given flow system when the pertinent bulk and wall temperatures are specified.

Effect of Reynolds number on heat transfer and pressure drop in fully established flow. For a given fluid the Nusselt number depends primarily on the flow conditions, which can be characterized by the Reynolds number Re. For flow in long conduits the characteristic length in the Reynolds number, as in the Nusselt number, is the hydraulic diameter, or

$$\mathrm{Re}_{D_H} = \frac{V D_H \rho}{\mu} = V D_H/\nu$$

In long ducts, where the entrance effects are not important, the flow is laminar when the Reynolds number is below 2100. In the range of Reynolds numbers between 2100 and 10,000, the transition from laminar to turbulent flow takes place. The flow in this regime is called transitional. At a Reynolds number of about 10,000, the flow becomes fully turbulent.

In laminar flow through a duct, just as for laminar flow over a plate, there is no mixing of warmer and colder fluid particles by eddy motion and the heat transfer takes place solely by conduction. Since all fluids with the exception of liquid metals have small thermal conductivities, the heat-transfer coefficients in laminar flow are relatively small. In transitional flow a certain amount of mixing occurs by means of eddies which carry warmer fluid into cooler regions, and vice versa. Since the mixing motion, even if it is only on a small scale, accelerates the transfer of heat considerably, a marked increase in the heat-transfer coefficient occurs above Re = 2100. This is illustrated in Fig. 8-2 where experimentally measured values of the average Nusselt number for atmospheric air flowing through a 60-in.-long, 1-in.-ID heated tube are plotted as a function of the Reynolds number. Since the Prandtl number for air does not vary appreciably, Eq. 8-2 reduces to Nu = ϕ(Re), and the curve drawn through the experimental points shows the dependence of Nu on the flow conditions. We note that, in the laminar regime, the Nusselt number remains small, increasing from about 2.2 at Re = 200 to 5.0 at Re = 2100. Above a Reynolds number of 2100, the Nusselt number begins to increase rapidly until the Reynolds number reaches about 8000. As the Reynolds number is further increased, the Nusselt number continues to increase, but at a slower rate.

A qualitative explanation for this behavior can be given by observing the fluid-flow field shown schematically in Fig. 8-3. At Reynolds numbers above 8,000, the flow inside the conduit is fully turbulent except for a very thin layer of fluid adjacent to the wall. In this layer turbulent eddies are damped out as a result of the viscous forces which predominate near the surface, and therefore heat flows through it mainly by conduction. The edge of this so-called laminar sublayer is indicated by a dotted line in Fig. 8-3. The flow beyond it is turbulent and the circular arrows in the turbulent-flow regime represent the eddies which sweep the edge of the laminar layer, probably penetrate it, and carry along with

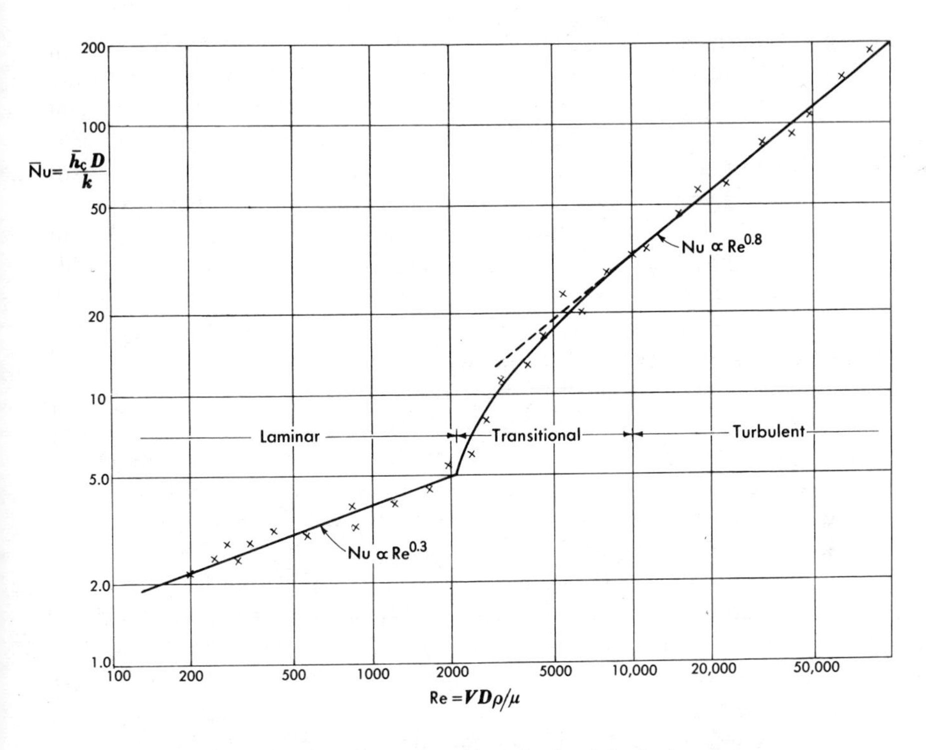

Fig. 8-2. Nusselt number vs. Reynolds number for air flowing in a pipe.

them fluid at the temperature prevailing there. The eddies mix the warmer and cooler fluids so effectively that heat is transferred very rapidly between the edge of the laminar boundary layer and the turbulent bulk of the fluid. It is thus apparent that, except for fluids of high thermal conductivity (e.g., liquid metals), the thermal resistance of the laminar layer controls the rate of heat transfer, and that most of the temperature drop between the bulk of the fluid and the surface of the conduit occurs in this layer. The turbulent portion of the flow field, on the other hand, offers little resistance to the flow of heat. The only effective method of increasing the heat-transfer coefficient is therefore to decrease the thermal

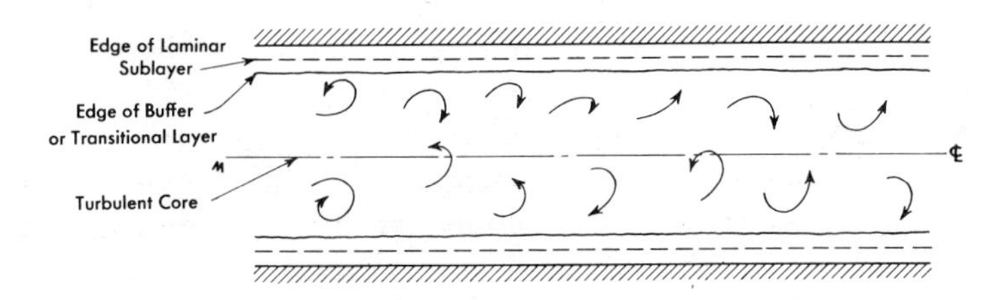

Fig. 8-3. Flow pattern for a fluid flowing turbulently through a pipe.

resistance of the laminar boundary layer. This can be accomplished by increasing the turbulence in the main stream so that the turbulent eddies can penetrate deeper into the laminar layer. An increase in turbulence, however, is accompanied by large energy losses which increase the frictional pressure drop in the conduit. In the design and selection of industrial heat exchangers, where not only the initial cost but also the operating expenses must be considered, the pressure drop is an important factor. An increase of the flow velocity yields higher heat-transfer coefficients which, in accordance with Eq. 1-13, decrease the size and consequently also the initial cost of the equipment for a specified heat-transfer rate. At the same time, however, the pumping cost increases. The optimum design therefore requires a compromise between the initial and operating costs. In practice it has been found that increases in pumping costs and operating expenses often outweigh the saving in the initial cost of heat-transfer equipment under continuous operating conditions. As a result, the velocities used in a majority of commercial heat-exchange equipment are relatively low, corresponding to Reynolds numbers of no more than 50,000. Whenever possible, laminar flow is avoided in heat-exchange equipment because of the low heat-transfer coefficients obtained. However, in the chemical industry, where frequently very viscous liquids must be handled, laminar flow sometimes cannot be avoided without producing undesirably large pressure losses.

It was shown in Sec. 6-10 that for turbulent flow of liquids and gases over a flat plate, the Nusselt number is proportional to the Reynolds number raised to the 0.8 power. Since in turbulent forced convection the laminar sublayer generally controls the rate of heat flow irrespective of the geometry of the system, it is not surprising that also for turbulent forced convection in conduits the Nusselt number is related to the Reynolds number by the same type of power law. For the case of air flowing in a pipe, this relation is illustrated in the graph of Fig. 8-2.

Effect of Prandtl number. The Prandtl number Pr is a function of the fluid properties alone. It has been defined previously as the ratio of the kinematic viscosity of the fluid to the thermal diffusivity of the fluid, that is,

$$\mathrm{Pr} = \frac{\nu}{a} = \frac{c_p \mu}{k}$$

The kinematic viscosity ν, or μ/ρ, is often referred to as the molecular diffusivity of momentum because it is a measure of the rate of momentum transfer between the molecules. The thermal diffusivity of a fluid $k/c_p\rho$ is often called the molecular diffusivity of heat. It is a measure of the ratio of the heat transmission and energy storage capacities of the molecules.

The Prandtl number relates the temperature distribution to the velocity distribution, as shown in Secs. 6-7 and 6-10 for flow over a flat plate. For flow in a pipe, just as over a flat plate, the velocity and temperature profiles are similar for fluids having a Prandtl number of unity. When the Prandtl number is smaller, the temperature gradient near a surface is less steep than the velocity gradient, and for fluids whose Prandtl number is larger than one, the temperature gradient is steeper than the velocity gradient. The effect of the Prandtl number on the temperature gradient in turbulent flow at a given Reynolds number in tubes is illustrated schematically in Fig. 8-4, where temperature profiles at different Prandtl numbers are shown at $Re_D = 10{,}000$. These curves reveal that, at a specified Reynolds number, the temperature gradient at the wall is steeper in a fluid having a large Prandtl number than in a fluid having a small Prandtl number. Consequently, at a given Reynolds number fluids with larger Prandtl numbers have larger Nusselt numbers.

Liquid metals generally have a high thermal conductivity and a small specific heat; their Prandtl numbers are therefore small, ranging from

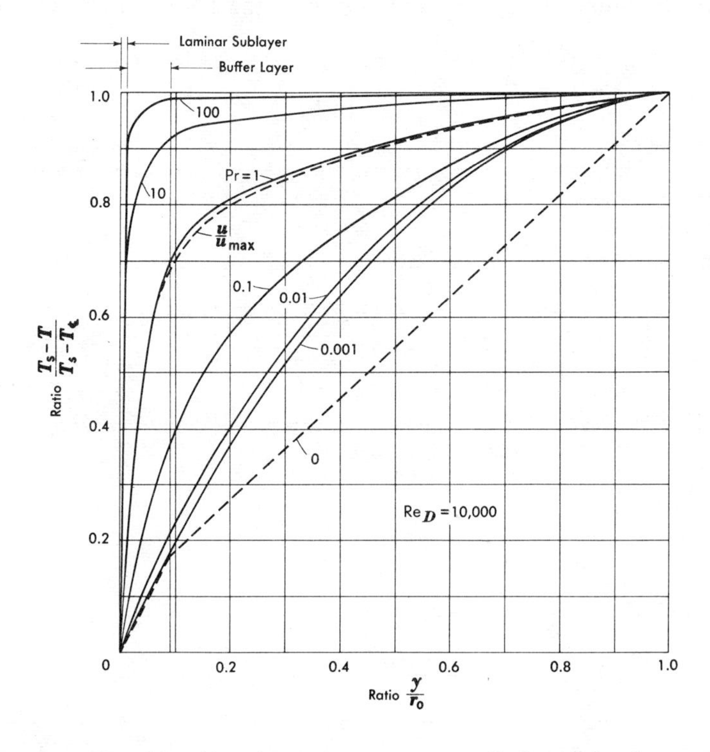

Fig. 8-4. Effect of Prandtl number on the temperature profile for turbulent flow in a long pipe. (Extracted from "Heat Transfer to Molten Metals," by R. C. Martinelli, *Trans. ASME*, Vol. 69, 1947, with permission of the publishers, The American Society of Mechanical Engineers).

0.005 to 0.01. The Prandtl numbers of gases range from 0.6 to 0.9. Most oils, on the other hand, have large Prandtl numbers because their viscosity is large and their thermal conductivity is small.

Entrance effects. In addition to the Reynolds number and the Prandtl number, several other factors can influence the conditions of heat transfer by forced convection. For example, when the conduit is short ($L/D_H < 50$), entrance effects are important. As a fluid enters a duct with a uniform velocity, the fluid immediately adjacent to the tube wall is brought to rest. For a short distance from the entrance a laminar boundary layer is formed along the tube wall. If the turbulence in the entering fluid stream is high, the boundary layer will quickly become turbulent. Irrespective of whether the boundary layer remains laminar or becomes turbulent, it will increase in thickness until it fills the entire duct. From this point on, the velocity profile across the duct remains essentially unchanged.

The development of the thermal boundary layer in a fluid which is heated or cooled in a duct is qualitatively similar to that of the hydrodynamic boundary layer. At the entrance, the temperature is generally

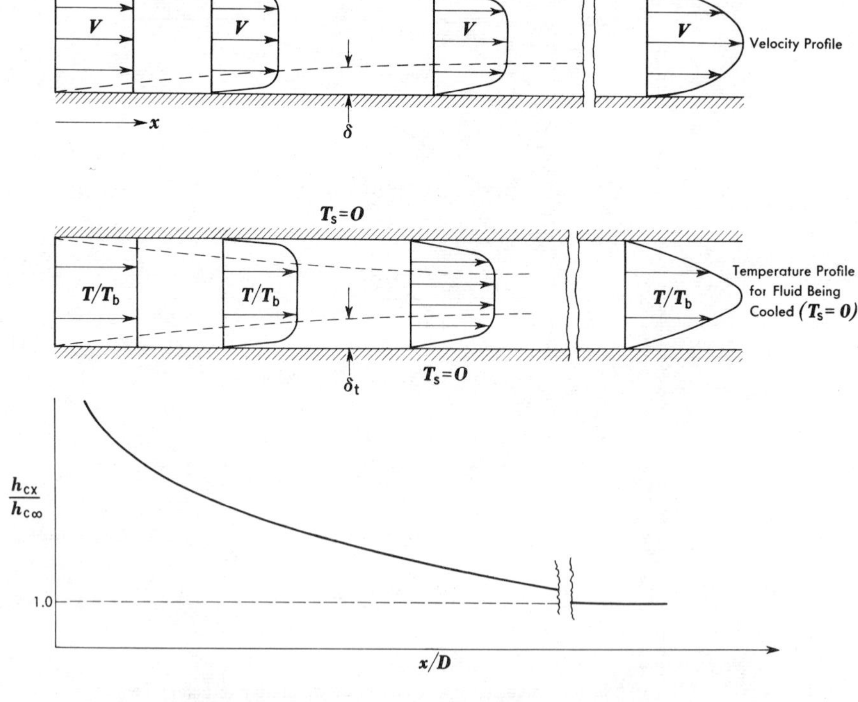

Fig. 8-5. Velocity distributions, temperature profiles and variation of the unit-convective conductance near the inlet of a tube for air being cooled in laminar flow.

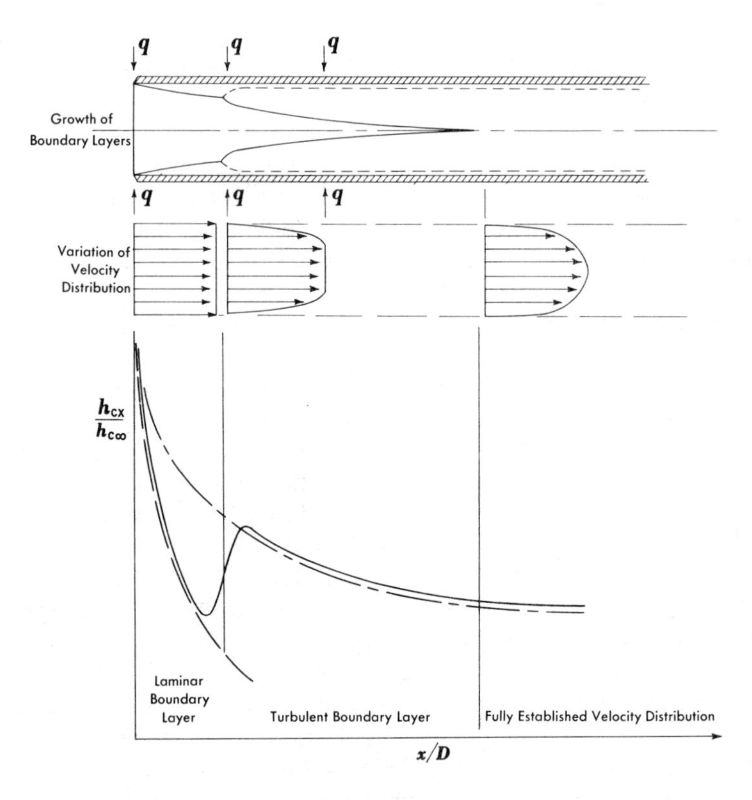

Fig. 8-6. Velocity distribution and variation of unit-convective conductance near the entrance of a tube for a fluid in turbulent flow.

uniform transversely, but as the fluid flows along the duct, the heated or cooled layer increases in thickness until heat is transferred to or from the fluid in the center of the duct. Beyond this point the temperature profile remains essentially constant if the velocity profile is fully established.

The final shapes of the velocity and temperature profiles depend on whether the fully developed flow is laminar or turbulent. Figures 8-5 and 8-6 illustrate qualitatively the growths of the boundary layers as well as the variations in the local unit-convective conductance near the entrance of a tube for laminar and turbulent conditions respectively. An inspection of these figures shows that the unit-thermal conductance varies considerably near the entrance. If the entrance is square-edged, as in most heat exchangers, the initial development of the hydrodynamic and thermal boundary layers along the walls of the tube is quite similar to that along a flat plate. Consequently, the conductance is largest near the entrance and decreases along the duct until both the velocity and the temperature profiles for the fully developed flow have been established. If the pipe Reynolds number for the fully developed flow $VD\rho/\mu$ is below 2100, the entrance effects may be appreciable for a length as much as 50 diameters from the entrance. For velocities corresponding to turbulent-pipe Reynolds numbers, the entrance effects disappear about 10 diameters from the entrance.

Variation of physical properties. Another factor which can influence the heat transfer and friction considerably is the variation of physical properties with temperature. When a fluid flowing in a duct is heated or cooled, its temperature, and consequently also its physical properties, vary along the duct as well as over any given cross section. For liquids, only the temperature dependence of the viscosity is of major importance. For gases, on the other hand, the temperature effect on the physical properties is more complicated than for liquids because the thermal conductivity and the density, in addition to the viscosity, vary significantly with temperature. In either case, the numerical value of the Reynolds number depends on the location at which the viscosity is evaluated. It is believed that the Reynolds number based on the bulk temperature is the significant parameter to describe the flow conditions. However, considerable success in the empirical correlation of experimental heat-transfer data has been achieved by evaluating the viscosity at an *average film* temperature, defined as a temperature approximately halfway between the wall and the bulk mean temperatures. Another method of taking account of the variation of physical properties with temperature is to evaluate all properties at the bulk mean temperature and to correct for the thermal effects by multiplying the right-hand side of Eq. 8-2 by a function proportional to the ratio of bulk to wall temperatures or viscosities. The latter is preferred because it is simpler to apply in practice and can also be justified on the basis of more advanced boundary-layer theory.

Thermal boundary conditions and compressibility effects. For fluids having a Prandtl number of unity or less, the heat-transfer coefficient also depends on the thermal-boundary condition. For example, in geometrically similar liquid metal heat-transfer systems a uniform wall temperature yields smaller convective conductances than a uniform heat input at the same Reynolds and Prandtl numbers (26,27,28). When heat is transferred to or from gases flowing at very high velocities, compressibility effects influence the flow and the heat transfer. Problems associated with heat transfer to or from fluids at high Mach numbers are discussed in Refs. 30 to 32.

Limits of accuracy in predicted values of convective heat-transfer coefficients. In the application of any empirical equation for forced convection to practical problems it is important to bear in mind that the predicted values of the heat-transfer coefficient are not exact. The results obtained by various experimenters, even under carefully controlled conditions, differ appreciably. In turbulent and in laminar flow the accuracy of a heat-transfer coefficient predicted from any available equation or graph may be no better than 30 percent. In the transition region, where experimental data are scant, the accuracy of the Nusselt number predicted from available information may be even lower.

8-2. Analogy between heat and momentum transfer

To illustrate the most important physical variables affecting heat transfer by turbulent forced convection to or from fluids flowing in a long tube or duct, we shall apply the analogy between heat and momentum transfer. The basic concepts of this analogy, introduced by Osborn Reynolds in 1874 (1), have been discussed in Sec. 6-9. The basic analogy was later improved by Prandtl (2), and additional refinements, particularly applicable to forced convection in circular ducts, were made over the years by von Karman (3), Boelter et al. (4), Martinelli (5), and most recently by Deissler (6,7). In this section we shall develop the analogy for pipe flow only in its simplest form and then present some of the important practical results of the more advanced refinements.

The assumptions necessary for the simple analogy are valid only for fluids having a Prandtl number of unity, but the fundamental relation between heat transfer and fluid friction for flow in ducts can be illustrated for this case without introducing mathematical difficulties. The results of the simple analysis can also be extended to other fluids by means of empirical correction factors, as will be shown in Sec. 8-3.

The rate of heat flow per unit area in a fluid can be related to the temperature gradient by the equation developed in Chapter 6

$$\frac{q}{A\rho c_p} = -\left(\frac{k}{\rho c_p} + \epsilon_H\right)\frac{dT}{dy} \tag{6-56}$$

This relation, as shown in Sec. 6-9, takes into account the heat flow by conduction as well as by eddy convection. In purely laminar flow $\epsilon_H = 0$, and, except for liquid metals, the term $k/\rho c_p$ is negligible in highly turbulent motion. Similarly, the shearing stress caused by the combined action of the viscous forces and the turbulent momentum transfer is given by

$$\frac{\tau g_c}{\rho} = \left(\frac{\mu}{\rho} + \epsilon_M\right)\frac{du}{dy} \tag{6-50}$$

According to the Reynolds analogy, heat and momentum are transferred by analogous processes in turbulent flow. Consequently, both q and τ vary with y, the distance from the surface, in the same manner. For fully developed turbulent flow in a pipe, the local shearing stress decreases linearly with the radial distance r. Hence we can write

$$\frac{\tau}{\tau_s} = \frac{r}{r_s} = 1 - \frac{y}{r_s} \tag{8-3}$$

and

$$\frac{q/A}{(q/A)_s} = \frac{r}{r_s} = 1 - \frac{y}{r_s} \tag{8-4}$$

where the subscript s denotes conditions at the inner surface of the pipe. Introducing Eqs. 8-3 and 8-4 into Eqs. 6-50 and 6-56 respectively yields

$$\frac{\tau_s g_c}{\rho}\left(1 - \frac{y}{r_s}\right) = \left(\frac{\mu}{\rho} + \epsilon_M\right)\frac{du}{dy} \tag{8-5}$$

and

$$\frac{q_s}{A_s \rho c_p}\left(1 - \frac{y}{r_s}\right) = -\left(\frac{k}{\rho c_p} + \epsilon_H\right)\frac{dT}{dy} \tag{8-6}$$

If $\epsilon_H = \epsilon_M$, the brackets on the right-hand side of Eqs. 8-5 and 8-6 are equal, provided the molecular diffusivity of momentum μ/ρ equals the molecular diffusivity of heat $k/\rho c_p$, that is, when the Prandtl number is unity. Dividing Eq. 8-6 by Eq. 8-5 yields under these restrictions

$$\frac{q_s}{A_s c_p g_c \tau_s} du = -dT \tag{8-7}$$

Equation 8-7 can be integrated between the wall where $u = 0$ and $T = T_s$, and the bulk of the fluid where $u = V$ and $T = T_b$. The integration then yields

$$\frac{q_s V}{A_s c_p g_c \tau_s} = T_s - T_b \tag{8-8}$$

which can also be written in the form

$$\frac{\tau_s g_c}{\rho V^2} = \frac{q_s}{A_s(T_s - T_b)}\frac{1}{c_p \rho V} = \frac{\bar{h}_c}{c_p \rho V} \tag{8-9}$$

since $\bar{h}_c$ is by definition equal to $q_s/A_s(T_s - T_b)$. Multiplying the numerator and the denominator of the right-hand side of Eq. 8-9 by $D_H \mu k$ and regrouping yields

$$\frac{\bar{h}_c}{c_p \rho V} \frac{D_H \mu k}{D_H \mu k} = \left(\frac{\bar{h}_c D_H}{k}\right)\left(\frac{k}{c_p \mu}\right)\left(\frac{\mu}{V D_H \rho}\right) = \frac{\text{Nu}}{\text{Re}\,\text{Pr}}$$

which we recognize as the Stanton number, St. To bring the left-hand side of Eq. 8-9 into a more convenient form, we make a force balance on a cylindrical mass of fluid as shown in Fig. 8-7. The pressure difference $p_1 - p_2$ exerts the force $(p_1 - p_2)\,\pi D^2/4$, which is balanced in steady flow by the shear at the wall, or

$$(p_1 - p_2)\frac{\pi D^2}{4} = \tau_s \pi D L \tag{8-10}$$

Solving for the wall shear per unit area yields

$$\tau_s = \frac{(p_1 - p_2)\,D}{4L} \tag{8-11}$$

In fluid mechanics the pressure drop is usually expressed in terms of a drag-friction coefficient f as

$$p_1 - p_2 = f\frac{L}{D}\frac{\rho V^2}{2g_c} \tag{8-12}$$

Substituting Eq. 8-12 for $p_1 - p_2$ in Eq. 8-11 gives

$$\tau_s = f\frac{\rho V^2}{8g_c} \tag{8-13}$$

Substituting Eq. 8-13 for τ_s in Eq. 8-9 finally yields the equation

$$\text{St} = \frac{\text{Nu}}{\text{Re}\,\text{Pr}} = \frac{f}{8} \tag{8-14}$$

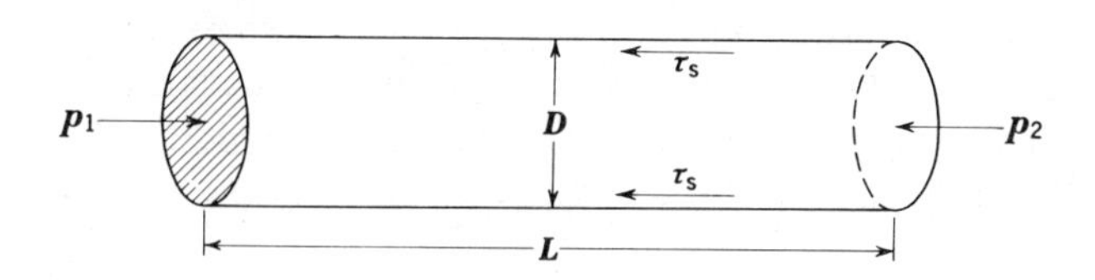

Fig. 8-7. Sketch illustrating nomenclature for force balance on a fluid element in a tube.

which is known as the *Reynolds analogy*.[1] It agrees fairly well with experimental data for heat transfer in gases whose Prandtl number is nearly unity.

According to experimental data for fluids flowing in smooth tubes in the range of Reynolds numbers from 10,000 to 120,000, the friction coefficient f is given by the empirical relation

$$f = 0.184 \, \mathrm{Re}_D^{-0.2} \tag{8-15}$$

Using this relation, Eq. 8-14 can be written as

$$\mathrm{St} = \frac{\mathrm{Nu}}{\mathrm{Re}\,\mathrm{Pr}} = 0.023 \, \mathrm{Re}_D^{-0.2} \tag{8-16}$$

or, since Pr was assumed unity, as

$$\mathrm{Nu} = 0.023 \, \mathrm{Re}_D^{0.8} \tag{8-17}$$

or

$$\bar{h}_c = 0.023 \, V^{0.8} D^{-0.2} k \left(\frac{\mu}{\rho}\right)^{-0.8}$$

We observe that, in fully established turbulent flow, the convective-unit conductance is directly proportional to the velocity raised to the 0.8 power and inversely proportional to the tube diameter raised to the 0.2 power. For a given flow rate, an increase in the tube diameter reduces the velocity and thereby causes a decrease in $\bar{h}_c$ proportional to $1/D^{1.8}$. The use of small tubes and high velocities is therefore conducive to large heat-transfer coefficients, but at the same time the power required to overcome the frictional resistance is increased. In the design of heat-exchange equipment it is therefore necessary to strike a balance between the gain in heat-transfer rates achieved by the use of ducts having small cross-sectional areas, and the accompanying increase in pumping requirements.

Figure 8-8 shows the effect of surface roughness on the friction coefficient. We observe that the friction coefficient increases appreciably with the relative roughness, defined as ratio of the average asperity height ϵ to the diameter D. According to Eq. 8-14 one would expect that roughening the surface, which increases the friction coefficient, also increases the convective conductance. Experiments performed by Cope (8) are qualitatively in agreement with this prediction, but even a considerable

---

[1] The Reynolds analogy can be extended to mass transfer. The analogies among mass, heat, and momentum transfer will be discussed in Chapter 12.

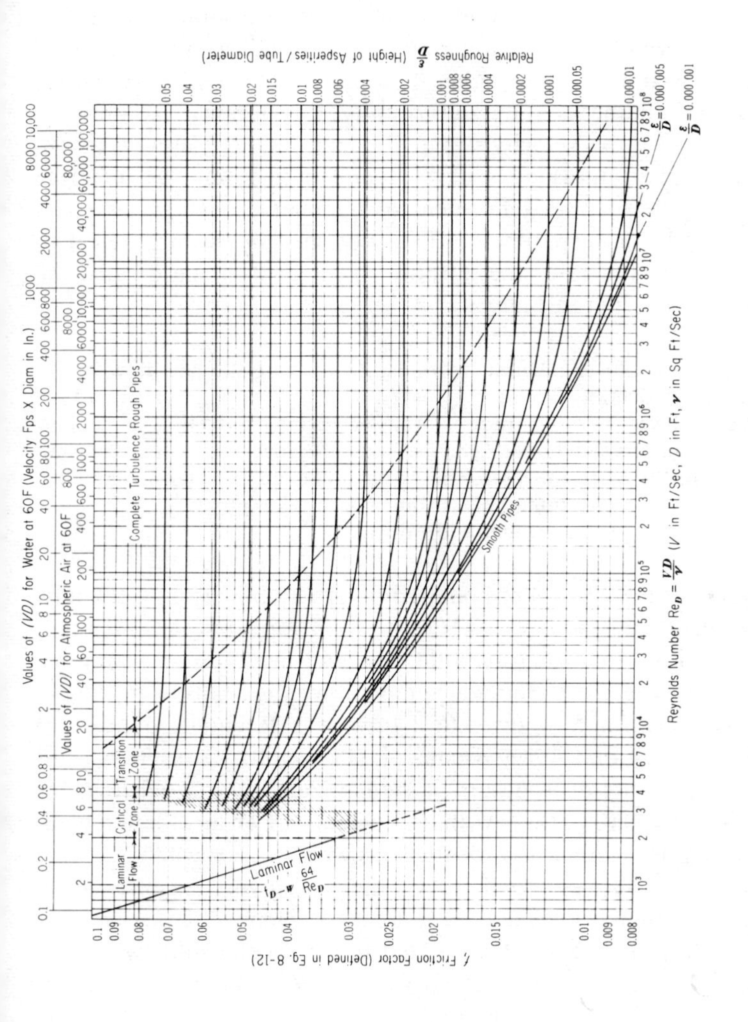

Fig. 8-8. Friction factor vs. Reynolds number for laminar and turbulent flow in tubes of various surface roughnesses. (Extracted from "Friction Factor for Pipe Flow," by L. F. Moody, published in *Trans. ASME*. Vol. 66, 1944, with permission of the publishers, The American Society of Mechanical Engineers)

increase in surface roughness improves the rate of heat transfer only very little.[2] Since an increase in the surface roughness causes a substantial increase in the frictional resistance, Cope found that, for the same pressure drop, the rate of heat transfer obtained from a smooth tube is larger than from a rough one.

The equations relating the Nusselt number to the flow conditions have been developed above for fluids having a Prandtl number of unity. The analogy between heat and momentum transfer has also been applied to fluids having Prandtl numbers other than unity (2,3,4,5,6,7). However, when the Prandtl number is not equal to unity, it is necessary to obtain a relationship between the velocity u and the coordinate y, as well as between the molecular diffusivities of heat and momentum, to integrate Eq. 8-6. Martinelli (5), in a refinement of the simple analogy, assumed that $\epsilon_M = \epsilon_H$ and used experimental data obtained by Nikuradse (9) to perform the integration. For the purpose of his analysis he divided the flow field into three separate regions:

1. A laminar sublayer adjacent to the surface where the heat-flow mechanism is conduction alone and $\epsilon_M = \epsilon_H = 0$.
2. A buffer layer in which heat is transferred by conduction as well as convection. In this buffer layer between the edge of the laminar sublayer and the turbulent core, the eddies build up in intensity and the transition between laminar and turbulent flow takes place.
3. A turbulent core in the center of the duct.

Although, as Deissler (7) has pointed out, the subdivision of the flow field is somewhat artificial and ceases to be valid for large Prandtl numbers, it is helpful in visualizing the fluid flow and heat-flow mechanisms. Figure 8-4 shows the cross-sectional temperature distribution in dimensionless coordinates for flow through a tube at a Reynolds number of 10,000. We observed that, for a viscous oil (Pr = 100), about 95 percent of the total temperature drop occurs in the laminar sublayer, whereas for a liquid metal (Pr = 0.01), it is less than 5 percent. For air, the temperature and velocity fields are nearly identical, as would be expected from the Reynolds analogy. As mentioned earlier, the reason why the thermal resistance of the laminar sublayer is only a small fraction of the total resistance in the case of a liquid metal is that the molecular diffusivity term $k/\rho c_p$ in Eq. 8-6 is much larger than ϵ_H when the thermal conductivity of the liquid is large. Hence, the main contribution to the total heat transfer comes from the conduction mechanism when the Prandtl number is small, whereas for fluids having a large Prandtl number the conduction is negligible compared to mixing in the bulk of the fluid.

---

[2]The effect of surface roughness on heat transfer to air has been investigated recently in detail by Nunner (23). The results of his theory and experiments corroborate Cope's conclusion qualitatively.

8-3. Heat-transfer coefficients for turbulent flow

The final expressions obtained from more advanced analogies are very complicated and the evaluation of the Nusselt number under given flow and thermal-boundary conditions requires usually a numerical integration. For this reason it is more convenient for engineering purposes to use semi-empirical equations, or graphs. In this section we shall present some of the engineering equations and graphs relating the Nusselt number to the Reynolds number, Prandtl number, the geometrical configuration of the system, the temperature gradient, and the thermal boundary condition.

For fluids having Prandtl numbers in the range from 0.5 to 100, Colburn (10) recommends, on the basis of experimental data, that the Stanton number in Eq. 8-16 be multiplied by $\mathrm{Pr}^{2/3}$, or

$$\mathrm{St}\mathrm{Pr}^{2/3} = j = 0.023\,\mathrm{Re}^{-0.2} = \frac{f}{8} \qquad \text{AT } T_f \qquad (8\text{-}18)$$

SEE (8-20)

The term $\mathrm{St}\mathrm{Pr}^{2/3}$ is usually called the Colburn j-factor in the heat-transfer literature.

To account for the variation in physical properties due to the temperature gradient, all of the physical properties in Eq. 8-18 should be evaluated at the average film temperature of the fluid T_f defined as

$$T_f = 0.5(T_s + T_b) \qquad (8\text{-}19)$$

where T_s is the temperature of the heat-transfer surface, or the wall temperature (11).

Denoting properties evaluated at T_f by the subscript f, Eq. 8-18 can be written as

$$\frac{\bar{h}_c}{c_p G} = 0.023 \left(\frac{\mu_f}{D_H G}\right)^{0.2} \mathrm{Pr}_f^{-2/3} \qquad (8\text{-}20)$$

where $G = \rho V$, i.e., the mass velocity per square foot of cross section in $\mathrm{lb_m/hr\ sq\ ft}$. Equation 8-20 has been found to correlate the results of numerous experimenters for moderate temperature differences, $T_s - T_b$, within 30 percent. In many practical situations the wall temperature and the bulk temperature are unfortunately not directly available, and then a trial-and-error solution becomes necessary. For this type of problem the Stanton number can often be evaluated more conveniently by a method which was originally suggested by Sieder and Tate (12) and later improved by Kays and London (13). This method uses, for gases flowing in long ducts, an equation of the type

$$\mathrm{StPr}^{2/3} = C\,\mathrm{Re}^{-0.2}\left(\frac{T_b}{T_s}\right)^n \tag{8-21}$$

and, for liquids, an equation of the type

$$\mathrm{St} = \phi(\mathrm{Re})\,\psi(\mathrm{Pr})\left(\frac{\mu_b}{\mu_s}\right)^n \tag{8-22}$$

In both of these equations all of the physical properties are evaluated at the *average fluid bulk temperature* T_b, and the variations in physical properties caused by the temperature gradient are accounted for either by the temperature or by the viscosity correction factor. The constant C in Eq. 8-21 which gives the best correlation with the available data for gases is

$C = 0.020$ for a constant duct-wall temperature

and

$C = 0.021$ for constant heat input per unit tube length or constant temperature difference in the flow direction

The exponent of the temperature-correction factor n in Eq. 8-21 is

$n = 0.575$ for gas heating

$n = 0.15$ for gas cooling

For liquids having Prandtl numbers larger than 1.0 the exponent n of the viscosity ratio (μ_s/μ_b) in Eq. 8-22 is

$n = 0.36$ for liquid heating

$n = 0.20$ for liquid cooling

The variation of the Stanton number with the Prandtl number in Eq. 8-22 is shown graphically in Fig. 8-9 for various values of the bulk Reynolds number GD_H/μ. This graph is based on an analysis by Deissler (7) for long circular tubes which is in excellent agreement with available experimental results. Its use is recommended to evaluate the Nusselt number for heating and cooling of liquids when large wall-to-fluid temperature differences exist. For liquids having Prandtl numbers larger than unity, Eq. 8-22 applies to any type of wall-temperature variation, so that no distinction between uniform heat input and uniform wall temperature is necessary.

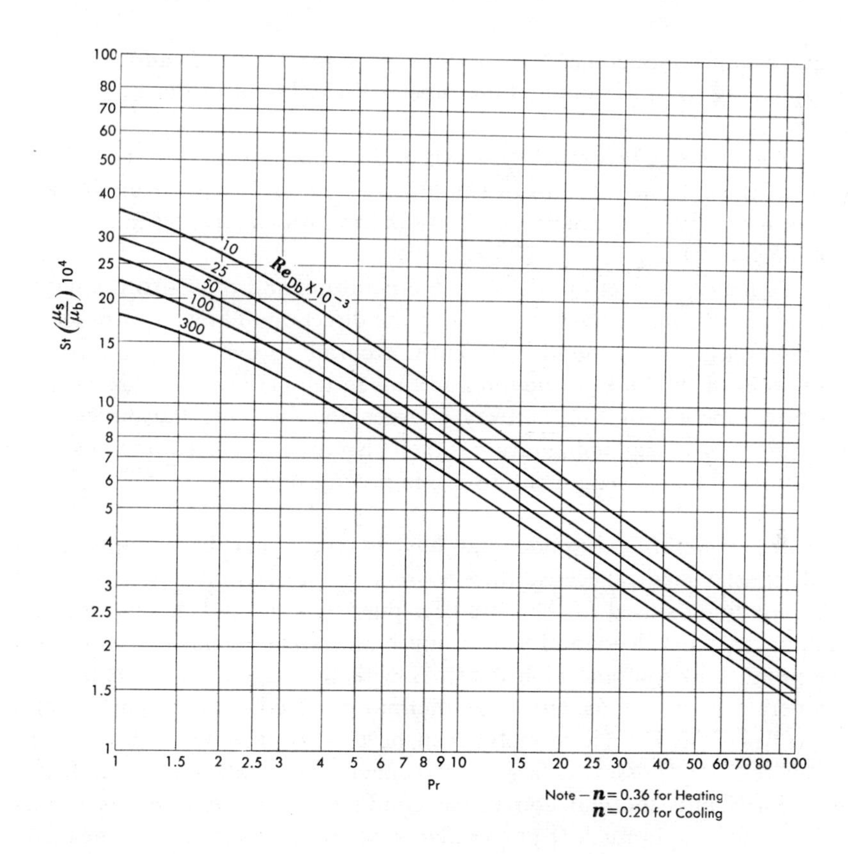

Fig. 8-9. Variation of the Stanton number with Prandtl number for various values of the bulk Reynolds number. (By permission from W. M. Kays and A. L. London, *Compact Heat Exchangers*, National Press, Palo Alto, 1955)

For gases and liquids flowing in *short* circular tubes ($2 < L/D < 60$) with abrupt contraction entrances, the entrance configuration of greatest interest in heat-exchanger design, the entrance effect for Reynolds numbers corresponding to turbulent flow (11) can be represented approximately by the equation

$$\frac{\bar{h}_{cL}}{\bar{h}_c} = 1 + (D/L)^{0.7} \tag{8-23}$$

when L/D is less than 20 but larger than 2, and by the equation

$$\frac{\bar{h}_{cL}}{\bar{h}_c} = 1 + 6\,D/L \tag{8-23a}$$

when L/D is larger than 20. In both of the above equations, $\bar{h}_{cL}$ is the average unit conductance for the tube of finite length L and $\bar{h}_c$ is the conductance for an infinitely long tube evaluated either from Eq. 8-21 or Eq. 8-22.

An extensive theoretical analysis of the heat transfer and the friction drop in the entrance regions of smooth passages is given in Ref. 14, and a complete survey of experimental results for various types of inlet condition in Refs. 21 and 22.

In many applications the fluid temperature, and thus also the physical properties, vary considerably along the direction of flow. For practical purposes it has been found sufficiently accurate to evaluate the physical properties of the bulk of the fluid at a mean temperature with respect to the flow-tube length, i.e., halfway between the inlet and the outlet temperature. This mean temperature is then also used to correct for property variations at a flow section, as discussed previously.

Liquid metals. Liquid metals have in recent years been employed as heat-transfer media because they possess certain advantages over other common liquids used for heat-transfer purposes. Liquid metals, such as sodium, mercury, lead, and lead-bismuth alloys, have relatively low melting points and combine high densities with low vapor pressures at high temperatures as well as with large thermal conductivities, ranging from 5 to 50 Btu/hr ft F. These metals can be used over wide ranges of temperatures, they possess a large heat capacity per unit volume, and also have large unit thermal convective conductances. They are especially suitable for use in nuclear power plants where large amounts of heat are liberated and must be removed in a small volume. Liquid metals pose some difficulties in handling and pumping, but the development of electromagnetic pumps has eliminated most of these problems.

A comprehensive summary of the available information on liquid-metal heat transfer is contained in Refs. 15, 20, and 26. The material presented here has been taken mainly from these references.

In liquid metals the most important heat-transfer mechanism is conduction, and even in a highly turbulent stream the effect of eddying is of secondary importance. As a result, the empirical equations for gases and liquids do not apply. Several theoretical analyses for the evaluation of the Nusselt number are available, but there still exist some unexplained discrepancies between many of the experimental data and the analytic results. This is illustrated in Fig. 8-10 where the experimentally measured Nusselt numbers for heating of mercury in long tubes by various observers are compared with the analysis of Martinelli (5). The results of Martinelli's analysis were simplified by Lyon (15) and Dwyer (26) who found that the equation

$$\mathrm{Nu} = 5.0 + 0.025\,(\mathrm{Re}_D\,\mathrm{Pr})^{0.8} \tag{8-24}$$

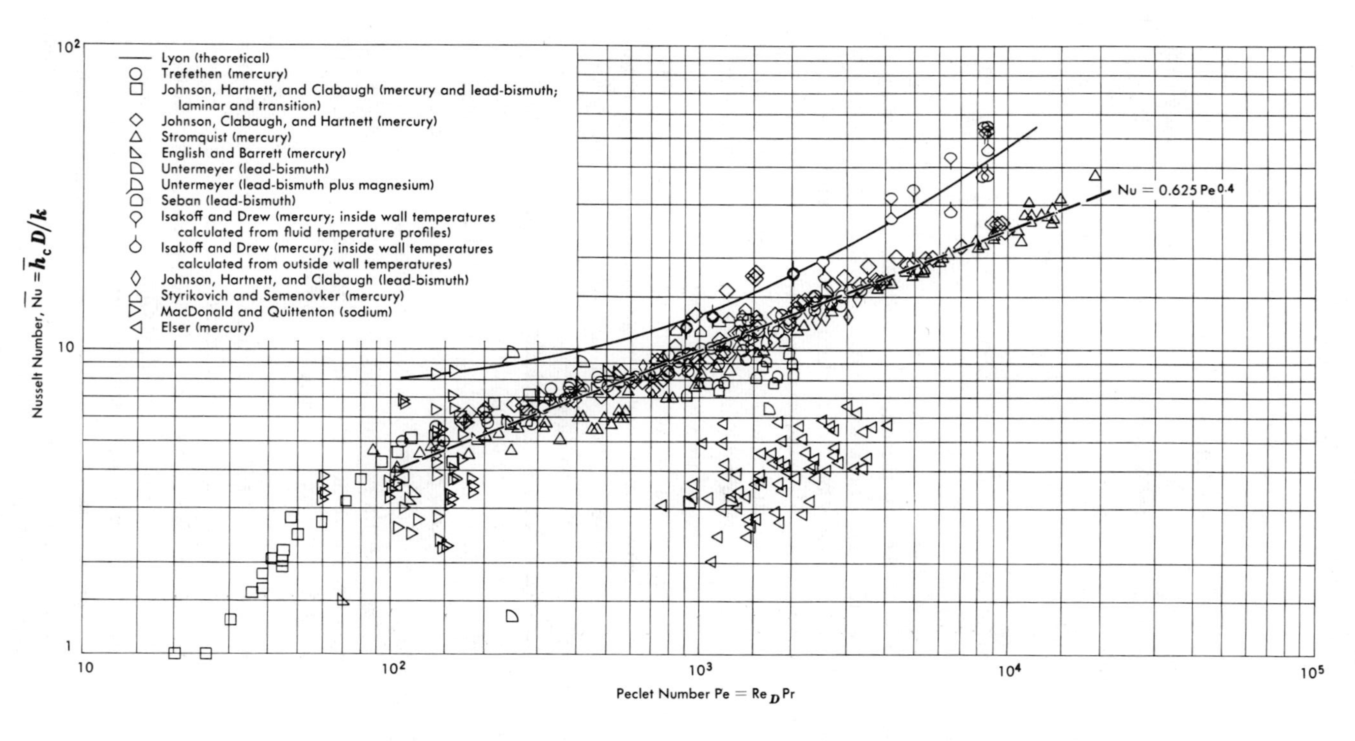

Fig. 8-10. Comparison of measured and predicted Nusselt number for liquid metals heated in long tubes with constant heat input. (Courtesy of National Advisory Committee for Aeronautics, *NACTA TN* 3336)

approximates the available data with satisfactory accuracy for a uniform heat input along the tube, whereas the equation

$$\mathrm{Nu} = 4.0 + 0.025\,(\mathrm{Re}_D\,\mathrm{Pr})^{0.8} \tag{8-25}$$

applies when the tube-wall temperature is constant. An inspection of the experimental data obtained by several independent investigations shows that they fall within 60 to 80 percent of their predicted values. Lubarsky and Kaufman (20) found that the empirical equation

$$\mathrm{Nu} = 0.625\,(\mathrm{Re}_D\,\mathrm{Pr})^{0.4} \tag{8-26}$$

correlates most constant-heat-flux data in the fully developed turbulent flow regime.

Physical properties do not vary significantly in liquid metals because the unit-surface conductance is usually quite large and the difference between the wall and the bulk temperature is, therefore, small. According to a theoretical analysis by Touloukian and Viscanta (25), the effect of property variation can, however, be taken into account by multiplying the Nusselt number evaluated with properties at the wall temperature by a_w/a_b, the ratio of the thermal diffusivity evaluated at the wall temperature to the diffusivity based on the bulk temperature.

Those points in Fig. 8-10 that fall very far below the average are believed to have been obtained in systems where the liquid metal did not wet the surface. However, no final conclusions regarding the effect of wetting have been reached to date.

The entrance corrections obtained from Eqs. 8-23 and 8-23a do not apply to liquid metals. The conditions in the entrance regions for fluids with small Prandtl numbers have been investigated analytically by Deissler (14) and experimental data supporting the analysis are summarized in Refs. 21 and 26. In turbulent flow the thermal entry length, $(L/D_H)_{\mathrm{entry}}$, is approximately 10 equivalent diameters when the velocity profile is already developed, and 30 equivalent diameters when it is developing simultaneously with the temperature profile.

Calculation of the heat-transfer coefficient. The application of the formulas for convective heat transfer requires a knowledge of the physical properties of the medium. As every practicing engineer knows from experience, the physical properties of fluids have only been measured accurately for some of the more common fluids. In many instances it is necessary to estimate the value of properties, especially at elevated temperatures. Since physical properties have been published by investigators of many countries, it is always important to note carefully the units used. The use of dimensionless numbers in the presentation of experimental data has eliminated many of the difficulties involved in the practical application. However, it cannot be overemphasized that a careful checking of the units of the quantities used in building the dimensionless numbers is a prerequisite to obtaining correct results. It is also suggested that a common-sense order-of-magnitude check be applied to the final result. The order of magnitudes for heat-transfer coefficients under typical conditions in Table 1-2 will aid in this task.

EXAMPLE 8-1. Determine the unit thermal convective conductance for water flowing at a velocity of 10 fps in an annulus formed between a 1-in.-OD tube and a $1\frac{1}{2}$-in.-ID tube. The water is at 180 F and is being cooled. The temperature of the inner wall is 100 F, and the outer wall of the annulus is insulated. Neglect entrance effects and compare the results of Eqs. 8-20 and 8-22. The properties of water are given in the accompanying tabulation.

T (F)	μ (lb_m/hr ft)	k (Btu/hr ft F)	ρ (lb_m/cu ft)	c (Btu/lb_m F)
100	1.67	0.36	62.0	1.0
140	1.14	0.38	61.3	1.0
180	0.75	0.39	60.8	1.0

Solution: The hydraulic diameter D_H for this geometry is 0.5 in. The Reynolds number based on the hydraulic diameter and the bulk temperature properties is

$$\text{Re}_{Db} = \frac{VD_H\rho}{\mu} = \frac{(10\ \text{ft/sec})(0.5/12\ \text{ft})(60.8\ \text{lb}_\text{m}/\text{cu ft})(3600\ \text{sec/hr})}{0.75\ \text{lb}_\text{m}/\text{hr ft}}$$

$$= 125{,}000$$

Based on the mean film temperature T_f, the Reynolds number is $\text{Re}_{Df} = 82{,}000$. The Prandtl number at the bulk temperature is

$$\text{Pr}_b = \frac{c\mu}{k} = \frac{(1.0\ \text{Btu/lb}_\text{m}\ \text{F})(0.75\ \text{lb}_\text{m}/\text{hr ft})}{0.39\ \text{Btu/hr ft F}} = 1.92$$

and at T_f, we find that $\text{Pr}_f = 3.0$. According to Eq. 8-20 we have

$$\text{St} = \frac{\bar{h}_c}{c\rho V} = 0.023\,\text{Re}_{Df}^{-0.2}\,\text{Pr}_f^{-2/3}$$

$$= 0.023/(9.6 \times 2.08) = 0.00115$$

so that $\bar{h}_c = (1\ \text{Btu/lb}_\text{m}\ \text{F})(61.3\ \text{lb}_\text{m}/\text{cu ft})(10\ \text{ft/sec})(3600\ \text{sec/hr})(0.00115)$

$$= 2570\ \text{Btu/hr sq ft F}$$

Using Fig. 8-9, we get

$$\text{St}(\mu_s/\mu_b)^n = 16 \times 10^{-4}$$

with $n = 0.20$ for cooling. Therefore, the unit-convective conductance is

$$\bar{h}_c = (16 \times 10^{-4}) \frac{k_b}{D_H} \mathrm{Re}_{Db} \mathrm{Pr}_b (\mu_b/\mu_s)^{0.2}$$

$$= (16 \times 10^{-4}) \left(\frac{0.39}{0.5/12}\right) (125{,}000)(1.92)(0.85)$$

$$= 3060 \text{ Btu/hr sq ft F} \qquad \textit{Ans.}$$

We see that the results obtained by two different methods of calculating $\bar{h}_c$ for turbulent flow in ducts agree within 16 percent.

8-4. Forced convection in laminar flow

Although heat-transfer coefficients for laminar flow are considerably smaller than for turbulent flow, in the design of heat-exchange equipment for very viscous liquids, it is sometimes economically necessary to accept a lower unit-surface conductance in order to reduce the pumping-power requirements. In recent years, laminar gas flow has also been considered for high-temperature, compact heat exchangers, where tube diameters are very small and gas densities very low. Another potential application of laminar-flow forced convection lies in the atomic-power field, where liquid metals are used as heat-transfer mediums. Since most liquid metals have a high thermal conductivity, their heat-transfer coefficients are relatively large even in laminar flow.

The heat-flow mechanism in purely laminar flow is conduction. The rate of heat flow between the walls of a conduit and the fluid flowing in it can be obtained analytically by solving the equations of motion and of conduction heat flow simultaneously. But to obtain a solution it is necessary to know or assume the velocity distribution in the duct. In fully developed laminar flow without heat transfer, the velocity distribution at any cross section has the shape of a parabola. The velocity profile for high-Prandtl-number fluids, such as oils, becomes fully established much more rapidly than the temperature profile, usually within 20 to 80 diameters from the entrance.[3] Heat-transfer equations based on the assumption of a parabolic velocity distribution will therefore not introduce serious errors for oils and other viscous fluids flowing in long ducts, if they are modified to account for effects caused by the variation of the viscosity due to the temperature gradient. For liquid metals, on the other hand, the temperature profile is established much more rapidly than the

[3] According to Langhaar (*Jour. Appl. Mech.*, Vol. 64, 1942, p. A-55) the length required to establish a parabolic velocity distribution in isothermal flow is $0.05\ D\ \mathrm{Re}_D$.

velocity profile as a result of the metals' high thermal conductivity, and the assumption of a uniform velocity profile may not involve large errors for many applications (26,34,35). For gases, the temperature and velocity profiles develop nearly at equal rates along the tube, and the actual behavior of both must be considered in a heat-transfer analysis.

Effect of free convection. An additional complication in the determination of a heat-transfer coefficient in laminar flow arises when the buoyancy forces are of the same order of magnitude as the external forces due to the forced circulation. Such a condition may arise in oil coolers when low flow velocities are employed. Also, in the cooling of rotating parts, such as rotor blades of gas turbines and ramjets attached to the propellers of helicopters, the free-convection forces may be so large that their effect on the velocity pattern cannot be neglected even in high-velocity flow. When the buoyancy forces are in the same direction as the external forces, e.g., the gravitational forces superimposed on upward flow, they increase the rate of heat transfer. When the external and buoyancy forces act in opposite direction, the heat transfer is reduced. Eckert (16,38) studied heat transfer in mixed flow, and his results are shown qualitatively in Figs. 8-11a and b. In the darkly shaded area, the contribution of free convection to the total heat transfer is less than 10 percent, whereas in the lightly shaded area, forced-convection effects are less than 10 percent and free convection predominates. In the unshaded area, both free and forced convection are of the same order of magnitude. In practice, free-convection effects are hardly ever significant in turbulent flow (37). In cases where it is doubtful whether forced- or free-convection flow applies, the heat-transfer coefficient is generally calculated by using forced- and free-convection relations separately, and the larger one is used (11). The accuracy of this rule of thumb is estimated to be about 25 percent.

Correlations and empirical equations. The details of the mathematical solutions for purely laminar flow are beyond the scope of this text. References listed at the end of this chapter, especially Refs. 17 and 18, contain the mathematical background for the engineering equations and graphs which are presented and discussed in this section.

For engineering applications it is most convenient to present the results of analytical and experimental investigations in terms of a Nusselt number defined in the conventional manner as

$$\overline{\mathrm{Nu}}_D = \frac{\bar{h}_c D}{k}$$

It was pointed out in Sec. 8-1 that the unit-convective conductance $\bar{h}_c$ varies along the tube. For practical applications the average value of the conductance is most important, and for the equations and charts pre-

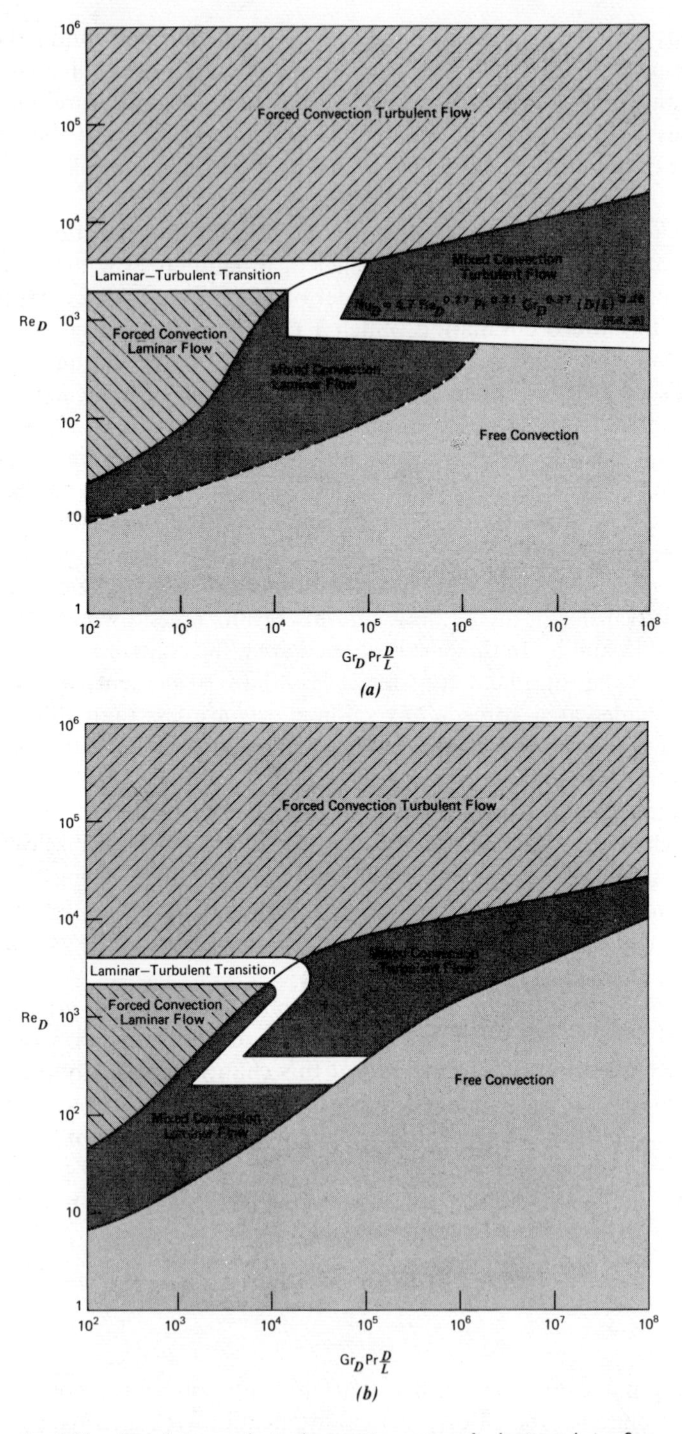

Fig. 8-11. Forced, free, and mixed convection regimes for horizontal pipe flow.

sented in this section we shall use a mean Nusselt number $\overline{\mathrm{Nu}}_D = \bar{h}_c D/k$, averaged with respect to the length of the duct L, or

$$\overline{\mathrm{Nu}}_D = \frac{1}{L}\int_0^L \mathrm{Nu}_x dx$$

where the subscript x refers to local conditions at x. This mean Nusselt number is often termed the log-mean Nusselt number because it can be used directly in the log-mean-rate equations for heat exchangers presented in Chapter 11.

The mean Nusselt numbers for laminar flow in tubes at a uniform wall temperature have been calculated analytically by various investigators. Their results are shown in Fig. 8-12 for several velocity distributions. All of these solutions are based on the idealizations of a constant tube-wall temperature and a uniform temperature distribution at the tube inlet and apply strictly only when the physical properties are independent of temperature. The abscissa is the dimensionless quantity $\mathrm{Re}_D \mathrm{Pr} D/L$, usually called the Graetz number, Gz. To determine the mean value of the Nusselt number for a given tube of length L and diameter D, one evalu-

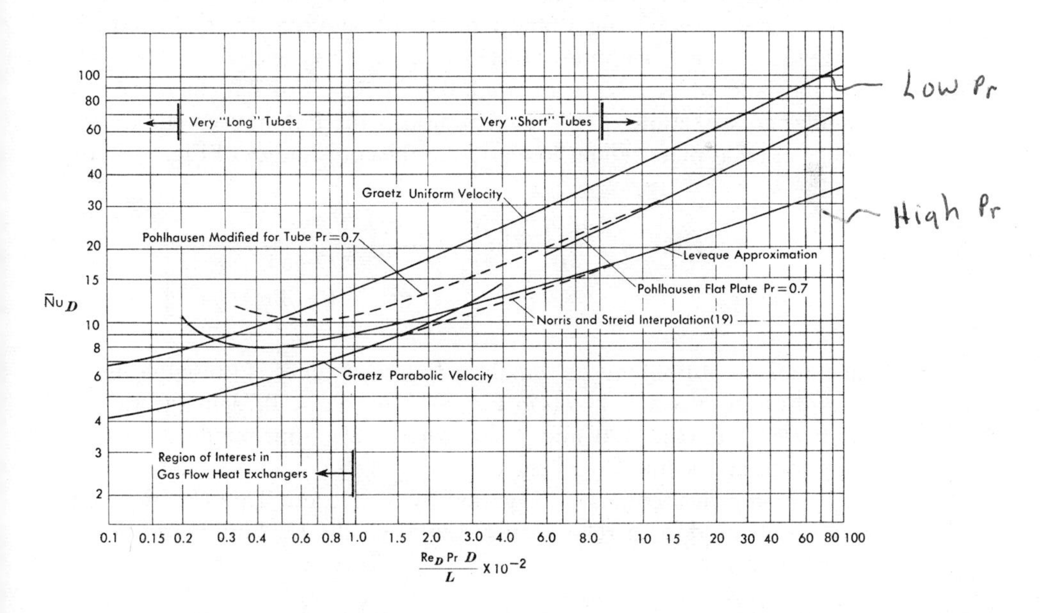

Fig. 8-12. Curves illustrating solutions for laminar-flow heat transfer at constant wall temperature. (Extracted from "Numerical Solutions for Laminar Flow Heat Transfer in Circular Tubes," by W. M. Kays, published in *Trans. ASME*, Vol. 77, 1955, with permission of the publishers, The American Society of Mechanical Engineers)

ates the Reynolds number Re_D, the Prandtl number Pr, forms the dimensionless parameter $Re_D Pr D/L$, and enters the curve of Fig. 8-12. The selection of the curve representing the conditions which most nearly correspond to the physical conditions depends on the nature of the fluid and the geometry of the system. For high-Prandtl-number fluids, such as oils, the velocity profile is established much more rapidly than the temperature profile. Consequently the application of the curve labeled "parabolic velocity" does not lead to a serious error in long tubes when $Re_D Pr D/L$ is less than 100. For very long tubes the Nusselt number approaches a limiting minimum value of 3.66 when the tube temperature is uniform. When the heat rate instead of the tube temperature is uniform, the limiting value of $\overline{Nu}_D$ is 4.36.

For very low-Prandtl-number fluids, such as liquid metals, the temperature profile is established much more rapidly than the velocity profile. For typical applications the assumption of a uniform velocity profile may give satisfactory results, although experimental evidence is insufficient for a quantitative evaluation of the possible deviation from the analytical solution for slug flow. For very short tubes or rectangular ducts with initially uniform velocity and temperature distribution, the flow conditions along the wall approximate those along a flat plate, and the Pohlhausen analysis presented in Sec. 6-7 is expected to yield satisfactory results for liquids having Prandtl numbers between 1.0 and 15.0 The Pohlhausen solution applies (18,19) when L/D is less than 0.0048 Re_D for tubes and when L/D is less than 0.0021 Re_{D_H} for flat ducts of a rectangular cross section. For these conditions the Pohlhausen equation for flow over a flat plate can be converted to the coordinates of Fig. 8-12, or

$$\overline{Nu}_D = \frac{Re_D Pr D}{4L} \ln \left[\frac{1}{1 - \dfrac{2.654}{Pr^{0.167} (Re_D Pr D/L)^{0.5}}} \right] \tag{8-27}$$

An extension of Pohlhausen's analysis to longer tubes is presented in Ref. 18, and the results are shown in Fig. 8-12 for Pr = 0.73 in the range of $Re_D Pr D/L$ between 100 and 1500, where this approximation is most likely to be applicable.

An empirical equation suggested by Sieder and Tate (12) which has been widely used to correlate experimental results for liquids can be written in the form

$$\overline{Nu}_D = 1.86 (Re_D Pr D/L)^{0.33} \left(\frac{\mu_b}{\mu_s} \right)^{0.14} \tag{8-28}$$

where the empirical correction factor $(\mu_b/\mu_s)^{0.14}$ is introduced to account for the effect of the temperature variation on the physical properties.

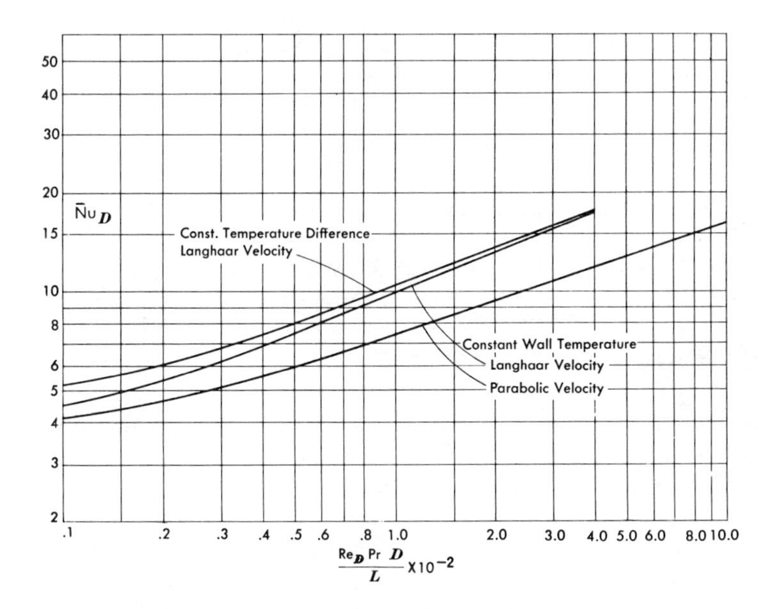

Fig. 8-13. Mean Nusselt number with respect to tube length for gases in laminar flow. (Extracted from "Numerical Solutions for Laminar Flow Heat Transfer in Circular Tubes," by W. M. Kays, published in *Trans. ASME*, Vol. 77, 1955, with permission of the publishers, The American Society of Mechanical Engineers)

In liquids the viscosity decreases with increasing temperature, while in gases the reverse trend is observed. When a liquid is heated, the fluid near the wall is less viscous than the fluid in the center. Consequently, the velocity of the heated fluid near the wall is larger than for an unheated fluid, but less in the center. The distortion of the parabolic velocity profile for liquids when heating or cooling is shown in Fig. 8-14. For gases the conditions are reversed, but the variation of density with temperature introduces additional complications.

The empirical viscosity correction factor is merely an approximate rule of thumb, and recent data indicate that it may not be satisfactory when large temperature gradients exist. As an approximation in the absence of a more satisfactory method, it is suggested that, for liquids, the Nusselt number obtained from analytic solutions presented in Figs. 8-12 and 8-13 also be multiplied by $(\mu_s/\mu_b)^{0.14}$ to correct for the variation of properties due to the temperature gradient. For gases Kays and London (13) suggest that the Nusselt number from Fig. 8-13 be multiplied by a temperature-correction factor. If all fluid properties are evaluated at the average bulk temperature, the corrected Nusselt number is

$$\overline{\text{Nu}}_D = \overline{\text{Nu}}_{D_{\text{Fig. 8-13}}} \left(\frac{T_b}{T_s}\right)^n \tag{8-29}$$

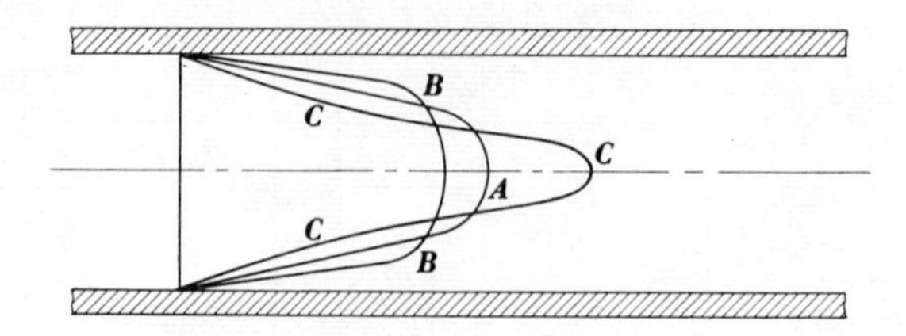

Fig. 8-14. Effect of heat transfer on velocity profiles in fully developed laminar flow. (Curve *A*, isothermal flow; curve *B*, heating of liquid or cooling of gas; curve *C*, cooling of liquid or heating of gas.)

where $n = 0.25$ for a gas heating in a tube, and 0.08 for a gas cooling in a tube.

For laminar flow of gases between two parallel heated plates a distance $2y_0$ apart, Swearingen and McEligot (33) have shown that gas property variations can be taken into account by the relation

$$\mathrm{Nu} = \mathrm{Nu}_{\text{constant properties}} + 0.024\, Q^{+0.3}\, \mathrm{Gz}_b^{0.75} \tag{8-30}$$

where $Q^+ = \dot{q}_w y_0/(kT)_{\text{entrance}}$
$\dot{q}_w$ = wall heat flux

The influence of free convection on the heat transfer to fluids in horizontal isothermal tubes has been investigated by Depew and August (29). They found that their own as well as previously available data for tubes with $L/D > 50$ could be correlated by the equation

$$\mathrm{Nu} = 1.75\left(\frac{\mu_b}{\mu_s}\right)^{0.14}[\mathrm{Gz} + 0.12(\mathrm{Gz}\,\mathrm{Gr}_D^{1/3}\,\mathrm{Pr}^{0.36})^{0.88}]^{1/3} \tag{8-31}$$

Correlations for vertical tubes and ducts are considerably more complicated because they depend on the direction of the heat flow, and the free convection. A summary of available information is given in Refs. 37 and 38.

Effect of heat transfer on the friction coefficient. The variation in physical properties also affects the friction coefficient. To evaluate the friction coefficient of fluids being heated or cooled it is suggested that, for liquids, one modifies the isothermal friction coefficient by

$$f_{\text{heat transfer}} = f_{\text{isothermal}}\left(\frac{\mu_s}{\mu_b}\right)^{0.14} \tag{8-32}$$

and for gases by

$$f_{\text{heat transfer}} = f_{\text{isothermal}} \left(\frac{T_s}{T_b}\right)^{0.14} \tag{8-33}$$

8-5. *Forced convection in transition flow*

The mechanisms of heat transfer and fluid flow in the transition region (Re_D between 2100 and 10,000) vary considerably from system to system. In this region the flow may be unstable, and fluctuations in pressure drop and heat transfer have been observed. There exists a large uncertainty in the basic heat-transfer and flow-friction performance, and consequently the designer is advised to design equipment, if possible, to operate outside this region. For the purpose of estimating the Nusselt number in the transition region, the curves of Fig. 8-15 may be used, but the actual performance may deviate considerably from that predicted on the basis of these curves.

8-6. *Closure*

To aid in the rapid selection of an appropriate relation to obtain the heat-transfer coefficient for flow in a duct, some of the most commonly

Table 8-1. Summary of useful equations for forced convection heat transfer inside tubes and ducts.

System	Equation	Eq. No. in Text
Long Ducts, Liquids and Gases, Laminar Flow ($\text{Re}_D < 2,100$, $\text{Pr} > 0.7$)	$\overline{\text{Nu}}_{D_H} = 1.86\,(\text{Re}_D \text{Pr} L/D_H)^{0.33} (\mu_b/\mu_s)^{0.14}$	8-28*
Short Ducts, Liquids and Gases Laminar Flow $100 < \text{Re}_D \text{Pr}\, D_H/L < 1500$ ($\text{Pr} > 0.7$)	$\overline{\text{Nu}}_{D_H} = \dfrac{\text{Re}_D \text{Pr}\, D_H}{4L} \ln \left[\dfrac{1}{1 - \dfrac{2.6}{\text{Pr}^{0.167} (\text{Re}_{D_H} \text{Pr}\, D_H/L)^{0.5}}}\right]$	8-27*
Long Ducts, Turbulent Flow, Liquid Metals Constant Heat Flux ($\text{Pr} < 0.1$)	$\overline{\text{Nu}}_{D_H} = 5.0 + 0.025\,(\text{Re}_{D_H} \text{Pr})^{0.8}$	8-24*
Long Ducts, Turbulent Flow Liquid Metals, Constant Wall Temperature ($\text{Pr} < 0.1$)	$\overline{\text{Nu}}_{D_H} = 4.0 + 0.025\,(\text{Re}_{D_H} \text{Pr})^{0.8}$	8-25*
Long Ducts, Liquids and Gases in Turbulent Flow ($\text{Re}_D > 6,000$, $\text{Pr} > 0.7$)	$\overline{\text{Nu}}_{D_H} = 0.023/(\text{Re}_{D_H}^{0.2} \text{Pr}^{0.67})$	8-20*
Short Ducts, Liquids and Gases in Turbulent Flow ($2 < L/D_H < 20$, $\text{Pr} > 0.7$)	$\overline{\text{Nu}}_{D_H} = \dfrac{0.023\,[1 + (D_H/L)^{0.7}]}{\text{Re}_{D_H}^{0.2} \text{Pr}^{0.67}}$	8-23*
Liquid Metals, Laminar Flow	See Reference 26.	

*All physical properties should be evaluated at the mean film temperature, $(T_s + T_b)/2$.

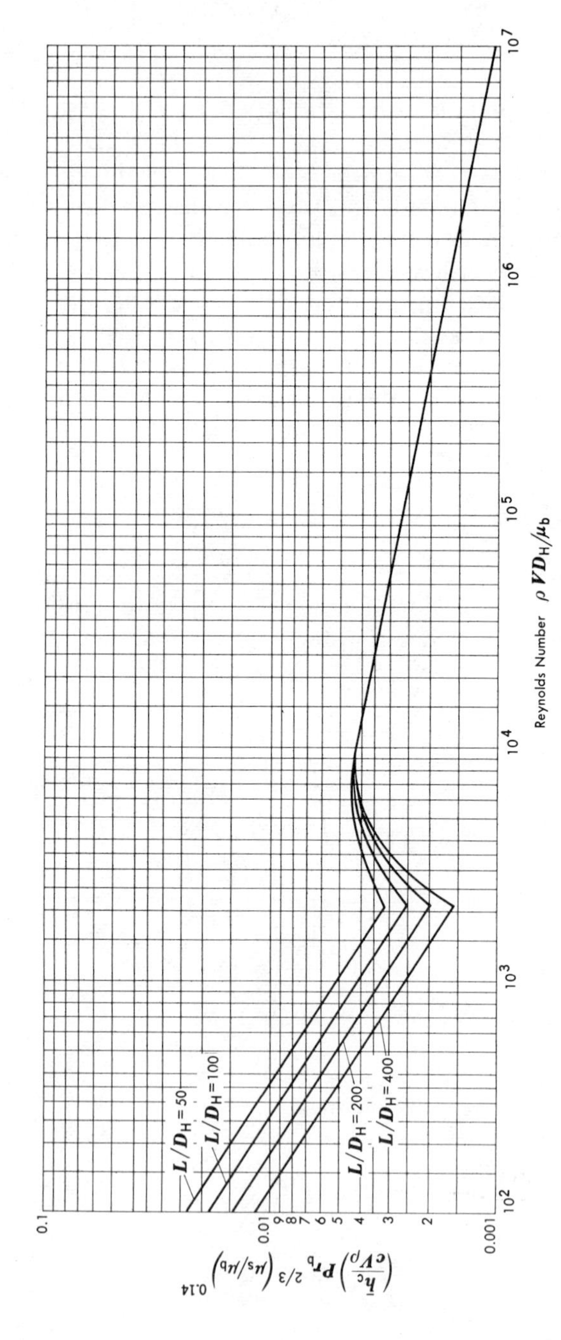

Fig. 8-15. Recommended curves for determining heat-transfer coefficient in the transition regime. (Reprinted from *Industrial and Engineering Chemistry*, Vol. 28, p. 1429, December 1936, with permission of the copyright owner, The American Chemical Society)

used empirical equations are summarized in Table 8-1. A complete and up to date survey of other useful correlation equations and their accuracy can be found in "Engineering Sciences Data" (37).

PROBLEMS

Problems 8-1 through 8-14 require only the evaluation of the heat-transfer coefficient and pressure loss. Problems 8-15 through 8-33 demand the analysis of a system in which the evaluation of the convection heat-transfer coefficient plays a major part.

8-1. Water at an average temperature of 80 F is flowing through a smooth 2-in.-ID pipe at a velocity of 3 fps. If the temperature at the inner surface of the pipe is 120 F, determine (a) the unit-surface conductance, (b) the rate of heat flow per foot of pipe, (c) the bulk-temperature rise per foot, and (d) the pressure drop per foot in psi. *Ans.* (a) $\bar{h}$ = 620 Btu/hr sq ft F ($\pm 10\%$); (b) q/ft = 13,000 Btu/hr; (c) ΔT/ft = 0.89 F; (d) Δp/ft $\simeq$ 0.008 psi (depending on roughness)

8-2. An aniline-alcohol solution is flowing at a velocity of 10 fps through a long 1-in.-ID thin-wall tube. On the outer surface of the tube, steam is condensing at atmospheric pressure, and the tube-wall temperature is 212 F. The tube is clean, and there is no thermal resistance due to a scale deposit on the inner surface. Using the physical properties tabulated below, estimate the unit-surface conductance between the fluid and the pipe by means of Eq. 8-20, as well as Eq. 8-22 and Fig. 8-9, and compare the results. Assume that the bulk temperature of the aniline solution is 68 F and neglect entrance effects.

Physical properties of the aniline solution:

Temp (F)	Viscosity (centipoises)	Thermal Conductivity (Btu/hr ft F)	Specific Gravity	Specific Heat (Btu/lb F)
68	5.1	0.100	1.03	0.50
140	1.4	0.098	0.98	0.53
212	0.6	0.095	....	0.56

8-3. For water at a bulk temperature of 90 F flowing at a velocity of 5 fps through a 1-in.-ID duct having a wall temperature of 110 F, calculate the Nusselt number and the convection heat-transfer coefficient by three different methods and compare the results. *Ans.* Nu = 230 ($\pm 10\%$)

8-4. Atmospheric air at a velocity of 200 fps and a temperature of 60 F enters a 2-ft-long square metal duct of 8- by 8-in. cross section. If the duct wall is at 300 F, determine the *average* unit-surface conductance. Comment briefly on the L/D_H effect. *Ans.* $\bar{h} \simeq$ 14 Btu/hr sq ft F ($\pm 15\%$)

8-5. Air at 60 F and atmospheric pressure enters a $\frac{1}{2}$-in.-ID tube at 100 fps. For an average wall temperature of 212 F, determine the discharge temperature of the air and the pressure drop in inches of water if the pipe is (a) 4 in. long, (b)

40 in. long. Use the average bulk temperature of the air between the inlet and the outlet to evaluate the rate of heat transfer between the wall and the air.

Ans. (a) $T_{out} = 82$ F, $\Delta p = 3.8$ in. H_2O; (b) $T_{out} = 175$ F, $\Delta p = 9.1$ in. H_2O

8-6. In a refrigeration system, brine (NaCl 10 percent by weight) having a viscosity of 4 lb/ft hr and a thermal conductivity of 0.15 Btu/hr ft F is flowing through a long 1-in.-ID pipe at 20 fps. Under these conditions the heat-transfer coefficient was found to be 200 Btu/hr sq ft F. For a brine temperature of 30 F and a pipe temperature of 65 F, determine the temperature rise of the brine per foot length of pipe if the velocity of the brine were doubled. Assume that the specific heat of the brine is 0.9 Btu/lb F and that its density is equal to that of water.

8-7. Determine the heat-transfer coefficient for liquid bismuth flowing through an annulus (2 in. ID, 2.5 in. OD) at a velocity of 15 fps. The wall temperature of the inner surface is 800 F and the bismuth is at 600 F. It may be assumed that heat losses from the outer surface are negligible.

8-8. Water at a bulk inlet temperature of 200 F is flowing with a velocity of 0.05 fps through a 0.05-ft-diam tube, 1 ft long. If the tube wall temperature is 400 F, determine the average heat-transfer coefficient and estimate the bulk temperature rise of the water. *Ans.* 60 Btu/hr sq ft F, 78 F

8-9. Air at an average temperature of 300 F flows through a short square duct (4 by 4 by 1 in.) at a rate of 116 lb/hr. The duct-wall temperature is 800 F. Determine the average heat-transfer coefficient using duct equation with appropriate L/D correction. Compare your results with flow-over-flat-plate relations.

8-10. In a long annulus (10 in. ID, 15 in. OD), atmospheric air is heated by steam condensing at 300 F on the inner surface. If the velocity of the air is 20 fps and its bulk temperature 100 F, calculate the heat-transfer coefficient.

8-11. The equation

$$\overline{\mathrm{Nu}} = 0.116(\mathrm{Re}^{2/3} - 125)\,\mathrm{Pr}^{1/3}[1 + (D/L)^{2/3}](\mu_b/\mu_s)^{0.14}$$

has been proposed by Hausen (Prob. 8-13) for the transition range ($2300 \le \mathrm{Re} < 8000$) as well as for higher Reynolds numbers. Compare the values of $\overline{\mathrm{Nu}}$ predicted by Hausen's equation for Re = 3000 and Re = 20,000 at D/L of 0.1 and 0.01 with those obtained from appropriate equations or charts in the text. Assume the fluid is water at 60 F flowing through a pipe at 200 F.

8-12. Compute the average unit-surface conductance, $\bar{h}_c$, for 50 F water flowing at 10 fps in a long 1-in.-ID pipe (surface temperature of 102 F) by three different equations and compare your results. Also determine the pressure drop per foot length of pipe.

8-13. The equation

$$\overline{\mathrm{Nu}} = \frac{\bar{h}_c D}{k} = \left\{ 3.65 + \frac{0.0668\,(D/L)\,\mathrm{Re}\,\mathrm{Pr}}{1 + 0.04\,[(D/L)\,\mathrm{Re}\,\mathrm{Pr}]^{2/3}} \right\} \left(\frac{\mu_b}{\mu_s} \right)^{0.14}$$

was recommended by H. Hausen (*Zeitschr. Ver. Deut. Ing., Beiheft* No. 4, 1943) for forced convection heat transfer in fully developed laminar flow through tubes. Compare the values of the Nusselt number predicted by Hausen's equation for Re = 1000, Pr = 1, and D/L = 2, 10, and 100 respectively with those obtained from appropriate equations or graphs in the text.

8-14. Water at 180 F is flowing through a thin copper tube (6 in. ID) at a velocity of 25 fps. The duct is located in a room at 60 F and the unit-surface conductance at the outer surface of the duct is 2.5 Btu/hr sq ft F. (a) Determine the heat-transfer coefficient at the inner surface. (b) Estimate the length of duct in which the water temperature drops 1 F.

Ans. (a) $\bar{h}_c \simeq$ 3800 Btu/hr sq ft F; (b) $L \simeq$ 1500 ft

8-15. The following thermal-resistance data were obtained on a 50,000 sq ft condenser constructed with 1-in.-OD brass tubes, 23¾ ft long, 0.049 in. wall thickness, at various water velocities inside the tubes [*Trans. ASME*, Vol. 58 (1936), p. 672].

$\frac{1}{U_o} \times 10^3$ (hr sq ft F/Btu)	Water Velocity (fps)	$\frac{1}{U_o} \times 10^3$ (hr sq ft F/Btu)	Water Velocity (fps)
2.060	6.91	3.076	2.95
2.113	6.35	2.743	4.12
2.212	5.68	2.498	6.76
2.374	4.90	3.356	2.86
3.001	2.93	2.209	6.27
2.081	7.01		

Assuming that the unit-surface conductance on the steam side is 2000 Btu/hr sq ft F, determine the scale resistance. HINT: Plot $(1/U_o)$ vs. $(1/V^{0.8})$. (This method is called the *Wilson plot.*)

8-16. A double-pipe heat exchanger is used to condense steam at 1 psia. Water at an average bulk temperature of 50 F flows at 10 fps through the inner pipe (copper, 1 in. ID, 1.2 in. OD). Steam at its saturation temperature flows in the annulus formed between the outer surface of the inner pipe and an outer pipe of 2 in. ID. The average unit-surface conductance of the condensing steam is 1000 Btu/hr sq ft F, and the thermal resistance of a surface scale on the outer surface of the copper pipe is 0.001 hr sq ft F/Btu. Determine (a) the over-all heat-transfer coefficient between the steam and the water based on the outer area of the copper pipe. Also sketch the thermal circuit and (b) evaluate the temperature at the inner surface of the pipe. (c) Estimate the length required to condense 1 lb of steam.

Ans. U = 350 Btu/hr sq ft F; T_{si} = 65 F

8-17. Determine the rate of heat transfer per foot length to a light oil flowing through a 1-in.-ID, 2-ft-long copper tube at a velocity of 6 fpm. The oil enters the tube at 60 F and the tube is heated by steam condensing on its outer surface at atmospheric pressure with a unit-surface conductance of 2000 Btu/hr sq ft F. The properties of the oil at various temperatures are listed in the accompanying tabulation:

T (F)	60	80	100	150	212
ρ (lb/cu ft)	57	57	56	55	54
c (Btu/lb F)	0.43	0.44	0.46	0.48	0.51
k (Btu/hr ft F)	0.077	0.077	0.076	0.075	0.074
μ (lb/hr ft)	215	100	55	19	8
Pr	1210	577	330	116	55

Ans. $q \simeq$ 1380 Btu/hr

8-18. Evaluate the rate of heat loss per foot from superheated steam flowing at 600 F and 250-psi pressure through schedule 80 4-in. pipe at a velocity of 100 fps. The pipe is lagged with a 2-in.-thick layer of asbestos. Heat is transferred to the surroundings by free convection and radiation.

8-19. Assume that the heat source in Prob. 8-7 is an aluminum-clad rod of uranium, 2 in. OD and 6 ft long. *Estimate* the heat flux that will raise the temperature of the bismuth 100 F and the maximum center and surface temperatures necessary to transfer heat at this rate.

8-20. If in the tube of Prob. 8-8 air instead of water were to be used, but the velocity of the air were to be increased until the heat-transfer coefficient with the air equals that obtained with water at 0.05 fps, determine the velocity required and the pressure drop in psi. *Ans.* 485 ft/sec

8-21. If the total resistance between the steam and the air (including the pipe wall and scale on the steam side) in Prob. 8-10 is 0.30 hr sq ft F/Btu, calculate the temperature difference between the outer surface of the inner pipe and the air. Show the thermal circuit.

8-22. A plastic tube 3 in. ID and $\frac{1}{2}$ in. thick having a thermal conductivity of 0.30 Btu/hr ft F, a density of 150 lb/cu ft, and a specific heat of 0.40 Btu/lb F is cooled from an initial temperature of 170 F by passing air at 70 F inside and outside of the tube parallel to its axis. The velocities of the two air streams are such that the coefficients of heat transfer are the same on the interior and exterior surfaces. Measurements show that at the end of 26 min, the temperature difference between the tube surfaces and the air is 10 percent of the initial temperature difference. It is proposed to cool a tube of a similar material having an inside diameter of 6 in. and a wall thickness of 1 in. from the same initial temperature also using air at 70 F and feeding to the inside of the tube the same number of pounds of air per hour as was used in the first experiment. The air-flow rate over the exterior surfaces will be adjusted to give a heat-transfer coefficient on the outside the same as on the inside of the tube. It may be assumed that the air-flow rate is so high that the temperature rise along the axis of the tube may be neglected. Using the experience gained initially with the 3-in. tube, estimate how long it will take to cool the surface of the larger tube to 80 F under the conditions described. Indicate all assumptions and approximations in your solution.

Ans. 174 min

8-23. Derive an equation of the form $\bar{h}_c = f(T, D, V)$ for turbulent flow of water through a long tube in the temperature range between 100 and 200 F.

8-24. A 1.0-in.-OD, 0.75-in.-ID steel pipe carries dry air at a velocity of 25 fps and a temperature of 20 F. Ambient air is at 70 F and has a dew point of 50 F. How much insulation with a conductivity of 0.15 Btu/hr ft F is needed to prevent condensation on the exterior of the insulation if $\bar{h}$ = 2.0 Btu/hr sq ft F on the outside.

8-25. A light oil with a 65 F inlet temperature flows at a rate of 1000 lb/min through a 2-in.-ID pipe which is enclosed by a jacket containing condensing steam at 300 F. If the pipe is 30 ft long, determine the outlet temperature of the oil.

8-26. Mercury at an inlet bulk temperature of 200 F flows through a $\frac{1}{2}$-in.-ID tube at a flow rate of 10,000 lb_m/hr. This tube is a part of a nuclear reactor in which heat can be generated uniformly at any desired rate by adjusting the neutron flux level. Determine the length of tube required to raise the bulk temperature of the mercury to 450 F without generating any mercury vapor.

8-27. Power generation in a nuclear reactor is limited principally by the ability of the coolant to absorb the heat generated. Compare the relative merits of water, liquid sodium, and carbon dioxide as coolants of a nuclear reactor, assuming that either a gas or a steam turbine may be used to generate electrical energy.

8-28. The energy conservation equation for steady laminar flow through a tube as shown in the accompanying sketch is

$$u \frac{\partial T}{\partial x} = a \frac{1}{r} \frac{\partial}{\partial r}\left(r \frac{\partial T}{\partial r}\right)$$

if the fluid properties are uniform. Under certain conditions it is quite reasonable to assume that the velocity profile is flat, i.e., $u(r) = U$, so that "rod" or "slug" flow prevails. For this type of flow, assuming that the fluid enters at $x = 0$ at a uniform temperature T_o and the wall of the tube is kept at temperature T_w, determine the temperature $T(r,x)$ as a function of relevant dimensionless parameters. (a) Letting $x' = x/L$ and $r' = r/r_o$, show that aL/Ur_o^2 is an appropriate dimensionless parameter for the independent variable. (b) Starting with the continuity equation in the form $w = \rho U \pi r_o^2$ show that $\pi k L/wc_p$ may also be used as a dimensionless parameter. (c) Derive an expression for the dimensionless temperature distribution $[T(r,x) - T_o]/(T_w - T_o)$ for slug flow. (d) Defining the "bulk mean" temperature at a given cross section as

$$\bar{T} = \frac{\int_0^{r_o} 2\pi u T(r)\, r dr}{\int_0^{r_o} 2\pi u r dr}$$

verify that for slug flow at a distance $x = L$ from the inlet

$$\frac{\overline{T} - T_o}{T_w - T_o} = 1 - 0.692e^{-5.78z} + 0.131\, e^{-30.4z}$$

where $z = aL/Ur_o^2$. (e) Derive an expression for the local Nusselt number based on the difference between T_w and the "bulk mean" temperature defined under part (d) and show that Nu $\rightarrow$ 5.78 as $x \rightarrow \infty$.

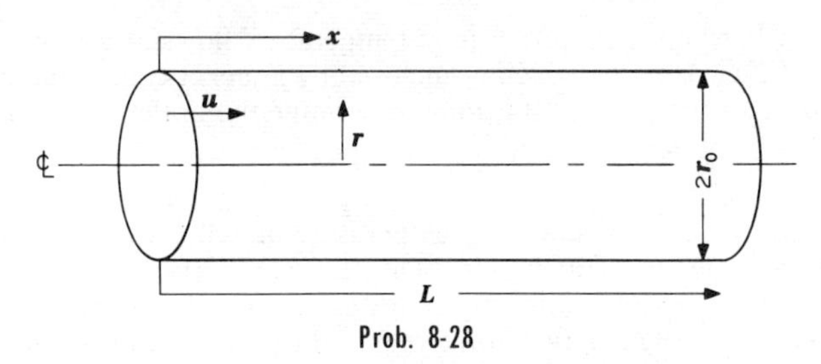

Prob. 8-28

8-29. It is proposed to heat dry sand by passing it steadily downward through a vertical pipe which is heated by a vapor condensing on the outside. The sand is assumed to flow through the pipe with a uniform velocity profile. At the exit of the pipe the sand flows into a mixer and its temperature becomes uniform. The inside wall of the pipe is maintained at 220 F, and the thermal resistance between the pipe wall and the sand is assumed to be negligible. The sand initially at a temperature of 120 F is fed to the pipe at a rate of 1.18 cu ft/hr. The pipe is 18 ft long and has an inside diameter of 1.2 in. Estimate the temperature of the mixed sand leaving the heater using the sand properties listed below:

Density = 100 lb/cu ft

Thermal conductivity = 0.20 Btu/hr sq ft F/ft

Specific heat = 0.24 Btu/lb F

HINT: See Prob. 8-28 for an approach.

8-30. Show that for fully developed laminar flow in a tube with a parabolic velocity profile, i.e.,

$$u(r) = \frac{2w}{\rho \pi r_o^2}\left[1 - \left(\frac{r}{r_o}\right)^2\right]$$

the Nusselt number $2h_c r_o/k$ is 48/11 if the wall temperature increases linearly with x, i.e., $\partial T/\partial x = C$. HINT: Assume a solution of the form $T = Cx + \overline{\theta}(r)$, where $\overline{\theta}$ represents the difference between the local fluid temperature and the wall temperature at the same location.

8-31. For fully turbulent flow in a long tube of diameter D, develop a relation between the ratio $L\Delta_T/D$ in terms of flow and heat-transfer parameters, where $L\Delta_T$ is the tube length required to raise the bulk temperature of the fluid by ΔT. Use Eq. 8-18 for fluids with Prandtl number of the order of unity or larger and Eq. 8-26 for liquid metals.

8-32. In a pipe within a pipe heat exchanger, water is flowing in the annulus and oil having the properties listed in Prob. 8-3 is flowing in the central pipe. The inner pipe is 0.527 in. ID, 0.625 in. OD, and the ID of the outer pipe is 0.750 in. For a water bulk temperature of 80 F and an oil bulk temperature of 175 F, determine the over-all heat-transfer coefficient based on the outer diameter of the central pipe and the frictional pressure drop per unit length of the water and the oil for the following velocities: (a) water rate 1 gpm, oil rate 1 gpm; (b) water rate 10 gpm, oil rate 1 gpm; (c) water rate 1 gpm, oil rate 10 gpm; and (d) water rate 10 gpm, oil rate 10 gpm ($L/D = 400$).

8-33. Water in *turbulent* flow is to be heated in a single-pass tubular heat exchanger by steam condensing on the outside of the tubes. The flow rate of the water, its pressure drop, its inlet and outlet temperatures, and the steam pressure are fixed. Assuming that the tube-wall temperature remains constant, determine the dependence of the total required heat-exchanger area on the inside diameter of the tubes. *Ans.* $A_{\text{total}} \sim (1/\text{ID})^{1/3}$

8-34. An incompressible fluid is flowing in steady laminar flow between two parallel infinite plates whose wetted surfaces are a distance $2a$ apart and are at uniform temperature T_w. Owing to a peculiar variation of viscosity with temperature, the velocity of the fluid is uniform, i.e., $u(y) = U$. Assuming that the physical properties c_p and k are constant, that the fluid enters the system ($x = 0$) at a uniform temperature T_o, and that conduction in the direction of flow is negligible, develop an expression for the temperature distribution in the fluid in terms of the spacing, $2a$; the fluid velocity, U; and the fluid properties, c_p, μ, and k. After having derived an expression for the temperature $T(x,y)$, find an expression for the local rate of heat transfer and the local Nusselt number $h(x)2a/k$.

8-35. Mercury flows inside a copper tube 30 ft long having 2-in. inside diameter at an average velocity of 24 ft per sec. The temperature at the inside surface of the tube is 100 F uniformly throughout the tube, and the arithmetic mean bulk temperature of the mercury is 150 F. Assuming the velocity and temperature profiles are fully developed, calculate the rate of heat transfer by convection for the 30-ft length by considering the mercury as (a) an ordinary liquid, and (b) liquid metal. Compare the results.

8-36. Show that for fully developed laminar flow between 2 flat plates spaced $2a$ apart, the Nusselt number based on the "bulk mean" temperature is 4.12 if the temperature of both walls varies linearly with the distance x, i.e., $\partial T/\partial x = C$. The "bulk mean" temperature is defined as

$$\overline{T} = \int_0^{2a} u(y)T(y)\,dy \bigg/ \int_0^{2a} u(y)\,dy$$

HINT: Assume a solution of the form $T(x,y) = Cx + \bar{\theta}(y)$, where $\bar{\theta}$ is the local temperature difference between the fluid at y and the wall.

8-37. Repeat Prob. 6-36, but assume that one wall is insulated while the temperature of the other wall increases linearly with x. *Ans.* Nu = 2.70

REFERENCES

1. O. Reynolds, "On the Extent and Action of the Heating Surface for Steam Boilers," *Proc. Manchester Lit. Phil. Soc.*, Vol. 8 (1874).
2. L. Prandtl, "Eine Beziehung zwischen Wärmeaustausch und Strömungswiederstand der Flüssigkeiten," *Phys. Zeit.*, Vol. 11 (1910), p. 1072.
3. T. von Karman, "The Analogy between Fluid Friction and Heat Transfer," *Trans. ASME*, Vol. 61 (1939), p. 705.
4. L. M. K. Boelter, R. C. Martinelli, and F. Jonassen, "Remarks on the Analogy Between Heat and Momentum Transfer," *Trans. ASME*, Vol. 63 (1941), pp. 447–455.
5. R. C. Martinelli, "Heat Transfer to Molten Metals," *Trans. ASME*, Vol. 69 (1947), p. 947.
6. R. G. Deissler, "Investigation of Turbulent Flow and Heat Transfer in Smooth Tubes Including the Effect of Variable Properties," *Trans. ASME*, Vol. 73 (1951), p. 101.
7. R. G. Deissler, "Analysis of Turbulent Heat Transfer, Mass Transfer and Friction in Smooth Tubes at High Prandtl and Schmidt Numbers," *NACA TN* 3145, May, 1954.
8. W. F. Cope, "The Friction and Heat Transmission Coefficients of Rough Pipes," *Proc. Inst. Mech. Engrs.*, Vol. 145 (1941), p. 99.
9. J. Nikuradse, "Wiederstandsgesetz und Geschwindigkeit von turbulenten Wasserströhmungen in glatten und rauhen Rohren," *Proc. 3rd Int. Cong. Appl. Mech.*, Vol. 1 (1930), p. 239.
10. A. P. Colburn, "A Method of Correlating Forced Convection Heat Transfer Data and a Comparison with Fluid Friction," *Trans. AIChE*, Vol. 29 (1933), p. 174.
11. W. M. McAdams, *Heat Transmission*, 3d ed. (New York: McGraw-Hill Book Company, Inc., 1954.)
12. E. N. Sieder and C. E. Tate, "Heat Transfer and Pressure Drop of Liquids in Tubes," *Ind. Eng. Chem.*, Vol. 28 (1936), p. 1429.
13. W. M. Kays and A. L. London, "Compact Heat Exchangers—A Summary of Basic Heat Transfer and Flow Friction Design Data," *Tech. Rep.* 23, Stanford University, 1954.
14. R. G. Deissler, "Turbulent Heat Transfer and Friction in the Entrance Regions of Smooth Passages," *Trans. ASME*, Vol. 77 (1955), pp. 1221–1234.
15. R. N. Lyon, Ed., *Liquid Metals Handbook*, 3d ed. (Washington, D.C.: Atomic Energy Commission and Department of the Navy, 1952.)
16. R. G. Eckert and A. J. Diaguila, "Convective Heat Transfer for Mixed Free and Forced Flow Through Tubes," *Trans. ASME*, Vol. 76 (1954), pp. 497–504.
17. T. B. Drew, "Mathematical Attacks on Forced Convection Problems: A Review," *Trans. AIChE*, Vol. 26 (1931), p. 26.
18. W. M. Kays, "Numerical Solution for Laminar Flow Heat Transfer in Circular Tubes," *Trans. ASME*, Vol. 77 (1955), pp. 1265–1274.

19. R. H. Norris and D. D. Streid, "Laminar-Flow Heat-Transfer Coefficients for Ducts," *Trans. ASME*, Vol. 62 (1940), p. 525.

20. B. Lubarsky and S. J. Kaufman, "Review of Experimental Investigations of Liquid-Metal Heat Transfer," *NACA TN* 3336, 1955.

21. J. P. Hartnett, "Experimental Determination of the Thermal Entrance Length for the Flow of Water and of Oil in Circular Pipes," *Trans. ASME*, Vol. 77 (1955), pp. 1211–1234.

22. L. M. K. Boelter, D. Young, and H. W. Iverson, "An Investigation of Aircraft Heaters—XXVII Distribution of Heat Transfer Rate in the Entrance Section of a Circular Tube," *NACA TN* 1451, 1948.

23. W. Nunner, "Wärmeübergang and Druckabfall in Rauhen Rohren," *VDI Forschungsheft* No. 455, *VDI Verlag GMBM*, Duesseldorf, 1956.

24. W. M. Rohsenow and H. Choi, *Heat, Mass, and Momentum Transfer*. (Englewood Cliffs, N.J.: Prentice Hall, Inc., 1961.)

25. R. Viskanta and Y. Touloukian, "Heat Transfer to Liquid Metals with Variable Properties," *Trans. ASME*, Ser. C, Vol. 82 (1960), pp. 333–340.

26. O. E. Dwyer, "Liquid-Metal Heat Transfer" Chapt. 5, Sodium and NaK Supplement to the *Liquid Metals Handbook*, 1970 ed., (Washington D.C.: Atomic Energy Commission).

27. J. R. Sellars, M. Tribus, and J. S. Klein, "Heat transfer to laminar flow in a round tube or flat conduit—the Graetz problem extended," *Trans. ASME*, Vol. 78 (1956), pp. 441–448.

28. C. A. Schleicher and M. Tribus, "Heat Transfer in a Pipe with Turbulent Flow and Arbitrary Wall-Temperature Distribution," *Trans. ASME*, Vol. 79 (1957), pp. 789–797.

29. C. A. Depew and S. E. August, "Heat Transfer Due to Combined Free and Forced Convection in a Horizontal and Isothermal Tube," *Trans. ASME*, ser. C., J. Heat Transfer, Vol. 93 (1971), pp. 380–384.

30. E. R. A. Eckert, "Engineering Relations for Heat Transfer and Friction in High Velocity Laminar and Turbulent Boundary Layer Flow over Surfaces with Constant Pressure and Temperature," *Trans. ASME*, Vol. 78 (1956), pp. 1273–1284.

31. W. D. Hayes and R. F. Probstein, *Hypersonic Flow Theory* (New York, N.Y.: Academic Press, Inc., 1959).

32. F. Kreith, *Principles of Heat Transfer*, 2nd ed. Chapt. 12, International Textbook Co., Scranton, Pa., 1965.

33. T. W. Swearingen and D. M. McEligot, "Internal Laminar Heat Transfer with Gas-Property Variation," *Trans. ASME*, ser. C, J. Heat Transfer, Vol. 93 (1971), pp. 432–440.

34. N. Z. Azer, "Thermal Entry Length for Turbulent Flow of Liquid Metals in Pipes With Constant Wall Heat Flux," *Trans. ASME*, ser. C, J. Heat Transfer, Vol. 90 (1968), pp. 483–485.

35. C. J. Hsu, "An Exact Mathematical Solution for Entrance-Region Laminar Heat Transfer with Axial Conduction," *Appl. Sci. Res.*, Vol. 17 (1967), pp. 359–376.

36. L. Duchatelle and L. Vautrey, "Détermination des coefficients de convection d'un alliage NaK en écoulement turbulent entre plaques planes parallèles," *Int. J. Heat-Mass Transfer*, Vol. 7 (1964), pp. 1017–1031.

37. "Engineering Sciences Data," Heat Transfer Subsciences, 1970, Technical Editing and Production Ltd., London.

38. B. Metais and E. R. G. Eckert, "Forced, Free, and Mixed Convection Regimes," *Trans. ASME*, ser. C., J. Heat Transfer, Vol. 86 (1964), pp. 295–296.

9
Forced convection over exterior surfaces

9-1. Flow over bluff bodies

In this chapter we shall consider heat transfer by forced convection between the exterior surface of bluff bodies, such as spheres, wires, tubes, and tube bundles, and fluids flowing perpendicularly to the axes of these bodies. The heat-transfer phenomena for these systems, as for those in which a fluid flows inside a duct or along a flat plate, are closely related to the nature of the flow. The most important difference between the flow over a bluff body and the flow over a flat plate or a streamlined body lies in the behavior of the boundary layer. We recall that the boundary layer of a fluid flowing over the surface of a streamlined body will separate when the pressure rise along the surface becomes too large. On a streamlined body the separation, if it takes place at all, occurs near the rear. On a bluff body, on the other hand, the point of separation often lies not far from the leading edge. Beyond the point of separation of the boundary layer, the fluid in a region near the surface flows in a direction opposite to the main stream, as shown in Fig. 9-1. The local reversal in the flow results in disturbances which produce turbulent eddies. This is illustrated in Fig. 9-2, which is a photograph of the flow pattern of a stream flowing at right angle to a cylinder. We can see that eddies from both sides of the cylinder extend downstream, so that a turbulent wake is formed in the rear of the cylinder.

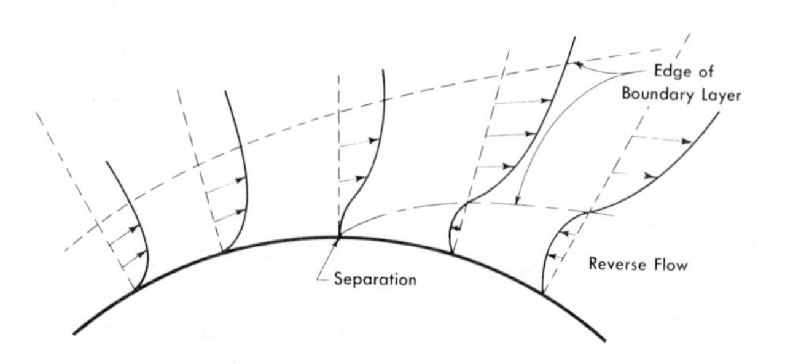

Fig. 9-1. Schematic sketch of boundary layer on a circular cylinder near separation point.

Associated with the separation of the flow are large pressure losses, since the kinetic energy of the eddies which pass off into the wake can not be regained. In flow over a streamlined body, the pressure drop is caused mainly by the skin-friction drag. For a bluff body, on the other hand, the skin-friction drag is small compared to the form drag in the Reynolds-number range of commercial interest. The form or pressure drag arises from the separation of the flow which prevents the closing of the streamlines and thereby induces a low-pressure region in the rear of the body. When the pressure over the rear of the body is lower than over the front, there exists a pressure difference which produces a drag force over and above that of the skin friction. The magnitude of the form drag decreases as the separation moves farther toward the rear.

The geometrical shapes which are most important for engineering work are the long circular cylinder and the sphere. The heat-transfer phenomena for these two shapes in crossflow have been studied by a num-

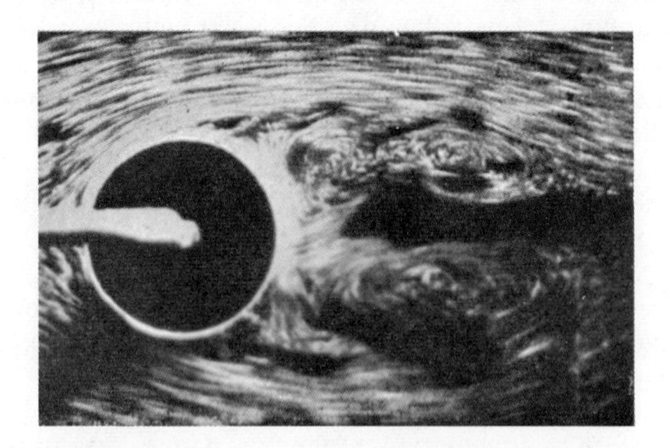

Fig. 9-2. Flow pattern in crossflow over a single horizontal cylinder. (Photograph by H. L. Rubach, *Mitt. Forschungsarb*, 185, 1916).

ber of investigators, and representative data are summarized in Sec. 9-2. In addition to the average surface conductance over a cylinder, the variation of conductance around the circumference will be considered. A knowledge of the peripheral variation of the heat transfer associated with flow over a cylinder is important for many practical problems such as heat-transfer calculations for airplane wings, whose leading-edge contours are approximately cylindrical. The interrelation between heat transfer and flow phenomena will also be stressed because it can be applied to the measurement of the velocity and its fluctuations in a turbulent stream by means of a hot-wire anemometer. Heat transfer to or from spherical bodies is of importance in systems where particles suspended in a fluid stream are heated or cooled. Examples of such systems are met in fluidization processes, settling operations, and cement preheaters.

Section 9-3 deals with heat transfer to or from bundles of tubes in crossflow, a configuration which is widely used in boilers, air-preheaters, and conventional shell-and-tube heat exchangers. Representative experimental data are presented and applied to typical engineering problems.

9-2 Cylinder and sphere in crossflow

Photographs of typical flow patterns for flow over a single cylinder and a sphere are shown in Figs. 9-2 and 9-3 respectively. The most forward points of these bodies are called stagnation points. Fluid particles striking there are brought to rest, and the pressure at the stagnation point p_o rises approximately one velocity head, i.e. $(\rho_\infty V_\infty^2/2g_c)$, above the pressure in the oncoming free stream p_∞. The flow divides at the stagnation point of the cylinder, and a boundary layer builds up along the surface. The fluid accelerates when it flows past the surface of the cylinder, as can be seen by the crowding of the streamlines shown in Fig. 9-4. This flow pattern for a nonviscous fluid in irrotational flow, a highly idealized case, is called *potential flow*. The velocity reaches a maximum at both sides of the cylinder, then falls again to zero at the stagnation point in the rear. The pressure distribution around the cylinder corresponding to this idealized flow pattern is shown by the solid line in Fig. 9-5. Since the pressure distribution is symmetrical about the vertical center plane of the cylinder, it is clear that there will be no pressure drag in irrotational flow. However, unless the Reynolds number is very low, a real fluid will not adhere to the entire surface of the cylinder but, as mentioned previously, the boundary layer in which the flow is not irrotational will separate from the sides of the cylinder as a result of the adverse pressure gradient. The separation of the boundary layer and the resultant wake in the rear of the cylinder give rise to pressure distributions shown for different Reynolds numbers by the dotted lines in Fig. 9-5. It can be seen that there is fair agreement between the ideal and actual pressure distribution in the neigh-

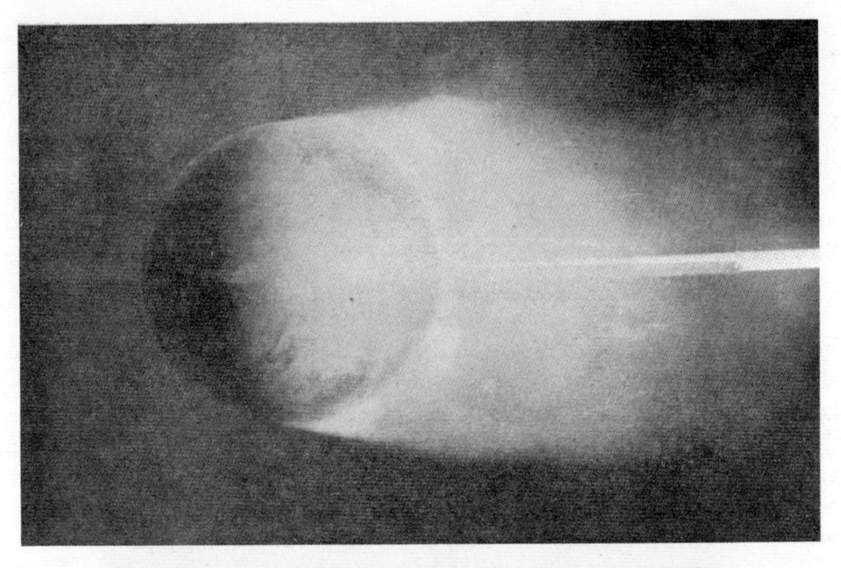

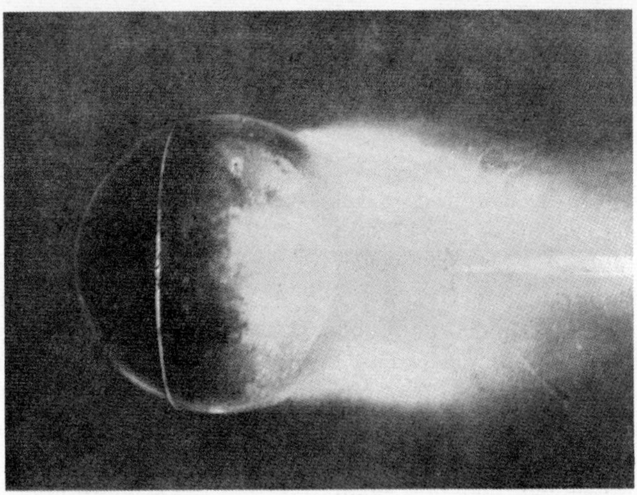

Fig. 9-3. Photographs of air flowing over a sphere. In lower picture a "tripping" wire induced early transition and delayed separation. (Courtesy of L. Prandtl and the *Journal of the Royal Aeronautical Society*)

borhood of the forward stagnation point. In the rear of the cylinder, however, the actual and the ideal distribution differ considerably. The characteristics of the flow pattern and of the boundary layer depend on the Reynolds number, $V_\infty D_o \rho/\mu$, which for flow over a cylinder or a sphere is based on the velocity of the oncoming free stream V_∞ and the outside diameter of the body D_o. The flow pattern around the cylinder undergoes a series of changes as the Reynolds number is increased, and

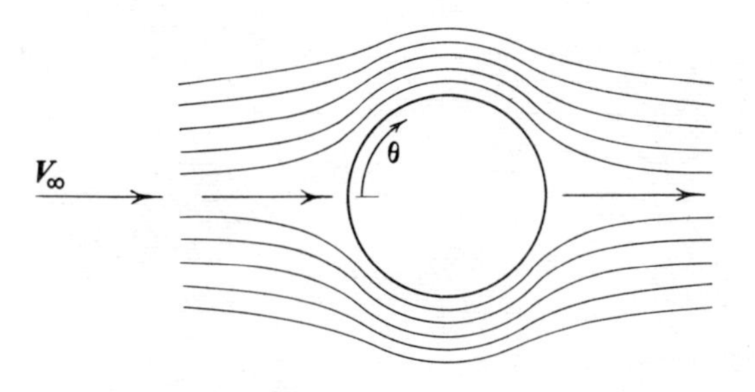

Fig. 9-4. Streamlines for potential flow over a circular cylinder.

since the heat transfer depends largely on the flow, we shall consider first the effect of Reynolds number of the flow and then interpret the heat-transfer data in the light of this information.

The sketches in Fig. 9-6 illustrate flow patterns typical of the characteristic ranges of Reynolds numbers. The letter symbols of the sketches in Fig. 9-6 correspond to the flow regimes indicated in the curve of Fig. 9-7 where the total drag coefficients of a cylinder and a sphere, C_D, are plotted as a function of the Reynolds number. The total drag coefficient is the sum of the pressure and frictional forces; it is defined by the following equation

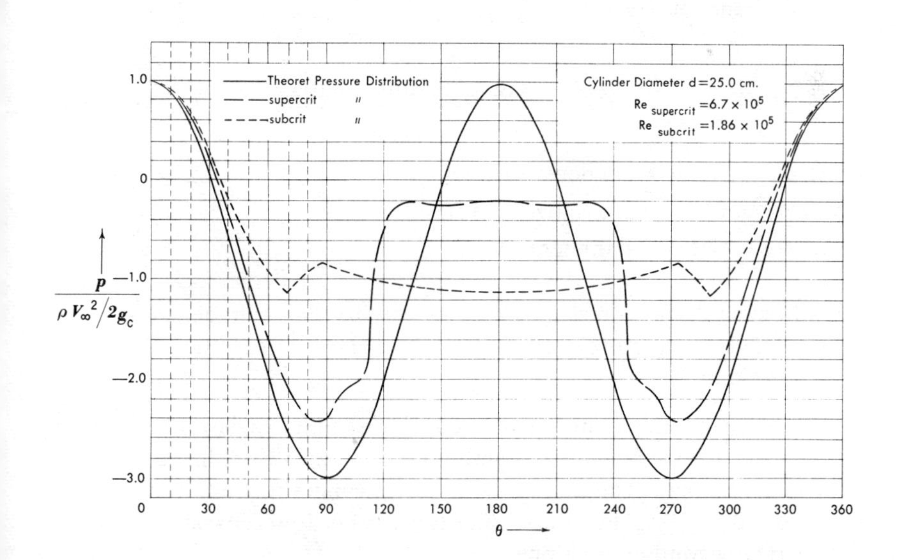

Fig. 9-5. Pressure distribution around circular cylinder in crossflow at various Reynolds numbers; p, local pressure; $\rho V_\infty^2/2g_c$, free-stream impact pressure; θ, angle measured from stagnation point. (By permission from L. Flachsbart, *Handbuch der Experimental Physik*, Vol. 4, Part 2)

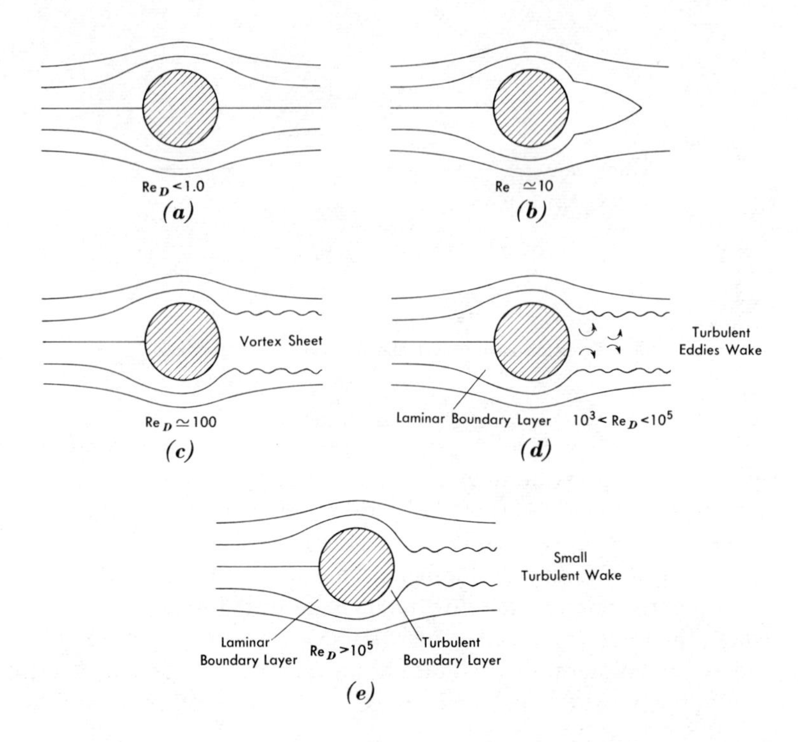

Fig. 9-6. Sketches illustrating flow pattern for crossflow over a circular cylinder at various Reynolds numbers.

$$C_D = \frac{\text{drag force/unit length}}{(\rho_\infty V_\infty^2/2g_c)\,D_o}$$

where ρ_∞ = free-stream density, in lb_m/cu ft;
V_∞ = free-stream velocity, in ft/sec;
D_o = outside diameter, in ft.

The following discussion strictly applies only to long cylinders, but it also gives a qualitative picture of the flow past a sphere. The letters a to e refer to Figs. 9-6 and 9-7.

a) At Reynolds numbers of the order of unity or less, the flow adheres to the surface and the streamlines follow those predicted from potential-flow theory. The inertia forces are negligibly small and the drag is caused only by viscous forces, since there is no flow separation. Heat is transferred by conduction alone.

b) At Reynolds numbers of the order of 10, the inertia forces become appreciable and two weak eddies stand in the rear of the cylinder. The pressure drag accounts now for about one-half of the total drag.

c) At a Reynolds number of the order of 100, vortices separate

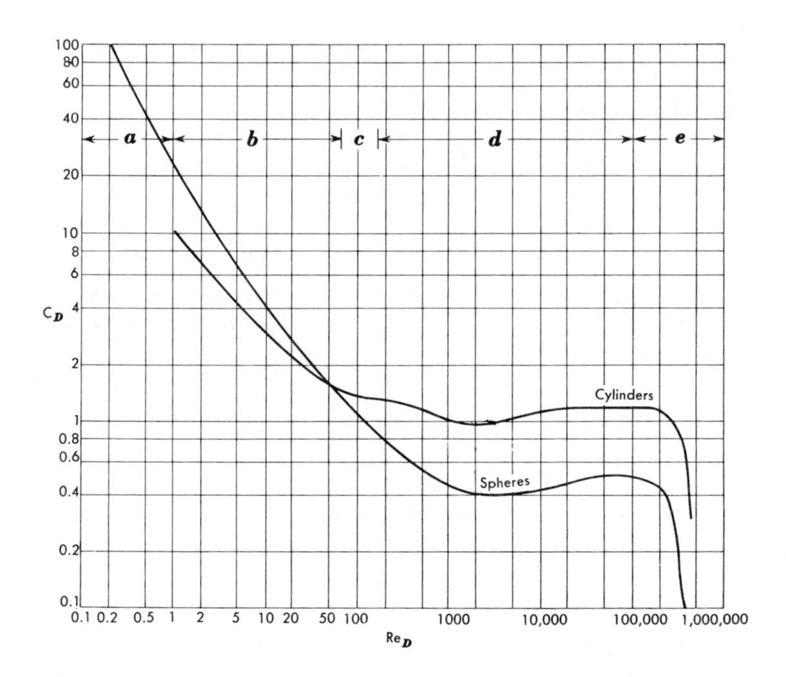

Fig. 9-7. Drag coefficient vs. Reynolds number for long circular cylinders and spheres in crossflow.

alternately from both sides of the cylinder and stretch a considerable distance downstream. These vortices are referred to as *von Karman vortex-streets* in honor of the scientist Theodore von Karman, who studied the shedding of vortices from bluff objects. The pressure drag now predominates.

d) In the Reynolds-number range between 10^3 and 10^5, the skin-friction drag becomes negligible compared to the pressure drag caused by turbulent eddies in the wake. The drag coefficient remains approximately constant because the boundary layer remains laminar from the leading edge to the point of separation, which lies throughout this Reynolds number range at an angular position, θ, between 80 and 85 deg measured from the direction of the flow.

e) At Reynolds numbers larger than about 10^5 (the exact value depends on the turbulence level of the stream) the kinetic energy of the fluid in the laminar boundary layer over the forward part of the cylinder is sufficient to overcome the unfavorable pressure gradient without separating. The flow in the boundary layer becomes turbulent while it is still attached, and the separation point moves toward the rear. The closing of the streamlines reduces the size of the wake, and the pressure drag is therefore also substantially reduced. Experiments by Fage and Falkner (1,2) indicate that, once the boundary layer has become turbulent, it will not separate before it has reached an angular position corresponding to a θ of about 130 deg.

Analyses of the boundary-layer growth and the variation of the local unit-surface conductances with angular position around circular cylinders and spheres have been only partially successful. Squire (3) has solved the equations of motion and energy for a cylinder at constant temperature in crossflow over that portion of the surface to which a laminar boundary layer adheres. He showed that, at the stagnation point and in its immediate neighborhood, the convective unit-surface conductance can be calculated from the equation

$$\mathrm{Nu}_D = \frac{h_c D_o}{k_f} = C\sqrt{V_\infty D_o/\nu_f} \tag{9-1}$$

where C is a constant whose numerical value at various Prandtl numbers is tabulated below:

Pr	0.7	0.8	1.0	5.0	10.0
C	1.0	1.05	1.14	2.1	1.7

Over the forward portion of the cylinder ($0 < \theta < 80$ deg), the empirical equation for $h_{c\theta}$, the local value of the unit-surface conductance at θ,

$$\frac{h_{c\theta} D_o}{k_f} = 1.14\left(\frac{V_\infty D_o}{\nu_f}\right)^{0.5} \mathrm{Pr}_f^{0.4}[1 - (\theta/90)^3] \tag{9-2}$$

has been found to agree satisfactorily (4) with experimental data. For air, Eq. 9-2 can be written in the form

$$h_{c\theta} = 0.194\, T_f^{0.49}(V_\infty \rho_\infty/D_o)^{0.5}[1 - (\theta/90)^3] \tag{9-2a}$$

where T_f is the arithmetic average of the absolute temperatures of the free stream and of the surface in degrees Rankine R. Giedt (5) has measured the local pressures and the local unit-convective conductances over the entire circumference of a long, 4-in.-OD cylinder in an air stream over a Reynolds-number range from 90,000 to 220,000. Giedt's results are shown in Fig. 9-8, and similar data for lower Reynolds numbers are shown in Fig. 9-9. If the data shown in Figs. 9-8 and 9-9 are compared at corresponding Reynolds numbers with the flow patterns and the boundary-layer characteristics described earlier, some important observations can be made.

At Reynolds numbers below 100,000, separation of the laminar boundary layer occurs at an angular position of about 80 deg. The heat transfer and the flow characteristics over the forward portion of the

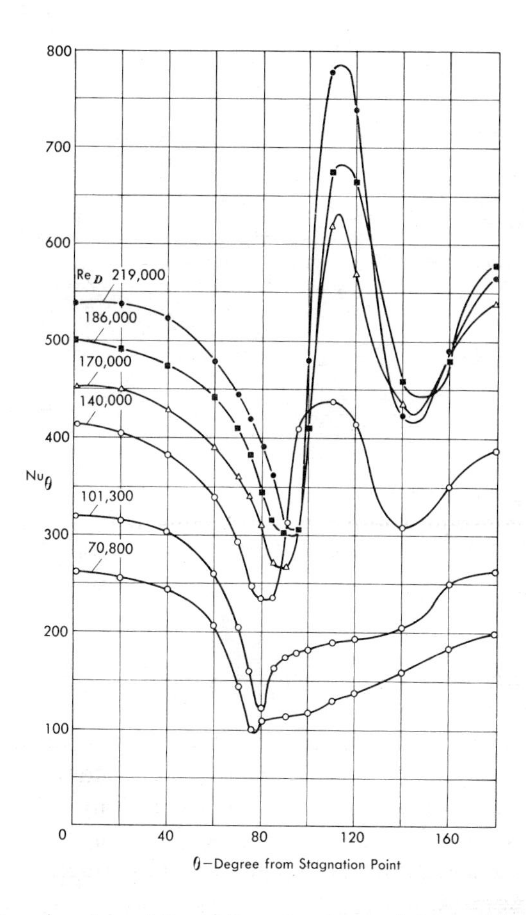

Fig. 9-8. Circumferential variation of the unit-surface conductance at high Reynolds numbers for a circular cylinder in crossflow. (Extracted from "Investigation of Variation of Point Unit-Heat-Transfer Coefficient Around a Cylinder Normal to an Air Stream," by W. H. Giedt, published in *Trans. ASME*, Vol. 71, 1949, with permission of the publishers, The American Society of Mechanical Engineers)

cylinder resemble those for laminar flow over a flat plate which were discussed earlier. The local conductance is largest at the stagnation point and decreases with distance along the surface as the boundary-layer thickness increases. The conductance reaches a minimum on the sides of the cylinder near the separation point. Beyond the separation point the local conductance increases because considerable turbulence exists over the rear portion of the cylinder where the eddies of the wake sweep the surface. However, the conductance over the rear is no larger than over the front, because the eddies recirculate part of the fluid and, despite their high turbulence, are not as effective in mixing the fluid in the vicinity of

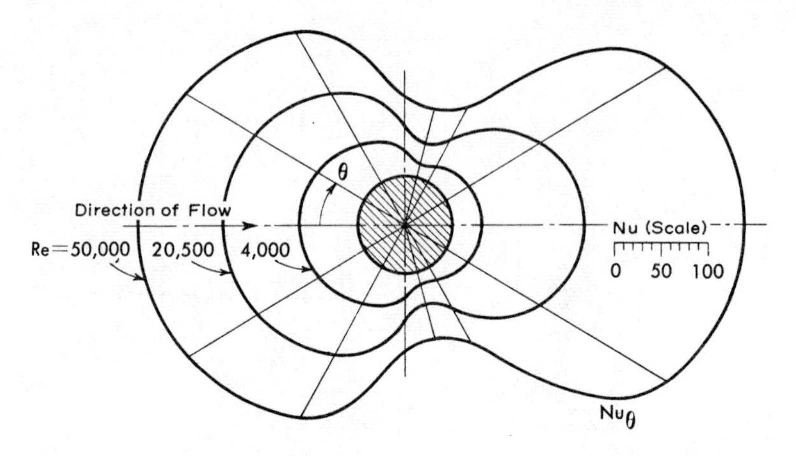

Fig. 9-9. Circumferential variation of the local Nusselt number $Nu_\theta = h_{c\theta}D_o/k_f$ at low Reynolds numbers for a circular cylinder in crossflow. (According to W. Lorisch from M. ten Bosch, "Die Wärmeübertragung," 3d ed., Springer Verlag, Berlin, 1936)

the surface with the fluid in the main stream as a turbulent boundary layer.

At Reynolds numbers large enough to permit transition from laminar to turbulent flow in the boundary layer without separation of the laminar boundary layer, the unit-surface conductance has two minima around the cylinder. The first minimum occurs at the point of transition. As the transition from laminar to turbulent flow progresses, the unit conductance increases and reaches a maximum approximately at the point where the boundary layer becomes fully turbulent. Then the unit-surface conductance begins to decrease again and reaches a second minimum at about 130 deg, the point at which the turbulent boundary layer separates from the cylinder. Over the rear of the cylinder the unit conductance increases to another maximum at the rear stagnation point.

EXAMPLE 9-1. To design a heating system for the purpose of preventing ice formation on an aircraft wing it is necessary to know the unit-surface conductance over the outer surface of the leading edge. The leading-edge contour may be approximated by a half cylinder of 12-in. diameter. The ambient air is at −30 F and the surface temperature is to be no less than 32 F. The plane is designed to fly at 25,000 ft altitude at a speed of 500 fps. Calculate the distribution of the convective unit-surface conductance over the forward portion of the wing.

Solution: At an altitude of 25,000 ft the standard atmospheric air pressure is 785 pounds per square foot (psf) and the density of the air is 0.034 lb_m/cu ft (see Table A-7 in the Appendix).

The unit-surface conductance at the stagnation point, i.e., at $\theta = 0$, is, according to Eq. 9-2a,

$$h_{c0} = 0.194 T_f^{0.49} \left(\frac{V_\infty \rho_\infty}{D_o} \right)^{0.5}$$

$$= 0.194 (461)^{0.49} \left(\frac{500 \times 0.034}{1.0} \right)^{0.5}$$

$$= 16.2 \text{ Btu/hr sq ft F}$$

The variation of h_c with θ is obtained by multiplying the value of the unit-surface conductance at the stagnation point by $1 - (\theta/90)^3$. The results are tabulated below.

θ (deg)	0	15	30	45	60	75
$h_{c\theta}$ (Btu/hr sq ft F)	16.2	16.1	15.6	14.2	11.4	6.88

Ans.

It is apparent from the foregoing discussion that the variation of the unit-surface conductance around a cylinder or a sphere is a very complex problem. For many practical applications it is fortunately not necessary to know the local value $h_{c\theta}$, but sufficient to evaluate the average value of the conductance around the body. A number of observers have measured mean conductances for flow over single cylinders and spheres. Hilpert (6) accurately measured the average conductances for air flowing over cylinders of diameters ranging from 0.008 to nearly 6 in. His results are shown in Fig. 9-10, where the average Nusselt number $\bar{h}_c D_o / k_f$ is plotted

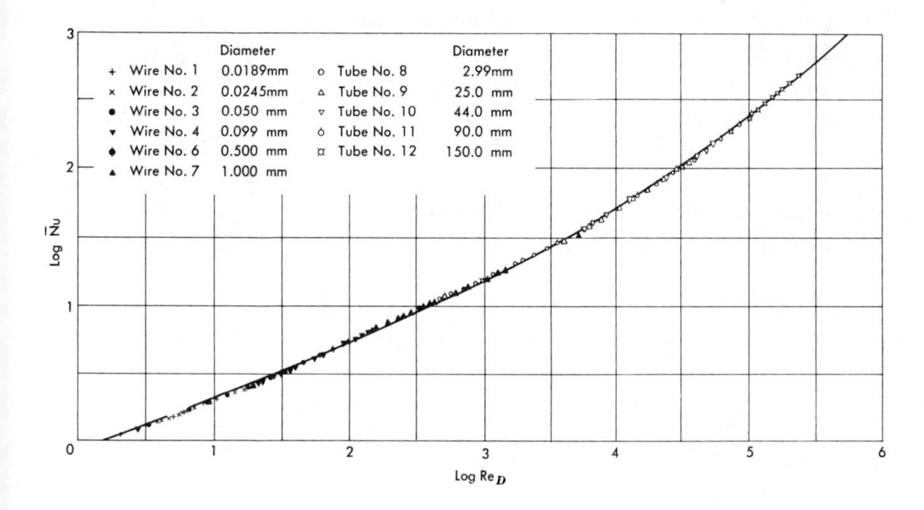

Fig. 9-10. Average heat-transfer coefficient vs. Reynolds number for a circular cylinder in crossflow with air. (After R. Hilbert, "Wärmeabgabe von geheizten Drähten und Rohren," *Forsch. Gebiete Ingenieurwesen*, Vol. 4, 1933, Ab. 9, page 220)

as a function of the Reynolds number $V_\infty D_o/\nu_f$. The data shown in Fig. 9-10 can be correlated by the equation

$$\frac{\bar{h}_c D_o}{k_f} = C\left(\frac{V_\infty D_o}{\nu_f}\right)^n \tag{9-3a}$$

where C and n are empirical constants whose numerical values vary with the Reynolds number as shown in Table 9-1. This empirical correlation agrees with the results shown in Fig. 6-7 to within 15 percent, which is within the accuracy of the experimental data. It should be noted that, if the turbulence level in the oncoming air is increased by placing a grid or some other type of turbulence promoter upstream of the cylinder, the surface conductance may increase by as much as 50 percent (33).

For liquids flowing over a single tube or wire, McAdams (9) suggests that the right-hand side of Eq. 9-3a be multiplied by the factor $1.1\ \mathrm{Pr}_f^{0.31}$, that is,

$$\frac{\bar{h}_c D_o}{k_f} = 1.1C\left(\frac{V_\infty D_o}{\nu_f}\right)^n \mathrm{Pr}_f^{0.31} \tag{9-3b}$$

A more sophisticated correlation has been developed by Whitaker (34) to correlate the available data for liquids as well as gases in the ranges of Reynolds numbers between 1.0 and 10^5 and Prandtl numbers between 0.67 and 300. This correlation takes account of the forward portion of the cylinder over which the boundary layer is attached, as well as of the wake region over the rear. In the boundary layer region the contribution to the Nusselt number should be proportional to $\mathrm{Re}^{0.5}$, as shown in Chapter 6, whereas in the wake region $\overline{\mathrm{Nu}}$ should be proportional to $\mathrm{Re}^{0.67}$. Using experimental data to evaluate appropriate coefficients, Whitaker showed that the average heat-transfer coefficient in flow over a single tube or wire can be determined from the relation

Table 9-1. Coefficients for calculation of average heat-transfer coefficient of a circular cylinder in a gas flowing normal to its axis, by eq. 9-3.

Re_{Df}	C	n
0.4—4	0.891	0.330
4—40	0.821	0.385
40—4,000	0.615	0.466
4,000—40,000	0.174	0.618
40,000—400,000	0.0239	0.805

$$\frac{\bar{h}_c D}{k} = (0.4\,\mathrm{Re}_D^{\,0.5} + 0.06\,\mathrm{Re}_D^{\,0.67})\,\mathrm{Pr}^{0.4}\,(\mu_s/\mu_\infty)^{0.25} \tag{9-3c}$$

where all physical properties should be evaluated at the free stream temperature T_∞, except μ_s which is the viscosity at the surface temperature T_s.

Hot-wire anemometer. The relationship between the velocity and the rate of heat transfer from a single cylinder in crossflow is used to measure velocity and velocity fluctuations in turbulent flow and in combustion processes by means of a hot-wire anemometer. This instrument consists basically of a thin (0.001 to 0.0001-in.-diameter) electrically heated wire stretched across the ends of two prongs. When the wire is exposed to a cooler fluid stream, it loses heat by convection. The temperature of the wire, and consequently its electrical resistance, depends on the temperature and the velocity of the fluid and the heating current. To determine the fluid velocity, the wire is either maintained at a constant temperature by adjusting the current and the fluid speed determined from the measured value of the current, or the wire is heated by a constant current and the speed deduced from a measurement of the electrical resistance or the voltage drop in the wire. In the first method the hot wire forms one arm in the circuit of a Wheatstone bridge as shown in Fig. 9-11a. The resistance of the rheostat arm R_e is adjusted to balance the bridge when the temperature, and consequently the resistance, of the wire has reached some desired value. When the fluid velocity increases, the current required to maintain the temperature and resistance of the wire constant also increases. This change in the current is accomplished by adjusting the rheostat in series

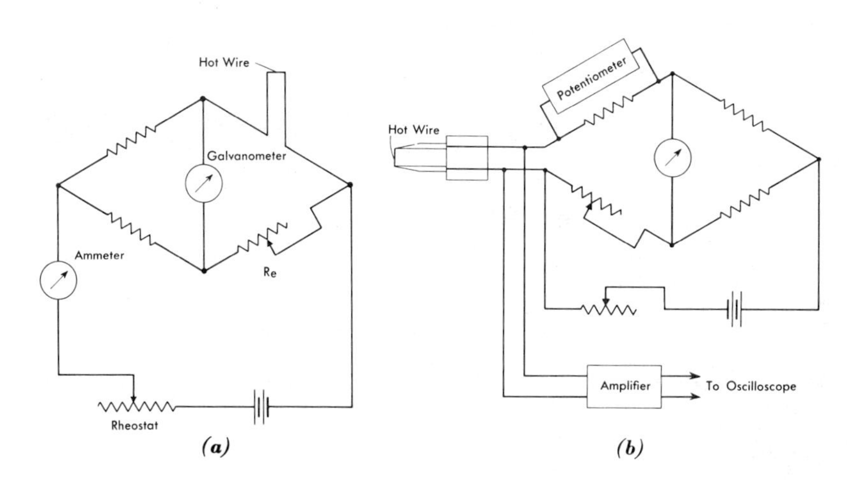

Fig. 9-11. Schematic circuits for hot-wire probes and associated equipment.

with the voltage supply. When the galvanometer indicates that the bridge is in balance again, the change in current, read on the ammeter, indicates the change in speed. In the other method the fluctuations in voltage drop caused by variations in the fluid velocity are impressed across the input of an amplifier, the output of which is connected to an oscilloscope. Figure 9-11b illustrates schematically an arrangement for the voltage measurement. Additional information on the hot-wire method is given in References 7 and 8.

EXAMPLE 9-2. A 0.005-in.-diameter polished-platinum wire 0.25 in. long is to be used for a hot-wire anemometer to measure the velocity of 70 F air in the range between 4 and 20 ft/sec. The wire is to be placed into the circuit of the Wheatstone bridge shown in Fig. 9-11a. Its temperature is to be maintained at 450 F by adjusting the current by means of the rheostat. To design the electric circuit it is necessary to know the required current as a function of air velocity. The electrical resistivity of platinum at 450 F is 17.1 microohms-cm.

Solution: Since the wire is very thin, conduction along the wire can be neglected; also, the temperature gradient in the wire at any cross section may be disregarded.

At the mean film temperature of 260 F the air has a thermal conductivity of 0.0195 Btu/hr ft F and a kinematic viscosity of 2.81×10^{-4} sq ft/sec. At a velocity of 4 ft/sec the Reynolds number is

$$\mathrm{Re} = \frac{(4\ \mathrm{ft/sec})\,(0.005/12\ \mathrm{ft})}{2.81 \times 10^{-4}\,\mathrm{sq\ ft/sec}} = 5.92$$

The Reynolds-number range of interest is therefore from 6 to 30. In this range the equation

$$\frac{\bar{h}_c D_o}{k_f} = 0.821\ \mathrm{Re}^{0.385}$$

applies according to Table 9-1 and Eq. 9-3. The average convective unit-surface conductance is therefore

$$\bar{h}_c = \left(\frac{0.0195\ \mathrm{Btu/hr\ ft\ F}}{0.005/12\ \mathrm{ft}}\right)(0.821)\left[\frac{0.005}{(12)\,(2.81 \times 10^{-4})}\right]^{0.385} V_\infty^{\ 0.385}$$

$$= 44.7\ V_\infty^{\ 0.385}\ \mathrm{Btu/hr\ sq\ ft\ F}$$

At this point it is necessary to estimate the unit-surface conductance for radiant heat flow. According to Eq. 1-6 we have

$$\bar{h}_r = \frac{q_r}{A\,(T_s - T_\infty)} = \frac{\sigma\epsilon\,(T_s^{\ 4} - T_\infty^{\ 4})}{(T_s - T_\infty)} = 0.173 \times 10^{-8}\epsilon\,(T_s^{\ 2} + T_\infty^{\ 2})(T_s + T_\infty)$$

or, since

$$(T_s^2 + T_\infty^2)(T_s + T_\infty) \simeq 4\left(\frac{T_s + T_\infty}{2}\right)^3$$

we have approximately

$$\bar{h}_r = 0.173 \times 10^{-8}\epsilon \times 4\left(\frac{T_s + T_\infty}{2}\right)^3$$

The emissivity of polished platinum from Table 5-1 is about 0.073, so that $\bar{h}_r$ is about 0.01 Btu/hr sq ft F. This shows that the amount of heat transferred by radiation is negligible compared to the heat transferred by forced convection. The rate at which heat is transferred from the wire is therefore

$$q = \bar{h}_c A (T_s - T_\infty) = 44.7\, V_\infty^{0.385} \frac{(\pi)(0.005)(0.25)}{144} 380$$

$$= 0.46\, V_\infty^{0.385} \text{ Btu/hr}$$

which is also the rate at which heat must be generated electrically to maintain equilibrium. The electrical resistance of the wire is

$$R_e = (17.1 \times 10^{-6} \text{ ohm cm})\left[\frac{0.25 \text{ in.}}{(\pi)\, 0.0025^2 \text{ sq in.}}\right]\left(\frac{1}{2.54}\,\frac{\text{in.}}{\text{cm}}\right)$$

$$= 0.0858 \text{ ohms}$$

A heat balance with the current i in amperes gives

$$i^2 R_e \text{ watts } (3.413 \text{ Btu/watt-hr}) = 0.46\, V_\infty^{0.385}$$

Solving for i we get the expression

$$i = \sqrt{\frac{0.46}{(0.0858)(3.413)}\, V_\infty^{0.1925}} = 1.25\, V_\infty^{0.1925} \text{ amp}$$

from which the current can be readily calculated for any velocity within the specified range. *Ans.*

Spheres. A knowledge of heat-transfer characteristics to or from spherical bodies is important to predict the thermal performance of systems where clouds of particles are heated or cooled in a stream of fluid. When the particles have an irregular shape, the data for spheres will yield satisfactory results if the sphere diameter is replaced by an equivalent diameter, i.e., if D_o is taken as the diameter of a spherical particle having the same surface area as the irregular particle.

The total drag coefficient of a sphere is shown as a function of the free-stream Reynolds number in Fig. 9-7[1] and corresponding data for heat transfer between a sphere and air are shown in Fig. 9-12. In the Reynolds-number range from about 25 to 100,000, the equation recommended by McAdams (9) for calculating the average unit-surface conductance for spheres heated or cooled by a gas is

$$\frac{\bar{h}_c D_o}{k_f} = 0.37 \left(\frac{V_\infty \rho_\infty D_o}{\mu_f} \right)^{0.6} = 0.37 \, \mathrm{Re}_D{}^{0.6} \tag{9-4a}$$

For Reynolds numbers between 1.0 and 25, the equation

$$\bar{h}_c = c_p V_\infty \rho_\infty \left(\frac{2.2}{\mathrm{Re}_D} + \frac{0.48}{\mathrm{Re}_D{}^{0.5}} \right) \tag{9-4b}$$

may be used for heat transfer in a gas. For heat transfer in liquid as well as gases, the equation

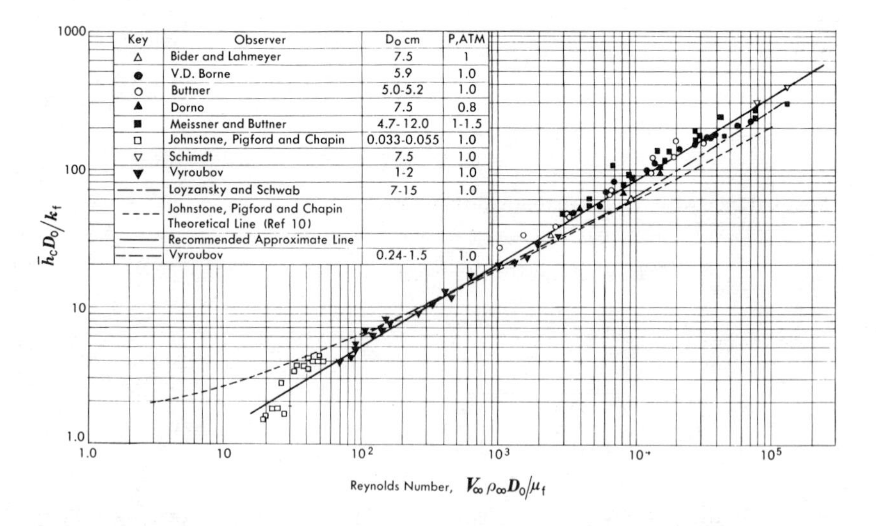

Fig. 9-12. Correlation of experimental average heat-transfer coefficients for flow over a sphere. (By permission from W. H. McAdams, *Heat Transmission*, 3d ed. New York: McGraw-Hill Book Company, Inc., 1954)

---

[1]When the sphere is dragged along by a stream, as for example, a liquid droplet in a gas stream, the pertinent velocity for the Reynolds number is the velocity difference between the stream and the body.

$$\frac{\bar{h}_c D}{k} = 2 + (0.4\ \mathrm{Re}_D^{0.5} + 0.06\ \mathrm{Re}_D^{0.67})\ \mathrm{Pr}^{0.4}\ (\mu_s/\mu_\infty)^{0.25} \qquad (9\text{-}4c)$$

correlates available data in the ranges of Reynolds numbers between 3.5 and 7.6×10^4 and Prandtl numbers between 0.7 and 380 (34).

In the limiting case when the Reynolds number is less than unity, Johnston, et al., (10) have shown from theoretical considerations that the Nusselt number approaches a constant value of two for a Prandtl number of unity unless the spheres have diameters of the order of the mean free path of the molecules in the gas.

Bluff objects. Sogin (31) determined experimentally the heat-transfer coefficient in the separated wake regions behind a flat plate of width D placed perpendicular to the flow and a half-round cylinder of diameter D over Reynolds numbers between 1 and 4×10^5 and found that the following equations correlated the mean heat-transfer results in air:

Normal flat plate:
$$\overline{\mathrm{Nu}}_D = \frac{\bar{h}_c D}{k} = 0.20\ \mathrm{Re}_D^{2/3} \qquad (9\text{-}5a)$$

Half-round cylinder with flat rear surface:
$$\overline{\mathrm{Nu}}_D = \frac{\bar{h}_c D}{k} = 0.16\ \mathrm{Re}_D^{2/3} \qquad (9\text{-}5b)$$

These results are in agreement with an analysis by Mitchell (32).

9-3. Tube bundles in crossflow

The evaluation of the convective conductance between a bank of tubes and a fluid flowing at right angles to the tubes is an important step in the design and performance analysis of many types of commercial heat exchangers. There are, for example, a large number of gas heaters in which a hot fluid inside the tubes heats a gas passing over the outside of the tubes. Figure 9-13 shows several arrangements of tubular air heaters in which the products of combustion, after they leave a boiler, economizer, or superheater, are used to preheat the air going to the steam-generating units. The shells of these gas heaters are usually rectangular and the shell-side gas flows in the space between the outside of the tubes and the shell. Since the flow cross-sectional area is continuously changing along the path, the shell-side gas speeds up and slows down periodically. A similar situation exists also in some unbaffled short-tube liquid-to-liquid heat exchangers in which the shell-side fluid flows over the tubes. In these units the tube arrangement is similar to that in a gas heater except that the shell cross-sectional area varies where a cylindrical shell is used.

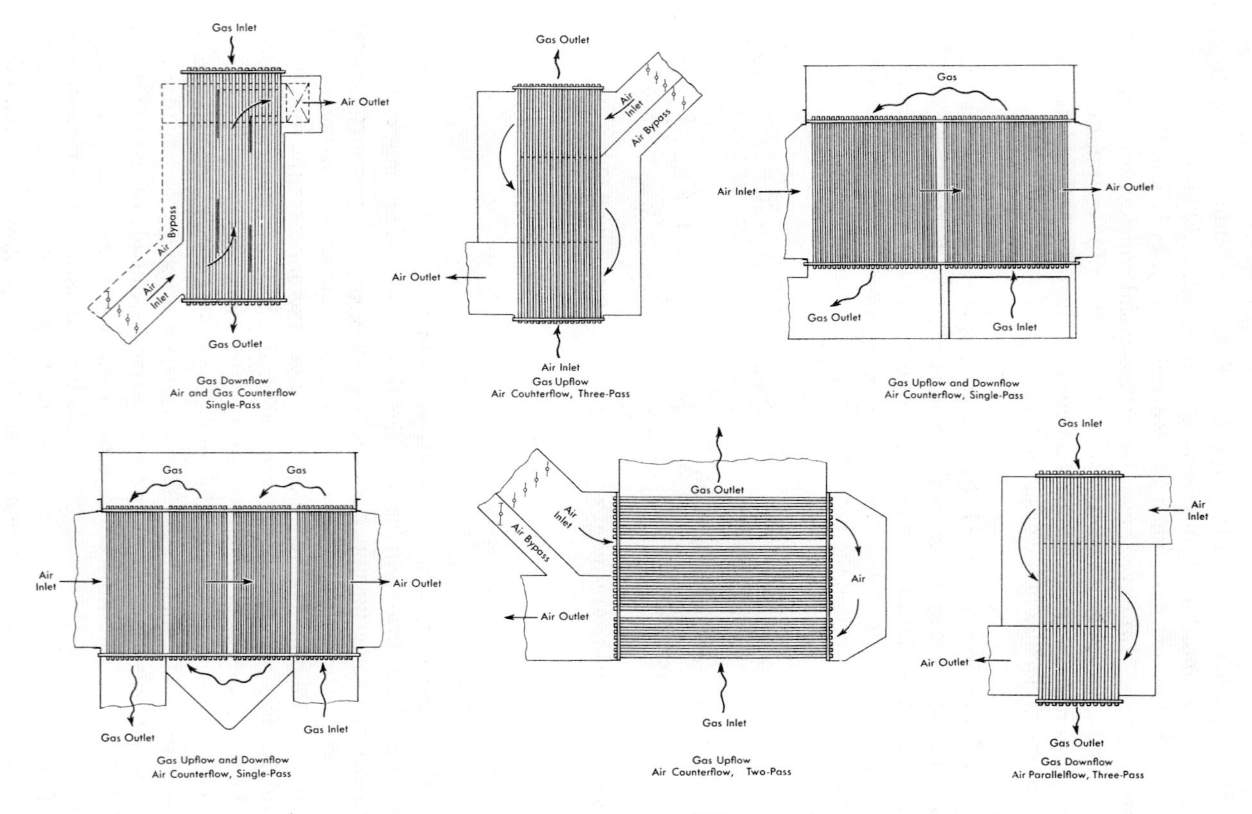

Fig. 9-13. Some arrangements for tubular air heaters. (Courtesy of The Babcock & Wilcox Company)

The heat transfer in flow over tube bundles depends largely on the flow pattern and the degree of turbulence, which in turn are functions of the velocity of the fluid and the size and arrangement of the tubes. The photographs of Figs. 9-14 and 9-15 illustrate the flow patterns for water flowing in the low turbulent range over tubes arranged *in line* and *staggered* respectively. The photographs were obtained (11) by sprinkling fine aluminum powder on the surface of water flowing perpendicularly to the axis of vertically placed tubes. We observe that the flow patterns around tubes in the first transverse rows are similar to those for flow around single tubes. Focusing our attention on a tube in the first row of the in-line arrangement, we see that the boundary layer separates from both sides of the tube and a wake forms behind it. The turbulent wake extends to the tube located in the second transverse row. As a result of the high turbulence in the wakes, the boundary layers around tubes in the second and subsequent rows become progressively thinner. It is therefore not unexpected that, in turbulent flow, the heat-transfer coefficients of tubes in the first row are smaller than the heat-transfer coefficients of tubes in subsequent rows. In laminar flow, on the other hand, the opposite trend has been observed (14).

For a closely spaced staggered-tube arrangement (Fig. 9-15), the size of the turbulent wake behind each tube is somewhat smaller than for similar in-line arrangements, but there is no appreciable reduction in the overall energy dissipation. Experiments on various types of tube arrangements (12) have shown that, for practical units, the relation between heat transfer and energy dissipation depends primarily on the velocity of the fluid, the size of the tubes, and the distance between the tubes. However, in the transition zone the performance of a closely spaced, staggered tube arrangement is somewhat superior to that of a similar in-line tube arrangement.

The equations available for the calculation of heat-transfer coefficients in flow over tube banks are based entirely on experimental data because the flow pattern is too complex to be treated analytically. Experiments have shown that, in flow over staggered-tube banks, the transition from laminar to turbulent flow is more gradual than in flow through a pipe, whereas for in-line tube bundles the transition phenomena resemble those observed in pipe flow. In either case the transition from laminar to turbulent flow begins at a Reynolds number based on the velocity at the minimum flow area, of about 200, and the flow becomes fully turbulent at a Reynolds number of about 6000.

For engineering calculations the average heat-transfer coefficient for the entire tube bundle is of primary interest. The experimental data for heat transfer in flow over banks of tubes are usually correlated by an equation of the form $\overline{\text{Nu}}_D = \text{const}\,(\text{Re})^m\,(\text{Pr})^n$, which has previously been used to correlate the data for flow over a single tube. To apply this equation to flow over tube bundles it is necessary to select a reference velocity,

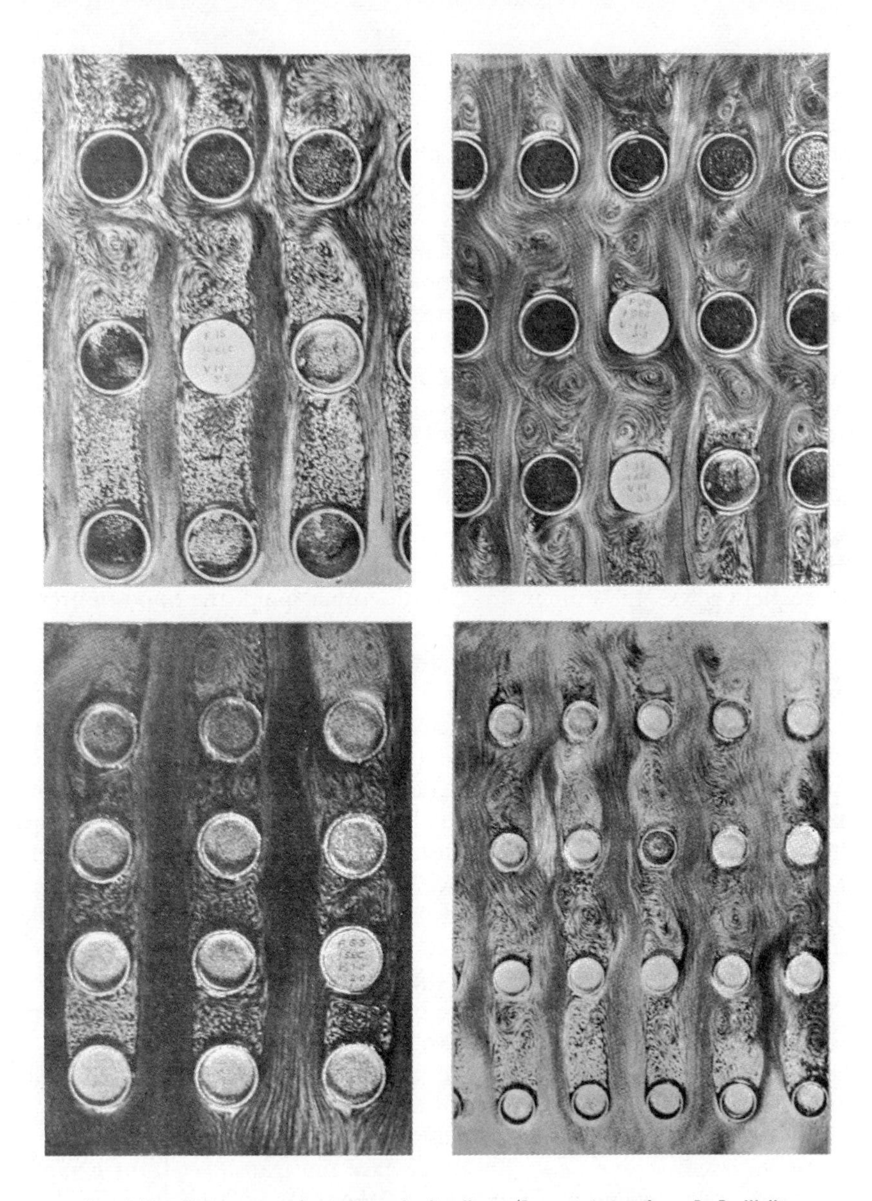

Fig. 9-14. Flow patterns for in-line tube bundles. (By permission from R. D. Wallis, "Photographic Study of Fluid Flow Between Banks of Tubes," *Engineering,* 148, 1933)

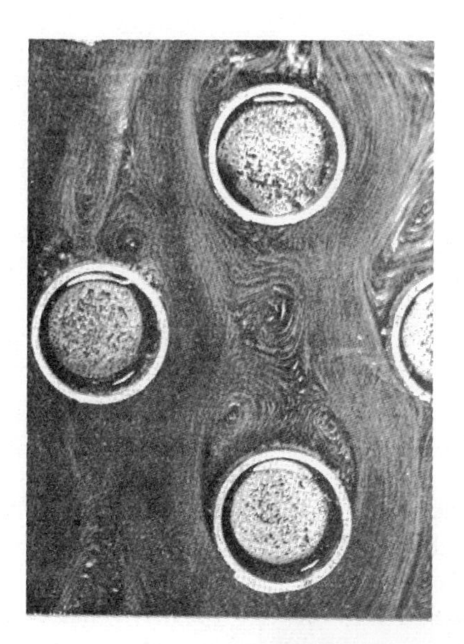

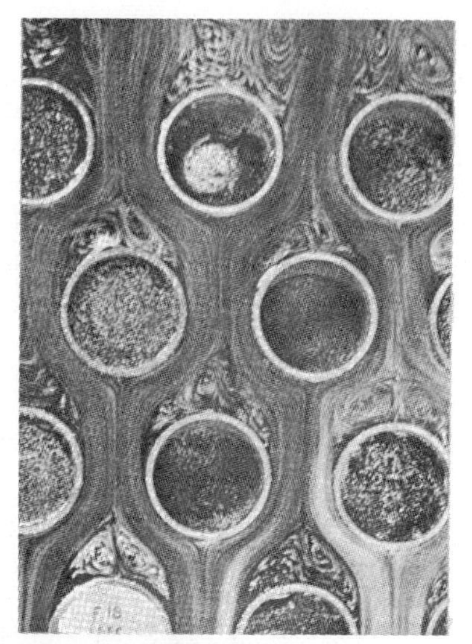

Fig. 9-15. Flow patterns for staggered tube bundles. (By permission from R. D. Wallis, "Photographic Study of Fluid Flow Between Banks of Tubes," *Engineering*, 148, 1933)

since the speed of the fluid varies along its path. The velocity used to build the Reynolds number for flow over tube bundles is based on the *minimum free area* available for fluid flow, regardless of whether the minimum area occurs in the transverse or diagonal openings. For in-line tube arrangements (Fig. 9-16), the minimum free-flow area per unit length of tube A_{min} is always $A_{min} = S_T - D_o$, where S_T is the distance between centers of the tubes in adjacent longitudinal rows (measured perpendicularly to the direction of flow), or the *transverse pitch*.

For staggered arrangements (Fig. 9-17) the minimum free-flow area may occur, as in the previous case, either between adjacent tubes in a row or if S_L/S_T is so small that $\sqrt{S_T^2 + S_L^2} < S_T + (D_o/2)$ between diagonally opposed tubes. In the latter case, the maximum velocity, V_{max}, is $S_T/(\sqrt{S_L^2 + S_T^2} - D_o)$ times the free-flow velocity based on the shell area without tubes. The symbol S_L denotes the center-to-center distance between adjacent transverse rows of tubes or pipes (measured in the direction of flow) and is called the *longitudinal pitch*.

To account for the effect of the tube arrangement on the heat-transfer coefficient, it is convenient to write a dimensionless correlation equation either in the form

$$\frac{\bar{h}_c D_o}{k_f} = 0.33 C_H \left(\frac{G_{max} D_o}{\mu_f}\right)^m \mathrm{Pr}_f^{1/3} \tag{9-7}$$

or in the form

$$j = \frac{\bar{h}_c}{c_p G_{max}} \mathrm{Pr}_b^{2/3} \left(\frac{\mu_s}{\mu_b}\right)^{0.14} = \phi \left(\frac{G_{max} D_o}{\mu_b}\right) \tag{9-8}$$

where $\bar{h}_c$ is the average heat-transfer coefficient for a tube bank of 10 or

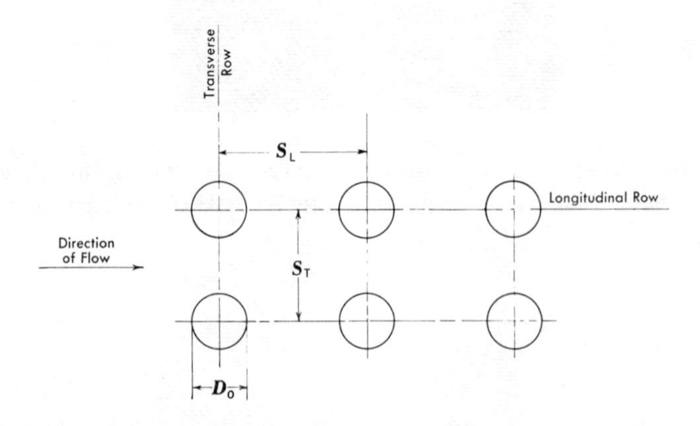

Fig. 9-16. Sketch illustrating nomenclature for in-line tube arrangements.

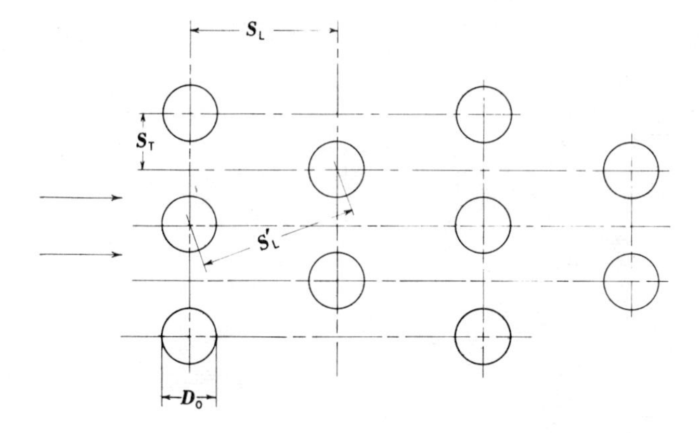

Fig. 9-17. Sketch illustrating nomenclature for staggered tube arrangements.

more transverse rows, G_{max} is the mass flow rate per unit of minimum free area, D_o is the outer diameter of the pipes or tubes, C_H and m are empirical coefficients whose value depends on the tube arrangement, ϕ is a functional relationship dependent on the tube arrangement, and the subscripts s, f, and b refer to conditions at the wall surface, the film, and the bulk respectively.

A relation of the type represented by Eq. 9-8 has been found to correlate data in the laminar-flow range ($G_{max} D_o / \mu_b < 200$) and in the transition-flow range ($200 < G_{max} D_o / \mu_b < 6000$), whereas Eq. 9-7 is used in the turbulent range. Experimental data for oil in laminar and transition flow over several different tube arrangements have been obtained by Bergelin et al. (12,13). The averaged results of this study are shown in Fig. 9-18 where the upper series of curves represent the friction data which will be discussed later, and the lower series of curves represent the heat-transfer data. The ordinate for the lower curves in Fig. 9-18 is the dimensionless Colburn j factor of Eq. 9-8 and the abscissa is the bulk Reynolds number, $G_{max} D_o / \mu_b$. The tubes in models 1 and 4 were arranged staggered in equilateral triangles; in model 3, the tubes were arranged in staggered squares; and the tubes of models 2 and 5 were arranged in in-line squares. The outside diameter of the tubes was $\frac{3}{8}$ in.; the pitch-to-diameter ratio for each model is shown for each curve in Fig. 9-18. Inspection of the curves shows that at a Reynolds number of 200 the experimental data begin to deviate markedly from the straight lines which represent the data in the viscous region. At a Reynolds number of about 5000 the heat-transfer curves for in-line and staggered tubes approach one another and the flow is assumed to be turbulent at higher values of the Reynolds number. It will be noted that the form of the curves for in-line and staggered tubes are different in the transition zone. There is a *dip region*, similar to that observed in pipe flow, for the curves of in-line tubes, but not

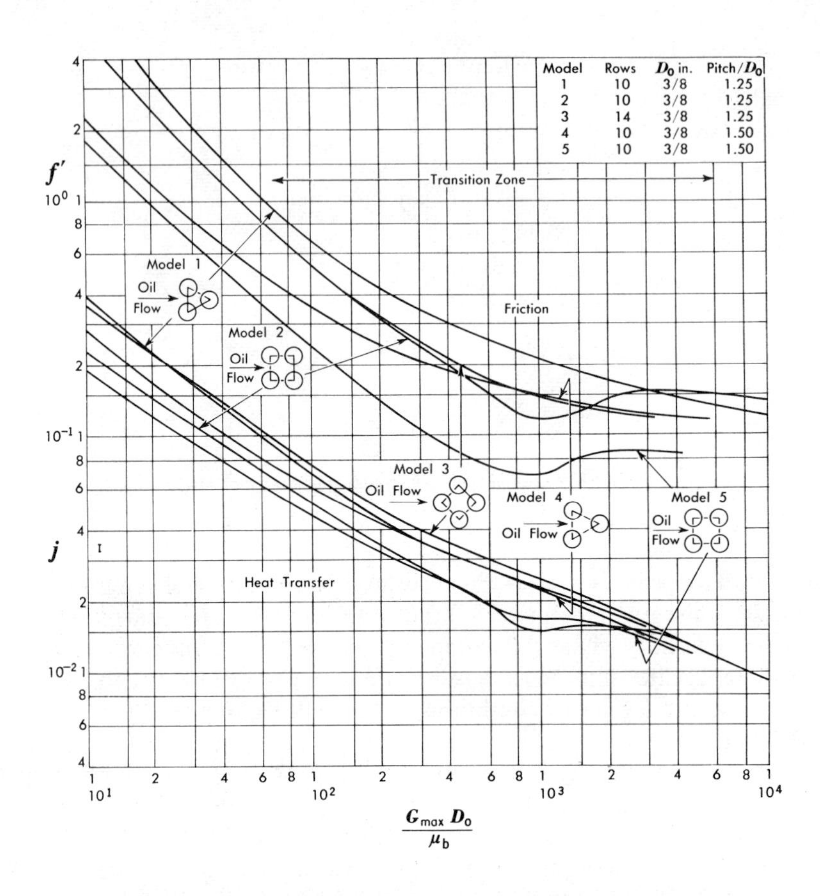

Fig. 9-18. Average friction and heat-transfer data for flow over five different arrangements of $\frac{3}{8}$-in.-diameter tube bundles in the laminar and transition regime. (Extracted from "Heat Transfer and Fluid Friction During Flow Across Banks of Tubes," by O. P. Bergelin, G. A. Brown, and S. C. Doberstein, published in *Trans. ASME*, Vol. 74, 1952, with permission of the publishers, The American Society of Mechanical Engineers)

for staggered tubes. It is believed that the flow in the free channels between the wakes (Fig. 9-14) of in-line tubes resembles flow in a pipe or a duct and the onset of turbulence occurs throughout the tube bank. In simple crossflow over staggered tubes, on the other hand, turbulence begins at the exit end, gradually works upstream as the flow is increased, and finally spreads throughout the tube bank. These general remarks apply only to simple crossflow and may not be true for baffled arrangements.

The effect of the number of transverse tube rows on the heat-transfer coefficient has been investigated for the laminar regime by Meece (14) with square in-line tube arrangements having one, two, four, six, eight, and ten rows of $\frac{3}{8}$-in. tubes with a pitch-to-diameter ratio of 1.25. Meece found that, for a given Reynolds number, the average heat-transfer coefficient for a single row of tubes was 50 per cent larger than for 10 rows. For the tube arrangements used, the variation of the heat-transfer coefficient with N, the number of tube rows in the direction of flow, can be generalized by the equation

$$j_{N\,\text{rows}}/j_{10\,\text{rows}} = (10/N)^{0.18} \tag{9-9}$$

when the flow is laminar. Figure 9-18 shows the results for 10 rows of tubes and can be combined with Eq. 9-9 to predict the average heat-transfer coefficient at Reynolds numbers below 1000 when the number of tube rows is less than 10. For more than 10 tube rows it is suggested that no correction be applied to the value of $\bar{h}_c$, obtained from Fig. 9-18.

Experiments similar to those described above have also been performed by Kays, et al., (15,16) with air flowing over banks of $\frac{1}{4}$- and $\frac{3}{8}$-in. tubes in various arrangements. The Reynolds numbers in these tests covered the transition regime and the low range of the turbulent regime, but did not extend into the laminar region. In the transition range the results obtained by Kays, et al., are in fairly good agreement with those shown in Fig. 9-18 for similar geometries. A summary of the results of these tests is presented in Reference 17.

For turbulent flow (i.e., $Re_{max} \geq 6000$) over banks of tubes or pipes, irrespective of whether they are staggered or arranged in line, the experimental heat-transfer data agree well with the equation

$$\frac{\bar{h}_c D_o}{k_f} = 0.33 C_H \left(\frac{G_{max} D_o}{\mu_f}\right)^{0.6} \Pr_f^{0.3} \tag{9-10}$$

if the tube bundle has 10 or more transverse rows. The value of the empirical coefficient C_H depends on the tube arrangement and the Reynolds number. Fishenden and Saunders (18) evaluated C_H from extensive experiments by Huge (19), Pierson (20), Grimison (21), Kuznetzkof and Lokshin (22) for longitudinal and transverse pitch-to-diameter ratios ranging from 1.25 to 3.0. They found that, for pitch-to-diameter ratios between 1.25 and 1.5, the range of practical interest for heat exchangers, the value of C_H did not deviate by more than 10 percent from unity for any of the tube arrangements tested. For preliminary calculations, Eq. 9-10 is therefore satisfactory. When high accuracy is desired, it is recommended that the correlation of Kays (17) be used in the Reynolds-number range between 1000 and about 15,000 and the correlation of Grimison (18 or 21) be used for the Reynolds-number range between 10,000 and 40,000.

The variation of the average heat-transfer coefficient of a tube bank with the number of transverse rows is shown in Table 9-2 for turbulent flow. To calculate the average heat-transfer coefficient for tube banks

Table 9-2. Ratio of $\bar{h}_c$ for N transverse rows to $\bar{h}_c$ for ten transverse rows in turbulent flow.

Ratio	N									
	1	2	3	4	5	6	7	8	9	10
Staggered tubes.....	0.68	0.75	0.83	0.89	0.92	0.95	0.97	0.98	0.99	1.0
In-line tubes.....	0.64	0.80	0.87	0.90	0.92	0.94	0.96	0.98	0.99	1.0

SOURCE: Ref. 15.

with less than 10 rows, the $\bar{h}_c$ obtained from Eq. 9-10 should be multiplied by the appropriate ratio $\bar{h}_{cN}/\bar{h}_c$.

The frictional pressure drop in lb_f/sq ft for flow over a bank of tubes Δp can be calculated from the equation

$$\Delta p = \frac{f' G_{max}^2 N}{\rho(2.09 \times 10^8)} \left(\frac{\mu_s}{\mu_b}\right)^{0.14} \tag{9-11}$$

where G_{max} = mass velocity at the minimum area, in lb_m/hr sq ft;
ρ = mass density, in lb_m/cu ft;
N = number of transverse rows;

and f' is an empirical friction factor which can be estimated, according to Jakob (23), for values of Reynolds number larger than 1000 from the equation

$$f' = \left[0.25 + \frac{0.118}{\left(\frac{S_T - D_o}{D_o}\right)^{1.08}}\right] \left(\frac{G_{max} D_o}{\mu_b}\right)^{-0.16} \tag{9-12}$$

for staggered tube arrangements and by the equation

$$f' = \left[0.044 + \frac{0.08 S_L/D_o}{\left(\frac{S_T - D_o}{D_o}\right)^{0.43 + 1.13 Do/S_L}}\right] \left(\frac{G_{max} D_o}{\mu_b}\right)^{-0.15} \tag{9-13}$$

for in-line tube arrangements.

For laminar flow the friction factors from the upper series of curves shown in Fig. 9-18 should be used in Eq. 9-11, but the exponent of 0.14 of the viscosity ratio (μ_s/μ_b) in Eq. 9-11 should be replaced by 0.25 (13).

Liquid metals. Experimental data for the heat-transfer characteristics of liquid metals in crossflow over a tube bank have been obtained at the Brookhaven National Laboratory (28 and 29). In these tests mercury ($Pr = 0.022$) was heated while flowing normal to a staggered tube bank consisting of 60 to 70 $\frac{1}{2}$-in. tubes, ten rows deep, arranged in an equilateral triangular array with a 1.375 pitch to diameter ratio. Both local and average heat-transfer coefficients were measured in turbulent flow. The average heat-transfer coefficients in the interior of the tube bank are well correlated by the equation

$$\overline{Nu}_D = 4.03 + 0.228\,(Re_{max}\,Pr)_f^{0.67} \tag{9-14}$$

in the Reynolds number range from 20,000 to 80,000. Additional data are presented in Reference 30.

The measurements of the distribution of the local unit-surface conductance around the circumference of a tube indicate that for a liquid metal the turbulent effects in the wake upon heat transfer are small compared to the heat transfer by conduction within the fluid. Whereas with air and water a marked increase in the local heat-transfer coefficient occurs in the wake region of the tube (see Fig. 9-8), with mercury the unit-surface conductance decreases continuously with increasing θ. At a Reynolds number of 83,000 the ratio $h_{c\theta}/\bar{h}_c$ was found to be 1.8 at the stagnation point, 1.0 at $\theta = 90$, 0.5 at $\theta = 145$, and 0.3 at $\theta = 180$ degrees.

9-4. *Application to heat-exchanger design*

In the design and selection of a stationary commercial heat exchanger, the power requirement and the initial cost of the unit must be considered. The results obtained by Pierson (20) show that the smallest possible pitch in each direction results in the lowest power requirement for a specified rate of heat transfer. Since smaller values of pitch also permit the use of a smaller shell, the cost of the unit is reduced when the tubes are closely packed. There is little difference in performance between in-line and staggered arrangements, but the former are easier to clean. The Tubular Exchanger Manufacturers Association recommends that tubes shall be spaced with a minimum center-to-center distance of 1.25 times the outside diameter of the tube and when tubes are on a square pitch, a minimum clearance lane of $\frac{1}{4}$ in. shall be provided.

EXAMPLE 9-3. Atmospheric air at 58 F is to be heated to 86 F by passing it over a bank of brass tubes inside which steam at 212 F is condensing. The unit-surface conductance on the inside of the tubes is about 1000 Btu/hr sq ft F. The tubes are 2 ft long, $\frac{1}{2}$ in. OD, BWG No. 18 (0.049 in. wall thickness). They are to be arranged in line in a square pattern with a pitch of $\frac{3}{4}$ in. inside a rectangular shell 2 ft wide and 15 in. high. If the total mass rate of flow of the air to be heated is 32,000 lb_m/hr, estimate (a) the number of transverse rows required, and (b) the pressure drop.

Solution: (a) Since the thermal resistance on the air side will be much larger than the combined resistance of the pipe wall and the steam, we shall first assume that the outside surface of the pipe is at the steam temperature. The mean film temperature of the air T_f will then be approximately equal to

$$\frac{1}{2}\left(\frac{58 + 86}{2} + 212\right) = 142 \text{ F}$$

The mass velocity at the minimum cross-sectional area, which is between adjacent tubes, is calculated next. The shell is 15 in. high and consequently holds 19 longitudinal rows of tubes. The minimum free area is

$$A_{min} = (20)(2)[(0.75 - 0.50)/12] = 0.792 \text{ sq ft}$$

and the maximum mass velocity is

$$G_{max} = 32{,}000/0.792 = 40{,}400 \text{ lb}_m/\text{hr sq ft}$$

Hence, the Reynolds number is

$$\text{Re}_{max} = \frac{G_{max} D_o}{\mu_f} = \frac{(40{,}400 \text{ lb/hr sq ft})(0.5/12 \text{ ft})}{0.0485 \text{ lb/hr ft}} = 34{,}600$$

Assuming that more than 10 rows will be required, the unit-surface conductance is calculated from Eq. 9-10, since the flow is turbulent. We get

$$\bar{h}_c = \left(\frac{k_f}{D_o}\right)(0.33)(\text{Re}_{max}{}^{0.6}\,\text{Pr}_f{}^{0.3})$$

$$= \left(\frac{0.016 \text{ Btu/hr ft F}}{0\ 5/12 \text{ ft}}\right)(0.33)(34{,}600)(0.905) = 61.2 \text{ Btu/hr sq ft F}$$

We can now determine the temperature at the outer tube wall, which was originally assumed equal to the steam temperature. There are three thermal resistances in series between the steam and the air. The resistance at the steam side per tube is

$$R_1 = \frac{1}{h_i}\Big/\pi D_i L = \left(\frac{1}{1000}\right)\Big/3.14\left(\frac{0.402}{12}\right)2 = 0.00474 \text{ hr F/Btu}$$

The resistance of the pipe wall ($k = 60$ Btu/hr ft F) is approximately

$$R_2 = \frac{0.049}{k}\Big/\pi\left(\frac{D_o + D_i}{2}\right)L = \left(\frac{0.049}{60}\right)\Big/$$

$$(3.14)(0.451)(2) = 0.000287 \text{ hr F/Btu}$$

The resistance at the outside of the tube is

$$R_3 = \frac{1}{h_o} \Big/ \pi D_o L = \left(\frac{1}{61.2}\right) \Big/ 3.14 \left(\frac{0.5}{12}\right) 2 = 0.0625 \text{ hr F / Btu}$$

The total resistance is then

$$R_1 + R_2 + R_3 = 0.0675 \text{ hr F / Btu}$$

Since the sum of the resistance at the steam side and the resistance of the tube wall is about 8 per cent of the total resistance, about 8 percent of the total temperature drop occurs between the steam and the outer tube wall. The mean film temperature can now be corrected and we get

$$T_f = 137 \text{ F}$$

This will not change the values of the physical properties appreciably, and no adjustment in the previously calculated value of $\bar{h}_c$ is necessary.

The mean temperature difference between the steam and the air can now be calculated. Using the arithmetic average we get

$$T_{\text{steam}} - T_{\text{air}} = 212 - \left(\frac{58 + 86}{2}\right) = 140 \text{ F}$$

The specific heat of air at constant pressure is 0.241 Btu/lb_mF. Equating the rate of heat flow from the steam to the air to the rate of enthalpy rise of the air gives

$$\frac{19 N \Delta T_{\text{avg}}}{R_1 + R_2 + R_3} = w c_p (T_{\text{out}} - T_{\text{in}})_{\text{air}}$$

Solving for N, the number of transverse rows, yields

$$N = \frac{(32{,}000)(0.24)(86 - 58)(0.0675)}{(19)(140)} = 5.5 \text{ or, say, } 6$$

Since the number of tubes is less than 10, it is necessary to correct $\bar{h}_c$ in accordance with Table 9-2, or

$$\bar{h}_{c6\text{ rows}} = 0.94\, \bar{h}_{c10\text{ rows}} = (0.94)(61.2) = 57.5 \text{ Btu/hr sq ft F}$$

Repeating the calculations with the corrected value of the average unit-surface conductance on the air side we find that six transverse rows are sufficient for heating the air according to the specifications. *Ans.*

b) The pressure drop is obtained from Eqs. 9-11 and 9-13. We first calculate the friction factor f. For the arrangement of the heater, $S_L = 1.5\, D_o$, and we get from Eq. 9-13

$$f' = \left[0.044 + \frac{(0.08)(1.5)}{(1.5 - 1)^{0.43 + 1.13/1.5}} \right] 33{,}000^{-0.15} = 0.067$$

Taking the density of the air ρ at 72 F as 0.075 lb_m/cu ft, the pressure drop is from Eq. 9-11

$$\Delta p = \frac{(0.067)(32{,}000^2)(6)}{(0.075)(2.09 \times 10^8)} \, 1.2^{0.14} = 26.8 \text{ lb}_f/\text{sq ft}$$

or about 5 in. of water. *Ans.*

In many commercial shell-and-tube heat exchangers, baffles are used to increase the velocity and consequently the heat-transfer coefficient on the shell side. Figure 9-19 is a photograph of a large baffled exchanger for vegetable-oil service. The flow of the shell-side fluid in baffled heat exchangers is partly perpendicular and partly parallel to the tubes. The heat-transfer coefficient on the shell side in this type of unit depends not only on the size and spacing of the tubes, the velocity and physical properties of the fluid, but also on the spacing and shape of the baffles. In addition, there is always leakage through the tube holes in the baffle and between the baffle and the inside of the shell, and there is bypassing between the tube bundle and the shell. Because of these complications, the heat-transfer coefficient can be estimated only by approximate methods or

Fig. 9-19. Heat exchanger tube bundle with baffles. (Courtesy of the Aluminum Company of America)

from experience with similar units. According to one approximate method which is widely used for design calculations (24), the average heat-transfer coefficient calculated for the corresponding tube arrangement in simple crossflow is multiplied by 0.6 to allow for leakage and other deviations from the simplified model. For additional information the reader is referred to References 24, 25, 26, and 27.

9-5. *Summary*

For the convenience of the reader useful correlation equations for the determination of the average value of the convection heat-transfer coefficients in crossflow over exterior surfaces are tabulated below.

Long Cylinders

$$\frac{\bar{h}_c D}{k} = (0.4\,\mathrm{Re}_D{}^{0.5} + 0.06\,\mathrm{Re}_D{}^{0.67})\,\mathrm{Pr}^{0.4}\,(\mu_s/\mu_\infty)^{0.25}$$

$$(1.0 > \mathrm{Re}_D > 10^5, 0.7 > \mathrm{Pr} > 300)$$

Spheres

$$\frac{\bar{h}_c D}{k} = (0.4\,\mathrm{Re}_D{}^{0.5} + 0.06\,\mathrm{Re}_D{}^{0.67})\,\mathrm{Pr}^{0.4}\,(\mu_s/\mu_\infty)^{0.25}$$

$$(3.5 > \mathrm{Re}_D > 7.6 \times 10^4, 0.7 > \mathrm{Pr} > 380)$$

Tube Bundles—gases and liquids, more than 10 rows

$$\frac{\bar{h}_c D}{k_b} = 0.33\left(\frac{\rho V_{\max} D}{\mu_f}\right)^{0.6} \mathrm{Pr}_f{}^{0.3}$$

$$(6{,}000 < \mathrm{Re}_D, 0.7 < \mathrm{Pr} < 300)$$

For $\mathrm{Re}_{\max}$ less than 6,000, $0.7 < \mathrm{Pr} < 300$ see Fig. 9-18.

Tube Bundles—liquid metals

$$\frac{\bar{h}_c D}{k_f} = 4.0 + 0.23\ \frac{\rho V_{\max} D}{\mu_f}^{0.67}\ \mathrm{Pr}_f{}^{0.67}$$

$$(20{,}000 < \mathrm{Re}_{\max}\,\mathrm{Pr} < 0.03)$$

If baffles are used, as in many heat exchangers, these equations must be modified as shown in Refs. 24, 25, 26, and 27.

PROBLEMS

9-1. Determine the unit-surface conductance at the stagnation point and average value of the conductance for a single 2-in. OD, 24-in.-long tube in cross-flow. The temperature of the tube surface is 500 F, the velocity of the fluid flowing perpendicularly to the tube axis is 20 fps, and its temperature is 100 F. The following fluids are to be considered: (a) air, (b) hydrogen, and (c) water.

9-2. A spherical water droplet of $\frac{1}{16}$-in. diameter is freely falling in atmospheric air. Calculate the average convection heat-transfer coefficient when the droplet has reached its terminal velocity. Assume that the water is at 130 F, the air is at 70 F and neglect mass transfer and radiation.

9-3. A mercury-in-glass thermometer at 100 F (OD = 0.35 in.) is inserted through the duct wall into a 100 fps air stream at 150 F. Estimate the unit-convective conductance between the air and the thermometer.

Ans. $\bar{h}_c \simeq 36$ Btu/hr sq ft F

9-4. Steam at 1 atm and 212 F is flowing across a 2-in. OD tube at a velocity of 20 fps. Estimate the Nusselt number, the heat-transfer coefficient, and the rate of heat transfer per foot length of pipe if the pipe is at 400 F.

9-5. Repeat Prob. 9-4 for a tube bank in which all of the tubes are spaced with their center lines 3.0 in. apart.

9-6. Determine the average unit-surface conductance for air at 142 F flowing at a velocity of 200 fpm over a bank of 2.375-in. OD tubes arranged as shown in the accompanying sketch. The tube-wall temperature is 242 F.

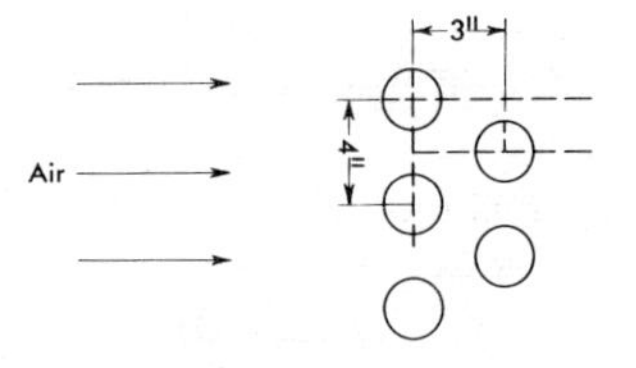

Prob. 9-6

9-7. A stainless-steel pin fin, 2 in. long, $\frac{1}{4}$-in. OD, extends from a flat plate into a 400 mph air stream as shown in the accompanying sketch. (a) Estimate the average heat-transfer coefficient between the air and the fin. (b) Estimate the temperature at the end of the fin. (c) Estimate the rate of heat flow from the fin.

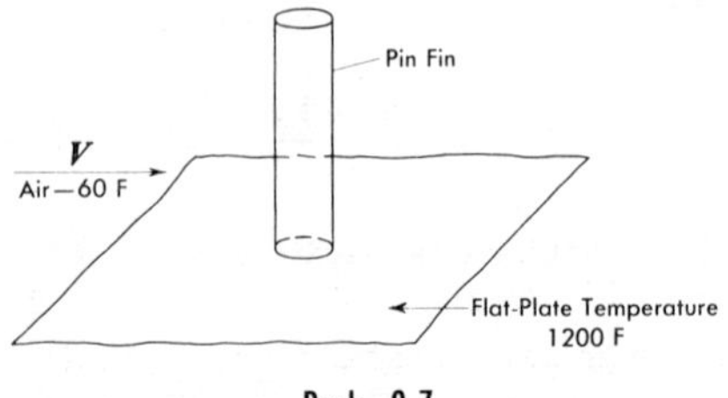

Prob. 9-7

9-8. Repeat Prob. 9-7 with glycerin flowing over the fin at 7 fps.

9-9. An inventor claims that pumping power can be reduced if the tubes in a bank in crossflow are replaced by hollow streamlined bodies whose cross sections have the shape of an ellipse. He claims that energy losses in the wake would be reduced without affecting the rate of heat transfer adversely. See the accompanying sketch. Present your evaluation of the inventor's claim in the form of a short report, and substantiate your conclusions by order-of-magnitude calculations. State all of your assumptions.

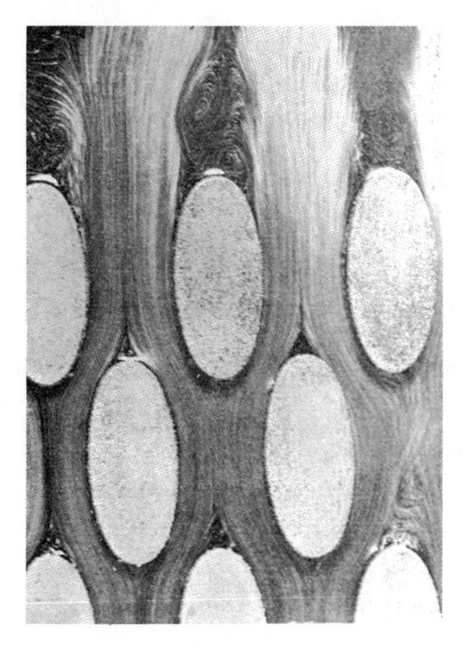

Prob. 9-9

9-10. The instruction manual for a hot-wire anemometer states that "Roughly speaking, the current varies as the fourth power of the average velocity at a fixed wire resistance." Check this statement, using the heat-transfer characteristics of a thin wire in air and water.

9-11. In a lead-shot tower, spherical $\frac{3}{8}$-in.-diam BB shots are formed by drops of molten lead which solidify as they descend in cooler air. At the terminal velocity, i.e., when the drag equals the gravitational force, estimate the total unit-surface conductance if the lead surface is at 340 F (ϵ = 0.63) and the air temperature is 60 F. Assume $C_D = 0.75$ for the first trial calculation.

9-12. Water at 350 F and at 10 ft/sec enters a bare, 50-ft-long, 1 in. wrought-iron pipe (1.05-in. ID, 1.32-in. OD). If air at 50 F flows perpendicular to the pipe at 40 ft/sec, determine the outlet temperature of the water. (Note that the temperature difference between the air and the water varies along the pipe.)

9-13. Estimate the unit-surface conductance for liquid sodium at 1000 F flowing over a 10-row staggered-tube bank arranged in an equilateral-triangular arrow with a 1.5 pitch-to-diameter ratio. The entering velocity is 10 ft/sec, based on the area of the shell, and the tube-surface temperature is 400 F. What is the outlet temperature of the sodium?

9-14. Estimate (a) the unit-surface conductance for a spherical fuel droplet injected into a diesel engine at 180 F and 300 ft/sec. The oil droplet is 0.001 in. in diameter, the cylinder pressure is 700 psia, and the gas temperature is 1700 R. (b) What is the time required to heat the droplet to its self-ignition temperature of 580 F?

9-15. An electrical transmission line of $\frac{1}{2}$-in. OD carries 200 amp and has a resistance of 1×10^{-4} ohm per foot of length. If the air around this line is at 60 F, determine the surface temperature on a windy day, assuming a wind blows across the line at 20 mph

9-16. Derive an equation in the form $\bar{h}_c = f(T, D, V_\infty)$ for flow of air over a long horizontal cylinder for the temperature range 0 to 200 F, using Eq. 9-3 as a basis.

9-17. Repeat Prob. 9-16 for water in the temperature range 50 to 100 F.

9-18. A copper sphere 1 in. in diameter is suspended by a fine wire in the center of an experimental hollow cylindrical furnace whose inside wall is maintained uniformly at 800 F. Dry air at a temperature of 200 F and a pressure of 1.2 atm is blown steadily through the furnace at a velocity of 45 fps. The inside diameter of the furnace is 8 in., the furnace is 32 in. long, and the emissivity of the interior surface of the furnace wall is 0.9. The copper is slightly oxidized, and its emissivity is 0.4. Assuming that the air is completely transparent to radiation, calculate for the steady state: (a) the overall heat-transfer coefficient between the copper sphere and the air, and (b) the temperature of the sphere.

Ans. (a) 28 Btu/hr sq ft F; (b) 260 F

9-19. The temperature of air flowing through a 10-in.-diameter duct whose inner walls are at 600 F is to be measured with a thermocouple soldered in a cylindrical steel well of $\frac{1}{2}$-in.-OD, whose exterior is oxidized as shown in the accompanying sketch. The air flows normal to the cylinder at a mass velocity of 3600 lb/hr sq ft. If the temperature indicated by the thermocouple is 400 F, estimate the actual temperature of the air. *Ans.* About 287 F

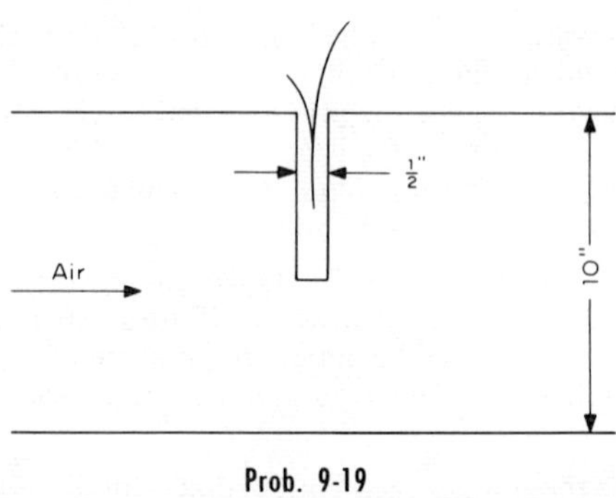

Prob. 9-19

9-20. Develop an expression for the ratio of the rate of heat transfer to water at 100 F from a thin flat strip of width πD and length L at zero angle of attack and a tube of the same length and diameter D in crossflow with its axis normal to the air stream in the Reynolds-number range between 50 and 4000. Assume both surfaces are at 200 F.

9-21. Repeat Prob. 9-20 for air flowing over the same two surfaces in the Reynolds-number range between 40,000 and 400,000, assuming the surface temperature is 100 F. Neglect radiation.

9-22. Liquid mercury at a temperature 600 F flows at a velocity of 3 fps over a staggered bank of $\frac{5}{8}$-in. 16 BWG stainless-steel tubes, arranged in an equilateral triangular array with a pitch-to-diameter ratio of 1.375. If water at 2 atm pressure is being evaporated inside the tubes, estimate the average rate of heat transfer to the water per foot length of the bank, if the bank is 10 rows deep and has 60 tubes in it.

9-23. Compare the rate of heat transfer and the pressure drop for an in-line and a staggered arrangement of a tube bank consisting of 300 tubes, 6 ft long and 1-in. OD. The tubes are to be arranged in 15 rows with normal and parallel spacing of 2 in. The tube-surface temperature is 200 F and water at 100 F is flowing at a mass rate of 12,000 lb/sec over the tubes.

REFERENCES

1. A. Fage, "The Air Flow Around a Circular Cylinder in the Region Where the Boundary Layer Separates from the Surface," Brit. Aero. Res. Comm., *R and M* 1179, 1929.
2. A. Fage and V. M. Falkner, "The Flow Around a Circular Cylinder," Brit. Aero Res. Comm., *R and M* 1369, 1931.
3. H. B. Squire, *Modern Developments in Fluid Dynamics*, 3d ed., Vol. 2. (Oxford: Clarendon Press, 1950.)
4. R. C. Martinelli, A. G. Guibert, E. H. Morin, and L. M. K. Boelter, "An Investigation of Aircraft Heaters VIII—A Simplified Method for Calculating the Unit-Surface Conductance over Wings," *NACA ARR*, March, 1943.
5. W. H. Giedt, "Investigation of Variation of Point Unit-Heat-Transfer Coefficient Around a Cylinder Normal to an Air Stream," *Trans. ASME*, Vol. 71 (1949), pp. 375–381.
6. R. Hilpert, "Wärmeabgabe von geheizten Drähten und Rohren," *Forsch. Gebiete Ingenieurw.*, Vol. 4 (1933), p. 215.
7. H. Dryden and A. N. Kuethe, "The Measurement of Fluctuations of Air Speed by the Hot-Wire Anemometer," *NACA Report* 320, 1929.
8. C. E. Pearson, "Measurement of Instantaneous Vector Air Velocity by Hot-Wire Methods," *J. Aero. Sci.*, Vol. 19 (1952), pp. 73–82.
9. W. H. McAdams, *Heat Transmission*, 3d ed. (New York: McGraw-Hill Book Company, Inc., 1953.)
10. H. F. Johnston, R. L. Pigford, and J. H. Chapin, "Heat Transfer to Clouds of Falling Particles," *Univ. of Ill. Bull.*, Vol. 38, No. 43 (1941).
11. R. D. Wallis, "Photographic Study of Fluid Flow Between Banks of Tubes," *Engineering*, Vol. 148 (1934), pp. 423–425.
12. O. P. Bergelin, G. A. Brown, and S. C. Doberstein, "Heat Transfer

and Fluid Friction During Flow Across Banks of Tubes," *Trans. ASME*, Vol. 74 (1952), pp. 953–959.

13. O. P. Bergelin, A. P. Colburn, and H. L. Hull, "Heat Transfer and Pressure Drop During Viscous Flow Across Unbaffled Tube Banks," *Bull.* 2, Univ. of Delaware Eng. Exp. Sta. (1950).

14. W. E. Meece, *The Effect of the Number of Tube Rows Upon Heat Transfer and Pressure Drop During Viscous Flow Across In-line Tube Banks*, M. S. Thesis, Univ. of Delaware, 1949.

15. W. M. Kays and R. K. Lo, "Basic Heat Transfer and Flow Friction Design Data for Gas Flow Normal to Banks of Staggered Tubes—Use of a Transient Technique," *Tech. Rep.* 15, Navy Contract N6-onr-251 T. O. 6, Stanford Univ., 1952.

16. W. M. Kays, "Basic Heat Transfer and Flow Friction Design Data for Flow Normal to Banks of In-line Circular Tubes—Use of a Transient Technique," *Tech. Rep.* 21, Navy Contract N6-onr-251 T. O. 6, Stanford Univ., 1954.

17. W. M. Kays and A. L. London, "Compact Heat Exchangers—A Summary of Basic Heat Transfer and Flow Friction Design Data," *Tech. Rep.* 23, Navy Contract N6-onr-251, T.O. 6, Stanford Univ., 1954. (Also published in book form under same title by National Press, Palo Alto, Calif., 1955.)

18. M. Fishenden and O. A. Saunders, *An Introduction to Heat Transfer.* (Oxford: Clarendon Press, 1950.)

19. E. C. Huge, "Experimental Investigation of Effects of Equipment Size on Convection Heat Transfer and Flow Resistance in Cross Flow of Gases over Tube Banks," *Trans. ASME*, Vol. 59 (1937), pp. 573–582.

20. O. L. Pierson, "Experimental Investigation of Influence of Tube Arrangement on Convection Heat Transfer and Flow Resistance in Cross Flow of Gases over Tube Banks," *Trans. ASME*, Vol. 59 (1937), pp. 563–572.

21. E. C. Grimison, "Correlation and Utilization of New Data on Flow Resistance and Heat Transfer for Cross Flow of Gases over Tube Banks," *Trans. ASME*, Vol. 59 (1937), pp. 583–594.

22. N. V. Kuznetzkoff and V. A. Lokshin, "Conventional Heat Transfer for the Cross Flow of a Fluid over Tube Banks," *Teplo i Sila*, Vol. 13, No. 10 (1937), p. 19.

23. M. Jakob, "Heat Transfer and Flow Resistance in Cross Flow of Gases over Tube Banks," *Trans. ASME*, Vol. 60 (1938), pp. 384–386.

24. T. Tinker, "Analysis of the Fluid Flow Pattern in Shell-and-Tube Heat Exchangers and the Effect Distribution on the Heat Exchanger Performance," *Inst. Mech. Eng. and ASME Proc. of the General Discussion on Heat Transfer*, September, 1951, pp. 89–115.

25. B. E. Short, "Heat Transfer and Pressure Drop in Heat Exchangers," *Bull.* 3819, Univ. of Texas, 1938. (See also revision, *Bull.* 4324, June, 1943.)

26. D. A. Donohue, "Heat Transfer and Pressure Drop in Heat Exchangers," *Ind. Eng. Chem.*, Vol. 41 (1949), pp. 2499–2511.

27. A. C. Mueller, "Thermal Design of Heat Exchangers," *Eng. Bull.* 121, Res. Series, Purdue Univ., 1954.

28. R. J. Hoe, D. Dropkin, and O. E. Dwyer, "Heat Transfer Rates to Crossflowing Mercury in a Staggered Tube Bank—I," *Trans. ASME*, Vol. 79 (1957), pp. 899–908.

29. C. L. Richards, O. E. Dwyer, and D. Dropkin, "Heat Transfer Rates to Crossflowing Mercury in a Staggered Tube Bank—II," *ASME—AIChE* Heat Transfer Conference Paper No. 57-HT-11, 1957.

30. S. Kalish and O. E. Dwyer, "Heat Transfer to NaK flowing through Unbaffled Rod Bundles," *Int. J. Heat Mass Transfer*, Vol. 10 (1967), pp. 1533–1558.

31. H. H. Sogin, "A Summary of Experiments on Local Heat Transfer from the Rear of Bluff Obstacles to a Lowspeed Airstream," Trans. ASME, ser. C, *J. Heat Transfer*, Vol. 86 (1964), pp. 200–202.

32. J. W. Mitchell, "Base Heat Transfer in Two-Dimensional Subsonic Fully Separated Flows," Trans. ASME, ser. C, *J. Heat Transfer*, Vol. 93 (1971), pp. 342–348.

33. J. Kestin and R. T. Wood, "The Influence of Turbulence on Mass Transfer from Cylinders," *Trans. ASME*, ser. C., J. Heat Transfer, Vol. 93 (1971), pp. 321–327.

34. S. Whitaker, "Forced Convection Heat Transfer Correlations for Flow in Pipes, Past Flat Plates, Single Cylinders, Single Spheres, and for Flow in Packed Beds and Tube Bundles," *A. I. Ch. E. Journal*, Vol. 18 (1972), pp. 361–371.

10
Heat transfer with change in phase

10-1. Fundamentals of boiling heat transfer

Heat transfer to boiling liquids is a convection process involving a change in phase from liquid to vapor. The phenomena of boiling heat transfer are considerably more complex than those of convection without phase change because, in addition to all of the variables associated with convection, those associated with the change in the phase are also relevant. Whereas in liquid-phase convection, the geometry of the system, the viscosity, the density, the thermal conductivity, the expansion coefficient, and the specific heat of the fluid are sufficient to describe the process, in boiling heat transfer, the surface characteristics, the surface tension, the latent heat of evaporation, the pressure, the density, and possibly other properties of the vapor play an important part. As a result of the large number of variables involved, neither general equations describing the boiling process nor general correlations of boiling-heat-transfer data are available to date. Considerable progress has been made, however, during the last few years in gaining a physical understanding of the boiling mechanism (1,67,68,69). By observing the boiling phenomena with the aid of high-speed photography, it has been found that there are various distinct regimes of boiling in which the heat-transfer mechanisms differ radically. To correlate the ɩ .perimental data it is therefore necessary to describe and analyze each of the boiling regimes separately.

To acquire a physical understanding of the phenomena which are characteristic of the various boiling regimes we shall first consider a simple system consisting of a heating surface, such as a flat plate or a wire, submerged in a pool of water at saturation temperature without external agitation. This is called *pool boiling*. A familiar example of such a system is the boiling of water in a kettle on a stove. As long as the temperature of the surface does not exceed the boiling point of the liquid by more than a few degrees, heat is transferred to liquid near the heating surface by free convection. The convection currents circulate the superheated liquid, and evaporation takes place at the free surface of the liquid. The heat-transfer mechanism in this process, although some evaporation occurs, is simply free convection, because only liquid is in contact with the heating surface.

As the temperature of the heating surface is increased, a point is reached where, in certain places, the energy level of the liquid adjacent to the surface becomes so high that some of the molecules break away from the surrounding molecules, are transformed from liquid into a vapor nucleus, and finally form a vapor bubble. This process occurs simultaneously at a number of favored spots on the heating surface. The vapor bubbles are at first small and condense before reaching the surface, but as the temperature is raised further, they become more numerous and larger until they finally rise to the free surface. These phenomena may be observed when boiling water in a kettle. They are also illustrated in Fig. 10-1 for a horizontal wire heated electrically in a pool of distilled water at atmospheric pressure and corresponding saturation temperature of 212 F (2,3). In this curve the heat flux is plotted as a function of the temperature difference between the surface and the saturation temperature. This temperature difference, ΔT_x, is called the excess temperature above the boiling point or *excess temperature* for short. We observe that, in regimes 2 and 3, the heat flux increases rapidly with increasing surface temperature. The process in these two regimes is called *nucleate boiling*. In the individual bubble regime most of the heat is transferred from the heating surface to the surrounding liquid by a vapor-liquid exchange action (4). As vapor bubbles form and grow on the heating surface, they push hot liquid from the vicinity of the surface into the colder bulk of the liquid. In addition, intense microconvection currents are set up as vapor bubbles are emitted and colder liquid from the bulk rushes toward the surface to fill the void. As the heat flux on the surface temperature is raised and the number of bubbles increases to the point where they begin to coalese, heat transer by evaporation becomes more important and eventually predominates at very large heat fluxes in regime 3 (63).

When the excess temperature is raised to about 100 F, we observe that the heat flux reaches a maximum (about 500,000 Btu/hr sq ft in a pool of water), and a further increase of the temperature causes a decrease in the rate of heat flow. This maximum heat flux is said to occur at the critical excess temperature.

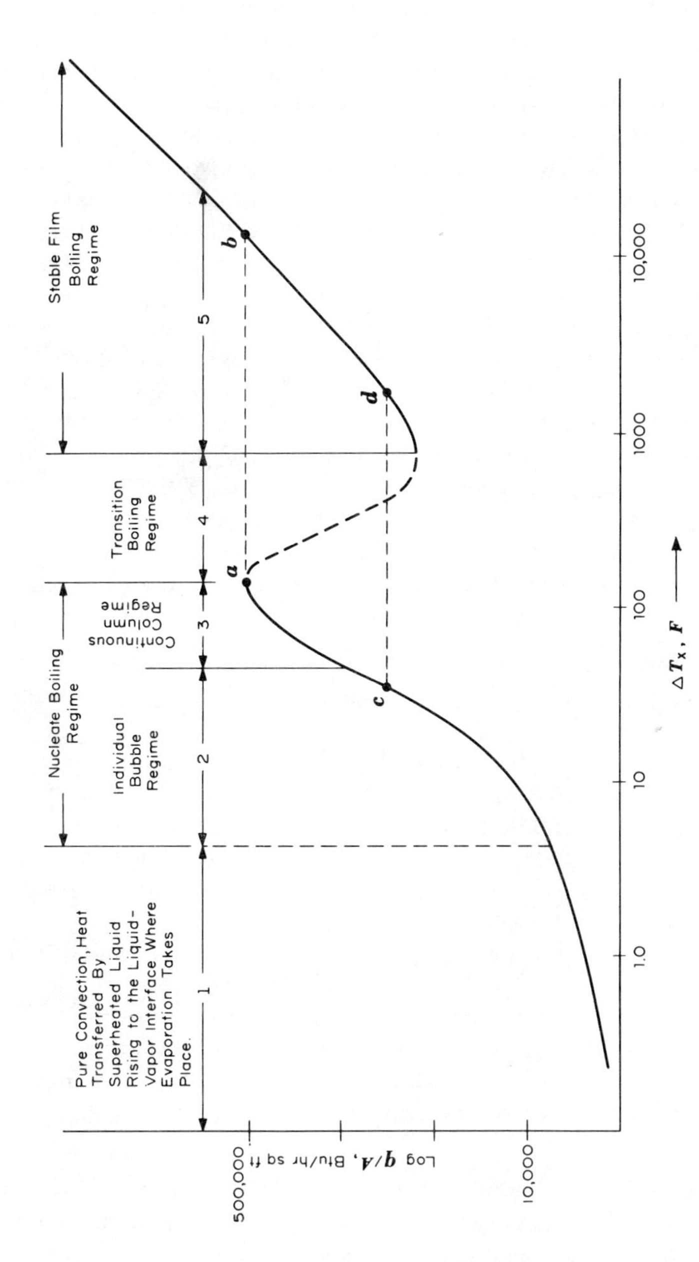

Fig. 10-1. Typical boiling curves for a wire, tube, or horizontal surface in a pool of water at atmospheric pressure.

The reason for the inflection point at about c in the curve may be found by examining the heat-transfer mechanism during boiling. At the onset of boiling, bubbles grow at certain favored spots on the surface until the buoyant force or currents of the surrounding liquid carry them away. But as the heat flux or the surface temperature is increased in nucleate boiling, the number of sites at which bubbles grow increases. Simultaneously, the rate of growth of the bubbles increases and so does the frequency of formation. As the rate of bubble emission from a site increases, bubbles collide and coalesce with their predecessors (5). This point is indicated as a transition from regime 2 to regime 3 in Fig. 10-1. Eventually successive bubbles merge into more or less continuous vapor columns and "mushrooms" (63, 67, 69).

As the maximum heat flux is approached, the number of vapor columns increases. But since each new column occupies space formerly occupied by liquid, there is a limit to the number of vapor columns that can be emitted from the surface. This limit is reached when the space between these columns is no longer sufficient to accommodate the streams of liquid which must move toward the hot surface to replace the liquid which evaporated to form the vapor columns.

If the surface temperature is raised further so that the ΔT_x at the maximum heat flux is exceeded, any of three things can occur, depending on the source of heat and the material of the heating surface (6):

1. If the heater surface temperature is the independent variable and the heat flux is controlled by it, the mechanism will change to transition boiling and the heat flux will decrease.

2. If the heat flux is controlled, as for example in an electrically heated wire, the surface temperature is dependent on it. Provided that the melting point of the heater material is sufficiently high, a transition from nucleate to film boiling will take place, and the heater will operate at a very much higher temperature. This case corresponds to a transition from point a to b in Fig. 10-1.

3. If the heat flux is independent, but the heater material has a low melting point, burnout occurs. For a very short time the heat supplied to the heater exceeds the amount of heat removed because when the peak heat flux is reached, an increase in heat generation is accompanied by a decrease in the rate of heat flow from the heater surface. Consequently, the temperature of the heater material will rise, reach the melting point, and the heater will melt.

In the stable film-boiling regime a vapor film blankets the entire heater surface, whereas in the transition film-boiling regime nucleate and stable film boiling occur alternately at a given location on the heater surface (7). The photographs in Figs. 10-2 and 10-3 illustrate the nucleate- and film-boiling mechanisms on a wire submerged in water at atmospheric pressure. Note the film of vapor which completely covers the wire in Fig.

Fig. 10-2. Photograph showing nucleate boiling on a wire in water. (Courtesy of J. T. Castles)

10-3. A phenomenon which closely resembles this condition is also observed when a drop of water falls on a red-hot stove. The drop does not evaporate immediately but dances on the stove because a steam film forms at the interface between the hot surface and the liquid and insulates the droplet.

Fig. 10-3. Photograph showing film boiling on a wire in water. (Courtesy of J. T. Castles)

When the surface temperature exceeds the saturation temperature, local boiling in the vicinity of the surface may take place even if the bulk temperature is below the boiling point. The boiling process in a liquid whose bulk temperature is below the saturation temperature but whose boundary layer is sufficiently superheated that bubbles form next to the heating surface is usually called *heat transfer to a subcooled boiling liquid* or *surface boiling*. The mechanisms of bubble formation and heat transfer are similar to those described for liquids at saturation temperature. However, the bubbles increase in number while their size and average lifetime decrease with decreasing bulk temperature at a given heat flux (8). As a result of the increase in the bubble population, the agitation of the liquid caused by the motion of the bubbles is more intense in a subcooled liquid than in a pool of saturated liquid, and much larger heat fluxes can be attained before burnout occurs. The mechanism by which a typical bubble transfers heat in subcooled and degassed water is illustrated by the sketches in Fig. 10-4 (9). The letters for the sequence of events described in the following paragraphs correspond to the designation of the sketches.

a) The liquid next to the wall is superheated.
b) A vapor nucleus of sufficient size to permit a bubble to grow has formed at a pit or scratch in the surface.
c) The bubble grows and pushes the layer of superheated liquid above it away from the wall into the cooler liquid above. The resulting motion of the liquid is indicated by arrows.
d) The top of the bubble surface extends into cooler liquid. The temperature in the bubble has dropped. The bubble continues to grow by virtue of the inertia of the liquid, but at a slower rate than during stage *c* because it receives less heat per unit volume.
e) The inertia of the liquid has caused the bubble to grow so large that its upper surface extends far into cooler liquid. It loses more heat by evapo-

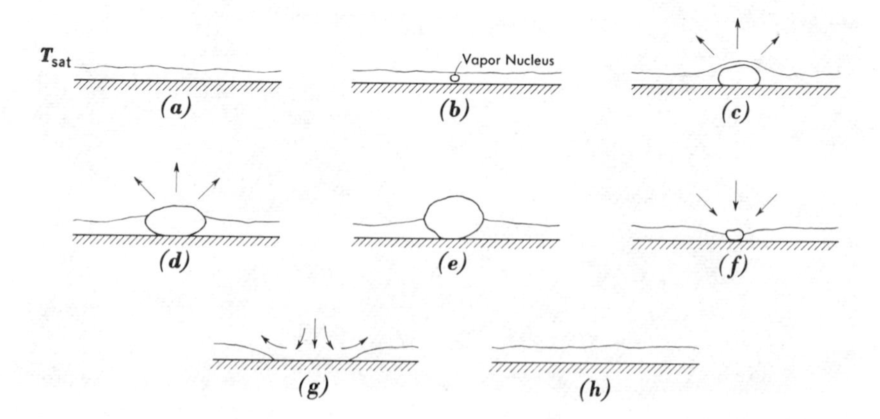

Fig. 10-4. Sketches illustrating the flow pattern induced by a bubble in a subcooled boiling liquid.

ration and convection than it received by conduction from the heating surface.

f) The inertia forces have been dissipated and the bubble begins to collapse. Cold liquid from above follows in its wake.

g) The vapor phase has been condensed, the bubble has disappeared, and the heat wall is splashed by a stream of cold liquid at high velocity.

h) The liquid film has settled and the cycle repeats.

The foregoing description of the life cycle of a typical bubble also applies qualitatively through stage *e* to liquids containing dissolved gases, to solutions of more than one liquid, and to saturated liquids. In these cases, however, the bubble does not collapse but is carried away from the surface by buoyant forces or convection currents. A void is created all the same and the surface is swept by cooler fluid rushing in from above. What eventually happens to the bubbles, whether they collapse on the surface or are swept away, has little influence on the heat-transfer mechanism, which depends mostly on the pumping action and liquid agitation.

The primary variable controlling the bubble mechanism is the excess temperature. It should be noted, however, that, in the nucleate boiling regime, the total variation of the excess temperature irrespective of the fluid bulk temperature is relatively small for a very large range of heat flux. For design purposes the conventional heat-transfer coefficient, which is based on the difference in temperature between the bulk of the fluid and the surface (Eq. 6-1), is therefore only of secondary interest as compared to the maximum heat flux attainable in nucleate boiling and the wall temperature at which boiling begins.

The generation of steam in the tubes of a boiler, the vaporization of fluids such as gasoline in the chemical industry, and the boiling of a refrigerant in the cooling coils of a refrigerator are processes which closely resemble those described above, except that in these industrial applications of boiling the fluid generally flows past the heating surface by forced convection. The heating surface is frequently the inside of a tube or a duct, and the fluid at the discharge end is a mixture of liquid and vapor. The foregoing descriptions of bubble formation and behavior also apply qualitatively to forced convection, but the heat-transfer mechanism is further complicated by the motion of the bulk of the fluid.

10-2. Correlation of boiling-heat-transfer data

The dominant mechanism by which heat is transferred in forced convection is the turbulent mixing of hot and cold fluid particles. As shown in Chapter 8 (Eq. 8-2), experimental data for forced convection without boiling can be correlated by a relation of the type

$$\mathrm{Nu} = \phi(\mathrm{Re})\,\psi\,(\mathrm{Pr})$$

where the Reynolds number Re is a measure of the turbulence and mixing motion associated with the flow. The increased heat-transfer rates attained with nucleate boiling are the result of the intense agitation of the fluid produced by the motion of vapor bubbles. To correlate experimental data in the nucleate-boiling regime the conventional Reynolds number in Eq. 8-2 is replaced by a modulus significant of the turbulence and mixing motion for the boiling process. A type of Reynolds number Re_b, which is a measure of the agitation of the liquid in nucleate-boiling heat transfer, is obtained by combining the average bubble diameter D_b, the mass velocity of the bubbles per unit area G_b, and the liquid viscosity μ_l to form the dimensionless modulus

$$\mathrm{Re}_b = \frac{D_b G_b}{\mu_l}$$

This parameter, often called the bubble Reynolds number, takes the place of the conventional Reynolds in nucleate boiling. If we use the bubble diameter D_b as the significant length in the Nusselt number, Eq. 8-2 can be modified for nucleate boiling into the form

$$\mathrm{Nu}_b = \frac{h_b D_b}{k_l} = \phi(\mathrm{Re}_b)\psi(\mathrm{Pr}_l) \tag{10-1}$$

where Pr_l is the Prandtl number of the saturated liquid and h_b is the *nucleate boiling heat-transfer coefficient* defined as

$$h_b = \frac{q/A}{\Delta T_x}$$

In nucleate boiling the excess temperature ΔT_x is the physically significant temperature potential. It replaces the temperature difference between the surface and the bulk of the fluid ΔT used in single-phase convection. Numerous experiments have shown the validity of this method, which obviates the need to know the exact temperature of the liquid and can therefore be applied to saturated as well as subcooled liquids.

Pool boiling. Using experimental data of pool boiling as a guide, Rohsenow (44) modified Eq. 10-1 by means of simplifying assumptions. An equation found convenient for the reduction and correlation of experimental data (70) is

$$\frac{c_l \Delta T_x}{h_{fg} \mathrm{Pr}_l^{1.7}} = C_{sf}\left[\frac{q/A}{\mu_l h_{fg}}\sqrt{\frac{g_c \sigma}{g(\rho_l - \rho_v)}}\right]^{0.33} \tag{10-2}$$

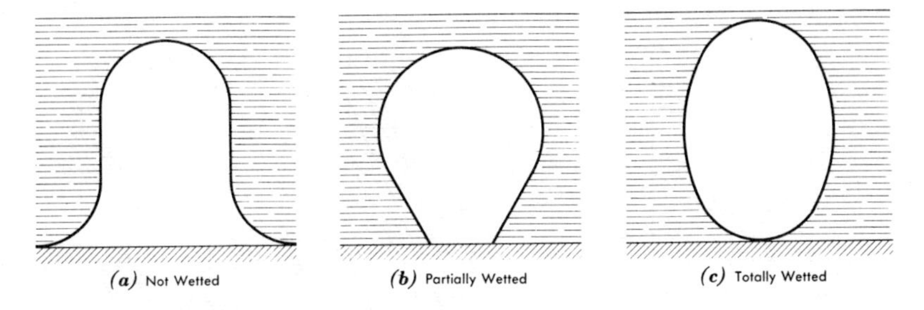

Fig. 10-5. Sketches illustrating the effect of surface wettability on the bubble contact angle.

where c_l = specific heat of saturated liquid, in Btu/lb_m F;
q/A = heat flux, in Btu/hr sq ft;
h_{fg} = latent heat of vaporization, in Btu/lb_m;
g_c = conversion factor, 4.17×10^8 lb_m ft/lb_f hr^2;
g = gravitational acceleration, in ft/hr^2;
ρ_l = density of the saturated liquid, in lb_m/cu ft;
ρ_v = density of the saturated vapor, in lb_m/cu ft;
σ = surface tension of the liquid-to-vapor interface, in lb_f/ft;
Pr_l = Prandtl number of the saturated liquid;
μ_l = viscosity of the liquid, in lb_m/hr ft;
C_{sf} = empirical constant which depends upon the nature of the the heating surface-fluid combination and whose numerical value varies from system to system.

The most important variables affecting C_{sf} are the surface roughness of the heater, which determines the number of nucleation sites at a given temperature (7), and the angle of contact between the bubble and the heating surface, which is a measure of the wettability of a surface with a particular fluid. The sketches of Fig. 10-5 show that the contact angle decreases with greater wetting. A totally wetted surface has the smallest area covered by vapor at a given excess temperature and consequently represents the most favorable condition for efficient heat transfer. In the absence of quantitative information on the effect of wettability and surface conditions on the constant C_{sf}, its value must be determined empirically for each fluid-surface combination.

Figure 10-6 shows experimental data obtained by Addoms (11) for pool boiling of water on a 0.024-in.-diameter platinum wire at various saturation pressures. These data, plotted as heat flux vs. excess temperature in Fig. 10-6, are replotted in Fig. 10-7 using

$$\frac{q/A}{\mu_l h_{fg}} \sqrt{\frac{g_c \sigma}{g(\rho_l - \rho_v)}}$$

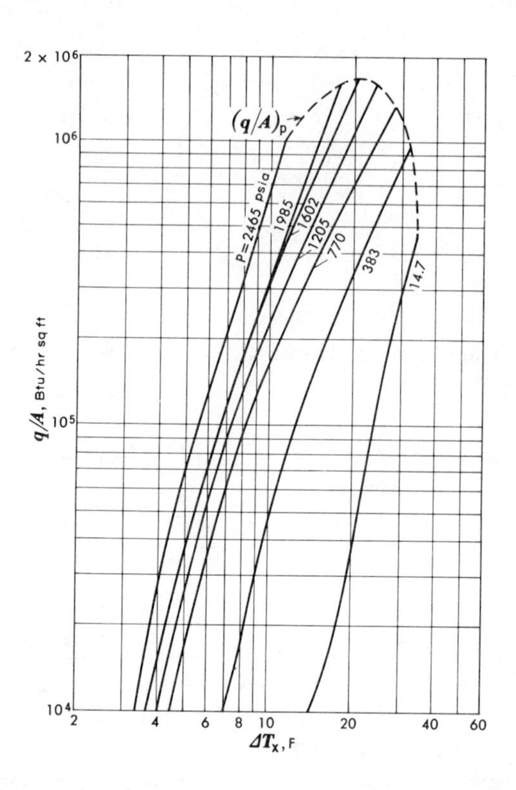

Fig. 10-6. Heat flux vs. excess temperature for nucleate boiling of water on a 0.024-in.-diam electrically-heated platinum wire. (Extracted from "A Method of Correlating Heat-Transfer Data from Surface Boiling Liquids," by W. M. Rohsenow, published in *Trans. ASME*, Vol. 74, 1955, with permission of the publishers, The American Society of Mechanical Engineers).

as the ordinate and $c_l \Delta T_x / h_{fg} \mathrm{Pr}_l^{1.7}$ as the abscissa. The slope of the straight line faired through the experimental points is 0.33; for water boiling on platinum, the value of C_{sf} is 0.013. For comparison the experimental values of C_{sf} for a number of other fluid-surface combinations are listed in Table 10-1.

Selected values of the vapor-liquid surface tension for water at various temperatures are shown in Table 10-2 for use in Eq. 10-2.

The principal advantage of the Rohsenow correlation is that the performance of a particular fluid-surface combination in nucleate boiling at any pressure and heat flux can be predicted from a single test. One value of the heat flux q/A and its corresponding value of the excess temperature difference ΔT_x are all that are required to evaluate C_{fs} in Eq. 10-12. It should be noted, however, that Eq. 10-2 applies only to clean surfaces. For contaminated surfaces the exponent of Pr_l has been found to vary be-

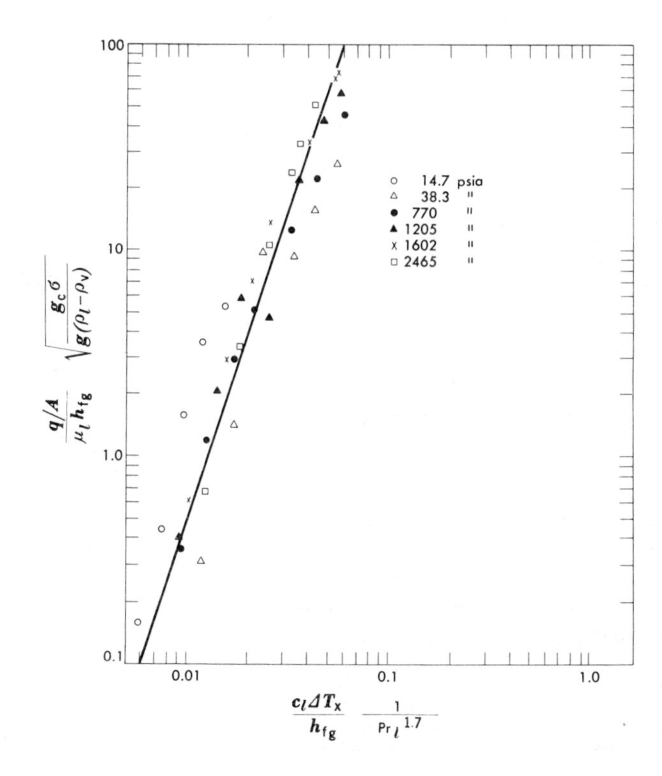

Fig. 10-7. Correlation of pool-boiling heat-transfer data by method of Rohsenow. (Extracted from "A Method of Correlating Heat-Transfer Data from Surface Boiling Liquids," by W. M. Rohsenow, published in *Trans. ASME*, Vol. 74, 1955, with permission of the publishers, The American Society of Mechanical Engineers).

tween 0.8 and 2.0. Contamination apparently does not influence the other exponent in Eq. 10-2.

The geometrical shape of the heating surface has no appreciable effect on the nucleate-boiling mechanism (15,16). This is not unexpected, since the influence of the bubble motion on the fluid conditions is limited to a region very near the surface.

Nucleate boiling with forced convection. The foregoing method of correlating data for nucleate pool boiling has also been applied successfully to boiling of fluids flowing inside tubes or ducts by forced (44) or natural convection (12).

Figure 10-8 shows curves faired through boiling data, typical of subcooled forced convection in tubes or ducts (17,18). The system in which these data were obtained consisted of a vertical annulus containing an electrically heated stainless-steel tube placed centrally in tubes of various diameters. The heater was cooled by degassed distilled water flowing up-

Table 10-1. Values of the coefficient C_{sf} in Eq. 10-2 for various liquid-surface combinations.

Fluid-Heating Surface Combination	C_{sf}
Water–copper (12)*	0.0130
Carbon tetrachloride–copper (12)	0.0130
35% K_2CO_3–copper (12)	0.0054
n-Butyl alcohol–copper (12)	0.00305
50% K_2CO_3–copper (12)	0.00275
Isopropyl alcohol–copper (12)	0.00225
n-Pentane–chromium (13)	0.0150
Water–platinum (11)	0.0130
Benzene–chromium (13)	0.0100
Water–brass (14)	0.0060
Ethyl alcohol–chromium (13)	0.0027
n-Pentane on Emery Polished Copper (70)	0.0154
n-Pentane on Emery Polished Nickel (70)	0.0127
Water on Emery Polished Copper (70)	0.0128
Carbon Tetrachloride on Emery Polished Copper (70)	0.0070
Water on Emery Polished, Paraffin-treated Copper (70)	0.0147
n-Pentane on Lapped Copper (70)	0.0049
n-Pentane on Emery Rubbed Copper (70)	0.0074
Water on Scored Copper (70)	0.0068
Water on Ground and Polished Stainless Steel (70)	0.0080
Water on Teflon Pitted Stainless Steel (70)	0.0058
Water on Chemically Etched Stainless Steel (70)	0.0133
Water on Mechanically Polished Stainless Steel (70)	0.0132

*Numbers in parentheses are those of references listed at end of chapter.

Table 10-2. Vapor-liquid surface tension for water.

Surface tension $\sigma \times 10^4$, lb_f/ft	Saturation temperature, *F*
51.8	32
50.2	60
47.8	100
45.2	140
41.2	200
40.3	212
31.6	320
21.9	440
11.1	560
1.0	680
0.0	705.4

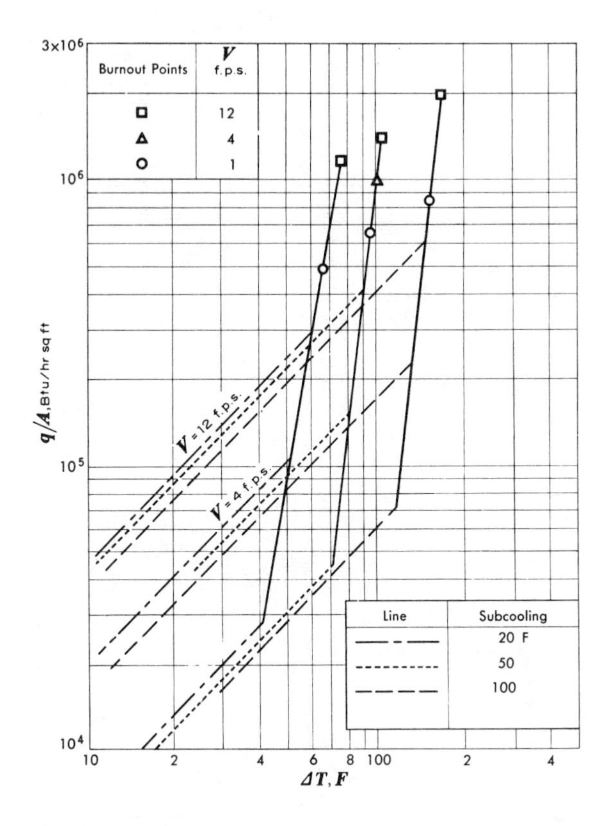

Fig. 10-8. Typical boiling data for subcooled forced convection—heat flux vs. temperature difference between surface and fluid bulk. (By permission from W. M. McAdams, W. E. Kennel, C. S. Minden, R. Carl, P. M. Picarnell, and J. E. Drew from "Heat Transfer at High Rates to Water with Surface Boiling," *Ind. Eng. Chem.*, Vol. 41, 1945).

ward at velocities from 1 to 12 fps and pressures from 30 to 90 psia. The scale of Fig. 10-8 is logarithmic. The ordinate is the heat flux q/A, and the abscissa is ΔT, the temperature difference between the heating surface and the bulk of the fluid. The dotted lines represent forced-convection conditions at various velocities and various degrees of subcooling. The solid lines indicate the deviation from forced convection caused by surface boiling. We note that the onset of boiling caused by increasing the heat flux depends on the velocity of the liquid and the degree of subcooling below its saturation temperature at the prevailing pressure. At lower pressures the boiling point at a given velocity is reached at lower heat fluxes. An increase in velocity increases the effectiveness of forced convection, decreases the surface temperature at a given heat flux, and thereby delays the onset of boiling. In the boiling region the curves are steep and the wall temperature is practically independent of the fluid ve-

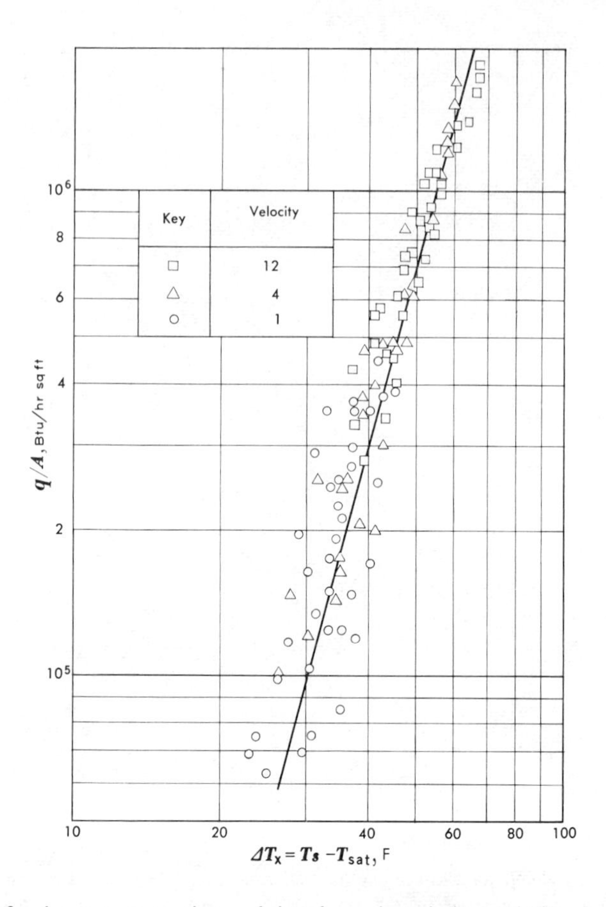

Fig. 10-9. Approximate correlation of data for nucleate boiling with forced-convection obtained by plotting heat flux vs. excess temperature. (By permission from W. M. McAdams, W. E. Kennel, C. S. Minden, R. Carl, P. M. Picarnell, and J. E. Drew from "Heat Transfer at High Rates to Water with Surface Boiling," *Ind. Eng. Chem.*, Vol. 41, 1945).

locity. This shows that the agitation caused by the bubbles is much more effective than turbulence in forced convection without boiling. The heat flux data with surface boiling are plotted separately in Fig. 10-9, vs. the excess temperature. The resulting curve is similar to that for nucleate boiling in a saturated pool shown in Fig. 10-1 and emphasizes the similarity of the boiling processes and their dependence on the excess temperature.

To apply the pool-boiling correlation to forced-convection boiling, the total heat flux must be separated into two parts, one a *boiling flux* q_b/A, the other a *convective flux* q_c/A, and

$$q_{total} = q_b + q_c$$

The boiling heat flux is determined by subtracting the heat-flow rate, accountable for by forced convection alone, from the total flux, or

$$q_b = q_{\text{total}} - A\bar{h}_c(T_s - T_b) \tag{10-3}$$

where $\bar{h}_c$ is determined from Eq. 8-18[1] or Eq. 8-22. This value of q_b is then used in Eq. 10-3 in the same manner as the total heat flux in nucleate pool boiling. The results of this method of correlating data for boiling superimposed on convection are shown in Fig. 10-10 for a number of

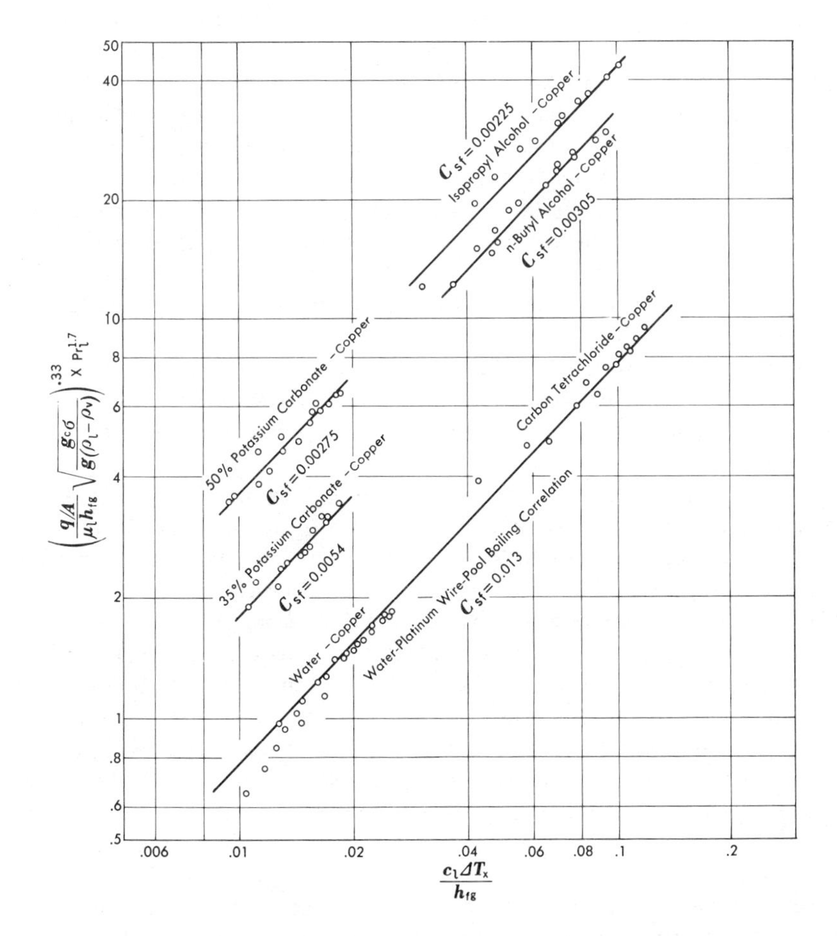

Fig. 10-10. Correlation of data for subcooled convection boiling by the Rohsenow method. (Extracted from "Recent Developments in Boiling Research," by W. H. Jens and G. Leppert, *J. Am. Soc. Naval Eng.*, Inc., Vol. 67, 1955, with permission of the publishers, The Society of Naval Engineers, Inc.).

---

[1] If Eq. 8-18 is used, Rohsenow (44) recommends that the coefficient 0.023 be replaced by 0.019.

fluid-surface combinations. Some of the data shown in Fig. 10-10 were obtained with subcooled liquids, others with saturated liquids containing various amounts of vapor.

Maximum heat flux in nucleate pool boiling. The Rohsenow method unifies the correlation of data for all types of nucleate-boiling processes, including pool boiling of saturated or subcooled liquids and boiling of subcooled or saturated liquids flowing by forced or natural convection in tubes or ducts. Specifically, the correlation equation (Eq. 10-2) relates the boiling heat flux to the excess temperature, provided the relevant fluid properties and the pertinent coefficient C_{fs} are available. The correlation is restricted to nucleate boiling and does not reveal the excess temperature at which the heat flux reaches a maximum or what the value of this flux is at which nucleate boiling breaks down, and an insulating vapor film forms. As mentioned earlier, the maximum heat flux attainable with nucleate boiling is sometimes of greater interest to the designer than the exact surface temperature, because for efficient heat transfer (19) and operating safety (1,17), particularly in high-performance constant-heat-input systems, operation in the film-boiling regime must be avoided.

Although there exists no satisfactory theory for predicting boiling heat-transfer coefficients, the maximum heat-flux condition in nucleate pool boiling, i.e., the *burnout point*, can be predicted with reasonable accuracy.

A close inspection of the nucleate-boiling regime (Fig. 10-1) shows (5) that it consists actually of at least two major subregimes. In the first region, which corresponds to low heat-flux densities, the bubbles behave as isolated entities and do not interfere with one another. But as the heat flux is increased, the process of vapor removal from the heating surface changes from an intermittent to a continuous one, and as the frequency of bubble emission from the surface increases, the isolated bubbles merge into continuous vapor columns.

The stages of the transition process from isolated bubbles to continuous vapor columns are shown schematically in Fig. 10-11a. The photographs in Fig. 10-11b and c show the two regimes for water boiling on a horizontal surface at atmospheric pressure (5). At the transition from the region of isolated bubbles to that of vapor columns, only a small portion of the heating surface is covered by vapor. But as the heat flux is increased, the column diameter increases and additional vapor columns form. As the fraction of a cross-sectional area parallel to the heating surface occupied by vapor increases, neighboring vapor columns and the enclosed liquid begin to interact. Eventually a vapor generation rate is attained where the close spacing between adjacent vapor columns leads to high relative velocities between the vapor moving away from the surface and the liquid streams flowing toward the surface to maintain continuity.

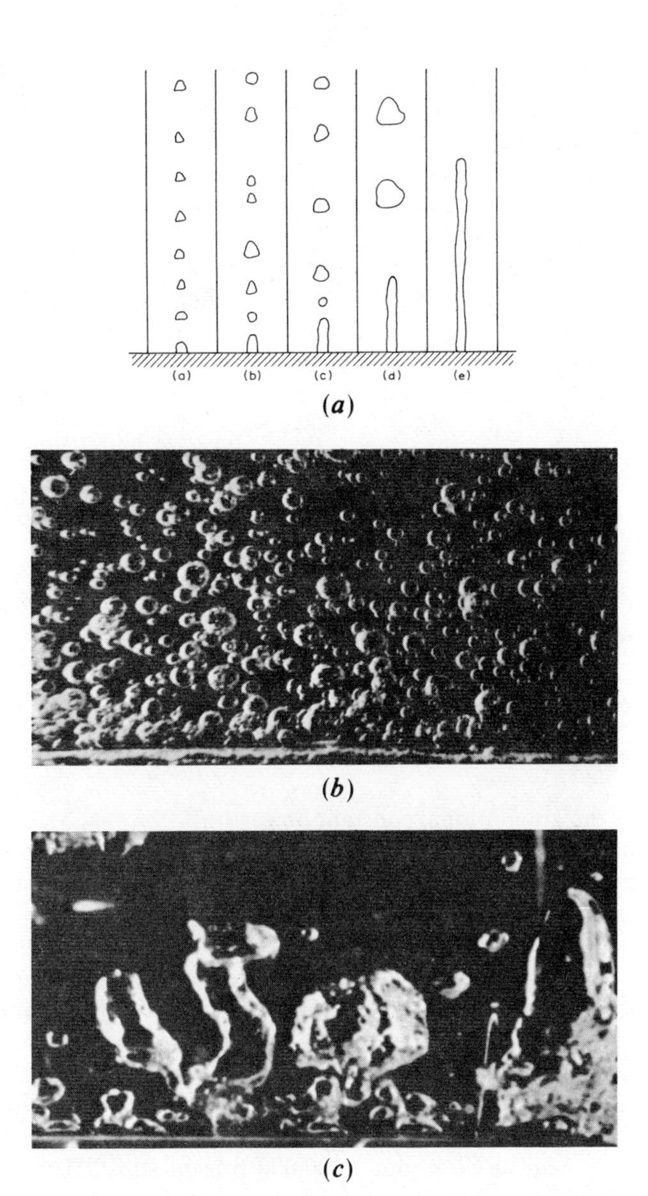

Fig. 10-11. Transition from isolated-bubble regime to continuous-column regime in nucleate boiling. (*a*) Schematic sketch of transition. (*b*) Photograph of isolated-bubble regime for water at atmospheric pressure and a heat flux of 38,400 Btu/hr sq ft. (*c*) Photograph of continuous-column regime for water at atmospheric pressure and a heat flux of 116,000 Btu/hr sq ft. (Extracted from "On the Hydrodynamic Transition in Nucleate Boiling," by Ralph Moissis and Paul J. Berenson, published in the *ASME Journal of Heat Transfer*, Aug. 1963, p. 222, with permission of the publishers, the American Society of Mechanical Engineers.)

The point of maximum heat flux occurs when the velocity of the liquid relative to the velocity of the vapor is so great that a further increase would either cause the vapor columns to drag the liquid away from the heating surface or the liquid streams to drag the vapor back towards the heating surface. Either case is obviously physically impossible without a decrease in the heat flux.

With this type of flow model as a guide, Zuber and Tribus (46) and Moissis and Berenson (5) have derived analytical relations for the maximum heat flux which are in essential agreement with an equation proposed earlier by Kutateladze (47) by empirical means. The recommended expression (45) for the peak flux in Btu/hr sq ft in saturated pool boiling is

$$\left(\frac{q}{A}\right)_{\max} = 0.13\,\rho_v h_{fg}\left[\frac{\sigma(\rho_l - \rho_v)\,g\,g_c}{\rho_v^{\,2}}\right]^{1/4}\left(\frac{\rho_l}{\rho_l + \rho_v}\right)^{1/2} \tag{10-4}$$

A simplified version of Eq. 10-4 proposed by Rohsenow and Griffith (21,40) is

$$\left(\frac{q}{A}\right)_{\max} = 143\,\rho_v h_{fg}\left(\frac{g}{g_o}\right)^{1/4}\left(\frac{\rho_l - \rho_v}{\rho_v}\right)^{0.6} \tag{10-4a}$$

where g_o is the gravitational constant on earth.

Equation 10-4a predicts that water will sustain a larger peak heat flux than any of the common liquids because water has such a large heat of vaporization. Further inspection of Eq. 10-4a suggests ways and means for increasing maximum heat flux. Pressure affects peak heat flux because it changes the vapor density and also the boiling point. Changes in the boiling point affect the heat of vaporization and the surface tension. For each liquid there exists, therefore, a certain pressure which yields the highest heat flux. This is illustrated in Fig. 10-12 where the peak nucleate-boiling heat flux is plotted as a function of the ratio of system pressure to critical pressure. For water the optimum pressure is about 1500 psi and the peak heat flux is about 1,200,000 Btu/hr sq ft. The quantity in the second bracket of Eq. 10-4a also shows that the gravitational field affects the peak heat flux. The reason for this behavior is that in a given field the liquid phase, by virtue of its higher density, is subject to a larger force per unit volume than the vapor phase. Since this difference in forces acting on the two phases brings about a separation of the two phases, an increase in the field strength as, for example, in a large centrifugal force field, increases the separating tendency and will also increase the peak flux. Conversely, experiments by Usiskin and Siegel (48) indicate that a reduced gravitational field decreases the peak heat flux in accordance with Eq. 10-4a; in a field of zero gravity, vapor does not leave the heated solid

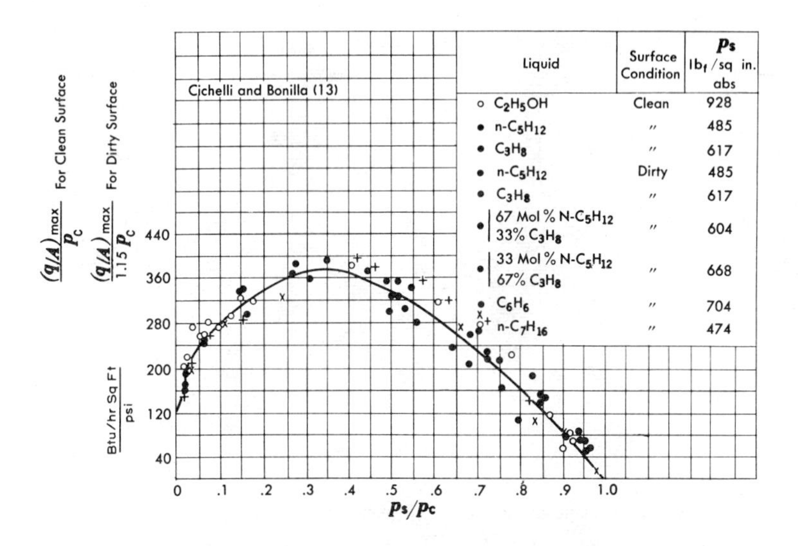

Fig. 10-12. Peak heat flux in nucleate boiling at various pressures—correlation of Ciechelli and Bonilla. (By permission from M. T. Ciechelli and C. F. Bonilla, "Heat Transfer to Liquids Boiling under Pressure," *AIChE Trans.*, Vol. 41, 1945).

and the heat flux tends towards zero. When the liquid bulk is subcooled, the maximum heat flux can be estimated (45) from the equation

$$\left(\frac{q}{A}\right)_{\text{max}} = \left(\frac{q}{A}\right)_{\text{max sat}} \left\{1 + \left[\frac{2k_l(T_{\text{sat}} - T_{\text{liquid}})}{\sqrt{\pi a_l \tau}}\right] \frac{24}{\pi h_{fg}\rho_v} \left[\frac{\rho_v^2}{\sigma g(\rho_l - \rho_v)}\right]^{1/4}\right\} \tag{10-5}$$

where

$$\tau = \frac{\pi}{3}\sqrt{2\pi}\left[\frac{\sigma}{g(\rho_l - \rho_v)}\right]^{1/2}\left[\frac{\rho_v^2}{\sigma g(\rho_l - \rho_v)}\right]^{1/4}$$

and $(q/A)_{\text{max sat}}$ can be determined from Eq. 10-4. Figure 10-13 illustrates the influence of the bulk temperature on the peak heat flux for distilled water and a 1 per cent aqueous solution of a surface-active agent boiling on a stainless-steel heater. The addition of the surface-active agent decreased the surface tension of water from 72 to 34 dynes/cm. This caused an appreciable decrease in the peak heat flux, an effect which is in agreement with Eq. 10-4. Non-condensable gases and non-wetting surfaces also reduce the peak heat flux at a given bulk temperature.

On the other hand, Westwater (6), Huber and Hoehne (60), and

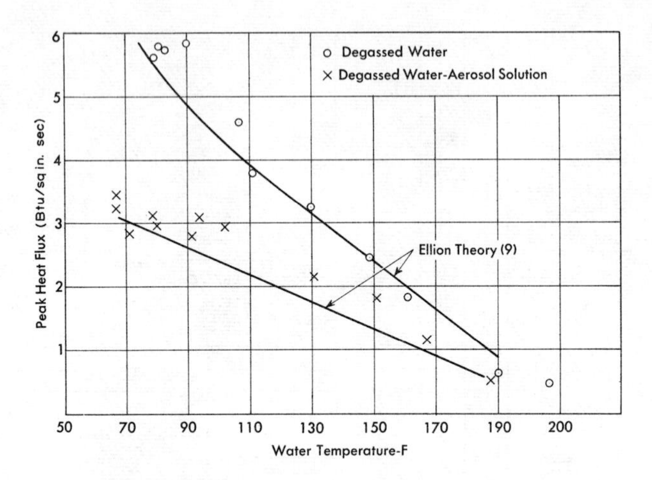

Fig. 10-13. Effect of bulk temperature on peak heat flux in pool boiling. (By permission from M. E. Ellion, "A Study of the Mechanism of Boiling Heat Transfer," Memo No. 20–28, Jet Propulsion Laboratory, Calif. Inst. of Tech., March, 1954).

others found that certain additives (e.g. small amounts of hyamine 1622) can increase the peak heat flux. Also, the presence of an ultrasonic or electrostatic field can increase the peak heat flux attainable in nucleate boiling.

EXAMPLE 10-1. Water at atmospheric pressure is boiling on a mechanically polished stainless steel surface which is heated electrically from below. Determine the heat flux from the surface to the water when the surface temperature is 222 F, and compare with the critical heat flux for nucleate boiling.

Solution: From Table 10-1, C_{sf} is 0.0132 for the fluid-surface combination. From steam tables, $h_{fg} = 970$ Btu/lb$_m$, $\rho_l = 60.04$ lb$_m$/ft^3, and $\rho_v = 0.0373$ lm$_m$/ft^3. From Table 10-2, the surface tension at 212 F is 40.3×10^{-4} lb$_f$/ft. Substituting these properties in Eq. 10-2 with $\Delta T_x = 222 - 212 = 10$ F gives

$$\frac{q}{A} = \left(\frac{c_l \Delta T_x}{C_{sf} h_{fg} \mathrm{Pr}_l^{1.7}}\right)^3 \mu_l h_{fg} \sqrt{\frac{g(\rho_l - \rho_v)}{g_c \sigma}}$$

$$= \left(\frac{1.0\ \mathrm{Btu/lb}_m F \times 10\ F}{0.0132 \times 970\ \mathrm{Btu/lb}_m \times 1.85^{1.7}}\right)^3 \times 0.19\ \mathrm{lb}_m/\mathrm{ft\ sec} \times 970\ \mathrm{Btu/lb}_m$$

$$\times \sqrt{\frac{32.2\ \mathrm{ft/sec}^2 \times 60.0\ \mathrm{lb}_m/\mathrm{ft}^3}{32.2\ \mathrm{lb}_m\ \mathrm{ft/lb}_f \mathrm{sec}^2 \times 40.3 \times 10^{-4}\ \mathrm{lb}_f/\mathrm{ft}}}$$

$$= 0.0172 \times 184\ \mathrm{Btu/ft\ sec} \times 1.22 \times 10^2\ \mathrm{ft}^{-1}$$

$$= 1.05 \times 10^3\ [\mathrm{Btu/sq\ ft\ sec}] \times 3600\ [\mathrm{sec/hr}] = 1.4 \times 10^5\ \mathrm{Btu/hr\ sq\ ft}$$

To determine the critical burnout heat flux use Eq. 10-4a, or

$$\left(\frac{q}{A}\right)_{max} = 143\,\rho_v h_{fg}(g/g_0)^{1/4}[(\rho_l - \rho_v)/\rho_v]^{0.6}$$

$$= 143 \times 0.0373\ \text{lb}_m/\text{ft}^3 \times 970\ \text{Btu/lb}_m \times (32.2/32.2)^{1/4} \times [(60.04 - 0.037)/0.037]^{0.6}$$

$$= 4.4 \times 10^5\ \text{Btu/hr sq ft}$$

Since at 10 F excess temperature the heat flux is less than the critical value, nucleate-pool boiling exists. If, on the other hand, the critical value had been larger than the value calculated from Eq. 10-2, film boiling would exist and the assumptions underlying the application of Eq. 10-2 would not be satisfied.

In applying the theoretical equations for the burnout heat flux in practice, a few words of caution are in order. Data have been presented in the literature indicating lower burnout heat fluxes than those predicted from Eq. 10-4 or 10-5. Berenson (7) explains this as follows: Boiling is a local phenomenon, while in most experiments and industrial installations an average heat flux is measured or specified. Therefore, if different locations of a heating surface have different heat fluxes or different nucleate-boiling curves, the measured result will represent an average. But the largest local heat flux at a given temperature difference will always be higher than the average measured value, and if the heat flux is not uniform, e.g., if considerable difference in subcooling or in surface condition exists or if gravity variations occur as around the periphery of a horizontal tube, a burnout may occur locally even if the average value of heat flux is below the critical value.

Boiling and vaporization in forced convection. The heat-transfer and pressure-drop characteristics of forced-convection vaporization play an important part in the design of boiling nuclear reactors, environmental control systems for spacecraft and space power plants, and other advanced power-production systems. Despite the large number of experimental and analytical investigations which have been conducted in the area of forced-convection vaporization, it is not yet possible to predict all of the characteristics of this process quantitatively. This is due to the great number of variables upon which the process depends and the complexity of the various two-phase flow patterns which occur as the quality of the vapor-liquid mixture, defined as the percentage of the total mass which is in the form of vapor at a given station, increases during vaporization. Recently, however, the forced-convection-vaporization process has been photographed (49,66), and based on these photographic observations, it is possible to give a qualitative description of the process.

In most practical situations, a fluid at a temperature below its boiling point at the system pressure enters a duct in which it is heated so that progressive vaporization occurs. Figure 10-14 shows schematically what happens in a duct in which a fluid is vaporized. Part e of the figure is a qualitative graph on which the heat-transfer coefficient at a specific location is plotted as a function of the local quality. In view of the fact that heat is continuously added to the fluid, the quality will increase with distance from the entrance.

The heat-transfer coefficient at the inlet can be predicted from Eq. 8-20 or 8-26 with satisfactory accuracy. However, as the fluid bulk temperature increases towards its saturation point, which occurs usually only a short distance from the inlet in a system designed to vaporize the fluid, bubbles will begin to form at nucleation sites and they will be carried into the main stream as in nucleate pool boiling. This regime, known as the *bubbly-flow regime*, is shown schematically in Fig. 10-14a. Bubbly flow occurs only a very low quality and consists of individual bubbles of vapor entrained in the main flow. In the very narrow quality range over which bubbly flow exists, the heat-transfer coefficient can be predicted by super-

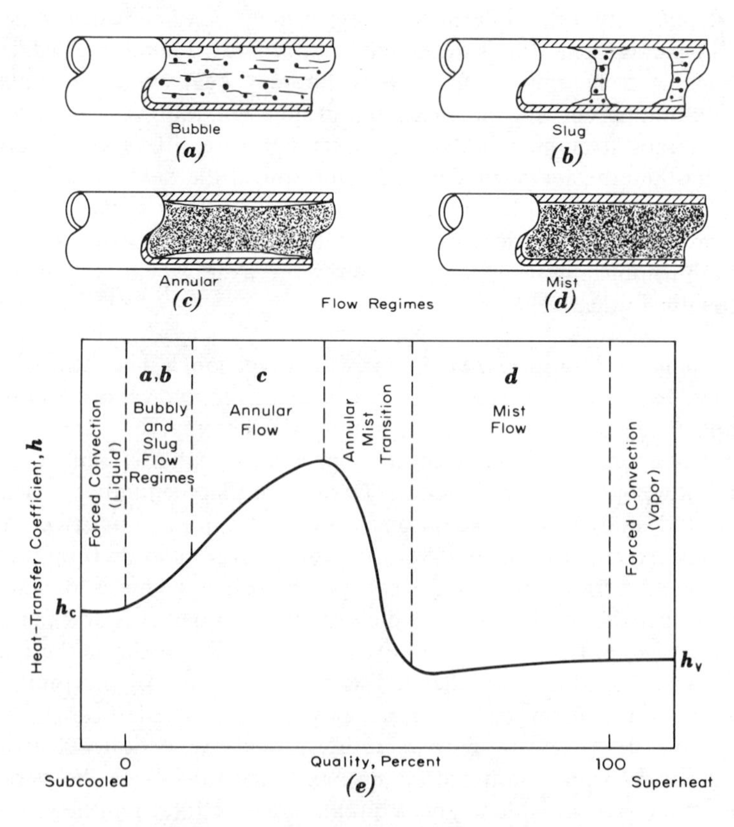

Fig. 10-14. Characteristics of forced-convection vaporization—heat-transfer coefficients vs. quality and types of flow regimes.

imposing liquid-forced-convection and nucleate-pool-boiling equations as long as the wall temperature is not so large as to produce film boiling. Satisfactory correlations for the heat-transfer coefficient and for the burnout heat flux in this regime have been proposed by Rohsenow (44) and Gambill (50) respectively.

As the vapor volume fraction increases, the individual bubbles begin to agglomerate and form plugs or slugs of vapor as shown in Fig. 10-14b. Although in this regime, known as *slug-flow regime*, the mass fraction of vapor is generally much less than 1 percent, as much as 50 percent of the volume fraction may be vapor and the fluid velocity in the slug-flow regime may increase appreciably. The plugs of vapor are compressible volumes which also produce flow oscillations within the duct even if the entering flow is steady. Bubbles may continue to nucleate at the wall, and it is probable that the heat-transfer mechanism in plug flow is the same as in the bubbly regime: a superposition of forced convection to a liquid and nucleate pool boiling. Owing to the increased liquid flow velocity the heat-transfer coefficient rises as can be seen in Fig. 10-14e.

While both the bubbly and slug-flow regimes are interesting, it should be noted that for density ratios of importance in forced-convection evaporators, the quality in these two regimes is too low to produce appreciable vaporization. These regimes become important in practice only if the temperature difference is so large as to cause film boiling, or if the flow oscillations produced in the slug-flow regime cause instability in a system.

As the fluid flows farther along in the tube and the quality increases, a third flow regime, commonly known as the *annular-flow regime*, appears. In this regime the wall of the tube is covered by a thin film of liquid and heat is transferred through this liquid film. In the center of the tube, vapor is flowing at a faster velocity and although there may be a number of active bubble nucleation sites at the wall, vapor is generated primarily by vaporization from the liquid-vapor interface inside the tube and not by the formation of bubbles inside the liquid annulus. In addition to the liquid in the annulus at the wall, there is a significant amount of liquid dispersed throughout the vapor core as droplets. The quality range for this type of flow is strongly affected by fluid properties and geometry, but it is generally believed that transition to the next flow regime, known as the *mist-flow regime*, occurs at qualities of about 25 percent or higher.

The transition from annular to mist flow is of great interest since this is presumably the point at which the heat-transfer coefficient experiences a sharp decrease as shown in Fig. 10-14e. Therefore, this transition point can be the cause of a burnout in forced-convection vaporization unless the heat flux is reduced appropriately before this condition is encountered. An important change takes place in the transition between annular and mist flow: In the former the wall is covered by a relatively high conductivity liquid, whereas in the latter the wall is covered by a low conductivity vapor. Berenson (49) observed that the wall-drying process occurs in the following manner: A small dry spot forms suddenly at the wall and grows in all directions as the liquid vaporizes because of conduction heat transfer through the liquid. The small strips of liquid remaining on the wall are almost stationary relative to the high velocity vapor and the liquid droplets in the vapor core. The dominant heat-transfer mechanism is conduction through the liquid film and although nucleation may produce the initial dry spot on the wall, it has only a small effect on the heat transfer. It thus appears that the drying process in transition to mist flow is similar to that which occurs with a thin film of liquid in a hot pan whose temperature is not great enough to cause nucleate boiling.

Most of the heat transfer in mist flow is from the hot wall to the vapor and after the heat has been transferred into the vapor core, it is transferred to the liquid droplets there. Vaporization in mist flow actually takes place in the interior of the duct, not at the wall. For this reason the temperature of the vapor in the mist-flow regime can be greater than the saturation temperature, and thermal equilibrium may not exist in the duct. While the volume fraction of the liquid droplets is small, they account for a substantial mass fraction because of the high rate of liquid to vapor density.

These observations are consistent with a theoretical stability analysis for a liquid film by Miles (51) which predicts that a liquid film is stable at sufficiently small Reynolds numbers irrespective of the vapor velocity. Since the Reynolds number of the liquid film in a forced-convection evaporator decreases as the quality increases, the liquid annulus will be stable at sufficiently high quality irrespective of the value of the vapor velocity.

For the flow of vapor-liquid mixtures through tubes Davis and David (64) found that as long as liquid wets the wall, the empirical equation

$$\left(\frac{\bar{h}D}{k_l}\right) = 0.06\left(\frac{\rho_l}{\rho_v}\right)^{0.28}\left(\frac{DG\chi}{\mu_l}\right)^{0.87}\mathrm{Pr}_l^{0.4} \tag{10-6}$$

where χ = the vapor mass fraction or quality, correlates the results of several investigations within about 20 percent.

Mist flow persists until the quality reaches 100 percent. Once this condition is reached, the heat-transfer coefficient can again be predicted by equations appropriate for forced convection of a vapor in a tube or a duct.

There have been several suggestions for the mechanism leading to transition from annular to mist flow and numerous correlations of the two-phase forced-convection heat-transfer coefficient have been proposed, but no agreement has yet been reached. Consequently, correlation equations which have been proposed to predict the burnout heat flux in the annular-flow and mist-flow regimes must be treated with considerable caution. These are empirical equations obtained under specific flow conditions and may not be applicable to other configurations and fluids. They are, however, typical of the manner in which attempts are being made at present to analyze and predict the behavior of forced-convection-vaporization systems. In view of the significance of this phenomenon, more reliable correlations will undoubtedly become available in the not-too-distant future. It is not expected, however, that a single model can serve for all the flow regimes and for this reason a familiarity with the characteristics of each flow pattern is important in the application of correlation equations.

Griffith (52) developed an empirical burnout correlation for forced convection covering a wide range of conditions. He correlated burnout data for water, benzene, *n*-heptane, *n*-pentane, and ethanol at pressures varying from 0.5 to 96 percent of critical pressure, at velocities from 0 to 100 fps, at subcooling from 0 to 280 F, and at qualities ranging from 0 up to 70 percent. The data used in this correlation were obtained in round tubes and rectangular channels. Figure 10-15 shows the correlated data and an inspection of this figure suggests that the burnout can apparently be predicted to within plus or minus 33 percent for the conditions used in this study. In Fig. 10-15, h_g is the saturated vapor enthalpy and h_b is the

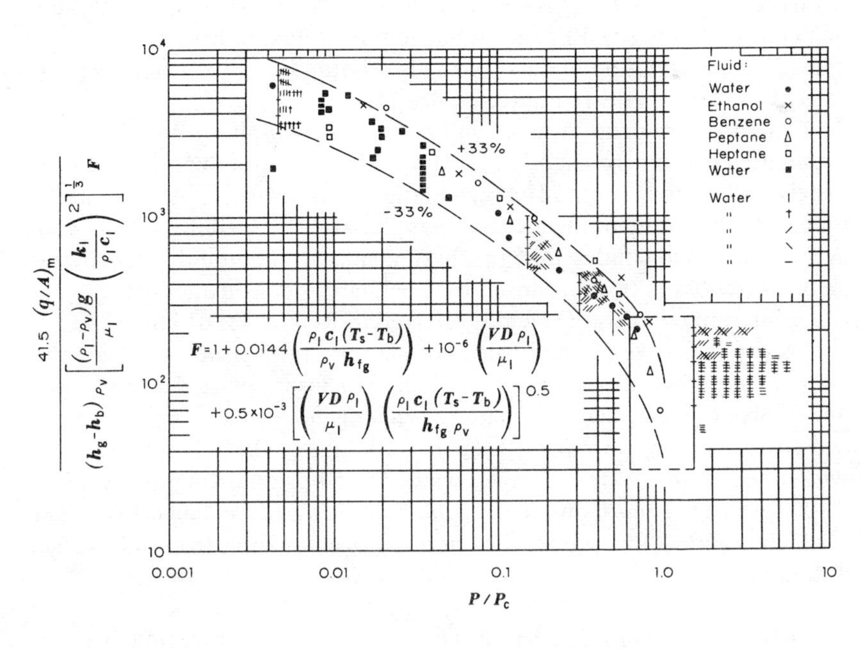

Fig. 10-15. Peak-heat-flux correlation for forced-convection boiling and vaporization. (Courtesy of Dr. P. Griffith and the American Society of Mechanical Engineers.)

bulk enthalpy of the fluid which may be subcooled liquid, saturated liquid, or a two-phase flow mixture at some quality less than 70 percent.

Lowdermilk, Lanzo, and Siegel (53) have reported measurements of burnout heat fluxes for water at pressures between 1 and 7 atm in tubes of 0.051- to 0.88-in.-diam length to diameter ratios varying between 25 and 250 and velocities between 0.1 and 98 fps. Inlet condition ranged from a liquid subcooled 140 F to a saturated liquid. Burnout heat fluxes were observed in the range between about 1×10^6 and 13×10^6 Btu/hr sq ft with net steam being generated from near zero to almost 100 percent quality. Under conditions combining a low velocity with high quality, the data were correlated by the equation

$$1 < \frac{G}{(L/D)^2} < 150: \left(\frac{q}{A}\right)_{\max} = \frac{270\, G^{0.85}}{D^{0.2}(L/D)^{0.85}} \tag{10-7}$$

whereas the data for conditions of high velocity and low quality were correlated by the equation

$$150 < \frac{G}{(L/D)^2} < 10{,}000: \left(\frac{q}{A}\right)_{\max} = \frac{1400\, G^{0.50}}{D^{0.2}(L/D)^{0.15}} \tag{10-7a}$$

In Eqs. 10-7 and 10-7a, G is the mass flux, in lb/hr; D is the inside-tube diameter, in ft; and $(q/A)_{\max}$ is the burnout heat flux, in Btu/hr sq ft. For an extensive correlation of data for the maximum best flux attainable with water the reader is referred to Reference 71.

The pressure drop in pipes and duct with two-phase flow has been investigated by numerous authors. The problem is quite complex and no entirely satisfactory method of calculation is available. A very useful summary of the state of the art has been prepared by Griffith (62) who, as several others, concludes that the best available method for predicting the pressure loss is that proposed by Lockhart and Martinelli (61). The reader interested in this problem is referred to References 61 and 62.

A very effective method for increasing the peak heat flux attainable in low-quality forced-convection boiling is to insert twisted tapes into a tube to produce a helical flow pattern which generates a centrifugal force field corresponding to many g_o's (54). Gambill, et al., (55) achieved in a swirl system with 110 F subcooled water at 850 psia flowing at a velocity of 100 fps in a 0.2-in.-diam tube, a peak heat flux of 55,000,000 Btu/hr sq ft, which is almost three times the energy flux emanating from the surface of the sun.

Transition boiling and film boiling. The transition region between the nucleate- and film-boiling regimes is difficult to characterize in a quantitative manner (67). Within this transition-boiling region the amount of vapor generated is not enough to support a stable vapor film, but it is too large to allow sufficient liquid to reach the surface to support nucleate boiling. Berenson (7) suggests, therefore, that nucleate and film boiling occur alternately at a given location. The process is unstable and photographs show that liquid surges sometimes towards the heating surface and sometimes away from it. At times, this turbulent liquid becomes so highly superheated that it explodes into vapor (6). From an industrial viewpoint, the transition-boiling regime is of little interest; equipment designed to operate in the nucleate-boiling region may be sized with more assurance and operate with more reproducible results.

Film boiling is characterized by a vapor film covering the heating surface. Since the vapor has a low thermal conductivity relative to the liquid, very large temperature differences are necessary to transfer heat at a rate approaching the nucleate-boiling regime. Film boiling is, therefore, used industrially only if circumstances make it unavoidable. Examples of such situations are when liquefied gases such as oxygen or hydrogen are boiling at ordinary temperatures. Film boiling may occur also if cryogenic fluids are used to cool rocket engines.

Film boiling requires a large temperature difference between the solid surface and the liquid, but it is not possible to predict exactly what the minimum excess temperature difference must be to sustain a stable film. For most organic liquids at atmospheric pressure the value is at least 200 F, but it is known that the lower limit is strongly influenced by pressure. There appears to be no upper limit to the temperature difference which will sustain a stable film, but at very high temperatures an appreciable amount of the heat transfer is due to radiation superimposed on the boiling.

The film-boiling process is classified according to whether the vapor film moves in viscous or turbulent flow. The flow is viscous if the flow path is short, e.g., on horizontal wires, small horizontal tubes, or short vertical surfaces. Figure 10-3 illustrates viscous film boiling on a wire.

When the flow is viscous it is possible to predict the thickness of the vapor film and to calculate the heat flux quite accurately. Bromley (26, 27) has studied stable film boiling on the outside of horizontal tubes of 0.19-, 0.24-, and 0.47-in.-diam both experimentally and analytically and his experimental results can be correlated with satisfactory accuracy by the equation

$$\bar{h}_b = 0.62\left[\frac{k_v^3 \rho_v(\rho_l - \rho_v) g \lambda'}{D_o \mu_v \Delta T_x}\right]^{1/4} \tag{10-8}$$

where k_v = thermal conductivity of saturated vapor, in Btu/hr ft F;
D_o = outside diameter of tube, in ft;
μ_v = viscosity of saturated vapor, in lb_m/hr ft;

while, except for λ', the other symbols are the same as those used in Eq. 10-2. The symbol λ' is defined as

$$\lambda' = h_{fg}\left(1 + \frac{0.4c_{pv}\Delta T_x}{h_{fg}}\right)$$

where c_{pv} is the specific heat of the saturated vapor.

The average unit-surface conductance $\bar{h}_b$ in Eq. 10-8 accounts only for the heat which is transferred by conduction through the vapor film and by boiling convection from the surface of the film to the surrounding liquid. Superimposed on this heat-flow path is the contribution of radiation to the total heat transfer. Since the heat transfer by radiation causes an increase in the thickness of the film, the coefficient h_b for conduction and convection in the presence of appreciable radiation is less than in the absence of radiation. The total surface conductance when radiation is appreciable can be estimated from the empirical relation

$$\bar{h} = \bar{h}_b\left(\frac{\bar{h}_b}{\bar{h}}\right)^{1/3} + \bar{h}_r$$

by trial and error. The radiation conductance $\bar{h}_r$ can be evaluated with the aid of Eq. 1-11. To determine the heat-transfer coefficient when the liquid is flowing past the surface of the tube, Bromley (27) suggests the equation

$$\bar{h}_b = 2.7\sqrt{\frac{V_\infty k_v \rho_v \lambda'}{D_o \Delta T_x}} \tag{10-8a}$$

if the velocity V_∞ is larger than $2\sqrt{gD_o}$. The total conductance, including radiation, is then

$$\bar{h} = \bar{h}_b + \tfrac{7}{8}h_r$$

under these conditions. At velocities less than $2\sqrt{gD_o}$, the flow is not fully developed turbulent and the conductance may be evaluated from data in Reference 27.

For laminar film boiling on the outside of small horizontal tubes and wires, Breen and Westwater (56) suggest the equation

$$\bar{h}_b = \left(0.59 + 0.069\frac{\lambda_c}{D_o}\right)\left(\frac{F}{\lambda_c}\right)^{1/4} \tag{10-9}$$

where

$$\lambda_c = 2\pi\left[\frac{g_c\sigma}{g(\rho_l - \rho_v)}\right]^{1/2}$$

and

$$F = \left[\frac{k_v^3\rho_v(\rho_l - \rho_v g h_{fg}\{1 + (0.34\, c_v\Delta T_x/h_{fg})\}}{\mu_v\Delta T_x}\right]^{1/4}$$

For horizontal surfaces facing upward, Berenson (59) derived analytically an expression similar to Eq. 10-9 which reads:

$$\bar{h}_c = 0.67\left(\frac{F}{\lambda_c}\right)^{1/4} \tag{10-10}$$

In the preceding two relations λ_c represents the wavelength of the smallest wave in the vapor film which can grow in amplitude when a flat layer of liquid lies above it.

For vertical surfaces of height L the equation

$$\bar{h}_b = 0.943\left[\frac{k_v^3\rho_v(\rho_l - \rho_v)g\lambda'}{L\mu_v\Delta T_x}\right]^{1/4} \tag{10-11}$$

is satisfactory as long as the vapor film remains laminar. Hsu and Westwater (57) have shown that as the vapor rises, a critical Reynolds number will be reached at some height L_{cr} which can be estimated from the relation

$$L_{cr} = \frac{100\,\mu_v\lambda'}{2k_v\Delta T_x}\left[\frac{200\,\mu_v^2}{g\rho_v(\rho_l - \rho_v)}\right]^{1/3}$$

For surfaces of height larger than L_{cr} Bankoff (58) showed that the equation

$$\bar{h}_b = 0.20\,\rho_v c_{pv}\left(\frac{g^2 L\mu_v}{\rho_v}\right)^{1/5}\left(\frac{\rho_l - \rho_v}{\rho_v}\right)^{2/5}\left(\frac{c_p\Delta T_x}{h_{fg}}\right)^{1/5} \tag{10-12}$$

can be used to predict the average heat-transfer coefficient in reasonably good agreement with a more complicated relation developed by Hsu and Westwater (57). Experimental data are available only for vertical heights up to 6.5 in. All of the equations for film boiling show that the rate of heat transfer is proportional to the gravitational constant raised to some fractional power. It is, therefore, anticipated that in a gravity free field the heat flux becomes zero, just as in nucleate boiling.

10-3. *Heat transfer in condensation*

When a saturated vapor comes in contact with a surface at a lower temperature, condensation occurs. Under normal conditions a continuous flow of liquid is formed over the surface and the condensate flows downward under the influence of gravity. Unless the velocity of the vapor is very high or the liquid film very thick, the motion of the condensate is laminar and heat is transferred from the vapor-liquid interface to the surface merely by conduction. The rate of heat flow depends, therefore, primarily on the thickness of the condensate film, which in turn depends on the rate at which vapor is condensed and the rate at which the condensate is removed. On a vertical surface the film thickness increases continuously from top to bottom, as shown in Fig. 10-16. As the plate is inclined from the vertical position, the drainage rate decreases and the liquid film becomes thicker. This, of course, causes a decrease in the rate of heat transfer.

Filmwise condensation. Theoretical relations for calculating the heat-transfer coefficients for filmwise condensation of pure vapors on

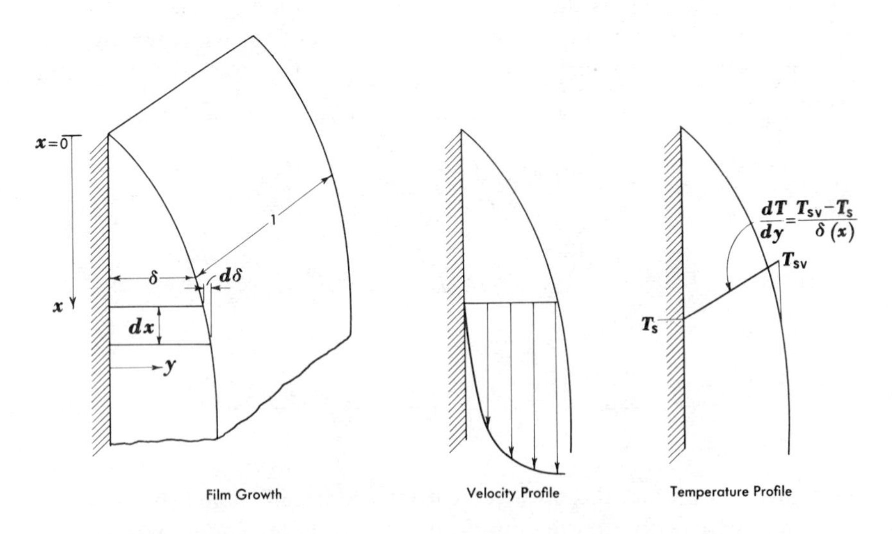

Fig. 10-16. Filmwise condensation on a vertical surface—film growth, velocity profile, and temperature distribution.

tubes and plates were first obtained by Nusselt (28), in 1916. To illustrate the classical Nusselt approach we shall consider a plane vertical surface at a constant temperature T_s on which a pure vapor at saturation temperature, T_{sv} is condensing. As shown in Fig. 10-16, a continuous film of liquid flows downward under the action of gravity, and its thickness increases as more and more vapor condenses at the liquid-vapor interface. At a distance x from the top of the plate the thickness of the film is δ. If the flow of the liquid is laminar and is caused by gravity alone, we can estimate the velocity of the liquid by means of a force balance on the element $dx\delta 1$. The downward force acting on the liquid at a distance greater than y from the surface is $(\delta - y)dx\rho_l g/g_c$. Assuming that the vapor outside the condensate layer is in hydrostatic balance ($dp/dx = \rho_v g/g_c$), a partially balancing force equal to $(\delta - y)dx\rho_v g/g_c$ will be present as a result of the pressure difference between the upper and lower faces of the element. The other forces retarding the downward motions consist of the drag at the inner boundary of the element. Unless the vapor flows at a very high velocity, the shear at the free surface is quite small and may be neglected. The remaining force will then simply be the viscous shear $(\mu_l du/dy)dx$ at the vertical plane y. Under steady-state conditions the upward and downward forces are equal, or

$$(\delta - y)(\rho_l - \rho_v)g = \mu_l \frac{du}{dy}$$

The velocity u at y is obtained by separating the variables and integrating. This yields the expression

$$u(y) = \frac{(\rho_l - \rho_v)g}{\mu_l}\left(\delta y - \frac{1}{2}y^2\right) + \text{const}$$

The constant of integration is zero because the velocity u is zero at the surface, i.e., $u = 0$ at $y = 0$.

The mass rate of flow of condensate per unit breadth Γ_c is obtained by integrating the local mass flow rate at the elevation x, $\rho u(y)$, between the limits $y = 0$ and $y = \delta$, or

$$\Gamma_c = \int_0^\delta \frac{\rho_l(\rho_l - \rho_v)g}{\mu_l}\left(\delta y - \frac{1}{2}y^2\right)dy = \frac{\rho_l(\rho_l - \rho_v)\delta^3}{3\mu_l} \tag{10-13}$$

The change in condensate flow rate Γ_c with the thickness of the condensate layer δ is

$$\frac{d\Gamma_c}{d\delta} = \frac{g\rho_l(\rho_l - \rho_v)}{\mu_l}\delta^2 \tag{10-14}$$

Heat is transferred through the condensate layer solely by conduction. Assuming that the temperature gradient is linear, the average enthalpy change of the vapor in condensing to liquid and subcooling to the average liquid temperature of the condensate film is

$$h_{fg} + \frac{1}{\Gamma_c} \int_0^{\delta} \rho_l u c_{pl}(T_{sv} - T)dy = h_{fg} + \frac{3}{8} c_{pl}(T_{sv} - T_s)$$

and the rate of heat transfer to the wall is $(k/\delta)(T_{sv} - T_s)$, where k is the thermal conductivity of the condensate. In the steady state the rate of enthalpy change of the condensing vapor must equal the rate of heat flow to the wall, or

$$\frac{q}{A} = k \frac{T_{sv} - T_s}{\delta} = \left[h_{fg} + \frac{3}{8} c_{pl}(T_{sv} - T_s)\right] \frac{d\Gamma_c}{dx} \tag{10-15}$$

Equating the expressions for $d\Gamma_c$ from Eqs. 10-14 and 10-15 gives

$$\delta^3 d\delta = \frac{k\mu_l(T_{sv} - T_s)}{g\rho_l(\rho_l - \rho_v)h'_{fg}}$$

where $h'_{fg} = h_{fg} + \frac{3}{8} c_{pl}(T_{sv} - T_s)$. Integrating between the limits $\delta = 0$ at $x = 0$ and $\delta = \delta$ at $x = x$ and solving for $\delta(x)$ yields

$$\delta = \left[\frac{4\mu_l kx(T_{sv} - T_s)}{g\rho_l(\rho_l - \rho_v)h'_{fg}}\right]^{1/4} \tag{10-16}$$

According to Eq. 6-6, the local heat-transfer coefficient h_x is k/δ. Substituting in Eq. 6-6 the expression for δ from Eq. 10-16 gives the unit-surface conductance as

$$h_x = \left[\frac{\rho_l(\rho_l - \rho_v)gh'_{fg}k^3}{4\mu_l x(T_{sv} - T_s)}\right]^{1/4} \tag{10-17}$$

and from Eq. 6-4 the dimensionless local Nusselt number at x is

$$\mathrm{Nu}_x = \frac{h_x x}{k} = \left[\frac{\rho_l(\rho_l - \rho_v)gh'_{fg}x^3}{4\mu_l k(T_{sv} - T_s)}\right]^{1/4} \tag{10-18}$$

Inspection of Eq. 10-17 shows that the unit conductance for condensation decreases with increasing distance from the top as the film thickens. The thickening of the condensate film is similar to the growth of a boundary layer over a flat plate in convection. At the same time it is also interesting to observe that an increase in the temperature difference $(T_{sv} - T_s)$ causes a decrease in the surface conductance. This is caused by the increase in the film thickness as a result of the increased rate of condensation. No comparable phenomenon occurs in simple convection.

The average value of the conductance $\bar{h}$ for a vapor condensing on a plate of height L is obtained by integrating the local value h_x over the plate and dividing by the area. For a vertical plate of unit width and height L we obtain by this operation

$$\bar{h}_c = \frac{1}{L}\int_0^L h_x dx = \frac{4}{3} h_{x=L} \tag{10-19}$$

or

$$\bar{h}_c = 0.943\left[\frac{\rho_l(\rho_l - \rho_v)gh'_{fg}k^3}{\mu_l L(T_{sv} - T_s)}\right]^{1/4} \tag{10-20}$$

It can easily be shown that, for a surface inclined by an angle ψ with the horizontal, the average conductance is

$$\bar{h}_c = 0.943\left[\frac{\rho_l(\rho_l - \rho_v)gh'_{fg}k^3 \sin\psi}{\mu_l L(T_{sv} - T_s)}\right]^{1/4} \tag{10-21}$$

A modified integral analysis for this problem by Rohsenow (41) which is in better agreement with experimental data if Pr > 0.5 and $c_{p_l}(T_{sv} - T_s)/h'_{fg} < 1.0$ yields results identical to Eqs. 10-17 through 10-21 except that h'_{fg} is replaced by $[h_{fg} + 0.68\, c_{p_l}(T_{sv} - T_s)]$. The effect of vapor shear stress on laminar film condensation is usually small, but it can be taken into account in the preceding analysis as shown by Rohsenow and Choi (40).

Although the foregoing analysis was made specifically for a vertical flat plate, the development is also valid for the inside and outside surfaces of vertical tubes if the tubes are large in diameter, compared with the film thickness. These results cannot be extended to inclined tubes, however. In such cases the film flow would not be parallel to the axis of the tube and the effective angle of inclination would vary with x.

The average unit conductance of a pure saturated vapor condensing on the outside of a horizontal tube can be evaluated by the same method which was used to obtain Eq. 10-21. For a tube of diameter D it leads to the equation

$$\bar{h}_c = 0.725\left[\frac{\rho_l(\rho_l - \rho_v)gh'_{fg}k^3}{D\mu_l(T_{sv} - T_s)}\right]^{1/4} \tag{10-22a}$$

If condensation occurs on N horizontal tubes so arranged that condensate from one tube flows directly onto the tube below, the average unit-surface conductance for the system can be estimated by replacing the tube diameter D in Eq. 10-22 by (DN). This method will in general yield conservative results because a certain amount of turbulence is unavoidable in this type of system (31).

An analysis which agrees better with experimental data was made by Chen (43) who suggested that, since the liquid film is subcooled, additional condensation occurs on the liquid layer between tubes. Assuming that all the subcooling is used for additional condensation, Chen's analysis yields the equation

$$\bar{h}_c = 0.728 \left[1 + 0.2 \frac{c_p(T_{sv} - T_s)}{h_{fg}} (N - 1)\right] \left[\frac{g\rho_l(\rho_l - \rho_v)k^3 h'_{fg}}{ND\mu_l(T_{sv} - T_s)}\right]^{1/4} \tag{10-22b}$$

which is in reasonably good agreement with experimental results, provided $[(N - 1) c_p(T_{sv} - T_s)/h_{fg}] < 2$.

In the preceding equations the unit-surface conductance will be in Btu/hr sq ft F if the other quantities are evaluated in the units listed below:

c_{pl} = specific heat of the liquid, in Btu/lb F.
k = thermal conductivity of liquid, in Btu/hr ft F.
ρ_l = density of liquid, in lb_m/cu ft.
ρ_v = density of the vapor, in lb_m/cu ft.
g = gravitational force, in ft/hr^2 (4.17×10^8 ft/hr^2 under normal conditions).
h_{fg} = latent heat of condensation or vaporization, in Btu/lb_m.
$h'_{fg} = h_{fg} + \frac{3}{8} c_p(T_{sv} - T_s)$
μ_l = viscosity of the liquid, in lb_m/hr ft.
D = tube diameter, in ft.
L = length of plane surface, in ft.
T_{sv} = temperature of saturated vapor, in F.
T_s = wall surface temperature, in F.

The physical properties of the liquid film in Eqs. 10-16 to 10-22 should be evaluated at the arithmetic average of the vapor and wall temperature. When used in this manner, Nusselt's equations are satisfactory for estimating surface conductances for condensing vapors. Experimental data are in general agreement with Nusselt's theory when the physical conditions comply with the assumptions inherent in the analysis. Deviations from Nusselt's film theory occur when the condensate flow becomes turbulent, when the vapor velocity is very high (42), or when a special effort is made to render the surface nonwettable. All of these factors tend to increase the surface conductance, and the Nusselt film theory will therefore always yield conservative results.

EXAMPLE 10-2. A $\frac{1}{2}$-in.-OD, 5-ft-long tube is to be used to condense steam at 6 psia. Estimate the unit-surface conductances for this tube in the (a) horizontal and (b) vertical positions. Assume that the average tube-wall temperature is 130 F.

Solution: (a) At the average temperature of the condensate film [$T_f = (170 + 130)/2 = 150$ F], the physical-property values pertinent to the problem are

$$\begin{aligned} k &= 0.383 \text{ Btu/hr ft F} \\ \rho &= 61.2 \text{ lb}_m\text{/cu ft} \\ h_{fg} &= 996.3 \text{ Btu/lb (from steam tables)} \\ \mu_f &= 1.06 \text{ lb/hr ft} \\ T_{sv} &= 170 \text{ F} \end{aligned}$$

For the tube in the horizontal position Eq. 10-22 applies and the unit-surface conductance is

$$\begin{aligned} \bar{h}_c &= 0.725\left[\frac{(0.383^3)(61.2^2)(4.17 \times 10^8)(996.3)}{(0.5/12)(1.06)(170 - 130)}\right]^{1/4} \\ &= 1920 \text{ Btu/hr sq ft F} \qquad \textit{Ans.} \end{aligned}$$

b) In the vertical position the tube may be treated as a vertical plate of area πDL and, according to Eq. 10-20, the average unit-surface conductance is

$$\begin{aligned} \bar{h}_c &= 0.94\left[\frac{(61.2^2)(4.17 \times 10^8)(996.3)(0.383^3)}{(1.06)(5)(170 - 130)}\right]^{1/4} \\ &= 730 \text{ Btu/hr sq ft F} \qquad \textit{Ans.} \end{aligned}$$

Effect of turbulence in the film. The results of the preceding calculations show that, for a given temperature difference, the average unit conductance is considerably larger when the tube is placed in a horizontal position where the path of the condensate is shorter and the film thinner than in the vertical position where the path is longer and the film thicker. This conclusion is generally valid when the length of the vertical tube is larger than 2.87 times the outer diameter, as can be seen by a comparison of Eqs. 10-21 and 10-22. However, both of these equations are based on the assumption that the flow of the condensate film is laminar and consequently do not apply when the flow of the condensate is turbulent. Turbulent flow is hardly ever reached on a horizontal tube but may be established over the lower portion of a vertical surface. When this occurs, the average heat-transfer coefficient becomes larger as the length of the condensing surface is increased because the condensate no longer offers as

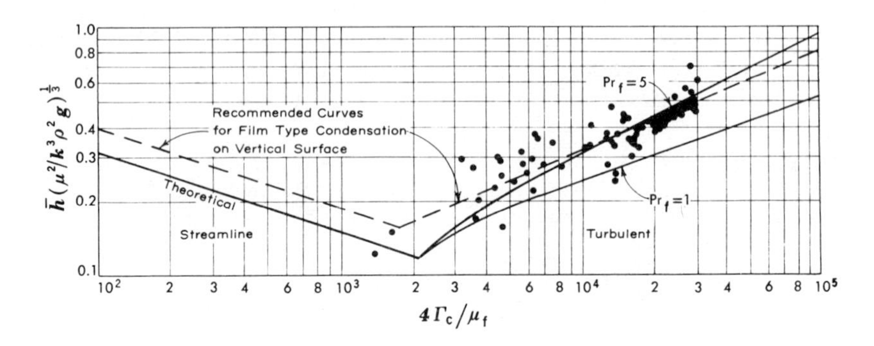

Fig. 10-17. Effect of turbulence in film on heat transfer with condensation.

high a thermal resistance as it does in laminar flow. This phenomenon is somewhat analogous to the behavior of a boundary layer.

Just as a fluid flowing over a surface undergoes a transition from laminar to turbulent flow, so the motion of the condensate becomes turbulent when its Reynolds number exceeds a critical value of about 2000. The Reynolds number of the condensate film Re_δ, when based on the hydraulic diameter (Eq. 8-1), can be written as $Re_\delta = 4A\Gamma_c/P\mu_f$, where P is the wetted perimeter equal to πD for a vertical tube and A is the flow cross-sectional area equal to $P\delta$. According to an analysis by Colburn (30) the local heat-transfer coefficient for turbulent flow of the condensate can be evaluated from the equation

$$h_x = 0.056 \left(\frac{4\Gamma_c}{\mu_f}\right)^{0.2} \left(\frac{k^3\rho^2 g}{\mu^2}\right)^{1/3} \mathrm{Pr}_f^{\,1/2} \tag{10-23}$$

To obtain average values of the conductance, integration of h_x over the surface by means of Eq. 10-17 for values of $(4\Gamma_c/P\mu_f)$ less than 2000 and Eq. 10-23 for values larger than 2000 is necessary. The results of such calculations for two values of the Prandtl number are plotted as solid lines in Fig. 10-17, where some experimental data obtained with diphenyl in turbulent flow are also shown (31). The heavy dashed line shown on the same graph is an empirical curve recommended by McAdams (16) for evaluating the average unit-surface conductance of single vapors condensing on vertical surfaces.

EXAMPLE 10-3. Determine whether or not the flow of the condensate in Example 10-2. is laminar or turbulent at the lower end of the tube.

Solution: The Reynolds number of the condensate at the lower end of the tube can be written with the aid of Eq. 10-13 as

$$\mathrm{Re}_\delta = \frac{4\Gamma_c}{P\mu_f^2} = \frac{4\rho^2 g\delta^3}{3\mu_f^2}$$

Substituting Eq. 10-16 for δ yields

$$\mathrm{Re}_\delta = \frac{4\rho^2 g}{3\mu_f^2}\left[\frac{4\mu kL(T_{sv} - T_s)}{gh_{fg}\rho^2}\right]^{3/4}$$

$$= \frac{4}{3}\left[\frac{4kL(T_{sv} - T_s)\rho^{2/3}g^{1/3}}{\mu^{5/3}h_{fg}}\right]^{3/4}$$

Inserting in the expression above the numerical values for the problem yields

$$\mathrm{Re}_\delta = \frac{4}{3}\left[\frac{(4 \times 0.383\ \mathrm{Btu/hr\ ft\ F})(10\ \mathrm{ft})(40\ \mathrm{F})\,(61.2\ \mathrm{lb_m/cu\ ft})^{2/3}(4.17 \times 10^8\ \mathrm{ft/hr^2})^{1/3}}{(1.06\ \mathrm{lb_m/hr\ ft})^{5/3}(996\ \mathrm{Btu/lb_m})}\right]^{3/4}$$

$$= 960\ \text{(dimensionless)}$$

Since the Reynolds number at the lower edge of the tube is below 2000, the flow of the condensate is laminar and the result obtained from Eq. 10-20 is valid. *Ans.*

Effect of high vapor velocity. One of the approximations made in Nusselt's film theory is that the frictional drag between the condensate and the vapor is negligible. This approximation ceases to be valid when the velocity of the uncondensed vapor is substantial compared with the velocity of the liquid at the vapor-condensate interface. When the vapor flows upward, it adds a retarding force to the viscous shear and causes the film thickness to increase. With downward flow of vapor, the film thickness decreases, and surface conductances substantially larger than those predicted from Eq. 10-20 can be obtained. In addition, the transition from laminar to turbulent flow occurs at condensate Reynolds numbers of the order of 300 when the vapor velocity is high. Carpenter and Colburn (33) determined the heat-transfer coefficients for condensation of pure vapors of steam and several hydrocarbons in a vertical tube, 8 ft long and $\frac{1}{2}$ in. ID, with inlet vapor velocities at the top up to 500 fps. Their data are correlated reasonably well by the equation

$$\frac{\bar{h}_c}{c_{pl}G_m}\mathrm{Pr}_l^{1/2} = 0.046\sqrt{\frac{\rho_l}{\rho_v}f} \tag{10-24}$$

where Pr_l = Prandtl number of liquid;
ρ_l = density of liquid, in $\mathrm{lb_m/cu\ ft}$;
ρ_v = density of vapor, in $\mathrm{lb_m/cu\ ft}$;
c_{pl} = specific heat of liquid, in $\mathrm{Btu/lb_m\ F}$;

$\bar{h}$ = average unit conductance, in Btu/hr sq ft F;
f = Fanning pipe-friction coefficient evaluated at the average vapor velocity;
G_m = mean value of the mass velocity of the vapor, in lb_m/hr sq ft.

The value of G_m in Eq. 10-24 can be taken as

$$G_m = \sqrt{\frac{G_1^2 + G_1 G_2 + G_2^2}{3}}$$

where G_1 = mass velocity at top of tube;
G_2 = mass velocity at bottom of tube.

All physical properties of the liquid in Eq. 10-24 are to be evaluated at a reference temperature equal to $0.25\ T_{sv} + 0.75\ T_s$. These results have not been verified on other systems but may be used generally as an indication of the influence of vapor velocity on the heat-transfer coefficient of condensing vapors when the vapor and the condensate flow in the same direction.

Condensation of superheated vapor. Although all of the preceding equations strictly apply only to saturated vapors, they can also be used with reasonable accuracy for condensation of superheated vapors. The rate of heat transfer from a superheated vapor to a wall at T_s will therefore be

$$q = A\bar{h}(T_{sv} - T_s) \tag{10-25}$$

where $\bar{h}$ = average value of unit conductance determined from equation appropriate to the geometrical configuration with same vapor at saturation conditions;
T_{sv} = *saturation temperature* corresponding to the prevailing system pressure.

Dropwise condensation. When a condensing surface is contaminated with a substance which prevents the condensate from wetting the surface, the vapor will condense in drops rather than as a continuous film (34). This is known as dropwise condensation. A large part of the surface is not covered by an insulating film under these conditions, and the heat-transfer coefficients are four to eight times as high as in filmwise condensation. So far, dropwise condensation has been reliably obtained only with steam. For the purpose of calculating the unit conductance in practice, it is recommended that filmwise condensation be assumed because, even with steam, dropwise condensation can be expected only under carefully controlled conditions which can not always be maintained in practice (76,77). Dropwise condensation of steam may, however, be a useful technique in experimental work when it is desirable to reduce the thermal resistance on one side of a surface to a negligible value.

Condenser design. The evaluation of the surface conductance of condensing vapors, as can be seen from Eqs. 10-20, 10-21, and 10-22, presupposes a knowledge of the temperature of the condensing surface. In practical problems this temperature is generally not known because its value depends on the relative order of magnitudes of the thermal resistances in the entire system. The type of problem usually encountered in practice, whether it be a performance calculation on an existing piece of equipment or the design of equipment for a specific process, requires simultaneous evaluation of thermal resistances at the inner and outer surfaces of a tube or the wall of a duct. In most cases the geometric configuration is either specified, as in the case of an existing piece of equipment, or assumed, as in the design of new equipment. When the desired rate of condensation is specified, the usual procedure is to estimate the total surface area required and then to select a suitable arrangement for a combination of size and number of tubes that meets the preliminary area specification. The performance calculation can then be made as though one were dealing with an existing piece of equipment, and the results can later be compared with the specifications. The flow rate of the coolant is usually determined by the allowable pressure drop or the allowable temperature rise. Once the flow rate is known, the thermal resistances of the coolant and the tube wall can be computed without difficulty. The unit-surface conductance of the condensing fluid, however, depends on the condensing-surface temperature, which can be computed only after the conductance is known. A trial-and-error solution is therefore necessary. One either assumes a surface temperature or, if more convenient, estimates the unit conductance on the condensing side and calculates the corresponding surface temperature. With this first approximation of the surface temperature, the unit-surface conductance is then recalculated and compared with the assumed value. A second approximation is usually sufficient for satisfactory accuracy.

The orders of magnitude of unit-thermal conductances for various vapors listed in Table 10-3 will aid in the initial estimates and reduce the amount of trial and error. We note that for steam the thermal resistance is very small, whereas for organic vapors it is of the same order of magnitude as the resistance offered to the flow of heat by water at a low turbulent Reynolds number. In the refrigeration industry and in some chemical processes, finned tubes have been used to reduce the thermal resistance on the condensing side. A method for dealing with condensation on finned tubes and tube banks is presented in Reference 32. We shall here consider only a simple example to illustrate the trial-and-error approach. When repeated calculations of the conductance for condensation of pure vapors are to be made, alignment charts devised by Chilton,

Table 10-3. Approximate values of unit-surface conductances for condensation of pure vapors

Vapor	System	Approximate Range of $T_{sv} - T_s$	Approximate Range of Average Unit Conductance (Btu/hr sq ft F)
Steam......	Horizontal tubes, 1–3 in. OD	5–40	2000–4000
Steam......	Vertical surface 10 ft high	5–40	1000–2000
Ethanol....	Vertical surface ½ ft high	20–100	200–340
Benzene....	Horizontal tube, 1 in. OD	30–80	250–350
Ethanol....	Horizontal tube, 2 in. OD	10–40	300–450
Ammonia...	Horizontal 2-to-3-in. annulus	2–7	250–450*

*Overall heat-transfer coefficient U for water velocities between 4 and 8 fps (35) inside the tube.

Colburn, Genereaux, and Vernon, reproduced in Reference 16, are convenient.

Mixtures of vapors and noncondensable gases. The analysis of a condensing system containing a mixture of vapors, or a pure vapor mixed with noncondensable gas, is considerably more complicated than the analysis of a pure-vapor system. The presence of appreciable quantities of a noncondensable gas will in general reduce the rate of heat transfer. If high rates of heat transfer are desired, it is considered good practice to vent the noncondensable gas, which otherwise will blanket the cooling surface and add considerably to the thermal resistance. It will also be shown in Chapter 12 that noncondensable gases inhibit the mass transfer by offering a diffusional resistance. A complete treatment of problems involving condensation of mixtures is beyond the scope of this text, and the reader is referred to References 16 and 30 for a comprehensive summary of available information on these topics.

10-4. Freezing and melting

Problems involving the solidification or melting of materials are of considerable importance in many technical fields. Typical examples in the the field of engineering are the making of ice, the freezing of foods, or the solidification and melting of metals in casting processes. In geology the solidification rate of the earth has been used to estimate the age of our planet. Whatever the field of application, the problem of central interest is the rate at which solidification or melting occurs.

We shall here consider only the problem of solidification, and it is

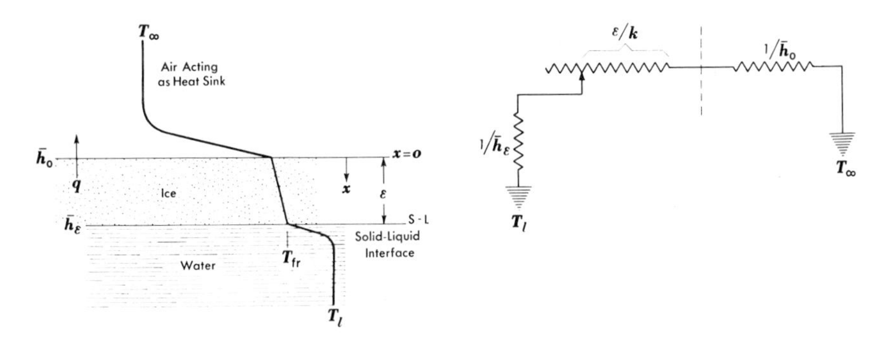

Fig. 10-18. Temperature distribution for ice forming on water with air acting as heat sink, and simplified thermal circuit for the system with heat capacity of solid considered to be negligible.

left for the reader as an exercise to show that a solution of this problem is also a solution to the corresponding problem in melting. Figure 10-18 shows the temperature distribution in an ice layer on the surface of a liquid. The upper face is exposed to air at subfreezing temperature. Ice formation occurs progressively at the solid-liquid interface as a result of heat transfer through the ice to the cold air. Heat flows by convection from the water to the ice, by conduction through the ice, and by convection to the sink. The ice layer is subcooled except for the interface in contact with the liquid, which is at the freezing point. A portion of the heat transferred to the sink is used to cool the liquid at the interface *SL* to the freezing point and to remove its latent heat of solidification. The other portion serves to subcool the ice. Cylindrical or spherical systems may be described in a similar manner, but solidification may proceed either inward (as for freezing of water inside a can) or outward (as for water freezing on the outside of a pipe).

The freezing of a slab can be formulated as a boundary-value problem in which the governing equation is the general conduction equation for the solid phase

$$\frac{\partial^2 T}{\partial x^2} = \frac{1}{a}\frac{\partial T}{\partial \theta}$$

subject to the boundary conditions that

$$\text{at } x = 0 \qquad -k\frac{\partial T}{\partial x} = \bar{h}_o(T_{x=0} - T_\infty)$$

$$\text{at } x = \epsilon \qquad -k\frac{\partial T}{\partial x} = \rho L_f \frac{d\epsilon}{d\theta} + \bar{h}_\epsilon(T_l - T_{fr})$$

where ϵ = distance to the solid-liquid interface which is a function of time, θ;
L_f = latent heat of fusion of the material;
a = thermal diffusivity of the solid phase $(c\rho/k)$;
ρ = density of the solid phase;
T_l = temperature of the liquid;
T_∞ = temperature of the heat sink;
T_{fr} = freezing point temperature;
$\bar{h}_o$ = unit conductance at $x = 0$, the air-ice interface;
$\bar{h}_\epsilon$ = unit conductance at $x = \epsilon$, the water-ice interface.

The analytic solution of this problem is very difficult and has only been obtained for special cases. The reason for the difficulty is that the governing equation is a partial-differential equation for which the particular solutions are unknown when physically realistic boundary conditions are imposed.

An approximate solution of practical value can however be obtained by considering the heat capacity of the subcooled solid phase as negligible relative to the latent heat of solidification. To simplify our analysis further we shall assume that the physical properties of the ice, ρ, k, and c, are uniform, that the liquid is at the solidification temperature (i.e., $T_l = T_{fr}$ and $1/\bar{h}_\epsilon = 0$), and that $\bar{h}_o$ and T_∞ are constant during the process.

The rate of heat flow per unit area through the resistances offered by the ice and the air, acting in series, as a result of the temperature potential $(T_{fr} - T_\infty)$ is

$$\frac{q}{A} = \frac{T_{fr} - T_\infty}{1/\bar{h}_o + \epsilon/k} \tag{10-26}$$

This is the heat-flow rate which removes the latent heat of fusion necessary for freezing at the surface $x = \epsilon$, or

$$\frac{q}{A} = \rho L_f \frac{d\epsilon}{d\theta} \tag{10-27}$$

where $(d\epsilon/d\theta)$ is the volume rate of ice formation per unit area at the growing surface in cu ft/hr sq ft, and ρL_f is the latent heat in Btu/cu ft. Combining of Eqs. 10-26 and 10-27 to eliminate the rate of heat flow yields the equation

$$\frac{T_{fr} - T_\infty}{1/\bar{h}_o + \epsilon/k} = \rho L_f \frac{d\epsilon}{d\theta} \tag{10-28}$$

which relates the depth of ice to the freezing time. The variables ϵ and θ

can now be separated and we get

$$d\epsilon\left(\frac{1}{\bar{h}_o} + \frac{\epsilon}{k}\right) = \frac{T_{fr} - T_\infty}{\rho L}\,d\theta \tag{10-29}$$

To make this equation dimensionless let

$$\epsilon^+ = \frac{\bar{h}_o \epsilon}{k}$$

and

$$\theta^+ = \theta \bar{h}_o^2 \frac{T_{fr} - T_\infty}{\rho L_f k}$$

Substituting these dimensionless parameters in Eq. 10-29 yields

$$d\epsilon^+(1 + \epsilon^+) = d\theta^+ \tag{10-30}$$

If the freezing process starts at $\theta = \theta^+ = 0$ and continues for a time θ, the solution of Eq. 10-30, obtained by integration between the specified limits, is

$$\epsilon^+ + (\epsilon^+)^2/2 = \theta^+ \tag{10-31}$$

or

$$\epsilon^+ = -1 + \sqrt{1 + 2\theta^+} \tag{10-32}$$

When the temperature of the liquids T_l is above the fusion temperature and the convective resistance at the liquid-to-solid interface is $\bar{h}_\epsilon$, the dimensionless equation corresponding to Eq. 10-30 in the foregoing simplified treatment becomes

$$\frac{(1 + \epsilon^+)\,d\epsilon^+}{1 + R^+T^+(1 + \epsilon^+)} = d\theta^+ \tag{10-33}$$

where $R^+ = \bar{h}_\epsilon / \bar{h}_o$
$T^+ = (T_l - T_{fr})/(T_{fr} - T_\infty)$

while the other symbols represent the same dimensionless quantities used previously in Eq. 10-30.

For the boundary conditions that, at $\theta^+ = 0$, $\epsilon^+ = 0$, and, at $\theta^+ = \theta^+$, $\epsilon^+ = \epsilon^+$, the solution of Eq. 10-33 becomes

$$\theta^+ = -\frac{1}{(R^+T^+)^2}\ln\left(1 + \frac{R^+T^+\epsilon^+}{1 + R^+T^+}\right) + \frac{\epsilon^+}{R^+T^+} \tag{10-34}$$

The results are shown graphically in Fig. 10-19 where the generalized

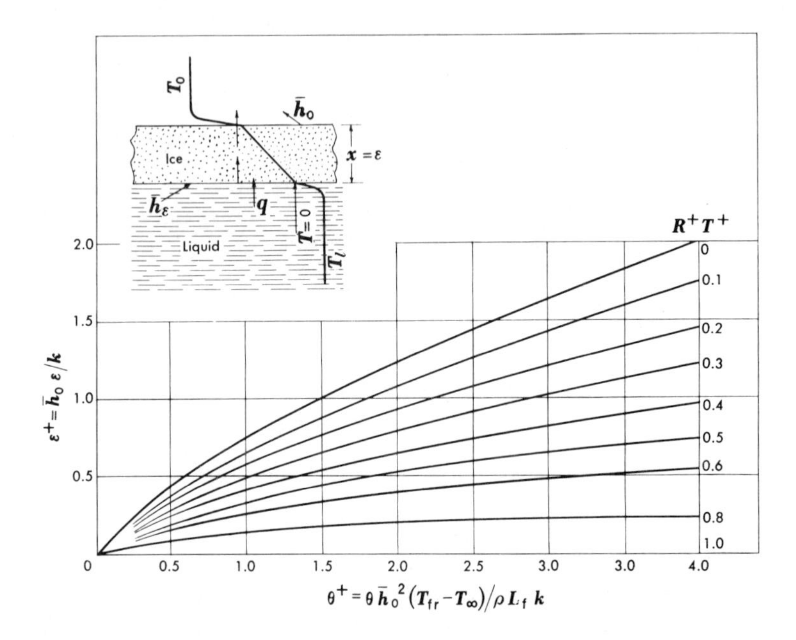

Fig. 10-19. Solidification of slab—thickness vs. time. (Extracted from "Rates of Ice Formation," by A. L. London and R. A. Seban, *Trans. ASME*, Vol. 65, 1943, with permission of the publishers, The American Society of Mechanical Engineers.)

thickness ϵ^+ is plotted vs. generalized time θ^+ with the generalized potential-resistance ratio R^+T^+ as parameter.

EXAMPLE 10-4. In the production of "Flakice," ice forms in thin layers on a horizontal rotating drum which is partly submerged in water. The cylinder is internally refrigerated with a brine spray at 12 F. Ice formed on the exterior surface is peeled off as the revolving-drum surface emerges from the water.

For the operating conditions listed below, estimate the time required to form an ice layer 0.1 in. thick.

Water liquid temperature . . . . . . 40 F
Liquid-surface conductance . . . . . . 10 Btu/hr sq ft F
Conductance between brine and
ice (including metal wall) . . . . . . 100 Btu/hr sq ft F

Solution: For the conditions stated above we have

$$R^+ = \frac{\bar{h}_\epsilon}{\bar{h}_o} = \frac{10}{100} = 0.1$$

$$T^+ = \frac{T_l - T_{fr}}{T_{fr} - T_\infty} = \frac{40 - 32}{32 - 12} = 0.4$$

$$\epsilon^+ = \frac{\bar{h}_o \epsilon}{k_{ice}} = \frac{(100 \text{ Btu/hr sq ft})(0.1/12 \text{ ft})}{1.34 \text{ Btu/hr sq ft F/ft}} = 0.622$$

We assume now that the ice is a sheet. This is justified because the thickness of the ice is very small compared to the radius of curvature of the drum. The boundary conditions of this problem are then the same as those assumed in the solution of Eq. 10-33. Hence, Eq. 10-34 is the solution to the problem at hand. Substituting numerical values for R^+, T^+, and ϵ^+ in Eq. 10-34 yields

$$\theta^+ = -\frac{1}{(0.04)^2} \ln\left(1 - \frac{0.0245}{1 - 0.04}\right) - \frac{0.622}{0.04} = 0.615$$

From the definition of θ^+, the time θ is

$$\begin{aligned}\theta &= 0.615 \times \rho L_f k / h_o^2 (T_{fr} - T_\infty) \\ &= (0.615)(57.3)(143.6)(1.34)/(10{,}000)(20) \\ &= 0.034 \text{ hr or about 2 min} \qquad \textit{Ans.}\end{aligned}$$

An estimate of the error caused by neglecting the heat capacity of the solidified portion has been obtained by means of an electrical network simulating the freezing of a slab originally at the fusion temperature (37). It was found that the error is not appreciable when $\epsilon \bar{h}_o / k$ is less than 0.1 (37) or when $[L_f/(T_{fr} - T_\infty)c]$ is larger than 1.5 (38). In the intermediate range, the freezing rates predicted by the simplified analysis are too large. The solutions presented here are valid for ice and other substances which have heats of fusion which are large compared to their specific heats. An approximate method for predicting the freezing rate of steel and other metals, where $[L_f/(T_{fr} - T_\infty)c]$ may be less than 1.5, is presented in Reference 38. Numerical methods of solution for systems involving a change of phase are presented in References 73 and 74. Melting and freezing in wedges and corners has been analyzed in Reference 75.

PROBLEMS

10-1. Show that the dimensionless equation for ice formation at the outside of a tube of radius r_o is

$$\theta^* = \frac{r^{*2}}{2} \ln r^* \left(\frac{1}{2R^*} + \frac{1}{4}\right)(r^{*2} - 1)$$

where $$r^* = \frac{\epsilon + r_o}{r_o} \qquad R^* = \frac{h_o r_o}{k} \qquad \theta^* = \frac{(T_f - T_\infty) k \theta}{\rho L r_o^2}$$

Assume that the water is originally at the freezing temperature T_f, that the cooling medium inside the tube surface is below the freezing temperature at a uniform temperature T_∞, and that h_o is the total conductance between the cooling medium and the pipe-ice interface. Also show the thermal circuit.

10-2. In the manufacture of can ice, cans having inside dimensions of 11 by 22 by 50 in. with 1-in. inside taper are filled with water and immersed in brine having a temperature of 10 F. [For details of the process see *The Refrigerating Data Book*, ASRE, Vol. II (1940), pp. 9,56]. For the purpose of a preliminary analysis, the actual ice can may be considered as an equivalent cylinder having the same cross-sectional area as the can, and end effects may be neglected. The overall conductance between the brine and the inner surface of the can is 40 Btu/hr sq ft F. Determine the time required to freeze the water and compare with the time necessary if the brine circulation rate would be increased to reduce the thermal resistance of the surface to one-tenth of the value specified above.

10-3. Estimate the time required to freeze vegetables in thin, tin cylindrical containers of 6-in. diameter. Air at 10 F is blowing at 15 fps over the cans, which are stacked to form one long cylinder. The physical properties of the vegetables may be taken as those of water and ice respectively.

10-4. Develop the Nusselt film-condensation relation for condensation inside small vertical tubes where the film builds up an annulus.

10-5. Consider a $\frac{1}{2}$-in.-ID vertical tube at a surface temperature of 150 F with atmospheric saturated steam inside. Determine the tube length at which the condensate fills the tube and chokes the flow.

10-6. Calculate the average heat-transfer coefficient for film-type condensation of water at pressures of 1 in. Hg abs and 14.7 psia for (a) a vertical surface 5 ft high; (b) the outside surface of a $\frac{5}{8}$-in.-OD vertical tube 5 ft long; (c) the outside surface of a $\frac{5}{8}$-in.-OD horizontal tube 5 ft long; and (d) a 10-tube vertical bank of $\frac{5}{8}$-in.-OD horizontal tubes 5 ft long. In all cases, assume that the vapor velocity is negligible and that the surface temperatures are constant at 20 F below saturation temperature.

10-7. Predict the nucleate-boiling heat-transfer coefficient for water boiling at atmospheric pressure on the outside surface of a $\frac{5}{8}$-in-.OD vertical tube 5 ft long. Assume the tube-surface temperature constant at 20 F above the saturation temperature.

10-8. Estimate the maximum heat flux obtainable with nucleate pool boiling on a clean surface for (a) water at 1 atm on brass, (b) water at 10 atm on brass, and (c) *n*-butyl alcohol at 3 atm on copper.

10-9. Determine the excess temperature at one-half of the maximum heat flux for the fluid-surface combinations in Prob. 10-8.

10-10. Estimate the time required to freeze a 1-in. thickness of water due to nocturnal radiation with ambient air and initial water temperatures at 40 F. Neglect evaporation effect.

10-11. For saturated pool boiling of water on a horizontal plate calculate the peak heat flux at pressures of 10, 20, 40, 60, and 80 percent of the critical pressure p_c and plot your results as q_{max} vs. p/p_c. The surface tension of water may be taken as $\sigma = 5.3 \times 10^{-3}(1 - 0.0014\,T)$ where σ is in pounds per foot and T in deg F.

10-12. A $\frac{1}{4}$-in.-thick flat plate of stainless steel, 3 in. high and 1 ft long, is immersed vertically at an initial temperature of 1800 F in a large water bath at 212 F and at atmospheric pressure. Determine how long it will take this plate to cool to 1000 F.

10-13. Calculate the maximum safe heat flux in the nucleate-boiling regime for water flowing at a velocity of 50 fps through a $\frac{1}{2}$-in.-ID tube 1 ft long if the water enters at 1 atm pressure and 212 F and the heat flux in the tube is uniform at a rate of 5×10^6 Btu/hr sq ft.

10-14. Compare the results obtained using Eqs. 10-7 and 10-7a with the correlation of Griffith shown in Fig. 10-15.

10-15. Calculate the maximum heat flux attainable in nucleate boiling with saturated water at 2 atm pressure in a gravitational field equivalent to one-tenth that of the earth.

10-16. Prepare a graph showing the effect of subcooling between 0 and 100 F on the maximum heat flux calculated in Prob. 10-15.

10-17. Calculate the heat-transfer coefficient for laminar film boiling of water in a $\frac{1}{2}$-in.-OD horizontal tube if the tube temperature is 1000 F and the system is placed under pressure of $\frac{1}{2}$ atm.

10-18. Calculate the critical height at which film-boiling water at atmospheric pressure will undergo a transition on a vertical surface at 1000 F. Calculate the average heat-transfer coefficient for the laminar regime.

10-19. A thin-walled horizontal copper tube of 0.2-in. OD is placed in a pool of water at atmospheric pressure and 212 F. Inside the tube an organic vapor is condensing and the outside surface temperature of the tube is uniform at 450 F. Calculate the average unit-surface conductance at the outside of the tube.

10-20. Estimate (a) the heat-transfer surface area required and (b) suggest a suitable arrangement for the condenser of a 10-ton refrigeration machine. The working fluid is ammonia condensing on the outside of horizontal pipes at a pressure of 170 psia. The condenser is to be constructed with 1-in. steel pipes (1.00-in. OD, 0.834-in. ID), cooling water is available at 79 F, and the average water velocity in the pipes is not to exceed 6 fps.

10-21. The inside surface of a 3-ft long vertical 2.0-in.-ID tube is maintained at 250 F. For saturated steam at 50 psia condensing inside estimate the average unit-surface conductance and the condensation rate assuming the steam velocity is small.

10-22. A horizontal 1.00-in.-OD tube is maintained at a temperature of 80 F on its outer surface. Calculate the average unit-surface conductance if saturated steam at 1.7 psia is condensing on this tube.

10-23. Repeat Prob. 10-22 for a tier of 6 horizontal 1.0-in.-OD tubes under similar thermal conditions.

10-24. Saturated steam at 5.0 psia condenses on a 3-ft-tall vertical plate whose surface temperature is uniform at 140 F. Compute the average unit-surface conductance and the value of the conductance 1 ft, 2 ft, and 3 ft from the top. Also, find the maximum height for which the condensate film will remain laminar.

10-25. At a pressure of 71 psia, the saturation temperature of sulfur dioxide (SO_2) is 90F, the density is 84 lb_m/ft^3, and the latent heat of vaporization 151 Btu/lb_m. If the SO_2 is to be condensed at 71 psia on an 8-inch flat surface, inclined at an angle of 45°, whose temperature is maintained uniformly at 75 F, calculate

a) the thickness of the condensate film 0.5 in. from the bottom,
b) the average heat transfer coefficient (*Ans.* 109 Btu/ft^2 hr F),
c) and the rate of condensation in lbs/hr.

10-26. Repeat Prob. 10-25, but assume that condensation occurs on a 2-in. OD horizontal tube.

10-27. Saturated methyl chloride at 62 psia condenses on a horizontal bank of tubes, ten-by-ten, 2-in, OD, equally spaced, 4 inches apart center-to-center on rows and columns. At 62 psia the latent heat of vaporization of methyl chloride is 167 Btu/lb and the specific volume of the saturated liquid is 0.017 ft^3/lb_m. If the surface temperature of the tubes is maintained at 45 F by water pumped through them, calculate the rate of condensation of methyl chloride in lb_m/hr.

10-28. Water at atmospheric pressure is boiling in a pot with a flat copper bottom on an electric range which maintains the surface temperature at 240 F. Calculate the boiling heat-transfer coefficient. *Ans.* 640 Btu/ft^2 hr F

10-29. The temperature of a cooling pond is 45 F on a winter day. If the air temperature suddenly drops to 20 F, calculate the thickness of ice formed after three hours.

REFERENCES

1. W. H. Jens and G. Leppert, "Recent Developments in Boiling Research," Parts I, II. *J. Am. Soc. Naval Engrs.*, Vol. 67 (1955), pp. 137–155; Vol. 66 (1955), pp. 437–456.
2. E. A. Farber and R. L. Scorah, "Heat Transfer to Water Boiling under Pressure," *Trans. ASME*, Vol. 70 (1948), pp. 369–384.
3. S. Nukyiyama, "Maximum and Minimum Values of Heat Transmitted From a Metal to Boiling Water Under Atmospheric Pressure," *J. Soc. Mech. Eng.*, Japan, Vol. 37, No. 206 (1934), pp. 367–394.

4. K. Engelberg-Forster and R. Greif, "Heat Transfer to a Boiling Liquid —Mechanism and Correlations," *Trans. ASME, J. of Heat Transfer*, Sec. C, Vol. 81 (February, 1959), pp. 43–53.

5. R. Moissis and P. J. Berenson, "On the Hydrodynamic Transitions in Nucleate Boiling," *Trans. ASME, J. of Heat Transfer*, Sec. C, Vol. 85 (August, 1963), pp. 221–229.

6. J. W. Westwater, "Boiling Heat Transfer," *Am. Scientist*, Vol. 47, No. 3 (September, 1959), pp. 427–446.

7. P. J. Berenson, "Experiments on Pool-Boiling Heat Transfer," *Int. J. of Heat Mass Transfer*, Vol. 5, Pergamon Press (1962), pp. 985–999.

8. F. C. Gunther, "Photographic Study of Surface Boiling Heat Transfer with Forced Convection," *Trans. ASME*, Vol. 73 (1951), pp. 115–123.

9. M. E. Ellion, "A Study of the Mechanism of Boiling Heat Transfer," *Memorandum 20–88*, Jet Propulsion Laboratory, Calif. Inst. of Tech., March, 1954.

10. F. C. Gunther and F. Kreith, "Photographic Study of Bubble Formation in Heat Transfer to Subcooled Water," *Prog. Rept.* 4–120, Jet Propulsion Lab., Calif. Inst. of Tech., March, 1950.

11. J. N. Addoms, *Heat Transfer at High Rates to Water Boiling Outside Cylinders, D.Sc. Thesis*, Dept. of Chem. Engrg., Massachusetts Institute of Technology, 1948.

12. E. L. Piret and H. S. Isbin, "Natural Circulation Evaporation Two-phase Heat Transfer," *Chem. Eng. Progress*, Vol. 50 (1954), p. 305.

13. M. T. Cichelli and C. F. Bonilla, "Heat Transfer to Liquids Boiling under Pressure," *Trans. AIChE*, Vol. 41 (1945), pp. 755–787.

14. D. S. Cryder and A. C. Finalbargo, "Heat Transmission from Metal Surfaces to Boiling Liquids: Effect of Temperature of the Liquid on Film Coefficient," *Trans. AiChE*, Vol. 33 (1937), pp. 346–362.

15. W. H. McAdams et al., "Heat Transfer from Single Horizontal Wires to Boiling Water," *Chem. Eng. Progress*, Vol. 44 (1948), pp. 639–646.

16. W. H. McAdams, *Heat Transmission*, 3d ed. (New York: McGraw-Hill Book Company, Inc., 1954.)

17. F. Kreith and M. J. Summerfield, "Heat Transfer to Water at High Flux Densities With and Without Surface Boiling," *Trans. ASME*, Vol. 71 (1949), pp. 805–815.

18. W. H. McAdams et al., "Heat Transfer at High Rates to Water with Surface Boiling," *Ind. Eng. Chem.*, Vol. 41 (1949), pp. 1945–1953.

19. *Steam—Its Generation and Use.* (New York: The Babcock & Wilcox Company, 1955.)

20. E. A. Kazekov, "Maximum Heat Transfer to Boiling Water at High Pressures," *Izvestia Adakmii Nauk USSR*, September, 1950, pp. 1377–1387. (Reviewed in *Engrg. Digest*, Vol. 12 (1951), pp. 81–85.)

21. W. Rohsenow and P. Griffith, "Correlation of Maximum Heat Flux Data for Boiling of Saturated Liquids." (Reprint, Heat Transfer Symposium, *Am. Inst. Chem. Engrs.*, Louisville, Ky., March, 1955.)

22. W. H. Jens and P. A. Lottes, "Analysis of Heat Transfer, Burnout, Pressure Drop and Density Data for High-Pressure Water," *Argonne Nat. Lab. Rpts.* ANL—4627, May, 1951.

23. W. H. Jens and P. A. Lottes, "Two Phase Pressure Drop and Burnout Using Water Flowing in Round and Rectangular Channels," *Argonne Nat. Lab. Rpts.* ANL—4915, October, 1952.

24. J. L. Schweppe and A. S. Foust, "The Effect of Forced Circulation Rate on Boiling Heat Transfer and Pressure Drop in a Short Vertical Tube," *Chem. Eng. Progress Symp.*, Series No. 5, Vol. 1944 (1953).

25. F. Kreith and A. S. Foust, "Remarks on the Stability and Mechanism of Surface Boiling Heat Transfer," *ASME Paper* 54-A-16, August, 1954.

26. L. A. Bromley, "Heat Transfer in Stable Film Boiling," *Chem. Eng. Progress*, Vol. 46 (1950), pp. 221–227.

27. L. A. Bromley et al., "Heat Transfer in Forced Convection Film Boiling," *Ind. Eng. Chem.*, Vol. 45 (1953), pp. 2639–2646.

28. W. Nusselt, "Die Oberflächenkondensation des Wasserdampfes," *Z. Ver. Deutsch. Ing.*, Vol. 60 (1916), pp. 541, 569.

29. A. P. Colburn, "Problems in Design on Research on Condensers of Vapours and Vapour Mixtures," Inst. Mech. Eng. and ASME, *Proc. General Discussion on Heat Transfer*, September, 1951, pp. 1–11.

30. A. P. Colburn, "The Calculation of Condensation where a Portion of the Condensate Layer is in Turbulent Flow," *Trans. Am. Inst. Chem. Engrs.*, Vol. 30 (1933), p. 187.

31. C. G. Kirkbridge, "Heat Transfer by Condensing Vapors on Vertical Tubes," *Trans. Am. Inst. Chem. Engrs.*, Vol. 30 (1933), p. 170.

32. D. L. Katz, E. H. Young, and G. Bolekjian, "Condensing Vapors on Finned Tubes," *Petroleum Refiner* (November, 1954), pp. 175–178.

33. E. F. Carpenter and A. P. Colburn, "The Effect of Vapor Velocity on Condensation-inside Tubes," Inst. Mech. Eng. ASME, *Proc. General Discussion on Heat Transfer*, 1951, pp. 20–26.

34. T. B. Drew, W. M. Nagle, and W. Q. Smith, "The Conditions for Dropwise Condensation of Steam," *Trans. Am. Inst. Chem. Engrs.*, Vol. 31 (1935), pp. 605–621.

35. A. P. Katz, H. J. Macintire, and R. E. Gould, "Heat Transfer in Ammonia Condensers," *Bull.* 209. Univ. Ill., Eng. Expt. Sta., 1930.

36. A. L. London and R. A. Seban, "Rate of Ice Formation," *Trans. ASME*, Vol. 65 (1943), pp. 771–778.

37. F. Kreith and F. E. Romie, "A Study of the Thermal Diffusion Equation with Boundary Conditions Corresponding to Freezing or Melting of Materials at the Fusion Temperature," *Proc. Phys. Soc.*, Vol. 68 (1955), pp. 277–291.

38. D. L. Cochran, "Solidification Application and Extension of Theory," *Tech. Rep.* 24, Navy Contract N6-onr-251, Stanford Univ., 1955.

39. W. H. McAdams, W. E. Kennel, C. S. Minden, R. Carl, P. M. Picornell, and J. E. Dew, "Heat Transfer at High Rates to Water with Surface Boiling," *Ind. Eng. Chem.*, Vol. 41 (1944), pp. 1945–1953.

40. W. M. Rohsenow and H. Choi, *Heat, Mass, and Momentum Transfer.* (Englewood Cliffs, N.J.: Prentice-Hall, Inc., 1961.)

41. W. M. Rohsenow, "Heat Transfer and Temperature Distribution in Laminar-Film Condensation," *Trans. ASME*, Vol. 78 (1956), pp. 1645–1648.

42. W. M. Rohsenow, J. M. Weber, and A. T. Ling, "Effect of Vapor Velocity on Laminar and Turbulent Film Condensation," *Trans. ASME*, Vol. 78 (1956), pp. 1637–1644.

43. M. M. Chen, "An Analytical Study of Laminar Film Condensation," Part 1, Flat Plates, and Part 2, Single and Multiple Horizontal Tubes, *Trans. ASME*, Sec. C, Vol. 83 (1961), pp. 48–60.

44. W. M. Rohsenow, "A Method of Correlating Heat-Transfer Data for Surface Boiling Liquids," *Trans. ASME*, Vol. 74 (1952), pp. 969–975.

45. N. Zuber, M. Tribus, J. W. Westwater, "The Hydrodynamic Crisis in Pool Boiling of Saturated and Subcooled Liquids," *Proceedings of the International Conference on Developments in Heat Transfer*, Am. Soc. of Mech. Engr., New York (1962), pp. 230–236.

46. N. Zuber and M. Tribus, "Further Remarks on the Stability of Boiling Heat Transfer," *Rep.* 58–5, Dept. of Eng., Univ. of Calif., Los Angeles, 1958.

47. S. S. Kutateladze, "A Hydrodynamic Theory of Changes in a Boiling Process Under Free Convection," *Izvestia Akademia Nauk Otdelenie Tekhnicheski Nauk*, No. 4 (1951), p. 524.

48. C. M. Usiskin and R. Siegel, "An Experimental Study of Boiling in Reduced and Zero Gravity Fields," *Trans. ASME*, Sec. C, Vol. 83 (1961), pp. 243–253.

49. P. J. Berenson and R. A. Stone, "A Photographic Study of the Mechanism of Forced-Convection Vaporization," *AIChE Reprint* No. 21, Symposium on Heat Transfer, San Juan, Puerto Rico, 1963.

50. W. R. Gambill, "Generalized Prediction of Burnout Heat Flux for Flowing, Subcooled, Wetting Liquids," *Chem. Eng. Progress*, Symp. Series, Vol. 59 (1963), pp. 71–87.

51. J. W. Miles, "The Hydrodynamic Stability of a Thin Film of Liquid in Uniform Shearing Motion," *J. of Fluid Mechanics*, Vol. 8 (1961), pp. 592–610.

52. P. Griffith, "Correlation of Nucleate-Boiling Burnout Data," *ASME Paper* 57-HT-21.

53. W. M. Lowdermilk, C. D. Lanzo, and B. L. Siegel, "Investigation of Boiling Burnout and Flow Stability for Water Flowing in Tubes," *N.A.C.A.T.N.* 4382, September, 1958.

54. F. Kreith and M. Margolis, "Heat Transfer and Friction in Turbulent Vortex Flow," *App. Sci. Res.*, Sec. A, Vol. 8 (1959), pp. 457–473.

55. W. R. Gambill, R. D. Bundy, and R. W. Wansbrough, "Heat Transfer, Burnout, and Pressure Drop for Water in Swirl Flow Through Tubes with Internal Twisted Tapes," *Chem. Eng. Prog.*, Symp. Ser. No. 32, Vol. 57 (1961), pp. 127–137.

56. J. W. Westwater and B. P. Breen, "Effect of Diameter of Horizontal Tubes on Film Boiling Heat Transfer," *Chem. Eng. Progress*, Vol. 58 (1962), pp. 67–72.

57. Y. Y. Hsu and J. W. Westwater, "Approximate Theory for Film Boiling on Vertical Surfaces," *Chem. Eng. Progress*, Symp. Series, Vol. 56, AIChE Heat Transfer Conference, Storrs, Conn., 1959, pp. 15–22.

58. S. G. Bankoff, "Discussion of Approximate Theory for Film Boiling on Vertical Surfaces," *Chem. Eng. Progress*, Symp. Series, AIChE Heat Transfer Conference, Storrs, Conn., 1959, pp. 22–24.

59. P. J. Berenson, "Film-Boiling Heat Transfer From a Horizontal Surface," *Trans. ASME*, Sec. C, Vol. 83 (1961), pp. 351–356.

60. D. A. Huber and J. C. Hoehne, "Pool Boiling of Benzene, Diphenyl, and Benzene-Diphenyl Mixtures Under Pressure," *Trans. ASME*, Sec. C, Vol. 85 (1963), pp. 215–220.

61. R. W. Lockhart and R. C. Martinelli, "Proposed Correlation of Data for Isothermal Two-Phase Two-Component Flow in Pipes," *Chem. Eng. Prog.*, Vol. 45 (1949), pp. 39–48.

62. P. Griffith, *Two Phase Flow in Pipes*, Course Notes, Mass. Inst. of Tech., Cambridge, Mass., 1964.

63. R. F. Gaertner, "Photographic Study of Nucleate Pool Boiling on a Horizontal Surface," *ASME Paper* 63—WA—76.

64. E. J. Davis and M. M. David, "Two-Phase Gas-Liquid Convection Heat Transfer," *I and E C Fundamentals*, Vol. 3 (1964), pp. 111–118.

65. B. B. Mikic and W. M. Rohsenow, "A New Correlation of Pool-Boiling Data Including the Effect of Heating Surface Characteristics," *Trans. ASME*, ser. C., *J. of Heat Transfer*, Vol. 91 (1969), pp. 245–250.

66. K. Konmutsos, R. Moissis, and A. Spyridonos, "A Study of Bubble Departure in Forced Convection Boiling," *Trans. ASME*, ser. C., *J. of Heat Transfer*, Vol. 90 (1968), pp. 223–230.

67. D. P. Jordan, "Film and Transition Boiling," *Advanced in Heat Transfer*, Vol. 5, T. F. Irvine, Jr., and J. P. Hartnett, eds. (New York: Academic Press, 1968, pp. 55–125.)

68. G. Leppert and C. C. Pitts, "Boiling," *Advances in Heat Transfer*, Vol. 1, T. F. Irvine, Jr., and J. P. Hartnett, eds. (New York: Academic Press, 1964, pp. 185–265.)

69. W. M. Rohsenow, "Boiling Heat Transfer," *Dev. in Heat Transfer*, W. M. Rohsenow, ed. (Cambridge, Massachusetts: MIT Press, 1964, pp. 169–260.)

70. R. I. Vachon, G. H. Nix, and G. E. Tanger, "Evaluation of Constants for the Rohsenow Pool-Boiling Correlation," *Trans. ASME*, ser. C, *J. of Heat Transfer*, Vol. 90 (1968), pp. 239–247.

71. R. V. Macbeth, "Burnout Analysis, Part 4: World Data for Uniformly Heated Round Tubes and Rectangular Channels," *AEEW*–R 267, Winfrith, (1963).

72. L. S. Tong, *Boiling Heat Transfer and Two-Phase Flow*. New York: John Wiley and Sons, Inc., 1965.)

73. W. D. Murray and F. Landis, "Numerical and Machine Solutions of Transient Heat Conduction Problems Involving Melting or Freezing," *Trans. ASME*, Vol. 81 (1959), pp. 106–112.

74. A. Lazaridis, "A Numerical Solution of the Multi-dimensional Solidification (or Melting) Problem," *Int. J. Heat Mass Transfer*, Vol. 13 (1970), pp. 1459–1477.

75. H. Budhia and F. Kreith, "Melting or Freezing in a Wedge," *Int. J. Heat Mass Transfer*, Vol. 15 (1972).

76. J. W. Rose, "On the Mechanism of Dropwise Condensation," *Int. J. Heat and Mass Transfer*, Vol. 10 (1967), pp. 755–762.

77. P. Griffith and M. S. Lee, "The Effect of Surface Thermal Properties and Finish on Dropwise Condensation," *Int. J. Heat and Mass Transfer*, Vol. 10 (1967), pp. 697–707.

11
Heat exchangers

11-1. Design and selection

A heat exchanger is a device which effects the transfer of heat from one fluid to another. The simplest type of heat exchanger is a container in which a hot and a cold fluid are mixed directly. In such a system both fluids will reach the same final temperature, and the amount of heat transferred can be estimated by equating the energy lost by the hotter fluid to the energy gained by the cooler one. Open feed-water heaters, desuperheaters, and jet condensers are examples of heat-transfer equipment employing direct mixing of fluids. More common, however, are heat exchangers in which one fluid is separated from the other by a wall or a partition through which the heat flows. These types of exchangers are called *recuperators*. There are many forms of such equipment ranging from a simple pipe-within-a-pipe with a few square feet of heat-transfer surface up to complex surface condensers and evaporators with many thousands of square feet of heat-transfer surface. In between these extremes is a broad field of common shell-and-tube exchangers. These units are widely used because they can be constructed with large heat-transfer surfaces in a relatively small volume, can be fabricated from alloys to resist corrosion, and are suitable for heating, cooling, evaporating, or condensing all kinds of fluids.

The complete design of a heat-exchanger can be broken down into three major phases:

1. The thermal analysis.
2. The preliminary mechanical design.
3. Design for manufacture.

The emphasis in this chapter will be on the thermal design. This phase of the design is primarily concerned with the determination of the heat-transfer surface area required to transfer heat at a specified rate for given flow rates and temperatures of the fluids.

The mechanical design involves considerations of the operating temperatures and pressures, the corrosive characteristics of one or both fluids, the relative thermal expansions and accompanying thermal stresses, and the relation of the heat exchanger to other equipment concerned.

The design for manufacture requires the translation of the physical characteristics and dimensions into a unit which can be built at a low cost. Selection of materials, seals, enclosures, and the optimum mechanical arrangement have to be made and the manufacturing procedures must be specified.

To achieve maximum economy the majority of manufacturers have adopted standard lines of heat exchangers. The standards establish tube diameters and pressure ratings and promote the use of standard drawings and standard fabrication procedures. Standardization does not mean, however, that heat exchangers can be delivered off the shelf, because service requirements vary too much. Some engineering design is necessary for almost every exchanger, but if service conditions permit, the use of exchangers built to standard lines saves money. The engineer concerned with the installation of heat exchangers in power plants and process equipment is therefore often called upon to select a heat-exchanger unit which is suitable for a particular application. The selection requires a thermal analysis to determine whether a standard unit of specified size and geometry can meet the requirements of heating or cooling a given fluid at a specified rate. In this type of analysis the initial cost must be weighed against such factors as life of equipment, ease of cleaning, and space required. It is also important that the requirements of the safety codes of ASME be met, and for this purpose the Standards of the Tubular Exchanger Manufacturers Association (TEMA) should be consulted.

11-2. Basic types of heat exchangers

The simplest type of shell-and-tube heat exchanger is shown in Fig. 11.-1. It consists of a tube or a pipe located concentrically inside another tube which forms the shell for this arrangement. One of the fluids flows through the inner tube, the other through the annulus formed between the inner and the outer tube. Since both fluid streams traverse the exchanger only once, this arrangement is called a *single-pass* heat ex-

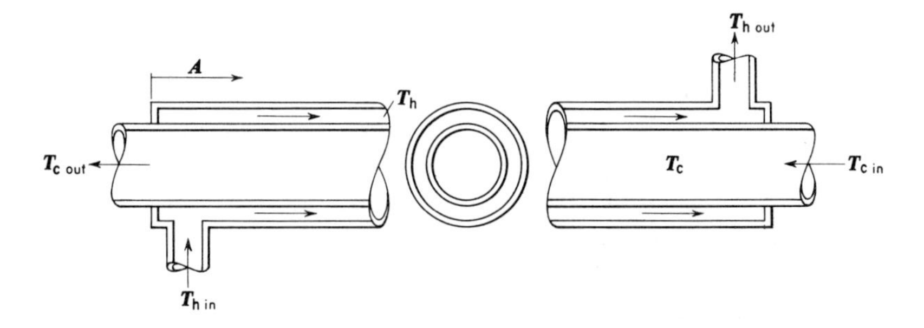

Fig. 11-1. Diagram of a simple tube-within-a-tube counterflow heat exchanger.

changer. If both fluids flow in the same direction, the exchanger is a *parallel-flow* type; if the fluids move in opposite directions, the exchanger is of the *counterflow* type. The temperature difference between the hot and the cold fluid is, in general, not constant along the tube, and the rate of heat flow will vary from section to section. To determine the rate of heat flow one must therefore use an appropriate mean-temperature difference, as shown in Sec. 11-3.

When the two fluids flowing along the heat-transfer surface move at right angles to each other, the heat exchanger is of the *crossflow* type. Three separate arrangements of this type of exchanger are possible. In the first case each of the fluids is *unmixed* as it passes through the exchanger and, therefore, the temperatures of the fluids leaving the heater section are not uniform, being hotter on one side than on the other. A flat-plate type heater (Fig. 11-2), a design used for turbine regenerators to reclaim the energy of the exhaust gases, or an automobile radiator approximates this type of exchanger. In the second case, one of the fluids is *unmixed* and the other is perfectly *mixed* as it flows through the exchanger.

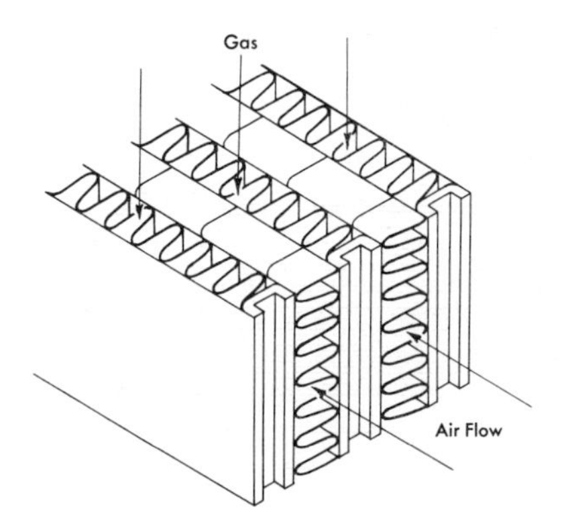

Fig. 11-2. Flat-plate type heat exchanger illustrating crossflow with both fluids unmixed.

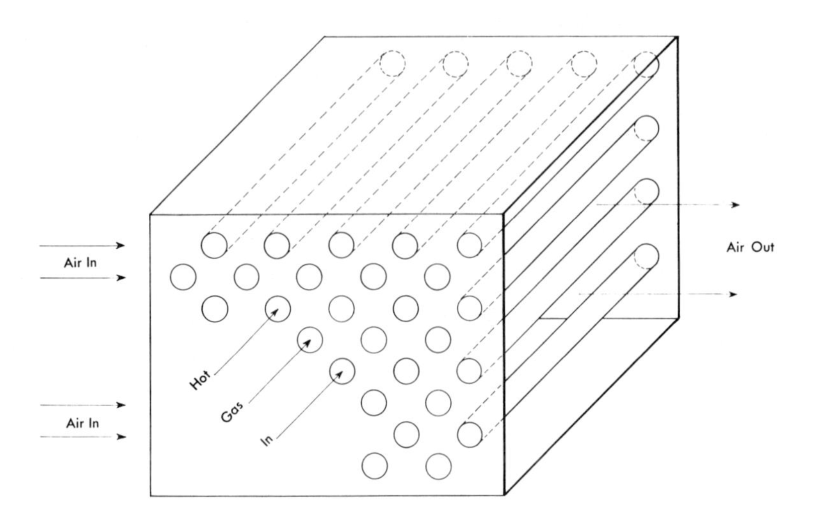

Fig. 11-3. Crossflow air heater illustrating crossflow with one fluid mixed, the other unmixed.

The temperature of the mixed fluid will be uniform across any section and will vary only in the direction of flow. An example of this type is the crossflow air heater shown schematically in Fig. 11-3. The air flowing over the bank of tubes is mixed, while the hot gases inside the tubes are confined and therefore do not mix. In the third case, both of the fluids are *mixed* as they flow through the exchanger; that is, the temperature of both fluids will be uniform across the section and will vary only in the direction of flow. This type of arrangement is less important than the other two and will not be discussed here.

In order to increase the effective heat-transfer surface area per unit volume, most commercial heat exchangers provide for more than a single pass through the tubes, and the fluid flowing outside the tubes in the shell

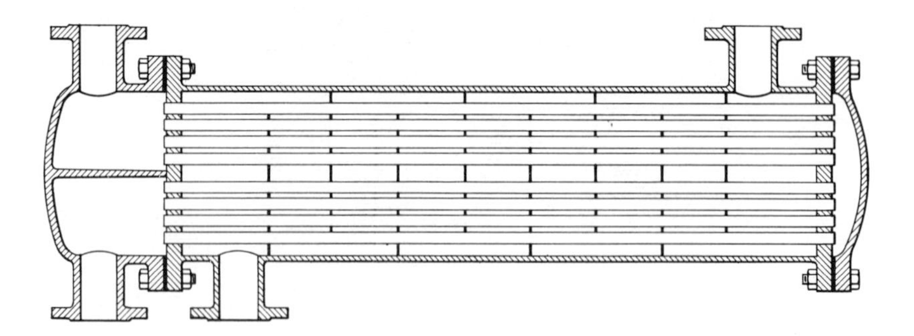

Fig. 11-4. Shell-and-tube heat exchanger with segmental baffles: two tube passes, one shell pass.

is routed back and forth by means of baffles. Figure 11-4 is a cross section of a heat exchanger with two tube passes and one cross-baffled shell pass. The baffles are of the segmental type. This and other typical types of baffles are shown in Fig. 11-5. In a baffled exchanger, the flow pattern on the shell side is complex. As shown by the arrows, part of the time the flow is perpendicular, and part of the time parallel, to the tube.

The heat exchanger illustrated in Fig. 11-4 has fixed tube plates at each end and the tubes are welded or expanded into the plates. This type of construction has the lowest initial cost but can only be used for small temperature differences between the hot and the cold fluid because no

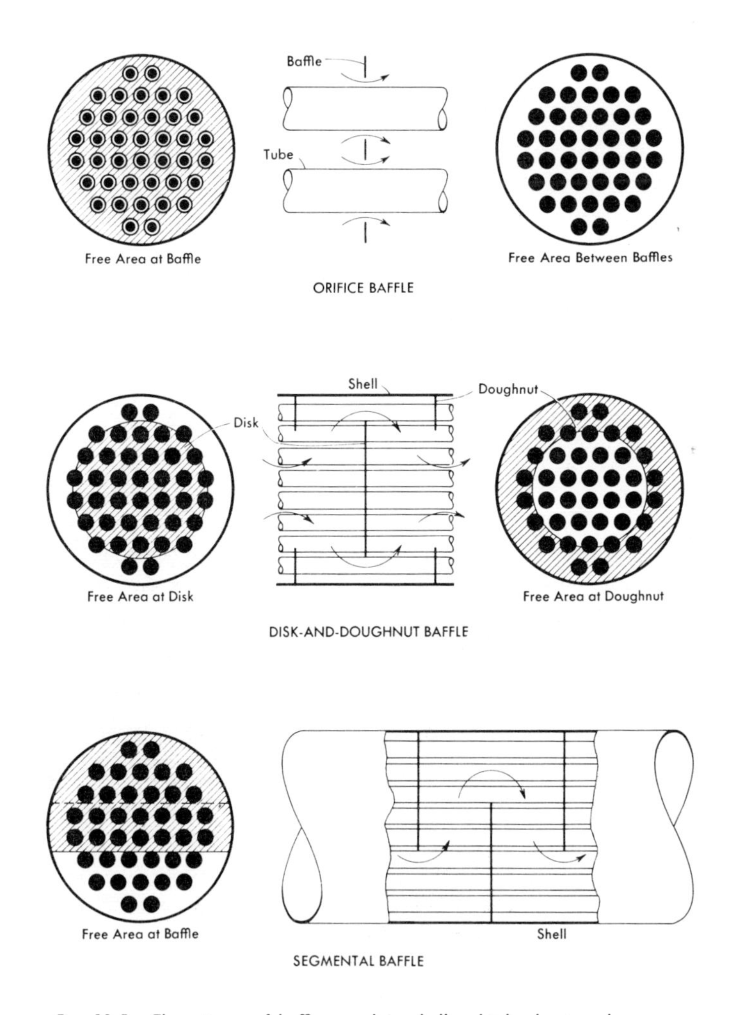

Fig. 11-5. Three types of baffles used in shell-and-tube heat exchangers. (After C. B. Cramer, *Heat Transfer*, 2d ed. International Textbook Company, Scranton, Pa.)

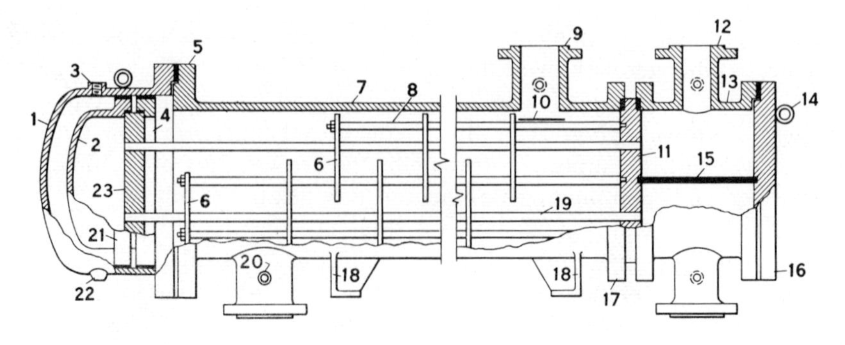

1. Shell cover
2. Floating head
3. Vent connection
4. Floating-head backing device
5. Shell cover—end flange
6. Transverse baffles or support plates
7. Shell
8. Tie rods and spacers
9. Shell nozzle
10. Impingement baffle
11. Stationary tube sheet
12. Channel nozzle
13. Channel
14. Lifting ring
15. Pass partition
16. Channel cover
17. Shell channel—end flange
18. Support saddles
19. Heat-transfer tube
20. Test connection
21. Floating-head flange
22. Drain connection
23. Floating tube sheet

Fig. 11-6. Shell-and-tube heat exchanger with floating head. (Courtesy of the Tubular Exchange Manufacturer's Association)

Fig. 11-7. Typical compact heat-exchanger section. (Courtesy of the Harrison Radiator Division, General Motors Corp.)

provision is made to prevent thermal stresses due to the differential expansion between the tubes and the shell. Another disadvantage is that the tube bundle cannot be removed for cleaning. These drawbacks can be overcome by the modification of the basic design as shown in Fig. 11-6. In this arrangement one tube plate is fixed but the other is bolted to a floating-head cover which permits the tube bundle to move relative to the shell. The floating tube sheet is clamped between the floating head and a flange so that it is possible to remove the tube bundle for cleaning. The heat exchanger shown in Fig. 11-6 has one shell pass and two tube passes.

For certain special applications such as regenerators for aircraft or automobile gas turbines, the rate of heat transfer per unit weight and unit volume is the prime consideration. Compact, lightweight heat exchangers for this type of service have been investigated by Kays and London (1). A typical design is shown in Fig. 11-7. For a complete description and analysis of compact heat exchangers, especially for the application of fins to increase the effectiveness of such units, the reader is referred to the original papers (1,2,3,4, and 5). The application of rough surfaces to heat exchanger design is discussed in Ref. 20.

11-3. Mean temperature difference

The temperatures of fluids in a heat exchanger are generally not constant, but vary from point to point as heat flows from the hotter to the colder fluid. Even for a constant thermal resistance, the rate of heat flow will therefore vary along the path of the exchangers because its value depends on the temperature difference between the hot and the cold fluid at the section. Figures 11-8, 11-9, 11-10, and 11-11 illustrate the changes in temperature that may occur in either or both fluids in a simple shell-and-tube exchanger (Fig. 11-1). The distances between the solid lines are proportional to the temperature differences ΔT between the two fluids.

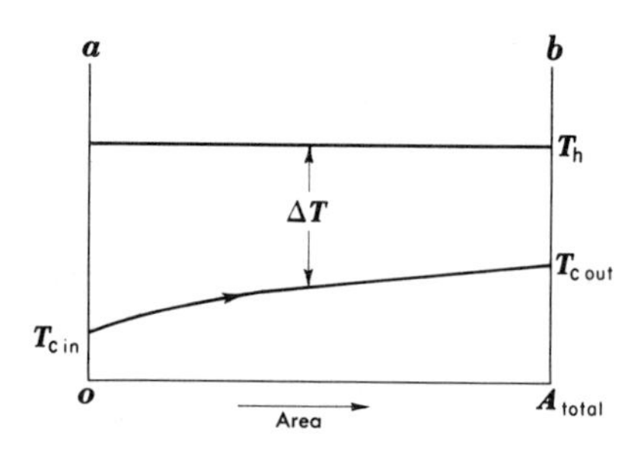

Fig. 11-8. Temperature distribution in single-pass condenser.

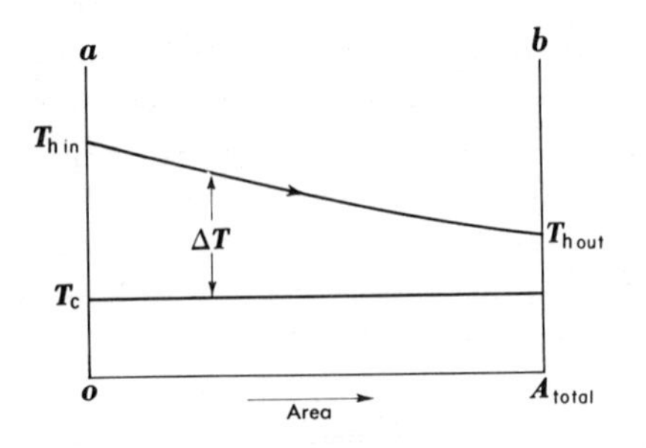

Fig. 11-9. Temperature distribution in single-pass evaporator.

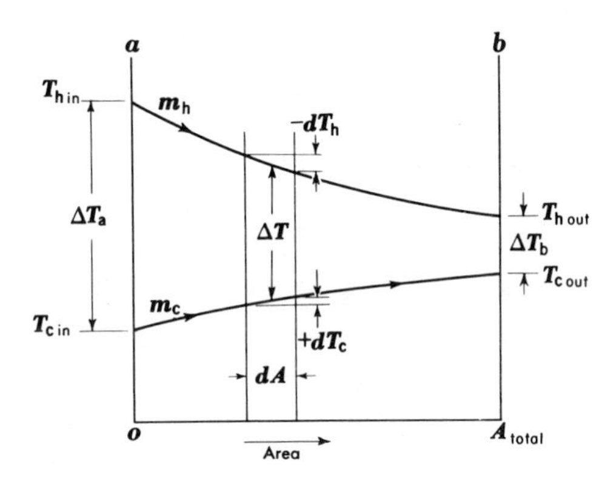

Fig. 11-10. Temperature distribution in single-pass parallel-flow heat exchanger.

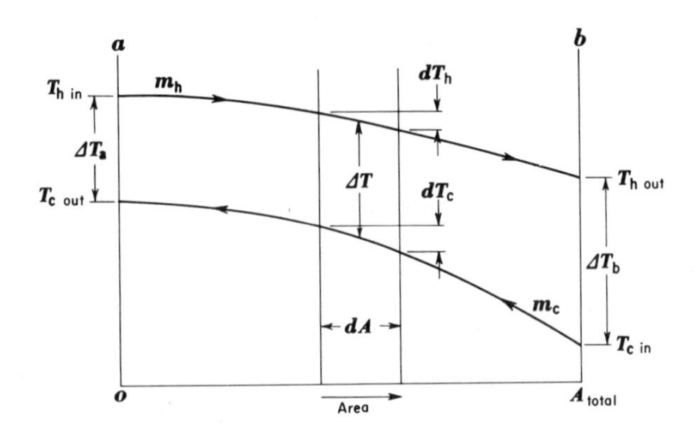

Fig. 11-11. Temperature distribution in single-pass counterflow heat exchanger.

Figure 11-8 illustrates the case where a vapor is condensing at a constant temperature while the other fluid is being heated. Figure 11-9 is representative of a case where a liquid is evaporated at constant temperature while heat is flowing from a warmer fluid whose temperature decreases as it passes through the heat exchanger. For both of these cases the direction of flow of either fluid is immaterial and the constant-temperature medium may also be at rest. Figure 11-10 represents conditions in a parallel-flow exchanger, and Fig. 11-11 applies to counterflow. No change of phase occurs in the latter two cases. Inspection of Fig. 11-10 shows that, no matter how long the exchanger is, the final temperature of the colder fluid can never reach the exit temperature of the hotter fluid in parallel flow. For counterflow, on the other hand, the final temperature of the cooler fluid may exceed the outlet temperature of the hotter fluid, since a favorable temperature gradient exists all along the heat exchanger. An additional advantage of the counterflow arrangement is that, for a given rate of heat flow, less surface area is required than in parallel flow.

To determine the rate of heat transfer in any of the aforementioned cases the equation

$$dq = U dA \Delta T \tag{11-1}$$

must be integrated over the heat-transfer area A along the length of the exchanger. If the overall unit conductance U is constant, if changes in kinetic energy are neglected, and if the shell of the exchanger is insulated, Eq. 11-1 can easily be integrated analytically for parallel or counterflow. An energy balance over a differential area dA yields

$$dq = -m_h c_{ph} dT_h = \pm m_c c_{pc} dT_c = U dA\,(T_h - T_c) \tag{11-2}$$

where m is the mass rate of flow in $\mathrm{lb_m/hr}$, c_p is the specific heat at constant pressure in $\mathrm{Btu/lb_m\,F}$, and T is the average bulk temperature of the fluid in F. The subscripts h and c refer to the hot and cold fluid respectively; the plus sign in the third term applies to parallel flow, and the minus sign to counterflow. If the specific heats of the fluids do not vary with temperature, we can write a heat balance from the inlet to an arbitrary cross section in the exchanger, or

$$-C_h(T_h - T_{h\,\mathrm{in}}) = C_c(T_c - T_{c\,\mathrm{in}}) \tag{11-3}$$

where $C_h = m_h c_{ph}$, the hourly heat-capacity flow rate of the hotter fluid in Btu/hr F;

$C_c = m_c c_{pc}$, the hourly heat-capacity flow rate of the colder fluid in Btu/hr F.

Solving Eq. 11-3 for T_h gives

$$T_h = T_{h\,\text{in}} - \frac{C_c}{C_h}(T_c - T_{c\,\text{in}}) \tag{11-4}$$

from which we obtain

$$T_h - T_c = -\left(1 + \frac{C_c}{C_h}\right)T_c + \frac{C_c}{C_h}T_{c\,\text{in}} + T_{h\,\text{in}} \tag{11-5}$$

Substituting Eq. 11-5 for $T_h - T_c$ in Eq. 11-2 yields after some rearrangement

$$\frac{dT_c}{-[1 + (C_c/C_h)]\,T_c + (C_c/C_h)\,T_{c\,\text{in}} + T_{h\,\text{in}}} = \frac{UdA}{C_c} \tag{11-6}$$

Integrating Eq. 11-6 over the entire length of the exchanger (i.e., from $A = 0$ to $A = A_{\text{total}}$) yields

$$\ln\left\{\frac{-[1 + (C_c/C_h)]\,T_{c\,\text{out}} + (C_c/C_h)\,T_{c\,\text{in}} + T_{h\,\text{in}}}{-[1 + (C_c/C_h)]\,T_{c\,\text{in}} + (C_c/C_h)\,T_{c\,\text{in}} + T_{h\,\text{in}}}\right\} = -\left(\frac{1}{C_c} + \frac{1}{C_h}\right)UA \tag{11-7}$$

Equation 11-7 can be simplified to read

$$\ln\left[\frac{(1 + C_c/C_h)(T_{c\,\text{in}} - T_{c\,\text{out}}) + T_{h\,\text{in}} - T_{c\,\text{in}}}{T_{h\,\text{in}} - T_{c\,\text{in}}}\right] = -\left(\frac{1}{C_c} + \frac{1}{C_h}\right)UA \tag{11-8}$$

From Eq. 11-3 we obtain for the total length of the exchanger

$$\frac{C_c}{C_h} = -\frac{T_{h\,\text{out}} - T_{h\,\text{in}}}{T_{c\,\text{out}} - T_{c\,\text{in}}} \tag{11-9}$$

which can be used to eliminate the hourly heat capacities in Eq. 11-8. After some rearrangement we get

$$\ln\left(\frac{T_{h\,\text{out}} - T_{c\,\text{out}}}{T_{h\,\text{in}} - T_{c\,\text{in}}}\right) = [(T_{h\,\text{out}} - T_{c\,\text{out}}) - (T_{h\,\text{in}} - T_{c\,\text{in}})]\,\frac{UA}{q} \tag{11-10}$$

since

$$q = C_c(T_{c\,\text{out}} - T_{c\,\text{in}}) = C_h(T_{h\,\text{in}} - T_{h\,\text{out}})$$

Letting $T_h - T_c = \Delta T$, Eq. 11-10 can be written

$$q = UA \frac{\Delta T_a - \Delta T_b}{\ln (\Delta T_a / \Delta T_b)} \tag{11-11}$$

where the subscripts a and b refer to the respective ends of the exchanger (see Figs. 11-10 and 11-11). In practice it is convenient to use an average effective temperature difference $\overline{\Delta T}$ for the entire heat exchanger defined by

$$q = UA\overline{\Delta T} \tag{11-12}$$

Comparing Eqs. 11-12 and 11-11, one finds that, for parallel or counterflow

$$\overline{\Delta T} = \frac{\Delta T_a - \Delta T_b}{\ln (\Delta T_a / \Delta T_b)} \tag{11-13}$$

which is called the logarithmic mean overall temperature difference often designated by LMTD. The LMTD also applies when the temperature of one of the fluids is constant, as shown in Figs. 11-8 and 11-9. When $m_h c_{ph} = m_c c_{pc}$, the temperature difference is constant in counterflow and $\overline{\Delta T} = \Delta T_a = \Delta T_b$. If the temperature difference ΔT_a is not more than 50 percent greater than ΔT_b, the arithmetic mean temperature difference will be within 1 percent of the LMTD and may be used to simplify calculations.

The use of the logarithmic mean temperature is only an approximation in practice because U is generally not constant. In design work, however, the overall conductance is usually evaluated at a mean section, usually halfway between ends, and treated as constant. If U varies considerably, a numerical step-by-step integration of Eq. 11-1 may be necessary, as shown in Sec. 11-6.

For more complex heat exchangers such as the shell-and-tube arrangements with several tube or shell passes and with crossflow exchangers having mixed and unmixed flow, the mathematical derivation of an expression for the mean temperature difference becomes quite complex. The usual procedure is to modify the simple LMTD by correction factors which have been published in chart form by Bowman, Mueller, and Nagle (6) and by the Tubular Exchanger Manufacturer's Association (7). Four of these graphs[1] are shown in Figs. 11-12, 11-13, 11-14, and 11-15. The ordinate of each is the correction factor F. To obtain the true mean temperature for any of these arrangements, the LMTD calculated for

[1] Correction factors for several other arrangements are presented in Reference 6.

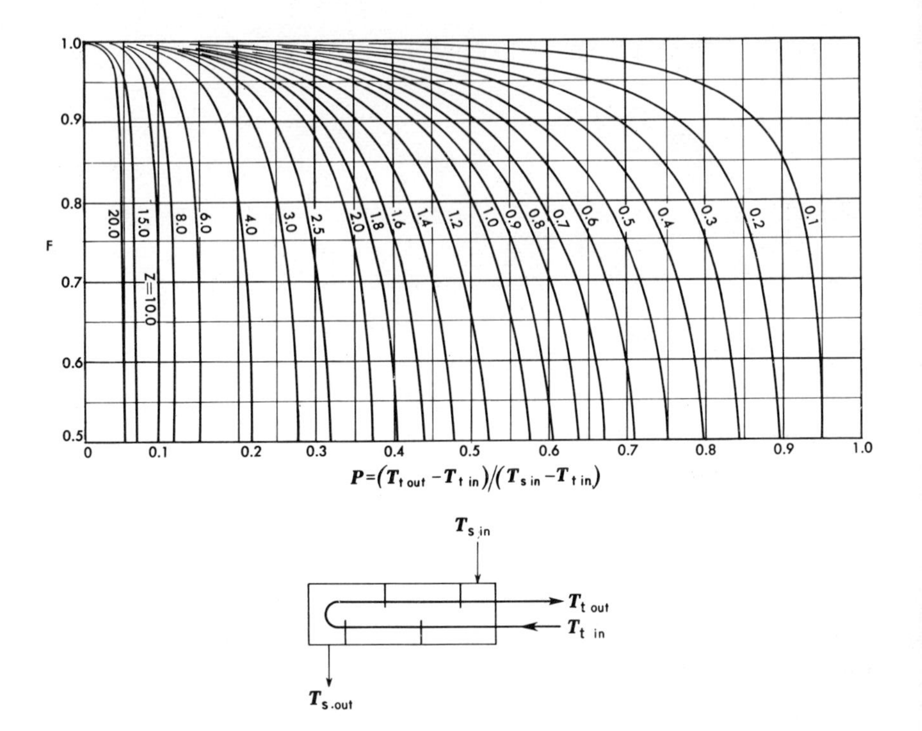

Fig. 11-12. Correction factor to counterflow LMTD for heat exchanger with one shell pass and two, or a multiple of two, tube passes. (Courtesy of the Tubular Exchange Manufacturer's Association)

counterflow must be multiplied by the appropriate correction factor, that is,

$$\Delta T_{\text{true mean}} = \text{LMTD} \times F \tag{11-14}$$

The values shown on the abscissa are for the dimensionless temperature-difference ratio

$$P = (T_{t\,\text{out}} - T_{t\,\text{in}})/(T_{s\,\text{in}} - T_{t\,\text{in}}) \tag{11-15}$$

where the subscripts t and s refer to the tube and shell fluid respectively, and the subscripts *in* and *out* refer to the inlet and outlet conditions respectively. The ratio P is an indication of the heating or cooling effectiveness and can vary from zero for a constant temperature of one of the fluids to unity for the case when inlet temperature of the hotter fluid equals the outlet temperature of the colder fluid. The parameter for each of the curves Z is equal to the ratio of the products of the mass-flow rate times the heat capacity of the two fluids $m_t c_{pt}/m_s c_{ps}$. This ratio is also

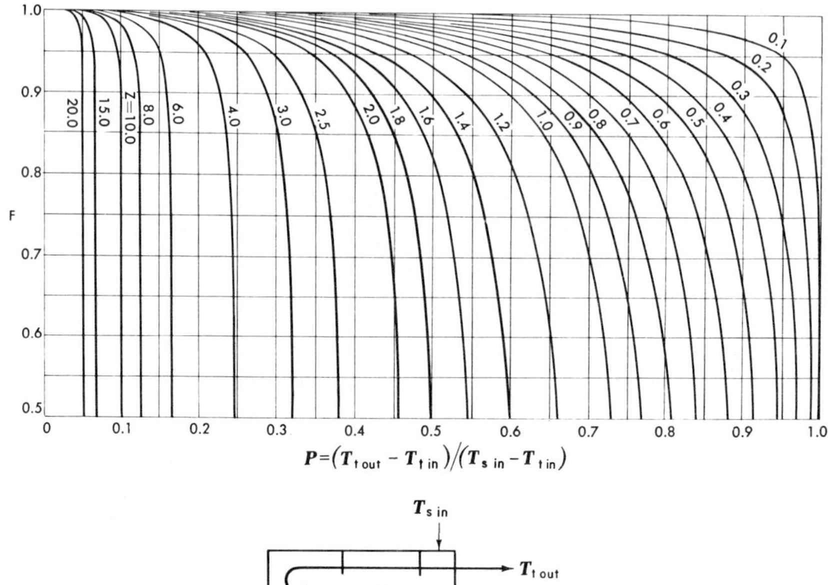

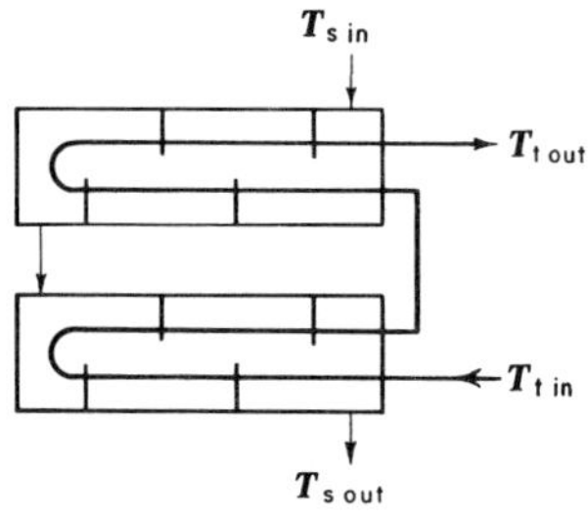

Fig. 11-13. Correction factor to counterflow LMTD for heat exchanger with two shell passes and a multiple of two tube passes. (Courtesy of the Tubular Exchange Manufacturer's Association)

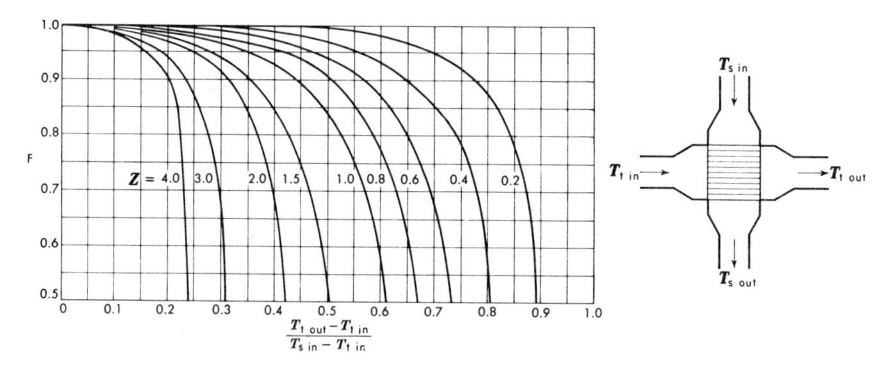

Fig. 11-14. Correction factor to counterflow LMTD for crossflow heat exchangers, fluid on shell side mixed, other fluid unmixed, one tube pass. (Extracted from "Mean Temperature Difference in Design," by R. A. Bowman, A. C. Mueller, and W. M. Nagel, published in *Trans. ASME*, Vol. 62, 1940, with permission of the publishers, The American Society of Mechanical Engineers)

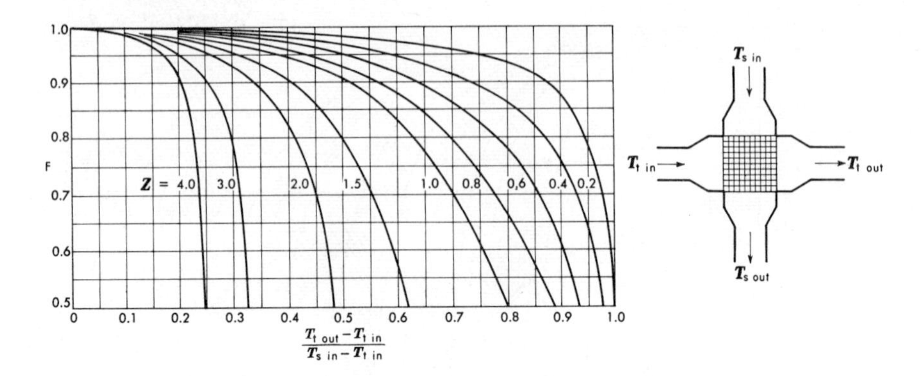

Fig. 11-15. Correction factor to counterflow LMTD for crossflow heat exchanger, both fluids unmixed, one tube pass. (Extracted from "Mean Temperature Difference in Design," by R. A. Bowman, A. C. Mueller, and W. M. Nagel, published in *Trans. ASME*, Vol. 62, 1940, with permission of the publishers, The American Society of Mechanical Engineers)

equal to the temperature change of the shell fluid divided by the temperature change of the fluid in the tubes, or

$$Z = \frac{m_t c_{pt}}{m_s c_{ps}} = \frac{T_{s\,\mathrm{in}} - T_{s\,\mathrm{out}}}{T_{t\,\mathrm{out}} - T_{t\,\mathrm{in}}} \tag{11-16}$$

In the application of the correction factors it is immaterial whether the warmer fluid flows through shell or tubes. If the temperature of either of the fluids remains constant, the direction of flow is also immaterial, since F equals 1 and the LMTD applies directly.

EXAMPLE 11-1. Determine the heat-transfer surface area required for a heat exchanger constructed from 1-in.-OD tube to cool 55,000 lb/hr of a 95 percent ethyl alcohol solution (c_p = 0.91 Btu/lb F) from 150 to 103 F, using 50,000 lb/hr of water available at 50 F. Assume that the overall coefficient of heat transfer based on the outer-tube area is 100 Btu/hr sq ft F and consider each of the following arrangements:

a) Parallel-flow tube and shell;
b) Counterflow tube and shell;
c) Reversed-current exchanger with two shell passes and 72 tube passes, the alcohol flowing through the shell and the water flowing through the tubes;
e) Crossflow, with one tube pass and one shell pass, shell-side fluid mixed.

Solution: (a) The outlet temperature of the water for any of the four arrangements can be obtained from an overall energy balance, assuming that the heat loss to the atmosphere is negligible. Writing the energy balance as

$$m_h c_{ph}(T_{h\,\text{in}} - T_{h\,\text{out}}) = m_c c_{pc}(T_{c\,\text{out}} - T_{c\,\text{in}})$$

and substituting the data in the above equation we obtain

$$(55{,}000)(0.91)(150 - 103) = (50{,}000)(1.0)(T_{c\,\text{out}} - 50)$$

from which the outlet temperature of the water is found to be 97 F. The rate of heat flow from the alcohol to the water is therefore

$$q = m_h c_{ph}(T_{h\,\text{in}} - T_{h\,\text{out}}) = (55{,}000)(0.91)(150 - 103) = 2{,}350{,}000 \text{ Btu/hr}$$

From Eq. 11-13, the LMTD for parallel flow is

$$\text{LMTD} = \frac{\Delta T_a - \Delta T_b}{\ln(\Delta T_a/\Delta T_b)} = \frac{100 - 6}{\ln(100/6)} = 33.4 \text{ F}$$

From Eq. 11-12 the heat-transfer surface area is

$$A = \frac{q}{(U)(\text{LMTD})} = \frac{2{,}350{,}000}{(100)(33.4)} = 703 \text{ sq ft}$$

The length of the exchanger for a 1-in.-OD tube would be too great to be practical. *Ans.*

b) For the counterflow arrangement, the appropriate mean temperature difference is 150 – 97 = 53 F because $m_c c_{pc} = m_h c_{ph}$. The required area is

$$A = \frac{q}{(U)(\text{LMTD})} = \frac{2{,}350{,}000}{(100)(53)} = 444 \text{ sq ft}$$

which is 40 percent less than the area necessary for parallel flow. *Ans.*

c) For the reversed-current arrangement, we determine the appropriate mean temperature difference by applying the correction factor found from Fig. 11-13 to the mean temperature for counterflow.

$$P = \frac{T_{c\,\text{out}} - T_{c\,\text{in}}}{T_{h\,\text{in}} - T_{c\,\text{in}}} = \frac{97 - 50}{150 - 50} = 0.47$$

and the hourly heat capacity ratio is

$$Z = \frac{m_c c_{pc}}{m_h c_{ph}} = 1$$

From the chart of Fig. 11-13, $F = 0.97$ and the heat-transfer area is

$$A = \frac{444}{0.97} = 460 \text{ sq ft} \qquad \textit{Ans.}$$

The length of the exchanger for 72 1-in.-OD tubes in parallel would be

$$L = \frac{A/72}{\pi D} = \frac{6.4}{(\pi)(1/12)} \simeq 24.4 \text{ ft}$$

This length is not unreasonable, but if it is desirable to shorten the exchanger, more tubes could be used.

d) For the crossflow arrangement (Fig. 11-3), the correction factor is found from the chart of Fig. 11-14 to be 0.88. The required surface area is thus 504 sq ft, about 10 percent larger than that for the reversed-current exchanger. *Ans.*

11-4. Heat-exchanger effectiveness

In the thermal analysis of the various types of heat exchanger presented in the preceding section, an equation (Eq. 11-12) of the type

$$q = UA\Delta T_{\text{mean}}$$

was used. This form will be found convenient when all of the terminal temperatures necessary for the evaluation of the appropriate mean temperature are known, and Eq. 11-12 is widely employed in the design of heat exchangers to given specifications. There are, however, numerous occasions when the performance of a heat exchanger (i.e., U) is known, or can at least be estimated, but the temperatures of the fluids leaving the exchanger are not known. This type of problem is encountered in the selection of a heat exchanger or when the unit has been tested at one flow rate but service conditions require different flow rates for one or both fluids. The outlet temperatures and the rate of heat flow can only be found by a rather tedious trial-and-error procedure if the charts presented in the preceding section are used. In such cases it is desirable to circumvent entirely any reference to the logarithmic or any other mean temperature difference. A method which accomplishes this has been proposed by Nusselt (8) and Ten Broeck (9).

To obtain an equation for the rate of heat transfer which does not involve any of the outlet temperatures, we introduce the *heat-exchanger effectiveness* $\mathcal{E}$. The heat-exchanger effectiveness is defined as the ratio of the actual rate of heat transfer in a given heat exchanger to the maximum possible rate of heat exchange. The latter would be obtained in a counterflow heat exchanger of infinite heat-transfer area. In this type of unit, if there are no external heat losses, the outlet temperature of the colder fluid equals the inlet temperature of the hotter fluid when $m_c c_{pc} < m_h c_{ph}$; when $m_h c_{ph} < m_c c_{pc}$, the outlet temperature of the warmer fluid equals the inlet temperature of the colder one. In other words, the effectiveness compares the actual heat-transfer rate to the maximum rate whose only limit is the second law of thermodynamics. Depending on which of the hourly heat capacities is smaller, the effectiveness is

$$\mathcal{E} = \frac{C_h(T_{h\,\text{in}} - T_{h\,\text{out}})}{C_{\min}(T_{h\,\text{in}} - T_{c\,\text{in}})} \tag{11-17a}$$

or

$$\mathcal{E} = \frac{C_c(T_{c\,\text{out}} - T_{c\,\text{in}})}{C_{\min}(T_{h\,\text{in}} - T_{c\,\text{in}})} \tag{11-17b}$$

where $C_{\min}$ is the smaller of the $m_h c_{ph}$ and $m_c c_{pc}$ magnitudes.

Once the effectiveness of a heat exchanger is known, the rate of heat transfer can be determined directly from the equation

$$q = \mathcal{E} C_{\min}(T_{h\,\text{in}} - T_{c\,\text{in}}) \tag{11-18}$$

since

$$\mathcal{E} C_{\min}(T_{h\,\text{in}} - T_{c\,\text{in}}) = C_h(T_{h\,\text{in}} - T_{h\,\text{out}}) = C_c(T_{c\,\text{out}} - T_{c\,\text{in}})$$

Equation 11-18 is the basic relation in this analysis because it expresses the rate of heat transfer in terms of the effectiveness, the smaller hourly heat capacity, and the difference between the inlet temperatures. It replaces Eq. 11-12 in the LMTD analysis but does not involve the outlet temperatures. Equation 11-18 is of course also suitable for design purposes instead of Eq. 11-12.

We shall illustrate the method of deriving an expression for the effectiveness of a heat exchanger by applying it to a parallel-flow arrangement. The effectiveness can be introduced into Eq. 11-8 by replacing $(T_{c\,\text{in}} - T_{c\,\text{out}})/(T_{h\,\text{in}} - T_{c\,\text{in}})$ by the effectiveness relation from Eq. 11-17. We obtain

$$\ln\left[1 - \mathcal{E}\left(\frac{C_{\min}}{C_h} + \frac{C_{\min}}{C_c}\right)\right] = -\left(\frac{1}{C_c} + \frac{1}{C_h}\right)UA$$

or

$$1 - \mathcal{E}\left(\frac{C_{\min}}{C_h} + \frac{C_{\min}}{C_c}\right) = e^{-(1/C_c + 1/C_h)UA}$$

Solving for $\mathcal{E}$ yields

$$\mathcal{E} = \frac{1 - e^{-[1+(C_h/C_c)]UA/C_h}}{(C_{\min}/C_h) + (C_{\min}/C_c)} \tag{11-19}$$

When C_h is less than C_c, the effectiveness becomes

$$\mathcal{E} = \frac{1 - e^{-[1+(C_h/C_c)]UA/C_h}}{1 + (C_h/C_c)} \tag{11-20}$$

and when $C_c < C_h$, we obtain

$$\mathcal{E} = \frac{1 - e^{-[1+(C_c/C_h)]UA/C_c}}{1 + (C_c/C_h)} \tag{11-20a}$$

The effectiveness for both cases can therefore be written in the form

$$\mathcal{E} = \frac{1 - e^{-[1+(C_{\min}/C_{\max})]UA/C_{\min}}}{1 + (C_{\min}/C_{\max})} \tag{11-21}$$

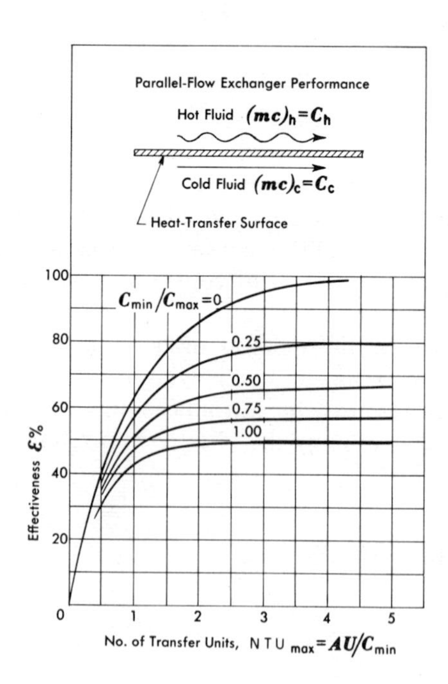

Fig. 11-16. Heat-exchanger effectiveness for parallel flow. (By permission from W. M. Kays and A. L. London, *Compact Heat Exchangers*, National Press, 1955)

The foregoing derivation illustrates how the effectiveness for a given flow arrangement can be expressed in terms of two dimensionless parameters, the hourly heat-capacity ratio C_{min}/C_{max} and the ratio of the overall conductance to the smaller hourly heat capacity, UA/C_{min}. The latter of the two parameters is called the *number of heat-transfer units*, or NTU for short. The number of heat-transfer units is a measure of the heat-transfer size of the exchanger. The larger the value of NTU, the closer the heat exchanger approaches its thermodynamic limit. By analyses which in principle are similar to the one presented here for parallel flow, effectivenesses may be evaluated for most flow arrangements of practical interest. The results have been put by Kays and London (1) into convenient graphs from which the effectiveness can be determined for given values of NTU and C_{min}/C_{max}. The effectiveness curves for some common flow arrangements are shown in Figs. 11-16 to 11-20. The abscissas of these figures are the NTU's of the heat exchangers. The constant parameter for each curve is the hourly heat capacity ratio C_{min}/C_{max}, and the effectiveness is read on the ordinate. Note that, for an evaporator or condenser, $C_{min}/C_{max} = 0$, because if one fluid remains at constant temperature

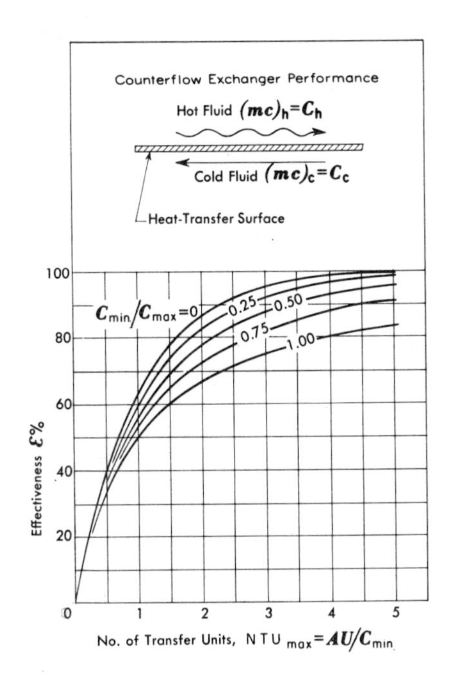

Fig. 11-17. Heat-exchanger effectiveness for counterflow. (By permission from W. M. Kays and A. L. London, *Compact Heat Exchangers*, National Press, 1955)

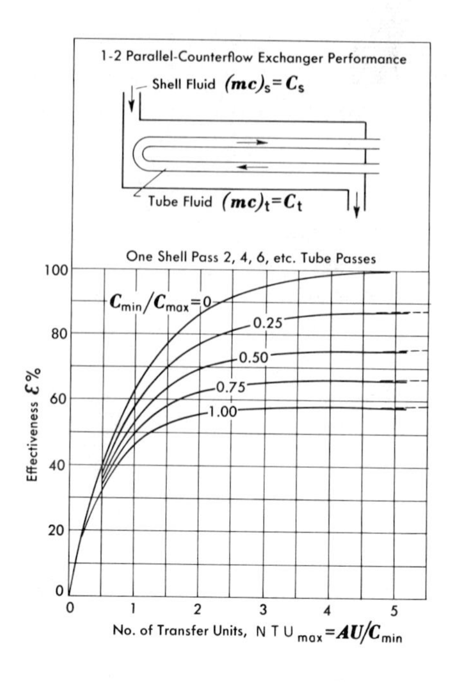

Fig. 11-18. Heat-exchanger effectiveness for shell-and-tube heat exchanger with one well-baffled shell pass and two, or a multiple of two, tube passes. (By permission from W. M. Kays and A. L. London, *Compact Heat Exchangers*, National Press, 1955)

throughout the exchanger, its effective specific heat, and thus its capacity rate, is by definition equal to infinity.

EXAMPLE 11-2. From a performance test on a well-baffled single-shell, two-tube-pass heat exchanger, the following data are available: oil (c_p = 0.5 Btu/lb F) in turbulent flow inside the tubes entered at 160 F at the rate of 5000 lb/hr and left at 100 F; water flowing on the shell side entered at 60 F and left at 80 F. A change in service conditions requires the cooling of a similar oil from an initial temperature of 200 F but at three fourths of the flow rate used in the performance test. Estimate the outlet temperature of the oil for the same water rate and inlet temperature as before.

Solution: The test data may be used to determine the hourly heat capacity of the water and the overall conductance of the exchanger. The hourly heat capacity of the water is from Eq. 11-9

$$C_c = C_h \frac{T_{h\,\mathrm{in}} - T_{h\,\mathrm{out}}}{T_{c\,\mathrm{out}} - T_{c\,\mathrm{in}}} = (5000)(0.5)\left(\frac{160 - 100}{80 - 60}\right) = 7500 \text{ Btu/hr F}$$

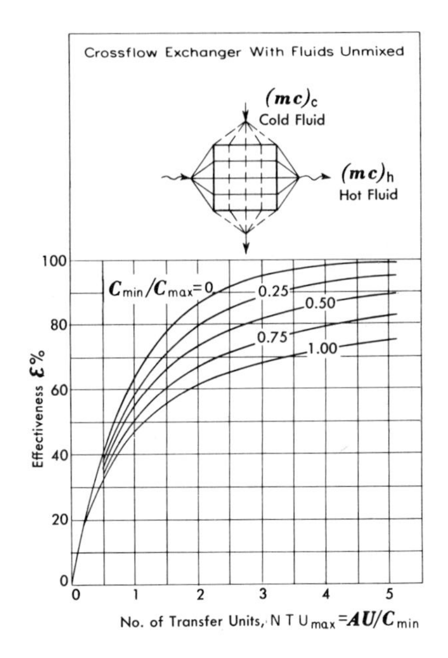

Fig. 11-19. Heat-exchanger effectiveness for crossflow with both fluids unmixed. (By permission from W. M. Kays and A. L. London, *Compact Heat Exchangers*, National Press, 1955)

and the temperature ratio P is, from Eq. 11-15,

$$P = \frac{T_{t\,\mathrm{out}} - T_{t\,\mathrm{in}}}{T_{s\,\mathrm{in}} - T_{t\,\mathrm{in}}} = \frac{60}{100} = 0.6 \qquad Z = \frac{20}{60} = 0.33$$

From Fig. 11-12, $F = 0.94$ and the mean temperature difference is

$$\overline{\Delta T} = F \times \mathrm{LMTD} = 0.94\,\frac{80 - 40}{\ln (80/40)} = 54.2\ \mathrm{F}$$

From Eq. 11-12, the overall conductance is

$$UA = q/\overline{\Delta T} = (7500)(20)/54.2 = 2760\ \mathrm{Btu/hr\ F}$$

Since the thermal resistance on the oil side is controlling, a decrease in velocity to 75 percent of the original value will increase the thermal resistance roughly by the velocity ratio raised to the 0.8 power. This can be verified by reference to Eq. 8-17. Under the new conditions the conductance, the NTU, and the hourly heat capacity ratio will therefore be approximately

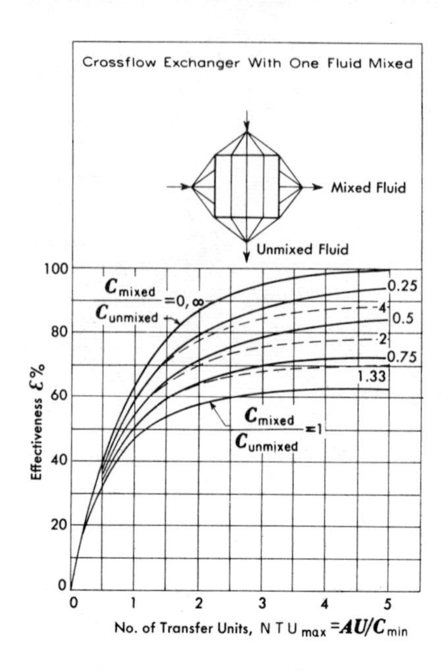

Fig. 11-20. Heat-exchanger effectiveness for crossflow with one fluid mixed, the other unmixed. When $C_{mixed}/C_{unmixed} > 1$, NTU_{max} is based on $C_{unmixed}$ (By permission from W. M. Kays and A. L. London, *Compact Heat Exchangers*, National Press, 1955)

$$UA \simeq (2760)(0.75^{0.8}) = 2190 \text{ Btu/hr F}$$

$$\text{NTU} = \frac{UA}{C_{oil}} = \frac{2190}{(0.75)(2500)} = 1.17$$

and $$C_{oil}/C_{water} = C_{min}/C_{max} = (0.75)(2500)/7500 = 0.25$$

From Fig. 11-18 the effectiveness is equal to 0.63. Hence from the definition of $\mathcal{E}$ in Eq. 11-17a, the oil outlet temperature is

$$T_{oil\,out} = T_{oil\,in} - \mathcal{E}\Delta T_{max} = 200 - [(0.63)(140)] = 112 \text{ F} \qquad \textit{Ans.}$$

EXAMPLE 11-3. A flat-plate-type heater (Fig. 11-21) is to be used to heat air with the hot exhaust gases from a turbine. The required air-flow rate is 6000 lb/hr, entering at 60 F; the hot gases are available at a temperature of 1600 F and at a rate of 5000 lb/hr. Determine the temperature of the air leaving the heat exchanger.

Solution: Inspection of Fig. 11-21 shows that the unit is of the crossflow

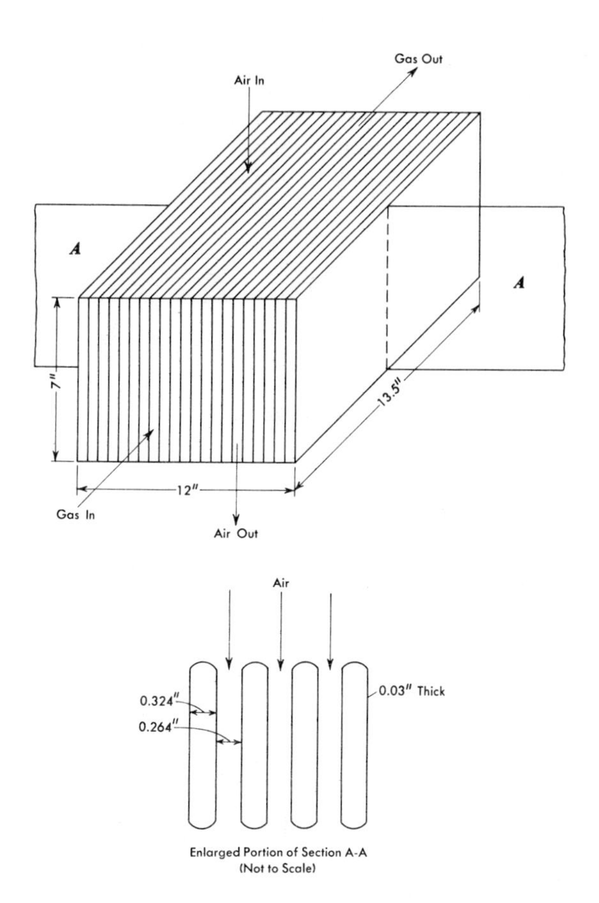

Dimensions at Section A–A

P_a =	Wetted perimeter on air side	2.3 ft
P_g =	Wetted perimeter on gas side	1.22 ft
A_a =	Cross-sectional area of air passage (per passage)	0.0246 sq ft
A_g =	Cross-sectional area of gas passage (per passage)	0.0158 sq ft
	Heat-transfer area	23.6 sq ft
	Air passages	19
	Gas passages	18

Fig. 11-21. Flat-plate-type heater.

type, both fluids unmixed. As a first approximation the end effects will be neglected. The flow systems for the air and gas streams are similar to flow in straight ducts having the following dimensions:

Length of air duct, L_a = 0.583 ft
Hydraulic diameter of air duct $D_{Ha} = 4A_a/P_a$ = 0.0427 ft
Length of gas duct L_g = 1.13 ft
Hydraulic diameter of gas duct $D_{Hg} = 4A_g/P_g$ = 0.0516 ft
Heat-transfer surface area A = 23.6 ft

The unit conductances may be evaluated from Eqs. 8-23a and 8-18 for flow in ducts ($L_a/D_{Ha} = 0.583/0.0427 = 13.7$, $L_g/D_{Hg} = 1.13/0.0516 = 21.9$). A difficulty arises, however, because the temperatures of both fluids vary along the duct. It is therefore necessary to estimate an average temperature and refine the calculations after the outlet temperatures have been found. Selecting the average air temperature at 160 F and the average gas temperature at 1500 F, the absolute viscosities at those temperatures are from Table A-3:

$$\mu_{\text{air}} = 1.4 \times 10^{-5} \text{lb}_\text{m}/\text{ft sec}$$
$$\mu_{\text{gas}} \text{ (assuming air properties)} = 3.0 \times 10^{-5} \text{lb}_\text{m}/\text{ft sec}$$

The mass rates per unit area are:

$$(m/A)_{\text{air}} = 6000/(19)(0.0246) = 12{,}840 \text{ lb/hr sq ft}$$
$$(m/A)_{\text{gas}} = 5000/(18))(0.0158) = 17{,}550 \text{ lb/hr sq ft}$$

The Reynolds numbers are:

$$\text{Re}_{\text{air}} = \frac{(m/A)_a D_{Ha}}{\mu_a} = \frac{(12{,}840 \text{ lb}_\text{m}/\text{hr sq ft})(0.0427 \text{ ft})}{(3600 \text{ sec/hr})(1.4 \times 10^{-5} \text{lb}_\text{m}/\text{ft sec})} = 11{,}100$$

$$\text{Re}_{\text{gas}} = \frac{(m/A)_g D_{Hg}}{\mu_g} = \frac{(17{,}550 \text{ lb}_\text{m}/\text{hr sq ft})(0.0516 \text{ ft})}{(3600 \text{ sec/hr})(3.0 \times 10^{-5} \text{lb}_\text{m}/\text{ft sec})} = 8200$$

Using Eqs. 8-18 and 8-23a the average unit conductances are:

$$\bar{h}_{\text{air}} = \left[(0.023)\left(\frac{k_a}{D_{Ha}}\right)(\text{Re}_D^{0.8}\,\text{Pr}^{0.33})\right]\left[1 + \left(\frac{D_{Ha}}{L}\right)^{0.7}\right]$$

$$= \left[(0.023)\left(\frac{0.0185}{0.0427}\right)(1720)(0.87)\right](1 + 0.16) = 17.3 \text{ Btu/hr sq ft F}$$

$$\bar{h}_{\text{gas}} = \left[(0.023)\left(\frac{0.048}{0.0516}\right)(1340)(0.84)\right](1 + 0.116) = 26.8 \text{ Btu/hr sq ft F}$$

If the thermal resistance of the metal wall is neglected, the overall conductance is

$$UA = \frac{1}{\dfrac{1}{\bar{h}_a A} + \dfrac{1}{\bar{h}_g A}} = \frac{1}{\left(\dfrac{1}{17.3 \times 23.6}\right) + \left(\dfrac{1}{26.8 \times 23.6}\right)} = 250 \text{ Btu/hr F}$$

The number of transfer units, based on the warmer fluid which has the smaller heat capacity rate, are

$$\text{NTU} = UA/C_{1\,\min} = 250/(5000 \times 0.24) = 0.208$$

the hourly heat-capacity ratio is

$$\frac{C_g}{C_a} = \frac{(5000)(0.24)}{(6000)(0.24)} = 0.833$$

and from Fig. 11-19 the effectiveness is 0.15. Finally, the average outlet temperature of the air is

$$T_{\text{air out}} = T_{\text{air in}} + \frac{C_g}{C_a} \mathcal{E} \Delta T_{\max} = 60 + (0.833)(0.15)(1540) = 247 \text{ F} \qquad \textit{Ans.}$$

A check on the mean air temperature gives

$$T_{\text{mean}} = \frac{247 + 60}{2} = 153.5 \text{ F}$$

which is sufficiently close to the assumed value of 160 F to make a second approximation unnecessary. To appreciate the usefulness of the approach based on the concept of heat-exchanger effectiveness, it is suggested that this same problem be worked out by trial and error, using Eq. 11-12 and the chart of Fig. 11-15.

The effectiveness of the heat exchanger in Example 11-3 is very low (15 percent) because the heat-transfer area is too small to utilize the available energy efficiently. The relative gain in heat-transfer performance which can be achieved by increasing the heat-transfer area is well represented on the effectiveness curves. A fivefold increase in area would raise the effectiveness to 60 percent. If, however, a particular design falls near or above the knee of these curves, increasing the surface area will not improve the performance appreciably, but may cause an undue increase in the frictional pressure drop.

11-5. Fouling factors

The performance of heat exchangers under service conditions, especially in the process industry, cannot often be predicted from a thermal analysis alone. During operation with most liquids and some gases, a dirt film gradually builds up on the heat-transfer surface. This deposit may be rust, boiler scale, silt, coke, or any number of things. Its effect, which is referred to as *fouling*, is to increase the thermal resistance. The manufacturer cannot usually predict the nature of the dirt deposit, nor the rate of fouling. Therefore, only the performance of clean exchangers can be guaranteed. The thermal resistance of the deposit can generally be obtained only from actual tests or from experience. If performance tests are made on a clean exchanger and repeated later after the unit has been in service for some time, the thermal resistance of the deposit can be determined from the relation

$$R_d = \frac{1}{U_a} - \frac{1}{U}$$

where U = unit conductance of the clean exchanger;
U_a = conductance after fouling has occurred;
R_d = unit thermal resistance of the scale.

Fouling factors for various applications have been compiled by the Tubular Exchanger Manufacturers Association and are available in their publication (7). A few samples are given in Table 11-1. The fouling factors should be applied as indicated in the following equation for the overall design heat-transfer coefficient U_d of *unfinned* tubes:

$$U_d = \frac{1}{\frac{1}{\bar{h}_o} + R_o + R_k + \frac{R_i A_o}{A_i} + \frac{A_o}{\bar{h}_i A_i}} \tag{11-22}$$

where U_d = design overall coefficient of heat transfer in Btu/hr sq ft F based on a unit area of the outside tube surface;
$\bar{h}_o$ = average unit-surface conductance of the fluid on the outside of tubing, in Btu/hr sq ft F;
$\bar{h}_i$ = average unit-surface conductance of fluid inside tubing, in Btu/hr sq ft F;
R_o = unit fouling resistance on outside of tubing, in hr sq ft F/Btu;
R_i = unit fouling resistance on inside of tubing, in hr sq ft F/Btu;
R_k = unit resistance of tubing in hr sq ft outside tube surface F/Btu;
A_o/A_i = ratio of outside tube surface to inside tube surface.

Table 11-1. Table of normal fouling factors.

Types of Fluid	Fouling Resistance (hr F sq ft/Btu)
Sea water below 125 F	0.0005
Sea water above 125 F	0.001
Treated boiler feed water above 125 F	0.001
East River water below 125 F	0.002–0.003
Fuel oil	0.005
Quenching oil	0.004
Alcohol vapors	0.0005
Steam, non-oil-bearing	0.0005
Industrial air	0.002
Refrigerating liquid	0.001

SOURCE: Ref. 7.

EXAMPLE 11-4. A heat exchanger (condenser) using steam from the exhaust of a turbine at a pressure of 4.0-in. Hg abs. is to be used to heat 25,000 lbs/hr of sea water ($c = 0.95$ Btu/lb F) from 60 to 110 F. The exchanger is to be sized for one shell pass and four tube passes with 20 parallel tube circuits of 0.995-in. ID and 1.125-in. OD brass tubing ($k = 60$ Btu/hr ft F). For the clean exchanger the average heat-transfer coefficients at the steam and water sides are estimated to be 600 and 300 Btu/hr ft^2 F, respectively. Calculate the tube length required for long-term service.

Solution: At 4.0-in. Hg abs. the temperature of condensing steam will be 125.4 F so that the required effectiveness of the exchanger is

$$\epsilon = \frac{T_{h\,\text{out}} - T_{c\,\text{in}}}{T_{h\,\text{in}} - T_{c\,\text{in}}} = \frac{110 - 60}{125.4 - 60} = 0.765$$

For a condenser $C_{\min}/C_{\max} = 0$ and from Fig. 11-18, NTU $= 1.4$. The fouling factors from Table 11-1 are 0.0005 for both sides of the tubes. The overall design heat-transfer coefficient per unit outside area of tube is from Eq. 11-22

$$U_d = \frac{1}{\dfrac{1}{600} + 0.005 + \dfrac{1.125}{2 \times 12 \times 60}\,\text{hr}\,\dfrac{1.125}{0.995} + \dfrac{0.005 \times 1.125}{0.995} + \dfrac{1.125}{300 \times 0.995}}$$

$$= 152 \text{ Btu/hr ft}^2\text{ F}$$

The total area Ao is $20\pi\, Do\, L$ and, since $U_d Ao/C_{\min} = 1.4$, the length of the tube is

$$L = \frac{1.4 \times 25{,}000 \times 0.95 \times 12}{20 \times \pi \times 1.125 \times 152} = 37 \text{ ft}$$

11-6. Analysis for variable heat-transfer coefficient

In Chapters 8 and 9 it was shown that the value of the convection-heat-transfer coefficient depends on the physical properties of the fluid which change with temperature. In flow through short tubes, especially in laminar flow, the heat-transfer coefficient varies also with the distance from the inlet. In the analysis of heat-exchanger performance in this chapter we have assumed so far that the value of the heat-transfer coefficient is uniform. This assumption is satisfactory for most situations if the tubes are long and if the physical properties are evaluated at the mean temperature between inlet and outlet. In some cases, however, the physical properties can vary substantially, e. g. with polimers or viscous oils, or heat exchangers can be so short that the heat-transfer coefficient cannot be assumed uniform. In this circumstance the thermal analysis must be performed numerically on a finite difference basis, preferably with a computer using the technique in Chapter 2.

To illustrate the application of numerical methods in heat-exchanger design, consider the simple tube-within-a-tube exchanger shown in Fig. 11-1. The exchanger is first divided into finite sections, each having a length Δx_i and internal surface area ΔA_i. For this incremental surface area the temperatures of the hot and the cold fluid are taken at their values at the inlet to ΔA_i, i. e. at T_{hi} and T_{ci} respectively. Then the overall heat-transfer coefficient can be expressed as a function of these fluid temperatures and the location from the inlet, or

$$U_i = U_i(T_{hi}, T_{ci}, A_i)$$

According to Eq. 11-2, for parallel flow, the rate of heat transfer over section ΔA_i in finite difference form is

$$\begin{aligned}\Delta q_i &= -(mc_p)_{h,i}(T_{h,i+1} - T_{h,i}) = (mc_p)_{c,i}(T_{c,i+1} - T_{c,i}) \\ &= U_i \Delta A_i (T_h - T_c)_i \end{aligned} \tag{11-23}$$

The change in the temperature difference between the hot and the cold fluid in section ΔA_i, $[(T_h - T_c)_{i+1} - (T_h - T_c)_i]$, can then be written in the finite difference form as

$$(T_h - T_c)_{i+1} - (T_h - T_c)_i = -\Delta q_i \left[\frac{1}{(mc_p)_{h,i}} + \frac{1}{(mc_p)_{c,i}}\right] \tag{11-24}$$

or

$$(T_h - T_c)_{i+1} - (T_h - T_c)_i$$
$$= -U_i(T_c, T_h, A_i)(T_h - T_c)_i \left[\frac{1}{(mc_p)_{h,i}} + \frac{1}{(mc_p)_{c,i}}\right] \Delta A_i$$

Letting $U_i(T_c, T_h, A_i)\left[\frac{1}{(mc_p)_{h,i}} + \frac{1}{(mc_p)_{c,i}}\right] = K_i$, the new temperature difference $(T_h - T_c)_{i+1}$ can be related to known parameters at section i, or

$$(T_h - T_c)_{i+1} = (T_h - T_c)_i [1 - K_i \Delta A_i] \tag{11-25}$$

The numerical procedure can therefore be cast into the same form as was used in Chapter 2 for Example 2-10.

1. Select convenient values for Δx_i and corresponding ΔA_i, subdividing the exchanger into N sections.
2. Calculate U_1 and $(T_h - T_c)_1$ at the inlet for A_1.
3. Calculate Δq_1 for the first section.
4. Calculate $(T_h - T_c)$, T_h, T_c, and U for the next section.

5. Repeat for N increments of ΔA_i.
6. Evaluate the total heat-transfer from

$$q_{\text{total}} = \sum_{i=1}^{N} \Delta q_i$$

Additional information on heat-exchanger analysis with variable properties can be found in the current literature. An analysis, which also includes economic factors, is presented in Ref. 19.

11-7. Closure

In this chapter we have studied the thermal design of heat exchangers in which two fluids at different temperatures flow in spaces separated by a wall and exchange heat by convection at and conduction through the wall. Such heat exchangers, sometimes called recuperators, are by far the most common and industrially important heat-transfer devices. In addition to recuperators there are, however, two other general types of heat exchangers in use. In both of these types the hot and cold fluid streams occupy the same space, a channel with or without solid inserts. In one type, the *regenerator*, the hot and the cold fluid pass alternately over the same heat-transfer surface. In the other type, the *cooling tower*, both fluids flow through the same passage simultaneously.

In a cooling tower the transfer of heat is accompanied by simultaneous transfer of mass. The discussion of the transfer mechanism will therefore be taken up in Chapter 12 in conjunction with the principles of mass transfer.

Periodic flow regenerators have been used in practice only with gases. The regenerator consists of one or more flow passages which are partially filled either with solid pellets or with metal matrix inserts. During one part of the cycle the inserts store internal energy as the warmer fluid flows over their surfaces. During the other part of the cycle internal energy is released as the colder fluid passes through the regenerator and is heated. Thus, heat is transferred in a cyclic process. The principal advantage of the regenerator is a high heat-transfer effectiveness per unit weight and space. The major problem is to prevent leakage between the warmer and cooler fluids at elevated pressures. Regenerators have been used successfully as air preheaters in open-hearth and blast furnaces and in gas liquefication processes.

The theories of the regenerators are very difficult and involved. The reader interested in the design and operation of these units is referred to References 14 to 16 for detailed information. Reference 14 contains a summary of the design theory with particular emphasis on the exhaust-

Table 11-2. Approximate overall coefficients for preliminary estimates.

Duty	Overall Coefficient (Btu/hr sq ft F)
Steam to water	
Instantaneous heater	400–600
Storage-tank heater	175–300
Steam to oil	
Heavy fuel	10–30
Light fuel	30–60
Light petroleum distillate	50–200
Steam to aqueous solutions	100–600
Steam to gases	5–50
Water to compressed air	10–30
Water to water, jacket water coolers	150–275
Water to lubricating oil	20–60
Water to condensing oil vapors	40–100
Water to condensing alcohol	45–120
Water to condensing Freon-12	80–150
Water to condensing ammonia	150–250
Water to organic solvents, alcohol	50–150
Water to boiling Freon-12	50–150
Water to gasoline	60–90
Water to gas oil or distillate	35–60
Water to brine	100–200
Light organics to light organics	40–75
Medium organics to medium organics	20–60
Heavy organics to heavy organics	10–40
Heavy organics to light organics	10–60
Crude oil to gas oil	30–55

SOURCE: Ref. 13.

gas thermal-energy regenerator in gas turbine power plants. Reference 15 presents calculated values for the effectiveness of regenerators and Reference 16 gives a complete and detailed treatment of regenerator theory and practice.

For preliminary estimates of heat-exchanger sizes and performance parameters, it is often sufficient to know the order of magnitude of the overall transmittance under average service conditions. Typical values of overall heat-transfer coefficients recommended for preliminary estimates by Mueller (13) are given in Table 11-2.

PROBLEMS

Note. The problems marked * require the direct or indirect evaluation of heat-transfer coefficients *before* the heat exchanger can be analyzed.

11-1. In a tubular heat exchanger with two shell passes and eight tube passes, 100,000 lb/hr of water are heated in the shell from 180 to 300 F. Hot exhaust gases having roughly the same physical properties as air enter the tubes at 650 F and leave at 350 F. The total surface, based on the outer tube surface, is 10,000 sq ft. Determine (a) the log-mean temperature if the heat exchanger were a simple counterflow type, (b) the correction factor F for the actual arrangement, (c) the effectiveness of the heat exchanger, (d) the average overall heat-transfer coefficient.

*11-2.** Design (i.e., determine the overall area and a suitable arrangement of shell and tube passes) for a tubular feed-water heater capable of heating 5000 lb/hr of water from 70 to 190 F. The following specifications are given: (a) saturated steam at 134 psia is condensing on the outer tube surface, (b) unit-surface conductance on steam side is 1200 Btu/hr sq ft F, (c) tubes are of copper, 1-in.-OD, 0.9-in.-ID, 8 ft long, and (d) water velocity is 3 fps.

*11-3.** Repeat Prob. 11-2, but assume that the design should contain a safety factor to allow for scale formation on the steam side which could add an additional thermal resistance of 0.002 hr sq ft F/Btu.

*11-4.** A small space heater is constructed of $\frac{1}{2}$-in., 18-gauge brass tubes, 2 ft long. The tubes are arranged in equilateral, staggered triangles on $1\frac{1}{2}$-in. centers, four rows of 15 tubes each. A fan blows 2000 cfm of atmospheric air at 70 F uniformly over the tubes (see sketch). Estimate: (a) heat-transfer rate; (b) exit temperature of the air; (c) rate of steam condensation, assuming that saturated steam at 2 psig inside the tubes as the heat source. State your assumptions. NOTE. Work parts a, b, and c of this problem by two methods. First use the LMTD, which requires a trial-and-error or graphical solution; then use the effectiveness method. (d) Also, estimate pressure drop of the air, in inches of water; (e) size motor required to drive the fan.

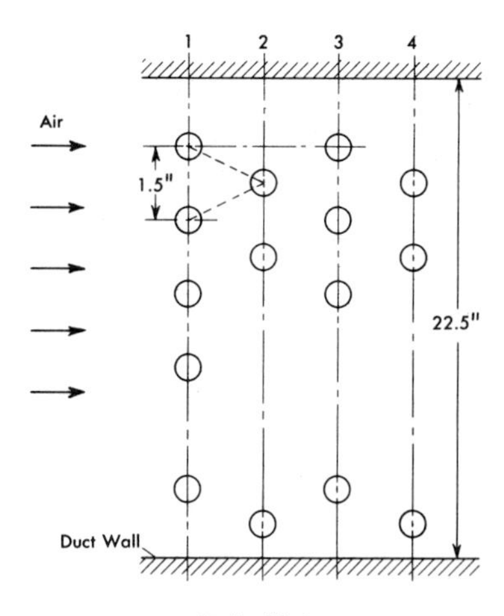

Prob. 11-4

*11-5.** Calculate the overall conductance and the rate of heat flow from the hot gases to the cold air in the crossflow tube-bank type of heat exchanger shown in the accompanying illustration for the following operating conditions:

Air flow rate = 3000 lb / hr.
Hot gas flow rate = 5000 lb / hr.
Temperature of hot gases entering exchanger = 1600 F.
Temperature of cold air entering exchanger = 100 F.
Both gases are approximately at atmospheric pressure.

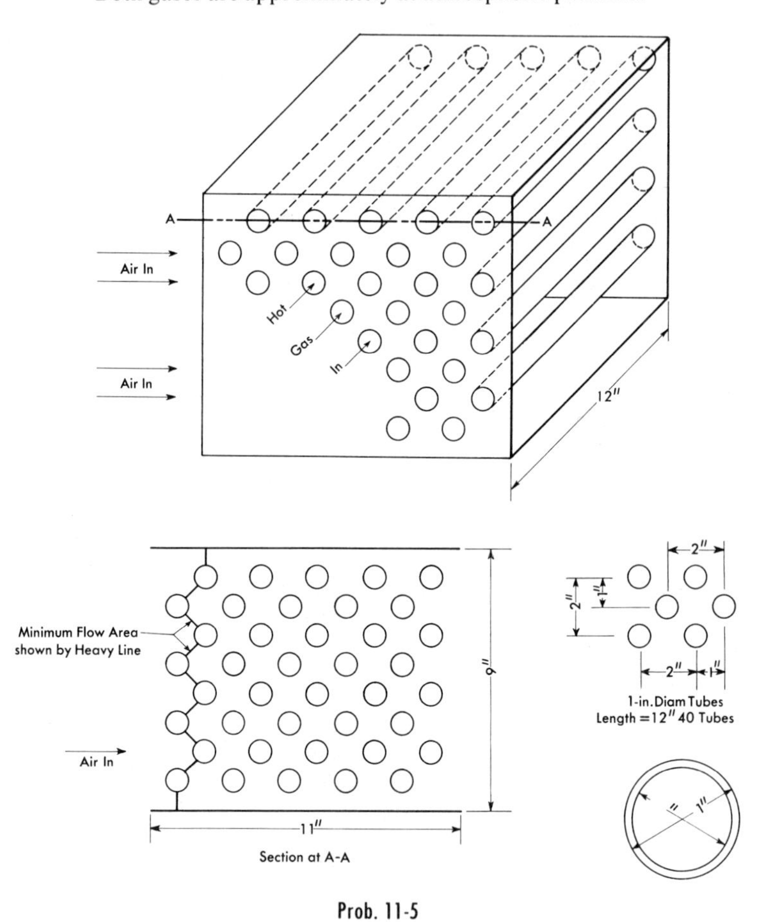

Prob. 11-5

11-6. In a single-pass counterflow heat exchanger, 10,000 lb / hr of water enter at 60 F and cool 20,000 lb / hr of an oil having a specific heat of 0.50 Btu/ lb F from 200 to 150 F. If the overall heat-transfer coefficient is 50 Btu / hr sq ft F, determine the surface area required.

11-7. Determine the outlet temperature of the oil in Prob. 11-6 for the same initial temperatures of the fluids if the flow arrangement is one shell pass and two tube passes, but with the same total area and average overall heat-transfer coefficient as the unit in Prob. 11-6.

11-8. Carbon dioxide at 800 F is to be used to heat 100,000 lb / hr of water from 100 F to 300 F while the gas temperature drops 400 F. For an overall heat-transfer coefficient of 10 Btu/hr sq ft F, compute the required area of the exchanger in square feet for (a) parallel flow, (b) counterflow, (c) a 2–4 reversed current exchanger, and (d) crossflow, gas mixed.

*11-9.** A double-pipe oil-water heat exchanger is constructed of a 10-ft-long brass tube, BWG No. 18 (0.527-in.-ID, 0.625-in.-OD), concentric within a well-insulated standard $\frac{3}{4}$-in. wrought-iron pipe (0.824 in. ID). Water flows in the annulus, entering at 60 F. A light oil flows in the tube, entering at 200 F. The flow arrangement is counterflow. For the velocities specified, determine (a) the exit temperature of the oil, (b) the rate of heat flow, and (c) the frictional pressure losses of the water and the oil.

The velocities specified are: (1) water rate = 1 gpm, oil rate = 1 gpm; (2) water rate = 10 gpm, oil rate = 1 gpm; (3) water rate = 1 gpm, oil rate = 20 gpm; (4) water rate = 10 gpm, oil rate = 20 gpm.

11-10. An economizer is to be purchased for a power plant. The unit is to be large enough to heat 60,000 lb/hr of water from 160 to 360 F. There are 100,000 lb/hr of flue gases (c_p = 0.24 Btu/lb F) available at 800 F. Estimate (a) the outlet temperature of the flue gases, (b) the heat-transfer area required for a counterflow arrangement if the overall heat-transfer coefficient is 10 Btu/hr sq ft F.

11-11. Saturated steam at 5 psi condenses on the outside of an 8.5-ft length of copper tubing heating 0.6 gpm of water flowing in the tube. The water temperatures, measured at 10 equally spaced stations along the tube length are:

Station	1	2	3	4	5	6	7	8	9	10	11
Temperature (F)	65	109	135	152	163	172	179	186	190	195	198

Calculate (a) average overall heat-transfer coefficient U_o based on the outside tube area; (b) average water-side heat-transfer coefficient $\bar{h}_w$ (assume steam-side coefficient at $\bar{h}_s$ = 2000 Btu/sq ft hr F), (c) local overall coefficient U_x based on the outside tube area for each of the 10 sections between temperature stations, and (d) local waterside coefficients h_{wx} for each of the 10 sections.

Plot all items vs. tube length. Tube dimensions: ID = 0.790 in., OD = 0.985 in., length = 8.5 ft. Temperature station 1 is at tube entrance and station 11 at tube exit.

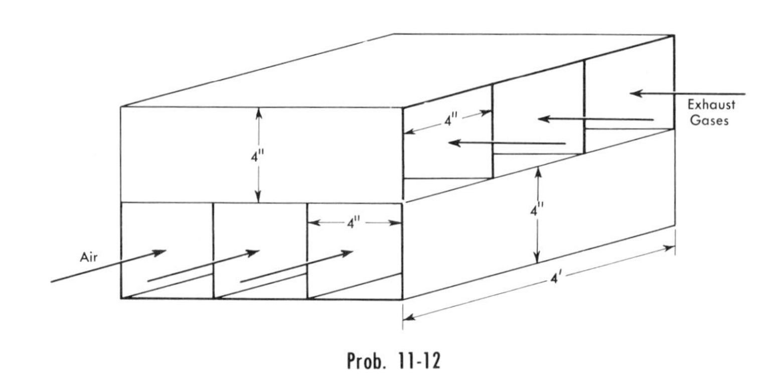

Prob. 11-12

*11-12.** A one-tube pass crossflow heat exchanger is considered for recovering energy from the exhaust gases of a turbine-driven engine. The heat exchanger is constructed of flat plates, forming an egg-crate pattern. The velocities of the entering air (50 F) and exhaust gases (800 F) are both equal to 200 fps. Assuming that the properties of the exhaust gases are the same as those of the air, estimate for a path length of 4 ft the overall heat-transfer coefficient U, neglecting the thermal resistance of the intermediate metal wall. Then determine the outlet temperature of the air, comment on the suitability of the proposed design, and if possible, suggest improvements.

11-13. Water is heated while flowing through a pipe by steam condensing on the outside of the pipe. (a) Assuming a uniform overall conductance along the pipe, derive an expression for the water temperature as a function of distance from the entrance. (b) For an overall conductance of 100 Btu/hr sq ft F, based on the inside diameter of 2 in., a steam temperature of 220 F, and a water-flow rate of 500 lb_m/min, calculate the length required to raise the water temperature from 60 F to 150 F. *Ans.* 475 ft

11-14. An oil having a specific heat of 0.50 Btu/lb_m F enters an oil cooler at 180 F at the rate of 20,000 lb_m/hr. The cooler is a counterflow unit with water as the coolant, the transfer area being 300 sq ft and the overall heat-transfer coefficient being 100 Btu/sq ft F hr. The water enters the exchanger at 80 F. Determine the water rate required if the oil is to leave the cooler at 100 F.

11-15. Steam is to be condensed at atmospheric pressure in a shell-and-tube heat exchanger consisting of 72 eight-foot lengths of standard 1-in. 18 BWG copper condenser tubing (0.902-in.-ID, 0.049 in. wall thickness). Water is available at a rate of 500,000 lb_m/hr; it flows inside the tubes, entering at 60 F. For the average unit-surface conductances listed below, estimate the pounds of steam per hour condensed. The unit-surface conductances (based on actual area) are:

Water side	800 Btu/hr sq ft F
Steam side	2000 Btu/hr sq ft F
Scale on steam side	1800 Btu/hr sq ft F
Scale of water side	2000 Btu/hr sq ft F

11-16. Show that the effectiveness for a counterflow arrangement is

$$\mathcal{E} = \frac{1 - e^{-[1-(C_{min}/C_{max})]NTU_{max}}}{1 - (C_{min}/C_{max})e^{-[1-(C_{min}/C_{max})]NTU_{max}}}$$

*11-17.** The following data were obtained with an experimental parallel-flow heat exchanger that consisted of a horizontal steel tube (0.53-in.-ID and 1.002-in.-OD) surrounded by a concentrically-arranged steel tube (1.263 in. ID) well insulated externally. A high pressure steam (2400 lb/hr) at an absolute pressure of 1643 lb/sq in. and at 821 F entered the inner tube and left with an absolute pressure of 1523 lb/sq in. and a temperature of 722 F; 953 lb/hr of low pressure steam at an absolute pressure of 189 lb/sq in. at 424 F entered the annular space and left at an absolute pressure of 122 lb/sq in. with a temperature of 744 F. Predict the heat-transfer rate if the steam flow in the inner tube were reversed, the entering temperatures remaining the same as before.

11-18. One hundred thousand lb/hr of benzene are to be cooled continuously from 180 to 130 F by 80,000 lb/hr of water available at 60 F. Using Table 11-2 estimate the surface area required for (a) crossflow, six tube passes, one-shell pass, neither of the fluids mixed; (b) reversed current exchanger, two-shell passes and eight tube passes, colder fluid inside of tubes.

*11-19.** An oil is being cooled by water in a double-pipe parallel-flow heat exchanger. The water enters the center pipe at a temperature of 60 F and is heated to 120 F. The oil which flows in the annulus is cooled from 270 to 150 F. It is proposed to cool the oil to a lower final temperature by increasing the length of the exchanger. Neglecting external heat loss from the exchanger, determine: (a) the minimum temperature to which the oil may be cooled; (b) the exit-oil temperature as a function of the fractional increase in the exchanger length; (c) the exit temperature of each stream if the existing exchanger were switched to counterflow operation; (d) the lowest temperature to which the oil could be cooled with counterflow operation; (e) the ratio of the required length for counterflow to that for parallel flow as a function of the exit-oil temperature.

Ans. (a) 130 F, (b) $270 - 140(1 - e^{5.84x})$,

$$\text{(e)}\ \frac{1}{3}\left(\ln \frac{75 + 0.5T_{h2}}{T_{h2} - 60} \Big/ \ln \frac{210}{75 + 1.5T_{h2}}\right)$$

11-20. In gas turbine recuperators the exhaust gases are used to heat the incoming air and C_{min}/C_{max} is therefore approximately equal to unity. Show that for this case $\mathcal{E} = 1 - e^{-NTU}$ for counterflow and $\mathcal{E} = \frac{1}{2}(1 - e^{-2NTU})$ for parallel flow.

11-21. In most gas turbine regenerators the heat capacity ratio C_{min}/C_{max} is near unity. For rapid cost estimates it is desirable to have performance curves for this condition showing the effectiveness, $\mathcal{E}$, as a function of the Number of Transfer Units, NTU. Prepare a series of such curves on one graph for counterflow, parallel-flow, 1 shell-2 tube passes, and unmixed crossflow. What conclusions can you draw?

11-22. It is proposed to preheat the water for a boiler with flue gases from the stack ($c_p = 0.24$ Btu/lb$_m$ F). The flue gases are available at 300 F, at the rate of 2000 lb$_m$/hr. The water entering the exchanger at 60 F at the rate of 400 lb$_m$/hr is to be heated to 200 F. The heat exchanger is to be of the reversed current type, one shell pass and 4 tube passes. The water flows inside the tubes which are made of copper (1-in.-ID, 1.25-in.-OD). The heat-transfer coefficient at the gas side is 20 Btu/hr sq ft F, while the heat-transfer coefficient on the water side is 200 Btu/hr sq ft F. A scale on the water side offers an additional thermal resistance of 0.01 hr sq ft F/Btu. (a) Determine the overall heat-transfer coefficient based on the *outer* tube diameter. (b) Determine the appropriate mean temperature difference for the heat exchanger. (c) Estimate the required tube length. (d) What would be the improvement in the effectiveness if the water flow rate would be doubled, giving an average-unit conductance of 320 Btu/hr sq ft F?

11-23. The heater arrangement of Prob. 9-6 is to heat air from 65 to 220 F with steam at 11 psig condensing inside the tubes. If the bank is 40 rows deep with 30 pipes in each row, estimate the capacity of the heater in pounds per hour of air and the pressure drop in inches of water. The pipes are 6 ft long and the free cross-sectional flow area is 60 ft. *Ans.* 83,000 lb_m/hr

11-24. At a rate of 5.43 gpm, water at 80 F enters a No. 18 BWG $\frac{5}{8}$-in. condenser tube made of nickel chromium steel (k = 15 Btu/hr ft F). The tube is 10 ft long and its outside is heated by steam condensing at 120 F. Under these conditions the average heat-transfer coefficient on the water side is 1750 Btu/hr sq ft F, and the heat-transfer coefficient on the steam side may be taken as 2000 Btu/hr sq ft F. On the interior of the tube, however, there is a scale having a thermal conductance equivalent to 1000 Btu/hr sq ft F. (a) Calculate the overall heat-transfer coefficient U per square foot of exterior surface area. (b) Calculate the exit temperature of the water. *Ans.* (a) 376; (b) 88

11-25. Dry air is cooled from 150 to 100 F, while flowing at the rate of 10,000 lb/hr in a simple adiabatic counterflow heat exchanger, by means of cold air which enters at 60 F and flows at a rate of 12,500 lb/hr. It is planned to lengthen the heat exchanger so that 10,000 lb/hr of air can be cooled from 150 to 80 F with a counterflow current of air at 12,500 lb/hr entering at 60 F. Assuming that the specific heat of the air is constant, calculate the ratio of the length of the new heat exchanger to the length of the original. *Ans.* 2.36

11-26. Water flowing at a rate of 100,000 lb/hr is to be cooled from 200 to 150 F by means of an equal flow rate of cold water entering at 100 F. The water velocity will be such that the overall coefficient of heat transfer U is 400 Btu/hr sq ft F. Calculate the square feet of heat-exchanger surface needed for each of the following arrangements: (a) parallel flow, (b) counterflow, (c) a multi-pass heat exchanger with the hot water making one pass through a well-baffled shell and the cold water making two passes through the tubes, and (d) a crossflow heat exchanger with the hot water making one pass through the shell and the cold water making one pass through the tubes. *Ans.* (b) 250; (c) 310; (d) 285

*11-27.** A shell-and-tube counterflow heat exchanger is to be designed for heating an oil from 80 to 180 F. The oil is to pass through $1\frac{1}{2}$-in. schedule 40 pipes at a velocity of 200 fpm and steam is to condense at 215 F on the outside of the pipes. The specific heat of the oil is 0.43 Btu/lb F and its mass density is 58 lb/cu ft. The steam-side heat-transfer coefficient is approximately 1800 Btu/hr sq ft, and the thermal conductivity of the metal of the tubes is 17 Btu/hr ft F. The results of previous experiments giving the oil-side heat-transfer coefficients for the same pipe size at the same oil velocity as those to be used in the exchanger are shown below:

ΔT (F)	135	115	95	75	35	
T_{oil} (F)	80	100	120	140	160	180
$\bar{h}_{ci}$ (Btu/hr sq ft F)	14	15	18	25	45	96

(a) Find the overall heat-transfer coefficient U, based on the outer surface area at the point where the oil is 100 F. (b) Find the temperature of the inside surface of the pipe when the oil temperature is 100 F. (c) Find the required length of the tube bundle. *Ans.* (a) 12.5 Btu/hr sq ft F; (b) 213 F; (c) 520 ft

11-28. A steam-heated single-pass tubular preheater is designed to raise 45,000 lb/hr of air from 70 to 170 F, using saturated steam at 20 psia. It is proposed to double the flow rate of air and, in order to be able to use the same heat exchanger and achieve the desired temperature rise, it is proposed to increase the steam pressure. Calculate the steam pressure necessary for the new conditions and comment on the design characteristics of the new arrangement. *Ans.* 25.4 psia

*11-29.** Water is to be heated from 50 to 90 F at the rate of 300,000 gal/hr in a single-pass shell-and-tube heat exchanger consisting of 1-in. schedule 40 steel pipe. The surface coefficient on the steam side is estimated to be 2000 Btu/hr sq ft F. A pump is available which can deliver the desired quantity of water provided the pressure drop through the pipes does not exceed 15 psi. Calculate the number of tubes in parallel and the length of each tube necessary to operate the heat exchanger with the available pump. *Ans.* 100 tubes of 27-ft length

*11-30.** In a shell-and-tube heat exchanger consisting of 1-in. schedule 40 steel pipe 10,000 standard cubic feet per minute of air are to be heated from 1 atm and 70 F to 205 F. On the outside of the tubes steam is condensing at 220 F. A fan is available which will deliver cold air to the entrance header at 70 F at a static pressure of 1.5 in. of water gauge. The hot air is to be discharged to atmosphere. The cross-sectional areas of the entrance and exit headers are twice the total internal cross-sectional area of the tubes. Assuming that the resistance on the steam side and the thermal resistance of the tube wall are negligible, calculate (a) the mass velocity of the air inside the tubes, (b) the heat-transfer coefficient from the air to the inside pipe wall, and (c) the number of tubes and the length of each tube.
Ans. (a) 9460 lb/hr sq ft; (b) 8.55 Btu/hr sq ft F; (c) 973 tubes of 13.4-ft length

*11-31.** A shell-and-tube heat exchanger in an ammonia plant is pre-heating 40,000 standard cubic feet of nitrogen per hour from 70 to 150 F, using steam condensing at 20 psia. The tubes in the heat exchanger have an inside diameter of 1 in. and the mass velocity of the nitrogen through the tubes is 10,000 lb/hr sq ft. In order to change from ammonia synthesis to methanol synthesis, the same heater is to be used to preheat carbon dioxide from 70 to 170 F, using steam condensing at 35 psia. Calculate the flow rate which can be anticipated from this heat exchanger in pounds of carbon dioxide per hour. *Ans.* About 5500 lb/hr

*11-32.** In an industrial plant a shell-and-tube heat exchanger is heating dirty water at the rate of 300,000 lb/hr from 140 to 235 F by means of steam condensing at 240 F on the outside of the tubes. The heat exchanger has 500 steel tubes (ID = 0.052 ft, OD = 0.700 ft) in a tube bundle which is 30 ft long. The water flows through the tubes while the steam condenses in the shell. If it may be assumed that the thermal resistance of the scale on the inside pipe wall is unaltered when the mass rate of flow is increased and that changes in water properties with temperature are negligible, estimate (a) the heat-transfer coefficient on the water side and (b) the exit temperature of the dirty water if its mass rate of flow is doubled. *Ans.* (a) About 600 Btu/hr sq ft F; (b) 226 F

*11-33.** A large surface condenser in a power plant was tested at three water velocities when new (i.e., with clean tubes), and again after considerable service. The results are tabulated below as overall coefficients of heat transfer, based on the outside surface of the tubes versus water velocity. The tubes have inside and outside diameters of 1.00 and 0.902 in. respectively, and the metal of which they are made has a thermal conductivity of 63 Btu/hr ft F.

Condition of Tubes	Clean			Dirty		
Water velocity (fps)	2.0	4.0	8.0	2.0	4.0	8.0
Overall heat-transfer coefficient (Btu/hr sq ft F)	357	550	795	293	410	534

Plot the data according to the Wilson method (see Prob. 8-15) and determine the following individual coefficients of heat transfer in Btu/hr sq ft F: (a) value of $\bar{h}_c$ on the steam side of the clean condenser, based on the *outside* surface; (b) value of $\bar{h}_c = k/x$ for the scale and slime deposited in the dirty tube, based on the *inside* surface; (c) value of $\bar{h}_c$ on the water side at a water velocity of 4 fps. (d) What overall coefficient would you expect with dirty tubes at a water velocity of 25 fps?

Ans. (a) 2300; (b) 1800; (c) 820; (d) 700 Btu/hr sq ft F

*11-34.** *N*-butyl alcohol is to be heated from 50 to 150 F at the rate of 33,000 lb/hr while passing through 1-in. 18 BWG standard copper condenser tubes on which steam at 250 F is condensing. The exchanger is to be designed for a pressure drop of 1 psi in the tubes. Neglecting the thermal resistance on the condensing steam side and the thermal resistance of the tubes, calculate (a) the required velocity through the tubes in fps, (b) the number of tubes in the tube bundle, and (c) the length of the tube bundle. *Ans.* (a) 4.2 fps; (b) 98 tubes; (c) 29 ft

*11-35.** Liquid benzene (specific gravity = 0.86) is to be heated in a counterflow concentric-pipe heat exchanger from 90 to 190 F. For a tentative design, the velocity of the benzene through the inside pipe (1-in., schedule 40) can be taken as 8 fps. *Saturated process steam at 200 psia* is available for heating. Two methods of using this steam are proposed: (a) Pass the process steam directly through the annular pipe of the exchanger; this would require that the latter be designed for the high pressure. (b) Throttle the steam adiabatically to 20 psia before passing it through the heater. In both cases the operation would be controlled so that *saturated water leaves the heater.* As an approximation, assume that for both cases the film coefficient for *condensing* steam remains constant at 2250 Btu/hr sq ft F, that the thermal resistance of the pipe wall is negligible, and that the pressure drop for the steam is negligible. If the inside diameter of the outer pipe is 2 in., calculate the mass rate of flow of steam (lb/hr per pipe) and the length of heater required for each arrangement.

Ans. (a) 485 lb/hr per pipe, 14.7 ft; (b) 407 lb/hr per pipe, 45 ft

11-36. An air cooler, comprising a tube bundle of 1-in. schedule 40 iron pipe enclosed in a well-baffled shell, is being built to cool air at a rate of 45,000 lb/hr from 200 to 90 F. The air flows in a single pass through the 1-in. pipe. Cooling water at 80 F, under sufficient pressure to force it through at a desired rate, flows counter-currently through the shell. The air flows inside the pipes at a mass velocity of 8600 lb/hr sq ft, which results in an overall coefficient U_i (based on the inside area) of 7.8 Btu/hr sq ft F. Cooling water costs $.20/1000 cu ft and the fixed charges on the cooler are $.50/sq ft of inside heating surface per year. It is proposed to operate 8400 hr per year. For the lowest total yearly cost, calculate (a) pounds of cooling water per pound of air, (b) the temperature difference at the hot end, (c) the length of each tube and the number of tubes in parallel, and (d) the total cost in cents per million Btu transferred.

Ans. (a) 0.66; (b) 80 F; (c) 18.7 ft; (d) $.31 per million Btu

REFERENCES

1. W. M. Kays and A. L. London, *Compact Heat Exchangers.* (Palo Alto, Calif.: National Press, 1955; 2nd ed., McGraw-Hill Book Company, New York, 1964.)

2. W. M. Kays, A. L. London, and D. W. Johnson, "Gas Turbine Plant Heat Exchangers," *ASME Research Publication*, April, 1951.

3. W. M. Kays and A. L. London, "Remarks on the Behavior and Application of Compact High-Performance Heat Transfer Surfaces," Inst. Mech. Eng. and ASME, *Proc. General Discussion on Heat Transfer*, 1951, pp. 127–132.

4. L. M. K. Boelter, R. C. Martinelli, F. E. Romie, and E. H. Morrin, "An Investigation of Aircraft Heaters XVIII—A Design Manual for Exhaust Gas and Air Heat Exchangers," *NACA Wartime Report*, ARR 5 AO6, August, 1945.

5. A. L. London and W. M. Kays, "The Gas Turbine Regenerator—the Use of Compact Heat Transfer Surfaces," *Trans. ASME*, Vol. 72 (1950), p. 611.

6. R. A. Bowman, A. C. Mueller, and W. M. Nagle, "Mean Temperature Difference in Design," *Trans. ASME*, Vol. 62 (1940), pp. 283–294.

7. Tubular Exchanger Manufacturers Association, *Standards TEMA*/ 3d ed. (New York, 1952.)

8. W. Nusselt, "A New Heat Transfer Formula for Cross-Flow," *Technische Mechanik and Thermodynamik*, Vol. 12 (1930).

9. H. Ten Broeck, "Multipass Exchanger Calculations," *Ind. Eng. Chem.*, Vol. 30 (1938), pp. 1041–1042.

10. K. A. Gardner, "Efficiency of Extended Surface," *Trans. ASME*, Vol. 67 (1945), pp. 621–631.

11. W. P. Harper and D. R. Brown, "Mathematical Equations for Heat Conduction in the Fins of Air Cooled Engines," *NACA Report* 158, 1922.

12. Townsend Tinker, "Shell Side Characteristics of Shell and Tube Heat Exchangers," Inst. Mech. Eng. and ASME, *Proc. General Discussion on Heat Transfer*, 1951, pp. 89–116.

13. A. C. Mueller, "Thermal Design of Shell-and-Tube Heat Exchangers for Liquid-to-Liquid Heat Transfer," *Eng. Bull.*, Res. Ser. 121, Purdue Univ. Eng. Exp. Sta., 1954.

14. J. E. Coppage and A. L. London, "The Periodic-Flow Regenerator—A Summary of Design Theory," *Trans. ASME*, Vol. 75, 1953, pp. 779–787.

15. T. J. Lambertson, "Performance Factors of a Periodic-Flow Heat Exchanger," M. S. Thesis, USN Postgraduate School, Monterey, Calif., 1957, also *ASME* Paper No. 57-SA-13, 1957.

16. M. Jakob, *Heat Transfer*, Vol. 2. (New York: John Wiley & Sons, Inc., 1957.)

17. R. Gregori, *Wärmeaustäuscher*, Aarau u. Frankfurt, a. M., 1959.

18. H. Hauser, *Wärmeübertragung im Gegenstrom, Gleichstrom, und Kreuzstrom*, Berlin/Göttingen/Heidelberg, 1950.

19. B. Gebhart, *Heat Transfer*, 2nd ed., McGraw Hill Book Co., New York, 1971.

20. R. L. Webb and E. R. G. Eckert, "Application of Rough Surfaces to Heat-exchanger Design," Int. J. Heat and Mass Transf., Vol. 15, 1972, pp. 1647–1658.

21. J. Taborek, T. Aoki, R. B. Ritter, J. W. Palen, and J. G. Knudsen, "Fouling: The Major Unresolved Problem in Heat Transfer," (Parts I and II), Chem. Eng. Prog., Vol. 68, 1972, No. 2, pp. 59–67 and No. 7, pp. 69–78.

12
Mass transfer

By L. Bryce Andersen

12-1. Introduction

The transport of one constituent of a fluid solution from a region of higher concentration to a region of lower concentration is called mass transfer. The mechanism of mass transfer can be most readily understood by drawing an analogy to heat transfer. Heat is transferred in a direction which reduces an existing concentration gradient; mass is transferred in a direction which reduces an existing concentration gradient. Heat transfer ceases when there is no longer a temperature difference; mass transfer ceases when the concentration gradient is reduced to zero. The rates of both heat and mass transfer depend on a driving potential and a resistance. Other similarities between heat and mass transfer will be discussed in connection with mass transfer theory. This chapter will develop the basic concepts of mass transfer and apply these concepts to a few typical problems. The detailed design of industrial mass transfer equipment will not be considered. Those readers who wish to pursue the subject of mass transfer further are referred to the references listed at the end of this chapter.

Mass transfer may occur either within the gas phase or within the liquid phase. In many chemical engineering unit operations, transfer of mass takes place between two different phases. In *gas absorption*, a soluble gas is removed from a gaseous mixture with an insoluble gas by transfer to a liquid phase. In *adsorption*, one constituent of a fluid phase is transferred to the surface of a solid adsorbent. In *distillation*, mass transfer takes place simultaneously in two directions: from the liquid to the vapor, and vice versa. The net effect is to increase the concentration of the more volatile constituent in the vapor phase and to deplete the liquid phase. *Liquid extraction* involves the transfer of a constituent from one liquid phase to another liquid phase. The two liquid phases must be immiscible to some extent, or no separation is possible. *Leaching*, or solid-liquid extraction, is an operation in which the soluble component of a solid phase is dissolved and transferred to a liquid solvent. An everyday example of leaching is the making of coffee, where the soluble component of the ground coffee is dissolved out by a hot-water phase.

In certain mass transfer operations, simultaneous heat transfer must be considered. For example, *humidification* is an operation in which a pure liquid is evaporated into a bulk gas phase. In the humidification of air, water is transferred from the liquid phase to the bulk air phase. Sufficient energy to supply the latent heat of vaporization of the water must be provided. This energy can be supplied by transferring heat from the gas to the liquid. Under this condition heat is transferred in a direction opposite to that of the mass transfer. Thermal effects are often important in distillation, since the liquid is continually vaporized and the vapor continually condensed. Other common mass-transfer operations are drying, evaporation, and condensation.

The mechanism of mass transfer, just as that of heat transfer, depends largely on the dynamics of the fluid phases. Mass can be transferred not only by random molecular motion in quiescent or laminar-flowing fluids, but also by eddy currents through fluids in turbulent motion. The former is analogous to conduction heat transfer, the latter, to convection. Before interphase mass transfer is considered, molecular and turbulent mass diffusion will be discussed.

12-2. Mass transfer by molecular diffusion

Mass transfer by molecular diffusion is directly analogous to conduction heat transfer or to momentum transfer in laminar flow. Mass transfer by molecular diffusion may occur in a stagnant fluid or in a fluid in laminar flow. The transient one-dimensional mass-transfer equation can be written in a form identical to the Fourier heat-transfer equation,

$$\frac{\partial c_A}{\partial \theta} = D_v \frac{\partial^2 c_A}{\partial y^2} \tag{12-1}$$

where c_A = concentration of component A in a mixture of A and B, in lb-moles/cu ft;

θ = time, in hr;

D_v = mass diffusivity, in sq ft / hr;
y = distance in the direction of diffusion, in ft.

In the steady state the concentration at any point does not vary with time, and

$$\frac{N_A}{A} = -D_v \frac{dc_A}{dy} \quad (12\text{-}2)$$

where N_A/A is the mass flux in lb-moles/hr sq ft. The negative sign appears because the concentration gradient is negative in the direction of mass transfer.

The equivalent expression for heat transfer is

$$\frac{q}{A} = -k\frac{dT}{dy} = -\frac{k}{c_p\rho}\frac{d(c_p\rho T)}{dy} = -a\frac{d(c_p\rho T)}{dy} \quad (12\text{-}3)$$

where q/A = heat flux, in Btu/hr sq ft;
k = thermal conductivity, in Btu/hr ft F;
c_p = heat capacity, in Btu/lb_m;
ρ = density, in lb_m/cu ft;
T = temperature, in F;
a = thermal diffusivity, in sq ft/hr.

Similarly, the equation for momentum transfer in laminar flow is

$$\tau g_c = -\mu\frac{du}{dy} = -\frac{\mu}{\rho}\frac{d(u\rho)}{dy} = -\nu\frac{d(u\rho)}{dy} \quad (12\text{-}4)$$

where τg_c = momentum flux, in ft lb_m/hr sq ft hr;
μ = absolute viscosity, in lb_m/ft hr;
u = velocity in the x-direction;
ν = momentum diffusivity (kinematic viscosity), in sq ft/hr.

An examination of Eqs. 12-2, 12-3, and 12-4 shows that they are all of the form: Flux = diffusivity × concentration gradient. Equation 12-2 is written for a mass concentration, c_A, Eq. 12-3 for a thermal concentration, $c_p\rho T$, and Eq. 12-4 for a momentum concentration, $u\rho$. All three diffusivities, D_v, a, ν, have the same dimensions and the concentration gradients are linear for a uniform medium at steady state.

Equation 12-2 states that mass will be transferred between two points in a fluid if a difference in concentration exists between the points. Mass transfer occurs at an appreciable rate only in gases and liquids. In solids, mass transfer is suppressed by the relative immobility of the molecules.

In the gas phase, concentrations are usually expressed as partial pressures. If the perfect gas law,

$$p_A = \frac{n_A \mathcal{R} T}{V} = c_A \mathcal{R} T \tag{12-5}$$

where p_A = partial pressure of gas A in a mixture, in atm;
n_A = number of moles of gas, in lb-moles;
$\mathcal{R}$ = gas constant, in cu ft atm / lb mole F;
V = gas volume, in cu ft;

is assumed to hold, Eq. 12-2 becomes

$$\frac{N_A}{A} = \frac{-D_v}{\mathcal{R} T} \frac{dp_A}{dy} \tag{12-6}$$

Integration of Eq. 12-6 between any two planes in the fluid gives

$$\frac{N_A}{A} = \frac{-D_v(p_{A_2} - p_{A_1})}{\mathcal{R} T(y_2 - y_1)} \tag{12-7}$$

where p_{A_1} is the partial pressure at y_1 and p_{A_2} is the partial pressure at y_2. Equation 12-7 is rigorously correct only for equimolar counterdiffusion. In equimolar counterdiffusion gases A and B diffuse simultaneously in opposite directions through each other. The rates of diffusion are equal but in opposite directions, i.e., $N_A = -N_B$. This situation has no counter-

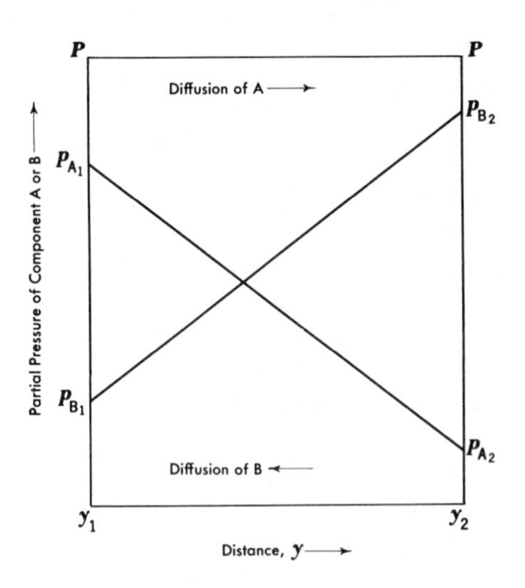

Fig. 12-1. Partial pressure gradients in equimolar counterdiffusion of two gases.

part in heat transfer, since heat can be transferred only in one direction at a time. Of course, gas B will be transferred only if a concentration gradient for B exists. This is shown schematically in Fig. 12-1. For equimolar counterdiffusion the partial pressure gradients must be equal but of opposite sign. Diffusion of this type can occur in distillation.

Diffusion of a gas through a second stationary gas often occurs in industrial mass-transfer equipment. For example, in the humidification of air, water vapor must diffuse from the air-water interface through an air layer which is stationary. Conversely, in the dehumidification of air, water vapor must diffuse from the bulk of the gas phase through stationary air to reach the surface at which it condenses.

Consider the case of gas A diffusing through a stationary gas B to a gas-liquid interface where gas A is absorbed but gas B is not (Fig. 12-2). Since gas A is diffusing toward the interface, there must be a partial-pressure gradient for A in the direction of diffusion. The rate of transfer of A is given by Eq. 12-6

$$\frac{N_A}{A} = \frac{-D_v}{\mathcal{R}T}\frac{dp_A}{dy}$$

Since there is a continuous gas phase, the total pressure P must be constant throughout the gas. Since $p_A + p_B = P$, a gradient in p_A will cause a gradient of p_B in the opposite direction. This gradient will force diffusion of gas B away from the interface, at the rate

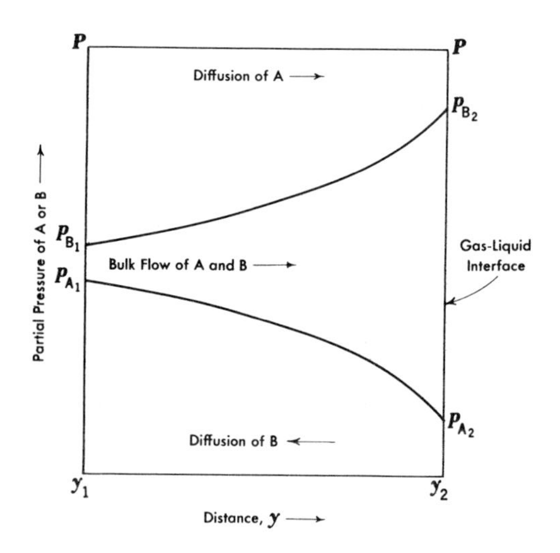

Fig. 12-2. Partial pressure gradients in the diffusion of a gas through a stationary gas.

$$\frac{N_B}{A} = \frac{-D_v}{\mathcal{R}T}\frac{dp_B}{dy} = \frac{D_v}{\mathcal{R}T}\frac{dp_A}{dy} \tag{12-8}$$

since $dp_B/dy = -dp_A/dy$. Since gas B is not being produced at the interface, even though it is diffusing away from the interface, some other mechanism must supply gas B to maintain a constant concentration of gas B at the interface. A bulk flow of gas toward the interface replenishes the gas B which is diffusing away. The bulk flow will consist of a mixture of A and B. The bulk flow of B toward the interface must equal $-N_B/A$ to balance the diffusion of B in the opposite direction. The presence of A in the bulk flow will effectively increase the rate of transfer of A toward the interface. Since the bulk flow rate of B toward the interface equals $-N_B/A$, the bulk flow rate of A toward the interface equals

$$\frac{\text{Moles } A \text{ in bulk flow}}{\text{Moles } B \text{ in bulk flow}} \times \text{bulk flow of } B = \frac{p_A}{p_B}\left(\frac{-N_B}{A}\right) = \frac{p_A}{P - p_A}\left(\frac{-N_B}{A}\right)$$

The total bulk flow rate equals the sum of bulk flow rates of A and B

$$\frac{-N_B}{A}\left(1 + \frac{p_A}{P - p_A}\right)$$

The total flux of A toward the interface is the sum of the diffusion of A and the bulk flow of A, or

$$\frac{N_{A_t}}{A} = \frac{-D_v}{\mathcal{R}T}\frac{dp_A}{dy} + \frac{p_A}{P - p_A}\left(\frac{-N_B}{A}\right) \tag{12-9}$$

Substitution for N_B from Eq. 12-8 yields

$$\frac{N_{A_t}}{A} = \frac{-D_v}{\mathcal{R}T}\left(1 + \frac{p_A}{P - p_A}\right)\frac{dp_A}{dy} \tag{12-10}$$

Integration gives

$$\frac{N_{A_t}}{A} = \frac{D_v}{\mathcal{R}T}\frac{P}{y_2 - y_1}\ln\frac{P - p_{A_2}}{P - p_{A_1}} \tag{12-11}$$

but since $p_B = P - p_A$,

$$\frac{N_{A_t}}{A} = \frac{D_v P}{\mathcal{R}T(y_2 - y_1)}\ln\frac{p_{B_2}}{p_{B_1}} \tag{12-12}$$

The definition of the logarithmic mean partial pressure of B is

$$p_{Bm} = \frac{p_{B_2} - p_{B_1}}{\ln \dfrac{p_{B_2}}{p_{B_1}}} \tag{12-13}$$

Since $p_{B_2} = P - p_{A_2}$ and $p_{B_1} = P - p_{A_1}$,

$$p_{B_2} - p_{B_1} = p_{A_1} - p_{A_2} \tag{12-14}$$

Combination of Eqs. 12-12, 12-13, and 12-14 yields

$$\frac{N_{A_t}}{A} = \frac{-D_v P(p_{A_2} - p_{A_1})}{\mathcal{R} T p_{Bm}(y_2 - y_1)} \tag{12-15}$$

Comparison of Eq. 12-15 with Eq. 12-7 shows that the factor P/p_{Bm} is introduced when diffusion through a stationary gas is considered. For a dilute mixture of A in B, p_{Bm} is approximately equal to P and Eq. 12-15 reduced to Eq. 12-7. It should be noted that the "stationary" characteristic of B does not imply that B is not moving, but refers to the *net* behavior of B. Since B is supplied by bulk flow at the same rate it diffuses away, there is no *net* movement of B. The partial-pressure gradients for diffusion through a stationary gas are not linear with distance, contrasted to the linear gradients in equimolar counterdiffusion.

Mass diffusivities must be evaluated experimentally. Selected values for gases and liquids are given in Table 12-1. The coefficients for gases

Table 12-1. Mass diffusivities for gases and liquids.

System	Diffusivity (sq ft/hr)
Gases at 77 F, 1 atm	
Ammonia-air	1.08
Water vapor-air	0.99
Ethanol-air	0.46
CO_2-air	0.64
O_2-air	0.80
H_2-air	1.60
Benzene-air	0.34
Liquid Phase at 68 F, Dilute Solutions	
Oxygen in water	7.0×10^{-5}
Ammonia in water	6.8×10^{-5}
Ethanol in water	3.8×10^{-5}
CO_2 in water	6.9×10^{-5}
H_2 in water	20.0×10^{-5}
HCl in water	10.0×10^{-5}
Sucrose in water	1.8×10^{-5}
NaCl in water	5.3×10^{-5}
CO_2 in ethanol	13.2×10^{-5}

in Table 12-1 are for either component diffusing through the other. The liquid-phase diffusivities are several orders of magnitude smaller than the gaseous diffusivities. This is due to the smaller molecular mobilities in liquids. Diffusivities for systems where no direct experimental data are available may be predicted by semi-empirical equations (see References 2, 6, and 7). Diffusion coefficients for gases and vapors vary approximately with the 3/2 power of the absolute temperature and inversely with the total pressure.

EXAMPLE 12-1. Calculate the rate of diffusion of water vapor from a pool of water at the bottom of a 20-ft well to dry air flowing over the top of the well. Assume the air in the well is stagnant and that the entire system is at 77 F and 1 atm.

Solution: This is the case of a gas diffusing through a second stationary gas. The bottom of the well is taken as point 1 and the top as point 2, and Eq. 13-15 is applied. Air is nearly insoluble in water, so diffusion of air into water can be ignored. The diffusivity of water vapor in air is taken from Table 12-1: D_v = 0.99 sq ft/hr. The partical pressure of water vapor at the water surface at the bottom of the well is equal to the saturated vapor pressure of water at 77 F. Therefore, from vapor pressure tables, $p_{A_1} = 0.031$ atm. Since the air at the top of the well is dry, $p_{A_2} = 0$. The gas constant $\mathcal{R}$ is 0.730 cu ft atm / lb-mole *R*.

$$p_{B_1} = P - p_{A_1} = 0.969 \qquad p_{B_2} = P - p_{A_2} = 1.0$$

$$p_{Bm} = \frac{p_{B_2} - p_{B_1}}{\ln p_{B_2}/p_{B_1}} = \frac{1 - 0.969}{\ln \dfrac{1}{0.969}} = 0.983$$

$$\frac{N_A}{A} = \frac{-(0.99)(1)(0 - 0.031)}{(0.730)(460 + 77)(0.983)(20 - 0)}$$

$$= 3.99 \times 10^{-6} \text{ lb moles/hr sq ft of well cross section} \qquad \textit{Ans.}$$

Since the water vapor partial pressure is small, Eq. 12-7 may be used as an approximation. It yields $N_A/A = 4.03 \times 10^{-6}$ lb moles/hr sq ft. The difference is only about 1 percent.

12-3. *Mass transfer by convection*

The mechanism of mass transfer in turbulent flow is similar to that of heat transfer in turbulent flow. Consider, for example, an air stream flowing over the surface of a pool of water. The velocity distribution in the air is the same as for flow over a plate (Chapter 6). Near the surface there is a laminar sublayer, followed by a buffer layer, and a turbulent main stream (Fig. 12-3). The rate of mass transfer of water vapor to the

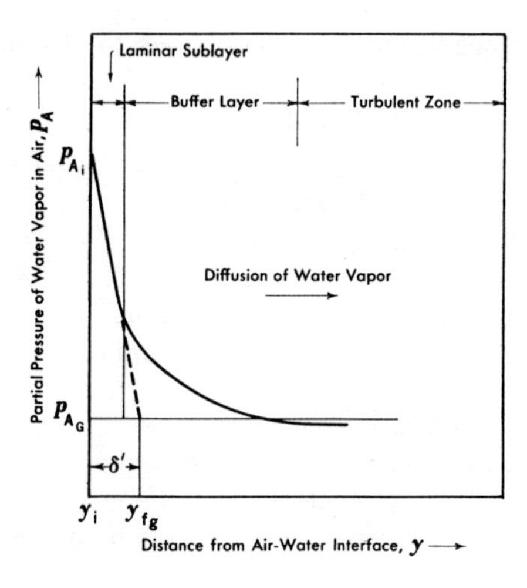

Fig. 12-3. Steady-state concentration gradient of water vapor in air flowing over a horizontal water surface.

air is given by the equation

$$\frac{N_A}{A} = \frac{-D_v P}{\mathcal{R} T p_{Bm}} \frac{(p_{A_G} - p_{A_i})}{(y_{fg} - y_i)} \tag{12-16}$$

where p_{A_G} = average partial pressure of the water vapor in the bulk gas phase (Fig. 12-3);
p_{A_i} = partial pressure of water vapor at the gas-liquid interface;
$y_{fg} - y_i$ = effective boundary-layer thickness for mass transfer, δ'_m, discussed previously in Sec. 6-3. Its resistance to molecular diffusion is the same as that offered to total diffusion by the sublayer, buffer layer, and turbulent region combined.

While simple theory predicts that the effective film thicknesses for mass and heat transfer should be the same, the experimental data available show that this is only an approximation.

Since the effective film thickness cannot be measured directly, Eq. 12-9 is rewritten

$$\frac{N_A}{A} = k_G(p_{A_i} - p_{A_G}) \tag{12-17}$$

where k_G is the gas-phase mass-transfer coefficient, defined by

$$k_G = \frac{D_v P}{\mathcal{R} T p_{Bm}(y_{fg} - y_i)} \text{ lb moles/hr sq ft atm} \tag{12-18}$$

The mass transfer coefficient k_G is analogous to the heat-transfer coefficient h_c.

For mass transfer in the liquid phase

$$\frac{N_A}{A} = \frac{-D_v c_t (c_{A_i} - c_{AL})}{c_{Bm}(y_i - y_{fl})} = k_L(c_{Ai} - c_{AL}) \tag{12-19}$$

where c_{A_i} = concentration of the diffusing component at the interface;
C_{A_L} = concentration of the diffusing component in the bulk liquid phase;
c_t = total concentration $(c_A + c_B)$;
c_{B_m} = log-mean concentration of component B;
$y_{fl} - y_i$ = thickness of the effective liquid film;
k_L = liquid-phase mass-transfer coefficient; defined by

$$k_L = \frac{D_v c_t}{c_{B_m}(y_{fl} - y_i)} \text{ lb moles/hr sq ft (lb mole/cu ft)} \tag{12-20}$$

The mass-transfer coefficients defined by Eqs. 12-18 and 12-20 apply to diffusion of one component through a second *stationary* component. Coefficients for equimolar counterdiffusion may be obtained similarly (see Prob. 12-9). In the humidification of air there is no resistance to diffusion of water in the liquid phase, since only water is present. Therefore, the liquid-phase mass-transfer coefficient is infinite, and only the gas-phase resistance need be considered.

12-4. Evaluation of mass-transfer coefficients

Mass-transfer coefficients must be evaluated experimentally. Where direct experimental data are lacking, empirical equations are available for predicting coefficients. These equations are quite similar to the equations derived for predicting heat-transfer coefficients. One would expect the mass-transfer equation to vary with the fluid properties and the physical characteristics of the equipment through which the fluid phases are flowing.

An equation relating the mass-transfer coefficient to the properties of the system may be derived by dimensional analysis. The coefficient will be a function of the velocity, density, viscosity, and mass diffusivity of the fluid, and some characteristic dimension, L, of the system, or

$$k_L = \phi(V, \rho, \mu, D_v, L) \tag{12-21}$$

Dimensional analysis yields the dimensionless equation

$$\frac{k_L L}{D_v} = \phi\left(\frac{LV\rho}{\mu}\right)\psi\left(\frac{\mu}{\rho D_v}\right) \tag{12-22a}$$

Equation 12-22a applies to diffusion through a stationary component if the ratio c_{B_m}/c_t is inserted, i.e.,

$$\frac{k_L L c_{B_m}}{D_v c_t} = \phi\left(\frac{LV\rho}{\mu}\right)\psi\left(\frac{\mu}{\rho D_v}\right) \tag{12-22b}$$

For the gas-phase mass-transfer coefficient, Eq. 12-22b becomes

$$\frac{k_G \mathcal{R} T p_{Bm} L}{D_v P} = \phi\left(\frac{LV\rho}{\mu}\right)\psi\left(\frac{\mu}{\rho D_v}\right) \tag{12-23}$$

The group on the left of Eq. 12-23 is the *Sherwood number*, Sh, which is analogous to the Nusselt number used in heat transfer. The functions ϕ and ψ are usually taken so that the dimensionless groups are related exponentially,in the same form as Eq. 8-2 for heat transfer.

From an equivalent analysis for heat transfer by convection it was found that

$$\frac{\bar{h}_c L}{k} = \phi\left(\frac{LV\rho}{\mu}\right)\psi\left(\frac{c_p \mu}{k}\right) \tag{8-2}$$

The heat-transfer coefficient is dependent on the Reynolds number, which characterizes flow conditions, and the Prandtl number, which is the ratio of the momentum diffusivity to the thermal diffusivity. Therefore, one would expect the mass-transfer coefficient to depend on the Reynolds number as well as on a dimensionless ratio of the momentum diffusivity and the mass diffusivity. This latter group, $\mu/\rho D_v$, is called the *Schmidt number* Sc. The Schmidt number characterizes mass transfer in the same manner as the Prandtl number characterizes heat transfer.

Since the mechanisms of mass, heat, and momentum transfer are closely related, one might expect data taken for one transfer operation to be useful in predicting the rate of transfer in the other operations. The interrelation of heat and momentum transfer was discussed earlier in connection with the Reynolds analogy, where

$$\frac{\mathrm{Nu}}{\mathrm{RePr}} = \frac{f}{2} = \frac{\bar{h}_c}{c_p \rho V} \tag{12-24}$$

A similar analysis for mass transfer yields

$$\frac{\text{Sh}}{\text{ReSc}} = \frac{f}{2} = \frac{k_L c_{B_m}}{V c_t} = \frac{k_G \mathcal{R} T p_{B_m}}{VP} \tag{12-25}$$

Combining Eq. 12-24 with Eq. 12-25 gives

$$\frac{\bar{h}_c}{c_p \rho V} = \frac{k_G \mathcal{R} T p_{B_m}}{VP}$$

or

$$k_G = \frac{\bar{h}_c}{c_p \rho} \frac{P}{\mathcal{R} T p_{B_m}} \tag{12-26}$$

Equation 12-26 is not generally applicable for prediction of mass-transfer coefficients from heat-transfer coefficients, since it is based on the Reynolds analogy which holds only for $\text{Pr} = 1$ and $\text{Sc} = 1$.

A more adequate correlation shows that for heat transfer to fluid flowing turbulently inside tubes

$$\text{Nu} = 0.023\, \text{Re}^{0.8} \text{Pr}^{0.33} \tag{12-27}$$

The analogous relation for mass transfer in a wetted-wall column (7) is

$$\text{Sh} = 0.023\, \text{Re}^{0.83} \text{Sc}^{0.33} \tag{12-28}$$

A wetted-wall column is a simple experimental mass-transfer device. It consists of a vertical tube with liquid flowing in a thin film down the inside wall of the tube and a gas flowing upward in the tube. Mass transfer takes place from the liquid film to the gas, or vice versa, depending on the characteristics of the system being studied. Equation 12-28 may be used to predict coefficients in wetted-wall columns. Data for industrial mass-transfer equipment may also be correlated by equations of the form of Eq. 12-28. A summary of these correlations can be found in Reference 7.

Colburn's *j* factors are now defined as

$$j_H = \frac{\text{Nu}}{\text{RePr}^{0.33}} = \left(\frac{\bar{h}_c}{c_p \rho V}\right) \left(\frac{c_p \mu}{k}\right)^{0.67} = \frac{1}{2} f \tag{12-29}$$

and

$$j_M = \frac{\text{Sh}}{\text{ReSc}^{0.33}} = \left(\frac{k_G \mathcal{R} T p_{Bm}}{VP}\right) \left(\frac{\mu}{\rho D_v}\right)^{0.67} = \frac{1}{2} f \tag{12-30}$$

where j_H is the j factor for heat transfer and j_M is the j factor for mass transfer. Combination of Eq. 12-27 with Eq. 12-29 gives

$$j_H = 0.023\,\mathrm{Re}^{-0.2} \tag{12-31}$$

and combination of Eq. 12-28 with Eq. 12-30 gives

$$j_M = 0.023\,\mathrm{Re}^{-0.17} \tag{12-32}$$

Experimental data for flow in tubes show that $j_H = \frac{1}{2}f = j_M$ within the accuracy of the data. This correlation considers only skin friction. In flow past blunt objects and in typical industrial mass-transfer equipment, separation of the boundary layer often induces additional pressure losses, and j_H and j_M are not equal to $\frac{1}{2}f$. However, in many cases, j_H is still approximately equal to j_M, and mass-transfer coefficients can be predicted by the relation

$$j_H = j_M$$

or

$$\left(\frac{\bar{h}_c}{V\rho c_p}\right)\left(\frac{c_p\mu}{k}\right)^{0.67} = \left(\frac{k_G \mathcal{R}T}{V}\right)\left(\frac{p_{Bm}}{P}\right)\left(\frac{\mu}{\rho D_v}\right)^{0.67}$$

So that

$$k_G = \left(\frac{\bar{h}_c}{c_p\rho}\right)\left(\frac{P}{\mathcal{R}Tp_{Bm}}\right)\left[\left(\frac{c_p\mu}{k}\quad\frac{\rho D_v}{\mu}\right)\right]^{0.67} \tag{12-33}$$

If the Prandtl and Schmidt numbers are equal, Eq. 12-33 reduces to Eq. 12-26. Where direct mass-transfer data for a new system are not available, Eq. 12-33 may be used to predict mass-transfer coefficients from heat-transfer data taken in a system of identical geometry and flow characteristics.

EXAMPLE 12-2. Predict the mass-transfer coefficient for liquid ammonia vaporizing into air at 77 F and 1 atm, knowing that the heat-transfer coefficient in the same equipment, at the same gas and liquid flow rates, is 800 Btu/hr sq ft F.

Solution: Equation 12-28 cannot be used, since equipment size and flow rates are not given. In any event, Eq. 12-28 is valid only for a wetted-wall column. Equation 12-33 will therefore be used.

For ammonia at 77 F, $D_v = 1.08$ sq ft/hr. The physical properties of the gas phase will be evaluated assuming a dilute mixture of ammonia in air. For air at

77 F and 1 atm:

$$\begin{aligned} \mu &= 0.018 \text{ centipoise} = 0.044 \text{ lb}_m/\text{ft hr} \\ \rho &= 0.074 \text{ lb/cu ft} \\ c_p &= 0.25 \text{ Btu/lb F} \\ k &= 0.015 \text{ Btu/hr ft F} \\ \mathcal{R} &= 0.730 \text{ cu ft atm/lb mole F} \\ T &= 460 + 77 = 537 \text{ R} \end{aligned}$$

For a dilute gas $p_{Bm} = P$, and

$$k_G = \left(\frac{800}{0.25 \times 0.074}\right)\left(\frac{1}{0.730 \times 537}\right)\left(\frac{0.25 \times 0.044}{0.015} \times \frac{0.074 \times 1.08}{0.044}\right)^{0.67}$$

$$= 134 \text{ lb moles/hr sq ft atm}$$

As an exercise, the dimensions of the above equation should be checked. If Eq. 12-26 is used

$$k_G = \frac{\bar{h}_c}{c_p \rho \mathcal{R} T} = \frac{800}{0.25 \times 0.074 \times 0.73 \times 537}$$

$$= 110 \text{ lb moles/hr sq ft atm} \qquad \textit{Ans.}$$

The difference in the two values of k_G is 18 percent.

12-5. Interphase mass transfer

All industrial mass-transfer operations involve the transfer of material from one phase to another. The total resistance to mass transfer in the two phases may be expressed in terms of an overall mass-transfer coefficient similar to an overall heat-transfer coefficient. There is, however, an important difference in the evaluation of overall mass- and heat-transfer coefficients.

To illustrate the difference between the two overall coefficients, consider first the transfer of heat from a hot gas to a cold liquid. The gas is insoluble in the liquid, and the liquid does not vaporize. The gas is flowing countercurrent to the liquid. The temperature gradient for this system is shown in Fig. 12-4. The driving potential for heat transfer in the gas phase is $(T_G - T_i)$, and in the liquid phase, $(T_i - T_L)$. The overall driving potential, $(T_G - T_L)$, is the sum of the two. The overall heat-transfer coefficient is given by

$$\left(\frac{1}{U} = \frac{1}{h_{\text{gas}}} + \frac{1}{h_{\text{liquid}}}\right) \qquad (12\text{-}34)$$

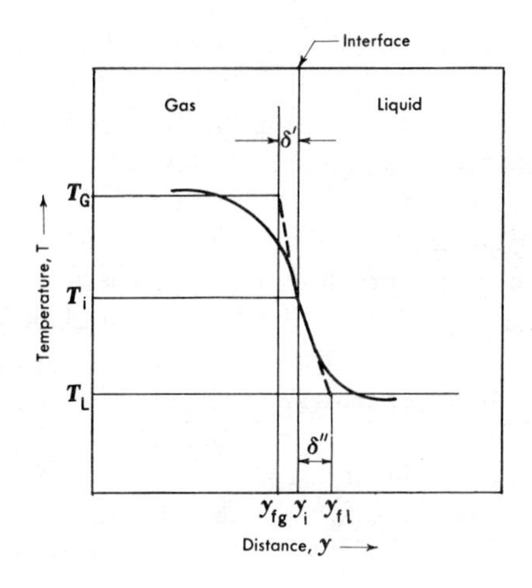

Fig. 12-4. Temperature gradient for heat transfer from a hot gas to a cold liquid.

and the heat flux by

$$q/A = U(T_G - T_L)$$

Now consider the transfer of mass from a gas to a liquid, as, for example, the absorption of ammonia from an air-ammonia mixture by water. In heat transfer the interfacial temperature T_i is identical for each phase, but as shown in Fig. 12-5, there is an apparent discontinuity in concentration at the gas-liquid interface which remains even when the concen-

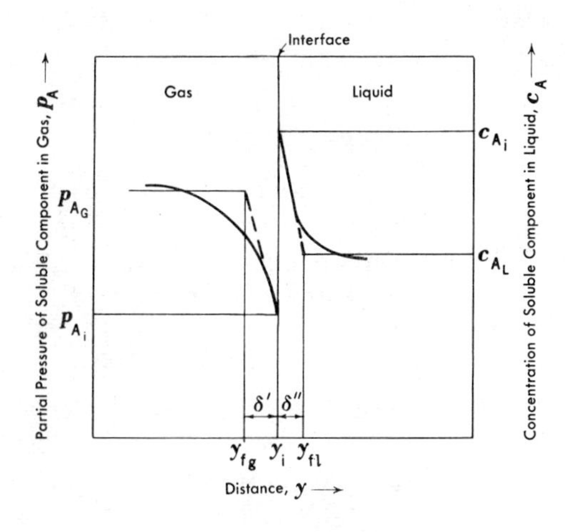

Fig. 12-5. Concentration gradient for mass transfer from a gas to a liquid.

trations in each phase are expressed in the same dimensions. If there is no resistance to heat transfer at the interface, the potentials are equal. In mass transfer, on the other hand, even if the two phases are assumed to be at equilibrium at the interface, the interfacial concentrations are not equal. The explanation of this apparent discrepancy lies in the choice of concentration as the driving potential for mass transfer. Strictly speaking the concentration is the driving potential for mass transfer within a phase, but not *between* phases. The correct driving potential between phases is a property called the *chemical potential.* In a single phase, the chemical potential is related to the concentration, but this relationship may change from one phase to another. Since the chemical potential is difficult to evaluate for industrial applications, it is seldom used in engineering calculations. It will not be considered in detail here, but it should be noted that when the chemical potentials of two phases are equal, they are in equilibrium. Thus, the chemical potentials at the interface in mass transfer are equal.

A simple illustration will show the possible great difference in mass concentration of two phases in thermodynamic equilibrium. Consider air at 77 F and 1 atm in equilibrium with water at the same temperature. If the air is saturated with water vapor, the partial pressure of the water vapor is 0.031 atm and the mole fraction of water vapor in the gas phase is 0.031. In the liquid phase the mole fraction of water is 1.0, since only water is present. (Actually a very small quantity of air would be dissolved in the water, but this will be ignored here.) Obviously, the concentrations of the two phases at equilibrium are not equal.

Experimental data for phase equilibria must be gathered for each system separately. Fortunately, groups of systems follow general laws, which facilitate the prediction of the equilibrium concentrations. For example, Henry's law adequately describes the equilibrium between a gas and a liquid phase for many gases and liquids by the expression,

$$p_A = mc_A \tag{12-35}$$

where p_A = partial pressure of component A in the gas phase, in atm;
c_A = concentration of component A in the liquid phase in equilibrium with the gas, in lb moles/cu ft;
m = Henry's law constant, experimentally determined, in atm cu ft/lb mole.

Values for the Henry's law constant for many gases, such as oxygen, nitrogen, and carbon dioxide dissolved in water, are given in Table 12-2. Gases such as sulfur dioxide and ammonia do not follow Henry's law, but data for these gases are tabulated in Reference 4.

An expression for the overall mass-transfer coefficient may now be derived for systems that follow Henry's law. Since it is not practical to

Table 12-2. Henry's law constants for various gases in water at moderate pressures.

$m \times 10^{-4}$, atm/(lb mole/cu ft)

T C	Air	O_2	N_2	CO_2
0	1.25	0.736	1.53	0.021
10	1.58	0.944	1.93	0.030
20	1.92	1.16	2.32	0.041
30	2.24	1.38	2.68	0.054
40	2.52	1.56	3.02	0.067
50	2.76	1.72	3.30	0.083
60	2.96	1.85	3.52	0.100
70	3.10	1.96	3.69	.....
80	3.18	2.04	3.74	.....
90	3.23	2.09	3.77	.....
100	3.22	2.11	3.79	.....

measure concentrations at the gas-liquid interface, p_{A_i} and c_{A_i} are unknown. However, Henry's law can be used to determine the partial pressure of a constituent of a gas in equilibrium with a given bulk liquid concentration

$$p_A^* = mc_{A_L} \tag{12-36}$$

where the * denotes that p_A is the gas-phase concentration in equilibrium with c_{A_L}. Similarly,

$$c_A^* = \frac{p_{A_G}}{m} \tag{12-37}$$

where c_A^* is the liquid-phase concentration in equilibrium with a gas of concentration p_{A_G}. It should be noted that c_{A_L} and p_{A_G} are actual concentrations (see Fig. 12-5), while p_A^* and c_A^* are fictitious concentrations, when mass transfer occurs.

Equations 12-17 and 12-19 may be written for mass transfer from the gas to liquid phase in the form

$$\frac{N_A}{A} = k_G(p_{A_G} - p_{A_i}) \qquad [12\text{-}17]$$

$$\frac{N_A}{A} = k_L(c_{A_i} - c_{A_L}) \qquad [12\text{-}19]$$

Since the interfacial concentrations cannot be evaluated, it is convenient to define overall coefficients

$$N_A = K_G(p_{A_G} - p_A^*) \tag{12-38}$$

and

$$N_A = K_L(c_A{}^* - c_{A_L}) \tag{12-39}$$

where K_G = overall mass-transfer coefficients based on the gas phase concentrations, in lb moles/hr sq ft atm;

K_L = overall mass-transfer coefficient based on the liquid-phase concentrations, lb moles/hr sq ft (lb mole/cu ft).

In the steady state the rate of mass transfer for 1 sq ft of transfer area N_A is the same in the gas and liquid films.

Solving Eq. 12-17 for p_{A_i} and Eq. 12-19 for c_{A_i} gives

$$p_{A_i} = p_{A_G} - \frac{N_A}{k_G} \tag{12-40}$$

and

$$c_{A_i} = c_{A_L} + \frac{N_A}{k_L} \tag{12-41}$$

or

$$mc_{A_i} = mc_{A_L} + \frac{mN_A}{k_L} \tag{12-42}$$

Since $p_{A_i} = mc_{A_i}$, Eq. 12-40 may be subtracted from Eq. 12-42, and

$$p_{A_G} - \frac{N_A}{k_G} = mc_{A_L} + \frac{mN_A}{k_L} \tag{12-43}$$

Since $mc_{A_L} = p_A{}^*$,

$$p_{A_G} - p_A{}^* = N_A\left(\frac{1}{k_G} + \frac{m}{k_L}\right) \tag{12-44}$$

Eliminating $p_{A_G} - p_A{}^*$ between Eq. 12-38 and Eq. 12-44 gives the relation

$$\frac{1}{K_G} = \frac{1}{k_G} + \frac{m}{k_L} \tag{12-45}$$

which is equivalent to Eq. 12-34 for heat transfer. In a similar manner, an expression for the overall liquid-phase mass-transfer coefficient can be derived:

$$\frac{1}{K_L} = \frac{1}{mk_G} + \frac{1}{k_L} \tag{12-46}$$

Combining Eq. 12-45 and Eq. 12-46 gives

$$K_L = mK_G \tag{12-47}$$

The overall coefficient based on either phase may be determined and used in calculations. The interrelation given in Eq. 12-47 is analogous to the interrelation of overall heat-transfer coefficients based on different areas. It is conventional to use the overall mass-transfer coefficient based on the phase where the major resistance to mass transfer lies.

In many cases the major resistance to mass transfer lies in one phase. For example, slightly soluble gases like oxygen and nitrogen have very large values of m. For systems having a large value of m, Eq. 12-46 reduces to $K_L \simeq k_L$. Absorption of a slightly soluble gas is therefore said to be "liquid-phase controlling." Similarly, for a very soluble gas, m is very small and by Eq. 12-45, $K_G \simeq k_G$. Therefore, the absorption of a very soluble gas is said to be "gas-phase controlling." Many systems are approximately either gas- or liquid-phase controlled. This approximation is made whenever possible, since calculations then require knowledge of only one individual phase coefficient. However, in certain systems resistance to mass transfer is appreciable in both phases, and both of the individual coefficients must be evaluated to calculate the mass-transfer rate. An example of a system where the resistance to mass transfer appears to be appreciable in both phases is the absorption of sulfur dioxide in water. The system has an additional complication of a chemical reaction between the sulfur dioxide and water in the liquid phase. The rate of such a chemical reaction may also influence the rate of absorption in such cases.

In cases of mass transfer where a pure phase is involved, no resistance to mass transfer exists in the pure phase. For example, in the humidification of air by water, the liquid is pure water and all resistance to mass transfer lies in the gas phase where there is a mixture of air and water vapor. If this concept is applied to mass transfer between two pure phases, one concludes that there is no resistance to mass transfer at all for such a case. An example of this would be the vaporization of water into pure steam, with no air present. In this case the rate of vaporization is determined by the rate at which heat is supplied to the liquid.

12-6. Simultaneous heat and mass transfer

Simultaneous heat and mass transfer must be considered in vaporization and condensation operations. Operations of particular interest to mechanical engineers are humidification and dehumidification of air. This discussion will be limited to the air-water system, but the theory also applies to any system of a condensing and a noncondensing gas.

When air is humidified in contact with liquid water, the latent heat of the water which is vaporized must be supplied by the gas, the liquid, or an outside source. Conversely, in condensation, heat must be removed by one of these agents. A number of cases will be considered from a theoretical viewpoint.

Humidification where equilibrium is established between water and air which is at a constant temperature. This is the case of water-cooling towers in power-plant installations. A limited quantity of liquid water is contacted with a stream of air. The quantity of air is sufficiently large so that the air temperature and humidity do not change appreciably, and it is assumed that no heat is supplied from the surroundings. If the air and water are initially at the same temperature, vaporization will tend to lower the temperature of the remaining water. This will establish a temperature gradient and heat will be transferred from the bulk gas phase to the water. The water will decrease in temperature until it reaches the temperature where the heat transferred to the water just balances the heat removed in vaporization. This temperature T_{wb} is known as the wet-bulb temperature of the gas. The expression for the equality of the heat transferred to the water and the heat supplied for vaporization is

$$\frac{q}{A} = \lambda_M N_A / A \tag{12-48}$$

where q/A = heat transferred per unit of interfacial area, in Btu/hr sq ft;
N_A/A = water vaporized, in lb moles/hr sq ft;
λ_M = molar latent heat of vaporization evaluated at T_{wb}, the wet-bulb temperature, in Btu/lb-mole.

At equilibrium, the liquid phase will be at a uniform temperature T_{wb}. The rate of heat transfer is

$$\frac{q}{A} = h_G(T_G - T_{wb}) \tag{12-49}$$

where h_G = gas-phase heat-transfer coefficient, in Btu/hr sq ft F;
T_G = dry-bulb temperature of the bulk of the gas, in F;
T_{wb} = temperature of the water at steady state, in F.

The resistance to heat and mass transfer lies only in the gas phase, since the liquid water is a pure phase, and the rate of mass transfer is

$$\frac{N_A}{A} = k_G(p_{wb} - p_G) \tag{12-50}$$

where k_G = gas-phase mass-transfer coefficient, in lb moles/hr sq ft atm;
p_{wb} = partial pressure of water vapor at the air-water interface; in this case it is the vapor pressure of water at temperature T_{wb}, in atm;
p_G = partial pressure of water vapor in the bulk gas phase, in atm.

Substitution of Eqs. 12-49 and 12-50 in Eq. 12-48 gives

$$p_{wb} - p_G = \frac{h_G}{\lambda_M k_G}(T_G - T_{wb}) \tag{12-51}$$

Since mass and heat are transferred by similar mechanisms, one might expect the ratio h_G/k_G to be constant. It is essentially constant for the conditions usually encountered in humidification. Equation 12-51 relates the wet- and dry-bulb temperatures for any mixture of water vapor and air. It is often written in terms of humidity

$$Y_{wb} - Y_G = \frac{1}{\lambda}\frac{h_G}{k_G M_G P}(T_G - T_{wb}) \tag{12-52}$$

where Y = absolute humidity of the air, in lb_m water vapor/lb_m dry air;
λ = specific latent heat of vaporization of water, in Btu/lb_m;
M_G = molecular weight of the gas phase—in this case air, in lb/lb mole;
P = total pressure, in atm.

Equation 12-52 follows from Eq. 12-51 when the humidity is defined as

$$Y = \frac{M_w}{M_G}\frac{p_w}{P - p_w} \tag{12-53}$$

where M_w = molecular weight of water;
p_w = partial pressure of water vapor in the gas phase.

Usually p_w is small compared to P and

$$Y = \frac{M_w p_w}{M_G P} \tag{12-54}$$

may be substituted in Eq. 12-51 to obtain Eq. 12-52.

The group $h_G/k_G M_G P$ has been evaluated for a limited number of systems. Selected values are given in Table 12-3. Where direct experi-

Table 12-3. Values of $h_G/k_G M_G P$ for various vapors in air.

Vapor	$h_G/k_G M_G P$
Water	0.26
Benzene	0.41
Carbon tetrachloride	0.44
Methyl alcohol	0.35

mental values are not available, the j factor relation can be used to calculate the group. Rearrangement of Eq. 12-33 gives

$$\left(\frac{h_G}{k_G \rho \mathcal{R} T}\right)\left(\frac{P}{p_{B_m}}\right) = c_p \left(\frac{\mu/\rho D_v}{c_p \mu/k}\right)^{0.67} \tag{12-55}$$

where the physical properties are those of the gas. For most humidification problems $p_{B_m}/P \cong 1$. From the perfect gas law, $\rho \mathcal{R} T = M_G P$. Equation 12-55 then becomes

$$\frac{h_G}{k_G M_G P} = c_p \left(\frac{\mu/\rho D_v}{c_p \mu/k}\right)^{0.67} = c_p \left(\frac{\text{Sc}}{\text{Pr}}\right)^{0.67} \tag{12-56}$$

Equation 12-56 gives a value of $h_G/k_G M_G P$ of 0.21 for the air-water system, compared to the experimental value of 0.26.

Humidification where equilibrium is established between water at a constant temperature and air. In this case the supply of air is limited and its temperature is lowered as heat is transferred to the liquid. This is the case usually approached in industrial air humidification towers when air is humidified and cooled. The final equilibrium temperature T_{as} is called the *adiabatic saturation temperature* and the operation is called *adiabatic humidification.* This implies that no heat is supplied to the air-water system from the surroundings.

An enthalpy balance can be written around the air and water, since adiabatic operation is assumed. Assume that the water enters at temperature T_{as}. The quantity of water supplied is large and the quantity vaporized is small so that its final temperature will also be T_{as}. Therefore, there is essentially no change in enthalpy of the water phase. An enthalpy balance for the gas phase gives semantically

$$\left(\begin{array}{c}\text{Enthalpy of entering dry air} \\ + \text{ water vapor}\end{array}\right) = \left(\begin{array}{c}\text{enthalpy of leaving dry air} \\ + \text{ water vapor}\end{array}\right)$$

or

$$\begin{pmatrix}\text{Enthalpy of entering dry air}\\ -\text{enthalpy of leaving dry air}\end{pmatrix} = \begin{pmatrix}\text{enthalpy of leaving water vapor}\\ -\text{enthalpy of entering water vapor}\end{pmatrix}$$

Then

$$c_a(T_G - T_{as}) = -Y_G c_w(T_G - T_{as}) - \lambda(Y_G - Y_{as}) \qquad \text{(12-57)}$$

where c_a = specific heat of air, in Btu/lb_m;
c_w = specific heat of water, in Btu/lb_m;
T_G = initial air temperature, in F;
T_{as} = final equilibrium air temperature, in F;
Y_G = absolute humidity of the initial air, in $lb_m\ H_2O/lb_m$ dry air;
Y_{as} = absolute humidity of the air at T_{as}, in $lb_m\ H_2O/lb_m$ dry air.

Rearrangement of Eq. 12-57 gives

$$Y_{as} - Y_G = \frac{1}{\lambda}(c_a + Y_G c_w)(T_G - T_{as}) \qquad \text{(12-58)}$$

Comparison of Eq. 12-58 with Eq. 12-52 shows that if $h_G/k_G M_G P$ = $(c_a + Y_G c_w)$, the adiabatic saturation temperature T_{as} is identical to the wet-bulb temperature T_{wb}. For air-water systems, the quantities are essentially equal and, therefore, $T_{as} = T_{wb}$. However, for any other vapor in air the temperatures are considerably different. The group $(c_a + Y_G c_w)$ is called the "humid heat," although a more appropriate term would be the humid heat capacity, which is designated by c_s. For the air-water vapor system $c_s = 0.24 + 0.45Y_G$.

EXAMPLE 12-3. A stream of air has a dry-bulb temperature of 120 F and a wet-bulb temperature of 90 F. What is the humidity of the air?

Solution: At 120 F, $Y_s = 0.08$ lb water vapor/lb dry air (from the saturation curve on a humidity chart, Reference 4) and $\lambda = 1025$ Btu/lb. From Table 13-2, $h_G/k_G M_G P = 0.26$. With Eq. 12-52,

$$Y_G = 0.080 - \frac{1\,(0.26)\,(120 - 90)}{1025} = 0.072 \text{ lb } H_2O\text{/lb dry air} \qquad \textit{Ans.}$$

If the air has a dry-bulb temperature of 120 F and an adiabatic saturation temperature of 90 F, Y_G may be evaluated from Eq. 12-58

$$Y_G = \frac{Y_{as} - c_a(T_G - T_{as})/\lambda}{1 + c_w(T_G - T_{as})/\lambda} = \frac{0.080 - 0.24(120 - 90)/1025}{1 + 0.45(120 - 90)/1025}$$

$$= 0.072 \text{ lb } H_2O/\text{lb dry air} \qquad \textit{Ans.}$$

Within the limits of accuracy of the calculation in Example 12-3, the answers are identical and $T_{as} = T_{wb}$ for water. This coincidence simplifies calculations for the air-water system. However, the adiabatic saturation temperature and the wet-bulb temperature are generally not equal for other systems.

Both of these cases are for conditions existing at the equilibrium of an air stream flowing past a water system. In actual humidification and water-cooling equipment, equilibrium is only approached. An infinitely high tower would be required to give true equilibrium between the two streams. The wet-bulb temperature and the adiabatic saturation temperature may be considered as limiting values beyond which no equipment can go.

Adiabatic humidification—cooling. The calculation of the size of industrial equipment for adiabatically humidifying and cooling an air stream requires integration of the rate equation over the height or length of the equipment. Usually in such equipment the quantity of water recirculated is large, so that the water remains constant at the adiabatic saturation temperature of the air, T_{as}.

The rate equation for mass transfer can be rewritten in terms of humidity

$$\frac{N'_A}{A} = k_Y(Y_{as} - Y) \qquad (12\text{-}59)$$

where k_Y = gas-phase mass-transfer coefficient, in lb of water transferred /hr sq ft unit ΔY;

N'_A = mass flux, in lb/hr sq ft of transfer area;

Y = absolute humidity of the air, in lb H_2O/lb dry air.

Since the water temperature is constant at T_{as}, and since there is no resistance to mass transfer in a pure water phase, the driving potential is $(Y_{as} - Y)$, in which Y_{as} is the saturated humidity of air at the water-air interface where the temperature is T_{as}.

To determine the rate of mass transfer, it is necessary to know the air-water interfacial area A. It is often impossible to estimate accurately the interfacial area available in industrial equipment. The water flows or is sprayed downward over wood slats or irregular packing to achieve a large interfacial area. This interfacial area may vary with liquid or gas flow rate. Because of the difficulty in estimating interfacial area for mass transfer, it is usually redefined as

$$A = aV = aSZ \tag{12-60}$$

where A = total interfacial area for mass transfer in the humidification tower, in sq ft;
a = interfacial area per unit volume of tower packing, in sq ft/cu ft;
S = cross-sectional area of the tower, in sq ft;
Z = height of the tower, in ft.

Combination of Eqs. 12-59 and 12-60 gives

$$N'_A = k_Y a (Y_{as} - Y) SZ \tag{12-61}$$

Since a is difficult to evaluate, it is combined with k_Y to form a new mass-transfer coefficient, $k_Y a$, which can be evaluated experimentally for a given tower packing and fluid flow rates.

Figure 12-6 is a schematic picture of a humidification tower. The water L_2 enters at the top of the tower and flows downward over packing to the bottom. The air G with a humidity Y_1 enters at the bottom of the tower, flows countercurrent to the water, and leaves the top of the tower at humidity Y_2. The mass velocity of air G is expressed as the lb dry air/hr sq ft of tower cross section; therefore it is constant through the tower, even though the humidity varies. The mass velocity of water L is given as lb water/hr sq ft of tower cross section, and it varies from L_2 at the top of the tower to L_1 at the bottom. A material balance on the water over the total height of the tower gives

$$L_2 - L_1 = G(Y_2 - Y_1) \tag{12-62}$$

i.e., the rate of vaporization from the liquid phase equals the rate of mass transfer to the gas.

Consider a differential height dZ as shown in Fig. 12-6. The change in humidity in the height dZ is dY and therefore the rate of transfer of water to the gas per square foot of tower cross-section is given by the equation

$$dL = GdY \tag{12-63}$$

Therefore, the rate of mass transfer across height dZ is

$$dN'_A = SGdY \tag{12-64}$$

since G is based on a square foot of tower cross section.

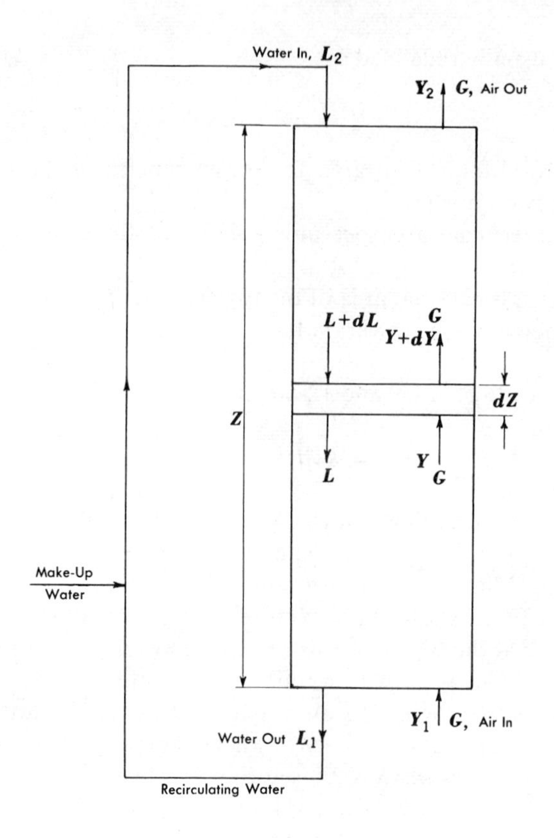

Fig. 12-6. Adiabatic humidification tower.

Combining Eqs. 12-61 and 12-64 yields

$$GdY = k_Y a(Y_{as} - Y)dZ \tag{12-65}$$

Integration from the bottom to the top of the tower, assuming $k_Y a$ is constant, gives

$$\int_0^Z dZ = \frac{G}{k_Y a} \int_{Y_1}^{Y_2} \frac{dY}{Y_{as} - Y}$$

$$Z = \frac{G}{k_Y a} \ln \frac{Y_{as} - Y_1}{Y_{as} - Y_2} \tag{12-66}$$

Equation 12-66 may be used to calculate the height of an adiabatic humidification tower required to humidify air from Y_1 to Y_2. The coefficient $k_Y a$ must be evaluated experimentally or by an empirical correlation.

EXAMPLE 12-4. Three thousand cu ft/min of air at 100 F and an absolute humidity of 0.003 lb water/lb dry air is to be adiabatically humidified and cooled in a packed tower by contacting it with 20 gal/min of recirculated water. The fluid flow rates dictate a cross-sectional area of 25 sq ft and the $k_Y a$ for the packing used has been determined as $k_Y a = 0.45\ GL^{0.2}$ (Reference 8). (a) Calculate the height required to cool the air to 70 F (corresponding to a humidity of 0.016 lb/lb). (b) Calculate the height of tower required to cool the gas of (a) to 62 F (corresponding to a humidity of 0.019).

Solution: (a) First calculate $k_Y a$. Neglecting the humidity of the incoming air,

$$G = \left(3000\ \frac{\text{cu ft}}{\text{min}}\right)\left(\frac{60\ \text{min}}{1\ \text{hr}}\right)\left(\frac{492\ \text{R}}{560\ \text{R}}\right)$$

$$\left(\frac{1\ \text{lb mole}}{359\ \text{cu ft at STP}}\right)\left(29\ \frac{\text{lb air}}{\text{lb mole air}}\right)\left(\frac{1}{25\ \text{sq ft}}\right)$$

$$= 513\ \text{lb dry air/hr sq ft of tower cross section}$$

$$L = \left(20\ \frac{\text{gal}}{\text{min}}\right)\left(\frac{60\ \text{min}}{1\ \text{hr}}\right)\left(\frac{8.34\ \text{lb}_\text{m}\ \text{H}_2\text{O}}{\text{gal}}\right)\left(\frac{1}{25\ \text{sq ft}}\right)$$

$$= 400\ \text{lb}_\text{m}/\text{hr sq ft of tower cross section}$$

Therefore

$$k_Y a = (0.45)(513)(400)^{0.2}$$

$$= 764\ \text{lb}_\text{m}/\text{hr cu ft unit}\ \Delta Y$$

The adiabatic saturation temperature T_{as} and humidity Y_{as} may be evaluated from Eq. 12-58 or from a humidity chart; $T_{as} = 62$ F and $Y_{as} = 0.019$. Use of Eq. 12-66 gives a height of

$$Z = \frac{513}{764} \ln \frac{0.019 - 0.003}{0.019 - 0.016} = 1.1\ \text{ft}$$

b) For cooling to 62 F,

$$Z = \frac{513}{764} \ln \frac{0.019 - 0.003}{0.019 - 0.019} = \infty \qquad \textit{Ans.}$$

This shows that an infinitely tall tower is required to reach the equilibrium condition of saturation.

Adiabatic humidification is a simple case of the more general humidification problem. In adiabatic humidification the enthalpies of both the liquid and gas are nearly constant; but, in general, this is not the case and energy transfer across an enthalpy potential must be considered. An equation similar to Eq. 12-65 can be written with enthalpy driving forces and it may be integrated graphically. A discussion of this method, with examples, is given in Reference 3. Use of the enthalpy potential is necessary whenever the enthalpy of either phase changes appreciably. For example, it would be required in the calculation of a water-cooling tower.

Dehumidification. Air conditioning often involves the removal of water vapor from air by direct cooling with cold water or by indirect cooling by contact with a cold metal wall. Although spray towers are widely used for direct cooling, little data have been published on their performance.

The contact of humid air with a cold metal wall results in mass and heat transfer from the air to the liquid layer flowing down the metal wall. The heat transferred across the liquid layer must equal the heat transferred across the gas film plus the latent heat given up at the gas-liquid interface on condensation of the mass transferred across the gas film. Figure 12-7 represents such a system. An expression for the heat transferred across a unit area is

$$h_L(T_i - T_L) = h_G(T_G - T_i) + \lambda k_Y(Y_G - Y_i) \tag{12-67}$$

where T_i = temperature at the gas-liquid interface;
T_G = temperature in the bulk-gas phase;
T_L = temperature in the liquid layer;
h_L = liquid-phase heat-transfer coefficient;

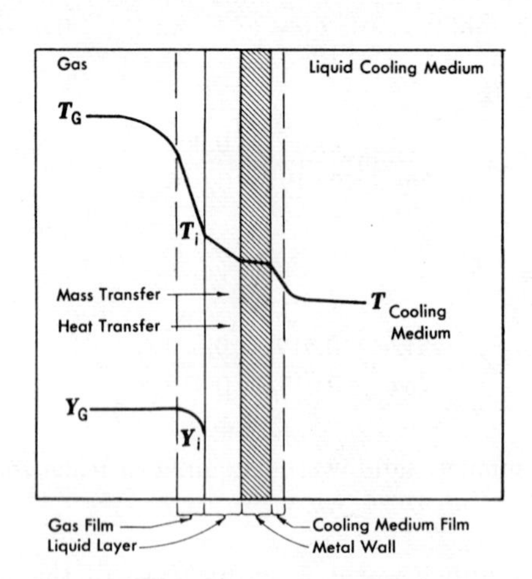

Fig. 12-7. Simultaneous heat and mass transfer in the dehumidification of air by indirect cooling.

h_G = gas-phase heat-transfer coefficient;
k_Y = gas-phase mass-transfer coefficient.

The use of this equation for dehumidification calculations involves a trial-and-error procedure, since the interface conditions are not known. The method of calculation is outlined in Reference 2.

12-7. Mass-transfer equipment

The theoretical relationships which have been discussed can be applied to the design of industrial equipment. However, the calculations are usually complex and are beyond the scope of this brief discussion. Detailed design methods and illustrations of industrial equipment can be found in References 4, 5, 6, and 7.

Mass-transfer equipment can be classified as batch or continuous flow. The tendency in industry has been toward continuous-flow equipment, where a steady state is reached and material is fed and withdrawn continuously. Calculation of batch equipment involves the consideration of transient mass transfer.

Continuous-flow equipment can be further classified as to whether it is stage-contact or continuous-contact. In stage-contacting, the two phases are brought together, mass is transferred between the phases, and finally the phases are mechanically separated. In continuous countercurrent stage-contacting, the resultant two phases are then sent in opposite directions to other stages for further contacting. Usually calculations are based on the assumption that the two phases leaving a stage are in equilibrium with each other. The number of *equilibrium stages* required to give the specified purity and recovery of product is determined from equilibrium and stoichiometric relationships. In such a calculation the *rate* of mass transfer is not considered, since it is assumed that transfer was rapid enough to establish equilibrium.

In an *actual* stage, for example a plate in a distillation column, the two phases are not usually in contact long enough to reach equilibrium. Therefore, more *actual* stages are required than *equilibrium* stages. A *stage efficiency* is applied to the number of equilibrium stages calculated to obtain the number of actual stages required. Stage efficiencies depend on many factors, including the physical configuration of the equipment, the phase-flow rates, and the rate of mass transfer. Experimental data on stage efficiencies for many systems are available. Correlations have been made on certain systems, such as petroleum distillation columns (see References 5 and 7).

An example of a continuous countercurrent stage-contacting device is a multiple-stage petroleum distillation column. A typical crude petroleum distillation column is shown with its accessory equipment in Fig. 12-8. A schematic diagram of the column, Fig. 12-9, shows the individual stages, the crude oil intake, and the points of withdrawal of the various products. The withdrawn products increase in volatility from the bottom to the top of the tower. Since the less volatile components may decom-

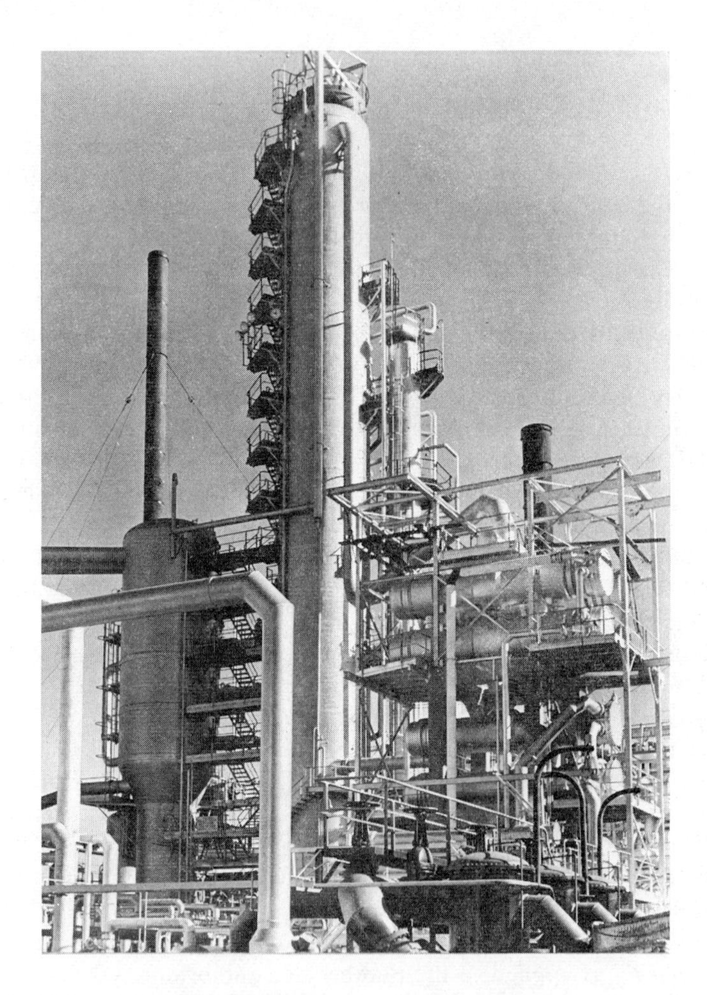

Fig. 12-8. Crude petroleum distillation unit. The taller tower in the center of the picture separates the lighter components of the crude petroleum. It is the upper part of the column shown in Fig. 12-9. The shorter tower at the left separates the heavier components and is represented by the lower part of the column shown in Fig. 12-9. Also of interest is the group of large shell-and-tube heat exchangers to the right of the taller tower. They are used to cool the product streams. (Courtesy of Standard Oil Company of California and California Research Corporation)

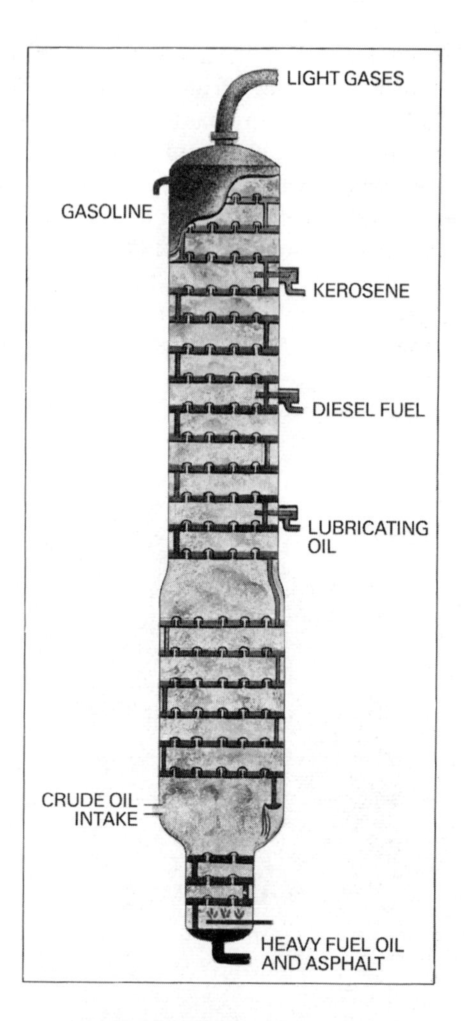

Fig. 12-9. Crude petroleum distillation column. (Courtesy of Standard Oil Company of California and California Research Corporation)

pose when heated to their boiling points at atmospheric pressure, the lower part of the column may be operated at less than atmospheric pressure to reduce the temperatures required for vaporization. The column shown in Fig. 12-8 has been split into two parts which are placed side by side to reduce the overall height of the unit.

In equipment such as packed absorption, distillation, or humidification towers the contact between the liquid and gas is *continuous* through the equipment. There is no mechanical separation of phases as occurs in stage equipment. For this case, differential forms of the rate equations are integrated over the height of the tower to determine the total mass transfer, as illustrated in Sec. 12-6. Problems encountered in calculating packed towers include variation in mass-transfer coefficients with flow rates and tower packing size and shape, unknown interfacial area for mass transfer, and variations in the flow pattern through the equipment. A natural-draft water-cooling tower is shown in Fig. 12-10. Water-cooling towers are used to conserve water by permitting reuse of cooling water. Warm water is distributed across the top of the tower. It flows downward through wood-slat gratings, continuously contacting air which is flowing upward by natural convection. As the water contacts the air, it humidifies the air and is cooled.

Fig. 12-10. Natural-draft water-cooling tower. (Courtesy of Standard Oil Company of California and California Research Corporation)

In many cases existing mass-transfer equipment is evaluated to determine its performance under new operating conditions or for a new separation. The principles involved are identical with those required for design of new equipment.

PROBLEMS

12.1. An open circular tank 10 ft in diameter which contains ethanol is exposed to the open air at 77 F and atmospheric pressure. Assuming that there is a stagnant layer of air 6 in. thick over the surface of the ethanol, calculate the weight of ethanol lost by evaporation in 24 hr. Vapor pressure of ethanol at 77 F = 58 mm Hg.

12-2. Calculate the rate of diffusion of ammonia across a water film 0.1 in. thick at 77 F. The concentration of ammonia is 2 percent (by weight) on one side of the film and 1 percent (by weight) on the other side.

12-3. Calculate the mass-transfer coefficient for the vaporization of water into air in a wetted-wall column under the following conditions.

Column diameter = 1.0 in.
Air and water temperature = 77 F
Average partial pressure of water vapor in air = 5 mm Hg
Total pressure = 753 mm Hg
Air flow rate = 2 cu ft/min

Calculate the effective film thickness for mass transfer.

12-4. Air at 100 C flows over a streamlined naphthalene body. Naphthalene sublimes into air and its vapor pressure at 100 C is 20 mm Hg. The heat-transfer coefficient for this system was previously found to be 3 Btu/hr sq ft F. The mass diffusivity of naphthalene vapor in air at 100 C is 0.32 sq ft/hr. The concentration of naphthalene in the bulk air stream is negligibly small. Calculate the mass-transfer coefficient and the mass flux for the system.

12-5. Ammonia is being absorbed from air by water at 77 F in an absorption column. At a point in the column the following conditions exist:

k_L = 0.95 lb mole/hr sq ft (lb mole/cu ft)
k_G = 0.15 lb mole/hr sq ft atm
Gas composition = 1 percent ammonia (by volume)
Total pressure = 1500 mm Hg
Liquid composition = 0.02 lb moles NH_3/cu ft
$p_{NH_3} = 0.38\, c_{NH_3}$ at 77 F for dilute solutions

Calculate: (a) overall gas-phase mass-transfer coefficient; (b) overall liquid-phase mass-transfer coefficient; (c) percentage of total resistance to mass transfer which lies in the gas phase; (d) interfacial composition of both phases; (e) mass flux of ammonia; (f) effective film thickness for each phase.

12-6. Air at 200 F and 1 atm has a humidity of 0.12 lb H_2O/lb dry air. Calculate: (a) the wet-bulb temperature; (b) the adiabatic saturation temperature.

12-7. Air at 100 F and 1 atm contains carbon tetrachloride vapor such that the wet-bulb temperature is 85 F. What is the adiabatic saturation temperature? At 100 F the vapor pressure of carbon tetrachloride is 200 mm Hg and the latent heat of vaporization is 83 Btu/lb.

12-8. Air at 120 F dry bulb and 70 F wet bulb is to be cooled and humidified adiabatically in a tower filled with packing which is the same as that in Example 12-4. The tower has a cross-sectional area of 17 sq ft. Air is supplied at a rate of 110 lb/min and water at 10 gal/min. Calculate and plot the height of tower required to cool the air to final temperatures between 70 F and 80 F.

12-9. Derive expressions for the liquid and gas phase mass-transfer coefficients for equimolar counterdiffusion. How do they differ from those for diffusion through a stagnant film?

12-10. Derive Eq. 12-22a by dimensional analysis.

12-11. Derive Eq. 12-23 by dimensional analysis.

12-12. Derive Eq. 12-46.

REFERENCES

1. W. L. Badger and J. T. Banchero, *Introduction to Chemical Engineering.* (New York: McGraw-Hill Book Company, Inc., 1955.)
2. A. P. Colburn and O. A. Hougen, *Ind. Eng. Chem.*, Vol. 26 (1934), pp. 1178–1182.
3. H. S. Mickley, *Chem. Eng. Prog.*, Vol. 45 (1949).
4. J. H. Perry, Ed., *Chemical Engineers' Handbook*, 3d ed. (New York: McGraw-Hill Book Company, Inc., 1950.)
5. C. S. Robinson and E. R. Gilliland, *Elements of Fractional Distillation*, 4th ed. (New York: McGraw-Hill Book Company, Inc. 1950.)
6. T. K. Sherwood and R. L. Pigford, *Absorption and Extraction*, 2d ed. (New York: McGraw-Hill Book Company, Inc., 1952.)
7. R. E. Treybal, *Mass Transfer Operations.* (New York: McGraw-Hill Book Company, Inc., 1955.)
8. F. Yoshida and T. Janaka, *Ind. Eng. Chem.*, Vol. 43 (1951), p. 1467.

Appendix I

Nomenclature

LETTER SYMBOLS

Symbol	Quantity	United States Engineering Units	International System of Units
a	velocity of sound	ft/sec	m/s
a	thermal diffusivity $= k/c\rho$	sq ft/hr	m^2/s
a	interfacial area per unit volume of tower packing in Chapter 12	sq ft/cu ft	m^2/m^3
A	area; A_c, cross-sectional area; A_p, projected area of a body normal to the direction of flow; A_q, area through which rate of heat flow is q; A_s, surface area; A_o, outside surface area; A_i, inside surface area; $\bar{A}$, logarithmic mean area defined by Eq. 2-8	sq ft	m^2
A	azimuth of the sun	deg	rad
b	breadth or width	ft	m
c	specific heat; c_p, specific heat at constant pressure; c_v, specific heat at constant volume; c_s, humid heat capacity in Chapter 12	$Btu/lb_m F$	J/kg K
c_A	concentration of component A in Chapter 12	lb moles/cu ft	kg moles/m^3
C	constant		
C	thermal capacity	Btu/F	J/K

Symbol	Quantity	United States Engineering Units	International System of Units
C	hourly heat capacity rate in Chapter 11; C_c, hourly heat capacity rate of colder fluid in a heat exchanger; C_h, hourly heat capacity rate of warmer fluid in a heat exchanger	Btu/hr F	w/K
C_e	electrical capacitance	farad	farad
C_D	total drag coefficient		
C_f	skin friction coefficient; C_{fx}, local value of C_f at distance x from leading edge; $\overline{C}_f$, average value of C_f defined by Eq. 6-20		
D	diameter; D_H, hydraulic diameter; D_o, outside diameter; D_i, inside diameter	ft	m
D_v	mass diffusivity	sq ft/hr	m^2/s
e	base of natural or Napierian logarithm		
E	electric potential	volt	volt
E	emissive power of a radiating body; E_b, emissive power of blackbody; E_λ, monochromatic emissive power per micron at wavelength λ	Btu/hr sq ft	w/m^2
$\mathcal{E}$	heat exchanger effectiveness defined by Eq. 11-17		
f	Fanning friction coefficient for flow through a pipe or a duct, defined by Eq. 8-12		
f'	friction coefficient for flow over banks of tubes, defined by Eq. 9-11		
F	force	lb_f	newton
F_T	temperature factor defined by Eq. 5-67		
$F_{1\text{-}2}$	geometrical shape factor for radiation from one blackbody to another defined by Eq. 5-35		
$\mathfrak{F}_{1\text{-}2}$	geometric shape and emissivity factor for radiation from one gray body to another		
g	acceleration of gravity	ft/sec^2 or ft/hr^2	m/s^2
g_c	dimensional conversion factor	32.2 ft $lb_m/lb_f sec^2$, or 4.18×10^8 ft $lb_m/lb_f hr^2$	1.0 kg m/N s^2
G	mass velocity or flow rate per unit area ($G = \rho V$)	lb_m/hr sq ft	$kg/m^2 s$
G	irradiation incident upon unit surface in unit time	Btu/hr sq ft	w/m^2
h	enthalpy per unit mass	Btu/lb_m	J/kg
$\bar{h}$	combined unit-surface conductance, $h = \bar{h}_c + \bar{h}_r$; h_b, unit-surface conductance of a boiling liquid, defined by Eq. 10-1; h_c, local unit convective conductance; $\bar{h}_c$, average unit convective conductance; $\bar{h}_r$, average unit conductance for radiation	Btu/hr sq ft F	w/m^2K
h_{fg}	latent heat of condensation or evaporation	Btu/lb_m	J/kg
h_G	gas-phase heat-transfer coefficient in Chapter 12	Btu/hr sq ft F	w/m^2K
h_L	liquid-phase heat-transfer coefficient in Chapter 12	Btu/hr sq ft F	w/m^2K
H	total hour angle from noon to sunrise or sunset	deg	rad
i	angle between sun direction and surface normal in Chapter 5	deg	rad
i	electric current flow rate	amp	amp (A)
I	intensity of radiation; I_λ, intensity per micron at wavelength λ	Btu/hr unit solid angle	w/sr
J	radiosity	Btu/hr sq ft	w/m^2
k	thermal conductivity; k_s, thermal conductivity of a solid; k_f, thermal conductivity of a fluid evaluated at the mean film temperature	Btu/hr ft F	w/m^2K
k_G	mass-transfer coefficient for the gas phase defined by Eq. 12-18	lb moles/hr sq ft atm	kg moles/m^2 s Pa

Symbol	Quantity	United States Engineering Units	International System of Units
k_L	mass-transfer coefficient for the liquid phase defined by Eq. 12-20	lb moles/hr sq ft (lb moles/cu ft)	kg moles/m^2 s (kg moles/m^3)
K	thermal conductance; K_k, thermal conductance for conduction heat transfer; K_c, thermal convective conductance; K_r, thermal conductance for radiation heat transfer	Btu/hr F	w/K
K_e	electrical conductance	amp/volt	amp/volt (A/V)
K_G	overall mass-transfer coefficient based on the gas phase	lb moles/hr sq ft atm	kg moles/s m^2 kg
K_L	overall mass-transfer coefficient based on the liquid phase	lb moles/hr sq ft (lb mole/cu ft)	kg moles/s m^2 (kg moles/m^3)
log	logarithm to the base 10		
ln	logarithm to the base e		
l	length, general	ft or in.	m
L	length along a heat flow path or characteristic length of a body	ft or in.	m
L_f	latent heat of solidification	Btu/lb	J/kg
m	mass flow rate	lb_m/sec or lb_m/hr	kg/s
m	Henry's law constant in Chapter 12	atm cu ft/lb mole	Pa m^3/kg mole
M	mass	lb_m	kg
m_A	mass of gas A in Chapter 12	lb mole	kg mole
M_G	molecular weight of the gas phase in Chapter 12	lb/lb mole	kg/kg mole
N	number in general; number of tubes, etc.		
p	static pressure; p_c, critical pressure; p_A, partial pressure of gas A in Chapter 12	psi/or lb_f/sq ft atm	pascal (Pa = N/m^2)
P	wetted perimeter	ft	m
P	total pressure in Chapter 12	atm	N/m^2 (pascal)
q	rate of heat flow; q_k, rate of heat flow by conduction; q_r, rate of heat flow by radiation; q_c, rate of heat flow by convection; q_b, rate of heat flow by nucleate boiling	Btu/hr	w (J/s)
$\bar{q}$	rate of heat flow per unit area or heat flux	Btu/hr sq ft	w/m^2
$\dot{q}$	rate of heat generation per unit volume	Btu/hr cu ft	w/m^3
Q	quantity of heat	Btu	J (joules)
$\dot{Q}$	volumetric rate of fluid flow	cu ft/hr	m^3/s
Q_e	electric charge of condenser	coulomb	C (coulomb)
r	radius; r_H, hydraulic radius; r_i, inner radius; r_o, outer radius	ft	m
R	thermal resistance; R_c, thermal resistance to convection heat transfer; R_k, thermal resistance to conduction heat transfer; R_r, thermal resistance to radiation heat transfer	hr F/Btu	s/w
R_e	electrical resistance	ohm	Ω (ohm)
$\mathcal{R}$	perfect gas constant	1545 ft lb/lb mole F, or 0.730 cu ft atm/lb mole F	8.314 J/K kg mole
s	molecular speed ratio		
S	cross-sectional area of mass-transfer tower in Chapter 12	sq ft	m^2
S	shape factor for conduction heat flow		

Symbol	Quantity	*United States Engineering Units*	*International System of Units*
S_L	distance between centerlines of tubes in adjacent longitudinal rows	ft	m
S_T	distance between centerlines of tubes in adjacent transverse rows	ft	m
T	temperature; T_b, temperature of bulk of fluid; T_f, mean film temperature; T_s, surface temperature; T_∞, temperature of fluid far removed from heat source or sink; T_m, mean bulk temperature of fluid flowing in a duct; T_{abs}, temperature on absolute scale; T_s, temperature at surface of a wall; T_{sv}, temperature of saturated vapor; T_{sl}, temperature of a saturated liquid; T_{fr}, freezing temperature; T_l, liquid temperature; T_o, total temperature; T_{as}, adiabatic wall temperature or adiabatic saturation temperature in Chapter 12; T_{wb} wet-bulb temperature	F or R	K
u	internal energy per unit mass	Btu/lb$_m$	J/kg
u	time average velocity in x direction; u', instantaneous fluctuating x component of velocity; u_∞, free stream velocity	ft/sec or ft/hr	m/s
U	overall unit conductance, overall heat-transfer coefficient, or overall transmittance	Btu/hr sq ft F	w/m^2 K
v	specific volume	cu ft/lb$_m$	m^3/kg
v	time average velocity in y direction; v', instantaneous fluctuating y component of velocity	ft/sec or ft/hr	m/s
V	volume	cu ft	m^3
V	average velocity; V_l, velocity of light; V_∞, free stream or flight velocity	ft/sec or ft/hr	m/s
x	distance from the leading edge; x_c, critical distance from the leading edge where flow becomes turbulent	ft	m
x	coordinate		
y	coordinate		
y	distance from a solid boundary measured in direction normal to surface	ft	m
Y	absolute humidity in Chapter 12	lb$_m$/lb$_m$	kg/kg
z	zenith distance	deg	rad
z	coordinate		
Z	ratio of hourly heat capacity rates in heat exchangers		
Z	height of mass-transfer equipment	ft	m

GREEK LETTERS

Symbol	*Quantity*	*United States Engineering Units*	*International System of Units*
α	absorptance for radiation; α_λ, monochromatic absportance at wavelength λ		
β	temperature coefficient of volume expansion	1/F	1/K
β_k	temperature coefficient of thermal conductivity	1/F	1/K
γ	specific heat ratio, c_p/c_r		
Γ	body force per unit mass	lb_f/lb_m	N/kg
Γ_c	mass rate of flow of condensate per unit breadth = $m/\pi D$ for a vertical tube	lb_m/hr ft	kg/s m
δ	boundary layer thickness; δ_h, hydrodynamic boundary layer thickness; δ_{th}, thermal boundary layer thickness; δ_m', effective boundary layer thickness for mass transfer	ft	m
δ	solar declination in Chapter 5	deg	rad
Δ	difference between values		
ϵ	emittance for radiation; ϵ_λ, monochromatic emittance at wavelength λ; ϵ_ϕ, emittance in direction of ϕ		
ϵ_H	thermal eddy diffusivity	sq ft/hr or sq ft/sec	m^2/s
ϵ_M	momentum eddy diffusivity	sq ft/hr or sq ft/sec	m^2/s
ζ	ratio of thermal to hydrodynamic boundary layer thickness, δ_{th}/δ_h		
η_f	fin efficiency		
θ	time	hr or sec	s
λ	wavelength; λ_{max}, wavelength at which monochromatic emissivity $E_{b\lambda}$ is a maximum (see Eq. 5-6)	micron	10^{-6} m
λ	latent heat of vaporization in Chapter 12; λ_M, molar latent heat of vaporization	Btu/lb_m or Btu/lb mole	J/kg
μ	absolute viscosity	lb_m/ft sec or lb_m/ft hr	Ns/m^2
ν	kinematic viscosity, μ/ρ	sq ft/hr or sq ft/sec	m^2/s
ν_r	frequency of radiation	1/sec	1/s
ρ	mass density, $1/v$; ρ_l, density of liquid; ρ_v, density of vapor	lb_m/cu ft	kg/m^3
ρ	reflectance for radiation		
τ	shearing stress; τ_s, shearing stress at surface; τ_w, shear at wall of a tube or a duct	lb_f/sq ft	Pa (N/m^2)
τ	transmittance for radiation		
σ	Stefan-Boltzmann constant	Btu/hr sq ft R^4	W/m^2 s
σ	surface tension	lb_f/ft	kg/kg
ϕ	phase lag angle	radians	rad
ϕ	latitude of location in Chapter 5	deg	rad
ψ	inclination from horizontal	deg	rad
ω	angular velocity	1/sec	1/s
ω	solid angle	steradian	sr
χ	quality	percent	percent

*DIMENSIONLESS GROUPS**

Bi	Biot number $= \bar{h}L/k_s$ or $\bar{h}r_o/k_s$
Fo	Fourier modulus $= a\theta/L^2$ or $a\theta/r_o^2$
Gz	Graetz number $= mc_p/k_f L$
Gr	Grashof number $= \beta g L^3 \Delta T/\nu^2$
j	Colburn j factor for heat transfer $=$ (Nu/Re Pr) $Pr^{2/3}$; j_M, j factor for mass transfer $=$ (Sh / Re Sc) $Sc^{2/3}$
Kn	Knudsen number
M	Mach number $= V/a$
Nu	Nusselt number $= h_c x/k_f$, Nu_x, local value of Nu at point x
$\overline{Nu}$	average value of Nu over surface $= \bar{h}_c\ L/k_f$; $\overline{Nu}_D$, diameter Nusselt number $= \bar{h}_c D/k_f$
Pe	Peclet number $=$ Re Pr
Pr	Prandtl number $= c_p\mu/k$ or ν/a
Re	Reynolds number $= V\rho L/\mu$; $Re_x = V\rho x/\mu$, local value of Re at a distance x from leading edge; Re_D, diameter Reynolds number; Re_b, bubble Reynolds number
θ	Boundary Fourier modulus $= \bar{h}^2 a\theta/k_s^2$
Sh	Sherwood number $= k_G \mathcal{R} T p_{Bm} L$
Sc	Schmidt number $= \mu/\rho D_v$
St	Stanton number $= h_c/\rho V c_p$ or Nu / RePr

MISCELLANEOUS

$a > b$	a greater than b	$\propto$	proportional sign
$a \gg b$	a much greater than b	$\cong$	approximately equal sign
$a < b$	a smaller than b	∞	infinity sign
$a \ll b$	a much smaller than b	Σ	summation sign

*The symbols used in this book for the dimensionless groups are generally in accordance with present day engineering usage, but differ slightly from those recommended recently by some committees of engineering societies who propose to use a capital N to denote any dimensionless group and then to identify the specific group by a subscript, e.g., N_{Nu} instead of Nu. It is, however, necessary to distinguish between local and average quantities and identify a significant length dimension as well as a temperature at which physical properties are to be evaluated. If these characteristics are indicated in the usual manner, that is, by sub- and superscripts attached to the symbol identifying the dimensionless group, combinations of symbols become clumsy and difficult to read. In order to avoid the use of a double subscript notation, the author decided, although not without serious misgivings, to omit the capital N from the symbols denoting the dimensionless groups.

Appendix II

Units, dimensions, and conversion factors

Numerical calculations in heat transfer, as in all other branches of engineering, require a consistent system of units. In the field of heat transfer a great variety of different units are encountered because contributions to this field have been made not only by engineers, but also by physicists and chemists of various countries. Many of the physical properties measured in the laboratory are in the International System (SI) of units, but engineers in this country generally use the engineering or technical system of units. Before one can proceed with numerical calculations, it is absolutely necessary to express all quantities in a consistent system. Several systems exist, each of them is equally correct. The choice is largely a matter of convenience but confusion between systems must be avoided.

A dimension is a name describing a geometrical or physical property which can be measured, observed or defined. It would be possible to assign a separate dimension to each property of interest, but it is more convenient to limit the number of dimensions to a few basic or primary dimensions and to express all other dimensions in terms of these fundamental quantities. The number of primary dimensions must of course be sufficient to express all derived or secondary dimensions in terms of them.

The physicist usually selects length, time, mass, and temperature as his primary set of dimensions. In heat transfer the dimension of the energy in transit due to a temperature difference, i.e. heat, is also included.

Dimensions differ from units of measurement. Dimensions describe a property *qualitatively* while units give a *quantitative* specification. For example, the length of a bar may be specified in feet, inches, or centimeters. All of these units are a quantitative specification of the primary dimension of length, L.

To familiarize the reader with the dimensions and units used in heat transfer, relations between some of the primary dimensions and the units associated with them will be briefly reviewed.

Time, θ, is the dimension of duration. The basic engineering unit is the second (sec), but in heat-transfer work the hour (hr) is frequently used. The SI unit is the second (s).

Length, L, is the dimension of distance. The basic engineering unit is the foot (ft). The SI unit is the meter (m).

Mass, M, is the dimension of quantity of matter. The basic engineering unit is the pound (lb_m). The SI unit is the kilogram (kg).

Temperature, T, is the dimension which describes the thermal potential of a system. It must be referred to an arbitrary datum, being somewhat analogous to the height above some reference level for gravitational action. The basic engineering unit is the degree Fahrenheit (F), which is closely equal to 1/180 of the temperature difference between the boiling and freezing temperature level of water at atmospheric pressure. For radiation phenomena the temperature is measured above absolute zero and is expressed in degrees Rankine (R). One Rankine degree equals one Fahrenheit degree, but the relation between the absolute values of the Fahrenheit and Rankine scales is

$$\text{Degrees Rankine} = 459.7 + \text{degrees Fahrenheit}$$

In SI units the absolute temperature is expressed in degrees Kelvin (K).

Force, F, is the dimension describing the action which tends to produce a change in the motion of a body. The basic engineering unit is the standard pound (lb_f) force, defined as the force necessary to support one pound mass under standard gravity conditions, corresponding to a gravitational force which accelerates one pound (lb_m) mass at the rate of 32.1739 ft/sec^2. Misunderstandings often arise because the word pound is used to denote the fundamental units of both mass and force. It is obvious of course that a pound of mass is an entirely different sort of thing from a pound force. In the SI system the unit of force is the newton (N), the force which accelerates 1 kg mass at 1 m/s^2. The international standard gravity is 9.80665 m/s^2.

Heat, Q, is the dimension of energy in transit by virtue of a temperature difference. The basic engineering unit is the British thermal unit or Btu, defined as the amount of heat required to raise the temperature of one pound mass of water at atmospheric pressure from 59.5 to 60.5 F.

Since heat is a form of energy it can be expressed in terms of its mechanical equivalent by means of the first law of thermodynamics. For a system whose state is not changed during a process the amount of heat added to the system Q must equal the work done by the system W, or

$$W = \mathrm{J}Q$$

where J is a dimensional conversion factor. Since work has the dimensions FL, J must have the dimensions FL/Q. For the system of units given here the experimentally measured value of the energy conversion factor J is

$$\mathrm{J} = 778.161 \text{ ft lb}_\mathrm{f}/\text{Btu}$$

which is often called "the mechanical equivalent of heat."

In the SI system the basic energy unit is the joule, J (newton meter), corresponding to the ft lb_f, but no separate unit for heat corresponding to the Btu is used. For the rate of heat flow, however, the SI system has a separate unit, the watt (w), corresponding to the Btu/hr in the engineering system.

Newton's second law of motion relates the independent physical quantities, force, mass, length and time, just as the first law of thermodynamics relates force, length, and heat. According to the second law of motion the net force F acting on a body of mass M is proportional to the product of the mass M and the acceleration a, or

$$F = \frac{1}{g_c} Ma$$

where g_c is an experimentally determined constant whose dimensions are always $ML/F\theta^2$, but whose magnitude depends on the units of force, mass, length, and time as shown in the following tabulation.

Mass	Length	Time	Force	g_c
lb_m	ft	sec	lb_f	32.1739 lb_m ft/lb_f sec^2
slug	ft	sec	lb_f	1.0 slug ft/lb_f sec^2
slug	ft	hr	lb_f	1.296×10^7 slug ft/lb_f hr^2
lb_m	ft	sec	poundal	1.0 lb_m ft/poundal sec^2
kg	m	sec	newton	1.0 kg m/N s^2

Table of conversion factors.

Length:	1 in. = 0.08333 ft
	1 cm = 0.03281 ft
	1 mile = 5280 ft
	1 μ (micron) = 3.281×10^{-6} ft
	1 A (angstrom unit) = 10^{-8} cm
Mass:	1 kg (kilogram) = 2.205 lb_m
	1 g (gram) = 2.205×10^{-3} lb_m
	1 slug = 32.1739 lb_m
Force:	1 poundal = 0.03108 lb_f
	1 dyne = 2.248×10^{-6} lb_f
	1 kg = 2.205 lb_f
Energy:	1 ft-lb_f = 0.001285 Btu
	1 kw-hr (kilowatt-hour) = 3413 Btu
	1 hp (horsepower) = 2544 Btu/hr
	1 kcal (kilocalorie) = 3.968 Btu
	1 joule = 9.478×10^{-4} Btu
Heat flow rate per unit area:	1 cal/sec sq cm = 13,272 Btu/hr sq ft
	1 watt/sq cm = 3171 Btu/hr sq ft
	1 cal/hr sq cm = 3.687 Btu/hr sq ft
Pressure:	1 atm = 2116 psf
	1 dyne/sq cm = 0.00209 psf
	1 cm Hg = 27.85 psf
	1 in. Hg = 70.73 psf
	1 in. water = 5.20 psf
	1 ft water = 62.43 psf
Density:	1 gm/cu cm = 62.43 lb_m/cu ft
	1 lb_m/gallon = 7.481 lb_m/cu ft
	1 lb_m/cu in. = 1728 lb_m/cu ft
Temperature:	1 R (degree Rankine) = 1 F (degree Fahrenheit)
	1 C (degree Centigrade) = 1.8 F
	1 K (degree Kelvin) = 1.8 F
Specific energy per degree:	1 cal/g C = 1 Btu/lb_m F
Thermal conductivity:	1 cal/sec sq cm (C/cm) = 241.9 Btu/hr sq ft (F/ft)
	1 watts/sq cm (C/cm) = 57.79 Btu/hr sq ft (F/ft)
	1 Btu/hr sq ft (F/in.) = 0.08333 Btu/hr sq ft (F/ft)
Unit thermal conductance:	1 cal/sec sq cm C = 7373 Btu/hr sq ft F
	1 watt/sq cm C = 1761 Btu/hr sq ft F
	1 cal/hr sq cm C = 2.048 Btu/hr sq ft F
Viscosity:	1 cp (centipoise) = 0.000672 lb_m/sec ft
	1 cp = 2.42 lb_m/hr ft
	1 lb_f sec/sq ft = 32.174 lb_m/sec ft
Volume:	1 gal (U.S.) = 0.1337 cu ft
	1 cu ft = 28.32 liters.

NOTE: To convert a given quantity from one set of units to another:

1. Write after the magnitude of the quantity the names of the units in which it is measured.
2. Replace each name by its equivalent in the new units, and arithmetically combine all numbers in the new expression.

For example, to change the density of water from slugs per cubic foot into pounds-mass per cubic foot, we have (to three significant figures)

$$\rho = 1.94 \text{ slugs/cu ft} = (1.94 \text{ \sout{slugs}/cu ft})(32.2 \text{ lb}_m/\text{\sout{slug}})$$
$$= 1.94 \times 32.2 \text{ lb}_m/\text{cu ft} = 62.4 \text{ lb}_m/\text{cu ft}$$

Table of conversion factor between U.S. engineering units and SI units.*

Physical Quantity	U.S. Engineering Unit	SI Unit	Conversion Constant
Length:	foot	m	1 ft = 0.3048 m
Area:	sq ft	m^2	1 sq ft = 0.0929 m^2
Volume:	cu ft	m^3	1 cu ft = 0.0283 m^3
	gallon		1 gallon = 0.004546 m^3
Mass:	pound	kg	1 lb_m = 0.4536 kg
Density:	lb_m/ft^3	kg m^{-3}	1 lb_m/ft^3 = 16.02 kg m^{-3}
Force:	pound (force)	N	1 lb_f = 4.448 N
Pressure:	psi	Nm^{-2}	1 psi = 6894.8 Nm^{-2}
	lb_f/ft^2		1 lb_f/ft^2 = 47.88 Nm^{-2}
	atmosphere		1 atm = 101.30 Nm^{-2}
Temperature:	deg R	K	1°R = (5/9) K
	deg F		t (°F) = (K − 273) 9/5 + 32
Energy:	Btu	J	1 Btu = 1055.1 J
	calorie		1 cal = 4.186 J
	ft-lb_f		1 ft-lb_f = 1.3558 J
Heat flow rate:	Btu/hr	w	1 Btu/hr = 0.293 w
Power:	HP	w	1 HP = 745.7 w
Heat Flux:	Btu/hr ft^2	wm^{-2}	1 Btu/hr ft^2 = 3.152 wm^{-2}
Spec. heat capacity:	Btu/lb F	J kg^{-1} K^{-1}	1 Btu/lb F = 4.184 J $kg^{-1}K^{-1}$
Thermal conductivity:	Btu/hr ft F	w m^{-1} K^{-1}	1 Btu/hr ft F = 1.731 wm^{-1} K^{-1}
Heat-transfer coefficient:	Btu/hr ft^2 F	w m^{-2} K^{-1}	1 Btu/hr ft^2 F = 5.67 wm^{-2} K^{-1}
Velocity:	ft/sec	m s^{-1}	1 ft/sec = 0.3048 m s^{-1}
Mass flow rate:	lb_m/hr	kg s^{-1}	1 lb/hr = 0.000126 kg s^{-1}
Viscosity:	lb_m/hr ft	Ns/m^2	1 lb_m/ hr ft = 2419.0 Ns/m^2
	lb_m/sec ft		1 lb_m/sec ft = 0.672 Ns/m^2
	ft^2/sec	m^2 s^{-1}	1 ft^2/sec = 0.1076 m^2 s^{-1}

*Abstracted from *Physical Measurements* by R. A. Ackley, Tech. Publications, San Diego, Calif., 1970, and *The International Metric System* by W. G. Canham, Chem. Eng. Prog., Vol. 68, July 1972, pp. 90–94.

In the engineering system, used in this text and shown in the fourth column of the table on page 629, g_c equals 32.1739 $lb_m ft/lb_f\ sec^2$, but a value of 32.2 is a satisfactory approximation in practice. In the SI system

$g_c = 1.0$ kg m/Ns^2. It is important to note that g_c is a universal constant entirely different from the acceleration of gravity which has the dimensions L/θ^2 and whose numerical value depends on the location.

When one is in doubt which system of dimensions and units is used in a reference, it is suggested that all of the dimensions and units in one of the equations be written out and the equation checked dimensionally. The procedure is illustrated in Sec. 6-6.

In heat transfer calculations it is most convenient to express all quantities in terms of feet, hours (or seconds), Btu, pound-mass, and degrees Fahrenheit. This choice of units does not cause trouble until one encounters problems in which fluid dynamics is involved. In fluid dynamics both force and mass are used as primary dimensions. The density of fluids is commonly expressed in pound-mass per cubit foot, but the viscosity is often given in pound-force-second per square foot. The pressure drop and the shear are always given in pound-force units. Since all physical properties in this book, including the viscosity, are expressed in pound-mass units, it is necessary to include the conversion factor g_c in equations derived from Newton's second law.

The Tables of Conversion Factors will be helpful in converting the units of a given quantity into the units used in this text.

Appendix III

Tables

The following tables have been compiled to facilitate the solution of the problems at the end of each chapter and are not intended to take the place of a handbook. Whenever answers to problems are given, they have been obtained with the aid of these tables.

Table A-1 gives the properties of metals and alloys. Table A-2 lists physical properties of nonmetals such as insulating and building materials. Table A-3 presents the property values of several gases at atmospheric pressure, of some liquids, and of three liquid metals. The property values have been extracted from various sources. The bibliography following Table A-3 lists these sources with the exception of some manufacturers' catalogs which may not be readily available. The reader interested in additional information on physical properties should consult the publications listed in the bibliography.

In Table A-4 the radiation functions described in Chapter 5 are tabulated. Tables A-5 and A-6 list the dimensions of tubes and steel pipes respectively. It should be noted that the schedule number is now used exclusively to characterize the pipe-wall thickness which was previously designated by "standard" or "extra strong."

Table A-7 contains selected physical properties of the atmosphere at altitudes up to 900,000 ft.

Table A-1. Thermal conductivity k, specific heat c, density ρ, and thermal diffusivity α of metals and alloys.

Material	k (Btu/hr ft F)				c (Btu/lb$_m$ F)	ρ (lb$_m$/cu ft)	α (sq ft/hr)
	32 F	212 F	572 F	932 F	32 F	32 F	32 F
Metals							
Aluminum	117	119	133	155	0.208	169	3.33
Bismuth	4.9	3.9	....	....	0.029	612	0.28
Copper, pure	224	218	212	207	0.091	558	4.42
Gold	169	170	....	....	0.030	1203	4.68
Iron, pure	35.8	36.6	....	....	0.104	491	0.70
Lead	20.1	19	18	....	0.030	705	0.95
Magnesium	91	92	....	....	0.232	109	3.60
Mercury	4.8	....	....	....	0.033	849	0.17
Nickel	34.5	34	32	....	0.103	555	0.60
Silver	242	238	....	....	0.056	655	6.6
Tin	36	34	....	....	0.054	456	1.46
Zinc	65	64	59	....	0.091	446	1.60
Alloys							
Admiralty metal	65	64					
Brass, 70% Cu, 30% Zn	56	60	66	....	0.092	532	1.14
Bronze, 75% Cu, 25% Sn	15	....	....	....	0.082	540	0.34
Cast iron							
Plain	33	31.8	27.7	24.8	0.11	474	0.63
Alloy	30	28.3	27	....	0.10	455	0.66
Constantan, 60% Cu, 40% Ni	12.4	12.8	....	....	0.10	557	0.22
18–8 stainless steel,							
Type 304	8.0	9.4	10.9	12.4	0.11	488	0.15
Type 347	8.0	9.3	11.0	12.8	0.11	488	0.15
Steel, mild, 1% C	26.5	26	25	22	0.11	490	0.49

Table A-2. Physical properties of some nonmetals.

Material	Average Temperature (F)	k (Btu/hr ft F)	c (Btu/lb_mF)	ρ (lb_m/cu ft)	a (sq ft/hr)
Insulating Materials					
Asbestos	32	0.087	0.25	36	~0.01
	392	0.12	....	36	~0.01
Cork	86	0.025	0.04	10	~0.006
Cotton, fabric	200	0.046			
Diatomaceous earth, powdered	100	0.030	0.21	14	~0.01
	300	0.036	....		
	600	0.046	....		
Molded pipe covering	400	0.051	....	26	
	1600	0.088	....		
Glass wool					
Fine	20	0.022	....		
	100	0.031	....	1.5	
	200	0.043	....		
Packed	20	0.016	....		
	100	0.022	....	6.0	
	200	0.029	....		
Hair felt	100	0.027	....	8.2	
Kaolin insulating brick	932	0.15	....	27	
	2102	0.26	....		
Kaolin insulating firebrick	392	0.05	....	19	
	1400	0.11	....		
85% magnesia	32	0.032	....	17	
	200	0.037	....	17	
Rock wool	20	0.017	....	8	
	200	0.030	....		
Rubber	32	0.087	0.48	75	0.0024
Building Materials					
Brick					
Fire-clay	392	0.58	0.20	144	0.02
	1832	0.95			
Masonry	70	0.38	0.20	106	0.018
Zirconia	392	0.84	....	304	
	1832	1.13	....		
Chrome brick	392	0.82	....	246	
	1832	0.96			
Concrete					
Stone	~70	0.54	0.20	144	0.019
10% moisture	~70	0.70	....	140	~0.025
Glass, window	~70	~0.45	0.2	170	0.013
Limestone, dry	70	0.40	0.22	105	0.017
Sand					
Dry	68	0.20	....	95	
10% H_2O	68	0.60	....	100	
Soil					
Dry	70	~0.20	0.44	....	~0.01
Wet	70	~1.5	....	......	~0.03
Wood					
Oak ⊥ to grain	70	0.12	0.57	51	0.0041
∥ to grain	70	0.20	0.57	51	0.0069
Pine ⊥ to grain	70	0.06	0.67	31	0.0029
∥ to grain	70	0.14	0.67	31	0.0067
Ice	32	1.28	0.46	57	0.048

Table A-3. Physical properties of gases, liquids, and liquid metals (all gas properties are for atmospheric pressure).

GASES

T (F)	ρ (lbm/cu ft)	c_p (Btu/lbm F)	$\mu \times 10^5$ (lbm/ft sec)	$\nu \times 10^3$ (sq ft/sec)	k (Btu/hr ft F)	Pr	a (sq ft/hr)	$\beta \times 10^3$ (1/F)	$\frac{g\beta\rho^2}{\mu^2}$ (1/F cu ft)
Air									
0	0.086	0.239	1.110	0.130	0.0133	0.73	0.646	2.18	4.2×10^6
32	0.081	0.240	1.165	0.145	0.0140	0.72	0.720	2.03	3.16
100	0.071	0.240	1.285	0.180	0.0154	0.72	0.905	1.79	1.76
200	0.060	0.241	1.440	0.239	0.0174	0.72	1.20	1.52	0.850
300	0.052	0.243	1.610	0.306	0.0193	0.71	1.53	1.32	0.444
400	0.046	0.245	1.750	0.378	0.0212	0.689	1.88	1.16	0.258
500	0.0412	0.247	1.890	0.455	0.0231	0.683	2.27	1.04	0.159
600	0.0373	0.250	2.000	0.540	0.0250	0.685	2.68	0.943	0.106
700	0.0341	0.253	2.14	0.625	0.0268	0.690	3.10	0.862	70.4×10^3
800	0.0314	0.256	2.25	0.717	0.0286	0.697	3.56	0.794	49.8
900	0.0291	0.259	2.36	0.815	0.0303	0.705	4.02	0.735	36.0
1000	0.0271	0.262	2.47	0.917	0.0319	0.713	4.50	0.685	26.5
1500	0.0202	0.276	3.00	1.47	0.0400	0.739	7.19	0.510	7.45
2000	0.0161	0.286	3.45	2.14	0.0471	0.753	10.2	0.406	2.84
2500	0.0133	0.292	3.69	2.80	0.051	0.763	13.1	0.338	1.41
3000	0.0114	0.297	3.86	3.39	0.054	0.765	16.0	0.289	0.815
Steam									
212	0.0372	0.451	0.870	0.234	0.0145	0.96	0.864	1.49	0.877×10^6
300	0.0328	0.456	1.000	0.303	0.0171	0.95	1.14	1.32	0.459
400	0.0288	0.462	1.130	0.395	0.0200	0.94	1.50	1.16	0.243
500	0.0258	0.470	1.265	0.490	0.0228	0.94	1.88	1.04	0.139
600	0.0233	0.477	1.420	0.610	0.0257	0.94	2.31	0.943	82×10^3
700	0.0213	0.485	1.555	0.725	0.0288	0.93	2.79	0.862	52.1
800	0.0196	0.494	1.700	0.855	0.0321	0.92	3.32	0.794	34.0
900	0.0181	0.50	1.810	0.987	0.0355	0.91	3.93	0.735	23.6
1000	0.0169	0.51	1.920	1.13	0.0388	0.91	4.50	0.685	17.1
1200	0.0149	0.53	2.14	1.44	0.0457	0.88	5.80	0.603	9.4
1400	0.0133	0.55	2.36	1.78	0.053	0.87	7.25	0.537	5.49
1600	0.0120	0.56	2.58	2.14	0.061	0.87	9.07	0.485	3.38
1800	0.0109	0.58	2.81	2.58	0.068	0.87	10.8	0.442	2.14
2000	0.0100	0.60	3.03	3.03	0.076	0.86	12.7	0.406	1.43
2500	0.0083	0.64	3.58	4.30	0.096	0.86	18.1	0.338	0.603
3000	0.0071	0.67	4.00	5.75	0.114	0.86	24.0	0.289	0.293
Oxygen									
0	0.0955	0.2185	1.215	0.127	0.0131	0.73	0.627	2.18	4.33×10^6
100	0.0785	0.2200	1.420	0.181	0.0159	0.71	0.880	1.79	1.76
200	0.0666	0.2228	1.610	0.242	0.0179	0.722	1.20	1.52	0.84
400	0.0511	0.2305	1.955	0.382	0.0228	0.710	1.94	1.16	0.256
600	0.0415	0.2390	2.26	0.545	0.0277	0.704	2.79	0.943	0.103
800	0.0349	0.2465	2.53	0.725	0.0324	0.695	3.76	0.794	48.5×10^3
1000	0.0301	0.2528	2.78	0.924	0.0366	0.690	4.80	0.685	25.8
1500	0.0224	0.2635	3.32	1.480	0.0465	0.677	7.88	0.510	7.50

Table A-3. Continued.

T (F)	ρ (lb_m/cu ft)	c_p (Btu/ lb_m F)	$\mu \times 10^5$ (lb_m/ ft sec)	$\nu \times 10^3$ (sq ft/ sec)	k (Btu/ hr ft F)	Pr	a (sq ft/hr)	$\beta \times 10^3$ (1/F)	$\frac{g\beta\rho^2}{\mu^2}$ (1/F cu ft)
				Nitrogen					
0	0.0840	0.2478	1.055	0.125	0.0132	0.713	0.635	2.18	4.55×10^6
100	0.0690	0.2484	1.222	0.177	0.0154	0.71	0.898	1.79	1.84
200	0.0585	0.2490	1.380	0.236	0.0174	0.71	1.20	1.52	0.876
400	0.0449	0.2515	1.660	0.370	0.0212	0.71	1.88	1.16	0.272
600	0.0364	0.2564	1.915	0.526	0.0252	0.70	2.70	0.943	0.110
800	0.0306	0.2623	2.145	0.702	0.0291	0.70	3.62	0.794	52.0×10^3
1000	0.0264	0.2689	2.355	0.891	0.0330	0.69	4.65	0.685	27.7
1500	0.0197	0.2835	2.800	1.420	0.0423	0.676	7.58	0.510	8.12
				Carbon Monoxide					
0	0.0835	0.2482	1.065	0.128	0.0129	0.75	0.621	2.18	4.32×10^6
200	0.0582	0.2496	1.390	0.239	0.0169	0.74	1.16	1.52	0.860
400	0.0446	0.2532	1.670	0.374	0.0208	0.73	1.84	1.16	0.268
600	0.0362	0.2592	1.910	0.527	0.0246	0.725	2.62	0.943	0.109
800	0.0305	0.2662	2.134	0.700	0.0285	0.72	3.50	0.794	52.1×10^3
1000	0.0263	0.2730	2.336	0.887	0.0322	0.71	4.50	0.685	28.0
1500	0.0196	0.2878	2.783	1.420	0.0414	0.70	7.33	0.510	8.13
				Helium					
0	0.012	1.24	1.140	0.950	0.078	0.67	5.25	2.18	77800
200	0.00835	1.24	1.480	1.77	0.097	0.686	9.36	1.52	15600
400	0.0064	1.24	1.780	2.78	0.115	0.70	14.5	1.16	4840
600	0.0052	1.24	2.02	3.89	0.129	0.715	20.0	0.943	2010
800	0.00436	1.24	2.285	5.24	0.138	0.73	25.5	0.794	932
1000	0.00377	1.24	2.520	6.69	......	.....	....	0.685	494
1500	0.0028	1.24	3.160	11.30	......	.....	....	0.510	129
				Hydrogen					
0	0.0060	3.39	0.540	0.89	0.094	0.70	4.62	2.18	86600
100	0.0049	3.42	0.620	1.26	0.110	0.695	6.56	1.79	36600
200	0.0042	3.44	0.692	1.65	0.122	0.69	8.45	1.52	18000
500	0.0028	3.47	0.884	3.12	0.160	0.69	16.5	1.04	3360
1000	0.0019	3.51	1.160	6.2	0.208	0.705	31.2	0.685	591
1500	0.0014	3.62	1.415	10.2	0.260	0.71	51.4	0.510	161
2000	0.0011	3.76	1.64	14.4	0.307	0.72	74.2	0.406	59
3000	0.0008	4.02	1.72	24.2	0.380	0.66	118.0	0.289	20
				Carbon Dioxide					
0	0.132	0.184	0.88	0.067	0.0076	0.77	0.313	2.18	15.8×10^6
100	0.108	0.203	1.05	0.098	0.0100	0.77	0.455	1.79	6.10
200	0.092	0.216	1.22	0.133	0.0125	0.76	0.63	1.52	2.78
500	0.063	0.247	1.67	0.266	0.0198	0.75	1.27	1.04	0.476
1000	0.0414	0.280	2.30	0.558	0.0318	0.73	2.75	0.685	71.4×10^3
1500	0.0308	0.298	2.86	0.925	0.0420	0.73	4.58	0.510	19.0
2000	0.0247	0.309	3.30	1.34	0.050	0.735	6.55	0.406	7.34
3000	0.0175	0.322	3.92	2.25	0.061	0.745	10.8	0.289	1.85

Table A-3. Continued.

LIQUIDS

T (F)	ρ (lb_m/cu ft)	c_p (Btu/lb_m F)	$\mu \times 10^3$ (lb_m/ft sec)	$\nu \times 10^5$ (sq ft/sec)	k (Btu/hr ft F)	Pr	$a \times 10^3$ (sq ft/hr)	$\beta_T \times 10^4$ (1/F)	$\frac{g\beta\rho^2}{\mu^2}$ (1/F cu ft)
Water									
32	62.4	1.01	1.20	1.93	0.319	13.7	5.07	−0.37	
40	62.4	1.00	1.04	1.67	0.325	11.6	5.21	0.20	2.3×10^6
50	62.4	1.00	0.88	1.40	0.332	9.55	5.33	0.49	8.0
60	62.3	0.999	0.76	1.22	0.340	8.03	5.47	0.85	18.4
70	62.3	0.998	0.658	1.06	0.347	6.82	5.57	1.2	34.6
80	62.2	0.998	0.578	0.93	0.353	5.89	5.68	1.5	56.0
90	62.1	0.997	0.514	0.825	0.359	5.13	5.79	1.8	85.0
100	62.0	0.998	0.458	0.740	0.364	4.52	5.88	2.0	118×10^6
150	61.2	1.00	0.292	0.477	0.384	2.74	6.27	3.1	440.0
200	60.1	1.00	0.205	0.341	0.394	1.88	6.55	4.0	1.11×10^9
250	58.8	1.01	0.158	0.269	0.396	1.45	6.69	4.8	2.14
300	57.3	1.03	0.126	0.220	0.395	1.18	6.70	6.0	4.00
350	55.6	1.05	0.105	0.189	0.391	1.02	6.69	6.9	6.24
400	53.6	1.08	0.091	0.170	0.381	0.927	6.57	8.0	8.95
450	51.6	1.12	0.080	0.155	0.367	0.876	6.34	9.0	12.1
500	49.0	1.19	0.071	0.145	0.349	0.87	5.99	10.0	15.3
550	45.9	1.31	0.064	0.139	0.325	0.93	5.05	11.0	17.8
600	42.4	1.51	0.058	0.137	0.292	1.09	4.57	12.0	20.6

T (F)	ρ (lb_m/cu ft)	c_p (Btu/lb_m F)	$\mu \times 10^5$ (lb_m/ft sec)	$\nu \times 10^5$ (sq ft/sec)	k (Btu/hr ft F)	Pr	$a \times 10^3$ (sq ft/hr)	$\beta_T \times 10^3$ (1/F)	$\frac{g\beta\rho^2}{\mu^2}$ (1/F cu ft)
Commercial Aniline									
60	64.0	0.48	325.0	5.08	0.10	56.0	3.25		
100	63.0	0.49	170.0	2.70	0.10	30.0	3.24	0.49	21.6×10^6
150	61.5	0.505	96.5	1.57	0.098	18.0	3.16	0.492	64.5
200	60.0	0.515	61.1	1.02	0.096	11.8	3.11		
300	57.5	0.54	32.5	0.565	0.093	6.8	3.00		
Ammonia (Saturated Liquid)									
−20	42.4	1.07	17.6	0.417	0.317	2.15	6.94		
0	41.6	1.08	17.1	0.410	0.316	2.09	7.04		
10	40.8	1.09	16.6	0.407	0.314	2.07	7.08		
32	40.0	1.11	16.1	0.402	0.312	2.05	7.03	1.2	238×10^6
50	39.1	1.13	15.5	0.396	0.307	2.04	6.95	1.3	266
80	37.2	1.17	14.5	0.386	0.293	2.01	6.73		
120	35.2	1.22	13.0	0.355	0.275	1.99	6.40		
Freon 12, CCl_2F_2, (Saturated Liquid)									
−40	94.8	0.211	28.4	0.300	0.040	5.4	2.00		
−20	93.0	0.214	25.0	0.272	0.040	4.8	2.01	1.03	4.6×10^9
0	91.2	0.217	23.1	0.253	0.041	4.4	2.07	1.05	5.27
20	89.2	0.220	21.0	0.238	0.042	4.0	2.14	1.34	7.80
32	87.2	0.223	20.0	0.230	0.042	3.8	2.16	1.72	10.5
60	83.0	0.231	18.0	0.213	0.042	3.5	2.19	2.1	14.4
100	78.5	0.240	16.0	0.206	0.040	3.5	2.12	2.5	19.4
120	75.9	0.244	15.5	0.204	0.039	3.5	2.12		

Table A-3. Continued.

T (F)	ρ (lb_m/cu ft)	c_p (Btu/lb_m F)	$\mu \times 10^5$ (lb_m/ft sec)	$\nu \times 10^5$ (sq ft/sec)	k (Btu/hr ft F)	Pr	$\alpha \times 10^3$ (sq ft/hr)	$\beta \times 10^3$ (1/F)	$\frac{g\beta\rho^2}{\mu^2}$ (1/F cu ft)
				n-Butyl Alcohol					
60	50.5	0.55	226	4.48	0.097	46.6	3.49		
100	49.7	0.61	129	2.60	0.096	29.5	3.16	0.45	21.5×10^6
150	48.5	0.68	67.5	1.39	0.095	17.4	2.88	0.48	80
200	47.2	0.77	38.6	0.815	0.094	11.3	2.58		
300	.....	.....	19.0						
				Benzene					
60	55.1	0.40	46.0	0.835	0.093	7.2	4.22	0.60	0.3×10^9
80	54.6	0.42	39.6	0.725	0.092	6.5	4.01		
100	54.0	0.44	35.1	0.650	0.087	5.1	3.53		
150	53.5	0.46	26.0	0.480	.....	4.5			
200	.....	.....	20.3	.....	.....	4.0			
				Light Oil					
60	57.0	0.43	5820	102	0.077	1170	3.14	0.38	1.17×10^4
80	56.8	0.44	2780	49	0.077	570	3.09	0.38	5.1
100	56.0	0.46	1530	27.4	0.076	340	2.95	0.39	16.7
150	54.3	0.48	530	9.8	0.075	122	2.88	0.40	1.34×10^6
200	54.0	0.51	250	4.6	0.074	62	2.69	0.42	6.4
250	53.0	0.52	139	2.6	0.074	35	2.67	0.44	21.0
300	51.8	0.54	83	1.6	0.073	22	2.62	0.45	56.5

T (F)	ρ (lb_m/cu ft)	c_p (Btu/lb_m F)	$\mu \times 10^2$ (lb_m/ft sec)	$\nu \times 10^2$ (sq ft/sec)	k (Btu/hr ft F)	Pr	$\alpha \times 10^3$ (sq ft/hr)	$\beta \times 10^3$ (1/F)	$\frac{g\beta\rho^2}{\mu^2}$ (1/F cu ft)
				Glycerin					
50	79.3	0.554	256	3.23	0.165	31×10^3	3.76		
70	78.9	0.570	100	1.27	0.165	12.5	3.67	0.28	56
85	78.5	0.584	42.4	0.54	0.164	5.4	3.58	0.30	332
100	78.2	0.600	18.8	0.24	0.163	2.5	3.45		
120	77.7	0.617	12.4	0.16	.....	$\simeq 1.6$			

LIQUID METALS

T (F)	ρ (lb_m/cu ft)	c_p (Btu/lb_m F)	$\mu \times 10^3$ (lb_m/ft sec)	$\nu \times 10^6$ (sq ft/sec)	k (Btu/hr ft F)	Pr	α (sq ft/hr)	$\beta \times 10^3$ (1/F)	$\frac{g\beta\rho^2}{\mu^2}$ (1/F cu ft)
				Bismuth					
600	625	0.0345	1.09	1.74	9.5	0.014	0.44	0.065	0.687×10^9
800	616	0.0357	0.90	1.5	9.0	0.013	0.41	0.068	
1000	608	0.0369	0.74	1.2	9.0	0.011	0.40	0.070	
1200	600	0.0381	0.62	1.0	9.0	0.009	0.39		
1400	591	0.0393	0.53	0.9	9.0	0.008	0.39		

Table A-3. Continued.

T (F)	ρ (lb_m/cu ft)	c_p (Btu/lb_m F)	$\mu \times 10^3$ (lb_m/ft sec)	$\nu \times 10^6$ (sq ft/sec)	k (Btu/hr ft F)	Pr	$a \times 10^3$ (sq ft/hr)	$\beta \times 10^3$ (1/F)	$\frac{g\beta\rho^2}{\mu^2}$ (1/F cu ft)
				Mercury					
50	847	0.033	1.07	1.2	4.7	0.027	0.17	0.1	2.02×10^9
200	834	0.033	0.84	1.0	6.0	0.016	0.22	0.1	2.02
300	826	0.033	0.74	0.9	6.7	0.012	0.25		
400	817	0.032	0.67	0.8	7.2	0.011	0.27		
600	802	0.032	0.58	0.7	8.1	0.008	0.31		
				Sodium					
200	58.0	0.33	0.47	8.1	49.8	0.011	2.6	0.150	73.5×10^6
400	56.3	0.32	0.29	5.1	46.4	0.007	2.6	0.20	243
700	53.7	0.31	0.19	3.5	41.8	0.005	2.5		
1000	51.2	0.30	0.14	2.7	37.8	0.004	2.4		
1300	48.6	0.30	0.12	2.5	34.5	0.004	2.4		

BIBLIOGRAPHY FOR PHYSICAL PROPERTIES

1. *International Critical Tables*, New York: McGraw-Hill Book Company, Inc., 1929.
2. L. S. Marks, et al., *Mechanical Engineers' Handbook*, ed. New York: McGraw-Hill Book Company, Inc.
3. J. H. Perry, ed., *Chemical Engineers' Handbook*, 3rd ed., McGraw-Hill Book Company, Inc., 1950.
4. J. L. Everhart, W. E. Lindlief, J. Kanegis, P. G. Weissler, and F. Siegel, *Mechanical Properties of Metals and Alloys*, Circular C447, U.S. Department of Commerce, National Bureau of Standards, Washington, D.C., 1943.
5. J. H. Keenan and F. G. Keyes, *Thermodynamic Properties of Steam*, New York: John Wiley & Sons, Inc., 1936.
6. J. H. Keenan and J. Kaye, *Gas Tables*, New York: John Wiley & Sons, Inc., 1945.
7. W. H. McAdams, *Heat Transmission*, 3rd ed., McGraw-Hill Book Company, Inc., 1954.
8. E. Schmidt, *Thermodynamics*, Oxford at the Claredon Press, 1949.
9. J. Hilsenrath, C. W. Beckett, W. S. Benedict, L. Fano, H. M. Hoge, J. F. Masi, R. L. Nuttall, Y. S. Touloukian, and H. W. Woolley, *Tables of the Thermal Properties of Gases*, National Bureau of Standards Circular 564, Washington, D.C., 1955.
10. F. B. Rowley and A. B. Algren, *Thermal Conductivity of Building Materials*, Bulletin No. 12, Eng. Exp. St., Univ. of Minnesota, 1937.
11. L. S. Kowalczyk, "Thermal Conductivity and its Variability with Temperature and Pressure," *Trans. ASME*, Vol. 77 (1955) p. 1021.
12. *Liquid-Metals Handbook*, 2d ed., U.S. Govt. Printing Office, Washington, D.C., 1952.
13. C. L. Mantell, ed., *Engineering Materials Handbook*, McGraw-Hill Book Company, Inc., 1958.

Table A-4. Radiation functions.*

λT	$\frac{E_{\lambda b} \times 10^5}{\sigma T^5}$	$\frac{E_{b(0-\lambda T)}}{\sigma T^4}$	λT	$\frac{E_{\lambda b} \times 10^5}{\sigma T^5}$	$\frac{E_{b(0-\lambda T)}}{\sigma T^4}$	λT	$\frac{E_{\lambda b} \times 10^5}{\sigma T^5}$	$\frac{E_{b(0-\lambda T)}}{\sigma T^4}$
1000	.0000394	0	7200	10.089	.4809	13400	2.714	.8317
1200	.001184	0	7400	9.723	.5007	13600	2.605	.8370
1400	.01194	0	7600	9.357	.5199	13800	2.502	.8421
1600	.0618	.0001	7800	8.997	.5381	14000	2.416	.8470
1800	.2070	.0003	8000	8.642	.5558	14200	2.309	.8517
2000	.5151	.0009	8200	8.293	.5727	14400	2.219	.8563
2200	1.0384	.0025	8400	7.954	.5890	14600	2.134	.8606
2400	1.791	.0053	8600	7.624	.6045	14800	2.052	.8648
2600	2.753	.0098	8800	7.304	.6195	15000	1.972	.8688
2800	3.872	.0164	9000	6.995	.6337	16000	1.633	.8868
3000	5.081	.0254	9200	6.697	.6474	17000	1.360	.9017
3200	6.312	.0368	9400	6.411	.6606	18000	1.140	.9142
3400	7.506	.0506	9600	6.136	.6731	19000	.962	.9247
3600	8.613	.0667	9800	5.872	.6851	20000	.817	.9335
3800	9.601	.0850	10000	5.619	.6966	21000	.702	.9411
4000	10.450	.1051	10200	5.378	.7076	22000	.599	.9475
4200	11.151	.1267	10400	5.146	.7181	23000	.516	.9531
4400	11.704	.1496	10600	4.925	.7282	24000	.448	.9589
4600	12.114	.1734	10800	4.714	.7378	25000	.390	.9621
4800	12.392	.1979	11000	4.512	.7474	26000	.341	.9657
5000	12.556	.2229	11200	4.320	.7559	27000	.300	.9689
5200	12.607	.2481	11400	4.137	.7643	28000	.265	.9718
5400	12.571	.2733	11600	3.962	.7724	29000	.234	.9742
5600	12.458	.2983	11800	3.795	.7802	30000	.208	.9765
5800	12.282	.3230	12000	3.637	.7876	40000	.0741	.9881
6000	12.053	.3474	12200	3.485	.7947	50000	.0326	.9941
6200	11.783	.3712	12400	3.341	.8015	60000	.0165	.9963
6400	11.480	.3945	12600	3.203	.8081	70000	.0092	.9981
6600	11.152	.4171	12800	3.071	.8144	80000	.0055	.9987
6800	10.808	.4391	13000	2.947	.8204	90000	.0035	.9990
7000	10.451	.4604	13200	2.827	.8262	100000	.0023	.9992
						∞	0	1.0000

* From Dunkle, R. V., *Trans. ASME*, **76**, 549 (1954).

Table A-5. Average properties of tubes.

Diameter		Thickness		External			Internal					Length of Tube Containing One Cu Ft
								Volume or Capacity per Lineal Foot				
External (in.)	Internal (in.)	BWG Gage	NOM Wall (in.)	Circumference (in.)	Surface per Lineal Foot (sq ft)	Lineal Feet of Tube per Square Foot of Surface	Transverse Area (sq in.)	Cu In.	Cu Ft	U. S. Gal		
	0.527	18	.049	1.9635	0.1636	6.1115	0.218	2.616	0.0015	0.011		661
⅝	0.495	16	.065				0.193	2.316	0.0013	0.010		746
	0.459	14	.083	↓	↓	↓	0.166	1.992	0.0011	0.009		867
	0.652	18	.049	2.3562	0.1963	5.0930	0.334	4.008	0.0023	0.017		431
	0.620	16	.065				0.302	3.624	0.0021	0.016		477
¾	0.584	14	.083				0.268	3.216	0.0019	0.014		537
	0.560	13	.095	↓	↓	↓	0.246	2.952	0.0017	0.013		585
	0.902	18	.049	3.1416	.2618	3.8197	0.639	7.668	0.0044	0.033		225
1	0.870	16	.065				0.595	7.140	0.0041	0.031		242
	0.834	14	.083				0.546	6.552	0.0038	0.028		264
	0.810	13	.095	↓	↓	↓	0.515	6.180	0.0036	0.027		280
	1.152	18	.049	3.9270	.3272	3.0558	1.075	12.90	0.0075	0.056		134
	1.120	16	.065				0.985	11.82	0.0068	0.051		146
1¼	1.084	14	.083				0.923	11.08	0.0064	0.048		156
	1.060	13	.095				0.882	10.58	0.0061	0.046		163
	1.032	12	.109	↓	↓	↓	0.836	10.03	0.0058	0.043		172
	1.402	18	.049	4.7124	.3927	2.5465	1.544	18.53	0.0107	0.080		93
	1.370	16	.065				1.474	17.69	0.0102	0.076		98
1½	1.334	14	.083				1.398	16.78	0.0097	0.073		103
	1.310	13	.095				1.343	16.12	0.0093	0.070		107
	1.282	12	.109	↓	↓	↓	1.292	15.50	0.0090	0.067		111
	1.620	16	.065	5.4978	.4581	2.1827	2.061	24.73	0.0143	0.107		70
	1.584	14	.083				1.971	23.65	0.0137	0.102		73
1¾	1.560	13	.095				1.911	22.94	0.0133	0.099		75
	1.532	12	.109				1.843	22.12	0.0128	0.096		78
	1.490	11	.120	↓	↓	↓	1.744	20.92	0.0121	0.090		83
	1.870	16	.065	6.2832	.5236	1.9099	2.746	32.96	0.0191	0.143		52
	1.834	14	.083				2.642	31.70	0.0183	0.137		55
2	1.810	13	.095				2.573	30.88	0.0179	0.134		56
	1.782	12	.109				2.489	29.87	0.0173	0.129		58
	1.760	11	.120	↓	↓	↓	2.433	29.20	0.0169	0.126		59

Table A-6. Steel-pipe dimensions.*

Nominal pipe size, in.	Outside diam, in.	Schedule No.	Wall thick-ness, in.	Inside diam, in.	Cross-sectional area metal, sq in.	Inside cross-sectional area sq ft
⅛	0.405	40†	0.068	0.269	0.072	0.00040
		80‡	0.095	0.215	0.093	0.00025
¼	0.540	40†	0.088	0.364	0.125	0.00072
		80‡	0.119	0.302	0.157	0.00050
⅜	0.675	40†	0.091	0.493	0.167	0.00133
		80‡	0.126	0.423	0.217	0.00098
½	0.840	40†	0.109	0.622	0.250	0.00211
		80‡	0.147	0.546	0.320	0.00163
		160	0.187	0.466	0.384	0.00118
¾	1.050	40†	0.113	0.824	0.333	0.00371
		80‡	0.154	0.742	0.433	0.00300
		160	0.218	0.614	0.570	0.00206
1	1.315	40†	0.133	1.049	0.494	0.00600
		80‡	0.179	0.957	0.639	0.00499
		160	0.250	0.815	0.837	0.00362
1¼	1.660	40†	0.140	1.380	0.699	0.01040
		80‡	0.191	1.278	0.881	0.00891
		160	0.250	1.160	1.107	0.00734
1½	1.900	40†	0.145	1.610	0.799	0.01414
		80‡	0.200	1.500	1.068	0.01225
		160	0.281	1.338	1.429	0.00976
2	2.375	40†	0.154	2.067	1.075	0.02330
		80‡	0.218	1.939	1.477	0.02050
		160	0.343	1.689	2.190	0.01556
2½	2.875	40†	0.203	2.469	1.704	0.03322
		80‡	0.276	2.323	2.254	0.02942
		160	0.375	2.125	2.945	0.02463
3	3.500	40†	0.216	3.068	2.228	0.05130
		80‡	0.300	2.900	3.016	0.04587
		160	0.437	2.626	4.205	0.03761
3½	4.000	40†	0.226	3.548	2.680	0.06870
		80‡	0.318	3.364	3.678	0.06170
4	4.500	40†	0.237	4.026	3.173	0.08840
		80‡	0.337	3.826	4.407	0.07986
		120	0.437	3.626	5.578	0.07170
		160	0.531	3.438	6.621	0.06447

* Based on A.S.A. Standards B36.10.
† Designates former "standard" sizes
‡ Former "extra strong."

Table A-6. Continued.

Nominal pipe size, in.	Outside diam, in.	Schedule No.	Wall thick-ness, in.	Inside diam, in.	Cross-sectional area metal, sq in.	Inside cross-sectional area, sq ft
5	5.563	40†	0.258	5.047	4.304	0.1390
		80‡	0.375	4.813	6.112	0.1263
		120	0.500	4.563	7.953	0.1136
		160	0.625	4.313	9.696	0.1015
6	6.625	40†	0.280	6.065	5.584	0.2006
		80‡	0.432	5.761	8.405	0.1810
		120	0.562	5.501	10.71	0.1650
		160	0.718	5.189	13.32	0.1469
8	8.625	20	0.250	8.125	6.570	0.3601
		30†	0.277	8.071	7.260	0.3553
		40†	0.322	7.981	8.396	0.3474
		60	0.406	7.813	10.48	0.3329
		80‡	0.500	7.625	12.76	0.3171
		100	0.593	7.439	14.96	0.3018
		120	0.718	7.189	17.84	0.2819
		140	0.812	7.001	19.93	0.2673
		160	0.906	6.813	21.97	0.2532
10	10.75	20	0.250	10.250	8.24	0.5731
		30†	0.307	10.136	10.07	0.5603
		40†	0.365	10.020	11.90	0.5475
		60‡	0.500	9.750	16.10	0.5185
		80	0.593	9.564	18.92	0.4989
		100	0.718	9.314	22.63	0.4732
		120	0.843	9.064	26.24	0.4481
		140	1.000	8.750	30.63	0.4176
		160	1.125	8.500	34.02	0.3941
12	12.75	20	0.250	12.250	9.82	0.8185
		30†	0.330	12.090	12.87	0.7972
		40	0.406	11.938	15.77	0.7773
		60	0.562	11.626	21.52	0.7372
		80	0.687	11.376	26.03	0.7058
		100	0.843	11.064	31.53	0.6677
		120	1.000	10.750	36.91	0.6303
		140	1.125	10.500	41.08	0.6013
		160	1.312	10.126	47.14	0.5592
14	14.0	10	0.250	13.500	10.80	0.9940
		20	0.312	13.376	13.42	0.9750
		30	0.375	13.250	16.05	0.9575
		40	0.437	13.126	18.61	0.9397
		60	0.593	12.814	24.98	0.8956
		80	0.750	12.500	31.22	0.8522
		100	0.937	12.126	38.45	0.8020
		120	1.062	11.876	43.17	0.7693
		140	1.250	11.500	50.07	0.7213
		160	1.406	11.188	55.63	0.6827

* Based on A.S.A. Standards B36.10.
† Designates former "standard" sizes.
‡ Former "extra strong."

Table A-7. Properties of the atmosphere.*

Altitude, ft	Altitude, miles	Absolute Temperature, R	Absolute Pressure, lb_f/sq ft	Pressure Ratio	Density, lb_m/cu ft	Density Ratio	Speed of Sound, ft/sec
0	0	518	2,116	1.00	7.65×10^{-2}	1.00	1,120
5,000	0.947	500	1,758	8.32×10^{-1}	6.60×10^{-2}	8.61×10^{-1}	1,100
10,000	1.894	483	1,456	6.87×10^{-1}	5.66×10^{-2}	7.38×10^{-1}	1,080
20,000	3.788	447	972	4.59×10^{-1}	4.08×10^{-2}	5.33×10^{-1}	1,040
30,000	5.682	411	628	2.97×10^{-1}	2.88×10^{-2}	3.76×10^{-1}	997
40,000	7.576	392	392	1.85×10^{-1}	1.88×10^{-2}	2.45×10^{-1}	973
50,000	9.470	392	243	1.15×10^{-1}	1.16×10^{-2}	1.52×10^{-1}	973
60,000	11.364	392	151	7.13×10^{-2}	7.32×10^{-3}	9.45×10^{-2}	973
70,000	13.258	392	94.5	4.47×10^{-2}	4.51×10^{-3}	5.90×10^{-2}	974
80,000	15.152	392	58.8	2.78×10^{-2}	2.80×10^{-3}	3.67×10^{-2}	974
90,000	17.045	392	36.6	1.73×10^{-2}	1.67×10^{-3}	2.28×10^{-2}	974
100,000	18.939	392	22.8	1.08×10^{-3}	1.1×10^{-3}	1.4×10^{-2}	975
150,000	28.409	575	3.2	1.5×10^{-3}	9.7×10^{-4}	1.3×10^{-3}	1,190
200,000	37.879	623	0.73	3.6×10^{-4}	2.2×10^{-5}	2.9×10^{-4}	1,240
300,000	56.818	487	0.017	9.0×10^{-6}	6.9×10^{-7}	9.0×10^{-6}	1,110
400,000	75.758	695	0.0011	5.2×10^{-7}	2.7×10^{-8}	3.5×10^{-7}	1,430
500,000	94.697	910	1.2×10^{-4}	8.5×10^{-8}	3.1×10^{-9}	4.1×10^{-8}	....
600,000	113.64	1,130	4.1×10^{-5}	1.9×10^{-8}	5.7×10^{-10}	7.5×10^{-9}	....
700,000	132.58	1,350	1.3×10^{-5}	6.2×10^{-9}	1.5×10^{-10}	1.9×10^{-9}	....
800,000	151.52	1,570	4.6×10^{-6}	2.2×10^{-9}	4.6×10^{-11}	6.0×10^{-10}	....
900,000	170.45	1,800	1.9×10^{-6}	9.0×10^{-10}	1.7×10^{-11}	2.2×10^{-10}	....

* Sources of Atmospheric Property Data:

1. C. N. Warfield, "Tentative Tables for the Properties of the Upper Atmosphere," *NACA TN* 1200, 1947.
2. H. A. Johnson, M. W. Rubesin, F. M. Sauer, E. G. Slack, and L. Fossner, "The Thermal Characteristics of High Speed Aircraft," AAF, AMC, Wright Field, TR 5632, 1947.
3. J. P. Sutton, *Rocket Propulsion Elements*, 2d ed., New York: McGraw-Hill Book Company Inc., 1957.

Table A-8. The error function (see Fig. 4-18 and Eq. 4-47).

$\frac{x}{2\sqrt{a\tau}}$	$f\left(\frac{x}{2\sqrt{a\tau}}\right)$	$\frac{x}{2\sqrt{a\tau}}$	$f\left(\frac{x}{2\sqrt{a\tau}}\right)$	$\frac{x}{2\sqrt{a\tau}}$	$f\left(\frac{x}{2\sqrt{a\tau}}\right)$
0.00	0.00000	0.76	0.71754	1.52	0.96841
0.02	0.02256	0.78	0.73001	1.54	0.97059
0.04	0.04511	0.80	0.74210	1.56	0.97263
0.06	0.06762	0.82	0.75381	1.58	0.97455
0.08	0.09008	0.84	0.76514	1.60	0.97635
0.10	0.11246	0.86	0.77610	1.62	0.97804
0.12	0.13476	0.88	0.78669	1.64	0.97962
0.14	0.15695	0.90	0.79691	1.66	0.98110
0.16	0.17901	0.92	0.80677	1.68	0.98249
0.18	0.20094	0.94	0.81627	1.70	0.98379
0.20	0.22270	0.96	0.82542	1.72	0.98500
0.22	0.24430	0.98	0.83423	1.74	0.98613
0.24	0.26570	1.00	0.84270	1.76	0.98719
0.26	0.28690	1.02	0.85084	1.78	0.98817
0.28	0.30788	1.04	0.85865	1.80	0.98909
0.30	0.32863	1.06	0.86614	1.82	0.98994
0.32	0.34913	1.08	0.87333	1.84	0.99074
0.34	0.36936	1.10	0.88020	1.86	0.99147
0.36	0.38933	1.12	0.88679	1.88	0.99216
0.38	0.40901	1.14	0.89308	1.90	0.99279
0.40	0.42839	1.16	0.89910	1.92	0.99338
0.42	0.44749	1.18	0.90484	1.94	0.99392
0.44	0.46622	1.20	0.91031	1.96	0.99443
0.46	0.48466	1.22	0.91553	1.98	0.99489
0.48	0.50275	1.24	0.92050	2.00	0.99532
0.50	0.52050	1.26	0.92524	2.10	0.997020
0.52	0.53790	1.28	0.92973	2.20	0.998137
0.54	0.55494	1.30	0.93401	2.30	0.998857
0.56	0.57162	1.32	0.93806	2.40	0.999311
0.58	0.58792	1.34	0.94191	2.50	0.999593
0.60	0.60386	1.36	0.94556	2.60	0.999764
0.62	0.61941	1.38	0.94902	2.70	0.999866
0.64	0.63459	1.40	0.95228	2.80	0.999925
0.66	0.64938	1.42	0.99538	2.90	0.999959
0.68	0.66378	1.44	0.95830	3.00	0.999978
0.70	0.67780	1.46	0.96105	3.20	0.999994
0.72	0.69143	1.48	0.96365	3.40	0.999998
0.74	0.70468	1.50	0.96610	3.60	1.000000

Index